8714

# World – Physical

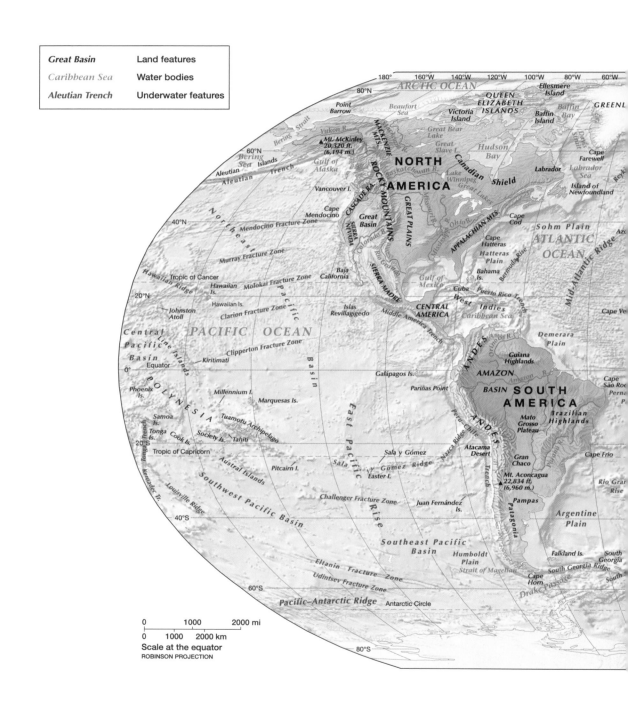

ARCTIC OCEAN

QUEEN ELIZABETH ISLANDS

Ellesmere Island

GREENL

80°N

Point Barrow

Beaufort Sea

Victoria Island

Baffin Island

Baffin Bay

Yukon R.

MACKENZIE MTS.

Great Bear Lake

Great Slave L.

Hudson Bay

Davis Strait

Cape Farewell

Reykj

60°N

Bering Strait

Mt. McKinley 20,320 ft. (6,194 m.)

NORTH AMERICA

Canadian Shield

Labrador

Labrador Sea

Island of Newfoundland

Bering Sea

Aleutian Islands

Gulf of Alaska

ROCKY MOUNTAINS

Saskatchewan R.

Lake Winnipeg

Aleutian Trench

Northeast

Vancouver I.

CASCADE RA.

GREAT PLAINS

Missouri R.

Great Lakes

Ohio R.

Cape Cod

Sohm Plain

Azo

40°N

Cape Mendocino

Great Basin

SIERRA NEVADA

Colorado R.

Rio Grande

APPALACHIAN MTS.

Cape Hatteras

ATLANTIC OCEAN

Mendocino Fracture Zone

Hatteras Plain

Bermuda Rise

Mid-Atlantic Ridge

Murray Fracture Zone

SIERRA MADRE

Hawaiian Ridge

Tropic of Cancer

Hawaiian Is.

Molokai Fracture Zone

Baja California

Gulf of Mexico

Bahama Is.

Cuba

Puerto Rico Trench

Cape Ve

20°N

Hawaiian Is.

Pacific

Islas Revillagigedo

CENTRAL AMERICA

West Indies

Caribbean Sea

Johnston Atoll

Clarion Fracture Zone

Middle America Trench

Cape

Central Pacific Basin

PACIFIC OCEAN

Line Islands

ANDES

Orinoco R.

Demerara Plain

Guiana Highlands

0° Equator

Kiritimati

Clipperton Fracture Zone

Basin

Galápagos Is.

AMAZON

Amazon R.

SOUTH AMERICA

Cape São Ro

Pern

Phoenix Is.

POLYNESIA

Millennium I.

Marquesas Is.

Pariñas Point

BASIN

Brazilian Highlands

Cape

Samoa Is.

Tonga Trench

Cook Is.

Tuamotu Archipelago

Society Is.

Tahiti

East Pacific

Mato Grosso Plateau

Peru-Chile

Tonga Is.

Tropic of Capricorn

20°S

Austral Islands

Pitcairn I.

Sala y Gómez

Atacama Desert

Gran Chaco

Cape Frio

Kermadec Tr.

Louisville Ridge

Southwest Pacific Basin

Sala y Gómez Ridge

Easter I.

Rise

Nazca Ridge

Mt. Aconcagua 22,834 ft (6,960 m.)

Rio Gran Rise

Challenger Fracture Zone

Juan Fernández Is.

Pampas

Patagonia

Argentine Plain

40°S

Southeast Pacific Basin

Humboldt Plain

Falkland Is.

South Georgia

Eltanin Fracture Zone

Udintsev Fracture Zone

Strait of Magellan

South Georgia Ridge

South

60°S

Pacific–Antarctic Ridge

Antarctic Circle

Cape Horn

Drake Passage

180°  160°W  140°W  120°W  100°W  80°W  60°W

80°S

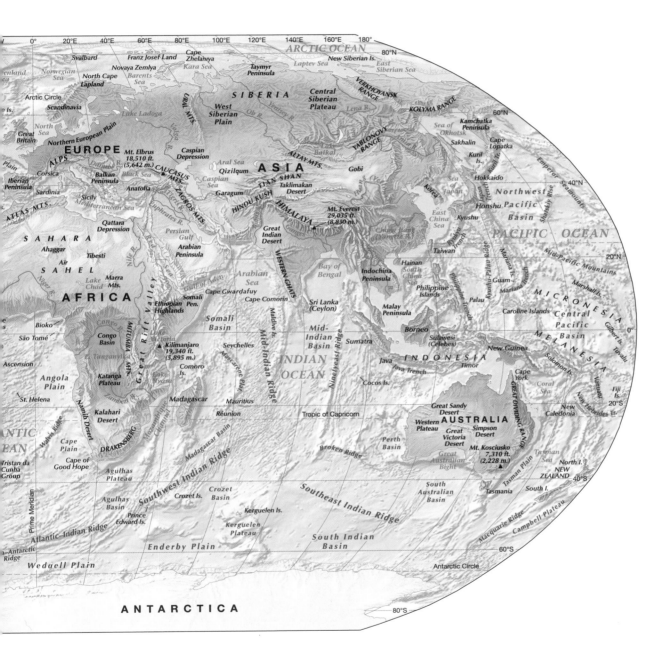

# Improve Your Grade!

## Access included with any new book.

**Registration will let you:**

- Prepare for exams by taking online quizzes.

- Watch videos and explore interactive maps to broaden your knowledge.

- Expand your understanding of important issues related to your world.

# www.mygeoscienceplace.com

## TO REGISTER

1. Go to www.mygeoscienceplace.com

2. Click "Register."

3. Follow the on-screen instructions to create your login name and password.

**Your Access Code is:**

*Note: If there is no silver foil covering the access code, it may already have been redeemed, and therefore may no longer be valid. In that case, you can purchase access online using a major credit card. To do so, go to www.mygeoscienceplace.com, click on "Buy Access," and follow the on-screen instructions.*

## TO LOG IN

1. Go to www.mygeoscienceplace.com

2. Click "Log In."

3. Pick your book cover.

4. Enter your login name and password.

**Hint:**
Remember to bookmark the site after you log in.

**Technical Support:**
http://247pearsoned.custhelp.com

# World Regions in Global Context

## Peoples, Places, and Environments

# World Regions in Global Context
## Peoples, Places, and Environments

FOURTH EDITION

## Sallie A. Marston
University of Arizona

## Paul L. Knox
Virginia Tech

## Diana M. Liverman
University of Arizona; Oxford University

## Vincent J. Del Casino, Jr.
California State University, Long Beach

## Paul F. Robbins
University of Arizona

**Prentice Hall**

Boston   Columbus   Indianapolis   New York   San Francisco   Upper Saddle River
Amsterdam   Cape Town   Dubai   London   Madrid   Milan   Munich   Paris   Montréal   Toronto
Delhi   Mexico City   São Paulo   Sydney   Hong Kong   Seoul   Singapore   Taipei   Tokyo

Geography Editor: Christian Botting
Editor in Chief, Geosciences and Chemistry: Nicole Folchetti
Marketing Manager: Maureen McLaughlin
Geography Editorial Project Editor: Tim Flem
Assistant Editor: Jennifer Aranda
Editorial Assistant: Christina Ferraro
Marketing Assistant: Nicola Houston
Managing Editor, Geosciences and Chemistry: Gina M. Cheselka
Senior Project Manager, Science: Beth Sweeten
Design Director: Mark Stuart Ong
Interior and Cover Design: Randall Goodall, Seventeenth Street Studios
Senior Technical Art Specialist: Connie Long
Art Studio: Precision Graphics

Senior Manufacturing and Operations Manager: Nick Sklitsis
Operations Supervisor: Maura Zaldivar
Senior Media Producer: Angela Bernhardt
Media Producer: Michelle Schreyer
Senior Media Production Supervisor: Liz Winer
Associate Media Project Managers: David Chavez and Ziki Dekel
Photo Research Manager: Elaine Soares
Photo Researcher: Clare Maxwell
Composition/Full Service: Laserwords Maine
Production Editor, Full Service: Patty Donovan
Cover Photograph: Crescent Moon Lake Oasis in Gobi Desert, near Dunhuang, China © Jose Fuste Raga / Corbis

**Library of Congress Cataloging-in-Publication Data**

World regions in global context : peoples, places, and environments / Sallie
A. Marston . . . [et al.]. — 4th ed.
      p. cm.
   Previous edition cataloged under Marston, Sallie A.
   ISBN-13: 978-0-321-65185-3
   ISBN-10: 0-321-65185-5
  1.   Geography. 2.   Globalization.   I.  Marston, Sallie A.   II. Marston,
Sallie A., World regions in global context.
   G116.W69 2011
   910—dc22

                                        2009047638

Printed in the United States
10 9 8 7 6 5 4 3 2 1

**Prentice Hall**
is an imprint of

www.pearsonhighered.com

ISBN 0-321-65185-5/978-0-321-65185-3 [Student]
ISBN 0-321-70700-1/978-0-321-70700-0 [Books á la Carte]

# BRIEF CONTENTS

# CONTENTS

# PREFACE

*. . . it takes a lot of things to change the world:*
*Anger and tenacity. Science and indignation,*
*The quick initiative, the long reflection,*
*The cold patience and the infinite perseverance,*
*The understanding of the particular case*
*and the understanding of the ensemble:*
*Only the lessons of reality can teach us to transform reality.*

Bertolt Brecht, Einverstandnis

In order to change the world, to make it a better place in which to live for all people, we need to understand not just our little corner of it, but the whole of it, the broad sweep of global geography that constitutes the larger world of which our small corners are just a part. *World Regions in Global Context* aims to provide an introduction to today's global geography set in the context of world regions. This goal is ambitious because our ability to trace the world's complex geographies in such a relatively short volume is impossible. That said, this textbook sheds some light on the dynamic and complex relationships between people and the worlds they inhabit. This book gives students the basic geographical tools and concepts needed to understand the complexity of today's global geography and the world regions that make up that geography, while asking students to appreciate the interconnections between their own lives and those of people in different parts of the world.

## NEW TO THE FOURTH EDITION

The fourth edition of *World Regions in Global Context* represents a thorough revision based on a number of substantive changes. These changes include a detailed review of the content and framework for the text. Below are a number of highlights that represent these broader changes.

- This edition introduces two new authors—Vincent Del Casino, Jr. and Paul Robbins—both of whom have extensive experience teaching world regional geography and conducting research in a number of different world regions.

- This edition includes a thorough engagement with the changing concepts being applied to geography today. In particular, this edition responds to the broader trend in geography, which is to incorporate a greater global sensibility into world regional geography. This is done with an emphasis on a global geography framework that moves toward de-centering a European and North American centered narrative about world history. Thus, every part of the book has been examined carefully with the dual goals of keeping topics current and improving the clarity of the conceptual framework, text, and graphics.

- This edition has also been shortened considerably. This has been done in several ways:

- First, we have reorganized the content of each regional chapter to meet the goals and expectations laid out in the significantly revised Chapter 1. This framework relies on a global geography approach that focuses attention on the world and its constitutive parts.

- Second, this means that, through reorganization, we have limited the redundancies in the chapters, while folding key elements of the "contemporary challenges in a globalizing world" sections into the chapters' new four-part construction. This construction highlights first the human-environment relationship, second economic and demographic conditions and linkages, third the significance of culture and politics, and fourth future geographies.

- Third, the "further readings" and "film, music, and popular culture" sections have moved to an online resource. This change will allow the authors and the production team to update these on a more regular basis.

- Fourth, we have also cut the "Geography Matters" boxed essays, folding some of the material from them into new "Signature Regions in Global Context" boxes or into the main text where appropriate.

- Even while shortening the text, we have provided a number of important innovations. Each chapter now opens with a narrative story designed to engage and draw the reader into the region. This new narrative vignette, along with the new "Key Facts" and "Main Points" sections that accompany the chapter opening regional map, help provide readers with both a local sense of place and a broad sense of each world region's environmental and demographic characteristics.

- We have also expanded key topics, including the conflicts and tensions in the Middle East; the growing concern and scientific evidence surrounding global climate change; the increasing political and economic importance of India and China; the aging of the world's population and the implications of global migration; shifts in politics in Latin America and Sub Saharan Africa; the spread of new technology systems; the changes in the European Union, North Korea, and Iran as political hot spots; minority politics and HIV in Southeast Asia and Peak Oil and food movements and issues around the world.

- The fourth edition of the book also incorporates a comprehensive updating of all the data, maps, and illustrative examples and striking new panoramic photographs in each chapter.

- Finally, the online Multimedia and Assessment for *World Regions in Global Context*, found at www.mygeoscienceplace.com, now includes a fully-integrated eBook, numerous geography video clips, interactive maps, activities, assessments, flash cards, RSS feeds, and additional resources such as further readings and film, music, and popular culture references.

# OBJECTIVES AND APPROACH

*World Regions in Global Context* has two primary objectives. The first is to provide a body of knowledge about how sets of global, natural, socio-cultural, and political–economic processes come together to produce world regions with distinctive political and economic territories, cultural and environmental landscapes, and socio-cultural attributes. The second is to emphasize that, although there is diversity among world regions, it is important to understand the current and historical interdependencies that exist among and between regions in order to build any real understanding of the modern world.

In an attempt to achieve these objectives, we have taken a fresh approach by constructing a global geography framework for studying world regions. This framework reflects the fact that world regions cannot be studied in isolation. For example, global changes in the flows of information (e.g., the World Wide Web) or the conditions affecting climate change (e.g., global warming) are impressed on regional and local landscapes and interpreted and taken up differently depending on where one might be located. The approach used in *World Regions in Global Context* thus provides access not only to the new ideas, concepts, and theories that address these historical global changes over time but also to the fundamentals of geography: the principles, concepts, theoretical frameworks, and basic knowledge that are necessary to build a geographic understanding of today's world.

A distinctive feature of our approach is that it emphasizes the short- and long-term interdependence of places and processes at different scales and how regions emerge from these processes. In overall terms, this approach provides an understanding of the relationships between widespread processes that are affecting places throughout the globe as well as the way these processes function in particular places. We are not only interested in understanding the internal dynamics of a world region; we are interested in that region's historical interconnections—through processes such as globalization—to other regions around the globe. We thus turn the student's attention to the three key conceptual points that guide the global geography framework in Chapter 1: world regions are produced through (1) global interdependence; (2) ongoing change; and (3) systemic relations.

This approach allows us to emphasize a number of important themes.

- *The connection between society and nature.*    Inherent to the basic geographic concepts of landscape, place, and region are the interactions between people and the natural environment that shape landscapes and give places and regions their distinctive characteristics. In this book, we explore the nature–society or human–environment relationships that assist in our understanding of regional geography. We emphasize that human use of Earth's physical environments has gone far beyond responses and adaptation to natural constraints to produce significant modifications of environments and landscapes, widespread environmental degradation and pollution and has produced global climate change.

- *The historical links between global and local processes.*    Throughout the book, we stress the interconnectedness of different parts of the world through common processes of economic, environmental, political, and cultural change. We approach the study of these processes through a historical framework that contextualizes the most recent phase of these interconnections, often defined as globalization. Different parts of the world have long been integrated into wider systems of exchange and control at what we might call the global scale. Each time the world has been reorganized, there have been major changes not only in global geography but also in the character and fortunes of individual regions. This approach demands, however, that we also explicitly recognize the underlying diversity of the world. While there are a range of processes that are likely to be common to most regions—urbanization, industrialization, and population distribution—the way these processes are manifested will vary from region to region and even within regions. In short, there are important variations within places and regions at every scale. For example, social well-being varies and there can be affluent enclaves in poor regions and pockets of poverty in rich regions.

- *The links among and between regions.*    While the book explores a set of coherent world regions, we also make it clear that regions are not isolated units but exist in complex relationship to other regions. The world, in short, is an integrated whole, and the concept of regions allows us to break it up into more manageable units. Yet, it is often the case that some regions have stronger and more long-standing connections to other regions or that some sub-areas of a region—certain key cities or industrial areas—may actually be more connected to outside regions than to their own. This emphasis on the links among and between regions, including flows of people and goods between them, enables us to demonstrate the interdependence of the world and how that interdependence is unevenly produced.

# THE GEOGRAPHY OF WORLD REGIONS

There is no standard way of dividing the world into regions. Textbooks, international organizations, and regional studies groups within universities, however, do divide up and make sense of the world in different ways, and many of these organizational frameworks are based on some broad arrangement of connections between a series of local places that make up wider world regions. In this text we utilize a schema that divides the world into ten major regions—Europe; The Russian Federation, Central Asia, and the Transcaucasus; the Middle East and North Africa; Sub-Saharan Africa; the United States and Canada; Latin America; East Asia; South Asia; Southeast Asia; and Australia, New Zealand, and the South Pacific. Although we review the distinctive characteristics of every region throughout each chapter and the futures of these regional formations at the end of each chapter, the changing and sometimes controversial process of defining world regions merits some further discussion here.

Over time, humans have divided the spaces of the planet into different regions, and these divisions have often depended on the local perceptions of people's place in the world. Early Greek geographers, for example, divided their known world into Europe, Africa, and Asia, with the boundaries defined by the Straits of Gibraltar (dividing Africa and Europe), the Red Sea (dividing Africa and Asia), and the Bosporus Strait (dividing Europe and Asia). As Europeans began to explore the world and come into contact with other world views, new regions were associated by them with major landmasses or continents, with the Americas usually split into North and South America and Australia and Antarctica added as the sixth and seventh continents. These divisions lumped together many different landscapes and cultures (especially in Asia) but served, in the minds of Europeans, to differentiate themselves from "others," and to provide a framework for organizing colonial exploration and administration. As a result, the European colonial period produced many new boundaries and nations and transformed cultures and landscapes in ways that produced more homogeneous regions. For example, 400 years of Spanish and Portuguese colonization of the region that stretches from Mexico to Argentina created a region of shared languages, religion, and political institutions that became known as Latin America. British colonization of what now constitutes Sri Lanka, India, Bangladesh, Pakistan, and Nepal interacted with local culture to produce a region frequently known as South Asia. In the Middle East and North Africa, the persistence of Muslim religion and tradition gave these regions an identity that separated them from Asia and from Africa south of the Sahara.

In the 20th century, new configurations of political power and economic alliances produced some reconfigurations of world regions. The most notable was the large block of Asia and eastern Europe associated with the socialist politics of the former Soviet Union centered on Russia, together with eastern European countries ranging from East Germany to Bulgaria.

In response to global conflicts and economic opportunities in the second half of the 20th century, governments and universities established programs and centers that focused on specific world areas and their languages. For example, in the United States, the Department of Education established university centers that focused on apparently coherent regions such as Latin America; the Caribbean; the Pacific; Europe; Africa; the Soviet Union and Eastern Europe; the Middle East; and East, South, and Southeast Asia.

At the beginning of the 21st century, these historical divisions of the world into regions have been challenged by events, critics, and the latest phases of globalization. When the Soviet bloc disintegrated in 1989, some states reoriented toward western Europe and to the economic alliance of the European Union, whereas others remained closer to Russia or looked eastward to an identity with countries such as Afghanistan as part of central Asia. As we note in the relevant chapters, regionalizations have been criticized for being based on race or religion (for example, the Middle East and Sub-Saharan Africa), for being remnants of colonial thinking (for example, Latin America or Southeast Asia), or for being grounded only in physical proximity or environmental characteristics rather than on cultural or other human commonalities (for example, Australia, New Zealand, and the Pacific islands clustered in Oceania). We will also discuss a number of countries or continents, such as Sudan, Cyprus, or Antarctica, that do not fit easily into the traditional regions or that fit into more than one region. Some scholars and institutions have proposed a dramatic rethinking of world regions. They suggest, for example, that all Islamic or oil-producing countries be treated together, or that countries be grouped according to their level of economic development or integration into the global economy. As another example, the World Bank commonly classifies nations into high-, middle-, and low-income countries, and this book identifies many regions and countries according to their relation to the core or periphery of the world-system.

Our own division of the world tries to take into account some of these changing ideas about world regions without deviating too radically from popular understandings or other texts or course outlines and by trying to create a manageable number and coherent set of chapters. Each chapter includes our rationale for treating the places in the chapter as a distinct region and a review of the limitations and debates about defining each region. In addition, each chapter emphasizes the links of the region under discussion to other regions and to processes of globalization that might be changing the nature and coherence of world regions.

# CHAPTER ORGANIZATION

One of the central challenges of writing a world regional geography text appropriate for the modern world involves balancing the emphasis between the study of dynamic global processes and regionally-specific generalizations, the latter having in the past been informed by detailed country-by-country descriptions. In order to provide a global geography approach to the study of world regions that balances the complexity of global interconnections with local approaches and interpretations of those interconnections, we divide each regional chapter into four major categories. These categories, which form the building blocks of our global geography, help us frame how world regions have emerged and been internally and externally organized over time.

## Environment and Society

We begin each chapter with a discussion of the physical and environmental context of the region, tying into this discussion a deeper examination of climate including adaptation and global change; geological resources, risks, and water; and ecology, land, and environmental management. Our aim here is to demonstrate the links between the natural environment and how that environment is shaped by, and shapes, the region's inhabitants over time.

## History, Economy, and Demographic Change

Consistent with our aim to highlight the enduring interdependence of the world's regions, we provide a section that places each of the regions within the larger context of global historical geography. This section emphasizes historical landscapes and legacies; economy, accumulation, development and the production of inequality; and demography, migration, and settlement. In addition, we provide wider historical context for the changes we see in the formation and organization of societies in world regional context.

## Culture and Politics

This newly developed section explores the complexity of cultural and political life in world regional context. It does so while emphasizing the broader contexts for the globalization of certain cultural forms and the geopolitical conflicts that are currently emerging across various places in the world. This section details these processes in three sub-sections, focusing on tradition, religion, and language; cultural practices, social differences, and identity; and territory and politics.

## Future Geographies

Keeping with the theme of this book, which emphasizes that a global geography framework is always informed by the study of ongoing change, each chapter concludes with a brief discussion of some of the key issues that might be faced within this region in the future. Moreover, in this section we consider the possibility of new regional formations at various scales moving forward.

The organization of the book is pedagogically useful in several ways. First, the conceptual framework of the book is built on a single opening chapter that describes the basics of a global geography perspective and introduces the key concepts that are deployed throughout the remaining regional chapters. In addition, Chapter 1 highlights the importance of a dynamic approach to the study of world regions that emphasizes both long-term and short-term global interconnections between and within world regions. The ten regional chapters that follow explore and elaborate the concepts and conceptual framework laid out in Chapter 1. We have also shortened each of the regional chapters and have instituted an even more consistent organizational framework to help guide students through the study of world regions in global context.

An important aspect of the book is the distinctive ordering of the chapters. The sequencing of the chapters is a deliberate move to avoid privileging any one region over any other or to cluster the regions according to any economic or political categorization. At the same time, we have decided to begin with the European region (Chapter 2) because that is the source of contemporary capitalism and many of the impulses for the contemporary world map—and a helpful place to begin our discussion of global connectivity in the modern era. Following the initial appearance of this region, however, we deliberately intersperse the other world regions in order to signal the interdependence of each. And, given the new conceptual framework at the beginning of the text, we encourage readers to explore the book in a different order than laid out here. This affords the reader the opportunity to think about the global geography framework offered in Chapter 1 and examine that framework in relation to the ten world regions presented in this text.

# FEATURES

This book takes a decidedly different approach to understanding world regions, and the features we use help to underscore that difference. The book employs a clean and accessible cartographic program and three different boxed essay features (Key Facts; Signature Regions in Global Context; and Geographies of Indulgence, Desire, & Addiction), as well as more familiar pedagogical devices such as end-of-chapter review questions for each region.

## Key Facts

At the beginning of each chapter, we include a section under the opening regional map called "Key Facts." This box provides some broad regional context—describing subregions and major physiogeographic features—along with some important statistics that help readers both understand the region and place it in relation to other world regions. In short, this new feature provides some basic background, while also allowing readers to see how the processes that connect the world also work out differently in different regional contexts. Throughout Chapter 1, the definitions of the terms outlined in the Key Facts box are highlighted and explained.

## Main Points and Main Points Revisited

On the second page of each chapter, we provide both a narrative as well as the main points covered in the chapter. These points signpost the major themes that readers should look for as they delve deeper into the chapter's content. These main points are intentionally broad, and the "answers" to

these main points do not lie in one particular section of the chapter, but draw from a number of different discussions throughout. At the end of the chapter, we return to these main points and offer some brief comment on them. Again, these are general in nature and cannot take the place of the complexity found throughout the chapter. But, the main points revisited section does help readers grapple with some of the larger conceptual material.

## Signature Regions in Global Context

This feature highlights specific subregions—places within the larger world region—with the intention of providing students with a more nuanced sense of what it is like to live in such a place, as well as how that place is globally interconnected. The "Signature Region" feature draws students closer to the textures of a specific regional geography.

## Geographies of Indulgence, Desire, and Addiction

This feature links people in one world region to people throughout the world through a discussion of the local production and global consumption of one of the region's primary commodities. The "Geographies of Indulgence, Desire, and Addiction" feature helps students to appreciate the links between producers and consumers around the world, as well as between people and the natural world.

# THE TEACHING AND LEARNING PACKAGE

In addition to the text itself, the authors and publisher have been pleased to work with a number of talented people to produce an excellent instructional package. This package includes the traditional supplements that students and professors have come to expect from authors and publishers, as well as new kinds of components that utilize electronic media.

# INSTRUCTOR RESOURCES

*Instructor Manual Download* (0-32-166795-6): The *Instructor Manual*, developed by one of the book's new co-authors Vincent Del Casino, Jr., follows the new organization of the main text and has been revitalized with new teaching tools. Strategies for Teaching Key Concepts provide instructors with a focused plan of action for every class session. Each chapter is broken down by means of how instructors can effectively teach the highlighted concepts presented in the Chapter Overview. The Web Exercises tie in with associated Interactive Maps, and Additional Resources such as journals and websites are provided.

*TestGen/Test Bank* (0-32-166774-3): TestGen is a computerized test generator that lets instructors view and edit *Test Bank* questions, transfer questions to tests, and print the test in a variety of customized formats. This *Test Bank* includes approximately 1000 multiple choice, true/false, and short answer/essay questions. Questions are correlated against the U.S. National Geography Standards and Bloom's Taxonomy to help instructors to better map the assessments against both broad and specific teaching and learning objectives. The *Test Bank* is also available in Microsoft Word®, and is importable into Blackboard and WebCT.

***Instructor Resource Center on DVD*** (0-32-166732-8): Everything instructors need where they want it. The Pearson Prentice Hall *Instructor Resource Center* helps make instructors more effective by saving them time and effort. All digital resources can be found in one, well-organized, easy-to-access place. The IRC on DVD includes:

- All textbook images as JPEGs, PDFs, and PowerPoint™ Presentations

- Pre-authored Lecture Outline PowerPoint™ Presentations, which outline the concepts of each chapter with embedded art and can be customized to fit instructors' lecture requirements

- CRS "Clicker" Questions in PowerPoint™ format, which correlate to the U.S. National Geography Standards and Bloom's Taxonomy

- The TestGen software, *Test Bank* questions, and answers for both MACs and PCs

- Electronic files of the *Instructor Manual* and *Test Bank*

- Over 120 Geography Video Clips

This Instructor Resource content is also available completely online via the Instructor Resources section of www.mygeoscienceplace.com and www.pearsonhighered.com/irc.

***Television for the Environment Earth Report Geography Videos on DVD*** (0-32-166298-9): This three-DVD set is designed to help students visualize how human decisions and behavior have affected the environment, and how individuals are taking steps toward recovery. With topics ranging from the poor land management promoting the devastation of river systems in Central America, to the struggles for electricity in China and Africa, these 13 videos from Television for the Environment's global *Earth Report* series recognize the efforts of individuals around the world to unite and protect the planet.

***Television for the Environment Life World Regional Geography Videos on DVD*** (0-13-159348-X): From the Television for the Environment's global *Life* series this two-DVD set brings globalization and the developing world to the attention of any world regional geography course. These 10 full-length video programs highlight matters such as the growing number of homeless children in Russia, the lives of immigrants living in the United States trying to aid family still living in their native countries, and the European conflict between commercial interests and environmental concerns.

***Television for the Environment Life Human Geography Videos on DVD*** (0-13-241656-5): This three-DVD set is designed to enhance any human geography course. These DVDs include 14 full-length video programs from Television for the Environment's global *Life* series, covering a wide array of issues affecting people and places in the contemporary world, including the serious health risks of pregnant women in Bangladesh, the social inequalities of the 'untouchables' in the Hindu caste system, and Ghana's struggle to compete in a global market.

***Aspiring Academics: A Resource Book for Graduate Students and Early Career Faculty*** (0-13-604891-9): Drawing on several years of research, this set of essays is designed to help graduate students and early career faculty start their careers in geography and related social and environmental sciences. This teaching aid stresses the interdependence of teaching, research, and service—and the importance of achieving a healthy balance in professional and personal life—in faculty work, and does not view it as a collection of unrelated tasks. Each chapter provides accessible, forward-looking advice on topics that often cause the most stress in the first years of a college or university appointment.

***Teaching College Geography: A Practical Guide for Graduate Students and Early Career Faculty*** (0-13-605447-1): Provides a starting point for becoming an effective geography teacher from the very first day of class. Divided into two parts, the first set of chapters addresses "nuts-and-bolts" teaching issues in the context of the new technologies, student demographics, and institutional expectations that are the hallmarks of higher education in the 21st century. The second part explores other important issues: effective teaching in the field; supporting critical thinking with GIS and mapping technologies; engaging learners in large geography classes; and promoting awareness of international perspectives and geographic issues.

***AAG Community Portal for Aspiring Academics & Teaching College Geography:*** This website is intended to support community-based professional development in geography and related disciplines. Here you will find activities providing extended treatment of the topics covered in both books. The activities can be used in workshops, graduate seminars, brown bags, and mentoring programs offered on campus or within an academic department. You can also use the discussion boards and contributions tool to share advice and materials with others. http://pearsonhighered.com/aag/

***Course Management Systems:*** Pearson Prentice Hall offers content specific to *World Regions in Global Context* within the BlackBoard and CourseCompass course management system platforms. Each of these platforms lets the instructor easily post his or her syllabus, communicate with students online or off-line, administer quizzes, record student results and track their progress. http://www.pearsonhighered.com/elearning/

# STUDENT RESOURCES

***Premium Website:*** A dedicated Premium Website with eText offers a variety of resources for students and professors, including an eText version of the textbook with linked/integrated multimedia, assignable Interactive Maps, geography videos with associated assessments, Flashcards, RSS Feeds, weblinks, annotated resources for further exploration, and Class Manager & GradeTracker Gradebook functionality for instructors. www.mygeoscienceplace.com

***Study Guide and Mapping Workbook*** (0-32-166775-1): This study guide/workbook provides students with basic map reading skills and offers exercises to further analyze the spatial dimensions of global and regional environmental, political–economic, and socio-cultural distributions, diffusions, and change. Furthermore, Suggested Paper Topics provide students with ideas on how to discuss or argue key chapter concepts for paper assignments.

*Goode's World Atlas* **22nd Edition** (0-32-165200-2): Goode's World Atlas has been the world's premiere educational atlas since 1923, and for good reason. It features over 250 pages of maps, from definitive physical and political maps to important thematic maps that illustrate the spatial aspects of many important topics. The 22nd edition includes 160 pages of new, digitally-produced reference maps, as well as new thematic maps on global climate change, sea level rise, $CO_2$ emissions, polar ice fluctuations, deforestation, extreme weather events, infectious diseases, water resources, and energy production.

*Encounter World Regional Geography Workbook & Website* (0-32-168175-4): *Encounter World Regional Geography* provides rich, interactive explorations of world regional geography concepts through Google Earth™ explorations. All chapter explorations are available in print format as well as online quizzes, accommodating different classroom needs. All worksheets are accompanied with corresponding Google Earth™ media files, available for download from www.mygeoscienceplace.com.

*Encounter Earth: Interactive Geoscience Explorations Workbook & Website* (0-32-158129-6): Ideal for professors who want to integrate Google Earth™ in their classrooms, *Encounter Earth* gives students a way to visualize key topics in their introductory geoscience courses. Each exploration consists of a worksheet, available in the workbook and as a PDF file, and a Google Earth™ KMZ file, containing placemarks, overlays, and annotations referred to in the worksheets. The accompanying *Encounter Earth* website is located at www.mygeoscienceplace.com.

*Dire Predictions* (0-13-604435-2): Periodic reports from the Intergovernmental Panel on Climate Change (IPCC) evaluate the risk of climate change brought on by humans. But the sheer volume of scientific data remains inscrutable to the general public, particularly to those who may still question the validity of climate change. In just over 200 pages, this practical text presents and expands upon the essential findings of the 4th Assessment Report in a visually stunning and undeniably powerful way to the lay reader. Scientific findings that provide validity to the implications of climate change are presented in clear-cut graphic elements, striking images, and understandable analogies.

# CONCLUSION

One important outcome of recent reforms in post-secondary education has been the inclusion of geography as a core subject in the Goals 2000: Educate America Act (Public Law 103-227). Another was the publication of a set of national geography standards for K 12 education (*Geography for Life,* published by National Geographic Research and Education for the American Geographical Society, the Association of American Geographers, the National Council for Geographic Education, and the National Geographic Society).

More broadly, this book is the product of conversations among the authors, colleagues, and students about how best to teach a course on world regional geography. In preparing the text, we have tried to help students make sense of the world by connecting our conceptual materials to the most compelling current events. We have also been careful to represent the best ideas and concepts the discipline of geography has to offer by mixing cutting-edge and innovative theories and concepts with more classical and proven approaches and tools. Finally, we have also tried to make it clear that no textbook is the product of its authors alone but is instead built from the intellectual and pedagogical toils and triumphs of thousands of colleagues around the world. Our aim has been to show how a geographical imagination is important, how it can lead to a greater understanding of the world and its constituent places and regions, and how it has practical relevance in our everyday and professional lives.

# ACKNOWLEDGMENTS

We are indebted to many people for their assistance, advice, and constructive criticism in the course of preparing this book. Among those who provided comments on various drafts and editions of this book are the following professors:

Donald Albert, *Sam Houston State University;* Brad Baltensperger, *Michigan Technological University;* Max Beavers, *University of Northern Colorado;* Richard Benfield, *Central Connecticut State University;* William H. Berentsen, *University of Connecticut;* Keshav Bhattarai, *Central Missouri State University;* Warren R. Bland, *California State University, Northridge;* Brian W. Blouet, *College of William and Mary* Pablo Bose, *University of Vermont;* Jean Ann Bowman, *Texas A & M University;* John Christopher Brown, *University of Kansas;* Stanley D. Brunn, *University of Kentucky;* Joe Bryan, *University of Colorado: Boulder;* Michelle Calvarese, *California State University, Fresno;* Craig Campbell, *Youngstown State University;* Xuwei Chen, *Northern Illinois University;* David B. Cole, *University of Northern Colorado;* Jose A. da Cruz, *Ozarks Technical Community College;* Tina Delahunty, *Texas Tech University;* Cary W. de Wit, *University of Alaska, Fairbanks;* Lorraine Dowler, *Pennsylvania State University;* Catie Finlayson, *Florida State University;* Ronald Foresta, *University of Tennessee;* Gary Gaile, *University of Colorado;* Roberto Garza, *University of Houston;* Jay Gatrell, *Indiana State University;* Mark Giordano, *Oregon State University;* Qian Guo, *San Francisco State University;* Devon A. Hansen, *University of North Dakota;* Julie E. Harris, *Harding University;* Russell Ivy, *Florida Atlantic University;* Kris Jones, *Saddleback College;* Tim Keirn, *California State University, Long Beach;* Lawrence M. Knopp, *University of Minnesota, Duluth;* Debbie Kreitzer, *Western Kentucky University;* Robert C. Larson, *Indiana State University;* Alan A. Lew, *Northern Arizona University;* John Liverman, *Independent Scholar;* Max Lu, *Kansas State University;* Donald Lyons, *University of North Texas;* Taylor Mack, *Mississippi State University;* Chris Mayda, *Eastern Michigan University;* Eugene McCann, *Simon Fraser University;* Tom L. McKnight, *University of California, Los Angeles;* M. David Meyer, *Central Michigan University;* Sherry D. Moorea-Oakes, *University of Colorado, Denver;* Barry Donald Mowell, *Broward Community College;* Tim Oakes, *University of Colorado;* Nancy Obermeyer, *Indiana State University;* J. Henry Owusu, *University of Northern Iowa;* Rosann Poltrone, *Arapahoe Community College;* Jeffrey E. Popke, *East Carolina University;* Henry O. Robertson, *Louisiana State University, Alexandria;* Robert Rundstrom, *University of Oklahoma;* Yda Schreuder, *University of Delaware;* Anna Secor, *University of Kentucky;* Daniel Selwa, *Coastal Carolina University;* Barry D. Solomon, *Michigan Technical University;* Joseph Spinelli, *Bowling Green State University;* Liem Tran, *Florida Atlantic University;* Samuel Wallace, *West Chester University;* Gerald R. Webster, *University of Alabama;* Mark Welford, *Georgia Southern University;* Keith Yearman, *College of Du Page; and* Anibal Yanez-Chavez, *California State University, San Marcos.*

Special thanks go to our editorial team at Pearson Prentice Hall, Christian Botting, Tim Flem, Jennifer Aranda, and Christina Ferraro; to our developmental editor, Anne Scanlan-Rohrer, and our Production Editor, Patty Donovan; to Clare Maxwell for photo research; and to art directors Maureen Eide and Mark Ong and their team for their creative work with the design of the text. We would also like to thank our excellent research assistant Ian G.R. Shaw. Finally, a number of colleagues gave generously of their time and expertise in guiding our thoughts, making valuable suggestions, and providing materials: David Lloyd, University of Southern California; Bruce Braun, University of Minnesota; Jennifer McCormack, University of Arizona; Stephen Cornell, Udall Center for Studies in Public Policy; Carolyn Gallaher, American University; Petr Pavlinek, University of Nebraska; and David Liverman, Memorial University, Newfoundland.

Sallie A. Marston
Paul L. Knox
Diana M. Liverman
Vincent J. Del Casino Jr.
Paul F. Robbins

# ABOUT THE AUTHORS

### SALLIE A. MARSTON

Sallie Marston received her Ph.D. in Geography from the University of Colorado, Boulder. She has been a faculty member at the University of Arizona since 1986. Her teaching focuses on the political and cultural aspects of social life, with particular emphasis on socio-spatial theory. She is the recipient of the College of Social and Behavioral Sciences Outstanding Teaching Award. Her teaching focuses on culture, politics, globalization, and methods. She is the author of over 70 journal articles, book chapters, and books and serves on the editorial board of several scientific journals. She has co-authored with Paul Knox, the introductory human geography textbook, *Places and Regions in Global Context*, also published by Pearson. She is currently a professor in the School of Geography and Development at the University of Arizona.

### PAUL L. KNOX

Paul Knox received his Ph.D. in Geography from the University of Sheffield, England. After teaching in the United Kingdom for several years, he moved to the United States to take a position as professor of urban affairs and planning at Virginia Tech. His teaching centers on urban and regional development, with an emphasis on comparative study. He has written several books on aspects of economic geography, social geography, and urbanization and he serves on the editorial board of several scientific journals. In 1996 he was appointed to the position of University Distinguished Professor at Virginia Tech, where he currently serves as Senior Fellow for International Advancement, and International Director of the Metropolitan Institute.

### DIANA M. LIVERMAN

Diana Liverman received her Ph.D. in Geography from the University of California, Los Angeles, and also studied at the University of Toronto, Canada, and University College London, England. Born in Accra, Ghana, she holds a joint appointment between the University of Arizona (where she co-directs the Institute of the Environment) and Oxford University. She has taught geography at Oxford University, the University of Arizona, Penn State, and the University of Wisconsin. Her teaching focuses on global environmental issues, climate and development, and on Latin America. She has served on several national and international advisory committees dealing with environmental issues and climate change and has written recent journal articles and book chapters on such topics as natural disasters, climate change, and environmental policy.

### VINCENT J. DEL CASINO JR.

Vincent J. Del Casino Jr. received his Ph.D. in Geography from the University of Kentucky in 2000. He is currently Professor and Chair of Geography at California State University, Long Beach, where he has taught since graduating from Kentucky. He has held a Visiting Research Fellow post at The Australian National University, and completed National Science Foundation supported research in Thailand. His current research reflects his ongoing interests in the areas of social and health geography, with a particular emphasis on HIV transmission, the care of people living with HIV and AIDS, and homelessness. He has published numerous articles and book chapters on his research, and he recently completed an upper-division textbook on social geography. His teaching focuses on social geography, geographic thought, and geographic methodology. He also teaches a number of general education courses in geography, including world regional geography, which he first began teaching as a graduate student in 1995.

### PAUL F. ROBBINS

Paul Robbins received his Ph.D. in Geography from Clark University in 1996. He has taught at the University of Arizona since 2005. Prior to this, he taught at Ohio State University, the University of Iowa, and Eastern Connecticut State University. His teaching and research focuses on the relationships between individuals (homeowners, hunters, professional foresters), environmental actors (lawns, elk, mesquite trees), and the institutions that connect them. He and his students seek to explain human environmental practices and knowledge, the influence the environment has on human behavior and organization, and the implications this holds for ecosystem health, local community, and social justice. Past projects have examined chemical use in the suburban United States, elk management in Montana, forest product collection in New England, and wolf conservation in India.

# ABOUT OUR SUSTAINABILITY INITIATIVES

This book is carefully crafted to minimize environmental impact. The materials used to manufacture this book originated from sources committed to responsible forestry practices. The paper is Forest Stewardship Council (FSC) certified. The binding, cover, and paper come from facilities that minimize waste, energy consumption, and the use of harmful chemicals.

Pearson closes the loop by recycling every out-of-date text returned to our warehouse. We pulp the books, and the pulp is used to produce items such as paper coffee cups and shopping bags. In addition, Pearson aims to become the first climate neutral educational publishing company.

The future holds great promise for reducing our impact on Earth's environment, and Pearson is proud to be leading the way. We strive to publish the best books with the most up-to-date and accurate content, and to do so in ways that minimize our impact on Earth.

**Mixed Sources**
Product group from well-managed forests, controlled sources and recycled wood or fiber
www.fsc.org   Cert no.  SW-COC-002985
©1996 Forest Stewardship Council

**Pearson Prentice Hall is an imprint of**

# World Regions in Global Context

## Peoples, Places, and Environments

▼ FIGURE 1.1

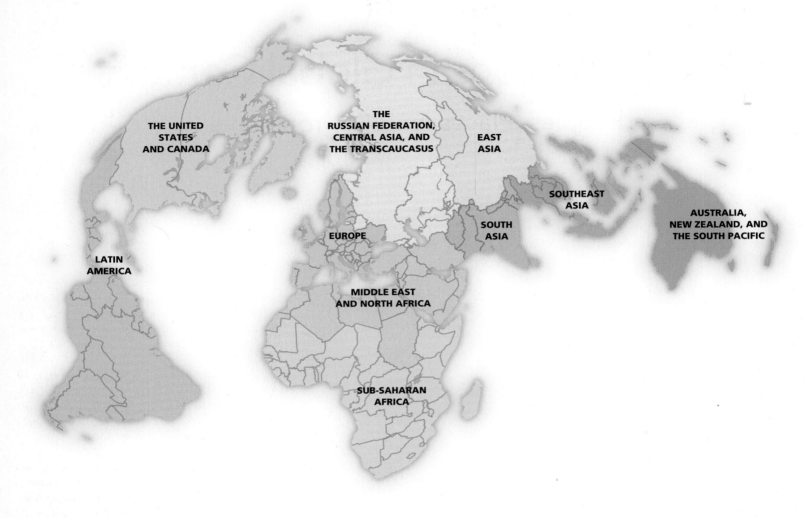

## WORLD REGIONS KEY FACTS

At the beginning of each chapter, readers are provided with a number of "Key Facts" about the region. Here, we provide world regional averages for all those Key Facts that are in numeric form. Readers can use these averages to compare the Key Facts in the region with the world regional averages.

- Population (2010): 6.83 billion

- Urbanization (2010): 57.90%

- GDP PPP (2010) Current International Dollars: 16,313

- Debt Service (total, % of GDP) (2005): 5.17

- Population in Poverty (%, < $2/day) (1990–2005): 33.9

- Gender-Related Development Index Value (2005): 0.76

- Average Life Expectancy, Overall (2008): 69.8

- Average Life Expectancy W (2008): 72.5

- Average Life Expectancy M (2008): 67.5

- Average Fertility Rate (2008): 2.68

- Population with Access to Internet (%, 2008): 33.7

- $CO_2$ Emission Per Capita, tonnes (2004): 7.14

- Access to Improved Drinking-Water Sources (%, 2006): 85

- Access to Improved Sanitation (% 2006): 73

# 1

# WORLD REGIONS
# IN GLOBAL CONTEXT

On the cover of this textbook is the image of the Crescent Moon Lake oasis, a place of apparent isolation in the heart of the Gobi Desert in China. As a possible stopping point along the old Silk Roads, however, this place has long been connected to the global flows of exchange and traffic that brought China into a relationship with India in South Asia, Iran (and the Persians) in the Middle East, and the Greeks and Romans in Europe. Today, this oasis and other spots like it are connected to the global flows of tourists, who seek to trek the old silk roads and experience the traffic in people and things that have been essential to the movement of history in this region.

It is not unusual to find other outposts like this one throughout the world, as people continue to migrate, trade, live, and survive in a variety of different places. A closer look at this cover image reveals that this oasis remains a vibrant space of social contact and interaction. People are seen moving from the oasis back into the desert, probably having stopped to take advantage of the clear blue groundwater that percolates up from underneath the surface of this vast, dry space, a geologic anomaly that makes this place so unique. The lush green spaces also suggest that this oasis is cultivated and well kept, a product of human intervention in this harsh landscape.

Although the Crescent Moon Lake oasis might seem a site of isolation, representing a regionally unique human adaptation, it reflects a much larger and more complex relationship between this place and others; indeed, its very origins and its ongoing existence are instead a product of *connections*. Where are the people going who are walking away from this oasis? How did they get there in the first place? What do people do in that structure located at the center of the oasis? Who tends those grounds and why? How did that water get there? It is these sorts of questions this textbook encourages. Our approach promotes asking questions about how places such as this oasis are both globally connected and locally situated—a product of their relationships to the wider world around them and a result of a unique set of sociocultural and political–economic processes that make them unique. In examining such places, this text critically analyzes today's world, asking and explaining how today's regional geographies have been shaped by longer-term relationships and wider-ranging connections.

It is in this spirit of critical inquiry that we ask readers to explore the images, maps, and text found throughout this book. More than this, though, we also ask readers to consider how we, as global citizens, have historically been connected to these broader global flows. Our goal is for readers to understand the historical and contemporary connections that distinguish our world's physical and built environments as well as its culture and politics. In short, in this book, we want to ask the obvious and not-so-obvious questions about why places like this oasis are where they are and what they mean to us both historically and today. To do this, we have to begin by developing a common conceptual framework for asking these questions. It is to this framework that we now turn.■

## MAIN POINTS

■ A global geography framework builds from the assumption that connections between and within world regions are part of a larger set of historical relations, which tie the world together environmentally, economically, politically, socially, and culturally.

■ The unique environmental characteristics of regions are formed through the connections that global climate and tectonic systems produce between regions. Human beings adapt to regional environmental conditions while remaking nature.

■ A historical geography of global relations suggests that economic connections and demographic patterns are best understood broadly by contextualizing those connections and patterns in comparative context.

■ Culture is a shared set of meanings manifest in both material and symbolic ways that connect people to places and to other people. Culture is dynamic and can be altered in both dramatic and subtle ways.

■ The state system—with some variations on it in the form of international and transnational organizations—coupled with capitalism, constitutes the foundation for contemporary political, cultural, and economic life around the globe.

■ Regional geography is an effective conceptual framework for studying the global processes tied to environmental, political–economic, and sociocultural change over time. It provides a lens for examining wider global processes in regional context, highlighting both similarities and differences between and within world regions.

# A GLOBAL GEOGRAPHY FRAMEWORK

The world has always been global. As a scientific discipline, geography has long understood this simple but important fact. For example, geographers study the long-term climate changes that affect our planet's temperature ranges and precipitation rates over time. And they examine the global exchange of peoples, practices, and ideas in a historical context, demonstrating how these movements have constructed a world defined by sociocultural and political–economic interaction. Geographers are also concerned with the connections that exist between places across the world and how those connections help redistribute energy, resources, and people, as well as flora and fauna. As a subject of scientific observation and study, geography has made important contributions to understanding a world that is more complex and fast changing than ever before. With basic foundational knowledge in geography, it is possible not only to appreciate the diversity and variety of the world's people and places, but also to be aware of their relationships to one another and to be able to make positive contributions to regional, national, and global development.

In this text, then, a *global geography* is about understanding the variety and distinctiveness of world regions without losing sight of the interdependence among them. Geographers learn about the world by finding out where things are and why they are there and by analyzing the spatial patterns and distributions that underpin regional differentiation and change. This chapter will introduce the basic tools and fundamental concepts that enable geographers to study the world in this way and describe the conceptual framework that informs each of the subsequent chapters. The power of geography comes from its integrative approach, which addresses global connections, historical trends, and systemic political–economic and sociocultural relations by drawing on an intellectual tradition in both the natural and social sciences. The integrative approach employed in this text stresses the following:

- *Global Interdependence.*—This reflects a key issue for geographers: documenting and respecting the individuality and uniqueness of world regions while explaining these global interdependencies. We want to understand broad geographic patterns without losing sight of the individuality and uniqueness of specific world regions and places within them. World regions are interdependent, and it is essential to know how places in and across world regions affect, and are affected by, one another.

- *Ongoing Change.*—World regions are the current configuration of an ongoing process of change. The sequence of change has not been the same everywhere. As political-economic and sociocultural connections within and between world regions change over time, the human character of places within those regions also changes. In the same way, the changing physical environments of the world are outcomes of interactions among complex environmental systems, as well as alterations and transformations made by human beings. Understanding the limits and resources that the natural world furnishes is just as important as understanding the strengths and weaknesses of the social worlds that organize the places in which humans live.

- *Systemic Relations.*—Changes and their consequences for different world regions are best understood by thinking in terms of the world as a changing, competitive, political–economic **world system.** World regions take on roles organized within larger arrangements of exchange and control. The parts of this system include local places where resources are harvested or where industries are located, and where the production or consumption of things occurs. By thinking systemically, geographers gain a better grasp of how and why world regions come to be.

The three pillars of this text's global geography are underpinned by the fields of both physical and human geography. The study of geography, in general, involves the study of Earth as created by natural forces and as modified by human action—an enormous amount of subject matter. *Physical geography* addresses Earth's natural processes and their outcomes. It is concerned with climate, weather patterns, landforms, soil formation, and plant and animal ecology. *Human geography* deals with the spatial organization of human activity and with people's relationships with their environments. This focus necessarily involves looking at physical environments insofar as they influence, and are influenced by, human activity. The result is that the study of human geography must cover a wide variety of phenomena. These phenomena include, for example, agricultural production and food security, population change, the ecology of human diseases, resource management, environmental pollution, regional planning, and the symbolism of places and landscapes. Most important, geographers are interested in studying physical geography and human geography conjointly through an examination of *environment–society relationships*.

The power of geography lies in the ability to explain and appreciate complex global geographic processes and their place-based effects. This text demonstrates that power, by advancing an argument about how global connections become the source of regional differences, as well as how existing differences help to fuel new global connections. In short, *the global processes that bind the world together also make places different*. Although this might seem counterintuitive, the reality is that the geographic processes at work in world regions serve to modify global climatic patterns on the one hand and global economic patterns on the other. Global warming does not affect all places in the same way, for example, whereas capitalism is practiced quite differently around the globe. The world can only be understood through its diverse places; to examine the world through a global framework, regions become the key. In short, world regions are increasingly important for organizing and understanding the world and we want to understand how and why this has become the case. This chapter builds toward a conceptual framework that can explain how world regions are *both* globally interconnected *and* locally differentiated (**Figure 1.1**). To do this, the text is organized around three broad interacting themes: *environment and society; history, economy, and demographic change;* and *culture and politics*. On the one hand, every region can be uniquely described using these elements, and their differences can be defined in terms of each. On the other hand, however, these regional differences, when examined more carefully, can themselves be explained by the connections between regions.

# ENVIRONMENT AND SOCIETY

Consider the case of global climate. Why is it persistently dry in the city of Phoenix, Arizona, in the United States southwest, but typically very wet in the city of Manaus in the Amazon jungle? What makes these two places so different? There are many components to Earth's climate system that explain this fact, including the shape of the North and South

American continents and the distance of these cities from the sea. The key explanation to this climatic difference, however, is that the two cities are *connected.* Air rising over the equator dumps its moisture over the Brazilian Amazon throughout the year. This same air spreads outward to the north and eventually descends over the deserts of the U.S. southwest, where it creates a band of pressure that locks out rainfall for much of the year. Phoenix is dry *because* Manaus is wet. The climate system makes both places globally interconnected and yet locally differentiated.

The physical and biological characteristics of these places, which include rainfall, temperature, vegetation, wildlife, and landforms, are also critical to the formation and transformation of these regions because they provide constraints on, and opportunities for, human activities. As people have settled these regions, they have acted to transform the conditions around them, building massive cities in the deserts of the United States and the jungle frontiers of Brazil. These regions are all influenced by the physical systems that help create them, even while people fully transform the environments in which they live.

For this reason, geographers view physical and environmental conditions as highly dynamic and best understood if one thinks of Earth as a system in which humans play an important role. **Earth system science** is an integrated approach to global patterns of geology, climate, and ecosystems and how they have changed over time and space, producing a physical geography that is dynamically shaped by both natural forces and human actions. Physical geographers work with other Earth scientists to understand the functioning of the Earth system and with human geographers to interpret the interactions between the Earth system and social, cultural, economic, and political circumstances. The fundamental Earth system processes that shape world regions are plate tectonics, atmospheric circulation, and ecosystem functioning. Plate tectonics explains the formation of continents and mountain ranges; atmospheric circulation shapes the pattern of world climates; and ecosystem characteristics affect the geography of vegetation, the cycling of key minerals, and the location of animals and other key organisms.

## Climate, Adaptation, and Global Change

**Atmospheric Circulation** Earth scientists have combined an understanding of basic physics with information about global patterns of temperature and precipitation to provide explanations of atmospheric circulation, the global movement of air that transports heat and moisture and explains the climates of different regions. Whereas **weather** is the instantaneous or immediate state of the atmosphere (for example, a rainy or freezing day) at a particular time and place, **climate** is the typical conditions of the weather expected at a place often measured by long-term averages of temperature and precipitation and at different seasons (for example, a place with wet, cool winters and hot, dry summers).

A simple model of atmospheric circulation is based on variations in the input of energy from the Sun and the configuration of the major

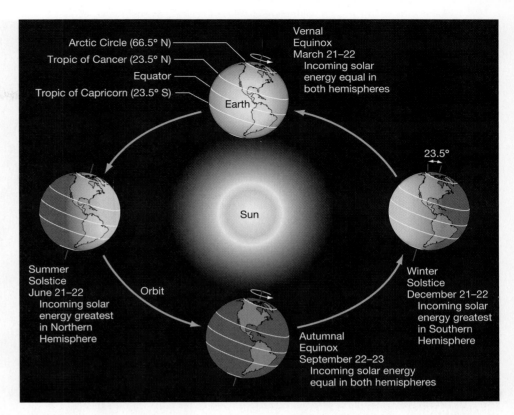

▲ **FIGURE 1.2 Seasonal incidence of Sun's rays by latitude** As Earth moves around the Sun, the angle at which sunlight hits Earth varies according to the seasons. The Sun's rays hit Earth most directly and focus the greatest solar energy and heat at the equator in March and September, at the latitude of the Tropic of Cancer (23.5°N) in the Northern Hemisphere in June, and at the latitude of the Tropic of Capricorn (23.5°S) in the Southern Hemisphere in December.

continents and mountain ranges. The spherical shape of Earth, the tilt of its axis, and its revolution around the Sun mean that the sunlight does not hit all parts of Earth's surface at the same angle (**Figure 1.2**). As Earth moves around the Sun, the angle at which sunlight hits Earth varies according to the seasons. The Sun's rays hit Earth most directly and focus the greatest solar energy and heat at the equator in March and October, at the latitude of the Tropic of Cancer (23.5°N) in the Northern Hemisphere in June, and at the latitude of the Tropic of Capricorn (23.5°S) in the Southern Hemisphere in December.

The constant high input of solar radiation at the equator produces warm temperatures throughout the year, and this warmer air has a tendency to rise into the atmosphere, creating low pressure at ground level, and cooling and condensing into clouds that eventually generate heavy rainfall. As in the case of Brazil described earlier, the ongoing heat associated with the region's location near the equator causes heavy rainfall. This process is called *convectional precipitation* and is typical of the equatorial climate with high temperatures and rainfall year-round.

The cooler air rises high into the atmosphere, moves out from the equator toward the poles, and eventually sinks over tropical latitudes (about 30° north and south latitude), creating a zone of high pressure (**Figure 1.3**). As the air moves toward the surface, it becomes warmer and drier, holding so little moisture by the time it reaches ground level that these regions are characterized by the very low rainfall, sparse vegetation, and warm, dry conditions of desert climates. As in the case of Phoenix described earlier, the sinking air creates a belt of pressure, locking out precipitation throughout much of the year.

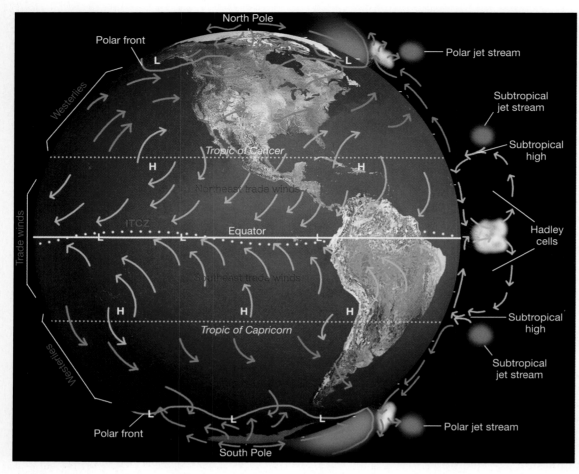

▲ **FIGURE 1.3 Atmospheric circulation** The general circulation of the atmosphere is based on air moving from regions of high to low pressure and on the vertical lift of air in the regions of highest heating over the equator at the intertropical convergence zone.

and when moisture-laden air encounters a landmass, especially coastal mountains, it condenses into rainfall or snow. Precipitation associated with mountains, called *orographic precipitation,* may result in the formation of a dry rainshadow region on the inland, or lee, side of the mountains, where sinking air that has lost its moisture becomes even drier. The trade winds flow across the oceans in tropical latitudes and frequently produce rain on east-facing coasts in what is sometimes called the *trade wind climates* (**Figure 1.4**).

Similarly, the westerly winds bring rain as they blow from the oceans onto western coasts. The regions on the margins of the trade winds, and on the margins of the equatorial rainfall zone, have highly seasonal climates with a distinct rainy season. Seasonal shifts in pressure and wind belts mean that the westerlies move nearer the equator in December and to the poles in June, resulting in distinct wet and dry seasons on the margins of the westerly circulation. When the global circulation shifts southward in December, storms spinning out of the Northern Hemisphere westerlies bring rain to the poleward margins of drier regions in the Northern Hemisphere.

This simple model of atmospheric circulation produces a pattern of global climate with warmer and wetter regions nearer the equator. East coasts in the trop-

When the sinking air reaches ground level, it diverges, and some of the air flows back toward the equator, where it converges with the heated air and rises again. This vertical circulation of air from the equator to the tropics back to the equator is called a *Hadley cell.* The **intertropical convergence zone (ITCZ)** is the region where air flows together and rises vertically as a result of intense solar heating at the equator, often with heavy rainfall. It shifts north and south with the seasons (see **Figure 1.3**).

The rotation of Earth tends to drag air flowing back from the tropical latitudes to the equator into a more east-to-west flow, creating a major wind belt called the *trade winds* that blow from east to west between the dry tropics and the equator. These are often called *easterly winds* or "easterlies," as they come *from* the east. Some of the air descending over the tropics also flows poleward and is pulled by Earth's spin into a major west-to-east flow called the *westerly winds,* as these winds flow *from* the west.

The seasonal variation associated with the tilt of Earth's axis and the changing orientation of the Northern and Southern Hemispheres toward the Sun (the Northern Hemisphere facing the Sun more directly in June than in December) means that the zones of rising and sinking air, and the major wind belts, move northward in June and southward in December, with corresponding shifts in the zones of rainfall and dry conditions. When winds blow across warmer oceans, they tend to pick up moisture,

ics and west coasts in temperate regions also tend to be wetter, as do the complex regions where warmer air flowing toward the poles meets colder air flowing toward the equator. This general pattern helps explain the climates and the associated interaction with human activity of each of the major world regions (**Figure 1.5**).

**Climate Zones**   The resulting zonation of precipitation and temperature creates a mosaic of climate regions across Earth. *Tropical climates* nearest the equator are characterized by heavy rainfall and persistent warm temperatures. These include those areas that have no dry season and are wet year-round. Also here in the wet tropics are areas that have a significantly uneven seasonal rainfall pattern. In *monsoonal climates,* much of the rain may come in a few months, with long periods with little or no rain (see Chapters 9 and 10 for examples from South and Southeast Asia).

Moving north and south of these tropical wet areas also moves toward the high pressure *arid and semiarid climates,* which typically occur between 28 and 30 degrees north and south of the equator. These dry zones include deep deserts but also open dry grassland. Arid climates may also occur in other areas and for other reasons, as in the case of the dry regions of western South America, for example, where cold water currents significantly affect regional climate. Similarly, regions in the rain shadow of mountains may be arid or semiarid, due to an oro-

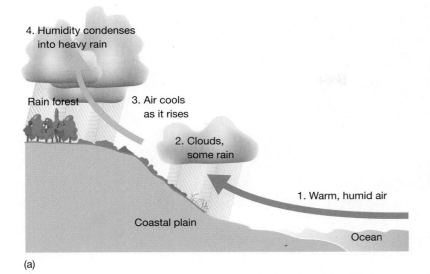

▲ **FIGURE 1.4 Orographic rainfall on a trade wind coast** (a) Winds blowing across the oceans pick up moisture. When they rise over coastal mountains, cooling causes the moisture to condense and fall as rainfall or snow. Orographic rainfall is common where winds, such as the trades, cross warm oceans, such as the Caribbean, and rise over mountains near the coasts of countries such as Costa Rica. Heavy orographic rainfall also occurs where the westerlies rise over coastal mountains, such as the Andes in Chile or where high mountains emerge from generally drier regions such as in East Africa and the western United States. (b) Mount Waialeale, on the Hawaiian Island of Kauai. One of the Earth's rainiest points, there are few other places on the globe that receive more rain than Waialeale. The mountain is frequently enshrouded in rain-dumping clouds, produced by the orographic effect of rising wet air along its slopes.

graphic effect (described earlier), as where the dry regions of Washington state in North America lie east of the Cascade Mountain Range. Between arid zones and wet tropics come *tropical savannas*, complex transition zones that span the areas between the desert edges up to the wet forests. The African Sahel (see Chapter 5, p. 156) is typical in this regard and describes a belt of semiarid savanna south of the Sahara desert and north of the deep wet tropical regions of Central Africa.

Further north and south beyond the tropical zone, climates are greatly affected by the side of the side of the continent on which they occur. *Mediterranean climates* typically occur on the western sides of continents between 30° and 45° (either north or south). These climates are warm, due to their proximity to the equator, but they have more moderate temperatures due to their proximity to the sea. They also tend to receive winter rains. *Humid subtropical climates*, on the other hand, tend to fall on the southeast side of continents in a similar range of latitudes. Because of their place on the receiving side of predominant easterlies or trade winds, however, they are far more wet and humid throughout the year than Mediterranean climates.

Further north and south still come more temperate climates. *Marine coast climates*, which occur between approximately 45° and 50° north and south, tend to be persistently wet and somewhat colder. Owing to westerly winds and oceanic currents, these are usually found on the west coasts of continents, just poleward from Mediterranean climates, although in Australia this climate occurs in the southeast. On the temperate interiors of the North American and Eurasian continents are *continental climates*, which can be somewhat drier and often have long cold winters.

In the extreme north and south are *polar climates*, which are dominated by the cold created by their distance from the equator. In rough mountainous areas, including the Rocky Mountains, the Himalayas, and the Andes, among others, the topography is too complex and variegated to be dominated by a single climate. Microclimates of the mountaintops in such places tend to be far different than the nearby valleys. The shorthand climate, *Highland*, is used to designate these important, but unpredictable, zones.

**Climate Change**  Although Earth system scientists believe that the general circulation of the atmosphere tends to remain the same over centuries, considerable evidence suggests that global and regional climates have varied over time. Landscapes show evidence of wetter or drier conditions in the past through remains of animals, vegetation, landforms, and lakes associated with a different climate. Human history records periods of hotter and cooler climates and their impact on harvests and migrations. Most dramatically, the landscapes of many regions show the marks of ice cover, erosion, and deposition from periods when it was so cold that rivers of ice (called *glaciers*) or even larger masses (*ice caps*) covered much of the world. Earth system scientists believe that the ice ages were caused by slight changes in the tilt of Earth's axis and its orbit around the Sun and associated changes in the amount of solar radiation reaching Earth (**Figure 1.6**).

Global and regional climates have also been affected by tectonic activity, especially volcanic eruptions. The most explosive volcanic eruptions can blast ash and gases high into the atmosphere, blocking sunlight and changing the chemical composition of the air. For example, the explosion of the Krakatoa volcano near Java in 1883 caused a worldwide drop in temperatures and harvest failures in Europe during the "year without a summer."

Human activity can also change the climate through altering the composition of the atmosphere. Of greatest concern is the role that human activity is playing in causing **global warming**, an increase in world temperatures and change in climate associated with increasing levels of carbon dioxide and other trace gases resulting from human activities such as deforestation and fossil-fuel burning. Global warming is associated with the **greenhouse effect**, the trapping of heat within the atmosphere by water

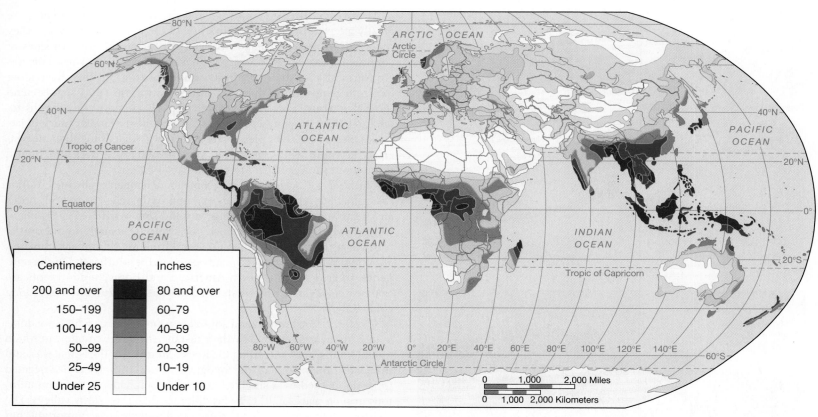

▲ **FIGURE 1.5  Major climate zones, including temperature and rainfall**  The world can be divided into very general climate zones based on temperature and on the total amount and seasonality of precipitation. (a) The global pattern of average annual rainfall indicates regions of low and high precipitation (averages are usually based on 30-year climate "normals"; in this case, 1961–90). (b) The climate system of the world can be generalized into ten major types based on temperature and precipitation. This map is based on a simplification of the climate classification of German climatologist Wladimir Köppen.

vapor and gases, such as carbon dioxide, resulting in the warming of the atmosphere and surface. Carbon dioxide ($CO_2$) is produced from burning fossil fuels, from cement production, and from deforestation, and the level of carbon dioxide in the atmosphere has increased dramatically with increases in human population and consumption over the last 100 years.

Under current science, the measure of a regional or national $CO_2$ emissions, therefore, is a strong (albeit imperfect) relative indicator of the degree to which that region or country impacts or accelerates global warming. Measured per capita (per each individual in a country or region), that number gives a sense of how great an impact the average person in that region or nation has on global warming. While the average person in the United States requires and emits around 19 metric tons per person every year, the average person in Switzerland emits less than 6 annually, for example.

Earth system scientists suggest that human-caused increases in carbon dioxide may result in an overall global temperature increase of about 3°C (5.5°F) over the next 50 years, as well as regional changes in the amount and distribution of precipitation. Projections vary, but worst-case scenarios are sobering (**Figure 1.7**). Evidence is accumulating that such changes are already occurring and include a warming of ground, ocean, and near-surface air temperatures; a decline in sea ice and snow cover in the Northern Hemisphere; a retreat of mountain glaciers; and an increase in precipitation in high latitudes.

## Geological Resources, Risk, and Water Management

Like the global climate system, the geologic system of the Earth helps to produce regions over long periods by uplifting mountains, forming water drainages and river systems, and presenting diverse resources and hazards around the world. At a global scale, the main driver of this system is plate tectonics.

**Plate Tectonics**   The theory of **plate tectonics** explains how Earth's crust is divided into large solid plates that move relative to each other and <u>cause mountain-building, volcanic, and earthquake activity when they separate or meet</u>. This layer, which is from 50 to 100 kilometers (31 to 62 miles) thick, is composed of about a dozen large "plates" of solid rocks floating on a layer of molten material. These plates move very slowly over the more fluid deeper layer and interact at their boundaries, where the resulting tensions are responsible for most of the world's volcanic, earthquake, and mountain-building activity (**Figure 1.8**). The

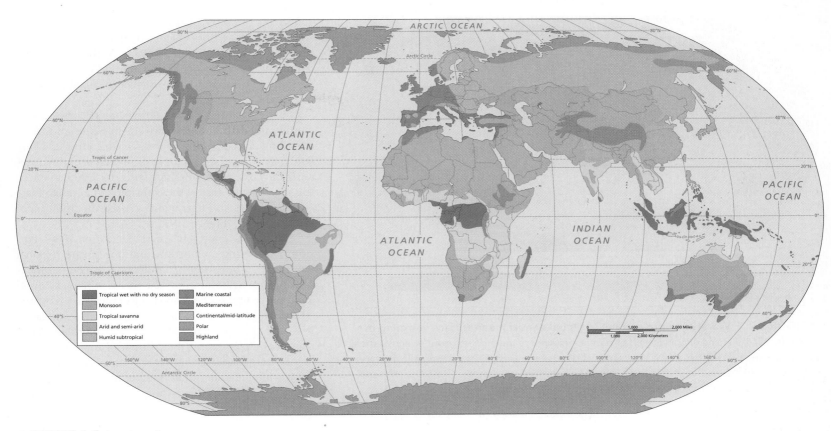

▲ **FIGURE 1.5** (continued)

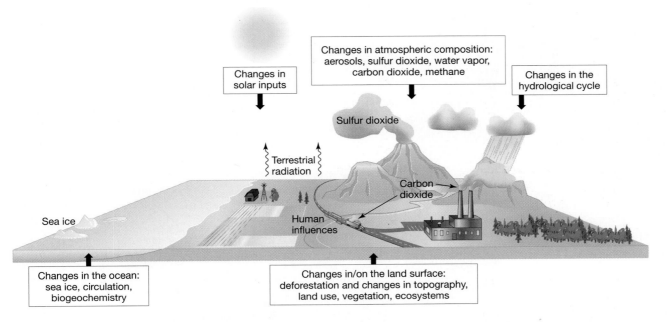

▲ **FIGURE 1.6 Major causes of climate changes** This diagram, based on the work of the Intergovernmental Panel on Climate Change, shows the major causes of climate change, including volcanic activity, changes in land use, and alterations in the composition of the atmosphere, such as increases in moisture, greenhouse gases including carbon dioxide, and pollutants such as sulfur dioxide. For example, scientists have shown that sulfur dioxide and dust from major volcanic eruptions and industrial pollution are likely to cool the climate and that deforestation and increased carbon dioxide from fossil-fuel burning are likely to warm the climate.

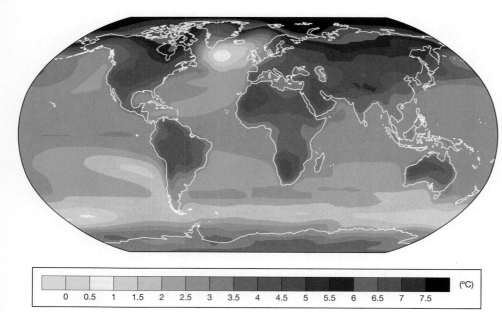

continents sit on the plates and emerge from the oceans where Earth's crust is thicker or where the rocks are lighter and therefore more buoyant. At the core of each continent is a region of old (500 million years or older) crystalline rocks called a *continental shield*. Shield areas often contain many minerals.

Plate tectonics builds on earlier theories about **continental drift**, the slow movement of the continents over long periods of time across Earth's surface, where land masses have split apart and collided over time, creating the current morphology of Earth. Emblematic in this regard is the case of the collision of the Indian Plate with the Eurasian Plate, which caused the uplifting of the Himalayas (**Figure 1.9**).

Within the continents, more local processes—erosion, weathering, and sedimentation—create regional landforms. The study of landforms in geography is called **geomorphology**. *Erosion* occurs when water and wind move across the land surface, picking up material and transporting it to other locations. In some cases, heat and the characteristics of water or rocks cause chemical changes and breakdown of material in the process called *weathering*. Erosion has affected many of the world's great mountain ranges, moving material to lower regions and depositing it in a process called

▲ **FIGURE 1.7 Worst-case projections of global annual mean surface warming in 2080 (in °C)** Assuming ongoing levels of current emissions of greenhouse gases, scientists predict sustained warming, with this model among the worst-case scenarios. Here, temperature increases between 4° and 8°C would result in the melting of glaciers and ice caps, the inundation of island nations, and the loss of large swaths of Earth's agricultural potential.

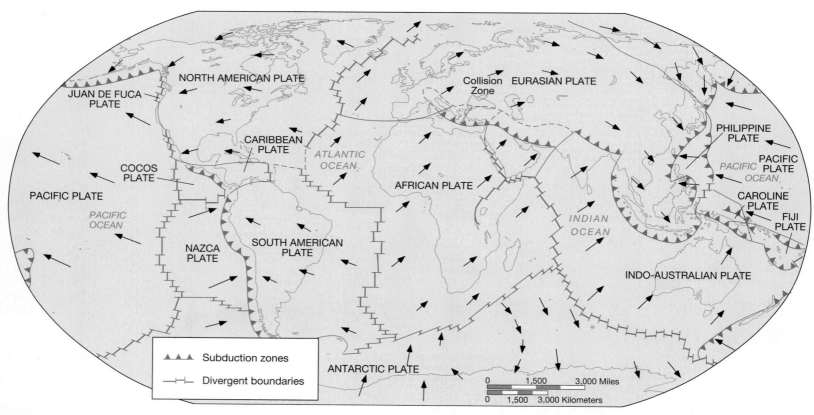

▲ **FIGURE 1.8 Major tectonic plates** Earth's crust is broken into a dozen or so rigid slabs or tectonic plates that are moving relative to one another. Each arrow represents 20 million years of movement and the direction of movement. Longer arrows indicate that the Pacific and Nazca plates are moving more rapidly than others. The long jagged boundaries indicate areas where plates are moving away from each other. The triangles show subduction zones, where one plate sinks underneath another.

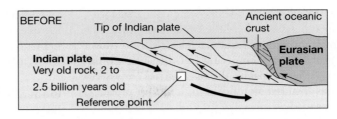

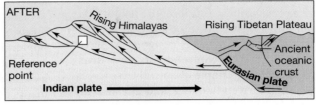

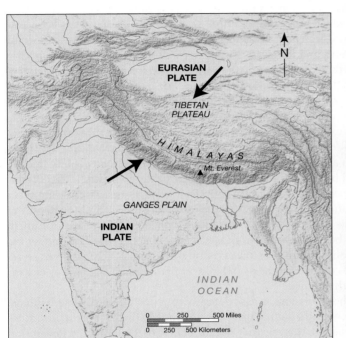

▲ **FIGURE 1.9  Plate tectonics and the creation of the Himalayas** (a) The collision between the Indian and Eurasian plates has pushed up the Himalayas and the Tibetan Plateau. (b) These cross sections show the orientation of the two plates before and after their collision. The reference points (small squares) show the amount of uplift of an imaginary point in Earth's crust during this mountain-building process. (c) The Himalayas are among the highest mountains in the world, resulting from very active mountain building in the region.

*sedimentation.* Extensive areas of deposited sediment occur in the large river basins, such as the Amazon, and across some of the vast plains, such as the North American prairies. Erosion and weathering have been critical in the development of better soils.

One of the central economic and geopolitical implications of these complex processes is the uneven distribution of Earth's resources. Of particular importance throughout the last 200 years is the availability of hydrocarbon deposits, including oil, natural gas, and coal, all of which have fueled industry since the 18th century. These deposits are the sedimented and crushed geological remains of prehistoric life, including the vast fern forests of the carboniferous period more than 300 million years ago. Due to the complexity and dynamism of plate activity and geomorphology, their availability is remarkably uneven. Though both the country of Jordan and the country of Iraq are located near the northern boundary of the shifting Arabian Plate, for example, Iraq is rich with oil, whereas Jordan has none (see Chapter 4, p. 116).

**Earthquakes and Volcanic Hazards**   Tectonic boundaries are also critical in determining the distribution of major hazards, especially including earthquakes and volcanic activity. Areas of plate convergence, as where

the Pacific Plate moves westward to collide with the several plates in the Western Pacific, give rise to massive volcanoes that have created the island chains of the region, including the islands of Japan. Here, tectonics have resulted in architectural innovation to sustain seismic earthquake shocks and evacuation planning to cope with sudden volcanic activity. In the complex undersea region where the Indian Plate converges on the Burma and Sunda Plates, a 2004 earthquake triggered a massive underwater pulse of energy, which came onshore in countries throughout South and Southeast Asia as a **tsunami**, an event that killed hundreds of thousands of people and left millions displaced (see Chapter 10, p. 337).

**River Formation and Water Management**   The uplift and folding of Earth's surface interacts with regional climate and rainfall to create the great river basins of the world. The formation of highlands and mountains through tectonic uplift provides a tilted surface over which rain falls and across which water flows to form river basins. These rivers carry fertile soils for agriculture and opportunities for traffic and trade. On the other hand, such rivers must be managed and maintained through straightening, damming, and drainage to carry barge traffic and protect areas of commerce along their length.

The Yangtze River provides a key example. Flowing out of the massive uplifted regions of the Himalayas, where the Indian and Eurasian Plates collide, the river flows with water from glacial melt from these high areas and is fed by rainfall before emptying into the East China Sea. Owing to its enormous size and capacity, the river carries a massive amount of sediment that has fostered significant agricultural productivity along its length and has made it an artery for global trade in grain and goods. These merits, however, cause problems, as the river is prone to catastrophic flooding and is not navigable along its entire length. The enormous **Three Gorges Dam** project (see Chapter 8, p. 277) is a feat of engineering by the government of China and is designed to maximize the benefits of the river's power while controlling its flow. River management is therefore a paradigmatic example of how diverse forces (tectonic uplift and erosion, rainfall driven flooding, and global commerce) act on one another and the way global circulations (of soil, water, and money) converge in locations to create regional difference. Humans occupy, and adapt to, differing conditions, like the complex flood hydrology of the Yangtze, while transforming and manipulating these landscapes into new forms.

## Ecology, Land, and Environmental Management

**Ecosystems**    The ecology of Earth's surface, like its hydrology, represents regionally diverse conditions and outcomes based on the interaction of climate, plate tectonics and geomorphology, and human activity. These together are also the major influences on the global distribution of **ecosystems**—the complexes of living organisms and their environments in particular places. For example, U.S. and Mexican ecosystems include the Sonoran Desert ecosystem, with its characteristic species such as the saguaro cactus adapted to dry conditions, and the prairie ecosystem of tall grasses once grazed by bison. When ecosystems are grouped into larger classifications, defined by major vegetation type such as desert, forest, or grassland, they are known as **biomes**, which are the largest geographic biotic units.

Warmer temperatures and higher rainfall are generally associated with more abundant vegetation. For example, the high rainfall and warm temperatures of tropical climate zones are associated with the lush vegetation of the tropical rainforest ecosystems or tropical forest biome. The hot, dry conditions of the arid climates are associated with sparser shrubs and drought-adapted plants and animals of the desert ecosystem. Colder deserts with slightly more rainfall produce the short grasses that are typical of the *steppe* or prairie ecosystems. The cold, drier conditions of regions nearer the poles correspond to a landscape of frost-resistant mosses, shrubs, and grasses called *tundra*. Just to the equator-ward side of the tundra, as temperatures and rainfall increase slightly, are the evergreen conifer boreal forest ecosystems (sometimes called the *taiga*). An ecosystem of scrub and dry forest is found in Mediterranean climate regions where the seasonal shift of the westerly winds brings rainfall only in the winter season. In marine coast climates, wet coastal rains are associated with towering conifer forest ecosystems. Similarly, the trade winds of humid subtropical climates produce rainforests on the southeast coasts of many continents. **Figure 1.10** shows the global distribution of ecosystems.

Geographers are also interested in the spatial distribution of plants, animals, and ecosystems, known as **biogeography**. They work with other scientists to examine patterns and to understand the relative roles of climate

and other factors in determining these patterns. Some of the most interesting questions in studying ecosystems and biogeography are those related to **biodiversity**—the differences in the types and numbers of species in different regions of the world.

**Innovating and Degrading Ecosystems**    This diversity has been a boon for human beings for more than 10,000 years, at least since the dawn of agriculture, when people began to manipulate the plants and animals around them to produce new environments and support new forms of social organization. More recently, this diversity has become a pool from which people have been able to extract genetic resources for better health and more productive agriculture using new biotechnologies. In other words, people have not only drawn on and adopted elements of the ecosystem around them, but have also innovated whole new ecosystems (like crop fields) and new life forms (like **transgenic crops**, which resist pests).

This extensive manipulation of Earth and its ongoing transformation with the extension of human settlement and agriculture, however, has led to a series of critical problems. First, there is an evident decline in overall biodiversity across the globe. Although possibly tens of millions of species exist on Earth, countless numbers are disappearing every day. It is impossible to estimate the exact rate of loss and the degree to which this decline is exclusively anthropogenic (the result of human activity), but of the number of species that have been evaluated by the International Union for Conservation of Nature (IUCN), 23% of vertebrates, 5% of invertebrates, and 70% of plants are designated as endangered or threatened.

Much of this loss is associated with a decline in diverse habitats, where the world's plants, insects, and animals thrive. Forest cover loss is emblematic in this regard. Over the course of the Industrial Revolution in Europe, most notably, forest cover declined from 230 million hectares in 1700 to 212 million hectares by 1980. **Deforestation** is ongoing elsewhere in the world (**Table 1.1**). Although there has been some modest return of forests to Europe, many world regions are in a state of critical land cover change and habitat decline.

**TABLE 1.1   Change in Cover of Forest and Wooded Land by World Region, Adapted Using Data from the United Nations Food and Agriculture Organization's 2005 Global Forest Resources Assessment**

|  | 1995–2005 (in '000 of Hectares) | Change (as % of 1995) |
|---|---|---|
| **Africa** | −63,949 | −9.14% |
| **Asia** | −2,910 | −0.51% |
| **Europe** | +12,074 | 1.22% |
| **North and Central America** | −4,941 | −0.70% |
| **Oceania** | −6,260 | −2.95% |
| **South America** | −59,278 | −6.65% |
| **Total World** | −125,264 | −3.07% |

*Source:* http://www.fao.org/forestry/32033/en/

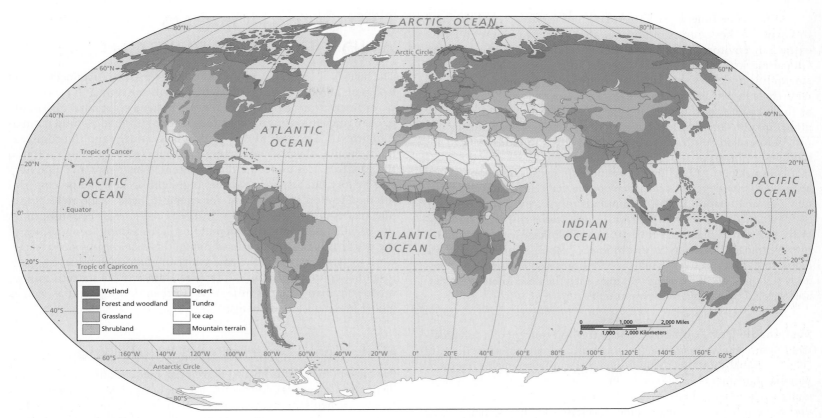

▲ **FIGURE 1.10 World ecosystems** This map shows the most general classification of global ecosystems into major biomes. Forests and woodlands are tree dominated communities, generally in areas of high precipitation. Shrublands of short woody plants and bushes are found in slightly drier regions, and grasslands occur where precipitation is even less plentiful. Deserts are found in the driest regions and include bare ground and a variety of scattered grasses, succulent plants, and bushes. Tundra is found in cooler regions and can include a variety of shorter plants, including grasses. Wetland vegetation is found in wet and flooded areas and includes mangroves and marshlands.

Ecosystem transformations extend beyond general deforestation, as seen in the chapters that follow, and include the destruction of tropical rainforests and the consequent loss of biodiversity; widespread, health-threatening pollution; the degradation of soil, water, and marine resources essential to food production; stratospheric ozone depletion; acid rain, and so on (even including the catastrophic destruction of a whole sea—see Chapter 3). Most of these threats are greatest in the world's periphery, where daily environmental pollution and degradation represent a catastrophe that will continue to unfold, in slow motion, in the coming years. Future trends will only intensify the contrasts between rich and poor regions.

**Commodifying Nature** One of the central drivers in this transformation has been the dramatic change in environmental use and management that has expanded in the period of economic transformation between the Industrial Revolution and the present. Essential to this economic transition has been the expansion of the idea that nature is a commodity to be exploited and produced. A **commodity** is anything that has a *use value* or that has some usefulness to someone. Sheep, rivers, and trees were all referred to as commodities because they could be used to sustain and shelter life. With the rise of capitalism, the popular understanding of this use value was replaced by that of *exchange value*, and a commodity came to mean anything useful that could be

bought or sold. Production for sale in the marketplace is what makes commodities out of things—whether foodstuffs, minerals, animals, even human beings. Capitalism has made it possible for everything and anything to be a commodity under certain conditions.

This commoditization of nature has had many effects, but chief among them is the increasing tendency for commodities to move globally through exchange, further connecting regions to one another; and the tendency for intensive extraction to lead to environmental damage and destruction. In terms of global exchange, the commoditization of nature that accompanied the rise of capitalism was feverishly pursued during the age of discovery as European explorers found new resources to exploit in new places (see **Historical Landscapes and Legacies** below). This period of increasing exchange linked regions to one another in new ways. Although wine, tobacco, sugar, oil, and diamonds had long been sources of exchange (see the history of those commodities in the following chapters), this period ushered in new methods of mining and cultivation, creating whole new regional landscapes of production, while opening new global markets and so inventing new regional cultures of consumption. This trend continues to the present as natural commodities increasingly tie places together through **commodity chains**, which are networks of labor and production processes that originate in the extraction or production of raw materials and end with the delivery or consumption of a finished commodity.

At the same time, the transformation of useful parts of the environment into key sources of profit has led in many cases to intensive extraction, environmental destruction, and exploitative labor systems. Global markets for diamonds over the 20th century led to massive stockpiling but also to brutal labor conditions for miners. The worldwide trade in petroleum has accelerated rampant speculation, as well as the incidence of oil spills and damage to the world's oceans. The global market for soybeans has encouraged widespread transformation of forests in South America to be converted to industrialized soy fields.

Following the global flow of commodities reveals how connections across and between regions causes environmental changes but also forges new regional relationships. As diamonds travel from South Africa to Saudi Arabia, petroleum flows from Saudi Arabia to Brazil, and soybeans move from Brazil to South Africa, each linkage leaves behind it a massive transformation to the environment (mining scars, oil spills, and deforestation, respectively), but these linkages also encourage increasingly tight global connections that may open the door to more sustainable ways of managing the environment.

**Sustainable Development**   Contesting the concept of a commoditized nature (or at least a *solely* commoditized nature) are a range of calls for global solutions to environmental problems, including international agreements to reduce pollution and protect species. A more benign relationship between nature and society has been proposed under the principle of **sustainable development**, a term that is now widely used but vaguely defined. One definition is that of the World Commission on Environment and Development, chaired by the former prime minister of Norway, Gro Brundtland, stating that sustainable development is "development that meets the needs of the present without compromising the ability of future generations to meet their own needs."[1] This definition incorporates the ethic of intergenerational equity, with its obligation to preserve resources and landscapes for future generations.

Geographers consider sustainable development to include the ecological, economic, and social goals of preventing environmental degradation while promoting economic growth and social equality. Sustainable development means that economic growth and change should occur only when the impacts on the environment are benign or manageable and the impacts (both costs and benefits) on society are fairly distributed across classes and regions. This means finding less-polluting technologies that use resources more efficiently and managing renewable resources (those that replenish themselves, such as water, fish, and forests) to ensure replacement and continued yield.

In practice, sustainable development policies of major international institutions, such as the World Bank, have promoted reforestation, energy efficiency and conservation, and birth control and poverty programs to reduce the environmental impact of rural populations. At the same time, however, the expansion and globalization of the world economy has resulted in increases in resource use and inequality that contradict many of the goals of sustainable development.

[1]World Commission on Environment and Development, *Our Common Future* (Brundtland Report) (New York: Oxford University Press, 1987), p. 43.

# HISTORY, ECONOMY, AND DEMOGRAPHIC CHANGE

The historical geography of the world is a story of interconnection and isolation, movement and settlement, growth and retreat. It is not a static or linear story, as places in the world have responded differently to changing global interconnections, economic changes, and population growth. The goal of this section is to understand the complexity of the global world, stretching our geographical imagination into the past to better understand the present. This is not a comprehensive history but a suggestive one, highlighting how the processes of interconnection and isolation, for example, have occurred in particular places at particular times. In the process, this section outlines the broader economic and population, for example geographies of the world as it exists today. In so doing, this section provides a framework for understanding the emergence of modern-day world regions in global context, placing within this framework a narrative about broader human patterns of rural and urban development as well as global integration and change.

## Historical Legacies and Landscapes

For most of the world's history, movement, not settlement, has been the distinguishing geographic feature for the human species. These long-term movements are an essential part of the human story (**Figure 1.11**). As people migrated from place to place, they developed and spread new knowledge systems—hunting-and-gathering techniques, language, religion, and later, animal and plant domestication—and created unique sets of cultural and social practices tied to their understandings and interpretations of the world around them. Some of the most important human societies historically have been migratory. Groups of animal herders—known as **pastoralists**—have played a significant role in world history, spreading knowledge systems as well as various cultural practices across vast spaces (**Figure 1.12**). Indo-Europeans in central Eurasia, for example, domesticated the horse and spread their diverse languages from Europe to South Asia. Hunting-and-gathering were often coupled with these movements, as people altered their environments to hunt—burning large swaths of land to target game animals. Massive environmental land-use change, however, became more dramatic with the advent of sedentary life.

From the archaeological evidence, it is known that some humans started to develop longer-term patterns of settlement during the Neolithic period. These settlements first emerged in places of abundant flora and fauna, which could serve as key food gathering and hunting sites (see Figure 1.12). Agricultural production most likely emerged thereafter not out of desire but out of necessity, as pressures from population and changing climate patterns produced the need to cultivate **staple crops** that could sustain growing populations. In many parts of the world where agriculture first took place, it was the conjoining of both plant and animal domestication that provided both the basic seed and fertilizer stock to sustain large-scale planting. Although the earliest agriculturalists practiced shifting or **swidden** agricultural practices—moving from field to field as nutrients were diminished—knowledge of fertilizers (both vegetative and animal) eventually slowed the need for such movements.

The establishment of villages during the Neolithic period and later urban life during the Ancient period also led to divisions of labor and the increasing complexity of social and cultural practice. A small specialist

# PERIODIZATION IN WORLD HISTORY AND GLOBAL GEOGRAPHY

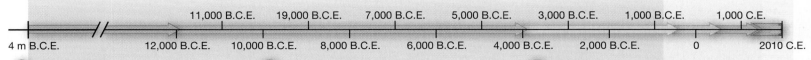

### Paleolithic
4 million B.C.E. – ~12,000 B.C.E.
*Old Stone Age*: This is the longest period in human history and includes human evolution to the present human species. Distinguished by the use of stone tools, peoples in this period relied almost exclusively on hunting and gathering. A distinguishing feature of this period is movement—both local and global—as peoples came to occupy all major landmasses with the exception of Antarctica.

### Neolithic
~12,000 B.C.E. – 4,000 B.C.E.
*New Stone Age*: This period is distinguished by changes in the social and spatial organization of humans, as some advanced techniques of animal husbandry (breeding and herding) and early agriculture were developed. While some peoples, particularly in harsher climates turned to pastoralism, others settled into small towns and villages, using agricultural practices to sustain growing kinship-based populations.

### Ancient
~4,000 B.C.E. – 500 B.C.E.
*Emerging Complex Urban Life*: During this period, urban life emerged as surplus food supplies were created in large river valleys throughout the world. The early adaptations of metallurgy facilitated this growth, allowing communities to take advantage of the rich soils of some of the world's largest river systems. Other technological advancements, such as writing and numerology, as well as the development of law and ethical codes, helped further advance the social organization of humans beyond kinship and into urban societies.

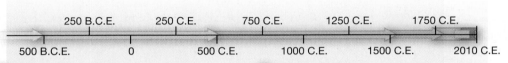

### Classical
~500 B.C.E. – 500 C.E.
*Expanding Global Empires*: As urban life took hold, competition for resources as well as the desire to control trade, people, and land pushed the expansion of large-scale globalizing empires across Eurasia and Africa. This intensified global connections as people, cultural practices, and technologies were further exchanged. This period also witnessed periods of increased violence, conflict, and disease at scales that expanded beyond the local.

### Post-Classical
~500 C.E. – 1500 C.E.
*Intensification of Global Connections*: The shift from the classical to the postclassical period is marked by increasing global connections across wider global spaces. Global trade networks expanded further, and commercial connections and relationships were developed that included the exchange of money, the rise of merchant classes, and the spread of global religions (e.g., Christianity, Islam, Hinduism, and Buddhism), which further facilitated economic and cultural connections.

### Early Modern
~1500 C.E. – ~1800 C.E.
*Resource Extraction and Direct Colonization*: Europeans driven by mercantilism and later capitalism expanded control over large territories in the Americas and parts of Asia, while also establishing themselves in coastal Africa. This resulted in massive demographic shift—movements of slaves from Africa to the Americas and the rapid decline of indigenous populations in the Caribbean. Global trade focused on the export of raw materials from newly established colonial holdings to Europe. Mainland parts of Asia remained relatively strong during this period and largely isolated from direct European colonization, with some powers, such as China, maintaining their hegemonic control in the region. Most of interior Africa, also remained isolated from European hegemony.

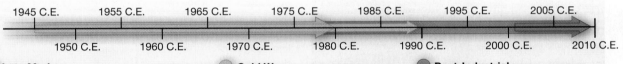

### Late Modern
~1800 C.E. – ~1980 C.E.
*Establishing Capitalism and the Nation-State*: The rise of capitalism helped further expand European hegemony as large parts of the Americas, Asia, Africa, and Oceania were brought into European systems of extraction, exploitation, and direct colonization. Other imperial powers, such as Japan, also emerged, while many parts of the Americas engaged in fights for decolonization. Cartographers carved the world up into discreet political units during the early part of this period and the Americas were brought into direct contact with Eurasia, Africa, and Australasia. Many of the colonial boundaries remain in place to this day, while others continue to be sites of struggle and contestation.

### Cold War
1945 C.E. – 1989 C.E.
*Global Conflict and Tension*: The post-World War II period is distinguished by the emergence of a "bipolar" world with two "superpowers" as well as a number of "nonaligned" nations. This was also a period of rapid decolonization, which facilitated the emergence of new "independent" states, some of which remained in a subordinate position to former colonizers. While the nation-state remained the "heart" of global geopolitics, supranational organizations (e.g., NATO, European Union, etc.) emerged as important global players as well.

### Post-Industrial
1980 C.E. – Present
*Global Realignment and Reorganization*: With the collapse of the Soviet Union and the Berlin Wall, realignments have taken place in the arena of economic, cultural, and social globalization. Far from reducing global tension, the end of the Cold War released new tensions, while other conflicts-based in ethnic, religious, and other social differences-escalated. This period is also distinguished by an increasing international division of labor, as manufacturing continues to move from the industrialized nations of North America and Europe to other world regions.

### Post-9/11
2001 C.E. – Present
*Global Tension and New/Old Local Struggles*: It is common to discuss the world as "post-9/11," although such a periodization centers global relations in the United States. Despite this caveat, it is clear that in the post-9/11 period, new challenges face the planet, as issues of security, global climate change, and cultural conflict continue to escalate. Radical and fundamental movements, based in religion, ethnicity, and race have created new visions about local autonomy and placed issues, such as immigration and national identity/borders, at the center-stage of numerous local and global conflicts.

▲ **FIGURE 1.11 Periodization in world history and global geography** Periods, like any human-constructed category, are representations of specific histories. Periods do provide a broad framework, however, to discuss geographic change and global connections over time. In this figure, certain periods overlap. This is a result of the fact that the end of one period and the start of a second is not a clear process in many cases and in most places. Throughout this text, we use the secular terms B.C.E., Before Common Era, and C.E., Common Era, to denote time. This is common practice today, as B.C.E. replaces B.C. and C.E. replaces A.D. in world history and global geography.

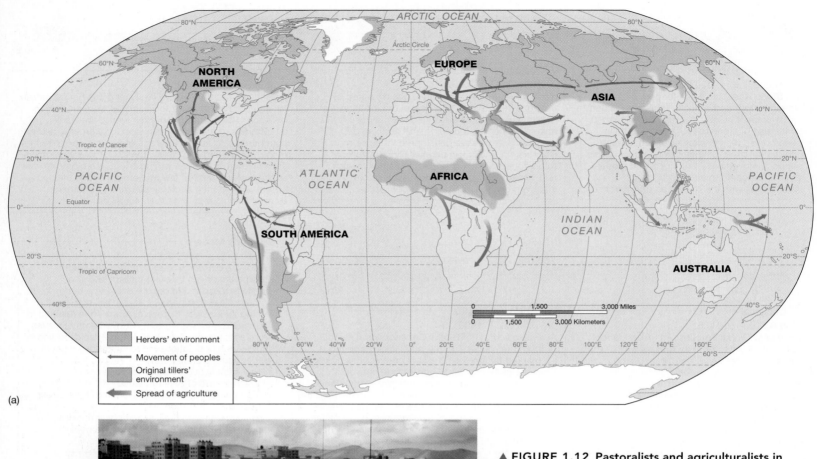

(a)

(b)

▲ **FIGURE 1.12 Pastoralists and agriculturalists in global context** (a) This map illustrates the emergence of agricultural production centers across the world, and the arrows demonstrate how these practices spread out from several core areas. Pastoral people were also important for the spread of knowledge systems, as they moved across large tracts of land in the more temperate climates in the Northern and Southern Hemispheres. This map also shows the important geographic relationship between rivers and agricultural production globally. (b) Migrant families newly arrived in Ulaanbaatar, Mongolia pitch their *gers* (felt nomadic tents) in the shadow of this growing metropolis. In the past, pastoralists constituted a majority of the world's population. Today, pastoral populations continue to dwindle. That said, their historical role cannot be underestimated.

population, which was not needed in agriculture, expanded craft production, religious systems, and other important cultural practices tied to their specific social group (**Figure 1.13**). It is clear from the archaeological record that sedentary life and agricultural expansion developed in different parts of the world and under varying environmental and social conditions during the Neolithic period, spreading from several core areas to other parts of the world as larger urban societies took hold during the Ancient period.

On the whole, larger populations were not possible until complex and intensive hydraulic infrastructures (such as large-scale irrigation) were developed during the Ancient period, which introduced control over river systems and provided water for agriculture (see Chapter 8,

p. 277). As irrigation systems became more sophisticated, urban life was also extended, and cities became more significant. With the development of metallurgy—first bronze and later iron—governing elites were soon able to control both agriculture and relatively cheap, strong weapons and agricultural tools. Written language and mathematical inventions also helped establish a social organizational structure for cities, providing mechanisms for regulating governance, trade, tribute, and ritualistic practice. These growing literate societies emerged across the world at slightly different times but many took hold between 4000 and 2000 B.C.E. (see Figure 1.11 for explanation of the term B.C.E.).

Although the Americas remained fairly autonomous from the rest of the world (after the receding ice shields cut them off from Eurasia and

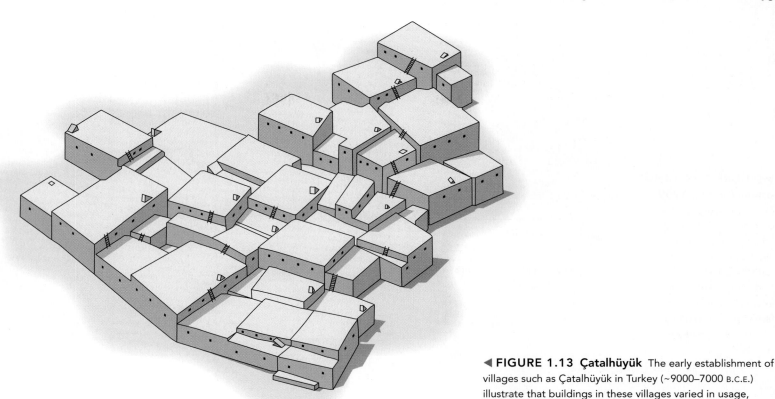

◀ **FIGURE 1.13 Çatalhüyük** The early establishment of villages such as Çatalhüyük in Turkey (~9000–7000 B.C.E.) illustrate that buildings in these villages varied in usage, suggesting that early social differences were emerging long before complex large-scale urban life developed.

Africa), parts of Africa, Europe, and Asia were intimately connected through systems of trade, migration, and eventually conflict and empire, beginning almost 4,000 years ago. During this period of expanding global connectivity, religions that had the potential to transcend locality also began to emerge (see Chapter 4, p. 123). As a result, over an extended period of time, ideas and practices moved back and forth across these trading routes, such as the Silk Roads (see Chapter 8, p. 284), as people developed various religious systems—such as Judaism (4,000 years ago), Hinduism (3,000 years ago), Buddhism (2,500 years ago), Christianity (2,000 years ago), and Islam (1,500 years ago)—and exchanged technological practices (such as rice agriculture [7,000 years ago in southern China and northern Vietnam], cotton weaving [5,000 years ago in the Indus River of Pakistan], and silk production [7,000 to 3,000 years ago in China], as examples).

In the classical periods, large-scale "empires" emerged, stretching across vast amounts of global space. As populations increased, connections became more important in different parts of the world—with the establishment of urban spaces, road networks, strong centralized states, and increasingly complex economic systems. The Romans, for example, following the example of the Persians, established a complex road network to link its empire (see Chapter 2, p. 53). This system of roads facilitated the spread of Christianity for years before it was officially "taken up" as the religion of the empire. River and oceanic systems also served to connect people and empires, as Egypt and Sudan, particularly through travel on the Nile River, served as a cultural conduit between sub-Saharan Africa and the Mediterranean, whereas the Chinese spread technological and cultural practices throughout the Indian Ocean, from the east coast of Africa to Japan. Indeed, for much of the world's history during the Classical and then the Postclassical period, the Indian Ocean and the Mediterranean

were the most globally significant water bodies in the world, facilitating trade and conquest, cultural exchange, and economic integration.

As **Figure 1.14** illustrates, by the height of the Postclassical period, the world of Eurasia and Africa was connected through land and sea networks, suggesting that the world has long been distinguished by transregional connections—trade and conquest brought together the ideas, practices, and objects of Asia and the Mediterranean, for example. Islam emerged during this period as well, with adherents covering a significant part of the world by 1491 (**Figure 1.15**). Large areas of the globe also remained either marginal to these connections (particularly in interior rural spaces) or were only partially integrated through systems of exchange and tribute. In fact, centralized authority and power was less clearly defined in many parts of the world at this time, and many localities were relatively autonomous, even as imperial powers sought economic, political, and sociocultural control. In the Americas, major **cultural hearth areas**—centers of sociocultural and political–economic production—in Central America/Mexico and South America, particularly the Andes Mountains, had spread technological practices, such as agriculture, throughout the wider region—with the knowledge of maize agriculture, for example, making its way to eastern North America by 1000 C.E.

By the end of the Postclassical period, geographic knowledge of the world was extensive, although quite different depending on one's perspective (see Figure 1.15), as certain places in these world regions established the Early Modern period as one of massive exploration. In global terms, the Atlantic and Pacific Oceans became more central to global movement and connectivity, as Europeans circumnavigated the globe and expanded their reach through the **colonization** (that is, the physical settlement of people from the colonizing state) of the Americas as

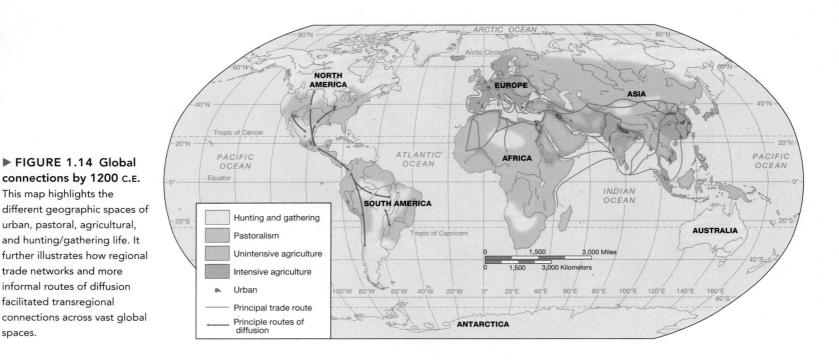

▶ **FIGURE 1.14 Global connections by 1200 C.E.** This map highlights the different geographic spaces of urban, pastoral, agricultural, and hunting/gathering life. It further illustrates how regional trade networks and more informal routes of diffusion facilitated transregional connections across vast global spaces.

well as parts of Africa and Asia. In fact, European **colonialism**—an established political–economic and sociocultural domination by a state over a separate society—took place in two majors waves.

The first wave, which took place between the late 1400s through the early 1800s, was facilitated by **mercantilism**, an economic policy prevailing in Europe during the 16th, 17th, and 18th centuries, wherein the government controlled industry and trade. During this period, European colonization was invested heavily in port cities and trading networks, with investment moving slowly toward the development of **plantation economies**, the classic institution of colonial agriculture (**Figure 1.16**). These were extensive, European-owned, operated, and financed enterprises where single crops were produced by local or imported labor for a world market.

The second wave, which marks the shift from the Early Modern to the Late Modern period, began in the early 1800s and lasted through the mid-20th century and incorporated the majority of the African continent as well as large parts of Asia, Australasia, and the South Pacific (see Figure 1.16). The direct colonization of non-European regions and people facilitated the further global expansion of complex innovations and institutions that stabilized and enabled the spread and evolution of **capitalism**, a form of economic and social organization characterized by the profit motive and the control of the means of production, distribution, and exchange of goods by private ownership. Colonization during this period was centered on the exploitation of **primary commodities**—such as agricultural products and mineral wealth—that could be transferred to the growing **secondary commodity** production complexes of the manufacturing sector in Europe. During this period, former colonies, such as the United States, also established themselves as colonial and imperial powers. The United States, for example, pushed its way across North America, marginalizing indigenous people while building its economy on slave labor.

Capitalist productivity and European colonial expansion in the Late Modern period were also tied to the Industrial Revolution—a period of manufacturing expansion in parts of Europe. As soon as the Industrial

Revolution had gathered momentum in the early 19th century, the industrial nations of Europe embarked on the inland penetration of the world's continents to exploit them for agricultural and mineral production. In the second half of the 19th century, and especially after 1870, there was a vast increase in the number of colonies and the number of people under colonial rule (**Figure 1.17**).

The colonization that accompanied European colonial expansion is tied to the evolution of world leadership cycles. **Leadership cycles** are periods of international power established by individual states through economic, political, and military competition. With a combination of economic, political, and military power, individual states can dominate globally, setting the terms for many economic and cultural practices and imposing their particular ideologies by virtue of their preeminence. This kind of dominance is known as hegemony (**Figure 1.18**). **Hegemony** refers to domination over the world economy, exercised—through a combination of economic, military, financial, and cultural means—by one national state in a particular historical epoch. Over the long run, the costs of maintaining this kind of power and influence tend to weaken the hegemon. This phase of the cycle is known as *imperial overstretch*. (**Imperialism** is the extension of the power of a state through the direct or indirect control of the economic and political life of other territories.) It is followed by another period of competitive struggle, which brings the possibility of a new dominant world power. As part of this imperial process, a growing European presence throughout the world restructured the political geographies of many regions, as colonial powers established defined and demarcated boundaries around their colonial holdings. In fact, many of the world's political boundaries today have their seeds in the period of European colonial expansion.

At the same time, it is unfair to say that Europeans were the only ones to take advantage of the mechanisms of colonialism and imperialism. Throughout the first half of the 20th century, Japan also established itself as a regional and eventually global power, expanding its control over large parts of East Asia, Southeast Asia, and the Pacific through a

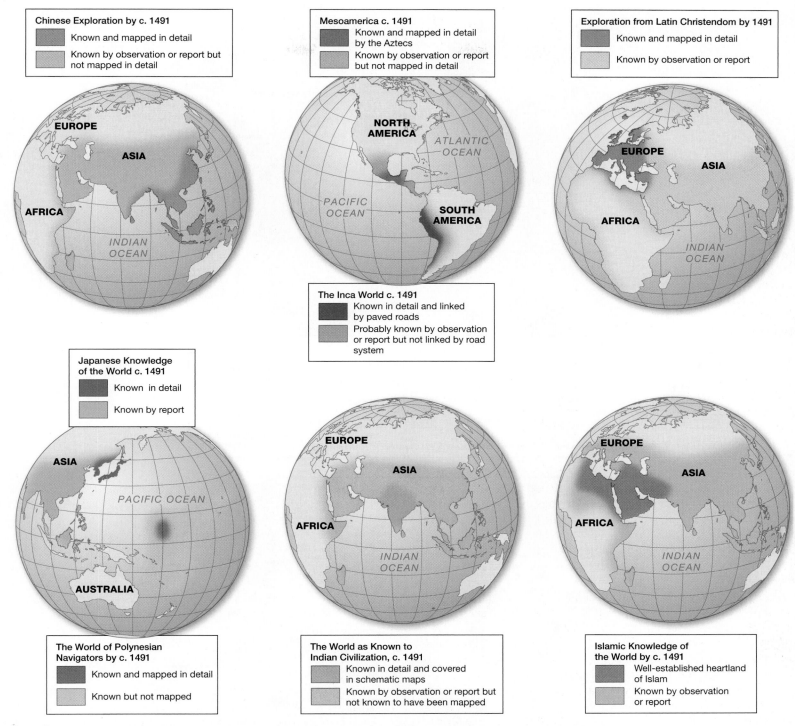

Chinese Exploration by c. 1491
- Known and mapped in detail
- Known by observation or report but not mapped in detail

Mesoamerica c. 1491
- Known and mapped in detail by the Aztecs
- Known by observation or report but not mapped in detail

Exploration from Latin Christendom by 1491
- Known and mapped in detail
- Known by observation or report

The Inca World c. 1491
- Known in detail and linked by paved roads
- Probably known by observation or report but not linked by road system

Japanese Knowledge of the World c. 1491
- Known in detail
- Known by report

The World of Polynesian Navigators by c. 1491
- Known and mapped in detail
- Known but not mapped

The World as Known to Indian Civilization, c. 1491
- Known in detail and covered in schematic maps
- Known by observation or report but not known to have been mapped

Islamic Knowledge of the World by c. 1491
- Well-established heartland of Islam
- Known by observation or report

▲ **FIGURE 1.15 Perspectives on the world in 1491 C.E.** As it does today, knowledge of the world varies depending on where one lives. Traditions of cartography, the mapping of the "known world," were not a European invention, and all cultures have traditions that represent and express their "place" in the world. The global representation presented here reflects some of these diverse perspectives, illustrating how complex the perspectives on the world were at this point in time. There is clearly overlap in some knowledge of the world, whereas in other contexts connections are scant.

massive wave of industrialization and military expansion (Chapters 8 and 10, pp. 285 and 350). Additionally, alternative political and economic philosophies, such as communism, emerged in the 19th and 20th centuries, establishing an alternative to the imperial expansion of Europe's capitalist states.

**Communism**, a form of economic and social organization characterized by the common ownership of the means of production, distribution, and exchange, developed in Europe and eventually spread in different forms to all parts of the globe, influencing the establishment of a variety of alternative economic systems in countries as diverse as China,

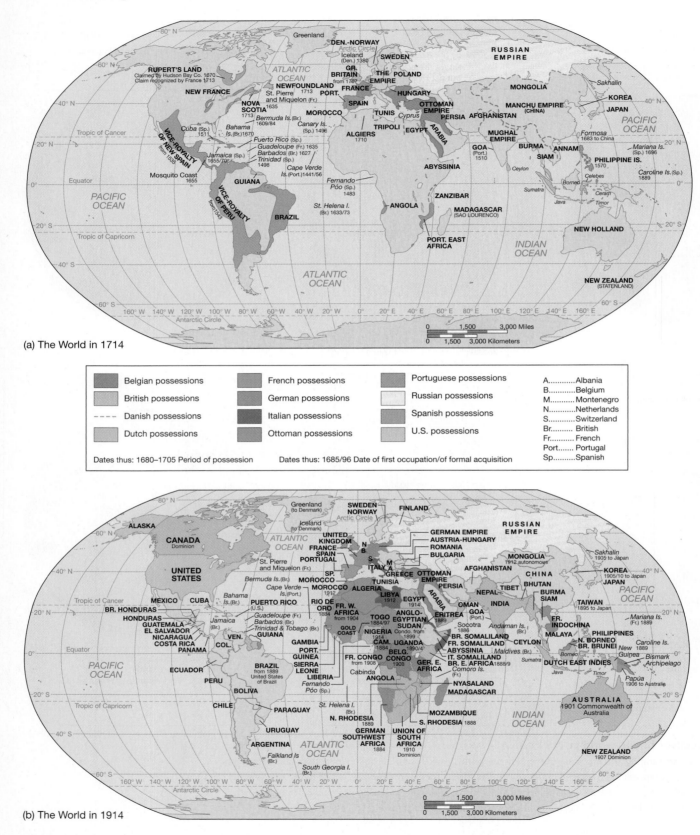

(a) The World in 1714

| | Belgian possessions | | French possessions | | Portuguese possessions | A............Albania |
|---|---|---|---|---|---|---|
| | British possessions | | German possessions | | Russian possessions | B............Belgium |
| - - - | Danish possessions | | Italian possessions | | Spanish possessions | M...........Montenegro |
| | Dutch possessions | | Ottoman possessions | | U.S. possessions | N............Netherlands |

S............Switzerland
Br.......... British
Fr........... French
Port....... Portugal
Sp..........Spanish

Dates thus: 1680–1705 Period of possession        Dates thus: 1685/96 Date of first occupation/of formal acquisition

(b) The World in 1914

▲ FIGURE 1.16 European colonialism in 1714 C.E. and 1914 C.E.  These two maps illustrate the transformation in the political geography of the globe that occurred during the 200 years that mark the most intense period of European global imperialism and colonialism. (a) By the middle of the first wave, European possessions amounted to less than 10% of the world's land area and only 2% of the world's population. (b) By the middle of the second wave, European colonies amounted to more than 55% of the world's land area and 34% of the world's population.

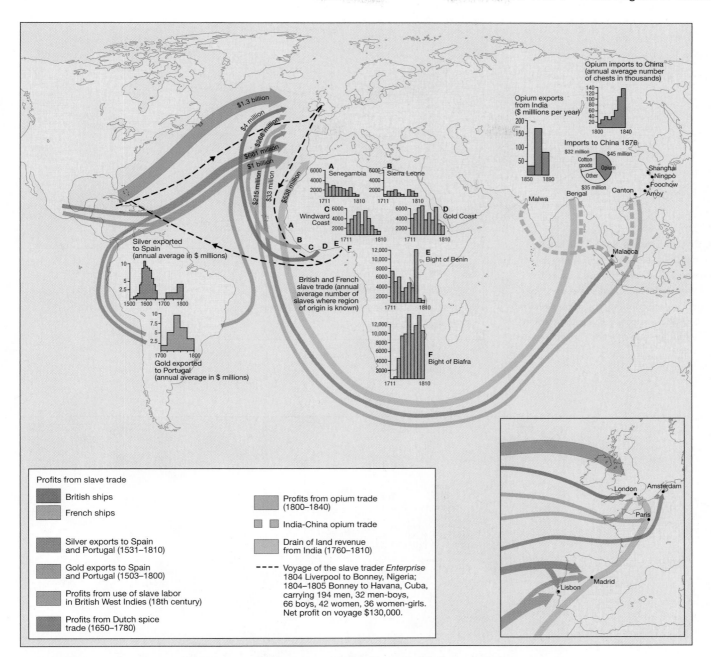

Profits from slave trade

■ British ships
■ French ships

■ Silver exports to Spain
and Portugal (1531–1810)

■ Gold exports to Spain
and Portugal (1503–1800)

■ Profits from use of slave labor
in British West Indies (18th century)

■ Profits from Dutch spice
trade (1650–1780)

■ Profits from opium trade
(1800–1840)

■ ■ India-China opium trade

■ Drain of land revenue
from India (1760–1810)

- - - Voyage of the slave trader *Enterprise*
1804 Liverpool to Bonney, Nigeria;
1804–1805 Bonney to Havana, Cuba,
carrying 194 men, 32 men-boys,
66 boys, 42 women, 36 women-girls.
Net profit on voyage $130,000.

▲ **FIGURE 1.17 Profits from European global expansion, 1500–1876 C.E.** This map illustrates the profits
generated through European plunder of global minerals, spices, and human beings over more than three centuries.
Silver and gold from the Americas, opium and tea from China, spices from the East Indies (Indonesia, Malaysia, and the
Philippines), and slaves from Africa are represented.

Cambodia, and Cuba. The emergence of communism along with the massive destruction of much of Europe during World War II facilitated the rapid decolonization of much of the world in the 1950s, 1960s, and 1970s. The challenge of communism also established what many political scientists called a bipolar world in the post–World War II period, as the United States and the Soviet Union established their role as global hegemons during a period defined as "**the Cold War**"—a period in which struggles between these two powers took place through a variety of proxy conflicts (**Figure 1.19**). At the same time, a nonaligned movement of states as diverse as India and Cuba tried to maintain a sense of autonomy,

despite enormous pressures to come under the political–economic and sociocultural control of one of these "poles." There have been many instances of resistance and adaptation, with some colonial countries (Tanzania, for example) attempting to become self-sufficient and others (Cuba, for example) seeking to opt out of the system altogether to pursue a different path. In either case, these regions on the edge of the world political system sought to respond, in one way or another, to the center.

This long history of world regional transformation until the end of the 20th century recalls several general themes highlighted previously. First, the world system we know today, with its specific centers of power

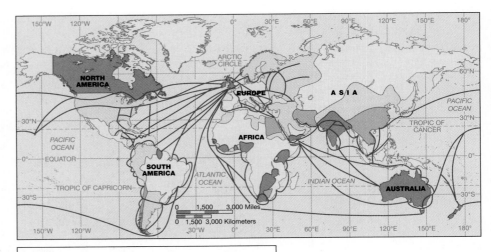

▶ **FIGURE 1.18 The British Empire, late 1800s** C.E.,
Protected by the all-powerful Royal Navy, the British merchant navy
established a web of commerce that collected food for British
industrial workers and raw materials for its industries, much of it
from colonies and dependencies appropriated by imperial might
and developed by British capital. So successful was the trading
empire that Britain also became the hub of trade for other states.

| | |
|---|---|
| ■ | British Empire |
| ■ | Over 50% of total imports to and from Britain |
| ■ | 25–49% of total imports and exports to and from Britain |
| — | Principal routes of commerce |

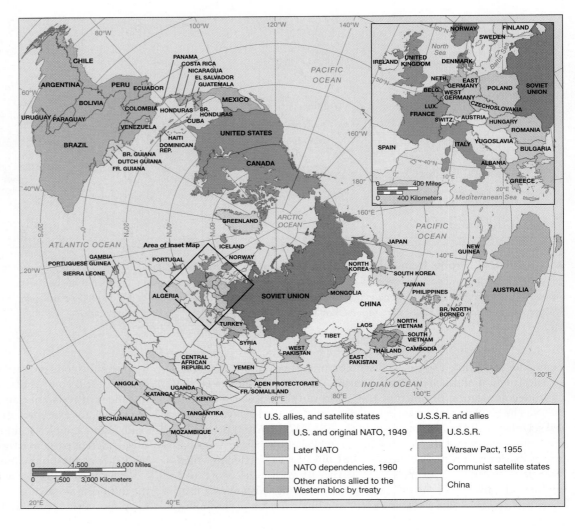

▶ **FIGURE 1.19 The alliances of the Cold
War** This map depicts the complex global
geography of the Cold War, illustrating how various
"blocs" emerged around different "poles" in this
global system of conflict and tension.

**U.S. allies, and satellite states**

- U.S. and original NATO, 1949
- Later NATO
- NATO dependencies, 1960
- Other nations allied to the Western bloc by treaty

**U.S.S.R. and allies**

- U.S.S.R.
- Warsaw Pact, 1955
- Communist satellite states
- China

and its peculiar well-connected and poorly connected peripheries, is not the first such system of regions to have existed in the world. At other moments in the past, different regions held the center of the world stage. Indeed, today's centers of economic power were marginal areas in other periods. Second, it emphasizes that changes in the regions of the world were predicated on changes in their *relations*. The rise of Europe, for example, was not a product of something unique to European culture, resources, or politics. Instead, it was Europe's relationship to its colonial periphery that allowed it to leverage a central place in recent history. Nor are the current positions of the world's powerful regions uncontested, especially as relationships between regions have been transformed since the end of the 20th century, and new global economic systems and regional patterns of accumulation have emerged.

## Economy, Accumulation, and the Production of Inequality

Toward the end of the Cold War period, a majority of the world's colonies had been emancipated, and many countries began to move away from communist forms of economic organization. Even prior to the "collapse" of the Soviet Union, there were clear shifts in the global economy, and some scholars have suggested that the latest period, the Postindustrial, actually began in the early 1980s after a series of "oil crises" changed how manufacturing happened and where. In fact, since the early 1980s, manufacturing production has moved dramatically from the imperial centers of the Cold War period—North America, Europe, and Japan—and toward mainland East Asia, Southeast Asia, Africa, and Latin America. Some have also suggested that although direct colonialism diminished in the Cold War and Postindustrial periods, a new set of **neocolonial** relationships took their place—with former colonies remaining in highly dependent relationships to their former colonizers. As a result, an **international division of labor** has emerged over time with the specialization of different people, regions, and countries concentrated in certain kinds of economic activities. Thus, although geographers often argue that the world is dominated by a core of economic powers—North America, Europe, and Japan—the world's economy continues to be complicated by the diversification and spatial distribution of economic practices. This is evinced, for example, by the growth of China as a significant economic player in the world (**Figure 1.20**). In considering our global geography, it is therefore important to understand some of the basic core tenets of this division of labor and of the global economy more generally.

**Sectors of the Global Economy**   The contemporary economic structure of a country or region is often described in terms of the rel-

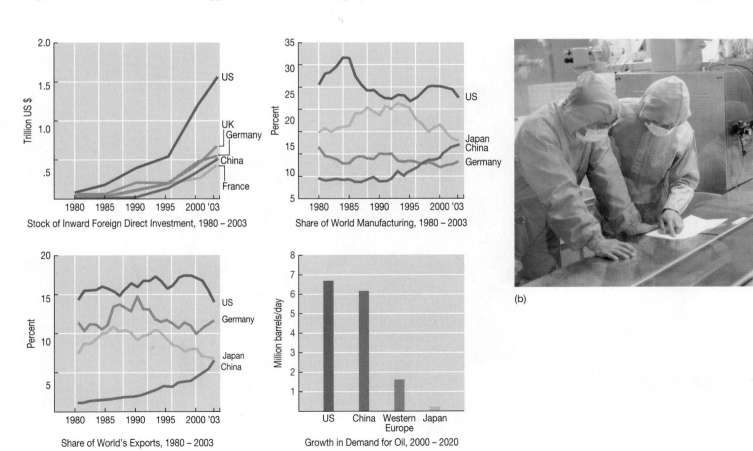

(a)

Stock of Inward Foreign Direct Investment, 1980 – 2003

Share of World Manufacturing, 1980 – 2003

Share of World's Exports, 1980 – 2003

Growth in Demand for Oil, 2000 – 2020

(b)

▲ **FIGURE 1.20  China's changing role in the world economy**  (a) Most forecasts indicate that by 2020 China's economy will exceed those of the world's core economic powers except the United States. China's reemergence as a significant player in the world economy is reflected in trends of inward investment, manufacturing output, and oil consumption. (b) Workers at a pharmaceutical company in discussion in Kunming, China.

# SIX KEY FACTORS OF GLOBALIZATION

1. **The "Newest" International Division of Labor** The newest international division of labor has resulted from the fact that the United States has declined as an industrial producer, while global manufacturing production has been globally decentralized. Tied to this, new specializations have emerged within North America, Europe, and Japan. These specializations include high-tech manufacturing and producer services—information services, insurance, and market research.

2. **The Internationalization of Finance** Banking and integrated financial markets are tied into the new international division of labor. Massive increases in levels of international direct investment from larger economies to smaller ones have risen dramatically over the past 30 years, thanks to computers and information systems. The nerve centers of the new system are located in just a few places—London, Frankfurt, New York, and Tokyo, in particular. Their activities are interconnected around the clock, and their networks penetrate every corner of the globe.

3. **New Technology Systems** A combination of technological innovations, including solar energy, robotics, microelectronics, biotechnology, digital telecommunications, and computerized information systems, has required the geographical reorganization of major postindustrial economies. It has also extended the global reach of finance and industry and made for a more flexible approach to investment and trade. Especially important in this regard have been new and improved technologies in transport and communications.

4. **Homogenization of International Consumer Markets** A new and materialistic international culture has taken root in which people save less, borrow more, defer parenthood, and indulge in affordable luxuries that are marketed as symbols of style and distinctiveness. This culture is easily transmitted through the new telecommunications media and has been easily reinforced through other aspects of globalization, including the internationalization of television—especially CNN, MTV, and Star Television—and the syndication of TV movies and light entertainment series.

5. **The Transnational Corporation** Business organizations that operate as "independent" actors in the global economy, separate from governments and stockholders, and with headquarters and holdings in multiple countries—known as transnational corporations (TNCs)—have been central to a major new phase of geographical restructuring that has been under way for the last 30 years. TNCs often work beyond the geographic and institutional boundaries of states, facilitating a much fuller integration of the economies of the worldwide system of states and a much greater interdependence of individual places and regions.

6. **Transnational Economic Integration** A series of global economic crises, as in 1997 and 2008, indicates that national economies are increasingly integrating into global financial networks. What has happened is that the logic of the world economy has in many ways transcended, and in some ways undermined, nation–states. The rationale and institutions of contemporary states are not conducive to transnational integration, whether economic or political, but the outcomes of the new international division of labor have forced many states to explore cooperative strategies of various kinds.

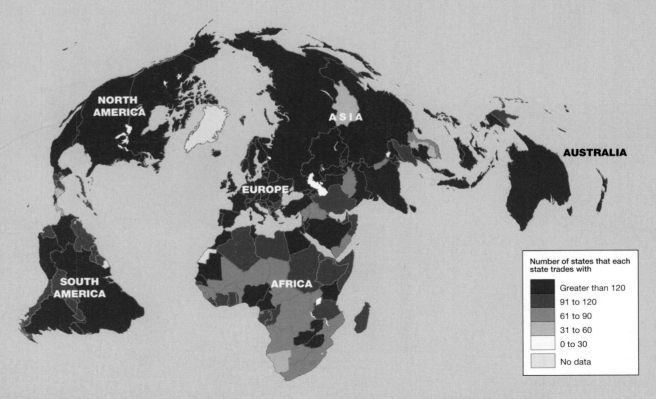

**Number of states that each state trades with**

- Greater than 120
- 91 to 120
- 61 to 90
- 31 to 60
- 0 to 30
- No data

▲ **FIGURE 1 World trade interconnectedness** Since WWII, several global trade agreements have been put in place to help foster the multilateral trading patterns shown in this map. In 1995, the World Trade Organization (WTO) established itself as the dominant body governing global trade.

ative share of primary, secondary, tertiary, and quaternary economic activities. **Primary activities** are those that are concerned directly with natural resources of any kind. These include agriculture, mining, fishing, and forestry. **Secondary activities** are those concerned with manufacturing or processing. They involve processing, transforming, fabricating, or assembling the raw materials derived from primary activities, or reassembling, refinishing, or packaging manufactured goods. These include, for example, steelmaking, food processing, furniture making, textile manufacturing, and garment manufacturing. **Tertiary activities** are those that involve the sale and exchange of goods and services. These include warehousing, retail stores, personal services such as hairdressing, commercial services such as accounting and advertising, and entertainment. **Quaternary activities** include handling and processing knowledge and information. Examples include data processing, information retrieval, education, and research and development (R & D).

**Understanding the Postindustrial Global Economy**  The main effect of the post-1980 shifts in the global economy, particularly the end of the Cold War, has been the consolidation of the world into the capitalist-driven system. This consolidation has sometimes been defined by the term **globalization,** a contested concept that has been applied to both the economy and to global culture (see **Culture and Politics,** p. 31). Globalization suggests that interconnectivity is a relatively new and unique function of the Modern and Postindustrial periods, a notion already rebuffed in the earlier historical discussion. That said, there is no doubt that the intensification of global economic relations makes the term *globalization* an appealing one because it suggests just how important our global geography is today (see **Six Key Factors of Globalization,** p. 22). It is also a term that appeals because within it is the suggestion that the consolidation of the global economy has intensified the differences between the rich and the poor over the recent past.

According to the United Nations Development Programme, the gap between the poorest fifth of the world's population and the wealthiest fifth increased more than threefold between 1960 and the turn of the new century. In 55 countries, per capita incomes actually fell during the 1990s. In sub-Saharan Africa, economic output fell by one-third during the 1980s and stayed low during the 1990s, so that people's standard of living there is now, on average, lower than it was in the early 1960s. At the beginning of the 21st century, the fifth of the world's population living in the highest-income countries had 74% of world income while the bottom fifth had just 1%. Such enormous differences have led many people to question the fairness of geographical variations in people's levels of affluence and well-being (**Figure 1.21**). The concept of **spatial justice** is important in this context, because it requires us to consider the distribution of society's benefits and burdens, taking into account both variations in people's needs and in their contribution to the production of wealth and social well-being.

This inequality resulting from the global economy is reflected—and reinforced—by many aspects of human well-being. Levels of infant mortality, a reliable indicator of social well-being, are far higher in the most economically marginal parts of the world. Likewise, for adults in the industrial core countries, life expectancy is high and continues to increase.

**Measuring Economic Development**  Understanding the structure of the world's economies tells only part of the story of their levels of development. At the global scale, levels of economic development are usually measured by national economic indicators such as gross domestic prod-

uct and gross national product (though national statistics tend to obscure within-country social and spatial inequalities). **Gross domestic product (GDP)** is an estimate of the total value of all materials, foodstuffs, goods, and services produced by a country in a particular year. To standardize for countries' varying sizes, the statistic is normally divided by total population, which gives an indicator, *per capita* GDP, that provides a good measure of relative levels of economic development. **Gross national product (GNP)** also includes the net value of income from abroad—flows of profits or losses from overseas investments, for example.

In making international comparisons, GDP and GNP can be problematic. They are calculated using national currencies (the value of which can be distorted by speculation and government policies), and they do not include nonmarket goods and services. (A nonmarket good or service does not have an observable monetary value. Examples of a nonmarket good include viewing wildlife or snorkeling at a coral reef. An example of a nonmarket service is the preparation of food for a family done by a parent.) Recently, it has become possible to compare national currencies based on **purchasing power parity (PPP)**. In effect, PPP measures how much of a common "market basket" of goods and services each currency can purchase locally, including goods and services that are not traded internationally. Using PPP-based currency values to compare levels of economic prosperity usually produces lower GDP figures in wealthy countries and higher GDP figures in poorer countries, compared with market-based exchange rates. Nevertheless, economic prosperity is very unevenly distributed across countries. As **Figure 1.22** shows, most of the highest levels of economic development are found in northern latitudes (very roughly, north of 30°N), which has given rise to another popular shorthand for the world's economic geography: the division between the **global north** and the **global south,** what some now refer to as the minority world (the global north) and the majority world (the global south) because, by population, most people live in the global south. In the countries of Japan, the United States, and Canada, and almost all of northwestern Europe, annual per capita GDP (in PPP) in 2003 exceeded $20,000. The only other countries that matched these levels were Australia, Hong Kong, Singapore, and the United Arab Emirates, where annual per capita GDP in 2003 was $29,632, $27,179, $24,481, and $22,420, respectively.

**Neoliberalism and Development**  Neoliberal policies of capitalist economic development—what is often referred to as "development"—have emanated from postindustrial countries and have been associated with the increasing influence of the World Bank and the International Monetary Fund over the last 25 years. The **World Bank**—whose chief is appointed by the United States—is a development bank and the largest source of development assistance in the world. Its goal is to help countries strengthen and sustain the fundamental conditions that will attract and retain private investment. The **International Monetary Fund (IMF)**—whose chief is appointed by the European Union (EU)—provides loans to governments throughout the world. To obtain these loans, governments must submit to IMF conditions. This often means rewriting laws so that they are more favorable to foreign investment. Along with the World Bank, the IMF aims to help countries strengthen their banking systems.

There have been several means by which the World Bank and the IMF have attempted to shape and assist economic development in poor countries. These include neoliberal policies that emphasize privatization, export production, and limited restrictions on imports. The term *neoliberal* (new liberal) refers to the revival of ideas popular at the end of the 19th century promoting free trade and economic integration. **Neoliberalism** promotes a reduction in the role and budget of govern-

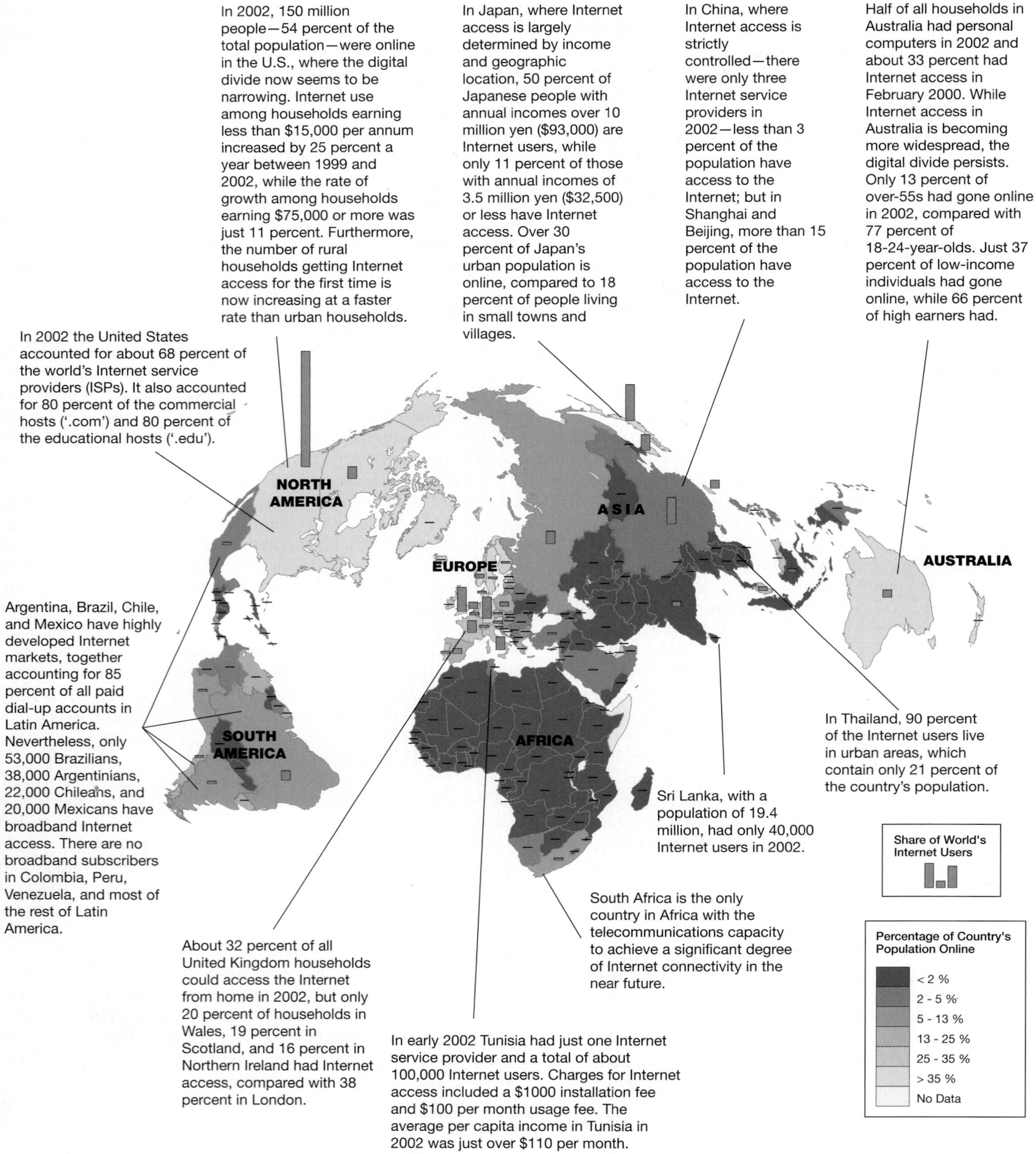

In 2002, 150 million people—54 percent of the total population—were online in the U.S., where the digital divide now seems to be narrowing. Internet use among households earning less than $15,000 per annum increased by 25 percent a year between 1999 and 2002, while the rate of growth among households earning $75,000 or more was just 11 percent. Furthermore, the number of rural households getting Internet access for the first time is now increasing at a faster rate than urban households.

In Japan, where Internet access is largely determined by income and geographic location, 50 percent of Japanese people with annual incomes over 10 million yen ($93,000) are Internet users, while only 11 percent of those with annual incomes of 3.5 million yen ($32,500) or less have Internet access. Over 30 percent of Japan's urban population is online, compared to 18 percent of people living in small towns and villages.

In China, where Internet access is strictly controlled—there were only three Internet service providers in 2002—less than 3 percent of the population have access to the Internet; but in Shanghai and Beijing, more than 15 percent of the population have access to the Internet.

Half of all households in Australia had personal computers in 2002 and about 33 percent had Internet access in February 2000. While Internet access in Australia is becoming more widespread, the digital divide persists. Only 13 percent of over-55s had gone online in 2002, compared with 77 percent of 18-24-year-olds. Just 37 percent of low-income individuals had gone online, while 66 percent of high earners had.

In 2002 the United States accounted for about 68 percent of the world's Internet service providers (ISPs). It also accounted for 80 percent of the commercial hosts ('.com') and 80 percent of the educational hosts ('.edu').

Argentina, Brazil, Chile, and Mexico have highly developed Internet markets, together accounting for 85 percent of all paid dial-up accounts in Latin America. Nevertheless, only 53,000 Brazilians, 38,000 Argentinians, 22,000 Chileans, and 20,000 Mexicans have broadband Internet access. There are no broadband subscribers in Colombia, Peru, Venezuela, and most of the rest of Latin America.

In Thailand, 90 percent of the Internet users live in urban areas, which contain only 21 percent of the country's population.

Sri Lanka, with a population of 19.4 million, had only 40,000 Internet users in 2002.

About 32 percent of all United Kingdom households could access the Internet from home in 2002, but only 20 percent of households in Wales, 19 percent in Scotland, and 16 percent in Northern Ireland had Internet access, compared with 38 percent in London.

South Africa is the only country in Africa with the telecommunications capacity to achieve a significant degree of Internet connectivity in the near future.

In early 2002 Tunisia had just one Internet service provider and a total of about 100,000 Internet users. Charges for Internet access included a $1000 installation fee and a $100 per month usage fee. The average per capita income in Tunisia in 2002 was just over $110 per month.

**Share of World's Internet Users**

**Percentage of Country's Population Online**

< 2 %
2 - 5 %
5 - 13 %
13 - 25 %
25 - 35 %
> 35 %
No Data

▲ **FIGURE 1.21  Global Internet connectivity** Like all previous revolutions in transportation and communications, the Internet is effectively reorganizing space. Although often spoken of as "shrinking" the world and "eliminating" geography, the Internet is highly uneven in its availability and use. This map shows the percentage of the total population in each country with access to the Internet (indicated by the density of shading) and the relative share of the world's Internet users accounted for by each country (indicated by the vertical bars).

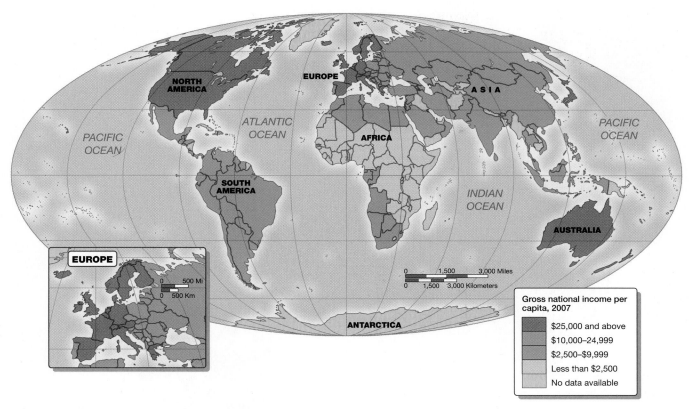

▲ **FIGURE 1.22 Gross domestic product (GDP) Per capita** GDP per capita is one of the best single measures of economic development. This map, based on 2003 data, shows the tremendous gulf in affluence among the countries of the world. The world's average annual per capita GDP (in PPP) is $8,229. The gap between the higher per capita GDPs ($62,298 in Luxembourg, $37,738 in Ireland, and $37,670 in Norway) and the lower ones ($548 in Sierra Leone and $608 in Malawi) is huge: In 2000, the average of the bottom ten was about 1/50th of the average of the top ten.

ment, including reduced subsidies and the privatization of formerly publicly owned and operated concerns such as utilities. The goal of neoliberal development policies is to enable poor countries to achieve higher economic standards of wealth and prosperity, while recognizing that preexisting conditions will have to be taken into account to construct a place-specific development path. Mostly associated with the IMF, **structural adjustment policies** require governments to cut budgets and liberalize trade in return for debt relief.

Since the 1980s, the apparent failure of existing neoliberal theory has generated radical criticism by feminist, postcolonial, and other perspectives. As a result of this radical refiguring of neoliberal theory and policy, contemporary theorists have called into question the whole grand notion of development. They argue instead for indigenous alternatives that empower grassroots movements and promote local knowledge and that can repair the damage done by neoliberal development projects. They have shown how the push to modernize is just one way to think about a very complicated reality, although one that has come to dominate as the only solution to social, cultural, economic, and political crises in the peripheral spaces of the global economy.

The contemporary critique of mainstream development practice challenges the position that institutions, based in the global north, know best how to solve the problems of the global south, and further argues that the neoliberal goal of economic development for all states and nations of the world might even be called into question. These challenges are part of a broader critique that argues for the need to recognize the importance of social and cultural factors in the process of economic and political development. Focusing on women and other underrepresented groups, this critique of neoliberalism contends that different groups in different places have different access to the power and resources that shape their daily lives. There has also been a growing understanding that development dramatically shapes the environment as well as people.

**Patterns of Social Well-Being** Inequality is reflected—and reinforced—by many aspects of human well-being. Patterns of infant mortality, a reliable indicator of social well-being, show a steep north–south gradient. For adults in the industrial countries, life expectancy is high and continues to increase. In contrast, life expectancy in the poorest countries is dramatically shorter.

The United Nations Development Programme (UNDP) has devised an overall index of human development based on measures of life expectancy, educational attainment, and personal income. **Figure 1.23** shows the international map of human development in 2003. Norway, Iceland, and Australia had the highest overall levels of human development (0.96, 0.96, and 0.95, respectively), whereas Niger and Sierra Leone (0.30 and 0.28) had the lowest levels. The same fundamental pattern is repeated in terms of the entire array of indicators of human development: adult literacy, poverty, malnutrition, access to physicians, public expenditure on higher education, telephone lines, Internet users, and so on. Inequality on this scale poses the most pressing, as well as the most intractable, questions of national and international policymaking.

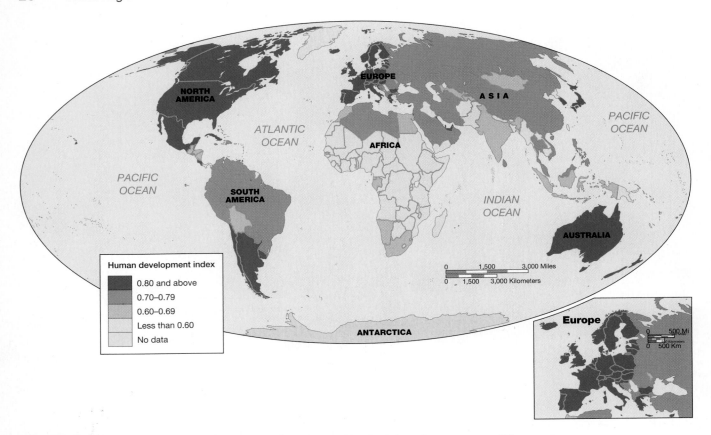

▲ **FIGURE 1.23 The Human Development Index** This UNDP index is based on measures of life expectancy, educational attainment, and personal income. A country that had the best scores among all of the countries in the world on all three measures would have a perfect index score of 1.0, whereas a country that ranked worst in the world on all three indicators would have an index score of 0. Most of the affluent countries have index scores of 0.9 or more, whereas the worst scores—those less than 0.4—are concentrated in Africa.

In addition to the HDI, the United Nations has also developed the **Gender-Related Development Index (GDI)**, which is a composite indicator of gender equality assessing the standard of living in a country. It aims to show the inequalities between men and women in the following areas: long and healthy life, knowledge, and a decent standard of living. The GDI is one of the five indicators used by the UNDP in its annual Human Development Report (HDR). The latest report from 2007/2008 shows Iceland, Norway and Australia in places one, two and three respectively, on the GDI. They also score in the top three places in the HDI. Niger, Guinea-Bissau, and Sierra Leone are in the bottom of the GDI where other African countries are numerous. Not surprisingly, rich countries tend to score higher on this index than poor countries, suggesting that in poor countries women are among the poorest of the poor.

The UNDP has also calculated that the annual cost of providing a basic education for all children in less-developed countries would be roughly $6 billion, which is less than the annual sales of cosmetics in the United States. Providing water and sanitation for everyone in less-developed countries is estimated to cost $9 billion per year, which is less than Europeans' annual expenditure on ice cream. Providing for basic health and nutrition for everyone in less-developed countries would cost an estimated $13 billion per year, which is less than the annual expenditure on pet foods in Europe and the United States. Reducing the military expenditures of affluent countries (at approxi-

mately $500 billion per year) by less than 10% each year would pay for basic education, water and sanitation, basic health and nutrition, and reproductive health programs for everyone in less-developed countries. Although most of these children live in poor countries, hundreds of thousands of underage workers are exploited in sweatshops, farm fields, and other workplaces in rich countries as well.

## Demography, Migration, and Settlement

The historical and economic geographies of the world are also intimately connected to the broader population geographies found today. As the world population density map demonstrates, some areas of the world are very heavily inhabited, others only sparsely (**Figure 1.24**). Almost all the world's inhabitants live on 10% of the world's land. Most live near the edges of land masses—near the oceans or seas or along rivers with easy access to a navigable waterway. Approximately 90% live north of the equator, where the largest proportion of the total land area (63%) is located. Finally, most of the world's population lives in temperate, low-lying areas with fertile soils.

In mid-2000 the world contained just over 6 billion people. The population division of the United Nations Department of Social and Economic

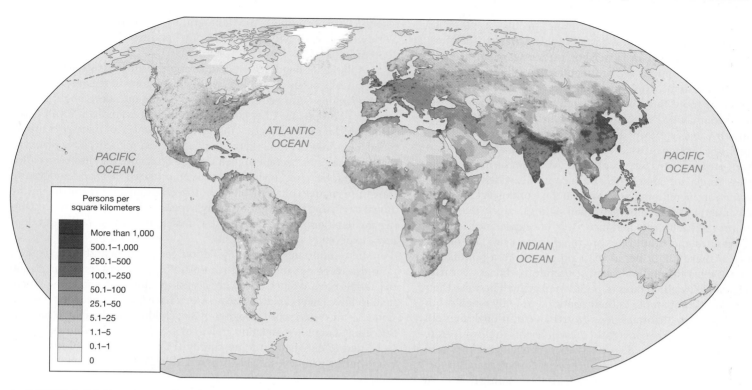

▲ **FIGURE 1.24 Population distribution** As this map shows, the world's population is not evenly distributed across the globe. Such maps are useful in understanding the relationship between population distributions and the national contexts in which they occur.

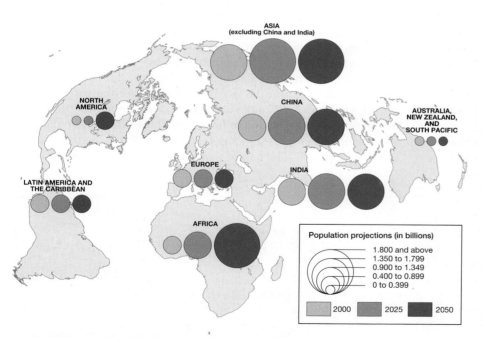

▲ **FIGURE 1.25 Population geography of the future** Population projections for 2025 and 2050 show very marked disparities among world regions, with parts of the world growing very little, whereas other parts continue to grow rapidly, suggesting a complicated future geography of resource needs and demands, among other issues.

Affairs projects that the world's population will increase by 1.2% annually to midcentury. This means that by the year 2050, the world is projected to contain nearly 9.3 billion people (**Figure 1.25**). The distribution of this projected population growth is noteworthy. Over the next half-century, population growth is predicted to occur overwhelmingly in regions least able to support it. Just six countries will account for half the increase in the world's population: Bangladesh, China, India, Indonesia, Nigeria, and Pakistan. Meanwhile, Europe and North America will experience very low and in some cases zero population growth.

Nevertheless, it is increasingly clear that human population will not continue to grow indefinitely, even in countries with already-high populations. The history of demographic change in the global north has prompted some analysts to suggest that many of the economic, political, social, and technological transformations associated with industrialization and urbanization lead to a "demographic transition"—a shift from a period of high growth, to one of low growth, no growth, or even population decline. The **demographic transition** is a model of population change when high birth and death rates are replaced by low birth and death rates.

Birthrates are a measure of the number of births in a population, usually expressed as births per 1,000 people per year, or as a percentage figure. High birthrates (as high as 40 births per thousand people or more) are associated with agrarian societies, where large families are critical for farm labor, gathering, or subsistence, and where birth control options are rudimentary. Death rates are measured with a similar statistic. High death rates (as high as 40 deaths per thousand people or more) are also associated with agrarian societies, where medical care may be rudimentary and where work is physically demanding and involves continued exposure to the elements.

Conversely, low birthrates (as low as perhaps 10 births per thousand people or fewer) are associated with contemporary urban societies, where children are less crucial for labor, where birth control is widely available, and where women are significant participants in the workforce. Low death rates are also common in contemporary urban settings, largely due to the ubiquity of improved health care and increasing life expectancy.

Where birthrates and death rates are roughly the same, no matter how high or low these might be, little or no growth in a population occurs. Most societies throughout history, therefore, did not actually experience high and sustained rates of population growth because their high birthrates were comparable to their death rates. Only when death rates become lower than birthrates does growth occur. If birthrates fall to match levels of death rates, conversely, growth begins to slow or stop.

Demographic transition predicts just such a change over time: from a period of no growth to a period of high growth and back again. In theory, once a society has moved from a preindustrial economic base to an industrial one, population growth slows. According to the demographic transition model, the slowing of population growth is attributable to improved economic production and higher standards of living brought about by changes in medicine, education, and sanitation.

As **Figure 1.26** illustrates, the high birth and death rates of the preindustrial phase (Phase 1) are replaced by the low birth and death rates of the industrial phase (Phase 4) only after passing through the critical transitional phase (Phase 2) and then more moderate rates (Phase 3) of natural increase and growth. The transitional phase of rapid growth is the direct result of early and steep declines in mortality while fertility remains at high, preindustrial levels. Recent demographic trends, particularly in countries such as Italy, where population growth is actually negative, has led some scholars to suggest a Phase 5, where deaths outstrip births, leading to a decrease in overall population. In this way of thinking, the relatively recent period of explosive population growth around the world, now followed in some regions by a period of decline, is simply a reflection of radical modern economic change. While the demands of growing populations are daunting (in terms of demands for

resources, for example), the demographic transition model suggests worldwide growth will "play itself out," ending altogether as soon as 2050 or 2060.

On the other hand, recent demographic history has shown that this transition may occur very differently in different world regions, countries, or even regions within countries if at all, raising questions about its universal application (see Chapter 9 for the startling example of South Asia). Although the model suggests that countries will pass through a transitional high-growth phase (between phases 2 and 3), many have observed that some regions become stalled there, a quagmire that has been called a "demographic trap." It should also be emphasized that the demographic transition model is based solely on the experience of countries in the global north and is thought by many experts to be less useful in explaining the demographic trends in the global south, whose entire development experience is quite different.

Most importantly, both population geographers and policymakers now also recognize that a close relationship exists between women's status and fertility. Women who have access to education and employment tend to have fewer children because they have less of a need for the economic security and social recognition that children are thought to provide. Success at damping population growth in less-developed countries appears to be tied to enhancing the possibility for a good quality of life and to empowering people, especially women, to make informed choices.

## CULTURE AND POLITICS

The relevance of culture and politics to the global geography of complex and extensive connections cannot be overstated. The world—whether seen as a constellation of interconnected world regions or a set of smaller networks of cities, neighborhoods and homes—is deeply affected by the circulation of cultural ideas and practices as well as the impact of political commitments and ideals. It is often argued that instantaneous communication is making the world a uniform place as people everywhere watch the same blockbuster films or operate within a largely comparable democratic political sphere. But these arguments don't take into account the fact that the dynamic and particular historical, environmental, and social context within which people live their daily lives provides a significant mediating influence on the globally extensive linkages—like hip-hop or the ballot box—that connect the world's people. The aim of this section is to identify some of the

 **FIGURE 1.26 Demographic transition** This model of economic development is based on the idea of successive stages of growth and change. Each stage is seen as leading to the next, though different regions or countries may require different periods of time to make the transition from one stage to the next.

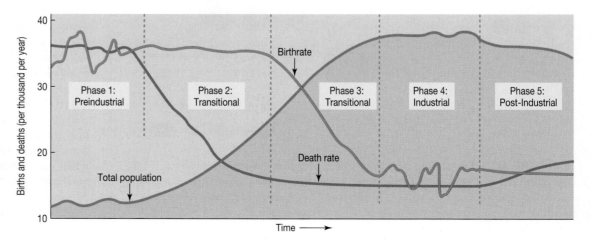

many key cultural and political forces that bind the world, not into a homogenous global space but one in which difference and similarity operate hand in hand. This aim is realized through the development of a conceptual framework for comprehending the cultural and political forces that produce both consistencies and rich variations among and between today's world regions.

## Tradition, Religion, and Language

**Tradition and Culture**    The term *culture* is often used to describe the range of activities that characterize a particular group, such as working-class culture, corporate culture, or teenage culture. Although this understanding of culture is accurate, it is incomplete. Broadly speaking, **culture** is a shared set of meanings that are lived through the material (dress or house style, for example) and symbolic (objects that represent something else, such as a cross as a symbol of Christianity) practices of everyday life. Culture is not something that is necessarily tied to a place and thus a fact to be discovered. Rather, it is the material and symbolic connections among people and places that can be altered and are therefore always changing, sometimes in subtle and other times in more dramatic ways (**Figure 1.27**). The "shared set of meanings" that constitutes culture can include values, beliefs, practices, and ideas about religion, language, family, food, gender, sexuality, and other important identities. These values, beliefs, ideas, and practices are, because of globalization, increasingly subject to reevaluation and redefinition and can frequently be transformed from both within and outside a particular group.

Culture is manifested in a wide range of forms from practices that persist over millennia or generations to ones that emerge seemingly overnight. Importantly, endurance does not necessarily mean consistent and unswerving reproduction. In fact, tradition is often the recreation of the spirit of a group belief, idea, or practice, rather than a slavish duplication. For example, the ceremony of the mass constitutes a 2,000-year-old tradition in Catholicism whether it's Roman, Coptic, or Greek Orthodox. But the actual performance of the mass by a priest and participation in the mass by a congregation can vary quite dramatically, sometimes even within the same church. One point to keep in mind about tradition and culture is the role the former plays in helping to maintain group identification and cohesion. Another is to recognize the rich variation that exists and to attempt to understand the ways that geography shapes it.

**Geographies of Religion**    The two most widely shared components of culture for most of the world's people are religion and language. **Religion** is a belief system and a set of practices that recognizes the existence of a power higher than humankind. Although religious affiliation is on the decline in some parts of the world, particularly among middle-class people, it still acts as a powerful shaper of daily life, from eating habits and dress codes to coming-of-age rituals and death ceremonies, to holiday celebrations and family practices. And, like language, religious beliefs and practices change as new interpretations are advanced or new spiritual influences are adopted.

The most important influence on religious change in the history of the world has been conversion from one set of beliefs to another. From the Arab invasions following Muhammad's death in 632 C.E., to the Christian Crusades of the Middle Ages, to the onset of modern period in the 15th century C.E., and into the present, religious missionizing—propagandizing and persuasion—as well as forceful and sometimes violent conversion have been key elements in changing geographies of

▲ **FIGURE 1.27  Belly dancers at the Doo Dah Parade, Pasadena, California**    The traditional dance form of the Middle East known in Egypt as *raqs sharki* or *rasbaladi* and in Turkey as *çiftetelli* came to be known in the United States, where it was first performed at the 1893 World's Fair, as belly dancing. This Arabic dance form is ancient, having originated in Babylon (present-day Iraq). Today the dance is performed around the world in both its traditional form as well as in a wide range of variations from Goth and hip-hop–inspired moves to reggae and rock to tribal fusion.

religion. Especially in the 500 years since the onset of the Columbian Exchange, traditional religions have become dramatically dislocated from their sites of origin not only through missionizing and conversion but also by way of diaspora and emigration. **Diaspora**—the spatial dispersion of a previously homogeneous group—and emigration involve the involuntary and voluntary movement of people who bring their religious beliefs and practices to their new locales. Missionizing and conversion are deliberate efforts to change the religious views of a person or people.

The processes of global political and economic change that have led to the massive movement of the world's populations over the last five centuries have also meant the dislodging and spread of the world's many religions from their historical sites of practice. Religious practices have become so spatially mixed that it is a challenge to present a map of the contemporary global distribution of religion. **Figure 1.28** identifies the contemporary distribution of what religious scholars consider to be the world's major religions because they contain the largest number of practitioners. As with other global representations, the map is useful in that it helps to present a generalized picture.

**Figure 1.29** identifies the source areas of four of the world's major religions and their diffusion from those sites over time. The map illustrates that the world's major religions originated and diffused from two fairly small areas of the globe. The first, where Hinduism and Buddhism (as well as Sikhism) originated, is an area of lowlands in the subcontinent of India drained by the Indus and Ganges rivers (Punjab on the map). The second area, where Christianity and Islam (as well as Judaism) originated, is in the deserts of the Middle East.

A notable observation about the present state of the global geography of religion is how during the colonial period religious missionizing and conversion flowed from Europe to the rest of the world. In the postcolonial period, however, an opposite trend has occurred. For example, the fastest-growing religion in the world today is Islam. And it is in world regions like Europe and the United States and Canada where Buddhism is making the greatest numbers of converts.

An important force in global religious practice today is fundamentalism, which is belief in or strict adherence to a set of basic principles, often in reaction to perceived doctrinal compromises with modern social and political life. Fundamentalism occurs across all religions and is an important force in global geographies as it is taken up as a political cause in regions as different as the United States and the Middle East.

Equally important, there is an historically significant and growing global population of agnostics, who remains unsure of the reality of divine beings and typically do not adhere to any particular religious faith, as well as atheists, who have no belief in god/gods. Among agnostics and atheists—who perhaps make up more than 15% of the global population—is a strong belief in the maintenance of a strict division between church and state, religion and politics. These practices tend to have less coalescence at the global scale, although organizations do provide networking around the world.

**Geographies of Language**   Language, like religion, is a key focus in understanding world regions because it is a central aspect of cultural identity. Without language, cultural accomplishments could not be transmitted from one generation to the next. And language itself reflects the ways that different groups regard and interpret the world around them. The distribution and diffusion of languages tell much about the changing history of human geography and the impact of globalization on culture. Before looking more closely at the geography of language and the impacts of globalization on the changing distribution of languages, however, it is necessary to become familiar with some basic vocabulary.

**Language** is a way of communicating ideas or feelings by means of a conventionalized system of signs, gestures, marks, or articulated vocal sounds. In short, communication is symbolic, based on commonly

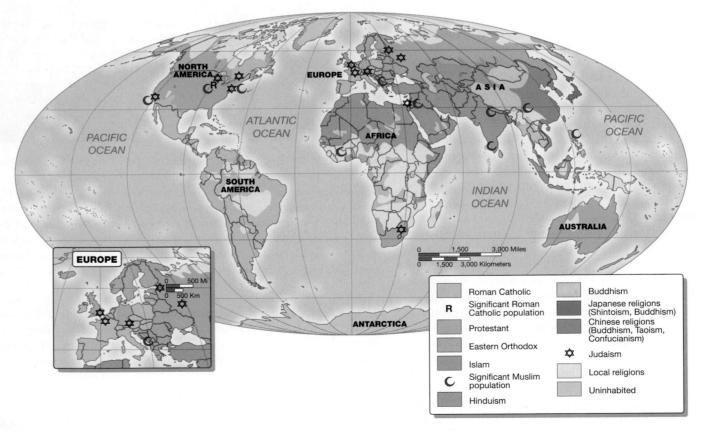

▲ **FIGURE 1.28 World distribution of major religions** Most of the world's people are members of these religions. Not evident on this map are the local variations in practices, as well as the many smaller religions that are practiced worldwide.

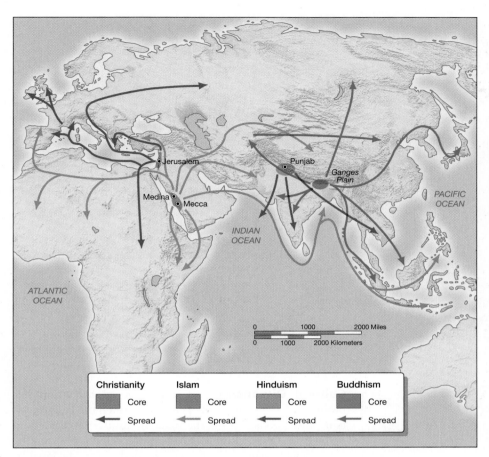

▲ **FIGURE 1.29 Origin areas and diffusion of four major religions** The world's major religions originated in the Middle East and South Asia. Christianity began in present-day Israel and Jordan. Islam emerged from western Arabia. Buddhism originated in India and Hinduism in the Indus region of Pakistan. These places are also the source areas of agriculture, urbanization, and key aspect of human development.

understood meanings of signs or sounds. Within standard languages (also known as official languages because they are maintained by offices of government such as educational institutions and the courts), regional variations, known as *dialects*, exist. Dialects emerge and are distinguishable through differences in pronunciation, grammar, and vocabulary that are place-based in nature. At the most general level, languages are grouped into families, which are collections of languages believed to be related in their prehistorical origins. These collections are further divided and subdivided.

Traditional approaches in cultural geography have identified the source areas of the world's languages and the paths of diffusion of those languages from their places of origin. Carl Sauer identified the origins of certain cultural practices with the label "cultural hearth." *Cultural hearths* are the geographic origins or sources of innovations, ideas, or ideologies. Language hearths are a subset of cultural hearths; they are the source areas of languages. **Figure 1.30** shows the locations of the world's indigenous languages or language families.

With globalization, the geography of language has become even more dynamic. For instance, the plethora of languages and dialects in a region like South Asia makes communication and commerce among different language speakers difficult. It may create problems for governing a population. That is why developing states often create one national language to facilitate communication and enable the efficient conduct of state busi-

ness. Unfortunately, where official languages are put into place, indigenous languages may eventually be lost (**Figure 1.31**). Yet the actual unfolding of globalizing forces—such as official languages—works differently in different places and in different times. Although the overall trend appears to be toward the loss of indigenous language (and other forms of culture), there have also been movements around the world to recuperate dying languages by introducing them into schools and offering courses for adults. In this way, language has become an important means of challenging the political, economic, cultural, and social forces of globalization as they occur in places like Belgium (among the Flemish people), Spain (with the Basque separatist movement), Canada (around the Québecois movement), and in other countries.

## Cultural Practices, Social Differences, and Identity

**Globalization and Culture** Anyone who has ever traveled between major world cities—or, for that matter, anyone who has been attentive to the backdrops of movies, television news stories, and magazine photojournalism—will have noticed the many familiar aspects of contemporary life in settings that, until recently, were thought of as being quite different from one another. Airports, offices, and international hotels have become notoriously alike, and their similarities of architecture and interior design have become reinforced by near-universal dress codes of the people who frequent them. The business suit, especially for males, has become the norm for office workers throughout much of the world. Meanwhile, jeans, T-shirts, and sneakers have become the universal attire for both young people and those in lower-wage jobs. The same automobiles can be seen on the streets of cities throughout the world (though sometimes they are given different names by their manufacturers); the same popular music is played on local radio stations; the same movies shown in local theaters; and the same brand names show up in stores and restaurants.

As a result, popular commentators have observed that cultures around the world are being Americanized, or "McDonaldized," which represents the beginnings of a single global culture that will be based on material consumption, with the English language as its medium. Yet, neither the widespread consumption of U.S. and U.S.-style products, nor the increasing familiarity of people around the world with global media and international brand names, adds up to the emergence of a single global culture (**Figure 1.32**). Rather, what is happening is that processes of globalization are exposing the globe's inhabitants to a common set of products, symbols, myths, memories, events, cult figures, landscapes, and traditions. People living in Tokyo or Tucson, Turin or Timbuktu, may be perfectly familiar with these commonalities without necessarily using or responding to them in uniform ways.

It is equally important to recognize that cultural flows take place in all directions, not just outward from the United States. Think, for example, of European fashions in U.S. stores; of Chinese, Indian, Italian, Mexican, and Thai restaurants in U.S. towns and cities; and of U.S. and European stores selling exotic craft goods from less-developed countries.

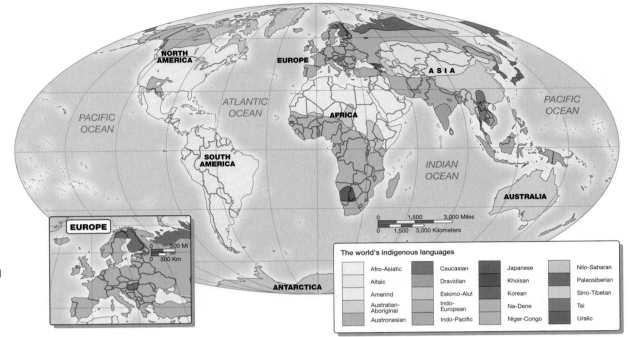

► FIGURE 1.30 World distribution of major language families Classifying languages by family and mapping their occurrence across the globe provides insights about important connections across space and time. For example, it is possible through mapping to discover linkages between seemingly unrelated cultures that are widely separated. This also provides understanding of the movement of populations across the globe from their areas of origin.

The world's indigenous languages

| | | | |
|---|---|---|---|
| Afro-Asiatic | Caucasian | Japanese | Nilo-Saharan |
| Altaic | Dravidian | Khoisan | Paleosiberian |
| Amerind | Eskimo-Alut | Korean | Sino-Tibetan |
| Australian-Aboriginal | Indo-European | Na-Dene | Tai |
| Austronesian | Indo-Pacific | Niger-Congo | Uralic |

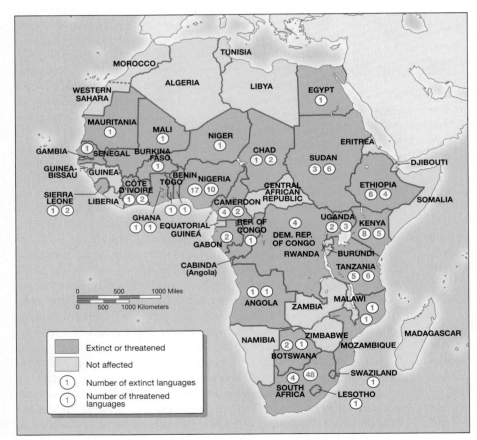

▲ FIGURE 1.31 African countries with extinct and threatened languages It is not absolutely certain how many languages are currently being spoken worldwide, but the estimates range between 4,200 and 5,600. Although some languages are being created through the fusion of an indigenous language with a colonial language, such as English or Portuguese, indigenous languages are mostly dying out. Although only Africa is shown in this map, indigenous languages are dying out throughout the Americas and Asia as well.

**Cultural Dissonance** At one level, globalization has brought a homogenization of culture through the language of consumer goods. This is the material culture predicated on Airbus jets, CNN, music video channels, cell phones, and the Internet and swamped by Coca-Cola, Budweiser, McDonald's, GAP clothing, Nikes, Walkmans, Nintendos, Toyotas, Disney franchising, and formula-driven Hollywood movies. These trends are transcending some of the traditional cultural differences around the world. It may make it easier to identify with people who use the same products, listen to the same music, and appreciate the same sports or music stars. At the same time, however, deep cultural differences are opening up between elites and marginalized people. By considering only superficial material consumption (clothing and food, for example), a focus on homogenization risks ignoring the emergence of new cultural gaps, often opening between previously compatible cultural groups and between ideologically divergent societies.

Several reasons account for the appearance of these new differences or "fault lines." One is the release of pressure brought about by the end of the Cold War. The evaporation of external threats has allowed people to focus on other perceived threats and intrusions. Another is the globalization of culture itself. The more people's lives are homogenized through their jobs and their material culture, the more many of them want to revive we/us feelings and reestablish a distinctive cultural identity. For the marginalized, a different set of processes is at work, however. The juxtaposition of poverty, environmental stress, and crowded living conditions alongside the materialism of wealthy elites creates a fertile climate for gangsterism. The same juxtaposition also provides the ideal circumstance for the spread and intensification of religious fundamentalism and for fundamentalist-inspired terrorism.

▲ **FIGURE 1.32 Michael Jackson impersonator in Ho Chi Minh City, Vietnam** Cult figures like Elvis Presley and Michael Jackson circulate widely around the world, even to the Socialist Republic of Vietnam, as shown here by way of an impersonator. But the ways that the Vietnamese audience understand and respond to Jackson, both at the height of his popularity and in response to his recent death, vary widely from the U.S. audience.

The overall result is that cultural fault lines are opening up everywhere. Metropolitan areas face a prospect of fragmented and polarized communities, with outright cultural conflict suppressed only through electronic surveillance (**Figure 1.33**) and the "militarization" of urban space via security posts and "hardened" urban design using fences and gated streets. This scenario presupposes a certain level of affluence to meet the costs of keeping the peace across economic and cultural fault lines. In the unintended metropolises of less-affluent world regions, where unprecedented numbers of migrants and refugees are thrown together, more grim prospects of anarchy and intercommunal violence exist.

**Culture and Identity**   Culture has an important impact on society through its influence on social organization. Social categories like kinship, gang, or generation, or some combination of categories, may figure more or less prominently in a group, depending on geography. Moreover, the salience of these social categories may change over time as the group interacts with people and forces outside its boundaries. Global media technologies, such as satellite television and the Internet, are increasingly shaping the potential for new social forms to emerge and old ones to be reconfigured. Generally, predominant forms of social organization in any of the world regions are likely to have persisted for hundreds of years, if not longer. It would be incorrect, however, to assume that both subtle and dramatic changes within these forms have not already occurred.

For example, kinship is a form of social organization that is particularly central to the culture of a wide range of world regions. **Kinship** is normally thought of as a relationship based on blood, marriage, or adoption. This definition can also include a *shared notion of relationship* among members of a group. The point is that not all kinship relations are understood by social groups to be exclusively based on biological or marriage ties. Although biological ties, usually determined through the father, are important in the Middle East, for instance, they are not the only important ties that link individuals and families. Though kinship is often expressed as a "blood" tie among social groups, neighbors,

▲ **FIGURE 1.33 Surveillance Cameras, Tianamen Square, China** Following the student democracy uprising 20 years ago here, the Chinese government has dramatically increased security in this important public place. There are now more than one hundred closed circuit television cameras in operation in the square keeping watchful eyes on those who traverse it.

friends, even individuals with common economic or political interests are often considered kin. Fraternities and sororities are examples in the United States of kinship systems—where the terms *brother* and *sister* are used to express solidarity.

Beyond cultural forms, such as religion and language, and traditional social groupings like kinship and tribe, many long-established and some more recently self-conscious cultural groups have begun to use their identities to assert political, economic, social, and cultural claims. Among these identities that are explored further in specific world regions are sexuality, ethnicity, race, and gender.

**Sexual Geographies**   Sexual identities are understood to be learned and expressed, or "performed," through standard, taken-for-granted practices that most people tend to view as "normal" or "natural." **Sexuality** is a set of practices and identities related to sexual acts and desires. Research on sexuality and the body in geography recognizes that space is central to our understanding of them. Where gay or straight identities can be performed plays a central role in who occupies those spaces, how they are occupied, and even what sexual identity you might be performing if you occupy one space or another. But issues of sexuality and space go beyond the performance of sexual identity to the emergence of new political cultures that are being constructed to protect the rights of lesbian, gay, and transgendered people in particular national and international spaces. For example, the increasing attention being paid to human rights issues around the globe has included sexual rights as a political cause (**Figure 1.34**).

**Ethnic Geographies**   Ethnicity is another area in which geographers are exploring cultural identity. **Ethnicity** is a socially created system of rules about who belongs to a particular group based on actual or perceived commonalities, such as language or religion. A geographic focus on ethnicity is an attempt to understand how it shapes and is shaped by space, and how ethnic groups use space with respect to mainstream culture. For cultural geographers, territory is also a basis for ethnic group cohesion. For example, cultural groups—ethnically identified or otherwise—may be spatially segregated from the wider society in ghettos or ethnic enclaves.

**Geographies of Race and Racialization**   Geographers also use prevailing ideas and practices with respect to race to understand the shaping of places. **Race** is a problematic and illusory classification of human beings based on skin color and other physical characteristics. **Racialization** is the practice of creating unequal castes where whiteness is considered the norm. Biologically speaking, however, no such thing as race exists within the human species.

Even without a biological reality, people's cultures often come to insist on the reality of racial differences. Cultural perceptions and practices can therefore create racialized places—such as the homelands of South Africa or the ghettos of Europe and the Russian Federation—and

the taken-for-grantedness of whiteness. In many places, whiteness is seen to be the norm or the standard against which all other visible differences are compared. This presents highly problematic opportunities for discrimination, unequal access to resources, and conflict or violence. But whiteness is itself a category of difference that depends on arbitrary visible distinctions and not biological ones and is always constructed in relation to other categories.

**Geographies of Gender**   Gender is another identity around which groups form. **Gender** is a category reflecting the social differences between men and women. Gender is not something people essentially are (because of a given set of physical characteristics), but something they do (something enacted by the presentation of bodily selves to the world). People may be born as biological males (a physically given condition), but learn how to act as men (a culturally imposed condition). As with other forms of identity, gender typically implies a socially created difference in power between groups. In the case of gender, the power difference gives an advantage to males over females and is not biologically determined, but socially and culturally created. Gender interacts with other forms of identity and can intensify power differences among and between groups. Furthermore, the implications of these differences are played out differently in different parts of the world. There are numerous examples that occur around the world.

For instance, women's subservience to men is deeply ingrained within South Asian cultures, and it is manifest most clearly in the cultural practices attached to family life, such as the custom of providing a dowry to daughters at marriage. The preference for male children in parts of East Asia is reflected in the widespread (but illegal) practice of selective abortion and female infanticide. Within marriages, many (but by no means all) poor women are routinely neglected and maltreated across all world regions. Women in rich countries, on average, earn less money than men do in the same jobs. But women have also organized for their rights and have secured improvements to their daily lives in terms of access to funding, education, safe drinking water, land, and birth control through personal sacrifice as well as support from government programs.

## Territory and Politics

**States and Nations**   Perhaps the most consequential political transformations for modern global geographies was the Peace of Westphalia in 1648, which created the concept of a Europe organized around sovereign states, and the overthrow or decline of monarchies in Europe in the late 18th to mid-19th centuries. A **sovereign state** exercises power over a territory and people and is recognized by other states with its independent power codified in international law. The result of these political innovations was the emergence and stabilization of a republican form of government. As distinct from monarchy, republican government requires the democratic participation and support of its population. Monarchical political power is derived from force and subjugation; republican political power derives from the support of the governed. By creating a sense of nationhood in tandem with republicanism, the newly emerging states of Europe were attempting to homogenize their multiple, and sometimes conflicting, constituencies so that they could govern with their active cooperation according to a sense of a common purpose.

▲ **FIGURE 1.34 Same-sex couple pose for a wedding photo in Beijing, China** Campaigns for the legalization of gay marriage are linked across the globe. Here a gay couple have their wedding photos taken on Valentine's Day, a time not only to celebrate loving relationships but also to advocate for gay rights and public approval of homosexuality.

In addition to enabling the creation of a stable democracy, the construction of a **nation**—a group of people sharing common elements of culture, such as language and religion—also facilitates the organization of a more extensive and coherent market where buyers and sellers all speak the same language and all have an investment in the success of the economic enterprise. Ultimately, a national identity is built not only on a common language but also on a common sense of history, geography, and purpose such that individuals feel compelled to defend the nation and to further the objectives of the state.

Over the last three and a half centuries, the republican form of government premised on a sovereign state has become the norm across the globe. Today, 195 states make up the totality of global sovereign entities. The state system—with some variations on it in the form of international and transnational organizations—constitutes the foundation for contemporary political, cultural, and economic life. And it is states and nations that are the building blocks of the world regions described in this book. States facilitate the flow of people, goods, and ideas across the globe in general, as well as across particular national boundaries. So, although the world has for thousands of years been a highly interconnected place, today's connectivities are negotiated through the complex architecture of the international state system, which not only controls all kinds of movements but also marks its citizens as members of a national community. An important observation to make about the international state system is the number of contemporary states that are at risk of failing. A **failed state** is one in which the central government is so weak that it has little control over its territory. In 2009, there were 20 such states, most of them in Africa.

The interconnectedness of the world has meant that now, more than ever, populations are highly mobile, and there has been a widespread dispersal of the world's people. As a result, contemporary members of a nation recognize a common identity, but they need not reside within a common geographical area. The term **nation–state** is an ideal form consisting of a homogeneous group of people, living in the same territory, and governed by their own state. In a true nation–state, no significant group exists that is not part of the nation. Today, there are very few true nation–states because mobility and history have meant that states govern highly multinational populations. **Nationalism** is the feeling of belonging to a nation as well as the belief that a nation has a natural right to determine its own affairs (**Figure 1.35**). Note that contemporary globalization has been accompanied by a rise in nationalism that has been manifested in both peaceful and conflictual ways. In the case of the latter, these conflicts are often over territory.

**Regionalism** often involves ethnic groups whose aims include autonomy from a national state and development of their own political power. An example is Basque regionalism, which has roots in the period at the turn of the 19th century. The Basque people of northeastern Spain and the southern part of Aquitaine in southwestern France (**Figure 1.36**) feared that cultural forces accompanying industrialization would undermine their traditions. To counter this trend, the Basque provinces of Spain and France sought autonomy from those states for most of the 20th century. Since the 1950s, agitation for political independence has occurred through terrorist acts—especially for the Basques in Spain. For more than 25 years, the French, Spanish, and more recently the Basque regional police have attempted to undermine the Basque Homeland and Freedom movement through arrests and imprisonments. However, not even the Spanish move to parliamentary democracy and the granting of autonomy to the Basque provinces has been able to slake

▲ **FIGURE 1.35 British National Party** Extremist forms of nationalism exist throughout the globe. One such right-wing manifestation of nationalism is the whites-only British National Party, which according to its constitution is "committed to stemming and reversing the tide of non-white immigration and to restoring, by legal changes, negotiation and consent, the overwhelmingly white makeup of the British population that existed in Britain prior to 1948." The BNP also proposes "firm but voluntary incentives for immigrants and their descendants to return home."

the thirst for self-determination among the Basques in Spain. Elections in summer 2006 in Catalonia (a semiautonomous region in Spain) point to the possibility of increasing political independence from Spain for them as well as the Basques. Meanwhile, on the French side of the Pyrenees, although a Basque separatist movement does exist, it is neither as violent nor as active as the movement in Spain.

▲ **FIGURE 1.36 Basque independence march** For decades, the Basque people of northwestern Spain have campaigned, peacefully and more violently, for complete autonomy and political independence. One of the Spanish provinces, the Basque region is highly industrialized with the highest GDP in Spain, which is also 40% higher than the European Union.

In certain cases, enclaves of ethnic minorities are claimed by the government of a country other than the one in which they reside. Such is the case, for example, of Serbian enclaves in Croatia, claimed by nationalist Serbs. When the government of a country asserts that a minority living outside its formal borders belongs to it historically and culturally, it is known as **irredentism**. In some circumstances, as with Serbia's claims on Serbian enclaves in Croatia in the early 1990s, irredentism can lead to war.

Not to be confused with regionalism or irredentism is **sectionalism**, an extreme devotion to regional interests and customs. Sectionalism has been identified as an overarching explanation for the U.S. Civil War. It was an attachment to the institution of slavery and the political and economic way of life that slavery enabled that prompted the southern states to secede from the Union. The Civil War was fought to ensure that sectional interests would not take priority over the unity of the whole—that is, that states' rights would not undermine the power of the federal government. Although the Civil War was waged around the real issue of permitting or prohibiting slavery, it was also fought at another level, a level that dealt with issues of the power of the federal state. The election of Abraham Lincoln to the presidency in 1860 reflected the sectionalism that dominated the country: He received no support from slave states.

**Political Globalization**   Political and cultural globalization have not been evenly matched by the growth of political globalization or a comprehensive system of government that can cope with these other powerful forces. Policymakers, for the most part, lack an adequate framework for coping with the consequences of globalization. Although institutions such as the World Criminal Court, the World Trade Organization, and the Kyoto Agreement constitute the beginnings of a global governance framework for crime, the economy, and the environment, their effectiveness is somewhat limited by the persistence of individual states. Further, trade policy has come to be governed by powerful transnational corporations, whereas national governments are unable to addess transnational environmental issues.

Instead of a coordinated system of global governance to accompany economic globalization, **supranational organizations**—collections of individual states with a common goal that may be economic or political and that diminish individual state sovereignty in favor of member states' interests—have emerged (**Figure 1.37**). Examples of supranational organizations include the European Union (EU), the North American Free Trade Agreement (NAFTA), and the Association of South East Asian Nations (ASEAN). In the absence of a deliberately organized global governance system, the United Nations has become something of a de facto replacement. Founded in 1945, the United Nations is an international organization aimed at facili-tating cooperation in international law, security, economic development, human rights, and world peace.

With headquarters in New York City, the United Nations is composed of five administrative organs. The first is the Security Council, which includes as permanent members the United States, Great Britain, China, France, and Russia. A second component of the organization is the General Assembly, which includes every one of the world's internationally recognized sovereign states (**Figure 1.38**). Absent from the General Assembly are Western Sahara and Kosovo, whose sovereignty is contested, and Vatican City, which is a nonmember observer. In addition to these two administrative bodies, the United Nations also includes the Economic and Social Council, the Secretariat, and the International Court of Justice, the latter of which is located in The Hague in the Netherlands and began operating in 1946.

At the same time that formal political organizations like the United Nations have been expanding their purview, informal political organizations and movements have also increased in overall numbers and membership and have become more globally connected in the last 50 years. Known as **social movements**—large, informal groupings of individuals and organizations that focus on social or political issues—these groups have come to constitute an arena known as global civil society. **Global civil society** is the worldwide array of voluntary civic and social organizations and institutions—also called nongovernmental organizations, or NGOs—that operate to represent the interests of citizens against the power of formal states and markets. Social movements embrace a spectrum of issues including women's rights, the environment, human rights, climate change, and tens of thousands of other issues that affect people very directly and around which people organize to secure formal political responses. Greenpeace (headquartered in Great Britain) is an example of an international social movement, as is Genocide Watch (headquartered in the United States) and the Green Belt Movement (headquartered in Kenya).

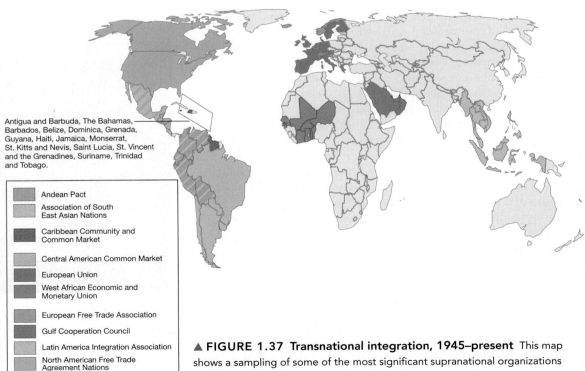

Antigua and Barbuda, The Bahamas, Barbados, Belize, Dominica, Grenada, Guyana, Haiti, Jamaica, Monserrat, St. Kitts and Nevis, Saint Lucia, St. Vincent and the Grenadines, Suriname, Trinidad and Tobago.

- Andean Pact
- Association of South East Asian Nations
- Caribbean Community and Common Market
- Central American Common Market
- European Union
- West African Economic and Monetary Union
- European Free Trade Association
- Gulf Cooperation Council
- Latin America Integration Association
- North American Free Trade Agreement Nations

▲ **FIGURE 1.37 Transnational integration, 1945–present** This map shows a sampling of some of the most significant supranational organizations around the globe.

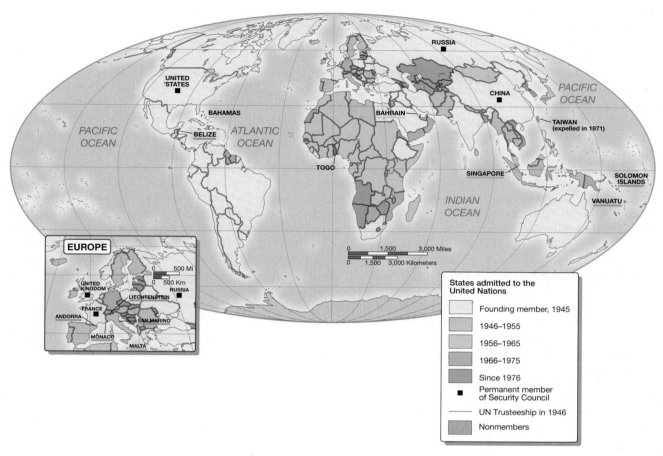

▲ **FIGURE 1.38 The United Nations member countries** Nearly all the sovereign states of the globe are members of the United Nations. This international organization, though powerful, does not supersede the sovereignty of its individual members.

# A WORLD OF REGIONS

To this point, the text has focused almost exclusively on building a *global geography* framework. But the remainder of this textbook is organized regionally. This is not simply because of convention but because a **regional geography approach** to global geography, with a focus on world regions in particular, affords an opportunity to examine global and regional complexity and difference in a comparative context. World regions are dynamic, with changing properties and fluid boundaries that are the product of the interplay of a wide variety of environmental and human factors. This dynamism and complexity is what makes travel so fascinating for many people. It is also what makes regions so influential in shaping people's lives. A regional geography approach to studying global geography allows us, as geographers, to combine elements of both physical and human geography, studying how the unique combinations of environmental and human factors produce world regions with distinctive landscapes and cultural attributes. What is distinctive about the study of regional geography is not so much the phenomena that are studied as the way they are approached. The contribution of regional geography is to reveal how natural, social, economic, political, and cultural phenomena come together to produce distinctive geographic settings.

One of the most important tenets of regional geography is that regions are not just distinctive outcomes of geographic processes; they are part of the processes themselves. They are created by people responding to the opportunities and constraints presented by their environments. As people live and work in particular geographic settings, they gradually impose themselves on their environments, modifying and adjusting the environment to meet their needs and express their values. At the same time, people gradually accommodate to both their physical environments and to the people around them. There is a continuous two-way process, in which people create and modify regions, while at the same time being influenced by the settings in which they live and work. Processes of geographic change are constantly modifying and reshaping regions, and the region's inhabitants are constantly coping with change. It is often useful to think of regions as representing the cumulative imprint of successive periods of change. For example, present-day Turkey embodies elements of Roman, medieval, Arabic, Persian, Ottoman, and modern European influences, among others.

Investigations of specific places must be framed within the compass of the entire globe, however, for two reasons. First, the world consists of a mosaic of places and regions that are interrelated and interdependent in many ways. Second, place-making forces—especially economic, cultural, and political forces that influence the distribution of human activities and the character of places and regions—have long operated globally. The interdependence of regions means that individual places are tied to wider processes of change that are reflected in broader geographic patterns.

Although the regional approach provides a rich and intuitively appealing way of organizing knowledge about the world, there are various methods of regional analysis and different ways of identifying and defining regions. In this book, this emphasis is on the interdependence of regions, explaining and analyzing them as the outcomes, in different physical environments, of successive eras of human activity that have been organized on the basis of different economic, cultural, and political systems and successive phases of economic and technological development. And while recognizing that integration does tend to blur some national and regional differences as the global marketplace brings a dispersion of people, tastes, and ideas, the overall result has often been an intensification of the differences.

**The Conceptual Language of a Regional Approach**  At the heart of geographers' concern with understanding how combinations of environmental and human factors produce regions with distinctive landscapes and cultural attributes is the belief that this "regional approach" is one of the best ways of organizing knowledge about the world. This, as with other branches of geography, depends on a working knowledge of one of geographers' basic tools: maps (see Appendix 1). However, there are different methods of regional analysis and different ways of identifying and defining regions. At the same time, place-making and regional differentiation carry over to affect landscapes and people's **sense of place**, which are the feelings evoked as a result of the experiences and memories that people associate with a place.

**Regionalization** is the geographer's classification of individual places or areal units; it is the geographer's equivalent of scientific classification. The purpose of regionalization is to identify "regions" of one kind or another. There are several ways individual areal units can be assigned to classes (regions). One is that of *logical division,* or "classification from above." This involves partitioning a universal set of areal units into successively larger numbers of classes, using more specific criteria at every stage. So a world regional classification of national states might be achieved by first differentiating between rich and poor countries, then dividing both rich and poor countries into those countries that have a trade surplus and those that have a deficit, and so on. A second way in which individual areal units can be assigned to classes (regions) is that of *grouping,* or "classification from below." This involves searching for regularities or significant relationships among areal units and grouping them together in successively smaller numbers of classes, using a broader measure of similarity at each stage.

**World regions** are extensive geographic divisions based on both physiographic *and* human historical processes. These regions are constructed, unraveled, and reconstructed as the realities of natural resources and technologies form a framework of opportunities and constraints to which particular cultures and societies respond. Superimposed on these regions, sometimes with only an approximate fit, are the formal, *de jure* territories of national states (*de jure* means "legally recognized"). Because of the inherent power of national governments, especially in relation to the flows of the goods, money, and information, national states represent a geographic scale that is often very significant. Once their boundaries are set, principles of national sovereignty mean that these boundaries tend increasingly to become regarded by

their inhabitants as somehow natural or immutable. National political boundaries are not fixed and unchanging, however. When economic circumstances change, national states may feel the need to adjust their boundaries or seek other means of accommodating economic reality, such as joining supranational organizations.

In the end, the regions introduced in this book are the *product* of ongoing physical and human geographic global dynamics. World regions are, after all, connected to one another through global systems (ecological, climatic, economic, and political); they are also transformed by these relations. In this sense, regions are the outcomes of a set of world-spanning systems. Each regional chapter stresses the global systems of connection that drive unique regional processes, thus making regions different. Studying regions promotes not only learning the critical elements of different places, but also understanding the fundamental processes that drive change. In sum, each regional chapter helps establish and explore the central theses of the book, in that world regions must always be thought of as *connected* and *related*. To understand how the world comes to be the way it is, world regions must be understood in a global context.

Finally, such an approach is necessary to anticipate or suggest how and in what ways global geographies may change in the future. By forecasting the changing relationships *between* regions, we can potentially anticipate the future character of those regions. Similarly, by thinking about the direction of change *within* regions, we can imagine how new connections between regions might emerge. In this way, this chapter concludes as do all our regional chapters, with a set of conceptual questions and real-world problems or issues that are likely to be important to reimagine the world and its regional differences and distinctions.

# FUTURE GEOGRAPHIES

At present, much of the established familiarity of the modern world and its geographies seems to be disappearing. The world has entered a period of transition, triggered by the end of the Cold War in 1989 and rendered more complex by the geopolitical and cultural repercussions of the terrorist attacks of September 11, 2001. Each of the regional chapters ahead contains a section that speculates about the way that events such as these or other current trends are likely to shape the future of that region. How those changes might affect other regions or global geography more generally is impossible to predict. What is knowable is the base from which the world is currently operating. That is, there are extensive data on human and physical resources, climate change, economic growth trends, technological innovation, and other key components of the global geography of the early 21st century. Extrapolating from what is known, it is possible to make some predictions about what might emerge in the future. Each global trend portends very different implications for the future of specific regions. Consider radically higher or dramatically lower oil prices on world markets, and what that might mean for countries like Saudi Arabia, the United States, or China.

**Resources and Development**  The expansion of the global economy and the globalization of industry will undoubtedly boost the overall demand for raw materials of various kinds, and this will spur the development of some previously underexploited but resource-rich regions in Africa, Europe, and Asia. Raw materials, however, will be only a fraction of future resource needs. The main issue, by far, will be energy resources. World energy consumption has been increasing

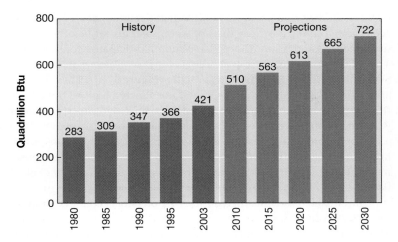

▲ **FIGURE 1.39 Trends in energy consumption** Global commercial energy consumption is expected to continue to rise steadily over the next two decades as peripheral countries continue to develop economically and require a larger share of the global energy pie.

steadily over the recent past (**Figure 1.39**). As the global south is industrialized and its population increases further, the demand for energy will expand rapidly. Basic industrial development tends to be highly energy intensive. The International Energy Agency, assuming (fairly optimistically) that energy in developing countries will be generated in the future as efficiently as it is in developed ones now, estimates that developing-country energy consumption will more than double by 2025, lifting total world energy demand by almost 50%.

Much of this demand will be driven by industrialization geared to meet the growing worldwide market for consumer goods, such as private automobiles, air conditioners, refrigerators, televisions, and household appliances. Without higher rates of investment in exploration and extraction than at present, production will be slow to meet the escalating demand. Many experts believe that current levels of production, in fact, represent "peak oil," that the world is at the halfway point of depleting its finite reserves of crude oil.

In countries that can afford research and development, new materials will reduce the growth of demand both for energy and for traditional raw materials such as aluminum, copper, and tin. Improved engineering and product design will also make it possible to reduce the need for input of some resources. In addition, the future may well bring technological breakthroughs that dramatically improve energy efficiency or make renewable energy sources (such as wind, tidal, and solar power) commercially viable. As with earlier breakthroughs that produced steam energy, electricity, gasoline engines, and nuclear power, such events would catalyze a major reorganization of the world's economic geographies.

**Economic Globalization and Global Governance**   Globalization is as much about restructuring geoeconomics as it has been about reshaping geopolitics. In the 20th century, from the end of World War II until 1989, when the Berlin Wall was dismantled, world politics was organized around two superpowers. The capitalist West rallied around the United States and the communist East gathered around the Soviet Union. But with the collapse of a bipolar world order, a new world order, organized around global

capitalism, emerged and has solidified around a new set of political powers and institutions. The new world order enabled by the technological (particularly communication and transportation) changes of the last 25 years is restructuring the architecture as well as the conduct of contemporary politics at both the international and domestic levels.

Regional and supranational organizations such as the EU, the North American Free Trade Agreement (NAFTA), the Association of Southeast Asian Nations (ASEAN), Organization of Petroleum Exporting Countries (OPEC), and the World Trade Organization (WTO) are increasingly constituting the world or respective regions as seamless trading areas not hindered by the rules that ordinarily regulate national economies. The increasing importance of these trade-facilitating organizations is the most telling indicator that the world, besides being transformed into one global economic space, is also experiencing global geopolitical transformations. Rather than disappearing altogether, the powers and roles of the modern state are changing, as it is forced to interact with these sorts of organizations as well as with a whole range of other political institutions, associations, and networks.

The involvement of the state in new global activities, the growth of supranational and regional institutions and organizations, the critical significance of transnational corporations to global capital, and the proliferation of transnational social movements and professional organizations is captured by the term **international regime**. The term is meant to convey the idea that the orientation of contemporary politics is now directed to the international arena, rather than the national one. This is so much the case that even city governments and local interest groups—from sister city organizations to car clubs—are making connections and conducting their activities beyond as well as within the boundaries of their own states.

**Security**   In some regions, it is quite possible that weak governments, lagging economies, religious extremism, and increasing numbers of young people will align to create a "perfect storm" for internal conflict, with far-reaching repercussions for security elsewhere. Such internal conflicts, particularly those that involve ethnic groups straddling national boundaries, risk escalating into regional conflicts and the failure of states, with expanses of territory and populations devoid of effective governmental control. These territories can become sanctuaries for transnational terrorists (such as Al-Qaeda in Afghanistan) or for criminals and drug cartels (such as in Colombia). According to the U.S. National Intelligence Council, within the next 15 years Al-Qaeda itself will likely be superseded by similarly inspired Islamic extremist groups, and there is a substantial risk that broad Islamic movements akin to Al-Qaeda will merge with local separatist movements. Information technology, allowing for instant connectivity, communication, and learning, will enable the terrorist threat to become increasingly decentralized, evolving into an eclectic array of groups, cells, and individuals that do not need a stationary headquarters to plan and carry out operations. Training materials, targeting guidance, weapons know-how, and fundraising will become virtual (existing online). As biotechnology information becomes more widely available, the number of people who can potentially misuse such information and wreak widespread loss of life will increase. Moreover, as biotechnology advances become more ubiquitous, stopping the progress of offensive biological warfare programs will become increasingly difficult. Over the next 10 to 20 years, there is a risk that advances in biotechnology will allow the creation of advanced biological agents designed to target specific systems—human, animal, or crop.

Economic and political turbulence and instability will also be conducive to transnational crime. Transnational crime syndicates pose a considerable threat to global security. They distribute harmful materials, weapons, and drugs; exploit local communities; disrupt fragile ecosystems; and control significant economic resources. In 2003, transnational crime syndicates may have grossed up to $2 trillion—more than all national economies except the United States, Japan, and Germany.[2] The bulk of crime syndicates' revenue comes from drug trafficking, but other significant sources include environmental products—everything from protected plants and animals to hazardous waste and banned chemicals. Trafficking in humans—for labor, sex work, and even for the removal of kidneys or other organs for transplant purposes—is another aspect of transnational crime. The U.S. State Department estimates that at least 600,000 to 800,000 people are sold internationally each year.

**Future Environmental Threats and Global Sustainability**   The world currently faces a daunting list of environmental threats: the destruction of tropical rainforests and the consequent loss of biodiversity; widespread, health-threatening pollution; the degradation of soil, water, and marine resources essential to food production; stratospheric ozone depletion; acid rain; and so on. Most of these threats are greatest in the world's poorest regions, where daily environmental pollution and degradation represent a catastrophe that will continue to unfold, in slow motion, in the coming years (see The Asian Brown Cloud, Chapter 8, p. 282).

Future trends will only intensify the contrasts between rich and poor regions. Environmental problems will be inseparable from processes of demographic change, economic development, and human welfare. In addition, it is becoming clear that environmental problems are going to be increasingly enmeshed in matters of national security and regional conflict. The spatial *interdependence* of economic, environmental, and social problems means that some parts of the world are ecological time bombs. The prospect of civil unrest and mass migrations resulting from the pressures of rapidly growing populations, deforestation, soil erosion, water depletion, air pollution, disease epidemics, and intractable poverty is real. Another dimension of the environmental issues is health issues that have emerged as a result of increases in global commerce and tourism. As people and goods move around the world, so do other species, pathogens, and zoonotic diseases (transmitted

between animals and people). New plant and animal species can invade ecosystems, choke pastures, disrupt water systems, drive other species to extinction, and result in expensive and unanticipated consequences. The total costs of losses from all kinds of invasive species in the United States now exceeds $138 billion annually (**Figure 1.40**). These specters are alarming not only for the people of the affected regions but also for the people of rich and faraway countries, whose continued prosperity will depend on processes of globalization that are not disrupted by large-scale environmental disasters, unmanageable mass migrations, or the breakdown of stability in the world-system as a whole. These pose a much greater security risk as a result of the speed and intensity of global flows.

At the same time, the interconnectivity of global economies and politics offers avenues for environmental innovation. The **Kyoto Protocol** is an agreement of international law that imposes reductions of greenhouse gas emission (including carbon) on countries that have ratified it. Though the United States announced in 2001 that it would not honor its commitment to the Kyoto agreement, it came into force for other signatory nations in February 2005. In negotiations since then, the United States and other important nations (including China and India) have continued to move toward some forms of greenhouse gas emission control. Such agreements might propel very different regional economies (fostering alternative energies, for example). In the process, these nations have begun to renegotiate their geopolitical commitments to one another with increased exchange of technology and ideas. Environmental problems open the possibility of new regions and new connections.

**Regional Integration and Fragmentation**   Will globalization bring about increased regional integration so that our current regions will remain the building blocks of the world order? Or will overall fragmentation lead to the emergence of new regions? The answers to these two questions are not easy, because for different parts of the world, there is evidence to support both successful integration and continued fragmentation. In Europe, for instance, regional integration has been remarkably successful. Since 1957, the various states of Europe have become increasingly integrated around political and economic concerns, so much so that the European Union (organized in 1957 as the European Economic Community, or EEC) is a shining global example of a coherent and prosperous world region. Elsewhere, however, especially in Africa and Latin America, attempts at regional integration have been far less successful, and in fact, subnational pressures have led to more fragmentation than integration. A helpful way to think about the possibilities for future regional integration is to consider why regional integration might be attractive for some states and troubling for others.

---

[2]E. Assadourian, "Transnational Crime," in Worldwatch Institute, *State of the World 2005* (New York: W. W. Norton, 2005), p. 20.

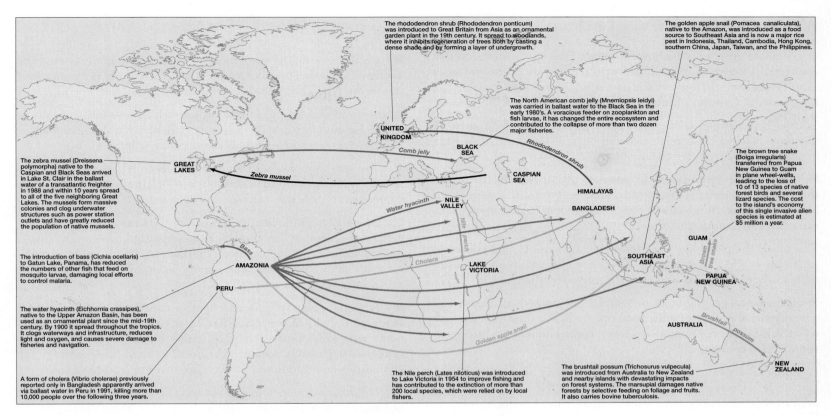

The rhododendron shrub (Rhododendron ponticum) was introduced to Great Britain from Asia as an ornamental garden plant in the 19th century. It spread to woodlands, where it inhibits regeneration of trees both by casting a dense shade and by forming a layer of undergrowth.

The golden apple snail (Pomacea canaliculata), native to the Amazon, was introduced as a food source to Southeast Asia and is now a major rice pest in Indonesia, Thailand, Cambodia, Hong Kong, southern China, Japan, Taiwan, and the Philippines.

The North American comb jelly (Mnemiopsis leidyi) was carried in ballast water to the Black Sea in the early 1980's. A voracious feeder on zooplankton and fish larvae, it has changed the entire ecosystem and contributed to the collapse of more than two dozen major fisheries.

The brown tree snake (Boiga irregularis) transferred from Papua New Guinea to Guam in plane wheel-wells, leading to the loss of 10 of 13 species of native forest birds and several lizard species. The cost to the island's economy of this single invasive alien species is estimated at $5 million a year.

The zebra mussel (Dreissena polymorpha) native to the Caspian and Black Seas arrived in Lake St. Clair in the ballast water of a transatlantic freighter in 1988 and within 10 years spread to all of the five neighboring Great Lakes. The mussels form massive colonies and clog underwater structures such as power station outlets and have greatly reduced the population of native mussels.

The introduction of bass (Cichla ocellaris) to Gatun Lake, Panama, has reduced the numbers of other fish that feed on mosquito larvae, damaging local efforts to control malaria.

The water hyacinth (Eichhornia crassipes), native to the Upper Amazon Basin, has been used as an ornamental plant since the mid-19th century. By 1900 it spread throughout the tropics. It clogs waterways and infrastructure, reduces light and oxygen, and causes severe damage to fisheries and navigation.

A form of cholera (Vibrio cholerae) previously reported only in Bangladesh apparently arrived via ballast water in Peru in 1991, killing more than 10,000 people over the following three years.

The Nile perch (Lates niloticus) was introduced to Lake Victoria in 1954 to improve fishing and has contributed to the extinction of more than 200 local species, which were relied on by local fishers.

The brushtail possum (Trichosurus vulpecula) was introduced from Australia to New Zealand and nearby islands with devastating impacts on forest systems. The marsupial damages native forests by selective feeding on foliage and fruits. It also carries bovine tuberculosis.

◀ **FIGURE 1.40 Worldwide invasive species** Shown is a sampling of the many invasive species that are changing the world's ecosystems.

Probably the main reasons to pursue regional integration involve strategic and security issues. As will be discussed in the regional chapters, regional trading blocks are created because they are understood to enable economic and trading advantages. The one variable that demonstrates the foundation of contemporary globalization is the increase in global trade linkages (see **Six Key Factors of Globalization**, p. 22). Regional trading blocks are also believed to provide security for weaker states against economic and political threats. Following the devastation wreaked on Europe by World War II and compounded by the loss of their colonies, the member states of the EEC viewed the regional trading block as a way to increase political and economic leverage in world affairs. It was widely understood that no single state could have exercised the power that a cooperative union could. The perceived military threat of the Soviet Union was also a factor in the decision to establish a regional bloc among the founding, Western European, members of the EEC. Finally, since its founding, the still-developing European Union has been seen by smaller and later-joining states—such as Denmark and Ireland—as a way to improve their political and market status, something they would lack if they remained isolated.

But just as there are strong incentives for forming regional blocks, there are also substantial obstacles that make it difficult for states to cooperate and intensify any blocks that already exist. These obstacles tend to revolve around difference and the desire by states to maintain control over their own affairs. The issue of difference centers on the fear that integration can undermine the cultural traditions of any particular nation because cooperation would necessarily require standardization of practices.

# Main Points Revisited

■ A global geography framework builds from the assumption that connections between and within world regions are part of a longer set of historical relations, which tie the world together environmentally, economically, politically, socially, and culturally.

The world has always been global, and a global geography framework illustrates how global connections and relationships have, over time, organized and reorganized the world.

■ The unique environmental characteristics of regions are formed through the connections that global climate and tectonic systems make between regions. Human beings adapt to regional environmental conditions while remaking nature.

This opens the possibility both for regional environmental crises and problems to affect the globe, but also for global environmental solutions to address potential and emerging regional problems.

■ A historical geography of global relations suggests that economic connections and demographic patterns are best understood through a broad lens that contextualizes those connections and patterns in comparative context.

Economic systems and flows have evolved over time, affecting the demographic patterns seen today. Although demographic change is not tied exclusively to economy, it is clear that as societies become more complex—politically, economically, socially, and culturally—their relationships to their own population practices change as well.

■ Culture is a shared set of meanings manifest in both material and symbolic ways that connect people to places. Culture is dynamic and can be altered in both dramatic and subtle ways.

At the same time that globalization has helped to transcend cultural divides among the world's people, new fault lines are also opening over perceived threats and intrusions over existing cultural norms.

■ The state system—with some variations on it in the form of international and transnational organizations—constitutes the foundation for contemporary political, cultural, and economic life around the globe.

But instead of a coordinated system of global governance to accompany economic globalization, supranational organizations—collections of individual states with a common goal that may be economic or political and that diminish individual state sovereignty in favor of member states' interests—have emerged.

■ Regional geography is an effective conceptual framework for studying the global processes tied to environmental, political–economic, and sociocultural change over time. It provides a lens for examining wider global processes in regional context, highlighting both similarities and differences between and within world regions.

The nature of world regions evolves and changes over time. To predict or imagine the future of the world requires a careful examination of the characteristics of world regions but also, therefore, an exploration of the changing relationship between regions. This text offers both.

# Key Terms

biodiversity (p. 10)

biogeography (p. 10)

biomes (p. 10)

capitalism (p. 16)

climate (p. 3)

Cold War (p. 19)

colonialism (p. 16)

colonization (p. 15)

commodity (p. 11)

commodity chains (p. 11)

communism (p. 17)

continental drift (p. 8)

cultural hearth areas (p. 15)

culture (p. 29)

deforestation (p. 10)

demographic transition (p. 27)

diaspora (p. 29)

earth system science (p. 3)

ecosystems (p. 10)

ethnicity (p. 34)

failed state (p. 35)

geomorphology (p. 8)

gender (p. 34)

gender-related development index (p. 26)

global civil society (p. 36)

global north (p. 23)

global south (p. 23)

global warming (p. 5)

globalization (p. 23)

greenhouse effect (p. 5)

gross domestic product (GDP) (p. 23)

gross national product (GNP) (p. 23)

hegemony (p. 16)

imperialism (p. 16)

international division of labor (p. 21)

International Monetary Fund (IMF) (p. 23)

international regime (p. 39)

intertropical convergence zone (ITCZ) (p. 4)

irredentism (p. 36)

kinship (p. 33)

Kyoto Protocol (p. 40)

language (p. 30)

leadership cycles (p. 16)

mercantilism (p. 16)

nation (p. 35)

nationalism (p. 35)

nation–state (p. 35)

neocolonial (p. 21)

neoliberalism (p. 23)

pastoralists (p. 12)

plantation economies (p. 16)

plate tectonics (p. 6)

primary activities (p. 23)

primary commodities (p. 16)

purchasing power parity (PPP) (p. 23)

quaternary activities (p. 23)

race (p. 34)

racialization (p. 34)

regional geography approach (p. 37)

regionalism (p. 35)

regionalization (p. 38)

religion (p. 29)

secondary activities (p. 23)

secondary commodities (p. 16)

sectionalism (p. 36)

sense of place (p. 38)

social movements (p. 36)

sovereign state (p. 34)

spatial justice (p. 23)

staple crops (p. 12)

structural adjustment policies (p. 25)

supranational organizations (p. 36)

sustainable development (p. 12)

swidden (p. 12)

tertiary activities (p. 23)

Three Gorges Dam (p. 10)

transgenic crops (p. 10)

tsunami (p. 9)

weather (p. 3)

World Bank (p. 23)

world regions (p. 38)

world systems (p. 2)

# Thinking Geographically

1. What is meant by the term "global geography," and how does a global geography framework relate to a regional geography approach?

2. What is global climate change, and what is the evidence that it may be happening? What might its impacts include?

3. What are the impacts of the global climate and tectonic system on human activities? How have human activities altered hydraulic and ecological systems?

4. How has commoditization of nature influenced human–environment relations in the last two centuries, and what does this suggest for the possibility of sustainable development?

5. How has the world changed, geopolitically, from the time of the Cold War to now? Are there any similarities between the Cold War period and the post Cold-War period?

6. As the global marketplace becomes more accessible to more people worldwide, why will regional geography remain relevant? List at least five reasons why people will maintain their sense of place in an increasingly global economy.

7. What are some of the differences in the geographic (or mapped) knowledge of the world for different peoples in 1491?

8. What may be some of the worldwide demographic trends over the next 25 years?

9. Globalization has a complex impact on cultural identity. Why do some groups seek to protect themselves from it while others embrace it?

10. What is global civil society, and how does it function with respect to formal states?

 Log in to www.mygeoscienceplace.com for Videos, Interactive Maps; RSS feeds; Further Readings; Suggestions for Films, Music and Popular Literature; and Self-Study Quizzes to enhance your study of World Regions in Global Context.

**▼ FIGURE 2.1**

Legend:
- ◉ More than 5 million
- ○ 1–5 million
- • Fewer than 1 million

Capital cities are underlined

# EUROPE KEY FACTS

- Major subegions: Eastern Europe, Northern Europe, Southern Europe, Western Europe

- Total land area: 10 million square kilometers/ 4 million square miles

- Major physiogeographic features: Alps, Pyrenees, and Carpathians mountains; longest river in Europe is Volga at 3.7 kilometers/ 2.9 miles; largest island in the region is Great Britain at 220,000 square kilometers/ 84,000 square miles. Temperate climate due to influence of Gulf Stream.

- Population (2010): 581 million, including 82 million in Germany, 64 million in France, 62 million in the U.K., and 60 million in Italy

- Urbanization (2010): 73%

- GDP PPP (2010) current international dollars: 26,523; highest = Luxembourg, 77,227; lowest = Moldova, 3108

- Debt service (total, % of GDP) (2005): Most countries that received aid were from Eastern Europe, raising the European average to 9.96% of GDP

- Population in poverty (% > $2/day) (1990–2005): 1.8

- Gender-related development index value: 0.89; highest = Iceland, 0.962; lowest = Moldova, 0.70

- Average life expectancy/fertility: 75 (M72, F79)/1.5

- Population with access to Internet (2008): 48.9%

- Major language families: Most languages in Europe stem from the Indo-European family, which includes Slavic, Germanic, and Romance branches. Other families include Finno-Urgic and Tukic.

- Major Religions: Roman Catholicism, Orthodox Christianity, Protestantism, Islam. Atheism and secularism are also strongly represented in Europe.

- $CO_2$ emission per capita, tonnes (2004): 7.97

- Access to improved drinking-water sources/sanitation (2006): 97%/93%

# 2

# EUROPE

Lucca, in Tuscany, Italy, is famous for its regional cuisine: simple, traditional dishes using fresh local vegetables, olive oil, and wine. The town itself is set amid rolling hills and olive groves, its narrow streets and Renaissance churches contained within massively thick 16th-century city walls. But in 2009 Lucca's city council provoked widespread contention and accusations of racism by prohibiting new ethnic food restaurants from opening within its historic center.

The only city in Tuscany with a traditionally right-wing government, Lucca has a long and fierce tradition of self-preservation. Some locals proudly believe their ancestry can be traced to the Etruscans, who founded Lucca before the Romans took over about 180 B.C.E. Many of the shops in the town center have been in the same families for generations. The traditional cuisine of the region has long been an important part of the town's identity and self-image: Even Sicilian food is considered "ethnic" by traditionalists. Indeed, the region's excellent food and wine, together with Lucca's picture-postcard town center—with its medieval towers, an intimate elliptical piazza on the site of a Roman amphitheater, its traditional-style bars and osterias, and its paths on the tree-lined city walls—attract the half a million tourists a year that the city relies on to survive.

When the city council imposed its ban on new ethnic food restaurants within the historic center, there were already four kebab houses, testaments both to Italy's growing immigrant population and to the effects of cultural globalization. Under the new law, these four can stay. "Lucca is 'very closed,' said Rogda Gok, a native of Turkey and the co-owner of Mesopotamia, a kebab restaurant, in the heart of the historical center. 'In Istanbul there's other food, like German and Italian, it's no problem,' she added. 'But here in Lucca, they only want Luccan food.' . . . 'It's shameful,' said Renata Barbonchielli, a Lucca resident, as she stood outside a bakery downtown. 'Of course there should be kebabs. We have to have openness here. . . . I've never tried one, but my kids like them.' . . . All the fuss has been good for business, said Mohammad Riaz, the Pakistani owner of Doner Kebab Casalgrande in downtown Lucca. The shop is particularly popular with students. As he ordered a kebab, Alessandro Tucci, 22, said he opposed the ban. 'It's ridiculous,' he said. 'People should have the freedom to cook what they want.' "[1] The city council rejects accusations of covert racism, asserting that the town simply wants to preserve its cultural and historical identity—and needs to, if it is to continue to attract tourists. Thus, for a while, Lucca became the latest battleground in a tug of war between the traditional Europe of the popular imagination and the more complex reality of a multicultural Europe and a global economy. ■

## MAIN POINTS

■ Contemporary Europe is highly urbanized and is a cornerstone of the world economy with a complex, multilayered, and multifaceted regional geography.

■ Beyond Europe's core regions, major metropolitan areas, and specialized industrial districts, a mosaic of landscapes has developed around distinctive physiographic regions characterized by broad coherence of geology, relief, landforms, soils, and vegetation. These regions are the Northwestern Uplands, the Alpine System, the Central Plateaus, and the North European Lowlands.

■ The rise of Europe as a major world region had its origins in the emergence of a system of merchant capitalism in the 15th century, when advances in business practices, technology, and navigation made it possible for merchants to establish the basis of a worldwide economy in the space of less than 100 years.

■ The European Union emerged after World War II as a major factor in reestablishing Europe's role in the world.

■ The reintegration of Eastern Europe has added a potentially dynamic market of 150 million consumers to the European economy. In overall terms, Europe, with about 12% of the world's population, accounts for almost 35% of the world's exports, almost 43% of the world's imports, and 33% of the world's aggregate GNP.

■ The principal core region within Europe is the Golden Triangle, which stretches between London, Paris, and Berlin and includes the industrial heartlands of central England, northeastern France, and the Ruhr district of Germany.

---

[1]Rachel Donadio, "A Walled City in Tuscany Clings to Its Ancient Menu," *New York Times*, March 12, 2009, http://www.nytimes.com/2009/03/13/world/europe/13lucca.html?_r=2&ref=dining

# ENVIRONMENT AND SOCIETY

Two aspects of Europe's physical geography have been fundamental to its evolution as a world region and have influenced the evolution of regional geographies within Europe itself. First, as a world region, Europe is situated between the Americas, Africa, and the Middle East. Second, as a satellite photograph of Europe reveals, the region consists mainly of a collection of peninsulas and islands at the western extremity of the great Eurasian landmass (**Figure 2.2**).

The largest of the European peninsulas is the Scandinavian Peninsula, the prominent western mountains of which separate Atlantic-oriented Norway from continental-oriented Sweden. Equally striking are the Iberian Peninsula, a square mass that projects into the Atlantic, and the boot-shaped Italian Peninsula. In the southeast is the broad triangle of the Balkan Peninsula, which projects into the Mediterranean, terminating in the intricate coastlines of the Greek peninsulas and islands. In the northwest are Europe's two largest islands, Britain and Ireland.

The overall effect is that tongues of shallow seas penetrate deep into the European landmass. This was especially important in the pre-Modern period, when the only means of transporting goods were by sailing vessel and wagon. The Mediterranean and North seas, in particular, provided relatively sheltered sea lanes, fostering seafaring traditions in the people all around their coasts. The penetration of the seas deep into the European landmass provided numerous short land routes across the major peninsulas, making it easier for trade and communications to take place in the days of sail and wagon. As we shall see, Europeans' relationship to the surrounding seas has been a crucial factor in the evolution of European—and, indeed, world—geography.

Europe's navigable rivers also shaped the human geography of the region. Although small by comparison with major rivers in other regions of the world, some of the principal rivers of Europe—the Danube, the Dneiper, the Elbe, the Rhine, the Seine, and the Thames—played key roles as routes. Also, the low-lying **watersheds** (dividing ridges between drainage areas) between the major rivers of Europe's plains allowed

canal building to take place relatively easily, thereby increasing the mobility of river traffic.

Whereas the western, northern, and southern limits of Europe as a whole are clearly defined by surrounding seas—the Atlantic Ocean to the west, the Arctic Ocean to the north, and the Mediterranean Sea to the south—the eastern edge of Europe merges into the vastness of Asia and is less easily defined. Geographers sometimes use the mountain ranges of the Urals to mark the boundary between Europe and Asia, but the most significant factors separating Europe from Asia are human and relate to race, language, and a common set of ethical values that stem from Roman Catholic, Protestant, and Orthodox forms of Christianity. As a result, the eastern boundary of Europe is often demarcated through political and administrative boundaries, rather than physical features.

## Climate, Adaptation, and Global Change

The seas that surround Europe strongly influence the region's climate (**Figure 2.3**). In winter, seas cool more slowly than the land, whereas in summer they warm up more slowly than the land. As a result, the seas provide a warming effect in winter and a cooling effect in summer. Europe's arrangement of islands and peninsulas means that this moderating effect is particularly marked, contributing to an overall climate that does not have great seasonal extremes of heat and cold. The moderating effect is intensified by the North Atlantic Drift, which carries great quantities of warm water from the tropical Gulf Stream as far as the United Kingdom.

Given its latitude (Paris, at almost 49°N, is the same latitude as Winnipeg and Newfoundland in Canada), most of Europe is remarkably warm. It is continually crossed by moist, warm air masses that drift in from the Atlantic. The effects of these warm, wet, westerly winds are most pronounced in northwestern Europe, where squalls and showers accompany the passage of successive eastward-moving weather systems. Weather in northwestern Europe tends to be unpredictable, partly because of the swirling movement of air masses as they pass over the Atlantic and partly because of the complex effects of the widely varying temperatures of interpenetrating bodies of land and water. Farther east, in continental Europe, seasonal weather tends to be more settled, with more pronounced extremes of summer heat and winter cold. In these interior regions, local variations in weather are influenced a great deal by the direction in which a particular slope or land surface faces and its elevation above sea level.

The Mediterranean Basin has a different and quite distinctive climate (**Figure 2.3**). Winters are cool, with an Atlantic airstream that brings overcast skies and intermittent rain—though snow is unusual. Low-pressure systems along the northern Mediterranean draw in rain-bearing weather fronts from the Atlantic. When low pressure over the northern Mediterranean coincides with high pressure over continental Europe, southerly airflows spill over mountain ranges and down valleys, bringing cold blasts of air. These events have local names: the mistral, for example, which blows down the Rhône Valley in southern France, and the bora, which blows over the eastern Alps toward the Adriatic region of Italy. In spring the temperature rises rapidly, and rainfall is more abundant. Then summer bursts forth suddenly as dry, hot, Saharan air brings 3 months of hot, sunny weather. There is no rain save an occasional storm; the soil cracks and splits and is easily washed away in

▲ **FIGURE 2.2 Europe from space** This image underlines one of the key features of Europe: the many arms of the seas that penetrate deep into the western extremity of the great Eurasian landmass.

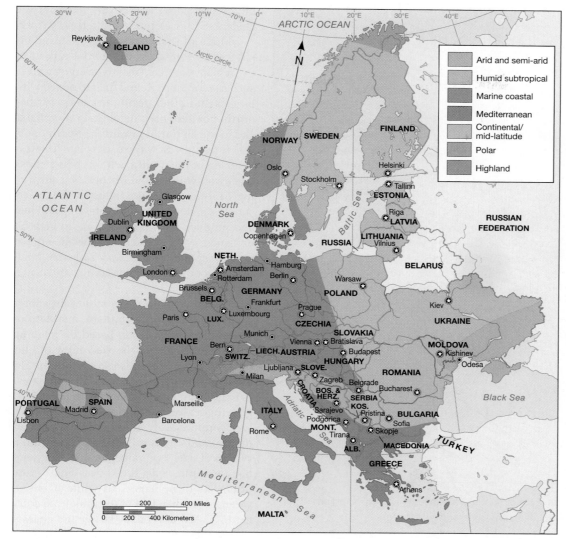

▲ **FIGURE 2.3  Climates of Europe.** Northern Europe is dominated by mid latitude climates with the influence of the westerly winds and storm tracks bringing Marine coastal wetter conditions to most of northwest Europe for most of the year whereas eastern Europe, farther from the moderating effects of the oceans, has the colder and drier conditions associated with the Continental mid latitude climate type. Southern Europe, along the shores of the Mediterranean, has the drier conditions and winter rains associated with a Mediterranean climate type with regions further from the sea in Spain and north of the Black Sea having dry conditions for most of the year (classified as semi-arid climates). Parts of Italy and the Balkans are wetter than the Mediterranean but warmer than climates to the north and are classified as Humid subtropical climates.

European land area in 2008 was 2.3°F (1.3°C) above preindustrial levels, whereas 9 out of the previous 12 years were among the warmest, on average, since 1850. High-temperature extremes and heat waves have become more frequent, whereas low-temperature extremes have become less frequent. The average length of summer heat waves over Western Europe doubled over the period 1880 to 2005, and the frequency of hot days has almost tripled.

Highly urbanized and industrialized, Europe is a major contributor to the carbon dioxide emissions that are a major contributor to global climate change. Under the United Nations Kyoto Protocol, a legally binding global agreement to reduce greenhouse gas emissions, industrialized nations are obliged to reduce the amount of greenhouse gases being released into the atmosphere. The European Union (27 countries that account for most of the region; see Figure 2.16, p. 59) is required to cut its emissions by 8% from 1990 levels by 2012. In an attempt to achieve this, the European Union has established an Emissions Trading Scheme (ETS), the world's largest emissions trading system and the first system to limit and to trade carbon dioxide emissions. Under the scheme, about 12,000 energy-intensive facilities (e.g., power generation plants and iron and steel, glass, and cement factories) have been able to buy and sell permits that allow them to emit carbon dioxide ($CO_2$) into the atmosphere. Overall, the ETS covers about 40% of the EU's total $CO_2$ emissions. Companies that exceed their individual limit are able to buy unused permits from firms that have taken steps to cut their emissions. Those that exceed their limit and are unable to buy spare permits are fined 40 euros (US$56) for every excess tonne of $CO_2$.

the occasional downpours. In October the temperature drops, and deluges of rain show that Atlantic air prevails once more.

In such conditions, delicate plants cannot survive. The Mediterranean climate precludes all plant species that cannot tolerate the range of conditions—cold as well as heat and drought as well as wet. The result is a distinctive natural landscape of dry terrain dotted with cypress trees, holm-oaks, cork oaks, parasol pines, and eucalyptus trees. These same conditions make agriculture a challenge. The crops that prosper best include olives, figs, almonds, vines, oranges, lemons, wheat, and barley. Sheep and goats graze on dry pastureland and stubble fields. Irrigation is often necessary, and in some localities it sustains high yields of fruit, vegetables, and rice.

Europe has warmed more than the global average as a result of changing global climate. The annual average temperature for the

## Geological Resources, Risk, and Water Management

The physical environments of Europe are complex and varied. It is impossible to travel far without encountering significant changes in physical landscapes. There is, however, a broad pattern to this variability, and it is based on four principal **physiographic regions** that are characterized by broad coherence of geology, relief, landforms, soils, and vegetation. These regions are the Northwestern Uplands, the Alpine System, the Central Plateaus, and the North European Lowlands (**Figure 2.4**).

**Northwestern Uplands**   The Northwestern Uplands are composed of the most ancient rocks in Europe, the product of the Caledonian mountain-building episode about 400 million years ago. Included in this region are the mountains of Norway and Scotland and the uplands of

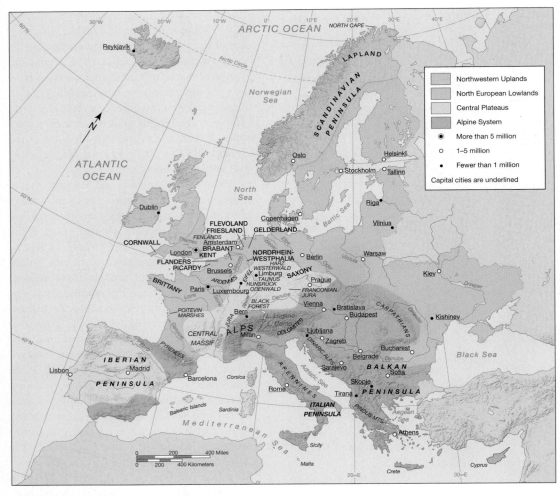

▲ **FIGURE 2.4 Europe's Physiographic Regions** Each of the four principal physiographic regions of Europe has a broad coherence in terms of geology, relief, landforms, soils, and vegetation.

dle of September and covers much of the landscape from October through early April.

Much of the Northwestern Uplands is covered by forests. In the southern parts of the region, conifers (mostly evergreen trees such as pine, spruce, and fir) are mixed with birches and other deciduous trees; further north, conifers become entirely dominant, and in the far north, toward the North Cape, the forest gives way to desolate treeless stretches of the **tundra**, with its gray lichens and dwarf willows and birches. Not surprisingly, forestry is a major industry. The forests supply timber for domestic building and fuel and produce the woods that are in greatest demand on world markets: pine and spruce for timber and for the paper industry, birch for plywood and cabinetmaking, and aspen for matches.

The farmers in these mountain subregions depend on **pastoralism**, a system of farming and way of life based on keeping herds of grazing animals, eked out with a little produce grown on the valley floors. In the less mountainous parts of Scandinavia, as in much of Baltic Europe, with their short growing season and cold, acid soils, agriculture supports only a low density of settlement. Landscapes reflect a mixed farming system of oats, rye, potatoes, and flax, with hay for cattle. Oats, the largest single crop, often has to be harvested while it is still green. More than half of the milk from dairy cattle is used to produce butter and cheese.

In these upland landscapes, farms and hamlets are casually situated on any habitable site, their buildings often widely scattered. The countryside is dotted with trim wooden houses, roofed with slate, tiles, shingles, or even sods of turf. Often, the buildings of a particular district are distinguished by some special stylistic feature. In some places, the old weatherboard houses that were used as shelters in bad weather still exist. They date back to the Middle Ages, as does the custom of storing reserve stocks of food or hay in a special isolated building, the *stabbur*, decorated with beautifully carved woodwork. The predominant color of the buildings is gray, the natural color that wood acquires with age. Nearer to towns, the houses are painted brighter colors: yellow, dark red, and mid-blue.

The more temperate subregions of the Northwestern Uplands (in Ireland and the United Kingdom) are dominated by dairy farming on meadowland, sheep farming on exposed uplands, and arable farming (mainly wheat, oats, potatoes, and barley) on drier lowland areas. In these areas, dispersed settlement, in the form of hamlets and scattered farms, is characteristic, and stone is more often the traditional building material.

**Alpine Europe**    The Alpine System occupies a vast area of Europe, stretching eastward for nearly 1290 kilometers (about 800 miles) across the southern part of Europe from the Pyrenees, which mark the

Iceland, Ireland, Wales, Cornwall (in England), and Brittany (in France). The original Caledonian mountain system was eroded and uplifted several times, and following the most recent uplift the Northwestern Uplands have been worn down again, molded by ice sheets and glaciers. Many valleys were deepened and straightened by ice, leaving spectacular glaciated landscapes. There are *cirques* (deep, bowl-shaped basins on mountainsides, shaped by ice action), glaciated valleys, and **fjords** (some as deep as 1200 meters—about 3900 feet) in Norway; countless lakes; lines of **moraines** that mark the ice sheet's final recession; extensive deposits of sand and gravel from ancient glacial deltas, and vast expanses of peat bogs that lie on the granite shield that forms the physiographic foundation of the region. Since the last glaciation, sea levels have risen, forming fjords and chains of offshore islands wherever these glaciated valleys have been flooded by the rising sea (**Figure 2.5**).

This formidable environment is rendered even more forbidding in much of the region as a result of climatic conditions. In the far north, the summer sun shines for 57 days without setting, but the winter nights are interminable. Oslo sees a *total* of only 17 hours of sunshine in the whole month of December. Rivers and lakes in the far north are frozen from mid-October on, whereas even farther south in Norway they freeze at the end of November and remain frozen until May. Snow, which is permanent in parts of Iceland and Lapland, begins to fall toward the mid-

▲ **FIGURE 2.5 The Northwestern Uplands** The northern parts of the region are characterized by high mountains and deeply eroded glacial valleys that have been drowned by the sea, creating distinctive fjord landscapes such as this one near Geiranger, Møre og Romsdal county, Norway.

border between Spain and France, through the Alps and the Dolomites and on to the Carpathians, the Dinaric Alps, and some ranges in the Balkan Peninsula. The Apennines of Italy and the Pindus Mountains of Greece are also part of the Alpine System. The Alpine System is the product of the most recent of Europe's mountain-building episodes, which occurred about 50 million years ago. Its relative youth explains the sharpness of the mountains and the boldness of their peaks. The Alpine landscape is characterized by jagged mountains with high, pyramidal peaks and deeply glaciated valleys (**Figure 2.6**). The highest peak, Mont Blanc, reaches 4,810 meters (15,781 feet). Most of the rest of the mountains are between 2,500 and 3,600 meters (about 8,200 and 11,800 feet) in height. Seven of the peaks in the western Alps exceed 4,000 meters (13,123 feet) in height. Although the Alps pose a formidable barrier between northwestern Europe and Italy and the Adriatic, a series of great passes—including the Brenner Pass, the Simplon Pass, the Saint Gotthard Pass, and the Great Saint Bernard Pass—and longitudinal valleys have always provided transalpine routes.

The dominant direction of the Alps and their parallel valleys is roughly southwest to northeast. The major Alpine valleys thus have one sunny, fully exposed slope that is suitable for vine growing and a shaded side rich with orchards, woods, and meadows. The mountains and valleys of the Alps proper are surrounded by glacial outwash deposits that provide rich farmland. The limestone of the Alpine region is widely quarried for cement, whereas mineral deposits—lead, copper, and iron—and

▼ **FIGURE 2.6 The Alpine System** Jagged peaks and glaciated valleys are typical of the Alpine ranges. This photograph is of the Geisler Peaks in the Dolomites, Italy.

small deposits of coal and salt have long been locally important throughout the region. In addition, the Alps are a valuable source of hydroelectric power: about 65 billion watt-hours in Switzerland (60% of the country's electricity consumption), about 72 billion watt-hours in France (15% of the country's electricity consumption), and about 45 billion watt-hours in Austria (85% of the country's electricity consumption).

The traditional staple of the economy, however, has been agriculture, and farming has given the Alpine region its distinctive human landscape: a patchwork of fields, orchards, vineyards, deciduous woodlands, pine groves, and meadows on the lower slopes of the valleys, with broad Alpine pasture above. In these pastures, which are dotted with wooden haylofts and summer chalets, dairy cattle wander far and wide. Farmers attach bells around the necks of their animals to be able to locate them, and the consequent effect is a resonant pastoral "soundscape" of clanking cowbells. Farms and hamlets tend to cling to lower elevations, the chalet-style architecture drawing on timber or rough-cast stone construction, with overhanging eaves, tiers of windows, and painted ornamentation.

The landscapes of the Alpine fringes are lusher. Lavender and fruit have been introduced to enrich and give variety to the mixed farming system of the Alpine fringes, which features vine and wheat growing, along with dairy cattle. The higher slopes, which receive more rainfall, provide lush pastures that have made the region famous for its rich cheeses, such as Gruyère.

The principal industry of the Alpine region is tourism. Attractive rural landscapes, together with magnificent mountain scenery, beautiful lakes, and first-class winter sports facilities, have attracted tourists to this region since the 1800s. Lakeside resorts such as Lucerne and Lugano, Switzerland; mountain resorts such as Chamonix, France, and Innsbruck, Austria; and winter sports resorts such as Val d'Isere, France, and Davos, St. Moritz, and Zermatt, Switzerland, are all well established, with an affluent clientele from across Europe. With the growth of the global tourist industry, Alpine resorts have attracted a new clientele from North America and Japan.

**Central Plateaus**   Between the Alpine System and the Northwestern Uplands are the landscapes of the Central Plateaus and the North European Lowlands. The Central Plateaus are formed from 250- to 300-million-year-old rocks that have been eroded down to broad tracts of uplands. Beneath the forest-clad slopes and fertile valleys of these plateaus lie many of Europe's major coalfields. For the most part, the plateaus reach between 500 and 800 meters (1,640 and 2,625 feet) in height, though they rise to more than 1,800 meters (5,905 feet) in the Central Massif of France. The Central Plateaus were generally too low to have been glaciated and too far south to have been covered by the great northern ice sheets of the last Ice Age. Rather, their landscape is characterized by rolling hills, steep slopes and dipping vales, and deeply carved river valleys.

In central Spain, the landscape is dominated by plateaus and high plains that go on for hundreds of miles, with long narrow mountain ranges—*cordillera*—stretched out like long cords along the edges. The flat landscape of the region is a result of immense horizontal sheets of sedimentary rock that cover a massif of ancient rock, with a general appearance of tables ending in ledges—hence the term *meseta*, which is generally used to designate the center of Spain (**Figure 2.7**). These dry, open lands are preeminently areas of grain crops and of flocks and herds of sheep and cattle that in summer are driven up to the cooler mountains. Olive trees dominate the shallow valley slopes, and in irrigated valley bottoms vines and prosperous market gardens flourish.

Further east—in the Massif Central in France and the Eifel, Westerwald, Taunus, Hunsruck, Odenwald, and Franconian Jura in Germany—where the climate is wetter, the landscape is dominated by gently rounded, well-wooded hills, with villages surrounded by neat fields and orchards in the vales. These hills rarely rise above 700 meters (2,300 feet), and the landscape includes many attractive and fertile subregions of low hills and rounded hillocks planted with vines and shallow valleys with orchards and meticulously maintained farms. The hills are covered with beech and oak forests, interspersed with growths of fir. The Black Forest is much higher, its bare granite summits reaching 1,493 meters (about 4,900 feet) in the south. It is scored by steep, narrow valleys with terraces of glacial outwash. The northeastern reaches of the Black Forest form an immense, silent, solid mass of fir forests. To the southwest, near the River Rhine, are small fields and meadows with prosperous farms at the edge of white fir forests and market towns nestling in tributary valleys. The Rhine has cut deep, scenic gorges through the higher plateau lands, but for much of its course across the central plateaus it is majestic and calm, with gentler slopes covered with vines and the bottomlands of the valley a busy corridor of prosperous towns surrounded by industrial crops and market gardens.

**North European Lowlands**   The North European Lowlands sweep in a broad crescent from southern France, through Belgium, the Netherlands, and southeastern England and into northern Germany, Denmark, and the southern tip of Sweden. Continuing eastward, they broaden into the immense European plain that extends through Poland, Czechia, Slovakia, and Hungary, all the way into Russia. Coal is found in quantity under the lowlands of England, France, Germany, and Poland and in smaller deposits in Belgium and the Netherlands. Oil and natural gas deposits are found beneath the North Sea and under the lowlands of southern England, the Netherlands, and northern Germany. Nearly all of this area lies below 200 meters (656 feet) in

▲ **FIGURE 2.7 The Central Plateaus** The *Meseta* of Spain covers thousands of square miles in the center of the Iberian Peninsula. This photograph shows the landscape near Consuegra, Toledo, Castile La Mancha, Spain.

▲ **FIGURE 2.8 The North European Lowlands** This photograph shows the landscape around Hieville, Pays d'Auge, Normandy, France.

elevation, and the topography everywhere is flat or gently undulating. As a result, the region has been particularly attractive to farming and settlement. The fertility of the soil varies, however, so that settlement patterns are uneven, and agriculture is finely tuned to the limits and opportunities of local soils, landscape, and climate.

The western parts of the North European Lowlands are densely settled and intensively farmed, with the moist Atlantic climate supporting lush agricultural landscapes (**Figure 2.8**). Surrounding the highly urbanized regions of the North European Lowlands are the fruit orchards and hop-growing fields of Kent, in southeastern England; the bulb fields of the Netherlands; the dikes and rectangular fields of the reclaimed marshland of North-Holland, Flevoland, and Friesland along the Dutch coastal plain (**Figure 2.9**); the pastures, woodlands, and forests of the upland plateaus of the Ardennes, the Eifel, the Westerwald, and the Harz; and the meadows and cultivated fields separated by hedgerows and patches of woodland that characterize most of the remaining countryside: Picardy in France; Flanders and Brabant in Belgium; Limburg, Nordrhein-Westphalia, and Saxony in Germany; and Gelderland in the Netherlands, for example.

Further east, the Lowlands are characterized by a drier Continental climate and lowland river basins with a rolling cover of sandy river deposits and **loess** (a fine-grained, extremely fertile soil). The hills are covered with oak and beech forests, but much of the region consists of broad loess plateaus where very irregular rainfall averages around 40 centimeters—approximately 16 inches—a year. There are no woods, and irrigation is often necessary to sustain the typical two-year rotation of corn and wheat. In some parts, several meters of loess and rich, sandy soil rest on the rocky substratum. Stone and trees are so scarce that houses are built of *pisé*, a kind of rammed-earth brick. The scarcity of trees forces storks to build nests atop chimneys and telegraph poles. The rich soils produce high yields of wheat and corn, together with hops, sugar beets, and forage crops for livestock.

The area also features residual regions of **steppe**: semiarid, treeless, grassland plains with landscapes that are infinitely monotonous. This land, which is too dry or too marshy to have invited cultivation, was once the domain of wild horses, cattle, and pigs, but today huge flocks of sheep find pasture there. The climate, though, is harsh, with seasonal extremes of burning hot and freezing cold, with the winter easterlies blowing down from midcontinent Russia. Population densities are low, and there are few villages.

It should be stressed that, within these broad physiographic divisions, marked variations exist. The mosaic of regions and landscapes within Europe is both rich and detailed. Physical differences are encountered over quite short distances, and numerous specialized farming regions exist, where agricultural conditions have influenced local ways of life to produce distinctive landscapes. In detail, these landscapes are a product of centuries of human adaptation to climate, soils, altitude, and **aspect** (exposure) and to changing economic and political circumstances. Farming practices, field patterns, settlement types, traditional building styles, and ways of life have all become attuned to the opportunities and constraints of regional physical environments, resulting in distinctive regional landscapes.

## Ecology, Land, and Environmental Management

Temperate forests originally covered about 95% of Europe, with a natural ecosystem dominated by oak, together with elm, beech, and lime (linden). By the end of the medieval period, Europe's forest cover had been reduced to about 20%, and today it is around 5%. Between C.E. 1000 and C.E. 1300, a period of warmer climate, together with advances in agricultural knowledge and practices, led to a significant transformation of the European landscape. The population more than doubled, from around 36 million to more than 80 million, and a vast amount of

▲ **FIGURE 2.9 Reclaimed landscape** This photograph shows the landscape around Kinderdijk, in the Netherlands.

land was brought under cultivation for the first time. By about 1200 most of the best soils of Western Europe had been cleared of forest, and new settlements were increasingly forced into the more marginal areas of heavy clays or thin sandy soils on higher ground and **heathlands.** (Heath is open land with coarse soil and poor drainage.) Many parts of Europe undertook large-scale drainage projects to reclaim marshlands.

The Romans had already demonstrated the effectiveness of drainage schemes by reclaiming parts of Italy and northwestern Europe. Under Roman colonization, land was often subdivided into a checkerboard pattern of rectilinear fields. This highly ordered system was known as *centuriation,* and the pattern can still be seen in some districts today—in parts of the Po valley, for example. Elsewhere, across large tracts of the Mediterranean, the soil can be cultivated only on a large scale, and the poor quality of pastureland necessitates vast untilled areas being left for flocks and herds. In these areas, successive conquerors, from the Greeks, Phoenicians, and Carthaginians to the Ottoman Turks and Christian Crusaders, carved out huge estates, known as **latifundia,** on which they set peasants to work. Land that did not belong to these big estates was often subdivided by independent peasant farmers into very small, intensively cultivated lots, or **minifundia,** most of which are barely able to support a family.

Another important legacy of the Romans was the concept of public good in relation to air and water, a concept that has endured as a legacy of Roman law and that is widely recognized in the doctrine of public trust, which asserts public rights in navigable waters, fisheries, and tidelands. It is reflected today in the European Union's approach to water management, which also features the most progressive overall system of water management, organized by river basin instead of according to administrative or political boundaries. On the other hand, the rising dominance of neoliberal policies since the 1980s has led to the privatization of water services and certain aspects of water management, especially in France and the United Kingdom.

In the 12th and 13th centuries extensive drainage and resettlement schemes were developed in Italy's Po Valley, in the Poitevin marshes of France, and in the Fenlands of eastern England. In Eastern Europe, forest clearances were organized by agents acting for various princes and bishops, who controlled extensive tracts of land. The agents would also arrange financing for settlers and develop villages and towns, often to standardized designs.

This great medieval colonization came to a halt nearly everywhere around 1300. One factor was the so-called Little Ice Age, a period of cooler climate that significantly reduced the growing season—perhaps by as much as 5 weeks. Another factor was the catastrophic loss of population during the period of the Black Death (1347–51) and the periodic recurrences of the plague that continued for the rest of the 14th century. The Little Ice Age lasted until the early 16th century, by which time many villages, and much of the more marginal land, had been abandoned.

The resurgence of European economies from the 16th century onward coincided with overseas exploration and trade, but domestic landscapes were significantly affected by repopulation, by reforms to land tenure systems, and by advances in science and technology that changed agricultural practices, allowing for a more intensive use of the land. In the Netherlands, a steadily growing population and the consequent requirement for more agricultural land led to increased efforts to reclaim land from the sea and to drain coastal marshlands. Hundreds of small estuarine and coastal barrier islands were slowly joined into larger units, and sea defense walls were constructed to protect low-lying land, which was drained by windmill-powered water pumps, the excess water being carried off into a web of drainage ditches and canals.

The resulting **polder** landscape provided excellent, flat, fertile, and stone-free soil. Between 1550 and 1650, 165,000 hectares (407,715 acres) of polderland was established in the Netherlands, and the sophisticated techniques developed by the Dutch began to be applied elsewhere in Europe—including eastern England and the Rhône estuary in southern France. Although most of these schemes resulted in improved farmland, the environmental consequences were often serious. In addition to the vulnerability of the polderlands to inundation by the sea, large-scale drainage schemes devastated the wetland habitat of many species, and some ill-conceived schemes simply ended in widespread flooding.

These environmental problems were but a prelude to the environmental changes and ecological disasters that accompanied the industrialization of Europe, beginning in the 18th century. Mining—especially coal mining—created derelict landscapes of spoil heaps; urbanization encroached on rural landscapes and generated unprecedented amounts and concentrations of human, domestic, and industrial waste; and manufacturing, unregulated at first, resulted in extremely unhealthy levels of air pollution. Much of Europe's forest cover was cleared, and remaining forests and woodlands suffered from the **acid rain** resulting from heavy doses of atmospheric pollution. Many streams and rivers also became polluted, and the landscape everywhere was scarred with quarries, pits, cuttings, dumps, and waste heaps.

# HISTORY, ECONOMY, AND DEMOGRAPHIC CHANGE

Perhaps more than that of any other world region, Europe's geography is the product of its role in world-spanning economic and demographic systems. Beginning in the 16th century, Europeans became, as Robert Reynolds put it, the "leaders, drivers, persuaders, shapers, crushers and builders"[2] of the rest of the world's economies and societies. It was as a result of these processes that the core areas of Europe emerged with a strong comparative advantage in the modern world-system. For several centuries, other world regions came to play a subordinate economic role to Europe. Nevertheless, the preindustrial trajectories of other parts of the world had often eclipsed that of Europe and were sometimes important in influencing events in Europe itself. Today, Europe is totally integrated into the world economy, with ethnically diverse populations that reflect the region's past economic and political history.

The foundations of Europe's human geography were laid by the Greek and Roman empires. Beginning around 750 B.C.E., the ancient Greeks developed a series of fortified city–states (called *poleis*) along the Mediterranean coast, and by 550 B.C.E. there were about 250 such trading colonies. **Figure 2.10** shows the location of the largest of these, some of which subsequently grew into thriving cities (for example, Athens and Corinth), whereas others remain as isolated

---

[2]R. Reynolds, *Europe Emerges: Transition Toward an Industrial World-Wide Society* (Madison: University of Wisconsin Press, 1961), p. vii.

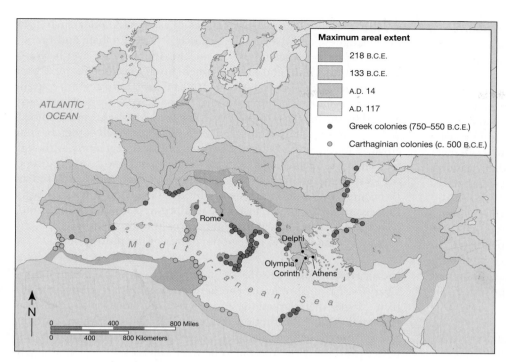

▲ **FIGURE 2.10  Greek Colonies and Extent of the Roman Empire**  This map shows the distribution of Greek *poleis* (city–states) and Carthaginian colonies and the spread of the Roman Empire from 218 B.C.E. to C.E. 117.

ruins or archaeological sites (for example, Delphi and Olympia). The Roman Republic was established in 509 B.C.E. and took almost 300 years to establish control over the Italian Peninsula. By C.E. 14, however, the Romans had conquered much of Europe, together with parts of North Africa and Asia Minor. Most of today's major European cities had their origin as Roman settlements. In quite a few of these cities, it is possible to find traces of the original Roman street layouts. In some, it is possible to glimpse remnants of defensive city walls, paved streets, aqueducts, viaducts, arenas, sewage systems, baths, and public buildings. In the modern countryside, the legacy of the Roman Empire is represented by arrow-straight roads, built by their engineers and maintained and improved by successive generations.

The decline of the Roman Empire, beginning in the 4th century C.E., was accompanied by a long period of rural reorganization and consolidation under feudal systems, a period often characterized as uneventful and stagnant. In fact, the roots of European regional differentiation can be traced to this long feudal era of slow change. **Feudal systems** were almost wholly agricultural, with 80% or 90% of the workforce engaged in farming and most of the rest occupied in basic craft work. Most production was for people's immediate needs, with very little of a community's output ever finding its way to wider markets.

By C.E. 1000 the countryside of most of Europe had been consolidated into a regional patchwork of feudal agricultural subsystems, each of which was more or less self-sufficient. For a long time, towns were small, their existence tied mainly to the castles, palaces, churches, and cathedrals of the upper ranks of the feudal hierarchy. These economic landscapes—inflexible,

slow-motion, and introverted—nevertheless contained the essential preconditions for the rise of Europe as the dynamic hub of the world economy.

## Historical Legacies and Landscapes

A key factor in the rise of Europe as a major world region was the emergence of a system of **merchant capitalism** in the 15th century. The immensely complex trading system that soon came to span Europe was based on long-standing trading patterns that had been developed from the 12th century by the merchants of Venice, Pisa, Genoa, Florence, Bruges, Antwerp, and the Hanseatic League (a federation of city–states around the North Sea and Baltic coasts that included Bremen, Hamburg, Lübeck, Rostock, and Danzig, as shown in **Figure 2.11**).

**Trade and the Age of Discovery**  In the 15th and 16th centuries a series of innovations in business and technology contributed to the consolidation of Europe's new merchant capitalist economy. These included several key innovations in the organization of business and finance: banking, loan systems, credit transfers, company partnerships, shares in stock, speculation in commodity futures, commercial insurance, and courier/news services. Meanwhile, technological innovations began to further strengthen Europe's economic advantages. Some of these innovations were adaptations and improvements of Oriental discoveries—the windmill, spinning wheels, paper manufacture, gunpowder, and the compass, for example. In Europe, however, there was a real passion for mechanizing the manufacturing process. Key engineering

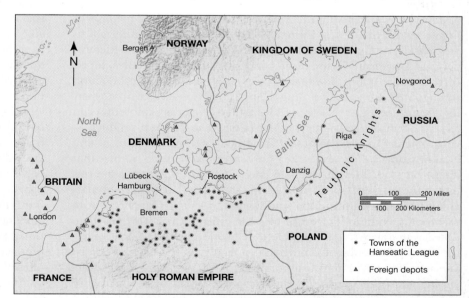

▲ **FIGURE 2.11  The Hanseatic League**  The Hanseatic League was a federation of city–states founded in the 13th century by north German towns and affiliated German merchant groups abroad to defend their mutual trading interests. The League, which remained an influential economic and political force until the 15th century, laid the foundations for the subsequent growth of merchant trade throughout Europe.

breakthroughs included the more efficient use of energy in water mills and blast furnaces, the design of reliable clocks and firearms, and the introduction of new methods of processing metals and manufacturing glass.

It was the combination of innovations in shipbuilding, navigation, and naval ordnance, however, that had the most far-reaching consequences for Europe's role in the world economy. In the course of the 15th century, the full-rigged sailing ship was developed, enabling faster voyages in larger and more maneuverable vessels that were less dependent on favorable winds. Meanwhile, Europeans developed navigational tools such as the quadrant (1450) and the astrolabe (1480) and acquired a systematic knowledge of Atlantic winds. By the mid-16th century, armorers in England, Holland, and Sweden had perfected the technique of casting iron guns, making it possible to replace bronze cannons with larger numbers of more effective guns at lower expense. Together, these advances made it possible for the merchants of Europe to establish the basis of a worldwide economy in less than 100 years after Portuguese explorer Bartholomeu Dias reached the Cape of Good Hope (the southern tip of Africa) in 1488.

**Changing Patterns of Advantage**   The geographical knowledge acquired during this Age of Discovery was crucial to the expansion of European political and economic power in the 16th century. In societies that were becoming more and more commercially oriented and profit conscious, geographical knowledge became a valuable commodity in itself. Information about overseas regions was a first step to controlling and influencing them; this in turn was a step to wealth and power. At the same time, every region began to open up to the influence of other regions because of the economic and political competition unleashed by geographical discovery. Not only was the New World affected by European colonists, missionaries, and adventurers, but the countries of the Old World also found themselves pitched into competition with one another for overseas resources.

Gold and silver from the Americas provided the first major economic transformation of Europe. The gold and silver bullion plundered by Spain and Portugal from the Americas created an effective demand for consumer and capital goods of all kinds—textiles, furniture, weapons, ships, food, and wine—thus stimulating production throughout Europe and creating the basis of a Golden Age of prosperity for most of the 16th century. Meanwhile, overseas exploration and expansion made available a variety of new and unusual products—cocoa, beans, maize, potatoes, tomatoes, sugarcane, tobacco, and vanilla from the Americas; tea and spices from the Orient—that opened up large new markets to enterprising merchants. Wine was one of the early luxury products that established the pattern of merchant trading within Europe, and when Europeans branched out to incorporate more of the world into the orbit of their world-system, they began organizing the production of wine wherever climatic conditions were encouraging: in warm temperate zones, roughly between latitudes 30° and 50° north and south (see Geographies of Indulgence, Desire, and Addiction: Wine, pp. 56–57). Not least, the emergence of a worldwide system of exploration and trade helped establish the foundations of modern academic geography.

These changes had a profound effect on the geography of Europe. Before the mid-15th century, Europe was organized around two sub-regional maritime economies—one based on the Mediterranean and

the other on the Baltic. The overseas expansions pioneered first by the Portuguese and then by the Spanish, Dutch, English, and French reoriented Europe's geography toward the Atlantic. The river basins of the Rhine, the Seine, and the Thames rapidly became focused on a thriving network of **entrepôt** seaports (intermediary centers of trade and transshipment) that transformed Europe. These three river basins, backed by the increasingly powerful states in which they were embedded—the Netherlands, France, and Britain, respectively—then became engaged in a struggle for economic and political hegemony. Although the Rhine was the principal natural routeway into the heart of Europe, the convoluted politics of the Netherlands allowed Britain and France to become the dominant powers by the late 1600s. Subsequently, France, under Napoleon, made the military error of attempting to pursue both maritime and continental power at once, allowing Britain to become the undisputed hegemonic power of the industrial era.

**Industrialization**   Europe's regional geographies were comprehensively recast once more by the new production and transportation technologies that marked the onset of the Industrial Revolution (from the late 1700s). Production technologies based on more efficient energy sources helped raise levels of productivity and create new and better products that stimulated demand, increased profits, and generated a pool of capital for further investment. Transportation technologies enabled successive phases of geographic expansion that completely reorganized the geography of Europe. As the application of new technologies altered the margins of profitability in different kinds of enterprise, so the fortunes of particular places and regions shifted.

There was in fact not a sudden, single Industrial Revolution but three distinctive transitional waves of industrialization, each with a different degree of impact on different regions and countries (**Figure 2.12**). The first, between about 1790 and 1850, was based on a cluster of early industrial technologies (steam engines, cotton textiles, and ironworking) and was highly localized. It was limited to a few regions in Britain where industrial entrepreneurs and workforces had first exploited key innovations and the availability of key resources (coal, iron ore, and water).

The second wave, between about 1850 and 1870, involved the diffusion of industrialization to most of the rest of Britain and to parts of northwest Europe, particularly the coalfield areas of northern France, Belgium, and Germany (see Figure 2.12). New opportunities were created as railroads and steamships made more places accessible, bringing their resources and their markets into the sphere of industrialization. New materials and new technologies (steel, machine tools) created opportunities to manufacture and market new products. These new activities prompted some significant changes in the logic of industrial location. Railway networks, for example, attracted industry away from smaller towns on the canal systems and toward larger towns with good rail connections. Steamships for carrying on coastal and international trade attracted industry to larger ports. At the same time, steel produced concentrations of heavy industry in places with nearby supplies of coal, iron ore, and limestone.

The third wave of industrialization saw a further reorganization of the geography of Europe as yet another cluster of technologies (including electricity, electrical engineering, and telecommunications) brought different resource needs and created additional investment opportunities. During this period, industrialization spread for the first time to remoter parts of the United Kingdom, France, and

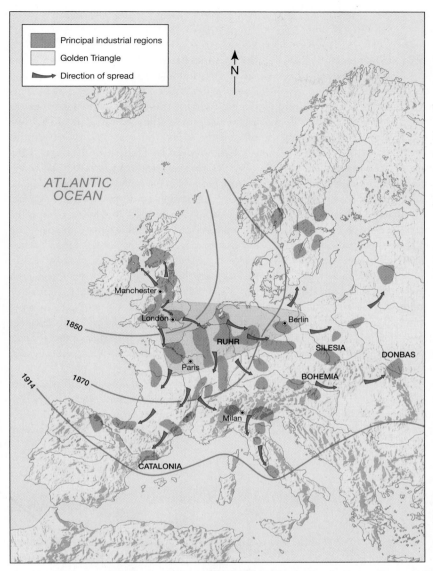

Legend:
- Principal industrial regions
- Golden Triangle
- Direction of spread

N

ATLANTIC OCEAN

Manchester
London
1850
RUHR
Berlin
Paris
SILESIA
DONBAS
1914
1870
BOHEMIA
Milan
CATALONIA

▲ **FIGURE 2.12 The Spread of Industrialization in Europe** European industrialization began with the emergence of small industrial regions in several parts of Britain. As new rounds of industrial and transportation technologies emerged, industrialization spread to other regions with the right attributes: access to raw materials and energy sources, good communications, and large labor markets.

Germany and to most of the Netherlands, southern Scandinavia, northern Italy, eastern Austria, Bohemia (in what was then Czechoslovakia), Silesia (in Poland), Catalonia (in Spain), and the Donbas region of Ukraine, then into Russia. The overall result was to create the foundations of a core-periphery structure within Europe (see Figure 2.12), with the heart of the core centered on the "Golden Triangle" stretching between London, Paris, and Berlin.

The peripheral territories of Europe—most of the Iberian peninsula, northern Scandinavia, Ireland, southern Italy, the Balkans, and east-central Europe—were slowly penetrated by industrialization over the next 50 years.

**Imperialism and War** Several of the most powerful and heavily industrialized European countries (notably the United Kingdom,

Germany, France, and the Netherlands) were by now competing for influence on a global scale. This competition developed into a scramble for territorial and commercial domination through **imperialism**—the deliberate exercise of military power and economic influence by core states to advance and secure their national interests. European countries engaged in preemptive geographic expansion to protect their established interests and to limit the opportunities of others. They also wanted to secure as much of the world as possible—through a combination of military oversight, administrative control, and economic regulations—to ensure stable and profitable environments for their traders and investors. This combination of circumstances defined a new era.

During the first half of the 20th century, the economic development of the whole of Europe was disrupted twice by major wars. The devastation of World War I was immense. The overall loss of life, including the victims of influenza epidemics and border conflicts that followed the war, amounted to between 50 and 60 million. About half as many again were permanently disabled. For some countries, this meant a loss of between 10% and 15% of the male workforce.

Just as European economies had adjusted to these dislocations, the Great Depression created a further phase of economic damage and reorganization throughout Europe. World War II resulted in yet another round of destruction and dislocation (**Figure 2.13**). The total loss of life in Europe this time was 42 million, two-thirds of whom were civilian casualties. Systematic German persecution of Jews—the Holocaust—resulted in approximately four million Jews being put to death in extermination camps such as Auschwitz and Treblinka, with up to two million more being exterminated elsewhere, along with gypsies and others. The German occupation of continental Europe also involved ruthless economic exploitation. By the end of the war, France was depressed to below 50% of its prewar standard of living and had lost 8% of its industrial assets. The United Kingdom lost 18% of its industrial assets (including

▲ **FIGURE 2.13 Dresden, Germany** Bomb damage during WWII.

# WINE

The production and consumption of wine reflects the evolution of the world-system. The original domestication of wine grapes (*Vitis vinifera*) seems to have taken place as early as 8000 B.C.E. along the mountain slopes of Georgia, eastern Turkey, and western Iran. Wine had symbolic and ritual significance in early civilizations of the eastern Mediterranean, partly because its ability to intoxicate and engender a sense of "other-worldliness" provided a means through which people could feel in contact with their gods, and partly because of the apparent death of the vine in winter and its dramatic growth and rebirth in the spring. Greek civilization established viticulture—the cultivation of grape vines for winemaking—as one of the staples of the Mediterranean agrarian economy, along with wheat and olives. By the 6th century B.C.E., Greek wine was being traded as far as Egypt, the shores of the Black Sea, and the southern regions of France.

Under the Roman Empire, viticulture spread west along the north shores of the Mediterranean and along the valleys of navigable rivers in France and Spain, while the wine trade extended north, to the North Sea and the Baltic. By the 1st century C.E., wine had become a commodity of indulgence, desire, and—for some—addiction throughout Europe. Viticulture and the art of winemaking survived (but did not prosper) through the Middle Ages, even in those parts of Mediterranean Europe that came under Islamic rule, where the consumption of alcohol was, theoretically, prohibited. Then, in the late medieval period, the growth of towns provided a substantial and increasingly affluent consumer market that led to the development of large, commercially oriented vineyards. Commercial viticulture spread to the southeast-facing slopes of the major river valleys in northern France and Germany, and merchant traders, drawing on innovations in finance, banking, and credit, facilitated the movement of vast quantities of wine, together with spices, perfumes, and silks, from the Mediterranean to England, Flanders, Scandinavia, and the Baltic. These northern European regions, meanwhile, paid for their luxury imports with the proceeds of exports of furs, fish, dairy produce, timber, and wool.

In the 16th century, Spanish and Portuguese overseas expansion saw the introduction of viticulture to the New World—to Mexico in the 1520s, Peru in the 1530s, Chile in the 1550s, and Florida in the 1560s. The British introduced viticulture to Virginia in the 1600s, and the Dutch established vineyards in the Cape Colony of southern Africa in the 1650s. The first vineyards in California were established by Franciscan missions in the 1770s, in southeastern Australia in the 1790s, and in New Zealand in the early 1800s. Meanwhile, in Europe, demographic growth and increasing prosperity rapidly expanded the market for wine. Winemakers developed new types of wine (including champagne, claret, and port), began to specialize in particular varieties of grapes (red grapes such as Cabernet Sauvignon, Nebbiolo, and Pinot Noir, and white grapes such as Chardonnay, Riesling, and Sauvignon Blanc), and found ways of storing wine, so that especially good quality wines could be aged without spoiling. Vintage wines, carefully aged and stored, acquired special value for connoisseurs.

Disaster hit European winemakers in the 1860s in the form of an aphid, *phylloxera*, which had somehow been brought to Europe on American vines. Though American vines were

▲ **FIGURE 1 Viticulture** This photograph shows the intensively cultivated landscape around Ladoix village, near Beaune, in Burgundy, France.

immune to *phylloxera*, the aphid killed European species. Many European vineyards were devastated before it was discovered, in 1881, that grafting European vines onto American rootstock would produce high-quality and *phylloxera*-resistant plants.

After recovering from the *phylloxera* episode, the makers (and consumers) of fine wines faced other problems: Unscrupulous merchants and foreign competitors sought to pass off lesser wines as prestigious wines, supplies of good wines were watered to stretch limited supplies, and poor wines were adulterated with chemicals to improve their color or their shelf life. In response, the exclusivity of wines was protected by new systems of regulation. In France, for example, the *Appellation Contrôlée* system was introduced to guarantee the authenticity of wines, district by district. In Germany, the classification system was based

not on geographic origin but on levels of quality and degrees of sweetness. Such regulations have been important in reinforcing the appeal of wine as a commodity of indulgence and desire.

Today, wine is one of the most widespread commodities of consumer indulgence. Fine wines such as those from Burgundy, France (**Figure 1**) denote affluence and distinction throughout the world's core regions and in the affluent enclaves of many of the metropolises of peripheral regions, whereas cheaper wines are consumed throughout much of the world. The globalization of viticulture and the more widespread consumption of wine have made branding an issue. Japanese supermarket shelves are lined with locally produced bottles with French-language labels that allude to nonexistent châteaus, whereas some California wineries produce "Chablis," "Champagne," and "Burgundy."

Two of the most important changes in relation to wine production date from the mid-20th century and have contributed a great deal to the globalization of wine as a commodity of indulgence and desire. First, the deployment of scientific approaches and new technologies, combined with large-scale capital investment in the industry, allowed the development of first-class wines in North America, Australia, New Zealand, and elsewhere (**Figure 2**). Second, the hedonistic cultural shift of the 1960s in Western societies brought the consumption of wine firmly into the routine practices of the middle classes. The result is that the production and retailing of wine is now a significant component of the activities of large, conglomerate, and transnational corporations. The most successful of these are able to exploit and manipulate changing patterns of consumption, introducing profitable new products such as wine coolers to broaden the market. ■

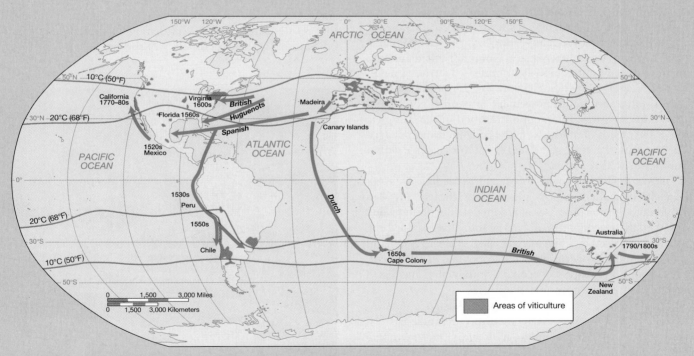

▲ **FIGURE 2 The Global Distribution of Viticulture** In general, the best areas for viticulture lie between the 10°C and 2°C annual isotherms, equating approximately to the warm temperate zones between latitudes 30° and 50° north and south. In detail, the geography of viticulture is heavily influenced by soils and microclimatic conditions. Where grapes are produced successfully nearer the equator, as in parts of Bolivia and Tanzania, it is usually because they are grown at higher altitudes.

overseas holdings), and the Soviet Union lost 25%. Germany lost 13% of its assets and ended the war with a level of income per capita that was less than 25% of the prewar figure. In addition to the millions killed and disabled during World War II, approximately 46 million people were displaced between 1938 and 1948 through flight, evacuation, resettlement, or forced labor. Some of these movements were temporary, but most were not.

After the war, the Cold War rift between Eastern and Western Europe resulted in a further handicap to the European economy and, indeed, to its economic geography. Ironically, this rift helped speed economic recovery in Western Europe. The United States, whose leaders believed that poverty and economic chaos in Western Europe would foster communism, embarked on a massive program of economic aid under the **Marshall Plan**. This pump-priming action, together with the backlog of demand in almost every sphere of production, provided the basis for a remarkable recovery. Meanwhile, Eastern Europe began an interlude of **state socialism**, a form of economy based on principles of collective ownership and administration of the means of production and distribution of goods dominated and directed by state bureaucracies.

### Eastern Europe's Legacy of State Socialism

After World War II, the leaders of the Soviet Union felt compelled to establish a **buffer zone** between their homeland and the major Western powers in Europe. The Soviet Union rapidly established its dominance throughout Eastern Europe: Estonia, Moldova, Latvia, and Lithuania were absorbed into the Soviet Union itself, and Soviet-style regimes were installed in Albania, Bulgaria, Czechoslovakia, East Germany, Hungary, Poland, Romania, and Yugoslavia. The result was what Winston Churchill called an "Iron Curtain" along the western frontier of Soviet-dominated territory: a militarized frontier zone across which Soviet and East European authorities allowed the absolute minimum movement of people, goods, and information. In addition, Soviet intervention resulted in the complete nationalization of the means of production, the collectivization of agriculture, and the imposition of rigid social and economic controls within the eastern European **satellite states.**

The economies of the former Soviet Union and its satellites were *not* based on true socialist or communist principles in which the working class had democratic control over the processes of production, distribution, and development. Rather, these economies evolved as something of a hybrid, in which state power was used by a bureaucratic class to create **command economies** in the pursuit of modernization and economic development. In a command economy, every aspect of economic production and distribution is controlled centrally by government agencies.

The Communist Council for Mutual Economic Assistance (CMEA, better known as COMECON) was established to reorganize eastern European economies in the Soviet mold—with individual members each pursuing independent, centralized plans designed to produce economic self-sufficiency. This quickly proved unsuccessful, however, and in 1958 COMECON was reorganized. The goal of economic self-sufficiency was abandoned, mutual trade among the *Soviet bloc*—the Soviet Union plus its eastern European satellite states—was fostered, and some trade with Western Europe was permitted. Meanwhile, Albania withdrew from the Soviet bloc in pursuit of a more authoritarian form of communism inspired by the Chinese revolution of 1949 (see Chapter 8), and Yugoslavia was expelled from the Soviet bloc (because of ideological differences over the interpretation of social-

ism) and allowed to pursue a more liberal, independent form of state socialism.

The experience of the east European countries under state socialism varied considerably, but, in general, rates of industrial growth were high. As in Western Europe, industrialization brought about radical changes in economic geography. In practice, however, the command economies of Eastern Europe did not result in any really distinctive forms of spatial organization. As in the industrial regions of the West, the industrialized landscapes of Eastern Europe came to be dominated by the localization of manufacturing activity, by regional specialization, and by core-periphery contrasts in levels of economic development.

The geography of industrial development under state socialism, as in democratic capitalism, was heavily influenced by the uneven distribution of natural resources and by the economic logic of initial advantage, specialization, and **agglomeration economies**—the cost advantages that accrue to individual firms because of their location among functionally related activities. The most distinctive landscapes of state socialism were those of urban residential areas, where mass-produced, system-built apartment blocks allowed impressive progress in eliminating urban slums and providing the physical framework for an **egalitarian society**—one based on belief in equal social, political, and economic rights and privileges—though at the price of uniformly modest dwellings and strikingly sterile cityscapes (**Figure 2.14**).

### The Reintegration of Eastern Europe

Eventually, the economic and social constraints imposed by excessive state control and the dissent that resulted from the lack of democracy under state socialism combined to bring the experiment to a sudden halt. By the time the Soviet

▲ **FIGURE 2.14 Socialist Housing** The socialist countries of Eastern Europe eradicated a great deal of substandard housing in the three decades following World War II, rehousing the population in mass-produced, system-built apartment blocks. Although this new housing provided adequate shelter and basic utilities at very low rents, space standards were extremely low, and housing projects were uniformly drab. This example is from the suburban district of Obuda in Budapest, Hungary.

bloc collapsed in 1989 (see Chapter 3), Poland and Hungary had already accomplished a modest degree of democratic and economic reform. By 1992, East Germany (the German Democratic Republic) had been reunited with West Germany (the German Federal Republic); Estonia, Latvia, and Lithuania had become independent states once more; and the whole of Eastern Europe had begun to be reintegrated with the rest of Europe. Only Kaliningrad, a small province on the Baltic between Poland and Lithuania, remains part of the Russian Federation, retained as Russian territory because of its warm-water naval port. In 1991 COMECON was abolished, and one by one, eastern European countries began a complex series of reforms. These included the abolition of controls on prices and wages, the removal of restrictions on trade and investment, the creation of a financial infrastructure to handle private investment, the creation of government fiscal systems to balance taxation and spending, and the **privatization** of state-owned industries and enterprises (**Figure 2.15**). After more than 40 years of state socialism, such reforms were difficult and painful. Indeed, the reforms are by no means complete in any of the countries, and economic and social dislocation is a continuing fact of life.

Nevertheless, the reintegration of Eastern Europe has added a potentially dynamic market of 130 million consumers to the European economy. Within a capitalist framework, Eastern Europe has the comparative advantage of relatively cheap land and labor. This has attracted a great deal of foreign investment, particularly from transnational corporations and from German firms and investors, many of whom have historic ties with parts of Eastern Europe. In some ways, the transition toward market economies has been remarkably swift. It did not take long for Western-style consumerism to appear on the streets and in many of the stores in larger eastern European cities. On the flip side, it also did not take long for inflation, unemployment, and homelessness to appear. Overall, Eastern Europe is increasingly reintegrated with the rest of Europe, but for the most part as a set of economically peripheral regions, with agriculture still geared to local markets and former COMECON trading opportunities and industry still geared more to heavy industry and standardized products than to competitive consumer products. Still very weakly developed are the service sector in general and knowledge-based industries in particular.

In detail, the pace and degree of reintegration vary considerably across Eastern Europe. Ethnic conflict in Bosnia and Herzegovina, Croatia, and the former republic of Yugoslavia has severely retarded reform and reintegration, whereas Albania, Bulgaria, Macedonia, Moldova, and Romania suffer from the combined disadvantages of having relatively poor resource bases, weakly developed communications and transportation infrastructures, and political regimes with little ability or inclination to press for economic and social reform. In Ukraine, which has a much better infrastructure, a significant industrial base, and the capacity for extensive trade in grain exports and in advanced technology, reintegration has been retarded by a combination of geographical isolation from Western Europe, continuing economic and political ties with Russia, and a surviving political elite that has little interest in economic and social reform.

The Baltic states of Estonia, Latvia, and Lithuania have been more successful in reintegrating with the rest of Europe. Their small size and relatively high levels of education have made them attractive as production subcontracting centers for western European high-technology industries. They are reviving old ties with neighboring Nordic countries, and in this regard Estonia is particularly well placed because its language

▲ **FIGURE 2.15 Privatized Industry** This aluminum truck wheel factory in Székesfehérvár, Hungary, was purchased from the state-owned Székesfehérvár Light Metal Works by Alcoa, the Pittsburgh-based global conglomerate, in 1996.

belongs to the Finnish family, and it was part of the Kingdom of Sweden when it was annexed by the Russians in 1710. The best-integrated states of Eastern Europe are Czechia, Hungary, Poland, and Slovenia. All have a relatively strong industrial base, and Hungary has a productive agricultural sector.

## Economy, Accumulation, and the Production of Inequality

Contemporary Europe is a cornerstone of the world economy with a complex, multilayered, and multifaceted regional geography. In overall terms, Europe, with about 12% of the world's population, accounts for almost 35% of the world's exports, almost 43% of the world's imports, and 33% of the world's aggregate GNP. Europe's inhabitants, on average, now consume about twice the quantity of goods and commercial services they did in 1975. Purchasing power has risen everywhere to the extent that basic items of food and clothing now account for only about 30% of household expenditures, leaving more resources for leisure and consumer durables. Levels of material consumption in much of Europe approach those of households in the United States.

Although Europe is a relatively affluent world region, there are in fact persistent and significant economic inequalities at every geographic scale. Annual per capita GDP (in PPP) in 2006 ranged from $2,396 in Moldova to $77,089 in Luxembourg. Regional income disparities within many European countries are increasing. In northwestern Europe this is generally a result of the declining fortunes of "rustbelt" regions and the relative prosperity of regions with high-tech industry and advanced business services. In southern and eastern Europe it is a result of differences between regions dominated by rural economies (generally poorer) and those dominated by metropolitan areas (generally more prosperous).

The resulting disparities are significant, with annual per capita GDP (in PPP) in 2009 ranging from over $100,000 in central London and more than $40,000 in many major metropolitan regions to between $12,000 and $18,000 in many peripheral rural regions. Poverty and homelessness exist in every European country, though poverty as measured on a global scale ($1 or $2 a day per person) is virtually unknown within Europe.

The development of European **welfare states** (institutions with the aim of distributing income and resources to the poorer members of society) has helped maintain households' purchasing power during periods of recession and ensured at least a tolerable level of living for most groups at all times. Levels of personal taxation are high, but all citizens receive a wide array of services and benefits in return. The most striking of these services are high-quality medical care, public transport systems, **social housing**, schools, and universities. The most significant benefits are pensions and unemployment benefits.

## The European Union: Coping with Uneven Development

Contemporary Europe is a dynamic region that embodies a great deal of change. Formerly prosperous industrial regions have suffered economic decline, but some places and regions have reinvented themselves to take advantage of new paths to economic development. Meanwhile, the former Soviet satellite states have been reintegrated into the European world region, and much of Europe has joined in the European Union (EU), a supranational organization founded to recapture prosperity and power through economic and political integration. The EU had its origins in the political and economic climate following World War II. The initial idea behind the EU was to ensure European autonomy from the United States and to recapture the prosperity Europe had forfeited as a result of the war. Part of the rationale for its creation was also to bring Germany and France together in a close association, which would prevent any repetition of the geopolitical problems in Western Europe that had led to two world wars. As the logic of the world economy has in many ways transcended the scale of nation–states the EU has grown in size and scope, becoming a major element in the reorganization of world economic geography.

The first stage in the evolution of the EU was the creation in the 1950s of several institutions to promote economic efficiency through integration. These were subsequently amalgamated to form the European Community (EC), which was in turn expanded in scope to form the European Union. EU membership has expanded from the six original members of the EC—Belgium, France, Italy, Luxembourg, the Netherlands, and West Germany—to 27 countries (**Figure 2.16**) with a population of nearly 492 million, and a combined gross domestic product (GDP) of $14.82 trillion in 2008, slightly larger than that of the United States. It has developed into a sophisticated and powerful institution with a pervasive influence on patterns of economic and social well-being within its member states. It also has a significant impact on certain aspects of economic development within some nonmember countries.

The origin of the organization that evolved into the EU was a compromise worked out between the strongest two of the original six members. West Germany wanted a larger but protected market for its industrial goods, whereas France wanted to continue to protect its highly inefficient but large and politically important agricultural sector from overseas competition. The result was the creation of a tariff-free market within the Community, the creation of a unified external tariff, and a Common Agricultural Policy (the CAP) to bolster the Community's agricultural sector.

Some of the most striking changes in the regional geography of the EU have been related to the operation of the CAP. Although agriculture accounts for less than 3% of the EU workforce, the CAP has dominated the EU budget from the beginning. For a long time, it accounted for more than 70% of the EU's total expenditures. Its operation has had a significant impact on rural economies, rural landscapes, and rural standards of living, and it has even influenced urban living through its effects on food prices.

The basis of the CAP is a system of EU support of wholesale prices for agricultural produce. This support has the dual effects of stabilizing the price of agricultural products and of subsidizing farmers' incomes. The overall result has been a realignment of agricultural production patterns, with a general withdrawal from mixed farming. Ireland, the United Kingdom, and Denmark, for example, have increased their specialization in the production of wheat, barley, poultry, and milk, whereas France and Germany have increased their specialization in the production of barley, maize, and sugar beets.

The reorganization of Europe's agricultural landscapes under the CAP brought some unwanted side effects, however, including environmental problems that have occurred as a result of the speed and scale of farm modernization, combined with farmers' desire to take advantage of generous levels of guaranteed prices for crops. Moorlands, woodlands, wetlands, and hedgerows have come under threat, and some traditional mixed-farming landscapes have been replaced by the prairie-style settings of specialized agribusiness.

Meanwhile, the rest of the world economy had changed significantly, intensifying the challenge to Europe. By the early 1980s, the U.S. and Japanese economies, having accomplished a large measure of restructuring, were becoming increasingly interdependent and prosperous on the basis of globalized producer services and new, high-tech industries. London's once preeminent financial services were losing ground to those of New York and Tokyo, and even the former West Germany, with Europe's healthiest economy at the time, faced the prospect of being left behind as a producer of obsolescent capital goods and consumer goods. In response, the European Community relaunched itself, beginning in the mid-1980s with the ratification of the Single European Act of 1985, which affirmed the ultimate aim of economic and political harmonization within a single supranational government. This relaunching represents an impressive achievement, particularly because it had to be undertaken at a time when there were major distractions: having to manage a changing relationship with the United States through trade renegotiations, having to cope with the reunification of Germany and the breakup of the former Soviet sphere of influence in Eastern Europe and, not least, having to cope with a widespread internal resurgence of nationalism.

Overall, EU membership has brought regional stresses as well as the prospect of overall economic gain. Existing member countries have found that the removal of internal barriers to labor, capital, and trade has worked to the clear disadvantage of peripheral regions and in particular to the disadvantage of those farthest from the Golden Triangle (between London, Paris, and Berlin), which is increasingly the European center of gravity in terms of both production and consumption. This regional imbalance was recognized by the Single European Act, which included "economic and social cohesion" as a

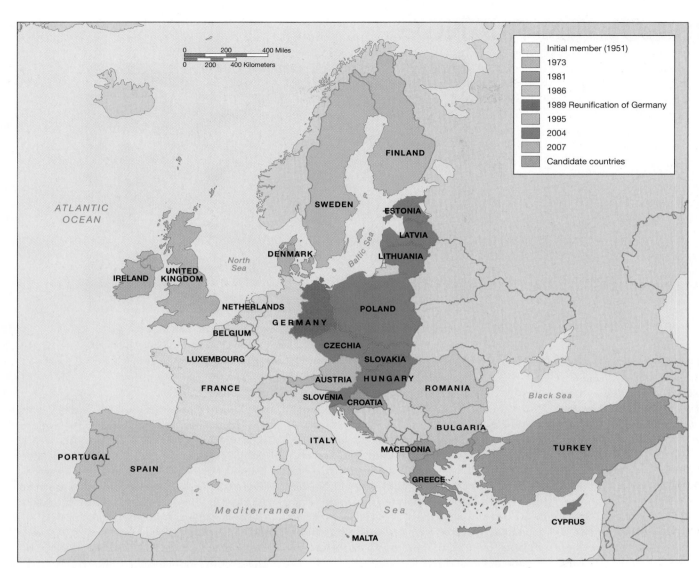

▲ **FIGURE 2.16 The Expansion of the European Union** The advantages of membership in the European Union have led to a dramatic growth in its size, transforming it into a major economic and political force in world affairs.

major policy. The SEA doubled its grant funding for regional development assistance and established a Cohesion Fund to help Greece, Ireland, Portugal, and Spain achieve levels of economic development comparable to those of the rest of the EU. Regions eligible for these funds are shown in **Figure 2.17**. In its 2007 to 2013 budget cycle, the EU has allocated $406 billion—about one-third of the total EU budget—to projects and policies designed to improve economic and social cohesion within and among its member countries.

## Growth, Deindustrialization, and Reinvestment

Europe provides a classic example of how long-term shifts in technology systems tend to lead to regional economic change (see Chapter 1). The innovations associated with new technology systems generate new industries that are not yet tied down by enormous investments in factories or tied to existing industrial agglomerations. Combined with innovations in transport and communications, this creates windows of opportunity that can result in new industrial districts and in some towns and cities growing into dominant metropolitan areas through new rounds of investment. Within Europe the regions that have prospered most through the onset of a new technology system are the Thames Valley to the west of London, the Île de France region around Paris, the Ruhr valley in northwestern Germany, and the metropolitan regions of Lyon–Grenoble (France), Amsterdam–Rotterdam (Netherlands), Milan and Turin (Italy), and Frankfurt, Munich, and Stuttgart (Germany).

Just as high-tech industries and regions have grown, the profitability of traditional industries in established regions has declined. Wherever the differential in profitability has been large enough, disinvestment has taken place in the less-profitable industries and regions, with production moving not so much to the growth regions of Europe but to emerging manufacturing regions in other world regions. **Disinvestment** means selling off assets such as factories and equipment. Widespread disinvestment leads to deindustrialization in formerly prosperous industrial regions. **Deindustrialization** involves a relative decline (and in extreme cases, an absolute decline) in indus-

# EUROPE'S GOLDEN TRIANGLE

The character and relative prosperity of Europe's chief core region, the Golden Triangle, stem from four advantages:

1. It is situated well for shipping and trade (**Figure 1**). Its geographic situation provides access to southern and central Europe by way of the Rhine and Rhône river systems and access to the sea lanes of the Baltic Sea, the North Sea, and, by way of the English Channel, the Atlantic Ocean.

2. Within it are the capital cities of the major former imperial powers of Europe.

3. It includes the industrial heartlands of central England, northeastern France, and the Ruhr district of Germany.

4. Its concentrated population provides both a skilled labor force and an affluent consumer market.

These advantages have been reinforced by the integrative policies of the European Union, whose administrative headquarters are situated squarely in the heart of the Golden Triangle, in Brussels, Belgium. They have also been reinforced by the emergence of Berlin, Paris, and, especially, London as world cities, in which a disproportionate share of the world's economic, political, and cultural business is conducted. As these world cities have come to play an increasingly central role in the world economy, so they have become home to a vast web of sophisticated financial, legal, marketing, and communications services. These services, in turn, have added to the wealth and cosmopolitanism of the region.

▲ **FIGURE 1 London's Container Port** Situated 25 miles downstream from central London on the River Thames, the container port at Tilbury is one of several, along with Southampton, Felixtowe, Hamburg, and Rotterdam, that handle the imports and exports of the Golden Triangle.

Agriculture within the Golden Triangle tends to be highly intensive and geared toward supplying the highly urbanized population with fresh dairy produce, vegetables, and flowers. Industry that remains within the Golden Triangle tends to be rather technical, drawing on the highly skilled and well-educated workforce. Most heavy industry and large-scale, routine manufacturing has relocated from the region in favor of cheaper land and labor found elsewhere in Europe or beyond.

In the Golden Triangle, a tightly knit network of towns and cities is linked by an elaborate infrastructure of canals, railways, and highways. The region has reached saturation levels of urbanization, as exemplified in Randstad Holland, the densely settled region of the western Netherlands that includes Dordrecht, Rotterdam (**Figure 2**), Delft, The Hague, Leiden, Haarlem, Amsterdam, Hilversum, and Utrecht. In addition to London, Paris, and Berlin, major cities of the Golden Triangle include Amsterdam, Antwerp, Birmingham, Brussels, Cologne, Dortmund, Düsseldorf, Frankfurt, Hamburg, Hannover, Lille, Portsmouth, and Rotterdam. Nevertheless, the Golden Triangle still contains fragments of

trial employment in core regions as firms scale back their activities in response to lower levels of profitability. This is what happened to the industrial regions of northern England, South Wales, and central Scotland in the early part of the 20th century; it is what happened to the industrial region of Alsace-Lorraine, in France, and to many other traditional manufacturing towns and regions within Western Europe in the 1960s and 1970s; and it is what has been happening to the industrial centers of Eastern Europe since the 1990s.

**Regional Development** The economic and political integration of Europe has intensified core-periphery differences within the region. The removal of internal barriers to flows of labor, capital, and trade has worked to the clear disadvantage of geographically peripheral regions within the EU, whereas core metropolitan regions have benefited. The principal core region within Europe is the Golden Triangle, centered on the area between London, Paris, and Berlin (see Signature Region in Global Context: Europe's Golden Triangle, above). A secondary emergent core is developing along a north–south crescent that straddles the Alps, stretching from Frankfurt, just to the south of the Golden Triangle, through Stuttgart, Zürich, and Munich to Milan and Turin. Both cores are linked by the new trans-European high-speed rail system.

**Europe's World Cities** In the 19th and early 20th centuries, London was the undisputed center of global economic and geopolitical power.

▲ **FIGURE 2 Rotterdam** Rotterdam is part of a metropolitan area called Rijnmond ("Mouth of the Rhine") with a total population of about 1.2 million. The port of Rotterdam is the largest in Europe, and until 2004, when it was superseded by Shanghai, it was the world's busiest port.

▲ **FIGURE 3 Rural England** The rolling countryside of Sussex, in southeastern England, is protected by strong planning regulations and environmental policies.

attractive rural landscapes (**Figure 3**), together with some unspoiled villages and small towns. These have survived partly because of market forces: They are very attractive to affluent commuters. Equally important to their survival, though, has been the relatively strong role of environmental, land use, and conservation planning in European countries.

Because of the affluence and dynamism of the region, it has attracted large numbers of immigrants. London, in particular, has become a city with a global mix of populations and subcultures.

Almost one-third of London's current residents—2.2 million people—were born outside England, and this total takes no account of the contribution of the city's second- and third-generation immigrants, many of whom have inherited the traditions of their parents and grandparents. Altogether, the people of London speak more than 300 languages, and the city has at least 50 nonindigenous communities with populations of 10,000 or more. Elsewhere in the Golden Triangle, immigrant groups have also added a new ethnic

dimension to city populations. Most have settled in distinctive enclaves, where they have retained powerful attachments to their cultural roots. In the United Kingdom, the Netherlands, and Germany, immigrant populations have tended to concentrate in older inner-city neighborhoods, whereas in France the pattern is one of concentration in suburban public housing projects. Many immigrant groups now face high unemployment and low wages as the postwar boom has leveled off. ■

Today, it remains one of the world's most important centers of finance and business (along with New York [Chapter 6] and Tokyo [Chapter 8]). It dominates the economic and political life of the whole of the United Kingdom and remains the single most cosmopolitan city in Europe. It is a vast, sprawling city that covers more than 3,900 square kilometers (about 1,500 square miles) of continuously built-up area, with a total population of just over 7 million (13 million, if the metropolitan fringe areas are included). London grew up around two core areas: a commercial core centered on its port and trading functions and an institutional core centered on its religious and governmental functions. The commercial core has Roman roots: *Londinium* was the fifth-largest Roman city north of the Alps, a major trading center that enjoyed the

advantages of a deepwater port (on the River Thames) and a key situation facing the continental North Sea and Baltic ports. These same advantages helped London prosper with the resurgence of trade in the medieval period, when the wealth accumulated by wool merchants provided the economic foundation for growth. From this commercial nucleus grew an extensive merchant and financial quarter (**Figure 2.18**). The docks spread eastward from the financial precinct (the City), but by the 1970s much of London's older Docklands had fallen into disuse. Then, between 1985 and 2005, they were redeveloped as a major financial center (**Figure 2.19**). London's institutional core, some 2.2 kilometers (2 miles) upstream from the commercial core, includes Westminster Abbey, the Houses of Parliament, St. James's Palace (built as a London

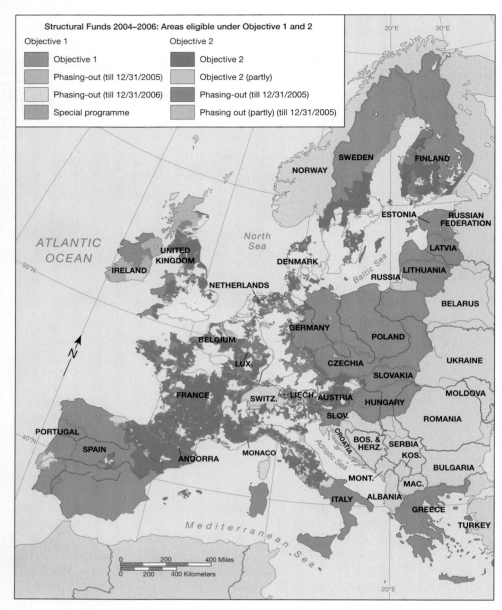

**Structural Funds 2004–2006: Areas eligible under Objective 1 and 2**

Objective 1
- Objective 1
- Phasing-out (till 12/31/2005)
- Phasing-out (till 12/31/2006)
- Special programme

Objective 2
- Objective 2
- Objective 2 (partly)
- Phasing-out (till 12/31/2005)
- Phasing out (partly) (till 12/31/2005)

◀ **FIGURE 2.17 European Union Regions Eligible for Aid, 2006** Just over half of the total population of the European Union lives in areas eligible for regional assistance from the European Regional Development Fund and Cohesion Fund. Most of the EU's regional aid is allocated to "lagging" regions (where per capita GDP is less than 75% of the EU average) and to declining industrial regions.

residence for the monarchy in the 16th century), Buckingham Palace (the royal residence), government offices, centers of culture (the Royal Academy, the National Gallery, and the Royal Opera House), the exclusive shops that cater to the city's elite, and the squares and townhouses of the rich and powerful.

Although London is the most cosmopolitan European city, Paris is the most urbane and the most spectacularly monumental. Paris is the unrivaled focus of political, economic, social, and cultural life in France. It is an industrial center as well as a major international financial center. It is, in short, *the* French city, and there are few cities in the rest of the world (perhaps only Tokyo) that so dominate their national urban systems, their national economy, their politics, and their culture. Within Paris itself, this dominance is reflected in the monumental buildings of the central core, or Ville de Paris. The nucleus of this core is the Île de la Cité (**Figure 2.20**), a boat-shaped island about ten blocks long and five blocks wide, the site of the great

cathedral of Notre Dame de Paris. French planners have sought to reduce the dominance of central Paris by establishing suburban new towns and encouraging nodes of commercial development outside the city center. The biggest and most successful of these suburban nodes is La Défense (**Figure 2.21**), a major example of modern planning and a key element of the city's standing as a world city. La Défense is a complex, multilevel, multiuse development with more than 1.39 million square meters (about 15 million square feet) of office space, 150,000 workers, and 20,000 residents.

If it were not for the geopolitical aftermath of World War II, Berlin would probably be a world city to rival Paris and London. Berlin was a cultural and intellectual center in the early part of the 20th century. It was a seedbed of avant-garde theater, film, cabaret, art, and architecture. World War II changed everything. By the end of the war, 34% of Berlin's housing had been destroyed, and another 54% was damaged. The city found itself embedded within the eastern, social-

▲ **FIGURE 2.18 London** London's original river port trade gave rise to a commercial core that developed into a major financial hub in the district of London known as the City. The skyline of this financial precinct is dominated by the distinctive "gherkin" building, seen here behind the historic Tower of London.

ist part of a partitioned nation and was itself partitioned into eastern and western sectors. West Berlin had to develop a new central business district, the old one having fallen within the eastern sector. In 1961 the division between the two half-cities was physically rein-forced when East Germany built the Berlin Wall to stem the flow of migrants to West Berlin. After the wall went up, both East and West Berlin remained highly militarized, with troops and their equipment a very visible part of the urban landscape. With the reunification of Germany in 1989, Berlin reassumed its prewar role as a national political and cultural center. The city experienced a surge of construction as the two parts of the city were reconnected, wired, and plumbed together again and as the federal government and investors raced to install new infrastructure, department stores, office blocks, hotels, and entertainment centers in keeping with the city's restored position in the world (**Figure 2.22**).

**The Southern Crescent** Stretching south from the Golden Triangle is a secondary, emergent, core region that straddles the Alps, running from Frankfurt in Germany through Stuttgart, Zürich, and Munich, and finally to Milan, Turin, and northern Italy. The prosperity of this Southern Crescent is in part a result of a general decentralization of industry from northwestern Europe and in part a result of the integrative effects of the European Union. Some of the capital freed up by the deindustrialization of traditional manufacturing regions in northwestern Europe has found its way to more southerly regions, where land is less expensive and labor is both less expensive and less unionized. The cities of the Southern Crescent have become key to the spatial reorganization of the whole region. Frankfurt (**Figure 2.23**) and Zürich are global-scale business and financial centers in their own right, whereas Milan (**Figure 2.24**) is a center of both finance and design, and Munich, Stuttgart, and Turin are important centers of industry and commerce.

This Southern Crescent stretches across a great variety of landscapes, from the plateaus of central Germany, across the Alps, and into northern Italy and the Apennines. Overall, these landscapes are much less urbanized than are those of the Golden Triangle. However, the *rate* of urbanization is much higher.

Much urban growth is taking place in smaller towns and cities that are part of new-style industrial subregions that have benefited from the deindustrialization of northwestern Europe. They represent a very different form of industrialization based on loose spatial

▲ **FIGURE 2.19 London Docklands** Once the commercial heart of Britain's empire, employing over 30,000 dockyard laborers, London's extensive Docklands fell into a sharp decline in the late 1960s because of competition from specialized ports using new container technologies. The remaking of the Docklands since the 1980s has seen a huge area of the Docklands converted into a mixture of residential, commercial, exhibition, and light industrial space, the largest single urban redevelopment scheme in the world.

▲ **FIGURE 2.20 Central Paris** Paris straddles the River Seine, and at the very center is the Île de la Cité, the site of the cathedral of Notre Dame and the city's Prefecture de Police, Palais de Justice, Hôtel-Dieu hospital, and Tribunal de Commerce.

▲ **FIGURE 2.22 Berlin** Since the reunification of Germany in 1989, Berlin has experienced a major construction boom, reflecting the city's restored role as a world city. Shown here is the redeveloped Potsdamer Platz, formerly a no-man's-land between East and West Berlin.

agglomerations of small firms that are part of one or more leading industries. Small firms using computerized control systems and an extensive subcontracting network have the advantage of being flexible in what they produce and when and how they produce it. Consequently, the new industrial districts with which they are associated are often referred to as **flexible production regions**. Within each of these regions, small firms tend to share a specific local indus-

trial culture that is characterized by technological dynamism and well-developed social and economic networks.

Northern Italy provides examples of a number of flexible production regions (**Figure 2.25**). Here, regional networks of innovative, flexible, and high-quality manufacturers make products that include textiles, knitwear, jewelry, shoes, ceramics, machinery, machine tools, and furniture. Other examples of flexible production regions based on a simi-

▼ **FIGURE 2.21 La Défense** One of the largest and most successful growth centers within metropolitan Paris, La Défense was designed by planners to attract office development from the central core of the city.

▼ **FIGURE 2.23 Frankfurt am Main, Germany** A city of global financial importance, Frankfurt is the seat of the European Central Bank and the Frankfurt Stock Exchange and is one of the most affluent cities in Europe.

▲ **FIGURE 2.24 Milan** Milan has long been a major regional center and today has developed into a prosperous city of global importance in finance, fashion, and industrial design.

lar mixture of design- and labor-intensive industries include the Baden-Württemberg region around Stuttgart in Germany (textiles, machine tools, auto parts, and clothing) and the Rhône-Alpes region around Lyon in France.

## Demography, Migration, and Settlement

A distinctive characteristic of Europe as a whole is the size and relative density of its population. With less than 7% of Earth's land surface, Europe contains about 13% of its population at an overall density of nearly 100 persons per square kilometer (260 per square mile). Within Europe, the highest national densities match those of Asian countries such as Japan, the Republic of Korea, and Sri Lanka. On the other hand, population density in Finland, Norway, and Sweden stands at about 15 persons per square kilometer, the same as in Kansas and Oklahoma (**Figure 2.26**). This reflects a fundamental feature of the human geography of Europe: the existence of a densely populated core and a sparsely populated periphery. We have already noted the economic roots of this core–periphery contrast.

Whereas the population of the world as a whole is increasing fast, the population of Europe is roughly stable. Europe's population boom coincided roughly with the Industrial Revolution of the late 18th to late 19th centuries. Today, Europe's population is growing slowly in some regions, while declining slightly in others. The main reason for Europe's slow population growth is a general decline in birthrates (though certain subgroups, especially immigrant groups, are an exception to this trend).

It seems that conditions of family life in Europe, including readily available contraception, have led to a widespread fall in birthrates. The average size of families has dropped well below the rate needed for replacement of the population (about 2.1 children per family), to about 1.5 per family in 2008. A "baby boom" after World War II was followed by a "baby bust." Meanwhile, life expectancy has increased because of improved health care, medical knowledge, and healthier lifestyles. The effect is not sufficient to outweigh falling birthrates, but it has meant a dramatic increase in the proportion of people over the age of 65, from 9% in 1950 to nearly 16% in 2008. Germany's population (**Figure 2.27**) reflects these trends and shows the impact of two world wars.

**The European Diaspora**  The upheavals associated with the transition to industrial societies, together with the opportunities presented by colonialism and imperialism and the dislocations of two world wars, have dispersed Europe's population around the globe. Beginning with the colonization of the Americas, vast numbers of people have left Europe for overseas. The full flood of emigration began in the early 19th century, partly in response to population pressure during the early phases of the demographic transition and partly in response to the poverty and squalor of the early phases of the Industrial Revolution. The main stream of migration was to the Americas, with people from northwestern and central Europe heading for North America and southern Europeans heading for destinations throughout the Americas. In addition, large numbers of British left for Australia and New Zealand and eastern and southern Africa. French and Italian emigrants traveled to North Africa, Ethiopia, and Eritrea, and the Dutch went to southern Africa and Indonesia. The final surge of emigration occurred just after World War II, when various relief agencies helped homeless and displaced persons move to Australia and New Zealand, North America, and South Africa, and large numbers of Jews settled in Israel.

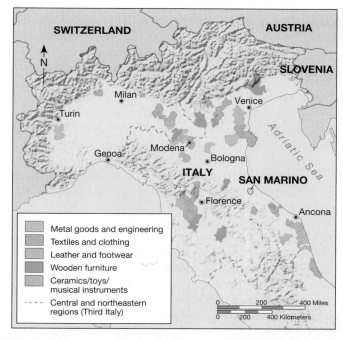

▲ **FIGURE 2.25 New industrial districts in Northern Italy** New industrial districts in the Southern Crescent are based on loose spatial agglomerations of small firms involved in a few key industries. There is a particular concentration of these new industrial districts in central and northeastern Italy—the so-called Third Italy.

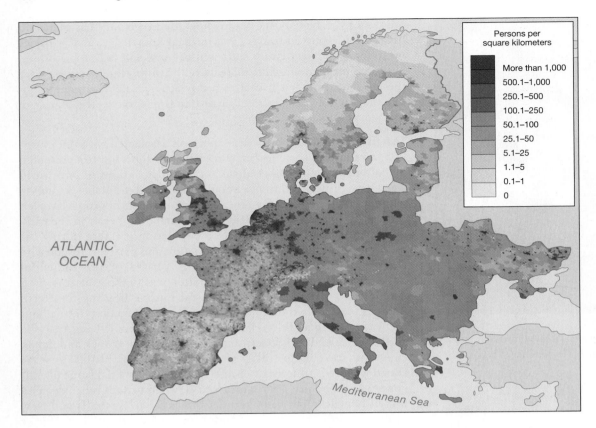

◀ **FIGURE 2.26 Population density in Europe** The distribution of population in Europe reflects the region's economic history, with the highest densities in the "Golden Triangle," the newer industrial regions of northern Italy, and the richer agricultural regions of the North European Lowlands.

**Migration within Europe**    Industrialization and geopolitical conflict have also resulted in a great deal of population movement within Europe. With the onset of industrialization, the regional redistribution of population within Europe followed the economic pattern. The three major waves of industrial development drew migrants from less-prosperous rural areas to a succession of industrial growth areas around coalfields.

As industrial capitalism evolved, the diversified economies of national metropolitan centers offered the most opportunities and the highest wages, thus prompting a further redistribution of population. In Britain this involved a drift of population from manufacturing towns southward to London and the southeast. In France, migration to Paris from towns all around France resulted in a polarization between Paris and the rest of the country. Some countries, developing an industrial base after the "coalfield" stage, experienced a more straightforward shift of population, directly from peripheral rural areas to prosperous metropolitan regions. In this way, Barcelona, Copenhagen, Madrid, Milan, Oslo, Stockholm, and Turin all emerged as regionally dominant metropolitan areas.

Wars and political crises have also led to significant redistributions of population within Europe. World War I forced about 7.7 million people to move. Another major transfer of population took place in the early 1920s, when more than one million Greeks were transferred from Turkey and a half million Turks were transferred from Greece in the aftermath of an unsuccessful Greek attempt to gain control over the eastern coast of the Aegean Sea. Soon afterward, more people were on the move, this time in the cause of ethnic and ideological purity, as the policies of Nazi Germany and fascist Italy began to bite. **Fascism**, of which Nazism was one variety, involves a cen-

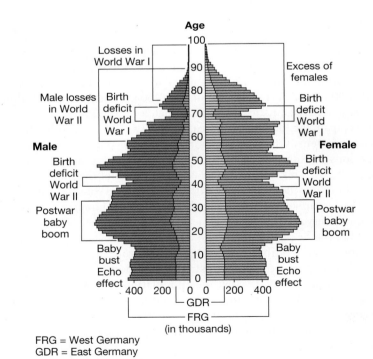

▲ **FIGURE 2.27 Population of Germany, by age and sex, 1989** Germany's population profile is that of a wealthy country that has passed through the postwar baby boom and currently possesses a low birthrate. It is also the profile of a country whose population has experienced the ravages of two world wars.

tralized, autocratic government and values nation and race over the individual. Jews, in particular, were squeezed out of Germany. With World War II, further forced migrations occurred that involved approximately 46 million people.

These migrations, together with mass exterminations undertaken by Nazi Germany, left large parts of west and central Europe with significantly fewer ethnic minorities than before the war. In Poland, for example, minorities constituted 32% of the population before the war but only 3% after the war. Similar changes occurred in Czechoslovakia—from 33% to 15%—and in Romania—from 28% to 12%. Southeast Europe did not experience such large-scale transfers, and as a result many ethnic minorities remained intermixed, or surrounded and isolated, as in the former republic of Yugoslavia. The geopolitical division of Europe after the war also resulted in significant transfers of population: West Germany, for example, had absorbed nearly 11 million refugees from Eastern Europe by 1961, when the Berlin Wall was built.

**Recent Migration Streams**   More recently, the main currents of migration within Europe have been a consequence of patterns of economic development. Rural–urban migration continues to empty the countryside of Mediterranean Europe as metropolitan regions become increasingly prosperous. Meanwhile, most metropolitan regions themselves have experienced a decentralization of population as factories, offices, and housing developments have moved out of congested central areas. Another stream of migration has involved better-off retired persons, who have tended to congregate in spas, coastal resorts, and picturesque rural regions.

The most striking of all recent streams of migration within Europe, however, have been those of migrant workers (**Figure 2.28**). These population movements were initially the result of Western Europe's postwar economic boom in the 1960s and early 1970s, which created labor shortages in Western Europe's industrial centers. The demand for labor represented welcome opportunities to many of the unemployed and poorly paid workers of Mediterranean Europe and of former European colonies. By the mid-1970s these migration streams had become an early component of the globalization of the world economy. By 1975, between 12 and 14 million immigrants had arrived in northwestern Europe. Most came from Mediterranean countries—Spain, Portugal, southern Italy, Greece, Yugoslavia, Turkey, Morocco, Algeria, and Tunisia. In Britain and France the majority of immigrants came from former colonies in Africa, the Caribbean, and Asia. In the Netherlands most came from former colonies in Indonesia. Most of these immigrants have stayed on, adding a striking new ethnic dimension to many of Europe's cities and regions.

Meanwhile, it is estimated that more than 18 million people moved within Europe during the 1980s and 1990s as refugees from war and persecution or in flight from economic collapse in Russia and Eastern Europe. Civil war and dislocation in the Balkans displaced more than four million people in the early 1990s. Wars in Iraq and Afghanistan have also resulted in significant numbers of immigrants to Europe, especially to Denmark and Sweden.

Until the global recession of 2008 to 2009, labor migration continued to expand throughout Europe. Economic growth and political stability for the countries that have recently joined the EU—Cyprus, Hungary, Czechia, Slovakia, and Slovenia—have become destination countries in their own right. Today, all European states are now net immigration countries, and within Europe as a whole there are more than 56 million migrants (compared to about 42 million in North America).

# CULTURE AND POLITICS

Although European culture and politics are distinctive at the global scale, they also bear the legacy of European interactions with other world regions through European colonialism, neocolonialism, emigration, and immigration. Europe's cultures and political systems are also characterized by some sharp internal regional variations. In the broadest terms, there is a significant north–south cultural divide, with a lingering legacy of an east–west political divide. Meanwhile, contemporary world-spanning processes of political and cultural change are beginning to modify some of the traditional patterns associated with European geography.

Southern Europe has always been more traditional in its religious affiliations—not just in terms of the dominance of Roman Catholicism over Protestantism, but of the prevalence of the conservative and more mystical forms of Catholicism. The Roman Catholic Church, still one of the most widespread within Europe, emerged in the 4th century under the bishop of Rome and spread quickly through the weakening Roman Empire. Missionaries helped spread not only the gospel but

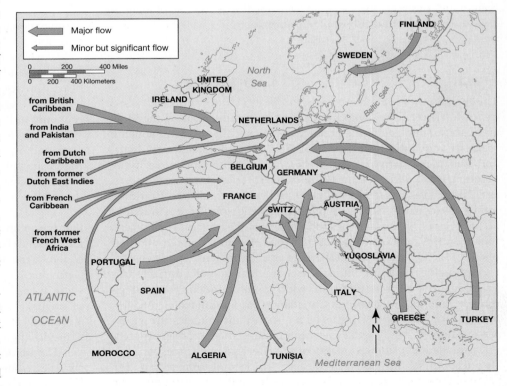

▲ **FIGURE 2.28 International labor migration** This map shows the main international labor migration flows to European countries between 1945 and 1972.

also the use of the Latin alphabet throughout most of Europe. The Eastern Orthodox Church, under the auspices of the Byzantine Empire centered in Constantinople (present-day Istanbul), dominated the eastern margins of Europe and much of the Balkans, whereas Islamic influence spread into parts of the Balkans (present-day Albania, the European part of Turkey, and parts of Bosnia-Herzegovina) and, for a while, southern Spain. With the religious upheavals of the 16th and 17th centuries, Protestant Christianity came to dominate much of northern Europe. More recently, immigrants from the Middle East, Africa, and South Asia have reintroduced Islam to Europe, adding an important dimension to contemporary politics as well as culture. Another distinctive aspect of southern European culture is its traditional patterns of family life, with larger, close-knit families that tend to stick together as a buffer against unemployment and poverty in societies with relatively underdeveloped social welfare systems.

As we have seen, for a few decades—between 1945 and 1989—there was also a significant geopolitical division within Europe into Eastern and Western Europe. These two subregions share much in terms of physical geography, racial characteristics, and cultural values, but for a while they were divided by the Cold War territorial boundary that separated the capitalist democracies of Western Europe from Soviet state socialism. Soviet influence to the east of the Iron Curtain created distinctive economic, social, and cultural conditions for a long enough period to have effected some modifications to the geography of Eastern Europe. Since the collapse of the Soviet Union in 1991, however, its former satellite states in Eastern Europe have reoriented themselves as part of the broader European world region. Meanwhile, most European states have joined together in the European Union, creating an extremely powerful economic and political force in world affairs. Europe as a whole still bears the legacy of a complex history of social and political development, but it has reemerged since World War II as a crucible of technological, economic, and cultural innovation.

## Tradition, Religion, and Language

The foundations of contemporary European culture were established by the ancient Greeks, who in turn had been influenced by the ideas and cultural and economic practices of ancient Egypt, Mesopotamia (centered along the Tigris and Euphrates rivers, in modern-day Iraq), and Phoenicia (centered along the coastal regions of modern-day Lebanon, Israel, Syria, and the Palestinian territories). Between 600 B.C.E. and 200 B.C.E. the Greeks built an intellectual tradition of rational inquiry into the causes of everything, along with a belief that individuals are free, self-understanding, and valuable in themselves. The Romans took over this intellectual tradition, added Roman law and a tradition of disciplined participation in the state as a central tenet of citizenship, and spread the resulting culture throughout their empire. From the Near East came the Hebrew tradition, which in conjunction with Greek thought produced Judaism and Christianity, religions in which the individual spirit is seen as having its own responsibility and destiny within the creation. At the heart of European culture, then, are the curiosity, open-mindedness, and rationality of the Greeks; the civic responsibility and political individualism of both Greeks and Romans; and the sense of the significance of the free individual spirit that is found in the main tradition of Christianity.

When Europeans pushed out into the rest of the world in their colonial and imperial ventures, they took these values with them, imposing them onto some cultures and grafting them into others. By the 18th century they also carried the idea of **Modernity**, the genesis of which was in the changing world geography of the Age of Discovery. Modernity emphasized innovation over tradition, rationality over mysticism, and utopianism over fatalism. As Europeans tried to make sense of their own ideas and values in the context of those they encountered in the East, in Africa, in Islamic regions, and among Native Americans during the 16th and 17th centuries, many certainties of traditional thinking were cracked open. In the 18th century this ferment of ideas culminated in the **Enlightenment** movement, which was based on the conviction that all of nature, as well as human beings and their societies, could be understood as a rational system. Politically, the Enlightenment reinforced the idea of human rights and democratic forms of government and society. Expanded into the fields of economics, social philosophy, art, and music, the Enlightenment gave rise to the cultural sensibility of Modernity.

By the late 20th century, after the decline of heavy industry and repeated episodes of economic recession; after two terrible world wars; after interludes of fascist dictatorships in Germany, Italy, Greece, Portugal, and Spain; after a protracted period of being on the front line of a Cold War that divided European geography in two; and after intermittent episodes of regional and ethnic conflict, it was not surprising that the culture of Europe had become a culture of doubt and criticism, heavily influenced by a search for radical rethinking. In this search, Europeans have not only established a new cultural sensibility for themselves, but they have also generated some powerful new ideas and philosophies that have begun to influence other cultures around the world. Dismay with the side effects of laissez-faire industrial capitalism and, later, horror at the results of fascism and Nazism gave a strong impetus to left-wing critiques that have been powerful enough to reshape entire national and regional cultures and, with them, some dimensions of regional geographies.

The most profound influence of all was Karl Marx, whose penetrating critique of industrial capitalism (written in London and drawing heavily on descriptions of conditions in Manchester, England, supplied by his colleague Friedrich Engels) inspired both a socialist political economy in Russia and a fascist countermovement in Germany. After World War II, western European left-wing critique portrayed both fascism and Soviet-style socialism as essentially imperialist, whereas American-style capitalism was critiqued as being intrinsically exploitative in privileging the individual and property over the community and the public good.

**Language and Ethnicity**    The western part of southern Europe shares the Romance family of languages, the development of which was fostered by the spread of the Roman Empire. A second major group of languages, Germanic languages, occupy northwestern Europe, extending as far south as the Alps (**Figure 2.29**). English is one of the Germanic family of languages, an amalgam of Anglo-Saxon and Norman French, with Scandinavian and Celtic traces. A third major language group consists of Slavic languages, which dominate Eastern Europe.

These broad geographic divisions of religion, language, and family life are reflected in other cultural traits: folk art, traditional costume, music, folklore, and cuisine. Thus there is a Scandinavian cultural subregion with a collection of related languages (except Finnish), a uniformity of

Protestant denominations, and a strong cultural affinity in art and music that reaches back to the Viking age and even to pre-Christian myths. A second distinctive subregion is constituted by the sphere of Romance languages in the south and west. A third is constituted by the British Isles, bound by language, history, art forms, and folk music, but with a religious divide between the Protestant Anglo-Saxon and Catholic Celtic spheres. A fourth clear cultural subregion is the Germanic sphere of central Europe, again with mixed religious patterns—Lutheran Protestantism in the north, Roman Catholicism in the south—but with a common bond of language, folklore, art, and music. The Slavic subregion of eastern and southeastern Europe forms another broad cultural subregion, though beyond the commonalities of related languages and certain physical traits among the general population, there is considerable diversity.

**Europe's Muslims** Europe is also home to approximately 15 to 20 million Muslims—between 4% and 5% of the total population of the region. The greatest concentrations of Muslims are in the Balkan countries, where Islam has been important for centuries, a legacy of the Turkish Ottoman Empire. Although the empire was dissolved at the end of World War I, Islamic culture has remained in place. In Albania, about 70% of the population is Muslim; in Bosnia-Herzegovina, the figure is 40%; in Macedonia, 30%; and in Serbia and Montenegro (including Kosovo), around 20%.

The majority of Europe's Muslims, however, are located in the industrial cities of Western Europe, and they are a relatively recent addition to the population of the region (**Figure 2.30**). The mass immigration of Muslims to Europe was a consequence of Europe's post–World War II economic boom, which created labor shortages in Western Europe's industrial centers in the 1960s and 1970s and spawned programs that encouraged migrant workers. The French Muslim population of between five and six million is the largest in Western Europe. About 70% have their heritage in France's former North African colonies of Algeria, Morocco, and Tunisia. By contrast, Germany's Muslim population of around three million is dominated by people of Turkish origin. In the United Kingdom, significant numbers of Muslims arrived in the 1960s as people from the former colonies took up offers of work. Some of the first were East African Asians, whereas many came from south Asia. Permanent communities formed, and at least 50% of the current population of 1.6 million was born in the United Kingdom, with one-third of the Muslim population currently aged 15 or less. Approximately one million Muslims live in the Netherlands, the majority with ties to the former Dutch colonies of Suriname and Indonesia. Other west European countries with significant Muslim populations include Austria (339,000, representing 4.1% of the total population), Belgium (400,000; 4%), Denmark (270,000; 5%), Italy (825,000; 1.4%), Spain (1 million; 2.3%), and Sweden (300,000; 3%).

Government policies in most European countries favor multiculturalism, an idea that, in general terms, accepts all cultures as having equal value. The growth of Muslim communities and their resistance to cultural assimilation, however, has challenged the European ideal of strict separation of religion and public life. Many of these immigrant groups, now facing high unemployment and low wages as the postwar boom has leveled off, have clustered together in distinctive urban neighborhoods. In Muslim neighborhoods with high concentrations of unemployment, heavy-handed policing and racial discrimination can easily trigger civil disorder. This is what happened in the fall of 2005 in the Paris suburbs of Clichy-sous-Bois (**Figure 2.31**), where days of rioting followed the deaths of two teenagers who had been chased by police. The riots quickly spread to other French cities, including Lille, Lyon, Marseille, St. Étienne, Strasbourg, and Toulouse; thousands of vehicles were burned, and the French government was forced to order a state of emergency that extended until January 2006.

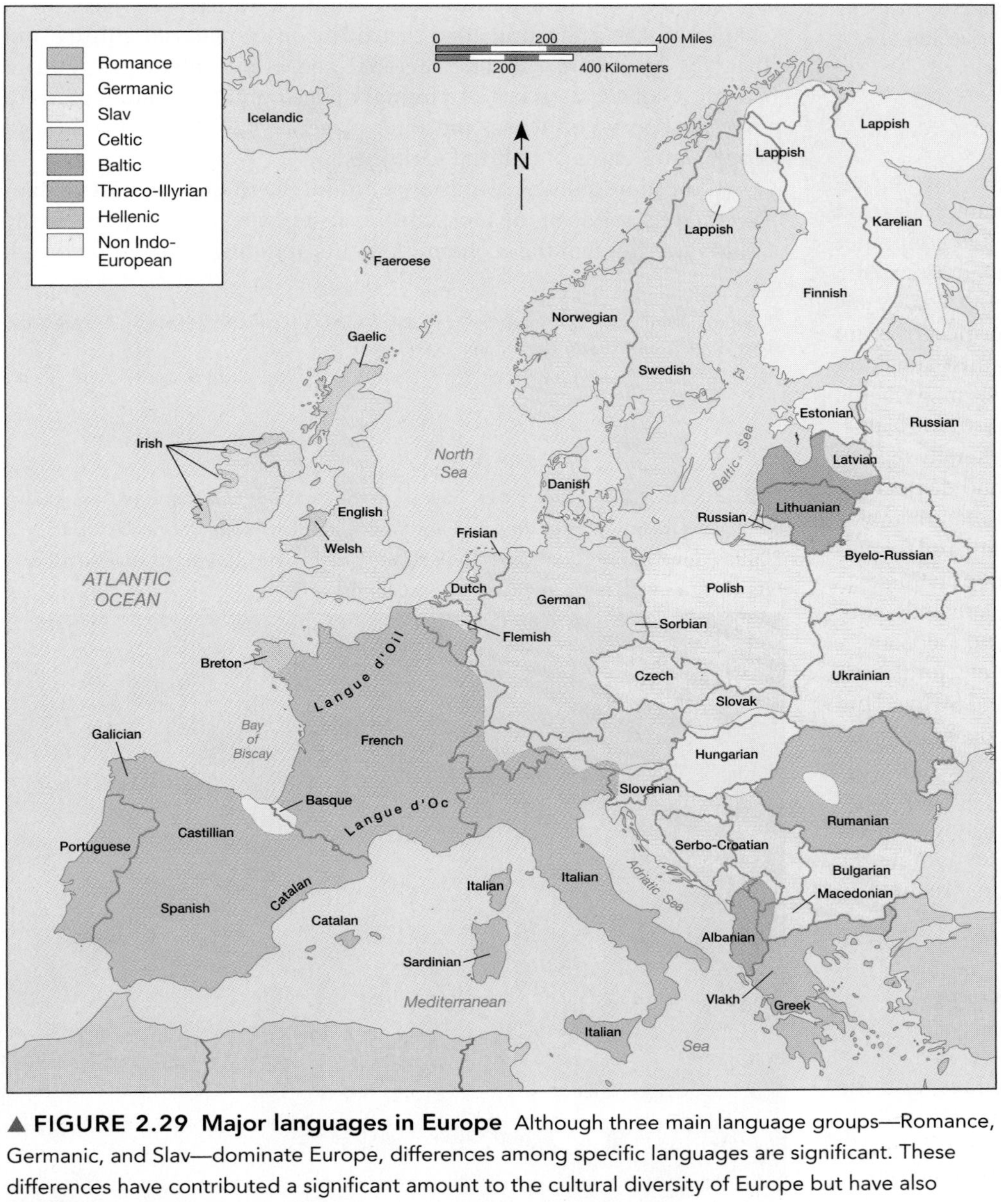

▲ **FIGURE 2.29 Major languages in Europe** Although three main language groups—Romance, Germanic, and Slav—dominate Europe, differences among specific languages are significant. These differences have contributed a significant amount to the cultural diversity of Europe but have also contributed a significant amount to ethnic and geopolitical tensions.

▲ **FIGURE 2.30 Europe's Muslims** Immigrants from former colonies, recruited to Europe to solve labor shortages after World War II, have added a new cultural dimension to many of Europe's cities. These women in Copenhagen, Denmark, are immigrants from North Africa.

Islamist terrorist attacks against west European targets have meanwhile heightened fears and tensions between Muslim communities and host populations. In March 2004, 13 bombs on four packed commuter trains killed 191 people and wounded more than 1,500 in Madrid, Spain. The attack was attributed to the Islamic militant group al-Qaeda. A few months later in the Netherlands a prominent filmmaker critical of Islam was murdered by a radical Islamist. In July 2005, an Islamist attack on buses and underground trains in London left 52 dead. Later in 2005, the Danish newspaper Jyllands-Posten published a series of cartoons featuring the prophet Mohammad. Whereas European culture generally regards satire and caricature as accepted elements of free speech and democracy, Muslims regard visual representations of the prophet Mohammad as a profanity. The result was that the cartoons became a lightning rod for cultural tensions. Muslim leaders in Denmark were able to internationalize the issue, and a few months later a sudden wave of anti-Danish demonstrations swept across the Middle East and in Indonesia and Pakistan.

In the past decade, immigration has become a major electoral issue in Austria, Belgium, Denmark, France, the Netherlands, and Switzerland (**Figure 2.32**), and Muslim communities across Europe have faced increased resentment and hostility from host populations. Tensions remain high, and the cultural issues associated with Europe's Muslim population are likely to continue to be an important dimension of European politics, especially with the possibility of Turkey joining the European Union, which would add around 83 million Muslims to Europe's population.

## Cultural Practices, Social Differences, and Identity

Contemporary Europe has a distinctive set of social values, and the "European Dream" is quite distinctive from the "American Dream":

The European Dream emphasizes community relationships over individual autonomy, cultural diversity over assimilation, quality of life over the

accumulation of wealth, sustainable development over unlimited material growth, deep play over unrelenting toil, universal human rights and the rights of nature over property rights, and global cooperation over the unilateral exercise of power.[3]

Present-day Europe also has a distinctive cultural cast. Intellectual debate—about the role of culture itself; about whether people's thoughts and lives should be understood in terms of the dynamics of the cultures in which they are embedded (structuralism) or in terms of individual consciousness (existentialism); and about whether any kind of single-viewpoint, big-picture understanding of the world is really possible (postmodernism)—has spilled over into literature, cinema, television, magazines, and newspapers. Thus contemporary European culture is marked by a critical awareness of the role of culture itself.

**A "European" Identity?**   Globalization has heightened people's awareness of cultural heritage and ethnic identities. As we saw in Chapter 1, the more universal the diffusion of material culture and lifestyles, the more valuable regional and ethnic identities tend to become. Globalization has also brought large numbers of immigrants to some European countries, and their presence has further heightened people's awareness of cultural identities.

In the more affluent countries of northwestern Europe, immigration has emerged as one of the most controversial issues since the end of the Cold War. Although the economic benefits of immigration far outweigh

---

[3]J. Rifkin, *The European Dream: How Europe's Vision of the Future Is Quietly Eclipsing the American Dream* (New York: Tarcher, 2004), p. 3.

▼ **FIGURE 2.31 Paris riots** Following the death of two teenage boys in an electrical substation after they had fled police in Clichy-sous-Bois in October 2005, extensive riots continued for 9 straight nights and spread to several other French cities with large immigrant populations.

▲ **FIGURE 2.32 The politics of immigration** A campaign poster from the right-wing Swiss People's Party (SVP), the top vote-getting party in the 2007 parliamentary elections in Switzerland. The poster was aimed at the SVP's proposal to deport foreigners—legal residents without Swiss citizenship—who commit crimes.

they have had to be accommodated in processing centers, have also provoked negative reactions (**Figure 2.33**).

All this raises the question, "How 'European' *are* the populations of Europe?" European history and ethnicity have resulted in a collection of national prides, prejudices, and stereotypes that are strongly resistant to the forces of cultural globalization. Germans continue to be seen by most other Europeans as a little overserious, preoccupied by work, and inclined to arrogance. Scots continue to carry the popular image of a dour, unimaginative, ginger-haired people who love bagpipe and accordion music, dress in kilts and sporrans, live on whisky and porridge, and generally spend as little as possible. The English are seen as a nation of lager-swilling hooligans, well-meaning middle classes, and out-of-touch aristocrats. Norwegians and Danes continue to resent the Swedes' "neutrality" during World War II, and so on. In reality, such stereotypes are, of course, exaggerations that stem from the behaviors of a relative minority, and opinion surveys show that these stereotypes, prejudices, and identities are steadily being countered by a growing sense of European identity, especially among younger and better-educated persons. Much of this can be attributed to the growing influence of the European Union.

**Women in European Society** Another powerful postwar movement deeply critical of the dominant structures of capitalist society was feminism, built on the ideas of Simone de Beauvoir in her 1949 book, *The Second Sex*. The strong liberal component of European culture has meant that, compared with peoples in most other world regions, Europeans have been more willing to address the deep inequalities between men and women that are rooted in both traditional societies and industrial capitalism. Still, patriarchal society and the culture of machismo remain strong in Mediterranean Europe—especially in rural areas—and working-class communities throughout Europe are still characterized by significant gender inequalities.

It is in northwestern Europe—and especially in Scandinavia—that gender equality has improved most, as a result of both the progressive social values of the "baby boom" generation and legislation that has translated these values into law. By the mid-1980s, younger men in

any additional demands that may be made on a country's health or welfare system, fears that unrestrained immigration might lead to cultural fragmentation and political tension have provoked some governments to propose new legislation to restrict immigration from the former communist states of Eastern Europe and from outside Europe. The same fears have been responsible for a resurgence of popular **xenophobia**—a hate, or fear, of foreigners—in some countries. In Germany, for example, right-wing nationalistic groups have attacked hostels housing immigrant families, whereas citizenship laws have prevented second-generation Gastarbeiter ("guest worker") families from obtaining German citizenship. In France, claims that immigrants from North Africa are a threat to the traditional French way of life have led to some success for the National Front Party. Asylum seekers, drawn to northwestern Europe from all parts of the globe in such large numbers that

▲ **FIGURE 2.33 Immigration concerns** Local residents of Lee-on-the-Solent, England, protest against a proposed asylum center for refugees.

much of northwestern Europe had acquired a new, progressive collective identity associated with ideals of gender equality—especially as they relate to men's domestic roles. More recently, however, a "men-behaving-badly" syndrome—known as "laddism" in the United Kingdom—has emerged in reaction.

In global context, Europe stands out as a region where women's representation in senior positions in industry and government is relatively high. Women in Europe generally have a significantly longer life expectancy than men and have comparable levels of adult literacy. Nevertheless, women in the European labor force tend to earn, on average, only 45% to 65% of what men earn. These statistics demonstrate a distinctly regional pattern. Broadly speaking, the gender gaps in education, employment, health, and legal standing are wider in southern and Eastern Europe and narrower in Scandinavia and northwestern Europe. In part, this reflects regional differences in social customs and ways of life; in part, it reflects regional differences in overall levels of affluence.

## Territory and Politics

Many of the countries of Europe are relatively new creations, and political boundaries within Europe have changed quite often, reflecting changing patterns of economic and geopolitical power and a continuous struggle to match territorial boundaries to cultural and ethnic identities. In 1500 the political map of Europe included scores of microstates (**Figure 2.34**), legacies of the feudal hierarchies of the Middle Ages. The idea of modern national states can be traced to the subsequent Enlightenment in Europe, when the ferment of ideas about human rights and democracy, together with widening horizons of literacy and communication, created new perspectives on allegiance, communality, and identity. In 1648 the Treaty of Westphalia, signed by most European powers, brought an end to Europe's seemingly interminable religious wars by making national states the principal actors in international politics and establishing the principle that no state has the right to interfere in the internal politics of any other state.

Gradually, these perspectives began to undermine the dominance of the great European continental empires controlled by family dynasties—the Bourbons, the Hapsburgs, the Hohenzollerns, the House of Savoy, and so on. After the French Revolution (1789–93) and the kaleidoscopic changes of the Napoleonic Wars (1800–15), Europe was reordered, in 1815, to be set in a pattern of modern states (**Figure 2.35**). Denmark, France, Portugal, Spain, and the United Kingdom had long existed as separate, independent states. The 19th century saw the unification of Italy (1861–70) and of Germany (1871) and the creation of Belgium, Bulgaria, Greece, Luxembourg, the Netherlands, Romania, Serbia, and Switzerland as independent national states. Early in the 20th century they were joined by Czechoslovakia, Estonia, Finland, Latvia, Lithuania, Norway, and Sweden. Austria was created in its present form in the aftermath of World War I, as part of the carving up of the German and Austro-Hungarian empires. In 1921 long-standing religious cleavages in Ireland resulted in the creation of the Irish Free State (now Ireland), with the six Protestant counties of Ulster remaining in the United Kingdom.

The European concept of the nation–state has immensely influenced the modern world. As we saw in Chapter 1, the idea of a nation–state is based on the concept of a homogeneous group of people governed by their own state. In a true nation–state, no significant group exists that is not part of the nation. In practice, most European states were established around the concept of a nation–state but with territorial boundaries that did in fact encompass substantial ethnic minorities (**Figure 2.36**). The result has been that the geography of Europe has been characterized by regionalism and irredentism throughout the 20th century and into the 21st.

We have already cited the example of Basque regionalism in Spain and France (see p. 35 in Chapter 1). Other examples of regionalism include regional independence movements in Catalonia (within Spain), Scotland (within the United Kingdom), and the Turkish Cypriots' determination to secede from Cyprus. Examples of *irredentism* include Ireland's claim on Northern Ireland (renounced in 1999), the claims of Nazi Germany on Austria and the German parts of Czechoslovakia and Poland, and the claims of Croatia and Serbia and Montenegro on various parts of Bosnia-Herzegovina. Some cases of regionalism have led to violence, social disorder, or even civil war, as in Cyprus. For the most part, however, regional ethnic separatism has been pursued within the framework of civil society, and the result has been that several regional minorities have achieved a degree of political autonomy. For example, the United Kingdom created regional parliaments for Scotland and Wales.

**Ethnic Conflict in the Balkans**   Most cases of irredentism, on the other hand, have contributed at some point in history to war or conflict. Nowhere has this been more evident than in the troubled region of the Balkans. When 19th-century empires were dismantled after World War I, an entirely new political geography was created in the Balkans. Those political boundaries survived until the 1990s, when the breakup of Yugoslavia marked the end of the Great Powers' attempt to unite Serbs, Croats, and Slovenes within a single territory. The repeated fragmentation and reorganization of ethnic groups into separate states within the region has given rise to the term **balkanization** in referring to any situation in which a larger territory is broken up into smaller units, and especially where territorial jealousies give rise to a degree of hostility. In the Balkans themselves, the geopolitical reorganizations of the 1990s have left significant **enclaves**, culturally distinct territories that are surrounded by the territory of a different cultural group, and **exclaves**, portions of a country or of a cultural group's territory that lie outside its contiguous land area. These enclaves and exclaves remain the focus of continued or potential hostility. In Romania, for example, there are more than 1.6 million Hungarians, whereas in Bulgaria there are more than 800,000 Turks. Serbian nationalism, however, has provided the principal catalyst for violence and conflict in the region. In the 1990s, Serbian nationalism led to attempts at **ethnic cleansing.**

The most extreme example of ethnic cleansing was in the Kosovo region. In 1998, Yugoslavia's Serbian leader, Slobodan Milosevic, initiated a brutal, premeditated, and systematic campaign of ethnic cleansing that was aimed at removing Kosovar Albanians from what had become their homeland. Serbian forces expelled Kosovar Albanians at gunpoint from villages and larger towns, looted and burned their homes, organized the systematic rape of young Kosovar Albanian women, and used Kosovar Albanians as human shields to escort Serbian military convoys. By systematically destroying schools, places of worship, and hospitals, Serbian forces sought to destroy social identity and the fabric of Kosovar Albanian society.

International outrage at these human rights violations finally led to the declaration of war against Yugoslavia by NATO in March 1999.

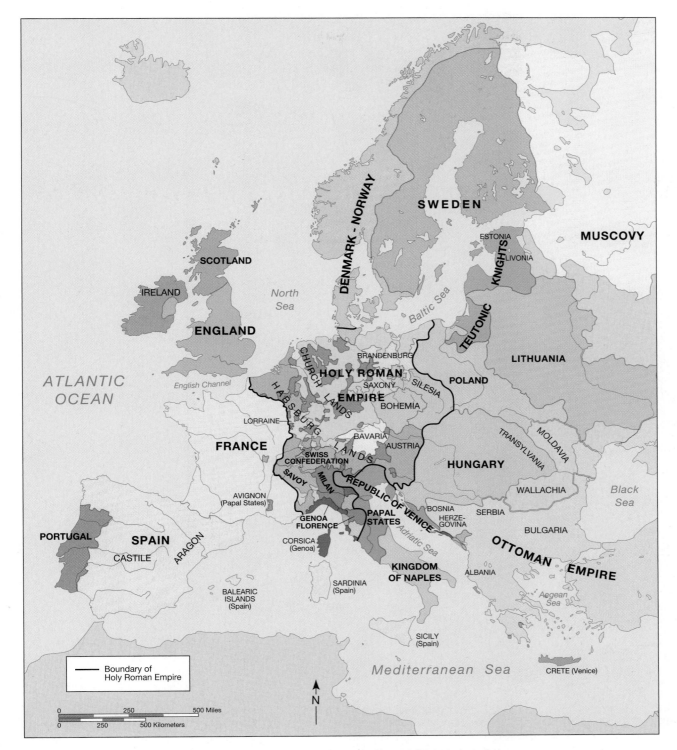

▲ FIGURE 2.34 Europe in 1500  In the late Middle Ages the political map of Europe included scores of microstates, legacies of the feudal system.

▲ **FIGURE 2.35 Europe in 1900** After the French Revolution at the end of the 18th century and the Napoleonic Wars at the beginning of the 19th century, Europe was reordered in a pattern of modern states. The 19th century saw the unification of Italy (1861–70) and Germany (1871) and the creation of Belgium, Bulgaria, Greece, Luxembourg, the Netherlands, Romania, Serbia, and Switzerland as independent national states.

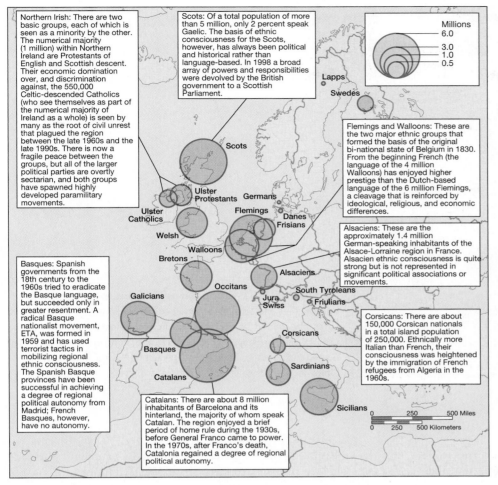

Northern Irish: There are two basic groups, each of which is seen as a minority by the other. The numerical majority (1 million) within Northern Ireland are Protestants of English and Scottish descent. Their economic domination over, and discrimination against, the 550,000 Celtic-descended Catholics (who see themselves as part of the numerical majority of Ireland as a whole) is seen by many as the root of civil unrest that plagued the region between the late 1960s and the late 1990s. There is now a fragile peace between the groups, but all of the larger political parties are overtly sectarian, and both groups have always spawned highly developed paramilitary movements.

Scots: Of a total population of more than 5 million, only 2 percent speak Gaelic. The basis of ethnic consciousness for the Scots, however, has always been political and historical rather than language-based. In 1998 a broad array of powers and responsibilities were devolved by the British government to a Scottish Parliament.

Flemings and Walloons: These are the two major ethnic groups that formed the basis of the original bi-national state of Belgium in 1830. From the beginning French (the language of the 4 million Walloons) has enjoyed higher prestige than the Dutch-based language of the 6 million Flemings, a cleavage that is reinforced by ideological, religious, and economic differences.

Alsaciens: These are the approximately 1.4 million German-speaking inhabitants of the Alsace-Lorraine region in France. Alsacien ethnic consciousness is quite strong but is not represented in significant political associations or movements.

Basques: Spanish governments from the 18th century to the 1960s tried to eradicate the Basque language, but succeeded only in greater resentment. A radical Basque nationalist movement, ETA, was formed in 1959 and has used terrorist tactics in mobilizing regional ethnic consciousness. The Spanish Basque provinces have been successful in achieving a degree of regional political autonomy from Madrid; French Basques, however, have no autonomy.

Corsicans: There are about 150,000 Corsican nationals in a total island population of 250,000. Ethnically more Italian than French, their consciousness was heightened by the immigration of French refugees from Algeria in the 1960s.

Catalans: There are about 8 million inhabitants of Barcelona and its hinterland, the majority of whom speak Catalan. The region enjoyed a brief period of home rule during the 1930s, before General Franco came to power. In the 1970s, after Franco's death, Catalonia regained a degree of regional political autonomy.

▲ **FIGURE 2.36 Minority ethnic subgroups in Western Europe** Regional and ethnic consciousness now represents a strong political factor in many European countries. Within Spain, for example, the dominant population—some 27 million—is Castilian, but there are between six and eight million Catalans, almost two million Basques, and about three million Galicians. Belgium is divided into four million French-speaking Walloons and six million Flemings, whose language is Dutch-based.

Within a few weeks Slobodan Milosevic and other Serbian leaders were indicted in the International Court of Justice for their roles in human rights violations. Since 2000, many Kosovar Albanian refugees have returned to their homeland in Kosovo to attempt to rebuild their lives, and in 2008 Kosovo declared its independence—recognized by 22 of the EU's 27 member states, but strongly opposed by Serbia.

# FUTURE GEOGRAPHIES

Europe will likely continue to develop and gain strength as the newer members of the EU become more economically strong and politically stable. And although there will still be an uneven distribution of wealth and development in the transforming EU, the new states will be better off than they were 10 years ago and Europe overall will be economically and politically stronger.

The realignment of subregional European geographies will be shaped in large part by changing patterns and technologies of transportation. With its relatively short distances between major cities, Europe is ideally suited for rail travel and less suited, because of population densities and traffic congestion around airports, to air traffic. Allowing for check-in times and accessibility to terminals, travel between many major European cities is already quicker by rail than by air. Improved locomotive technologies and specially engineered tracks and rolling stock will make it possible to offer passenger rail services at speeds of 275 to 350 kilometers per hour (180 to 250 miles per hour). New tilt-technology railway cars, which are designed to negotiate tight curves by tilting the train body into turns to counteract the effects of centrifugal force, are being introduced in many parts of Europe to raise maximum speeds on conventional rail tracks. These high-speed rail routes will cause some restructuring of the geography of Europe. They will have only a few time-tabled stops because the time penalties that result from deceleration and acceleration undermine the advantages of high-speed travel. Places that do not have scheduled stops will be less accessible and, then, less attractive for economic development. Places linked to the routes will be well situated to grow in future rounds of economic development. This high-speed rail network will be crucial to the success of what will likely remain one of the world's most vibrant and sophisticated industrial core regions that stretches from Southeast England to Northern Italy, with emerging axes of economic growth that stretch east–west and north–south (**Figure 2.37**).

There are, nevertheless, concerns that the geopolitical enlargement of the EU will risk destroying its own internal balance and cohesion. Europe's future international role also depends greatly on whether it undertakes major structural economic and social reforms to deal with the problem of its aging workforce. This will demand more immigration and better integration of workers, most of whom are likely to be coming from North Africa and the Middle East. Even if more migrant workers are not allowed in, Western Europe will have to integrate its growing Muslim population. Barring increased legal entry may only lead to more illegal migrants who will be harder to integrate, posing a long-term problem. It is possible to imagine European nations successfully adapting their workforces and social welfare systems to these new realities, but it is harder to see some countries—Germany, for example—successfully assimilating millions of new immigrant workers in a short period of time.

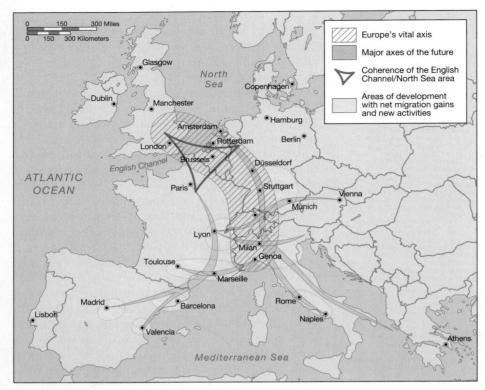

◀ **FIGURE 2.37 European growth areas** Most of Europe's major cities and advanced manufacturing regions lie along a crescent-shaped axis that extends from southeast England through southwest Germany to northern Italy, and along emerging axes that extend to southern France, Spain, eastern Austria, and southern Italy.

# Main Points Revisited

■    Contemporary Europe is highly urbanized and is a cornerstone of the world economy with a complex, multilayered, and multifaceted regional geography.

In overall terms, Europe accounts for almost two-fifths of world trade and about one-third of the world's aggregate GNP.

■    Beyond Europe's core regions, major metropolitan areas, and specialized industrial districts, a mosaic of landscapes has developed around distinctive physiographic regions characterized by broad coherence of geology, relief, landforms, soils, and vegetation. These regions are the Northwestern Uplands, the Alpine System, the Central Plateaus, and the North European Lowlands.

In detail, these landscapes are a product of centuries of human adaptation to climate, soils, altitude, and aspect and to changing economic and political circumstances. Farming practices, field patterns, settlement types, local architecture, and ways of life have all become attuned to the opportunities and constraints of regional physical environments, producing distinctive regional landscapes.

■    The rise of Europe as a major world region had its origins in the emergence of a system of merchant capitalism in the 15th century, when advances in business practices, technology, and navigation made it possible for merchants to establish the basis of a worldwide economy in the space of less than 100 years.

These changes also had a profound effect on Europe's geography, reorienting the region toward the Atlantic and away from the subregional maritime economies of the Mediterranean and the Baltic. Since then, Europe's regional geographies have been comprehensively recast several times: by the new production and transportation technologies that

marked the onset of the Industrial Revolution, by two world wars, and by the Cold War rift between Eastern and Western Europe.

■    The European Union (EU) emerged after World War II as a major factor in reestablishing Europe's role in the world. In overall terms, Europe, with about 12% of the world's population, accounts for almost 35% of the world's exports, almost 43% of the world's imports, and 33% of the world's aggregate GNP.

The EU itself is now a sophisticated and powerful institution with a pervasive influence on patterns of economic and social well-being within its member states. It has a population of more than 490 million, with a combined GDP slightly larger than that of the United States.

■    The reintegration of Eastern Europe has added a potentially dynamic market of 150 million consumers to the European economy.

Overall, Eastern Europe functions as a set of economically peripheral regions, with agriculture still geared to local markets and former COMECON trading opportunities and industry still geared more to heavy industry and standardized products than to competitive consumer products.

■    The principal core region within Europe is the Golden Triangle, which stretches between London, Paris, and Berlin and includes the industrial heartlands of central England, northeastern France, and the Ruhr district of Germany.

A secondary, emergent core is developing along a north–south crescent that straddles the Alps, stretching from Frankfurt, just to the south of the Golden Triangle, through Stuttgart, Zürich, and Munich, to Milan and Turin.

# Key Terms

acid rain (p. 52)

agglomeration economy
(p. 58)

aspect (p. 51)

balkanization (p. 74)

buffer zone (p. 58)

command economy (p. 58)

deindustrialization (p. 61)

disinvestment (p. 61)

egalitarian society (p. 58)

enclave (p. 74)

Enlightenment (p. 70)

entrepôt (p. 54)

ethnic cleansing (p. 74)

exclave (p. 74)

fascism (p. 68)

feudal system (p. 53)

fjord (p. 48)

flexible production region
(p. 66)

heathland (p. 52)

imperialism (p. 55)

latifundia (p. 52)

loess (p. 51)

Marshall Plan (p. 58)

merchant capitalism (p. 53)

minifundia (p. 52)

Modernity (p. 70)

moraine (p. 48)

pastoralism (p. 48)

physiographic region
(p. 47)

polder (p. 52)

privatization
(p. 59)

satellite state (p. 58)

social housing (p. 60)

state socialism
(p. 58)

steppe (p. 51)

tundra (p. 48)

watershed (p. 46)

welfare state (p. 60)

xenophobia (p. 73)

# Thinking Geographically

1. How has Europe benefited from its location and its major physical features?

2. Many European cultures have a strong history of seafaring. How did that become a crucial factor in European and world geography?

3. What key inventions during the period from 1400 to 1600 helped European merchants establish the basis of today's global economy? Why?

4. Which imports from the American colonies helped transform Europe? Focus on natural resources and new crops.

5. What factors led to the end of the European colonial era?

6. What was the Cold War? Why did the Soviet Union establish a buffer zone in Eastern Europe? Which two eastern European countries left the Soviet bloc to pursue alternate forms of state socialism?

7. How did aid from the Marshall Plan and COMECON help rebuild Europe after World War II? Which regions or economies benefited first?

8. How did the European Union (EU) develop? Why is the EU's Common Agricultural Policy (CAP) so important?

9. What migration patterns characterized Europe during the 19th and 20th centuries? Consider movement within Europe as well as movement to and from Europe.

Log in to www.mygeoscienceplace.com for Videos, Interactive Maps; RSS feeds; Further Readings; Suggestions for Films, Music and Popular Literature; and Self-Study Quizzes to enhance your study of Europe.

**▼ FIGURE 3.1**

# RUSSIAN FEDERATION, CENTRAL ASIA, AND THE TRANSCAUCASUS KEY FACTS

- Major subregions: The Russian Federation, Central Asia, and the Transcaucasus

- Total land area: 21 million square kilometers/ 8 million square miles

- Major physiogeographic features: the Russian Plain, the Central Siberian Plateau, the West Siberian Plain, Urals, desert plateaus of Central Asia. The tundra is an arctic wilderness that accounts for 13% of Russia. Continental climate with long and intense winters.

- Population (2010): 228 million, including 140 million in the Russian Federation, 28 million in Uzbekistan, and 15 million in Kazakhstan.

- Urbanization (2010): 52%

- GDP PPP (2010) current international dollars: 7,289; highest = Russia, 15,423; lowest = Kyrgyz Republic, 2,240.

- Debt service (total, % of GDP; 2005): 9.28%

- Population in Poverty (%< $2/day; 1990–2005): 25.2%

- Gender-related development index value: 0.75; highest = Belarus, 0.80; lowest = Tajikistan, 0.67

- Average life expectancy/fertility: 68 (M64, F73)/2.24

- Population with access to internet (2008): 13.2%

- Major language families: Indo-European (Slavic, Iranian, Armenian), Altaic, Caucasian, Uralic, Palea-Siberian

- Major religions: Orthodox Christianity, Protestantism, Islam, Roman Catholic

- $CO_2$ emission per capita, tonnes (2004): 5.23

- Access to improved drinking-water sources/sanitation (2006): 88%/91%

# 3

# THE RUSSIAN FEDERATION, CENTRAL ASIA, AND THE TRANSCAUCASUS

In 2008, a mysterious Russian multibillionaire broke a world record by agreeing to pay over $530 million for a luxury Belle Époque mansion on the heights of Villefrance, on the French Riviera. In 2009, having lost half his fortune as a result of the global banking crisis, he pulled out of the deal. Under French property law, he forfeited his 10% down payment. (The mysterious multibillionaire turned out to be Mikhail Prokhorov, former owner of Norilsk Nickel and Russia's fifth-richest man, with estimated wealth of $22.6 billion.)

Prokhorov is one of a number of Russian "oligarchs"—a small group of individuals who acquired tremendous wealth in the 1990s after the collapse of the Soviet Union. The rise of the oligarchs was one of the bizarre results of Russia's transformation to capitalism. In the turmoil that followed the collapse of the Soviet Union, the state-owned resources of the superpower—especially its mining and petroleum assets—were snapped up through legally questionable privatizations by a tiny group of smart, ruthless, ambitious, and well-connected men who quickly found themselves among the ranks of the world's super rich. The oligarchs are credited with having exercised significant political influence in the early years of the Russian Federation. They also became celebrity figures on the global stage, buying overseas companies and making splashy vanity purchases like racehorses, yachts, and planes. Like Prokhorov, they also bought villas in the French Riviera, townhomes in exclusive districts of London, and stately homes in the European countryside.

In addition to Prokhorov, the oligarchs included Oleg Deripaska (Russia's richest man, with interests in aluminum and automobile manufacturing), Mikhail Khodorkovsky (former head of oil company Yukos), Yuri Luzhkov (mayor of Moscow), and Roman Abramovich (former head of oil company Sibneft), now a resident of London and owner of English Premiership football club Chelsea. With the rise to power in Russia of populist Vladimir Putin, some of the oligarchs found themselves on the wrong side of domestic politics. Others, though, focused on global business expansion, taking out generous loans from Russian banks, buying shares, and then taking out more loans from Western banks against the value of these shares. But with the global financial crisis of 2008 to 2009, the overlapping bubbles of the Russian stock market, commodity prices, and easy credit burst, leaving the wealthiest 25 oligarchs with a collective loss of more than $230 billion in the 6 months following the market crash. Although the oligarchs have been able to hold on to most of their yachts and villas, the Russian economy as a whole has been seriously affected by the global economic downturn. After a decade of growth based on commodity exports and high levels of Western investment, Russia has seen a return of high levels of unemployment, with migrant workers from former Soviet republics with struggling economies among the hardest hit. ∎

## MAIN POINTS

■ The Russian Federation, Central Asia, and the Transcaucasus as a world region is very much in transition, struggling not only with the transition from socialism to capitalism but also with the shifting patterns of wealth and investment associated with capitalism.

■ These transitions are taking place at different speeds in different places and with rather uncertain outcomes.

■ Every country in the region has a legacy of serious environmental problems that stem from mismanagement of natural resources and failure to control pollution during the Soviet era.

■ The region itself, meanwhile, is still struggling to find its new place in the world economy.

■ Since the breakup of the Soviet Union, popular culture has gone through an enormous upheaval, thanks to the collapse of censorship, the pressures of market forces, and a flood of cultural imports from Europe, the United States, and elsewhere.

■ The Russian Federation, as the principal successor state to the Soviet Union, remains a nuclear power with a large standing army, but its future geopolitical standing remains uncertain.

■ Now freed from the economic constraints of state socialism, the Russian Federation stands to benefit a great deal by establishing economic linkages with the expanding world economy.

# ENVIRONMENT AND SOCIETY

This is a vast region (**Figure 3.1**) that encompasses a wide variety of climatic and physiographic settings. At the broadest scale, its physical geography means that, in global context, much of it has been cut off from the rest of the world for long periods of history. A satellite photograph (**Figure 3.2**) emphasizes the region's restricted access to the world's seas. It is bounded on the north by the icy seas of the Arctic and on the south by a mountain wall that stretches from the Elburz Mountains of northern Iran to the Altay and Sayan Mountains, which separate Siberia from Mongolia, and the Yablonovyy and Khingan ranges, which separate southeastern Siberia from northern China. The region stretches for more than 10,000 kilometers (6,200 miles) east–west and more than 2,500 kilometers (1,550 miles) north–south at its broadest. It takes a full week to traverse the region by train from Vladivostok in the east to St. Petersburg in the west. Nearly half the territory of the Russian Federation is north of 60°N. Moscow is approximately the same latitude as Juneau, Alaska, and Tbilisi, Georgia—one of the southernmost cities of the region—is approximately the same latitude as Chicago (42°N).

## Climate, Adaptation, and Global Change

The northerliness and vast size of the region exert strong influences on its climate (**Figure 3.3**). The absence of mountainous terrain, except in the far south and east, and the lack of any significant moderating influence of oceans and seas means that the prevailing climatic pattern is relatively simple. The region is dominated by a severe continental climate, with long, cold winters and relatively short, warm summers. The cold winters become colder eastward, as one moves away from the weak marine influence that carries over from the westerly weather systems that cross Europe from the Atlantic. Pronounced high-pressure systems develop over Siberia in winter, bringing clear skies and calm air. Average January temperatures in Verkhoyansk, a mining center in the middle of this high-pressure area, are in the region of −50°C (−58°F). The region's long and intense winters mean that the subsoil is permanently frozen—a condition known as **permafrost**—in more than two-thirds of the Russian Federation. In the extreme northeast, winter conditions can last for 10 months of the year.

The northerliness of most of the region means that most ports are icebound during the long winter. Murmansk, on the Kola Peninsula in the far north, is a major exception. It benefits from its location near the tail end of the warm Gulf Stream and is open year-round. Some ports, such as Vladivostok, on the Sea of Japan in the Far East, can be kept open by icebreakers. At least the Russian Federation has some warm-water ports, including Kaliningrad, a small province on the Baltic between Poland and Lithuania, retained as an exclave by the Russian Federation for its naval port. All the other countries in the region are landlocked, with the exception of Georgia, which has access to international sea-lanes by way of its Black Sea ports.

Summer comes quickly over most of Belarus and the Russian Federation, as spring is usually a brief interlude of dirty snow and much mud. Because many rural roads remain unpaved, they are typically impassable before the summer heat bakes the mud. As the landmass warms, low-pressure systems develop, drawing moist air across the western Russian Federation from Atlantic Europe and resulting in moderate summer rains. In late summer, the Chinese **monsoon** brings heavy rains to the southeastern corner of the Far East. Across much of Siberia, though, summer rainfall is quite low. The summers become hotter southward, and drought is a frequent problem in the southwestern and south-

▲ **FIGURE 3.2 The Russian Federation, Central Asia, and the Transcaucasus from space** This image underlines one of the key features of the Russian Federation, Central Asia, and the Transcaucasus: its sheer size.

▲ **FIGURE 3.3  Climate of Russia.** The Russian climate is dominated by the cold dry conditions of a Continental mid latitude climate where winters are very cold and summer droughts are common.  Northern Russia has a very cold and dry Polar climate around the Arctic Circle. Further south, warmer temperatures and distance from the sea mean that climate are dry but warmer, typical of semi arid climate types. The higher mountain regions have colder Highland climates.

ern parts of the Russian Federation. In Central Asia, aridity is a severe problem, with desert and semidesert covering much of Kazakhstan, Uzbekistan, and Turkmenistan. The climate in this subregion is harsh: Total annual precipitation in the deserts is less than 18 centimeters (7 inches), shade temperatures can reach 50°C (122°F), and ground surfaces can heat up to 80°C (176°F). The scorching heat is aggravated by strong drying winds, called *sukhovey*, that blow on more than half the days of summer, often causing dust storms. In late summer and fall, the increasing temperature range between the hot days and the longer, cooler nights becomes so extreme that rocks exfoliate, or "peel," leaving the debris to be blown away by the wind.

In the Transcaucasus, climatic patterns are mainly a result of the presence of massive mountain ranges to the north and south and substantial bodies of water to both east and west. A lot of precipitation falls on the windward side of the mountains, though the Transcaucasus is also influenced by the warm, dry air masses that originate over the deserts of Central Asia. The most distinctive feature of climatic patterns in the Transcaucasus, however, is the subtropical niche of western Georgia, on the shores of the Black Sea—a unique and striking feature in a region of otherwise severe climatic regimes.

The region is already experiencing the effects of global climate change. The implications for northern latitudes are mixed. Milder winters mean that, as growing seasons become longer and precipitation patterns change, it is becoming possible to use lands for agricultural purposes that previously have been inhospitable for too much of the year. It is also becoming possible to raise new crops and new varieties of crops in some regions. On the other hand, melting permafrost means that massive investments

will be required to stabilize roads, bridges, and the infrastructure of public utilities. Global warming also has important implications for the oil and gas industries that are so important to the region's economies. In addition to the infrastructure challenges associated with thawing permafrost, the warming climate is bringing milder and shorter heating seasons, which in turn lead to reduced energy demand within the region. Meanwhile, increased water availability—particularly along the Siberian rivers that can be used for hydroelectric power—should result in increased domestic power production.

In the Transcaucasus and the southern parts of European Russia, climate change means that agriculture is becoming more reliant on irrigation, pesticides, and herbicides (because of the spread of pests and disease) and, like many other world regions, more vulnerable to droughts and other extreme weather events. Some of the most affected regions are areas where socioeconomic and sociopolitical relations are unsettled. For example, the politically turbulent North Caucasus region (see page 108) is becoming drier and hotter and so is likely to become less prosperous as an agricultural region. In the Russian Far East, long-standing cross-border tensions are likely to intensify as water availability becomes an increasingly serious challenge in Central Asia, Mongolia, and northeastern China, prompting large-scale in-migration to the Russian Federation.

## Geological Resources, Risk, and Water Management

The most striking physical feature of the entire region is the monotony of its plains over thousands and thousands of kilometers. The reason for this monotony lies in the geological structure of the region. Two large and geologically ancient and stable shields of highly resistant crystalline rocks provide platforms for extensive plains of sedimentary material and glacial debris. In the west is the first shield, the Russian Plain, an extension of the Central European Plateau that runs from Belarus in the west to the Ural Mountains in the east and from the Kola Peninsula in the north to the Black Sea in the south (**Figure 3.4**). East of the Urals is a second shield that extends as far as the Lena River. It is so vast that it is conventionally divided into several physiographic subregions: the West Siberian Plain, the Central Siberian Plateau, and the desert plateaus of Central Asia. The Urals themselves represent a third distinctive physiographic region, and a fourth is the mountain wall that runs along the southern and eastern margins of the two shields.

The Russian Plain has a gently rolling topography, with the hard crystalline rock shield providing a flat platform that is covered by several meters or more of sedimentary deposits. Much of the Russian Plain is poorly drained, boggy, and marshy, though the major rivers that drain it—the Dnieper, the Don, and the Volga—have eroded the sedimentary layer in

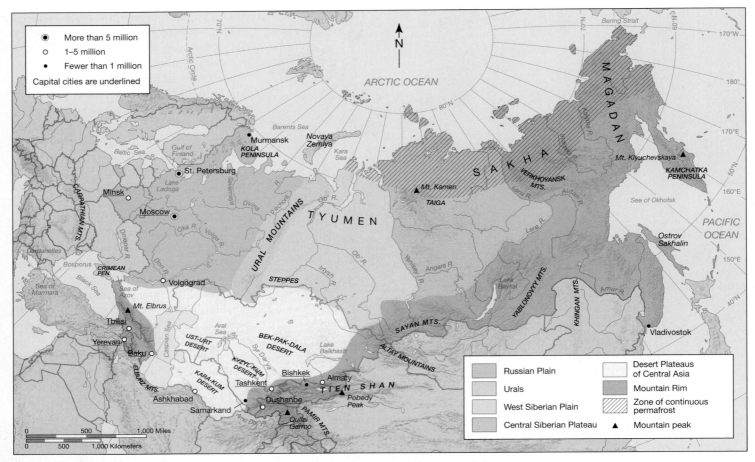

▲ **FIGURE 3.4 Physiographic regions of the Russian Federation, Central Asia, and the Transcaucasus**
The framework for the physical geography of the region consists of two stable shields on either side of the ancient Ural Mountains, with a wall of young mountains that runs along the southern and eastern margins of the shields.

places, resulting in more varied and attractive topography. The West Siberian Plain is even flatter and contains still more extensive wetlands and tens of thousands of small lakes. Poorly drained by the slow-moving Ob' and Irtysh rivers, the West Siberian Plain (**Figure 3.5**) is mostly inhospitable for settlement and agriculture, though it contains significant oil and natural gas reserves. The West Siberian Plain is distinctive for its absolute flatness: Across the whole broad expanse—more than 1,800 kilometers (1,116 miles) in each direction—relief is no more than 400 meters (1,312 feet). The monotony of the landscape is captured in this quote from Russian writer A. Bitov, describing a train journey:

> Once I was traveling through the Western Siberian Lowlands. I woke up and glanced out of the window—sparse woods, a swamp, level terrain. A cow standing knee-deep in the swamp and chewing, levelly moving her jaw. I fell asleep, woke up—sparse woods, a swamp, a cow chewing, knee deep. I woke up the second day—a swamp, a cow. . . ."[1]

[1]A. Bitov, *A Captive of the Caucasus* (Cambridge: Cambridge University Press, 1993, p. 50). Quoted in A. Novikov, "Between Space and Race: Rediscovering Russian Cultural Geography," in *Geography and Transition in the Post-Soviet Republics*, ed. M. J. Bradshaw (Chichester: John Wiley & Sons, 1997), p. 45.

▲ **FIGURE 3.5 The West Siberian Plain** This photograph of marshland near Primorye, western Siberia, shows very clearly the difficult, boggy conditions that prevail in much of western Siberia in summer.

▲ **FIGURE 3.6  The Central Siberian Plateau** The rock shield of the Central Siberian Plateau has been uplifted by geological movements; as a result, the land has been dissected by rivers into a hilly upland with occasional deep river gorges.

The Yenisey River marks the eastern boundary of this flat condition and the beginning of the Central Siberian Plateau, where the rock shield, having been uplifted by geological movements, averages about 700 meters (2,297 feet) in elevation. Stretching between 800 and 1,900 kilometers (496 to 1,178 miles) west–east, the Central Siberian Plateau (**Figure 3.6**) has been dissected by rivers into a hilly upland topography with occasional deep river gorges.

The Urals (**Figure 3.7**) consist of a once-great mountain range of ancient rocks that have been worn down over the ages. For the most part only 600 to 700 meters (1,969 to 2,297 feet) above sea level, and only in a few places rising above 2,000 meters (6,562 feet) in elevation, the Urals are penetrated by several broad valleys, and so they do not constitute a major barrier to transport. The Urals stretch for more than 3,000 kilometers (1,864 miles) from the northern frontier of Kazakhstan to the Arctic coast of the Russian Federation, the range reappearing across the Kara Sea in the form of the islands of Novaya Zemlya. The rocks of the Urals are heavily mineralized and contain significant quantities of chromite, copper,

▲ **FIGURE 3.7  Urals landscape** The mountains of the Urals are ancient and have been worn down to rounded landforms with thin soils that support small farms. Shown here is a small farming village near Kungur.

gold, graphite, iron ore, nickel, titanium, tungsten, and vanadium. As a result, a number of significant industrial cities, including Chelyabinsk, Magnitogorsk, Perm', Ufa, and Yekaterinburg have developed in the Urals, together forming a major industrial region.

The mountain wall that runs along the southern and eastern margins of the two stable shields on either side of the Urals is the product of geological instability. Younger, sedimentary rocks have been pushed up against the older and more stable shields in successive episodes of mountain-building, forming a series of mountain ranges of varying height, composition, and complexity. The highest ranges are those of the Caucasus (where Mt. Elbrus reaches 5,642 meters, or 18,510 feet) and the Pamirs and Tien Shan ranges along the borders with Iran, Afghanistan, and China, where many peaks reach 5,000 to 6,000 meters, and two—Pobedy and Qullai Garmo (formerly Communism Peak)—exceed 7,400 meters (24,278 feet). In the Far East, the ranges of the Kamchatka Peninsula contain numerous active volcanoes, including Mt. Klyuchevskaya (4,750 meters; 15,584 feet) and Mt. Kamen (4,632 meters; 15,197 feet). Only in the western part of Central Asia does the mountain wall fall outside the region, running along the southern side of Turkmenistan's border with Iran and Afghanistan. North of this border, extending through Uzbekistan into Kazakhstan and beyond is a huge geosyncline, a geological depression of sedimentary rocks. This syncline is of special importance as a source of energy resources, with oil reserves equivalent to between 15 and 31 billion barrels—about 2.7% of the world's proven reserves—plus significant deposits of coal and about 7% of the world's proven reserves of natural gas.

The northerliness of the region as a whole limits the use of its rivers for navigation and for generating hydroelectric power. Many rivers are frozen for much of the year, whereas the mouths of some remain frozen through the spring, causing backed-up meltwater to flood extensive areas of wetlands. Managing the increased flows resulting from climate change is becoming increasingly problematic, especially when they coincide with extreme weather events such as heavy downpours or springtime ice-clogged floods. Nevertheless, the sheer size of the territory sustains several rivers of considerable size, all of which were used historically as transport routes, allowing for conquest, colonization, and trade. It takes some of the longest rivers on Earth to drain the huge Siberian landmass. Rising in the southern mountains of Central Asia, the Lena, Kolyma, Ob', and Yenisey flow north to the Arctic Ocean, whereas the Amur flows north to the Pacific. In the western part of the Russian Federation, the rivers flow south from the Central Region occupied by Moscow and Nizhniy Novgorod. The Dnieper flows to the Black Sea, the Don to the Sea of Azov (which in turn connects to the Black Sea), and the Volga to the Caspian Sea. In the modern period, the Volga has become particularly important as a navigable waterway and source of hydropower, with huge reservoirs built during the Soviet era to regulate the flow of the river, conserving spring floodwaters for the dry summer months.

The Black Sea itself is an inland sea, connected to the Aegean Sea and the Mediterranean by way of the Bosporus, a narrow strait, the Sea of Marmara, and then another narrow strait, the Dardanelles. The many rivers that empty into the Black Sea give its surface waters a low salinity, but it is almost tideless, and below about 80 fathoms it is stagnant and lifeless. The Caspian Sea is the largest inland sea in the world, at 371,000 square kilometers (143,205 square miles—roughly the size of Germany). With the Black Sea and the Aral Sea, it once formed part of a much greater inland sea. Though perhaps dwarfed by the vastness of the region, there are several inland lakes of significant size, including Lake Balkhash (17,400 square kilometers; 6,715 square miles) and Lake Baykal (30,500 square kilometers; 11,775 square miles), which, with a depth of 1,615 meters (5,300 feet), is the deepest lake in the world.

### Resources and Political Realignment

One immediate consequence of the breakup of the Soviet Union in 1989 was that the collective natural resource base was fragmented among the new states. The states of Central Asia and the Transcaucasus were particularly affected because their smaller territories and less varied physiography left each of them with a relatively narrow resource base—though the oil and natural gas reserves of Central Asia are a major asset. Kazakhstan has the bulk of the oil reserves, whereas the natural gas fields are mainly to the south, in Turkmenistan and Uzbekistan. Proven oil reserves in the Central Asian geosyncline, around and beneath the Caspian Sea, amount to between 15 and 31 billion barrels, but estimates of the potential reserves run between 60 and 140 billion barrels. Only the oil fields of the Persian Gulf states and Siberia are larger. This represents a tremendous economic asset for the Central Asian states. Exploiting these assets is beset with difficulties, however. Most of the states involved are effectively landlocked, which means that expensive pipelines have to be constructed before the oil fields and gas fields can be fully developed. But pipeline construction and routing are both risky and contentious because of political tensions and instability in the region. The Russian Federation has lost free access to these oil and gas reserves and to the large uranium reserves of Tajikistan and Uzbekistan. Nevertheless, Russia still has a broad resource base, with huge reserves of coal, lignite, vanadium, manganese, and iron. It also has substantial reserves of oil and gas.

### Radioactivity

Contamination by radioactivity is seen by many to epitomize the consequences of Soviet attitudes toward the environment. Both the large-scale civilian nuclear energy program and the military nuclear capability of the Soviet Union were developed in ways that have resulted in an alarming incidence of radioactive pollution. The 1986 disaster at Chernobyl, in which a nuclear reactor exploded in a power plant in the former Soviet republic of Ukraine, became emblematic of the Soviet nuclear legacy. Stories of radiation sickness and eventual ghastly deaths of the facility workers, firefighters, medical personnel, and other volunteers filled international newspapers for weeks following the meltdown. Less graphic and less well remembered are the invisible and enduring effects on the population of the area surrounding the power plant as well as on the natural environment. Radiation particles entered the soil, the vegetation, the human population, and the rivers, effectively contaminating the entire food chain of the region. Secondary radiation continues to be a problem. In the immediate area surrounding Chernobyl, all the trees were contaminated. As the trees slowly die, rot, and decay, radioactive material enters the physical system as "hot" nutrients. More than 3,000 square kilometers (1,161 square miles) of trees turned brown from radiation immediately following the accident, and it is still not clear how to decontaminate such a large area. Entire towns remain abandoned (**Figure 3.8**).

Meanwhile, economic stress has led the Russian Federation to agree to become a dumping ground for other countries' nuclear waste. All over the industrialized world, atomic power plant construction is significantly slowing down because of unresolved safety considerations and the failure to develop a safe, permanent means of disposing long-lived nuclear waste. But the Russian Federation undertakes to store or reprocess highly radioactive nuclear waste from other countries at a relatively low price. Altogether, hundreds of thousands of tons of uranium waste have been

▲ **FIGURE 3.8 Pripyat, Ukraine** Established in 1970 about 1.5 km from the Chernobyl nuclear power plant to house the workers and their families, Pripyat was entirely evacuated on April 27, 1986, the day after the explosion of the number 4 reactor at Chernobyl, and remains abandoned still today.

▲ **FIGURE 3.9 Cotton cultivation** Manual labor remains important in the extensive cotton farming on irrigated lands in Turkmenistan.

shipped to Russia since the early 1990s. The risks, though, are appallingly high, given the Russian Federation's industrial inefficiency, corruption, and organized crime. The uranium shipments are made using conventional Russian transportation, without strong safety or security measures, along a route that passes through major cities like St. Petersburg and Tomsk and along the coasts of Belgium, the Netherlands, Denmark, Germany, Sweden, Norway, and Finland.

**The Aral Sea**   Radioactivity is only one of several major environmental problems that have left an enduring legacy to the Soviet Union's successor states. Soviet modernization programs brought large-scale irrigation schemes to the desert and semidesert regions of Central Asia, notably the Kara-Kum Canal, a 770-kilometer (478-mile) irrigation canal that diverts water from the Amu Dar'ya and irrigates 1.5 million hectares (4.7 million acres) of arable land and 5 million hectares (12.4 million acres) of pasture as it trails across southern Turkmenistan. Cotton was the dominant crop in these irrigated lands and remains so (**Figure 3.9**). In Turkmenistan, for example, more than half the arable land is devoted to cotton monoculture, and Uzbekistan is the world's fifth-largest producer of raw cotton and third-largest exporter of cotton.

In many ways, however, irrigated cotton cultivation has been harmful. Yields remain comparatively low, despite irrigation, because of soil exhaustion and salinization. **Salinization** is caused when water evaporates from the surface of the land and leaves behind salts that it has drawn up from the subsoil. An excess of salt in the soil seriously affects the yields of most crops. In addition to salt, residues of the huge doses of defoliants, pesticides, and fertilizers used on the cotton fields have found their

way into the drinking-water systems of the region. Meanwhile, cotton monoculture has rendered the countries of the region heavily dependent on food imports.

The worst consequence of the Soviet program of irrigated cotton cultivation has been the effects of excessive withdrawals of water from the main rivers that drain into the Aral Sea. The Kara-Kum Canal alone took away almost one-quarter of the Aral Sea's annual supply of water. The Aral Sea was the fourth-largest lake on the planet in 1960. By 2008 it had shrunk to 10% of its original size. The sea has shriveled into three major residual lakes, two of which are so salty that fish have disappeared. The level of the Aral Sea has already dropped by more than 10 meters (33 feet), and the desiccation of the former seabed, now littered with stranded ships (**Figure 3.10**), generates a constant series of dust storms that are

▲ **FIGURE 3.10 The Aral Sea** Like this area near Aralsk, Kazakhstan, some 24,000 square kilometers (11,000 square miles) of former seabed in the Aral Sea have become a desert of sand and salt.

thought to cause unusually high levels of respiratory ailments among the people of the region. The acute desiccation of the Aral Sea region has devastated its fishing industry and left ports such as Aralsk and Moynaq stranded more than 40 kilometers (25 miles) from the retreating lakeshore in the midst of a new "White Desert" of former lake bed sands.

**Lake Baykal**    Another notorious example of environmental disregard is Lake Baykal, where the unique ecosystem has been threatened by industrial pollution. Lake Baykal is the world's deepest lake at 1,615 meters (5,300 feet—more than a mile) and contains about 20% of all the freshwater on Earth—more than North America's five Great Lakes combined. It has a unique ecology, with more than 2,500 recorded plant and animal species, 75% of which are found nowhere else. These include the nerpa, Baykal's freshwater seal. Lake Baykal is also a place of incredible beauty—"The Pearl of Siberia"—that has become emblematic of the pristine wilderness of the region (**Figure 3.11**). But the lake's purity and unique ecosystem have been compromised by environmental mismanagement. In the 1960s, increasing levels of pollution were carried into the lake by the Selenga River, which supplies about half the water that flows into the lake. The Selenga rises in mountain ranges to the south but collects agricultural chemicals such as DDT and PCB, as well as human and industrial waste from several large cities before entering Lake Baykal.

Meanwhile, the purity of the lake's waters caught the attention of Soviet economic planners, who began to see the lake as a good location for factories that needed plentiful supplies of pure water. When the huge Baikalsk Pulp and Paper Mill, which produced high-quality cellulose for the Russian defense industry, was opened in the early 1960s, concern for Lake Baykal triggered the birth of Russia's environmental movement. It is estimated that the mill has spewed more than a billion tons of waste into the lake. When thousands of the lake's freshwater seals began dying in 1997, the lake's fragile ecology came under international scrutiny, and in 1998 the lake was designated a World Heritage Site by UNESCO, the UN cultural agency. As a result of unprecedented

▲ **FIGURE 3.11 Lake Baykal** The purity of the deep waters of Lake Baykal has been compromised by industrialization along its shores and along the rivers that feed the lake, but the unspoiled nature of much of the shoreline is beginning to attract tourists.

expressions of public concern, the Soviet government ordered the mill closed in 1986. But the government collapsed before the closure took effect, and in 1989 the mill was partially privatized, making pulp for low-quality paper, rather than cellulose, until it was closed in 2009 when its aging equipment was no longer profitable.

## Ecology, Land, and Environmental Management

The natural ecologies of the Russian Federation, Central Asia, and the Transcaucasus follow a strikingly straightforward pattern of seven long, latitudinal zones that run roughly from west to east. These zones are very closely related to global climatic patterns, glacial geomorphology, and soil type and remain easily recognizable to the modern traveler, despite centuries (or, in places, millennia) of human interference and modification. The northernmost zone is that of the tundra, which fringes the entire Arctic Ocean coastline and part of the Pacific (see Figure 3.4).

**The Tundra**    The tundra is an arctic wilderness where the climate precludes any agriculture or forestry. The tundra zone extends along the northern shores of the Russian Federation and includes its Arctic islands. Altogether, it amounts to 2.16 million square kilometers (833,760 square miles), representing almost 13% of the country. Almost all of it was sculpted by one or another of the great ice sheets of the Quaternary ice ages (between 1.7 million and 10,000 years ago). These, wrote geographer W. H. Parker, "scraped, polished, grooved, crushed or sheared the rocks in their advance, and dropped boulders and stones haphazardly on their retreat."[2] Frost action is still the main modifier of the landscape, as running water is almost entirely absent. For 9 months or more, the landscape is locked up by ice and covered by snow, whereas during the brief summer, much of the melted snow and ice is trapped in ponds, lakes, boggy depressions, and swamps by permafrost that extends, in places, to a recorded depth of 1,450 meters (4,757 feet). Water seeps slowly to streams that drain into the slow-moving rivers that cross the tundra and drain into the Arctic (**Figure 3.12**).

During winter, the tundra landscape has a uniformity that derives from snow cover and the somber effect of long nights and weak daylight, the sun remaining low in the sky. There is little sign of life, apart from herds of reindeer or occasional polar bear or fox. In summer, vegetation bursts into bloom, and animals, birds, and insects appear. The days are long—in June the sun circles the horizon, and there is no night at all. Mosses and tiny flowering plants provide color to the landscape, contrasting with the black, peaty soils and the luminous bright skies. Swans, geese, ducks, and snipe arrive on lakes and wetlands for their breeding season, as do seabirds and seals along the coast. From the forests to the south, wolves enter the tundra in search of prey that includes Alpine hare and lemmings (small, fat rodents that can produce five or six litters annually). Everywhere there are swarms of gnats and mosquitoes. Summer also brings a zonal differentiation to the region, with shrubs—blackberry, crowberry, and cowberry—and dwarf trees—birch, spruce, and willow—becoming more frequent as one moves south toward the taiga.

**The Taiga**    South of the tundra is the most extensive zone of all—a belt of coniferous forest known as the *taiga*. The term **taiga** originally referred

[2]W. H. Parker, *The Soviet Union* (Chicago: Aldine, 1969), p. 42.

▲ **FIGURE 3.12 Tundra landscape** The tundra landscape is bleak, with sparse vegetation and a surface strewn with rocks, which frost-heave action has arranged into geometric patterns. This photograph was taken in the Magadan region.

▲ **FIGURE 3.13 Taiga landscape** The taiga is a zone of boreal coniferous forest that stretches from the Gulf of Finland to the Kamchatka Peninsula.

to virgin forest, though it is now used to describe the entire zone of boreal (northern) coniferous forest (spruce, fir, and pines, for example) that stretches from the Gulf of Finland to the Kamchatka Peninsula—more than 4.4 million square kilometers (1.3 million square miles) of territory (**Figure 3.13**). The underlying rock shield, having been steadily lifted by geological movements, averages about 700 meters (2,296 feet) in elevation and has been dissected by rivers into a hilly upland topography with deep river gorges that are cut 300 meters (984 feet) or more below the general level of the plateau. The steep sides of the valleys are notched by numerous terraces, marking successive stages in the uplift of the ancient crystalline rock plateau. This topography, though, is given uniformity by the distinctive forest cover of the region. The characteristic forest is made up of larches: hardy, flat-rooted trees that can establish themselves above the permafrost. Slow-growing and long-lived, they are able to counter the upward encroachment of moss and peat by putting out fresh roots above the base of the trunk. The larches can grow to 18 meters (59 feet) or more, allowing an undergrowth of dwarf willow, juniper, dog-rose, and whortleberry.

The indigenous inhabitants of the taiga were hunters and gatherers, not farmers. Where the forest is cleared, some cultivation of hardy crops such as potatoes, beets, and cabbage is possible, but the poor, swampy soils and short growing season make agriculture chancy. Settlement here is sparse (**Figure 3.14**), the small populations of towns and villages engaged in mining, administration, construction, and local services. The central Siberian taiga is one of the richest timber regions in the world. Overall, close to 90% of the territory is covered with forest, and more than a quarter of the Russian Federation's lumber production comes from the region. More recently the taiga has become commercially important for the fur-bearing animals whose luxuriant pelts are well adapted to the bitter cold, and for the forest itself, whose timber is now methodically exploited and exported.

**Mixed Forest and Steppe Landscapes**   This zone is a continuation of the mixed forests of central Europe that extend through Belarus and into the Russian Federation as far as the Urals, with discontinuous patches in Siberia and the far east. Here, firs, pines, and larches are mixed with stands of birch and oak. It quickly shades into another relatively narrow zone of wooded steppe. In this zone the grasslands of the steppe are interspersed with less-extensive stands of mixed woodland, mostly in valley bottoms. Both the mixed forest and the wooded steppe were cleared and cultivated early in Russian history, providing both an agricultural heartland for the emerging Russian empire and a corridor along which Russian traders and colonists pushed eastward in the 16th and 17th centuries—through the middle Volga region to the southern Urals and eventually to the Pacific coast via the mixed forests of the Amur valley.

In recent decades, poor forest resource management has given cause for concern. After decades of relentless Soviet exploitation, large swathes of Siberia, once dense and practically impassable, have been cleared. Loggers are now moving farther and farther north to cut down century-old pines. The post-Soviet transition has intensified concern over forest resource management, as privatization has attracted U.S., Korean, and Japanese transnational corporations to invest in "slash-for-cash" logging operations (**Figure 3.15**) in a loosely regulated and increasingly corrupt business environment. The future of the central Siberian taiga has become an issue of international concern because the region accounts for a significant fraction of the world's temperate forests, which absorb huge amounts of carbon dioxide gas in the process of photosynthesis, thereby removing a main contributor to global warming from the atmosphere.

The wooded steppe, in turn, quickly shades into the steppe proper. The steppe belt stretches about 4,000 kilometers (2,486 miles) from the Carpathians to the Altay Mountains (see Figure 3.4), covering a total area of more than 4.25 million square kilometers (1.25 million square miles). The topography of the region is strikingly flat: Yellow loess, several meters thick, has blanketed the underlying geology, creating a rolling landscape of unbounded horizons, punctuated here and there by incised streams and river valleys. The natural vegetation of tall and luxuriant

▲ **FIGURE 3.14 Amguema, Chukotka** The most northeasterly region of Russia, Chukotka has a population just over 55,000. The principal town and administrative center is Anadyr. Chukotka has large reserves of oil, natural gas, coal, gold, and tungsten, which are slowly being exploited, but much of the rural population exists on subsistence reindeer herding, hunting, and fishing.

feather grasses and steppe fescue have matted roots that are able to trap whatever moisture is available in this rather arid region. The accumulated and decayed debris of these grasses has produced a rich dark soil, known as black earth, or **chernozem**. These soils, along with related brown and chestnut soils, have high natural fertility, but when they are plowed, they are vulnerable to the aridity of the region and can easily degenerate into wind-driven dustbowl conditions. Trees and shrubs are restricted to valleys, where broad-leaved woods of oak, ash, elm, and maple have established themselves; pinewoods and a low scrub of blackthorn, laburnum, dwarf cherry, and Siberian pea-trees grow in drier locations.

For centuries, the steppe region was the realm of nomadic people, but in the late 1700s, when the Turkish empire's hold on the steppes was broken, large numbers of colonists began to enter the western steppe. Wheat growing rapidly expanded wherever transportation was good enough to get the grain to the expanding world market, but large flocks of sheep dominated most of the colonized steppe until the railway arrived. Then, German Mennonites began mixed farming on the rich soils, Greeks began tobacco farming, and Armenians specialized in business and commerce. Farther east, the flat steppe of northern Kazakhstan remained largely untouched until the 19th century, save for Kazakh nomads and their herds and a few Russian forts and trading posts. During the 19th century, settlers came from the west in increasing numbers—a million or more by 1900—displacing the Kazakh nomads, thousands of whom died in famines or in unsuccessful uprisings against the Russians.

The Soviet period brought significant modifications to the region and its landscape. The colonists' small farms were merged into collectives, linear shelter belts of trees were planted in an attempt to modify climatic conditions, and rivers were dammed to provide hydroelectric power and irrigation for extensive farming of wheat, corn, and cotton (**Figure 3.16**). In the 1950s, Nikita Khruschev announced that the "virgin and idle lands" of the eastern steppe would be plowed and farmed for wheat. The Soviet government organized state farms and built large villages of new wooden houses to receive an army of 350,000 immigrants from European Russia. The eastern steppe was quickly transformed by modern machinery, fertilizers, and pesticides, initially producing as much wheat, on average, as the annual wheat harvest of Canada or France. This extensive wheat farming quickly led to dust-bowl conditions, however. In response, dry-farming techniques and irrigation have been introduced. **Dry-farming** techniques allow the cultivation of crops without irrigation in regions of limited moisture (50 centimeters, or 20 inches, of rain per year). Such techniques include keeping the land free from weeds and leaving stubble in the fields after harvest to trap snow. Together with irrigation schemes, dry-farming techniques now allow for the cultivation of not only wheat, but also millet and sunflowers, together with silage corn and fodder crops to support livestock.

**Semidesert and Desert**   South of the steppe are zones of semidesert and desert. They are largely a feature of Central Asia—covering most of Kazakhstan, Turkmenistan, and Uzbekistan—and they continue south of Siberia into Chinese and Mongolian territory. The semidesert is characterized by boulder-strewn wastes and salt pans (areas where salt has been deposited as water evaporated from short-lived lakes and ponds created by runoff from surrounding hills) and patches of rough vegetation used by nomadic pastoralists. The desert proper is characterized by bare rock and extensive sand dunes, though there are occasional oases and fertile river valleys. The landscapes of the northern zone of the desert region are arid plateaus with rocky outcrops, hillocks, and shal-

▲ **FIGURE 3.16 Extensive agriculture on the steppes** Beginning in the 1950s, the eastern steppes were transformed into an extensive wheat farming region through massive investments of modern machinery, fertilizers, and pesticides.

▲ **FIGURE 3.15 Illegal logging** More than 1.5 million cubic meters of oak, cedar, and ash are illegally logged in the far eastern Primorye region each year. Chinese traders pay $100 per cubic meter and resell the wood at prices of $400 to $500 per cubic meter.

low depressions that have become crusted with salty deposits as a result of the evaporation of runoff from surrounding hills.

The two principal deserts here are the Ust-Urt and Bek-pak-Dala. Farther south is a zone of sandy deserts: the Kara-Kum (Black Sands) and the Kyzyl-Kum (Red Sands). Here, the landscape is dominated by long ridges of sand in crescent-shaped dunes called *barchans* and by vast plains of level sand punctuated by patches of sand hills and by isolated remnants of worn-down mountain ranges called *inselbergs*. Several rivers—including the Amu Dar'ya, the Syr Dar'ya, and the Zeravshan—drain from the Tien Shan and Pamir mountain ranges across these deserts toward the Aral Sea. Dotted along their valleys are irrigated oases, whereas *tugay*—impenetrable thickets of hardy trees and thorny bushes—thrive in the rather salty soils of the valley floors.

**Environmental Legacies of Socialism** Meanwhile, every country in the region has a legacy of serious environmental problems that stem from mismanagement of natural resources and failure to control pollution during the Soviet era. Soviet central planning placed strong emphasis on industrial output, with very little regard for environmental protection. Early Soviet ideology had propagated the view that it would be feasible to harness and transform nature through the collective will

and effort of the people. Nature, it was asserted, is a dangerous force that needs to be subdued and transformed, and natural resources have no value in a socialist society until people's labor has been applied to them. As a result of this way of thinking, the tendency during the Soviet era was to squander natural resources and to "take on" and "conquer" nature through ambitious civil engineering projects. Soviet authorities saw problems of pollution and environmental degradation as an inevitable cost of modernization and industrialization, and the people most affected—the general public—had no political power or means to voice environmental concerns.

Today, serious environmental degradation affects all parts of the region (**Figure 3.17**), a legacy of Soviet problems that in many ways have been intensified by the transition to **market economies**, in which goods and services are produced and distributed through free markets. The ubiquitous corruption that has come to characterize the region during the post-Soviet transition means that environmental regulations are easily ignored or circumvented. In addition, economic problems in the Russian Federation have clearly limited the country's ability to address its environmental problems. The resulting legacy of environmental problems includes overcutting of forests, widespread overuse of pesticides, heavy pollution of many rivers and lakes, extensive problems of acid rain and soil erosion, and serious levels of air pollution in industrial towns and cities. Fragile environments at the margins of human settlement and on the peripheries of the Soviet empire have been among the worst affected. Across the far north, for example, air pollution from coal-burning industries in northern midlatitude countries produces a

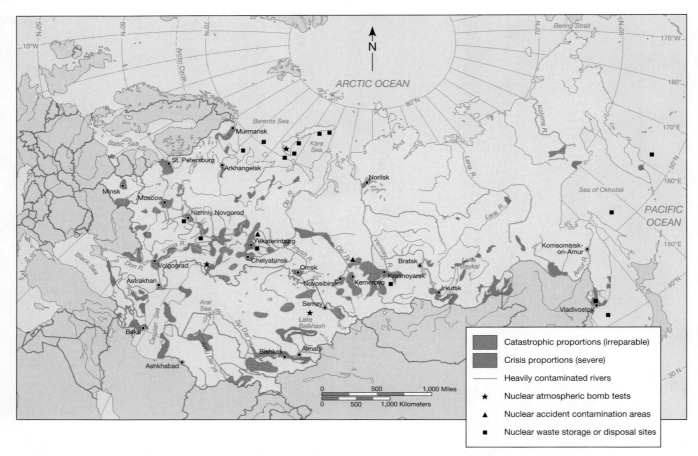

**▲ FIGURE 3.17 Environmental degradation** Serious environmental degradation afflicts all parts of the Russian Federation, Central Asia, and the Transcaucasus. The range of issues is enormous, and the problems are so intense that they are likely to persist for several generations.

phenomenon known as "Arctic haze," seriously reducing sunlight and destroying delicate vegetation complexes that underpin fragile ecosystems. In the semiarid and arid regions of the south, the diversion of rivers for irrigation schemes aimed at boosting agricultural productivity has depleted water resources in some areas and led to widespread soil erosion and desertification.

# HISTORY, ECONOMY, AND DEMOGRAPHIC CHANGE

Although relatively closed off from the people and economies of the rest of the world's regions for a good part of the 20th century as a result of Soviet policies, the character of this world region, like that of every other world region, derives in large part from world-spanning processes that intersect with its unique internal attributes and processes. As explained in the previous section, huge tracts of this world region are decidedly marginal. Agriculture and settlement have been greatly restricted by severe climatic conditions, highly acidic soils, poor drainage, and mountainous terrain. Even in the zone of rich chernozem soils, low and irregular rainfall rendered agriculture and settlement marginal until large-scale irrigation schemes were introduced in the 20th century. Only in the mixed forest and the wooded steppe west of the Urals were conditions suitable for the emergence of a more prosperous

and densely settled population. It was this area, in fact—from Smolensk in the west to Nizhniy Novgorod in the east, and from Tula in the south to Vologda and Velikiy Ustyug in the north—that was the Russian "homeland" that developed around the principality of Muscovy from late medieval times.

Long before then, however, the towns of Central Asia had become key nodes in the vast trading network known as the Silk Road. The **Silk Road** is the collective name given to a network of overland trade routes that connected China with Mediterranean Europe, facilitating the exchange of silk, spices, and porcelain from the East and gold, precious stones, and Venetian glass from the West. It had existed since Roman times and remained important until Portuguese navigators found their way around Africa and established the seaborne trade routes that exist to this day. Along the Silk Road stood the ancient cities of Samarkand, Bukhara (**Figure 3.18**), and Khiva, places of glory and wealth that astonished Western travelers such as Marco Polo in the 13th century. These cities were east–west meeting places for philosophies, knowledge, and religions. In their prime they were known for their leaders in mathematics, music, architecture, and astronomy: scholars such as Al Khoresm (780–847), Al Biruni (973–1048), and Ibn Sind (980–1037). The cities' prosperity was marked by impressive feats of Islamic architecture. Their civilization was overcome by Mongol Tatar horsemen, who ruled until the 14th century. Timur (Tamerlane), one of Genghis Khan's descendants and a convert to Islam, subsequently built up a vast Central Asian empire stretching from northern India to Syria, with its capital in Samarkand. The decline of

▲ **FIGURE 3.18 Bukhara, Uzbekistan** The historic center of Bukhara, an important center on the old Silk Road overland trade route.

Timur's empire in the 16th century saw the rise of nomadic people, who established three khanates, or kingdoms: in Bukhara, Khiva, and Kokand. They prospered as trading posts on the transdesert caravan routes until the late 19th century, when the three khanates fell to Russian troops.

## Historical Legacies and Landscapes

The spread of colonization and the extension of political control by the people of the Russian homeland is key to understanding the present-day geography of the entire region. In the mid-15th century, Muscovy was a principality of approximately 5,790 square kilometers (2,235 square miles) centered on the city of Moscow. Over a 400-year period, the Muscovite state expanded at a rate of about 135 square kilometers (52 square miles) per day so that by 1914, on the eve of the Russian Revolution, the empire occupied more than 22 million square kilometers (roughly 8.5 million square miles), or one-seventh of the land surface of Earth (**Figure 3.19**). At first, Muscovy formed part of the Mongol-Tatar Empire, whose armies were known as the Golden Horde, and Russian princes were obliged to pay homage to the Khan, the leader of the Golden Horde. In 1552, under Ivan the Terrible,

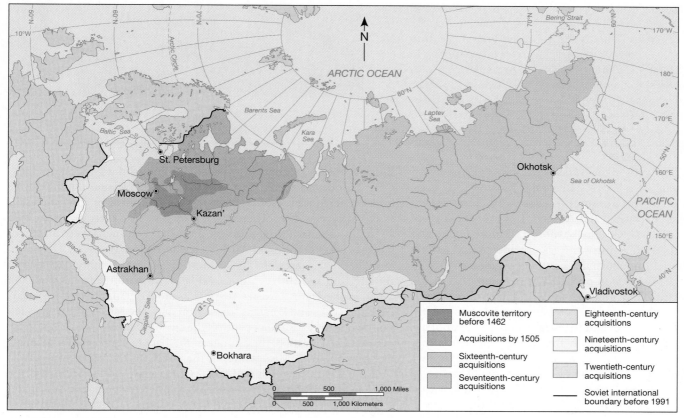

▲ **FIGURE 3.19 Territorial growth of the Muscovite/Russian state** The Muscovite empire was vast and was conquered over the same period (15th century to the late 20th century) that corresponds to the globalization of the world economy. What makes the Russian case unique is that the lands conquered were adjacent ones and not overseas. When the Bolsheviks came to power at the beginning of the 20th century, some of the territory was lost. Eventually, however, the Bolsheviks were able to control most of the territories formerly held by the tsars, and it was on this that they also built the soviet state.

# FURS

From earliest times, fur has been a prized commodity. In cold regions, fur coats, hats, and boots are valued for their warmth and as a practical investment that is also a portable form of wealth. In Russia and Europe, fur has long had royal and aristocratic connotations and, as a result, became a status symbol for all who could afford it. In the world of women's fashion, furs have become synonymous with haute couture, and in much of the world fur is seen by status-conscious consumers as a fashionable luxury good, a clear marker of material wealth. Furs have, however, become a controversial luxury item. In Europe and North America in particular, the market for furs has been significantly affected by people's concern that certain animal species might be threatened with extinction and the realization that fur trapping and fur farming can involve unnecessary cruelty to animals.

European merchants began trading fur in the Middle Ages, and fur was the commodity that drew Russian trappers and traders to Siberia in the 16th century. In 1581, under sponsorship of the rich merchant family of Stroganov, a military expedition opened a route to Siberia for promyshlenniks (fur hunters), who were drawn eastward in search of sable and sea otter. The pelts of these animals were exchanged for Chinese and Indian goods, and the tax revenues from the trade were the mainstay of the Russian imperial treasury for the next 300 years. It became government policy to encourage the fur trade and to support the promyshlenniks in their ruthless displacement of indigenous people and their sustainable economies. By the reign of Peter the Great (1682–1725), the promyshlenniks had reached the Sea of Okhotsk, and fur hunting was coming to a saturation point. In response, Peter the Great sponsored maritime expeditions to the Kamchatka Peninsula and to the offshore islands of the northeast. After Vitus Bering's expedition to the Northern Pacific in 1741–42 established that the islands had abundant populations of sea otter, foxes, seals, and walruses, a "fur rush" drew promyshlenniks all the way across the Bering Sea to Alaska.

Meanwhile, trade in fur pelts (beavers, muskrats, minks, and martens) had attracted Europeans' initial interest in North America. Beaver, trapped by Native Americans, was a main source of barter at trading posts that later grew into such cities as Chicago, Detroit, Montréal, New Orleans, Québec, St. Louis, St. Paul, and Spokane. The Hudson's Bay Company, founded in England in the mid-17th century to trade skins for guns, knives, and kettles, gained almost total control of the North American fur trade. For the company's first 200 years, its business consisted entirely of trading in furs. The company abandoned fur sales in 1991 but in 1997 started them again, largely because of a surge in demand from China and the other rapidly growing capitalist economies of Asia. Today, fur farming (raising animals in captivity under controlled conditions), rather than trapping, is the principal source of furs for the world market (**Figure 1**). Fur farming was started

▲ **FIGURE 1 Russian fur farm** Most of Russia's fur farms are located in remote settings, with few regulatory controls or inspections.

the Muscovites defeated the Tatars at the battle of Kazan'—a victory that prompted the commissioning of St. Basil's Cathedral in Moscow.

Desirous of more forest resources—especially furs (see Geographies of Indulgence, Desire, and Addiction: Furs, above)—Muscovy expanded into Siberia. Gradually, more and more territory was colonized. By the mid-17th century, the eastern and central parts of Ukraine had been wrested from Poland. The steppe regions, though, remained very much a frontier region of the Russian empire because of the constant threat of attack by nomads. Early in the 18th century, Peter the Great (1682–1725) founded St. Petersburg and developed it as the planned capital of Russia. Beyond the wealth and grandeur of a few cities, however, Russia was very much a rural, peasant economy. In the latter part of the 18th century, under Catherine the Great (1762–96), Russia secured the territory that would eventually become southern Latvia, Lithuania, Belarus, and western Ukraine. Then, with the defeat of the Crimean Tatars in the late 18th century, the steppes were opened to colonization by Russians and by ethnic and religious minorities—including Mennonites and Hutterites—from eastern and central Europe. It was during this period that Russia ousted the Ottoman Turks from the Crimean Peninsula and gained the warm-water port city of Odessa on the Black Sea.

**Empire** Russia's imperial expansion followed the same impulses as other European empires. The factors behind expansion were the drive for more territorial resources (especially a warm-water port) and additional subjects. The difference for Russia, however, was that vast stretches of adjacent land on the Eurasian continent were annexed, whereas other empires established new territories overseas.

The final phases of expansion of the Russian empire occurred in the late 18th and 19th centuries. Finland was acquired from Sweden in 1809. In the Transcaucasus, Georgians and Armenians were "rescued" from the Turks and Persians. In Central Asia, the Moslem Khanates fell one by one under Russian control: the city of Tashkent in 1865, the city of Samarkand in 1868, the Emirate of Bukhara in 1868, and the Khanate of Khiva in 1874. Meanwhile, in the Far East, the weakening of the Manchu dynasty, which had ruled China since 1664, prompted the Russian annexation of Chinese territory, where colonization and settlement was aided by the construction of the Trans-Siberian Railroad in the final years of the 19th century. By 1904, when Japanese victory in Manchuria brought a halt to Russian territorial expansion, the Russian empire contained about 130 million persons, only 56 million of whom were Russian. Of the rest, which included more than 170 distinct ethnic groups, some 23 million were Ukrainian, 6 million were Belorussian, in Canada in 1887 on Prince Edward Island. Through controlled breeding, animals with unique characteristics of size, color, or texture can pass on those characteristics to their offspring. The silver fox, developed from the red fox, was the first fur so produced.

Overall, the Scandinavian countries produce about 45% of the world supply of pelts, Russia 30%, the United States 10%, and Canada 3%. Retail sales of furs in the United States grew from less than $400 million in the early 1970s to $1.8 billion by the mid-1980s but fell off to between $1.0 billion and $1.2 billion annually in the 1990s. The reason for this leveling-off in sales during the sustained economic boom of that period was a shift in attitudes toward furs, led by animal rights activists. In the late 1980s, Greenpeace commissioned photographer David Bailey to work on an advertisement with the slogan, "It takes more than 100 dumb animals to make a fur coat, but only one to wear one." In the late 1990s, PETA (People for the Ethical Treatment of Animals) ran its "I'd Rather Go Naked Than Wear Fur" poster, featuring five supermodels—Naomi Campbell, Christy Turlington, Claudia Schiffer, Cindy Crawford, and Elle Macpherson—doing just that. Antifur demonstrations targeting designers, led by entertainment icons such as Charlize Theron, Ricky Gervais, Pamela Anderson, and Angelina Jolie, have captured widespread attention (**Figure 2**). Most leading designers dropped furs

from their fashion lines in the mid-1990s, but furs have made a fashion comeback in the 2000s, led by designer labels such as Dolce & Gabbana and Louis Vuitton and promoted by celebrities such as Jennifer Lopez, Halle Berry, Maggie Gyllenhaal, Kanye West, Victoria Beckham, and even some who once took part in antifur protests, such as Madonna, Cindy Crawford, Naomi Campbell, and Kate Moss.

Meanwhile, a number of well-publicized cases of maltreatment on fur farms have further reinforced the case of Western antifur activists. Some Russian fur farmers, faced with a combination of falling consumer demand because of economic recession, higher taxes, and widespread corruption, let their animals go hungry. Western visitors to Siberian fur farms found starving animals in tiny cages with no bedding, no protection against the elements, and no veterinary care. Many Russian fur farmers have slaughtered most of their animals rather than watch them starve. Of the 200 fur farms in Russia in the

early 1990s, less than 40 were still operating in 2006. The others have closed or were gradually phasing out production. Nevertheless, Russian consumers have not been affected by Western activists' concerns. Many middle-class Russians own fur coats, and most consider them necessities in the harsh winter. ■

▲ **FIGURE 2 Antifur protest** PETA (People for the Ethical Treatment of Animals) protesters hold signs protesting Jennifer Lopez's use of fur for her fashion line, May 2005.

more than 4 million were Kazakh or Kyrgyz, nearly 4 million were Jews, and nearly 3 million were Uzbek.

To meet the challenge of different ethnicities under one state, Russia needed to apply binding policies and practices. Russia's strategies to bind together the 100-plus "nationalities" (non-Russian ethnic people) into a unified Russian state were oftentimes punitive and unsuccessful. Non-Russian nations were simply expected to conform to Russian cultural norms. Those who did not were persecuted. The result was opposition and, sometimes, rebellion and stubborn refusal to bow to Russian cultural dominance.

**Revolution** Meanwhile, since the time of Peter the Great, Russia had been seeking to modernize. By 1861, when Tsar Alexander II decreed the abolition of serfdom, Russia had built up an internal core with a large bureaucracy, a substantial intelligentsia, and a sizable group of skilled workers. The abolition of feudal serfdom was designed to accelerate the industrialization of the economy by compelling the peasantry to raise crops on a commercial basis; the idea was that the profits from exporting grain would be used to import foreign technology and machinery. In many ways, the strategy seems to have been successful: Between 1860 and 1900 grain exports increased fivefold, whereas manufacturing activity expanded

rapidly. In 1906 further measures, known as the Stolypin Agricultural Reform, helped establish large, consolidated farms in place of some of the many small-scale peasant holdings. The consequent flood of dispossessed peasants to the cities created acute problems as housing conditions deteriorated and urban labor markets became inundated.

These problems, to which the tsars remained indifferent despite the petitions of desperate city governments, fueled deep discontent among the population. At the turn of the 20th century, Russia was in the grip of a severe economic recession. Inflation, with high prices for food and other basic commodities, led to famine and widespread hardship, but there was no real mechanism for legitimately voicing the concerns and aspirations of the majority of the population. Unions were illegal, as were strikes. Nevertheless, riots spread across the countryside and, in 1905, after an embarrassing military defeat by the Japanese in Manchuria the previous year, there was a revolutionary outbreak of strikes and mass demonstrations. A network of grassroots councils of workers—called *soviets*—emerged spontaneously not only to coordinate strikes but also to help maintain public order. The unrest was eventually subdued by brute force, and the soviets were abolished. But the discontent continued, intensified if anything by the flood of dispossessed peasants to cities after the Stolypin Agricultural Reform of 1906. World War I intensified the discontent of the population, as casualties mounted and the government's handling of both the armed forces and the domestic economy led to the socialist revolution of 1917.

**The Soviet Empire** From the beginning, the state socialism of the Soviet Union was based on a new kind of social contract between the state and the people. In exchange for people's compliance with the system, their housing, education, and health care were to be provided by state agencies at little or no cost. This new social contract, though, had its roots in the traditional Russian traits of collectivism and authoritarianism. It was not the exploited peasantry or the oppressed industrial proletariat that emerged from the chaos of revolution to take control of this new system. It was the Bolsheviks, a dissatisfied element drawn from the former middle classes, whose orientation from the beginning favored a strategy of economic development in which the intelligentsia and more highly skilled industrial workers would play the key roles.

By the early 1920s, Nikolai Lenin, whose real name was Vladimir Ilich Ulyanov, the revolutionary leader and head of state, was also able to focus attention on the more idealistic aspects of state socialism. The Bolsheviks were internationalists, believing in equal rights for all nations and wanting to break down national barriers and end ethnic rivalries. Lenin's solution was recognition of the many nationalities through the newly formed Union of Soviet Socialist Republics (USSR). Lenin believed that this federal system, with each republic defined according to the geographic extent of ethnonational communities, would provide different nationalities with a measure of political independence.

Lenin was optimistic that once international inequalities were diminished, and once the many nationalities became united as one Soviet people, the federated state would no longer be needed: Nationalism would be replaced by communism. Lenin's vision was short-lived, and following his death in 1924, the federal ideal faded. After eliminating several rivals, Joseph Stalin came to power in 1928 and enforced a new nationality policy, the aim of which was to construct a unified Soviet people whose interests transcended nationality. Although the federal administrative framework remained in place, nations increasingly lost their independence and by the 1930s were punished for displays of nationalism. **Figure 3.20** shows the administrative units and nationalities that were part of the USSR during Stalin's tenure as premier (1928–53). Figure 3.19 also shows how, during and immediately after World War II, Stalin expanded the power of the Soviet state westward to include Albania, Bulgaria, Czechoslovakia, the German Democratic Republic, Hungary, Poland, Romania, and Yugoslavia.

**Collectivization and Industrialization** Under Stalin's leadership, a major shift in power occurred within the Soviet Union. The New Economic Policy and its "bourgeois specialists" were replaced by a much more centralized allocation of resources: a command economy operated by engineers, managers, and *apparatchiks* (state bureaucrats) drawn from the membership of the Communist Party. With this shift, the Soviet Union chose to withdraw from the capitalist world economy as far as possible, relying on the capacity of its vast territories to produce the raw materials needed for rapid industrialization. The foundation of Stalin's industrialization drive was severe exploitation of the rural population. This involved the compulsory relocation of peasants into state or collective farms, where their labor was expected to produce bigger yields. The state would then purchase the harvest at relatively low prices so that, in effect, the collectivized peasant was to pay for industrialization by "gifts" of labor.

Severe exploitation required severe repression. Stalin employed police terror to compel the peasantry to comply with the requirements of the Five-Year Plans that provided the framework for his industrialization drive. Dissidents, along with enemies of the state uncovered by purges of the army, the bureaucracy, and the Communist Party, conveniently provided convict (*zek*) labor for infrastructure projects to support the industrialization drive. Altogether, some 10 million people were sentenced to serve in the zek workforce, to be imprisoned, or to be shot. The barbarization of Soviet society was the price paid for the modernization of the Soviet economy.

The Soviet economy *did* modernize, however. Between 1928 and 1940 the rate of industrial growth increased steadily, reaching levels of more than 10% per year in the late 1930s—growth rates that had never before been achieved and that since then have been equaled only by Japan (in the 1960s) and China since the 1990s. An industrial revolution in the Western sense was achieved in just over a decade. When the Germans attacked the Soviet Union in 1941, they took on an economy that in absolute terms (though not *per capita*) had industrial output figures comparable with their own.

**Growth, the Cold War, and Stagnation** Meanwhile, the whole Soviet bloc gave high priority to industrialization. Between 1950 and 1955, output in the Soviet Union grew at nearly 10% per year, though it subsequently fell away to more modest levels. In addition to their desire for rapid growth, Soviet economic planners sought to follow three broad criteria in shaping the economic geography of state socialism. First was the idea of technical optimization. Without free markets to provide competitive cost-minimization strategies, Soviet planners had to organize industry in ways that ensured both internal and external efficiencies. Perhaps the most striking result of this was the development of **territorial production complexes**, regional groupings of production facilities based on local resources that were suited to clusters of interdependent industries: petrochemical complexes, for example, or iron-and-steel complexes (**Figure 3.21**). Second was the idea of fostering industrialization in economically less-developed subregions, such as Central Asia and the Transcaucasus. A third consideration was secrecy and security from external military attack. This criterion led to some military–industrial development in Siberia and to the creation of scores of so-called secret cities—closed cities, where even the inhabitants' contacts with relatives and friends were strictly controlled because of the presence of military research and production facilities.

Soviet regional economic planners also sought to ameliorate many of the country's marginal environments through ambitious infrastructure schemes. Stalin insisted that it must be feasible to harness and transform nature through the collective will and effort of the people. As a result, plans were drawn up to reverse the flow of major rivers and divert them to feed irrigation schemes. There were also plans to ameliorate local climatic conditions in steppe regions through vast plantings of trees in shelter-belts. The prohibitive cost of these grandiose schemes kept most of them on the drawing board, but nevertheless the tendency to undertake civil engineering projects of heroic scale lasted through most of the Soviet era, resulting in some dramatic examples of the mismanagement of natural resources.

By the 1960s, the Soviet Union had clearly demonstrated its technological capabilities with its manned space program and the production of some of the world's most sophisticated military hardware. These successes were paralleled by the Soviet Union's geopolitical influence.

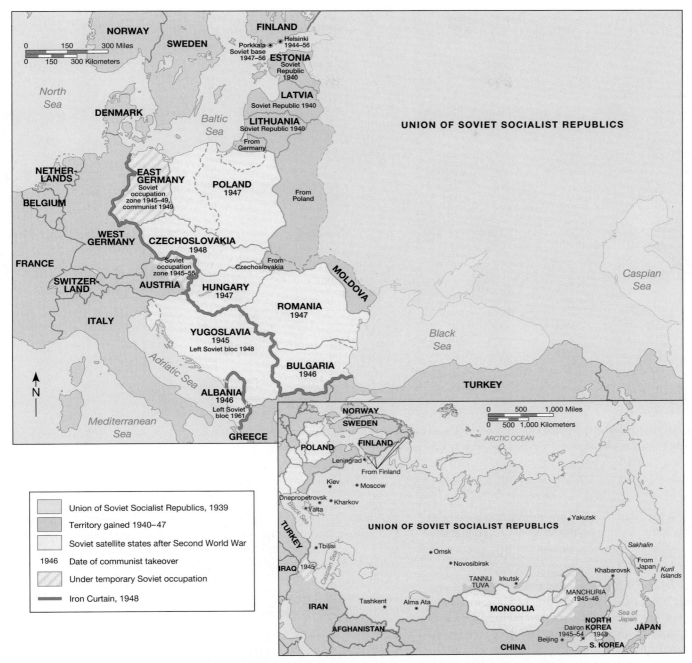

▲ **FIGURE 3.20 Soviet state expansionism, 1940s and 1950s** World War II gave the Soviet state the pretext for moving westward for additional territories. Insisting that these countries would never again be used as a base for aggression against the USSR, Stalin retained control over Poland, East Germany, Czechoslovakia, Hungary, Romania, Bulgaria, Albania, Yugoslavia, and eastern Austria. In 1945 Stalin promised democratic elections in these territories. After 1946, however, Soviet control over eastern and central Europe became complete as noncommunist parties were dissolved and Stalinist governments installed.

The Soviet Union not only had an extensive nuclear arsenal but also an ideological alternative to the capitalist and imperialist ideology that had created peripheral regions throughout much of the world. Armed with these, the Soviet Union posed a very real threat to U.S. hegemony, waging a Cold War that between 1950 and 1989 provided the principal framework for world affairs. In that period, Soviet influence caused significant tension and a succession of geopolitical crises in many regions of the world, including Cuba, much of the Middle East, South Asia, East Asia, Southeast Asia, and parts of South America (Chile),

Central America (Nicaragua and Panama), and Africa (Angola, Libya, and Egypt).

Yet throughout most of the Soviet Union itself, millions of peasants worked with primitive and obsolete equipment as they toiled to meet centrally planned production targets. Most nonmilitary industrial productivity was also constrained by technological backwardness and by cumbersome and bureaucratic management systems. A second economy—an informal or shadow economy—of private production, distribution, and sales emerged. It was largely tolerated by the government,

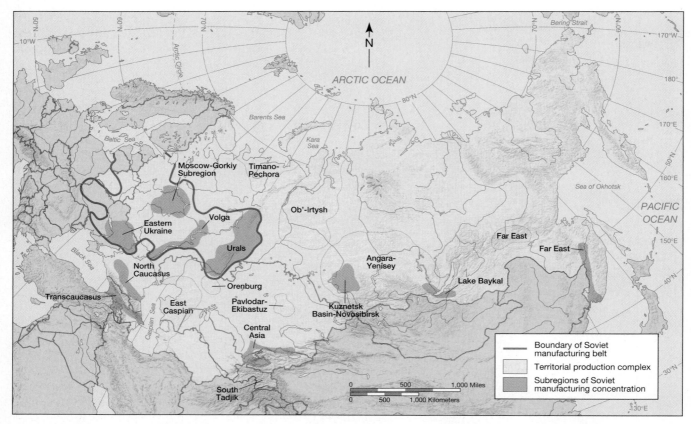

▲ FIGURE 3.21 Industrial regions of the Soviet Union Soviet planners gave a high priority to industrialization and sought to take advantage of agglomeration economies by establishing huge regional concentrations of heavy industry.

mainly because without it the formal economy would not have been able to function as well as it did. By the 1970s the Soviet economic system was steadily being enveloped by an era of stagnation.

**The Breakup of the Soviet Empire** By the 1980s the Soviet system was in crisis. In part, the crisis resulted from a failure to deliver consumer goods to a population that had become increasingly well informed about the consumer societies of their foreign enemies. Persistent regional inequalities also contributed to a loss of confidence in the Soviet system as an alternative mode of economic development. The cynical manipulation of power for personal gain by ruling elites and the drain on national resources from the arms race with the United States also undoubtedly played some role in undermining the Soviet model. The critical economic failure, however, was state socialism's inherent inflexibility and its consequent inability to take advantage of the new computerized information technologies that were emerging elsewhere.

Surprising even the most astute observers, the Soviet system unraveled rapidly between 1989 and 1991, leaving 15 independent countries as successors to the former USSR. The former states of Yugoslavia and Czechoslovakia were broken into smaller entities; East Germany was absorbed into Germany; Hungary, Poland, and the Baltic states (Estonia, Latvia, and Lithuania) were drawn rapidly into the European Union's sphere of influence; and Moldova and Ukraine began to show signs of a Western orientation. Belarus, the Russian Federation, and the states of Central Asia and the Transcaucasus continue to experience somewhat chaotic transitions, at different speeds, toward market economies. In the process, all local and regional economies have been disrupted, leaving many people to survive by supplementing their income with informal activities, such as street trading and domestic service.

## Economy, Accumulation, and the Production of Inequality

Restoring capitalism in countries where it had been suppressed for more than 70 years and reestablishing global connectivities has proven to be problematic. Although still a nuclear power with a large standing army and a vast territory containing a rich array of natural resources, the Russian Federation is economically weak and internally disorganized. The latter years of the Soviet system left industry in the Russian Federation with obsolete technology and low-grade product lines, epitomized by its automobiles and civilian aircraft. Similarly, the infrastructure inherited by the Russian Federation's economy was poorly developed, shoddy, and often downright dangerous. Investment in the development of computers and new information networks had been deliberately suppressed by Soviet authorities because, like photocopiers and fax machines, computers were seen as a threat to central control. As a result, the Russian Federation's economy faced a massive task of modernization before approaching its full potential.

**Building a Market Economy**   Overall, the economy of the Russian Federation shrank by 62% between 1991 and 1999. Global capital had begun to flow into the Russian Federation, but it was targeted mainly at the fuel and energy sector, natural resources, and raw materials (which now account for about half of the Russian Federation's total exports), rather than manufacturing industry. By the end of the 1990s, the Russian Federation's economy was in crisis. About one-half of the government's budget revenue was being absorbed by repaying debts to creditor nations; the rate of inflation had reached 100%; and economic output had plunged to about half that of 1989.

At the same time that there was the massive task of establishing the institutions of business and democracy after 70 years of state socialism, the Russian Federation found it difficult to create some of the essential pillars of a market economy. In the institutional vacuum that followed the breakup of the command economy, there were no accepted codes of business behavior, no civil code, no effective bank system, no effective accounting system, and no procedures for declaring bankruptcy. Security agencies were disorganized, bureaucratic lines of command were blurred, and border controls between the new post-Soviet states were nonexistent. The absence of these key economic elements has provided enormous scope for crime and corruption and fostered regional ethnic *mafiyas*, including Chechen, Azeri, and Georgian *mafiyas*. State assets were often sold off quickly and at an extremely undervalued price, so that some well-connected entrepreneurs managed to amass huge amounts of assets and wealth within the newly created liberal market environment. These new oligarchs quickly became extremely unpopular among the Russian public because of their political influence, extreme wealth, and control over media outlets. As a result, many have taken up residence outside Russia—particularly in upscale areas of London (humorously dubbed "Moscow on the Thames" or "Londongrad").

Eventually, Russia's economy grew in strength, largely because of the size and value of its commodities exports. Meanwhile, growing sectors of the economy include medicines, furniture, cosmetics, clothing, electrical appliances, and automobiles. Overall, the Russian economy grew between 5% and 10% each year between 2000 and 2008. In the same period, real incomes more than doubled while the proportion of population living below the poverty line decreased from 30% to 14%. The average wage increased from 2,200 rubles ($90) to 12,500 rubles ($500), and the average pension from 823 rubles ($33) to 3,500 rubles ($140). Russia's super rich prospered as never before. By early 2008, Moscow had 74 billionaires—more than New York and twice as many as London.

But the global financial crisis that began in fall 2008 hit Russia's economy—and the oligarchs—very hard. The crux of Russia's problem is that it depends on the outside world to provide much of the cash that keeps the financial system afloat. During the boom period of the early 2000s, foreigners purchased about two-thirds of the bonds issued by Russian companies. Foreign banks, meanwhile, put up roughly half of Russia's accumulated $900 billion in bank loans, including almost all long-term debt. Foreign investors had already started to slow their rate of investment as they began to worry about increasing risk in an emerging market that was not fully democratic; they were made still more nervous by Russia's invasion of Georgia (in August 2008). With the collapse of financial institutions in the United States and the United Kingdom the following month, the slowdown in investment turned into a rush to withdraw from Russian markets, resulting in a sudden drop in Russia's GDP and a sharp rise in unemployment.

**Regional Development**   Although the breakup of the Soviet Union and the transition toward market economies are beginning to modify patterns of regional development, at present the core regions of this world region remain broadly the same as under state socialism and, before that, Imperial Russia. The principal core region is the Central Region that extends for a radius of approximately 400 kilometers (248 miles) around Moscow. A secondary core area, developed around a long-standing industrial base, exists in the Urals. Between them, these core regions contain 11 of the 13 largest Russian Federation cities and about half of the total population of the Russian Federation.

The Central Region (**Figure 3.22**) was the hearth of the Muscovite state, the base from which its growing power thrust along the rivers in all directions toward distant seas. The industrial roots of the region go back to the 1600s, when various early industries, drawing on local resources of flax, hemp, hides, wool, and bog iron, developed to serve the needs of the growing capital, Moscow. By the 1700s, the region had developed a specialization in textiles, and with the onset of Russia's industrial revolution in the 1800s, the textile industry expanded and was joined by a broad range of engineering and manufacturing. Although the Central Region has no significant sources of energy (apart from low-grade lignite and peat deposits that can be converted into electricity in power stations), Soviet economic planners regarded the region as pivotal to their industrialization policies. Significant imports of coal, oil, gas, and electricity were used to develop a broad economic base. The region was also key to the Soviet Union's drive for technological supremacy, and a considerable proportion of the country's leading scientific research and development institutes were established as part of the Central Region's massive military–industrial complex.

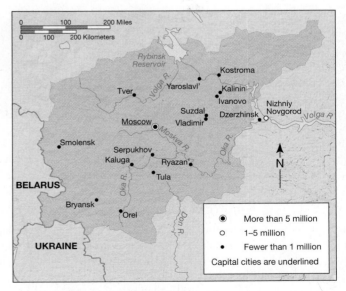

▲ **FIGURE 3.22 The Central Region**  Reference map showing principal physical features, political boundaries, and major cities of this region.

# THE MOSCOW REGION

The Moscow region itself has been significantly rewritten by the transition to an open, internationalized market economy. The effects of the transition are very uneven. On the positive side, the gateway situation of the Moscow region, with good transport and communication links to other regions and countries, is attracting a good deal of foreign direct investment, whereas the region's high-tech labor force and research institutes have also been attractive to investors. On the negative side, reduced domestic demand and competition from cheap imports have had severe adverse effects on the region's smokestack industries, especially textiles, machine building, and engineering. Among the places worst affected by deindustrialization are the eastern parts of the Moscow region, Bransk, Kostroma, Vladimir, and Yaroslavl'.

The transition to a market economy has already left its mark on Moscow in several ways. The combination of a newly emerging wealthy class, together with a chaotic planning situation, has sparked a spate of uncontrolled housing construction. Some has taken place within the borders of the city but a good deal of the recent growth has been in the forest protection belt, where speculative developments, mostly funded by foreign companies, have sprung up, providing expensive housing in community-style developments with tight security. New office buildings for transnational companies and new,

▲ **FIGURE 1 Traffic congestion in Moscow**

Western-style stores have begun to appear in the center of the city, consuming many of the parks and public spaces created by Soviet planners. Meanwhile, a sharp rise in the number of private automobiles (from approximately 0.5 million vehicles in 1985 to over 2 million by 2008) has led to an equally sharp rise in traffic congestion (**Figure 1**) and unprecedented strain on the existing road infrastructure.

The greatest effects of the transition to a market economy, however, are the social and economic consequences. Formerly a leading industrial center, by 2005 Moscow had slipped to 14th place among the Russian Federation's economic regions in terms of industrial production. The decline of the traditional industrial sector in Moscow has been mitigated largely by the rise of new sectors in the economy, particularly in tourism, retailing, and banking. Moscow has quickly developed as the Russian Federation's principal center for financial and business services, which in turn has attracted considerable foreign investment and many joint ventures. This has also led to the emergence of new culture and entertainment industries, whereas the overall climate of change has fostered a proliferation of small, private enterprises. The pace of change, however, has far outstripped the capacity of the city's authorities to regulate and control it, so that the positive aspects of transition have been accompanied by dramatic increases in social polarization and crime and a backlog in providing adequate infrastructure for ground and air transportation and telecommunications. ■

Today, the Central Region is highly urbanized, with about 85% of the population living in towns and cities. Moscow (**Figure 3.23**; population 10,400,000 in 2007) dominates the entire region, but other significant centers include Dzerzhinsk (285,000), Ivanovo (474,000), Nizhniy Novgorod (1,460,000), Ryazan' (536,000), Serpukhov (139,000), Smolensk (355,000), Tula (532,000), Vladimir (339,000), and Yaroslavl' (629,000). The region accounts for about 80% of the Russian Federation's textile manufactures. Cotton textiles are most important and are pro-

duced mainly in towns along the Klyazma valley between Moscow and Nizhniy Novgorod and at Ivanovo. Woolens are manufactured in and around Moscow; synthetic fibers are produced in Kalinin, Serpukhov, and Vladimir; and the ancient linen industry survives at Kostroma. Engineering, automobile and truck manufacture, machine tools, chemicals, electrical equipment, and food processing are also important. Overall the Central Region accounts for about 20% of the Russian Federation's industrial production.

Despite the high degree of urbanization and industrialization, much of the region has a rural flavor. About 25% of the Central Region remains forested, and there are numerous lakes and marshy areas. The traditional staple crop of the region was rye, but when the railways made it possible to import cheaper grain, farmers turned to industrial crops, such as flax, and to potatoes, sugar beets, fodder crops, dairying, and market gardening. Around rural settlements, there are orchards of apples, cherries, pears, and plums.

St. Petersburg (**Figure 3.24**) is a large metropolitan area (population 4.55 million in 2007), located some 650 kilometers (403 miles) from Moscow, at the eastern end of the Gulf of Finland. As such, it falls well outside the Central Region proper. Nevertheless, it must be considered as an extension of the Central Region, an industrial, cultural, and administrative metropolis, closely tied to the regional development of Moscow and the Central Region. The city was established as a new capital city by Peter the Great at the beginning of the 18th century. It was designed to be an imperial capital to rival those of continental Europe. Connections to Moscow—first by canal and later by railway—allowed St. Petersburg to flourish as Russia's chief trading port, and the city quickly became a cultural and intellectual center. With the Soviet revolution of 1917, however, Moscow was reinstated as the capital city, and St. Petersburg was renamed Petrograd. The city's trading function withered under state socialism's doctrine of national economic self-sufficiency, but Soviet planners quickly developed Petrograd—renamed again as Leningrad in 1924—into a key component of the Soviet military–industrial complex. During the Soviet era, the city was mostly a manufacturing center, the principal industries being electrical and power machinery, shipbuilding and repair, armaments, electronics,

▲ **FIGURE 3.23  Moscow**  The Kremlin, Red Square (center left), and the Moskva River.

▲ **FIGURE 3.24  St. Petersburg**  The city still bears the legacy of Catherine the Great (1762–96) and Alexander I (1801–25), who commissioned dozens of grand architectural projects in the center of the city in an attempt to make it Europe's most imposing capital. Today they are attractions for tourists and the backdrop for military ceremonies, like the parade shown here in front of the Winter Palace on the 60th anniversary of the Soviet Union's victory over Nazi Germany in World War Two.

chemicals, and high-quality engineering. The city suffered terribly in World War II, with an estimated 1 million of its residents dying from hunger or disease while the city was under siege by the German army for 872 days.

Today, St. Petersburg is once again taking advantage of its gateway situation, whereas its imperial legacy makes it an attractive international tourist destination. Already, the city handles about 35% of the Russian Federation's imports and about 30% of its exports. Although the city's infrastructure badly needs upgrading, its history and its European ambience are proving attractive not only to tourists but also to Western investors.

The transformation to market economies has intensified the unevenness of patterns of regional economic development. Market forces have introduced a much greater disparity between the economic well-being of regional winners and losers, while at the same time allowing for the more volatile spatial effects of the ebbs and flows of investment capital. After two decades of transition, many of the regional winners are the same as under state socialism. This is partly because of the natural advantages of certain regions and partly because of the initial advantage of economic development inherited from Soviet-era regional planning. Meanwhile, two different kinds of regions have experienced significantly decreasing levels of prosperity. The first consists of regions of armed territorial conflict (such as North Ossetia, Ingushetia, and Chechnya in the North Caucasus; Nagorno-Karabakh; and Tajikistan). The second consists of resource-poor peripheral regions—mainly in the European north, in Siberia, and in the Far East, where conditions are harsh in both rural and urban settings (**Figure 3.25**).

## Demography, Migration, and Settlement

A distinctive characteristic of this world region as a whole is the relatively low density of its population (**Figure 3.26a**). With a total population in 2008 of some 219 million and almost 14% of Earth's land surface, the Russian Federation, Central Asia, and the Transcaucasus contains about 4.8% of Earth's population at an overall density of only 11 persons per square kilometer (28 per square mile). The highest national densities—123 per square kilometer in Armenia, 87 per square kilometer in Azerbaijan, and 79 per square kilometer in Georgia— approximate the population densities of Colorado, Kansas, and Maine. Within the Russian Federation there is an area of relatively high population density (between 40 and 60 per square kilometer) that corresponds to the region of mixed forest and the wooded steppe west of the Urals. In contrast, population density in much of the far north, Siberia, and the Far East stands at less than 1 person per square kilometer, about the same as in the far north of Canada.

Levels of urbanization reflect this same broad pattern. Most large cities are in the European part of the Russian Federation and in the Urals. These include Moscow, Nizhniy Novgorod, St. Petersburg, Volgograd, and Yekaterinburg. Most of the other cities of any significant size are found in southern Siberia, on or near the Trans-Siberian Railway. Overall, both Belarus and the Russian Federation are quite highly urbanized, with 72% and 76% of their total populations living in cities, according to their respective census counts in the mid-1990s. The populations of the Transcaucasus are moderately urbanized (56% to 69% living in cities), whereas those of Central Asia are more rural (only 30% to 50% living in cities).

Overall, this is a world region with a relatively slow-growing population. Throughout the 20th century there was a general decline in both birth- and death rates (**Figure 3.26b**). In the 1990s the population began to register a decline as a result of more deaths than births. Viewed in greater detail, it is clear that this trend masks some important regional differences. In Belarus and the Russian Federation, population growth has for a long time been relatively modest, and it is in these countries that recent declines have been most pronounced. In contrast, in Central Asia and the Transcaucasus, birthrates have historically been relatively high, and rates of natural increase remain at a level comparable with those in South Asia and Southeast Asia.

Both World War I and World War II resulted in huge population losses that are still reflected in the age–sex profile of the Russian

◄ **FIGURE 3.25 Anadyr, Chukotka** One of the Soviet era's great achievements was construction of an adequate supply of sound housing for Soviet citizens. Most of the housing, however, came from a very restricted range of prefabricated apartment block designs; as a result, places throughout the Soviet empire acquired the same standardized appearance, sometimes broken up by bold color schemes.

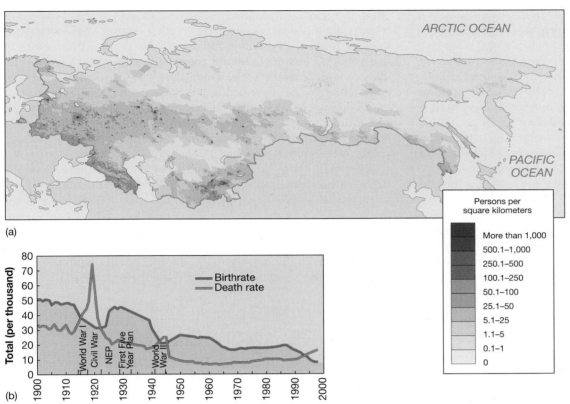

(a)

(b)

◀ **FIGURE 3.26 Population density and vital rates in the Russian Federation, Central Asia, and the Transcaucasus, 1995** (a) The distribution of population in the Russian Federation, Central Asia, and the Transcaucasus reflects the region's economic history, with the highest densities in the industrial regions of the western parts of the Russian Federation and the richer agricultural regions of the Transcaucasus. (b) This graph shows the dramatic drop in the birthrate in Russia that characterized the 1960s and 1990s. Note also the sharp rise in death rates since 1990.

Federation (**Figure 3.27**). It was not until the 1960s, however, that rates of natural population increase in the Soviet empire began to decrease significantly on a long-term basis. At the beginning of the 1960s, birthrates fell sharply as a result of a combination of the legalization of abortion, a greater propensity to divorce, planned deferral of marriage among the rapidly expanding urban population, and a growing preference to trade off parenthood for higher levels of material consumption. At about the same time, there began a steady rise in death rates, which increased sharply after the breakup of the Soviet empire. The reasons for this increase in death rates are several. Deteriorating health-care systems and the worsening health of mothers have contributed to an escalation of infant mortality rates. Meanwhile, public health standards have generally deteriorated, environmental degradation has intensified, and the rate of industrial accidents and alcohol-related illnesses has increased. By 2008 the average life expectancy of those born in the Russian Federation had slipped to 67 from the mid-1980s peak of 70 years.

**The Russian Diaspora and Migration Streams**   The spread of the Russian empire from its hearth in Muscovy took Russian colonists and traders to the Baltic, Finland, Ukraine, most of Siberia, the Far East, and parts of Central Asia and the Transcaucasus. In the late 19th and early 20th centuries many Russians joined the stream of emigrants headed toward North America. Concentrations of Russian immigrants developed in Chicago, New York, and San Francisco. They were joined by others who fled the civil war and Bolshevik revolution of 1917. More recently, in the first 5 years after the breakup of the Soviet Union, the United States resettled nearly 250,000 refugees from the former Soviet Union, mostly from Russia. Over a quarter of these immigrants have

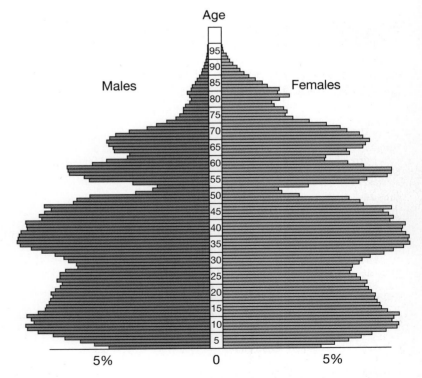

▲ **FIGURE 3.27 Age-sex pyramid for the Russian Federation**   This profile of the Russian Federation's population in the mid-1990s shows very clearly the effects of World War II (the relative lack of men and women in their early 50s and the reduced number of men aged 70 and older) and the reduced birthrates of the 1960s and 1990s.

settled in New York City, the majority in Brooklyn, where distinctive Russian exclaves, such as the Brighton Beach neighborhood of southern Brooklyn, have emerged as vital nodes in the Russian global diaspora.

With the rise of the Soviet empire, many Russians were directed and encouraged to settle in the Baltic, Ukraine, Siberia, the Far East, Central Asia, and the Transcaucasus—partly to further the Stalinist ideal of a transcendent Soviet people and partly to provide workers needed to run the mines, farms, and factories required by Soviet economic, strategic, and regional planners. By the time of the breakup of the Soviet Union, 80% or more of the population of Siberia and the Far East were Russian, and the Russian diaspora had become very pronounced in most of the Soviet Union's successor states beyond the borders of the Russian Federation.

In 1989, without any sense of ever having emigrated from their homeland, some 25 million Russians suddenly found themselves to be ethnic minorities in newly independent countries (**Table 3.1**). The largest number was in Ukraine, where more than 11.3 million Russians made up 22% of the population of the new state. In Kazakhstan, Russians represented nearly 38% of the population. Overall, the sudden collapse of the Soviet Union created havoc in the lives of many families, who suddenly found themselves living "abroad." During the 1990s, a good number of them decided to migrate back to the Russian Federation. In the Transcaucasus, where the proportion of Russians was generally lower than elsewhere, strongly nationalistic governments of the successor states quickly enacted policies that encouraged Russians to leave—reducing the number of Russian-language schools, for example. In Central Asia, too, nationalistic policies were enacted with similar

effect. Kyrgyzstan, Turkmenistan, and Uzbekistan dropped the use of the Cyrillic alphabet, deliberately creating institutional barriers for Russian speakers. Civil war in Tajikistan led to the departure of 80% of that country's Russian-speaking population within just 3 years of its independence from the Soviet Union. About 17% of the Russian population of Kyrgyzstan departed in that same period, mainly because of the withdrawal of the Russian Federation's defense industry enterprises and military installations. Altogether, almost four million ethnic Russians migrated to the Russian Federation from the other Soviet successor states between 1989 and 1999.

Meanwhile, an even greater number of people emigrated from the Russian Federation and the other successor states to countries elsewhere in the world. The annual loss, at about 100,000 per year, is not particularly significant in terms of raw numbers. What is significant, however, is the fact that most are well-educated individuals, and some are among the most talented. The countries of the former Soviet Union have thus been suffering something of a "brain drain," with the principal beneficiaries being Germany, Israel, and the United States.

# CULTURE AND POLITICS

Within the compass of this vast world region there exists a great deal of cultural and political diversity. This section first outlines the traditional religious and linguistic geographies of the region. They have been shaped by the influence of past movements of people into the region from Europe, the Middle East, South Asia, and East Asia. European influences have been particularly important in Belarus and the western parts of the Russian Federation, whereas the Arab conquests of the 8th century introduced Islam to the Transcaucasus and Central Asia. Since the region's exposure to global economic, political, and cultural systems in the post-Soviet period, traditional patterns of ethnicity and culture have become increasingly important as the basis for subregional political identity.

## Tradition, Religion, and Language

Patterns of religious adherence are a direct reflection of the past global connectivities of the region. The dominant religion in Russia, professed by about 75% of citizens who describe themselves as religious believers, is Eastern Orthodox Christianity. The emergence of Eastern Orthodox Christianity dates from the 11th century and the "Great Schism" between Rome and Constantinople, which led to the separation of the Roman Catholic Church and the Eastern Byzantine Churches, now the Eastern Orthodox. The schism was based in part on political factors and cultural differences between Latins and Greeks as well as doctrinal issues. Eastern Orthodoxy adheres to ancient traditions and practices rooted in Greek, Slavic, and Middle Eastern traditions. Russians of all religions have enjoyed freedom of worship since the collapse of the officially atheist regime of the Soviet Union, and large numbers of abandoned or converted religious buildings have been returned to active religious use. In Russia today about 10% of the population self-identify as Muslim. They are concentrated among the ethnic minority nationalities located to the north of the Caucasus, between the Black Sea and the Caspian Sea and in the larger cities of European Russia. Because of higher birthrates, Muslims account for the fastest-growing religious grouping in Russia. About 20% of the

### TABLE 3.1 The Russian Diaspora

| Republic | Number of Russians | Russians as % of Total Population of Republic |
|---|---|---|
| Ukraine | 11,356,000 | 22.1 |
| Belarus | 1,342,000 | 14.2 |
| Estonia | 475,000 | 30.3 |
| Latvia | 906,000 | 34.0 |
| Lithuania | 344,000 | 9.4 |
| Moldova | 562,000 | 14.0 |
| Georgia | 341,000 | 6.3 |
| Armenia | 51,600 | 1.6 |
| Azerbaijan | 392,000 | 5.6 |
| Kazakhstan | 6,228,000 | 37.8 |
| Uzbekistan | 1,653,000 | 8.3 |
| Kyrgyzstan | 917,000 | 21.5 |
| Turkmenistan | 334,000 | 9.5 |
| Tajikistan | 388,000 | 7.6 |

Source: D. B. Shaw, *Russia in the Modern World*. Malden, MA: Blackwell, 1999, p. 256.

population professes no religious affiliation—a similar figure to Germany, Italy, Spain, and the United Kingdom.

In Central Asia the Sunni branch of Islam is the dominant religious group. The dominance of Islam in the region dates from the Battle of Talas in 751 C.E., when the Arab Caliphate, based in Baghdad, defeated the Chinese Tang Dynasty, thereby gaining control of the region. A small minority group, the Pamiris, are adherents of Shia Islam, which first appeared in Central Asia in the early 10th century. The Arab Conquest also brought Islam to the Transcaucasus in the 8th century, displacing Zoroastrianism and various pagan cults. In Georgia, Eastern Orthodoxy was introduced after King David IV defeated Turkish rulers at the Battle of Didgori in 1121. Today Eastern orthodoxy is the dominant religion in Georgia, claimed by about 80% of the population. In Azerbaijan, on the other hand, about 95% of the population is nominally Islamic, mostly Shi'a Muslim.

Linguistic patterns and ethnicity are rather more complex. Dominant today throughout Belarus and the Russian Federation are Slavic people, among whom Russians represent one particular ethnic group. The Slavs are fundamentally defined by linguistic commonalities rather than territorial, racial, or other attributes. The Slavonic group of languages forms one of the major components of the great Indo-European language family, whose speakers range from north India (Hindi and Urdu) through Iran (Farsi) and parts of Middle Asia (Tajik) to virtually the whole of Europe. Written language came late to the Slavs, and when it did it was the deliberate effort of two missionaries—Constantine (later, as a monk, called Cyril) and Methodius—who were sent by the 9th-century Byzantine emperor Michael III to the Slavic nation of Greater Moravia (which occupied much of present-day Hungary, Germany, Slovakia, and Czechia) to spread the Scriptures. The new alphabet that Constantine/Cyril devised to accommodate Slavonic speech sounds became known as the Cyrillic alphabet. As **Figure 3.28** shows, Slavonic-speaking people correspond to the most densely settled parts of the region, extending eastward along the zone of wooded steppe and steppe to the Far East.

A second major language group is that of Turkic languages, which belong to the Altaic family of languages. These are spoken by the peoples of Central Asia and parts of the Transcaucasus and were spread into Russia itself through the Tatar invasion and period of rule (c. C.E. 1240–1480).

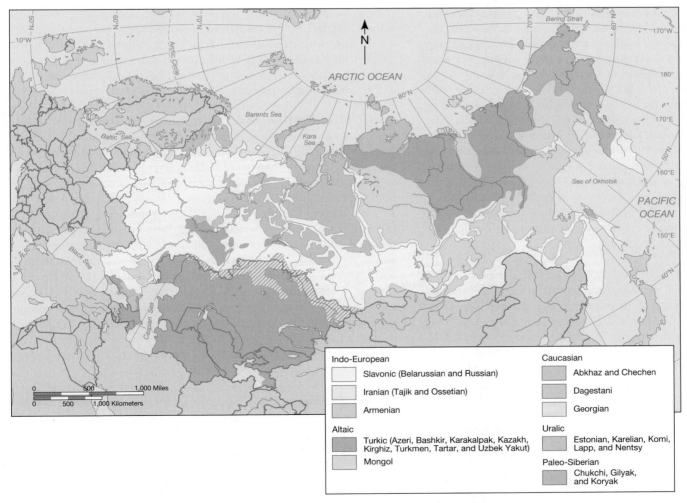

▲ **FIGURE 3.28 Languages of the Russian Federation, Central Asia, and the Transcaucasus** More than 100 languages are spoken in the region, the majority by very small ethnic groups and hence unrecordable on any but the most detailed maps. The greatest diversity is seen in the Caucasus, especially in Dagestan, on the northeastern flank of the range.

Much of northern and eastern Siberia is occupied by people who speak other branches of the Altaic language group, whereas in the Far East are people whose languages are part of the Paleo-Siberian language family, including Gilyak and Koryak. Finally, there are several smaller areas of Caucasian languages: Abkhaz and Chechen on the northern slopes of the Caucasus, and Georgian and Dagestani languages in the Transcaucasus.

# Cultural Practices, Social Differences, and Identity

Just as the economies of the Russian Federation and the Transcaucasus are in transition from state socialism to capitalism and are reestablishing global connectivities, so their cultural practices and identities are involved in complex dynamics, both adapting to and contributing to globalization. In some ways, this is simply a resumption of the emerging cultural practices of the pre-Soviet era, when Western, Byzantine, and Oriental influences all contributed to the cultural makeup of the region, and the folkways of Russia and the Transcaucasus found their way into the "high culture" of the West through art, literature, and symphonic music.

**High Culture**   The rising influence of European culture in Russia during the 17th and 18th centuries, strongly encouraged by Peter the Great and Catherine the Great, brought Russian high culture closer to the traditions of Western Europe. But by the end of the 19th century, uniquely Russian artistic styles had developed, some of them having developed in conjunction with liberal forces of social reform, and some having been inspired by Russian rural folk culture. The late 19th and early 20th centuries were a golden age for Russian high culture. In the performing arts there was the work of composers such as Borodin, Tchaikovsky, Mussorgsky, Rimsky-Korsakov, Rachmaninoff, Prokofiev, Stravinsky, and Shostakovich, together with the ballet impresario Sergei Diaghilev and the dancers Vaslav Nijinsky and Anna Pavlova. Equally influential in both Russia and the West were the novels of Dostoevsky and Tolstoy, the stories and plays of Chekhov, and the poetry of Pushkin. Meanwhile, in the cities and industrial regions of Russia and the Transcaucasus there developed an emergent modern mass culture of popular music, pulp fiction, variety stage acts, mass graphics, movies, and dances.

**Revolutionary Culture**   The 1917 Revolution halted these emergent cultural dynamics, while at the same time inspiring radical cultural and political movements in much of the rest of the world, eventually taking root in different forms in China, North Korea, and Cuba as well as being imposed on the Soviet satellite states of Eastern Europe after World War II. In Russia itself, the trajectories of both high culture and mass culture were repressed by revolutionary ideology. But to an emergent artistic avant-garde movement, the new Bolshevik regime seemed to promise just the sort of radicalism that they had been working toward for years. They produced political posters with a distinctive genre of graphic design and developed a distinctive form of modernist architecture—Constructivism—that drew on the new regime's emphasis on the importance of industrial power, rationality, and technology. Meanwhile, the Bolsheviks saw mass culture as a way of educating and transforming the population and therefore did their best to use film, fiction, radio, television, and poster art to help create the "new Soviet person."

After Stalin's death in 1953 the ideological controls on culture were somewhat relaxed, and this led to the development of subcultures, based largely on covert jokes, underground grassroots media such as cassette tapes, and "samizdat" publications (dissident or banned literature produced through systems of clandestine printing and distribution). Western rock music, in particular, became popular in the 1960s largely through illegal copies of albums that circulated from hand to hand.

**Popular Culture**   Since the breakup of the Soviet Union, popular culture has gone through an enormous upheaval, thanks to the collapse of censorship, the pressures of market forces, and a flood of cultural imports from Europe, the United States, and elsewhere. The advent of new television programming and different ways of thinking about issues like class, identity, and sexuality opened up post-Soviet society, especially in western Russia and the larger cities. While at first importing a great deal of popular culture from the West, it did not take long before a hybridized, home-grown popular culture began to emerge. No longer do American television shows and Brazilian and Mexican soap operas dominate Russian viewing figures. People now are more inclined to watch domestic programming, where their own lives are portrayed. Similarly, thousands of rock groups of all kinds—hard, soft, punk, art, folk, fusion, retro and heavy metal—now flourish in Russia and the Transcaucasus.

One distinctive outcome in post-Soviet Russia was the sensationally violent and abnormally graphic sexual nature of entertainment and mass culture that emerged. Free from censorship for the first time in Russia's history, the popular culture industry began to disseminate works that featured myriad depictions of deviance in pornographic and also detective fiction, with excessive and appalling details of social and moral decay. Cultural theorists have interpreted this as the popular culture industry's response to the scale of Russia's national collapse in the 1990s: a distraction from despair over economic woes and everyday threats.

**Social Stratification and Conspicuous Consumption**   Other aspects of life have also changed significantly. Restaurants, for example, were not highly developed under communism, but the post-Soviet period has seen an explosion of restaurants, cafés, and fast-food places in cities. The majority of people do not eat out often, mainly for economic reasons, but for the new business classes it is part of a new pattern of conspicuous and competitive consumption. This, in turn, reflects the emergence of new class factions with distinctive identities. Although they had special privileges, most officials in the Soviet system did not accrue wealth. Postsocialist privatization has allowed some of them to build large fortunes by taking advantage of insider status to acquire a share of direct ownership of state resources and industries. A new entrepreneurial class has also emerged, some of whose members have become significantly wealthy. More slowly, an affluent middle class is emerging in the cities, formed of an educated elite newly employed in business ventures and midlevel management. Most of the rest of the population, meanwhile, is relatively impoverished.

**Gender and Inequality**   When the Soviet system collapsed in 1989, economic liberalization produced chaos, hyperinflation, and industrial collapse; the usual social and economic safety nets were dismantled; and social and economic upheaval was accompanied by an intensification of inequality. In the early 1990s, industrial production in Russia fell by nearly half, whereas hyperinflation devalued people's savings, leaving many destitute. Then in the late 1990s there was a deep crisis of national finance as a result of Russia's weak trading record. The consequences for many parts of the Russian population included rising mortality rates, especially among older men, attributed in part to the stresses surrounding economic dislocation, in part to increasing poverty, and in part to the decline in the provision of health care in post-Soviet Russia. There was an increase in the rate of industrial accidents and alcohol-related illnesses and a rapid escalation of infant mortality rates. The modest levels of material welfare to which families became accustomed under state socialism were increasingly difficult to sustain.

Amid this upheaval, the role of women changed significantly. Women, together with all other groups discriminated against under the tsars, were "freed" by the 1917 Bolshevik revolution, which declared them equal and granted them all social and political rights. Women were trained for and encouraged to take up what was previously male-only labor, such as operating agricultural machinery, working in construction, and laying and maintaining roads and rail beds. Nurseries and day-care centers were established to free women from child rearing. Women's increased participation in medicine, engineering, the sciences, and other fields was encouraged. In practice, however, the Soviet political system became male-dominated, its legislative organs developing into a rigid structure based on proportional representation. After the collapse of the Soviet system, many women began to take up new opportunities presented by the transition to a market economy, venturing into small trade and opening their own businesses. Some women also took advantage of opportunities in newly established firms, quickly climbing through the ranks to become managers. Today, the number of women holding full-time jobs (about 45%) in Russia is on par with the developed countries. Nevertheless, where women's wages had been, on average, about 70% of men's during the 1980s, they had dropped to 52% by 1999, recovering only to 64% by 2005. The transition to new market economies has cost women many of the benefits they enjoyed under state socialism—such as child care, health care, equal pay, and political representation. Between 1985 and 2005, the number of working women in the Russian Federation fell by 24%. Another area in which the women of the region have clearly regressed is political representation. Under state socialism, quotas ensured that one-third of the seats in parliament went to women. In 2005, only 3.4% of the seats in the Russian Federation's upper parliament was held by women, and there were no women in government at the ministerial level.

Many women who might have hoped for a clerical or professional job under state socialism now find themselves forced into unskilled work to make ends meet while caring for children and keeping their family together. Often, this means working in the unprotected realm of the informal economy. For some, it means being drawn into the illegal activities of the informal sector. A great deal of media attention has been given to the fact that tens of thousands of women have been forced into prostitution, often after being trafficked abroad on the pretense that they would work as maids or waitresses.

## Territory and Politics

Within the Russian Federation, there are approximately 27 million non-Russians. This number encompasses 92 different ethnonational groups (though 25 of these groups include minority people of the north, who together number fewer than 200,000). Although most of the larger ethnonational groups enjoy a fair degree of administrative territorial autonomy within the Russian Federation, secessionist and irredentist claims are numerous (**Figure 3.29**). One of the most troubled regions

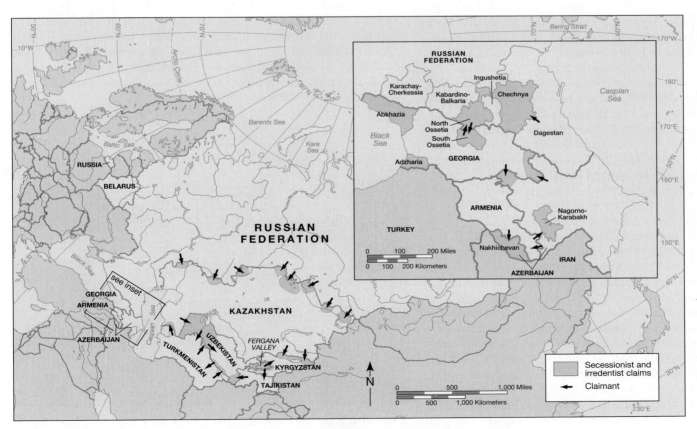

▲ **FIGURE 3.29 Secessionist and irredentist claims in the Post-Soviet states** The politics of multiculturalism is especially important in the Transcaucasus and Central Asia, where long-standing ethnic tensions, suppressed by the Soviet regime, have found renewed energy and expression.

is the North Caucasus, a complex mosaic of mountain people with strong territorial and ethnic identities. Soon after the breakup of the Soviet Union, Ingushetia broke away from the Chechen-Ingush Republic, and Chechnya promptly declared independence from the Russian Federation.

**Chechnya**  Of the many ethnonational movements that surfaced with the breakup of the Soviet Union, the Chechen independence movement has been the most bloody. In this region of the North Caucasus, clans, not territory, had been the traditional form of political organization. From the time that imperial Russia began its territorial expansion into the northern Caucasus in the late 1700s, the Sunni Muslim Chechens put up strong resistance, periodically waging holy wars against Christian Russia. When revolution came in 1917, Chechens scarcely looked on the Bolsheviks as a liberating force, not least because of the formal adoption of scientific atheism as the state religion of the newly created Soviet Union. Following a brief, failed attempt on the part of the people of the North Caucasus and Transcaucasus to resist Soviet domination, the Soviet strategy was to divide and conquer by creating administrative regions that encompassed a mixture of clans and ethnic groups. The anti-Soviet Chechens were put in the same region as the Ingush people.

The Chechens remained defiant but paid a terrible price for doing so. In the late 1930s, tens of thousands of Chechens were liquidated by Stalin in his purges against all suspected anti-Soviet elements. Then, in 1944, after invading German forces had been forced to retreat from the North Caucasus, Stalin accused the Chechens of having collaborated with the Nazis and ordered the entire Chechen population—then numbering about 700,000—to be exiled to Kazakhstan and Siberia. Brutal treatment during this mass deportation led to the death of more than 200,000 Chechens.

In 1957 Nikita Khruschev embarked on a program of de-Stalinization that included the rehabilitation of Chechens. But when Chechens returned, they found that newcomers had taken over many of their homes and possessions. Over the next 30 years, many of these newcomers withdrew, whereas the Chechen population consolidated and grew to almost 1 million. When Mikhail Gorbachev initiated his policy of glasnost in 1985, Chechens finally saw the possibility for self-determination, and with the breakup of the Soviet empire in 1989, Chechens wasted no time in unilaterally declaring their complete independence.

The Russian Federation chose at first to ignore Chechnya's declaration of independence but could not tolerate the possibility of the loss of the region, particularly because the area around Grozny is one of the Russian Federation's major oil-refining centers and has significant natural gas reserves. In December 1994, Russian troops invaded Chechnya. The ensuing conflict brought terrible suffering to the Chechen population and resulted in mass migrations away from the scene of the fighting. Chechen resistance continued, with increased popular support because of the invasion. Russian forces were disgraced in the fighting, and in 1996 the Russian Federation settled for peace, leaving Chechnya with de facto independence. For 3 years there were protracted negotiations over the nature of the peace settlement. Then, in the summer of 1999, after Chechen rebels had taken the fight to the neighboring republic of Dagestan and to the Russian heartland with a series of terrorist bombings of apartment blocks, the Russians renewed the military effort. After bitter and intense fighting, during which the Russian army suffered more than 400 deaths and nearly

1,500 wounded while hundreds of thousands of Chechens were made homeless and several thousand were dead or missing, Russian troops took the capital, Grozny, in February 2000. By that time, everything and everyone in Grozny had been brutalized. Between early 2000 and April 2009, Russian Federation troops maintained control of Grozny, with restrictions such as curfews, roadblocks, periodic searches and summary detention rules. The special regime, which included restricted access for journalists, encouraged massive rights violations in the region: human rights organizations documented patterns of abduction, detention, disappearances, collective punishment, extra-judicial executions, and the systematic use of torture by Russian and Chechen authorities. Chechen rebel guerillas, meanwhile, used terrorist attacks against the regime, including attacks on children.

**Ethnonationalism in the Transcaucasus and Central Asia**  In the Transcaucasus, one big trouble spot is the region of Nagorno-Karabakh, in Azerbaijan. For many years, this region was dominated by Armenians. At one time the region's population had been about 90% Armenian. By the mid-1980s the population of Nagorno-Karabakh was still more than 75% Armenian, and the breakup of the Soviet Union brought the opportunity for them to formally petition for secession from Azerbaijan. When the petition was refused, pent-up anger was unleashed in both Azerbaijan and Armenia against ethnic minorities from the other state. Riots, pogroms, and forced migrations quickly led to civil war, the outcome of which was that Armenian military forces secured Nagorno-Karabakh and established a militarized corridor as a lifeline to Armenia. Russian Federation armed forces were invited to serve a peacekeeping role in the region, and the United States, France, and Russia have been involved in mediating a peaceful settlement. The political situation in Azerbaijan is of broad international interest because of the 1760-kilometer (1094-mile) pipeline from Baku, Azerbaijan, to Ceyhan, Turkey, that opened in 2005 to carry oil from U.S. and European companies' oil fields in the Caspian Sea to Western markets via the Mediterranean (**Figure 3.30**).

▲ **FIGURE 3.30 Oil Pipeline**  The BTC (Baku-Tbilisi-Ceyhan) oil pipeline was opened in 2006. It was one of the most expensive oil projects in history, costing approximately $3.6 billion. The main backer was BP, in a consortium that included Unocal and Turkish Petroleum Inc.

Another flashpoint in the Transcaucasus is South Ossetia, which declared its independence from Georgia in 1991—mainly because of strong ethnic ties to North Ossetia-Alania, a republic of the Russian Federation. In 2008, Georgia launched a massive artillery attack on the separatists of South Ossetia, prompting a brief but full-scale war that drew Russian troops into South Ossetia, where they remained at observation and security posts for several months.

The breakup of the Soviet Union also led to an evanescence of ethnonational movements in Central Asia. Tajikistan has been beset by conflict between Tajik and Uzbek "patriots." A brief civil war ended in 1993 when the ruling government accepted intervention by the Russian Federation, acting to assert its claims to a special sphere of influence in the Near Abroad. Tajikistan now relies on the Russian Federation army to defend its borders. Since 1993 there have been frequent military skirmishes with a force of some 5,000 rebels who are based in Afghanistan with the tacit support of the Afghan government and the active support of Islamic fundamentalists. Elsewhere in Central Asia, one of the most complex regions of ethnonational movements and irredentism is in the Fergana Valley, where Tajik, Kyrgyz, and Uzbek nationalists living in areas outside, but adjacent to, their homeland states have called for the redrawing of international borders.

**Geopolitical Shifts**   In an attempt to counter some of the economic disruption caused by the political disintegration of the Soviet Union, several successor states agreed to form a loose association, known as the Commonwealth of Independent States (CIS). The CIS was designed to provide a forum for discussing the management of economic and political problems, including defense issues, cooperation in transport and communications, the creation of regional trade agreements, and environmental protection. The founding members were the Russian Federation, Belarus, and Ukraine; soon they were joined by the Central Asian states and some Transcaucasus states.

Meanwhile, however, the reorientation of the Baltic and eastern European states toward Europe and the imminent prospect of European Union and NATO membership for some of these states has not only undercut the economic prospects of the CIS (which has never really blossomed) but has also weakened the geopolitical security of the Russian Federation. In response, the leadership of the Russian Federation has asserted that country's claims to a special sphere of influence in what it calls the **Near Abroad,** the former components of the Soviet Union, particularly those countries that contain a large number of ethnic Russians.

The Russian Federation is clearly finding it problematic to adjust to a new role in the world. But although embarrassed by the disintegration of the Soviet Union and bankrupt by the subsequent dislocation to economic development, the Russian Federation is still accorded a great deal of influence in international affairs and may yet reemerge as a major contender for world power.

The new, post-Soviet states have joined the capitalist world system in semiperipheral roles, and all of them have had to find markets for uncompetitive products while at the same time engaging in domestic economic reform. Inevitably, patterns of regional interdependence have been disrupted and destabilized. The transition from state socialism to market economies involves a complex set of processes that requires more liberal and stable economic policies to encourage development of both private and international markets. All this has required radical changes in institutional, organizational, and technological structures and processes, along with equally radical changes in political

and cultural life and in the behavior and lifestyles of different socioeconomic groups.

**Toward Democracy?**   The ethnonational movements described in the previous section represent one of the major regional dimensions of the problems involved in the transition from state socialism to market economies. If democracy is to flourish, the new states must be able to guarantee territorial integrity, physical security, and effective governance. In some regions, secessionist and irredentist tensions are clearly undermining these preconditions for democracy. A second and more widespread problem concerns the vitality of civil society. **Civil society** involves the presence of a network of voluntary organizations, business organizations, pressure groups, and cultural traditions that operate independent of the state and its political institutions. A vibrant civil society is an essential precondition for **pluralist democracy**—a society in which members of diverse groups continue to participate in their traditional cultures and special interests. The Soviet state did not tolerate a civil society. Since the breakup of the Soviet Union, civil society has begun to flourish only in parts of the former Soviet empire that have reoriented themselves toward Europe. In the countries covered in this chapter, civil society is emerging only slowly, and in some regions—especially in Central Asia—democratic reform has been so limited that the emergence of civil society has been hard to detect.

# FUTURE GEOGRAPHIES

The range of possible futures for the region is starkly divergent. On the one hand are potentially very positive changes to be derived from integration with global market forces. This would mean significantly more open and progressive societies, though it would likely be accompanied by increasing social and spatial inequality. On the other hand, multiple constraints could limit the region's ability to achieve its full potential. Chief among them are shortfalls in investment in infrastructure and energy, decaying education and public health sectors, an underdeveloped banking sector, corruption, and organized crime.

Recent economic growth, particularly in Russia, has largely been the result of windfall profits from sustained increases in oil and commodity prices. Oil and gas currently account for roughly 20% of Russia's economy, 55% of its export earnings, and 40% of its total tax revenues. A generation of globally competitive companies in energy and metals has emerged, helping to further solidify Russia's position in the global marketplace. The export of primary commodities and raw materials is likely to remain the leading sector of economic development. Yet primary commodity markets are relatively more susceptible to fluctuations than are manufacturing markets. In addition, international investors may well be scared off by the region's record of lack of entrepreneurial investment, its endemic corruption, and fears of organized crime. Throughout the region, meanwhile, public investment in maintenance and new construction of infrastructure has fallen dramatically, so that deteriorating roads, bridges, railways, ports, housing, schools, and hospitals are likely to become an increasing deterrent to private investment. Much will depend on the effectiveness of the emerging systems of state capitalism as an approach to economic management, and in particular their ability to steer investment toward key infrastructure programs, to diversify the economic base and to engage with the world economy.

Demographic changes, already fairly clear-cut because of the existing composition and dynamics of regional populations, will also bring challenges. The populations of Russia and Belarus are aging dramatically. Between now and 2025, their populations are expected to decline by as much as 10%. The chances of stemming such a steep decline over this period are slim: The population of women in their 20s—their prime childbearing years—will be declining rapidly, falling to around 55% of today's total by 2025. As a result, the labor force may be insufficient for the size of the economy unless immigration is significantly increased, and meanwhile, the burden of dependency of the elderly on the young will intensify significantly. Within these overall trends, though, the Muslim minority population is projected to grow—from 14% in 2005 to 19% in 2030, and 23% in 2050, in Russia's case—something that is likely to provoke a nationalist backlash and even contribute to the emergence of a nationalistic, authoritarian, petro-state.

Finally, climate change will have an increasing influence, particularly on the northern regions of Russia. Russia has vast untapped reserves of natural gas and oil in Siberia and also offshore in the Arctic, and warmer temperatures should make the reserves considerably more accessible. The opening of an Arctic waterway could provide economic and commercial advantages. However, Russia could be also be hurt by damaged infrastructure as the Arctic tundra melts and will need expensive new technology to develop the region's fossil energy.

# Main Points Revisited

■ **The Russian Federation, Central Asia, and the Transcaucasus as a world region very much in transition, struggling not only with the transition from socialism to capitalism but also with the shifting patterns of wealth and investment associated with capitalism.**

After more than seven decades under the Soviet system, the Russian Federation, Belarus, and the Soviet Union's other successor states in Central Asia and the Transcaucasus are now experiencing transitions to new forms of economic organization and new ways of life.

■ **These transitions are taking place at different speeds in different places and with rather uncertain outcomes.**

It is clear, though, that the process is having a significant impact on local economies and ways of life throughout the region.

■ **Every country in the region has a legacy of serious environmental problems that stem from mismanagement of natural resources and failure to control pollution during the Soviet era.**

Soviet central planning placed strong emphasis on industrial output, with very little regard for environmental protection. The Chernobyl disaster was emblematic of the consequences of this attitude, whereas deforestation, the desertification of the Aral Sea and its surrounding region, and the pollution of Lake Baykal also represent large-scale environmental disasters.

■ **Since the breakup of the Soviet Union, popular culture has gone through an enormous upheaval, thanks to the collapse of censorship, the pressures of market forces, and a flood of cultural imports from Europe, the United States, and elsewhere.**

New television programming and different ways of thinking about issues like class, identity, and sexuality opened up post-Soviet society, especially in western Russia and the larger cities. But it did not take long before a hybridized, homegrown popular culture began to emerge.

■ **The region itself, meanwhile, is still struggling to find its new place in the world economy.**

Some of the old ties among the countries of the region have been weakened or reorganized, as have the interdependencies with Ukraine, Moldova, and the Baltic states. All of the new, post-Soviet states have joined the capitalist world system in semiperipheral roles, and all of them have to find markets for uncompetitive products while at the same time engaging in domestic economic reform. Inevitably, patterns of regional interdependence have been disrupted and destabilized.

■ **The Russian Federation, as the principal successor state to the Soviet Union, remains a nuclear power with a large standing army, but its future geopolitical standing remains uncertain.**

It has a formidable arsenal of sophisticated weaponry; a large, talented, and discontented population; a huge wealth of natural resources; and a pivotal strategic location in the center of the Eurasian landmass.

■ **Now freed from the economic constraints of state socialism, the Russian Federation stands to benefit a great deal by establishing economic linkages with the expanding world economy.**

Similarly, the collapse of the Communist Party has removed a major barrier to domestic economic and political development. The Russian Federation also has an ample labor force and a domestic market large enough to form the basis of a formidable economy. The Russian Federation is still accorded a great deal of influence in international affairs.

# Key Terms

chernozem (p. 90)

civil society (p. 109)

dry farming (p. 90)

market economy (p. 91)

monoculture (p. 87)

monsoon (p. 82)

Near Abroad (p. 109)

permafrost (p. 82)

pluralist democracy
(p. 109)

salinization (p. 87)

Silk Road (p. 92)

taiga (p. 88)

territorial production complex
(p. 96)

# Thinking Geographically

1. How does the Russian Federation suffer from its location, physical features, and climate? What is unique about the Transcaucasus area in terms of climate?

2. How did climate and physical geographic features spur Russia's imperial expansion?

3. Why is fur far more than an indulgence in Russia? What role did the fur trade play in the expansion of Russia?

4. How did the establishment of the Soviet bloc aid development of the Soviet Union following World War II? Discuss with regard to technical optimization, industrialization, and military security.

5. What factors led to the breakup of the Soviet empire?

6. In 1986, what happened at Chernobyl? Today, what policy does the Russian Federation have regarding the storage of nuclear waste?

7. Discuss the environmental degradation of Lake Baykal and the Aral Sea.

8. During the Soviet era, the human population of Siberia's tundra and taiga rose sharply. Currently people are leaving the area. Discuss migration in and out of Siberia with regard to natural resource development, industrialization, forced labor, military strategy, and free market forces.

9. Ethnic diversity in Central Asia contributed to the breakup of the Soviet Union. How have national identities been asserted in the decade since the Central Asian republics became independent countries? What cultural factors serve to unify or separate the states in this region?

Log in to www.mygeoscienceplace.com for Videos, Interactive Maps; RSS feeds; Further Readings; Suggestions for Films, Music and Popular Literature; and Self-Study Quizzes to enhance your study of The Russian Federation, Central Asia, and the Transcaucasus.

## ▼ FIGURE 4.1

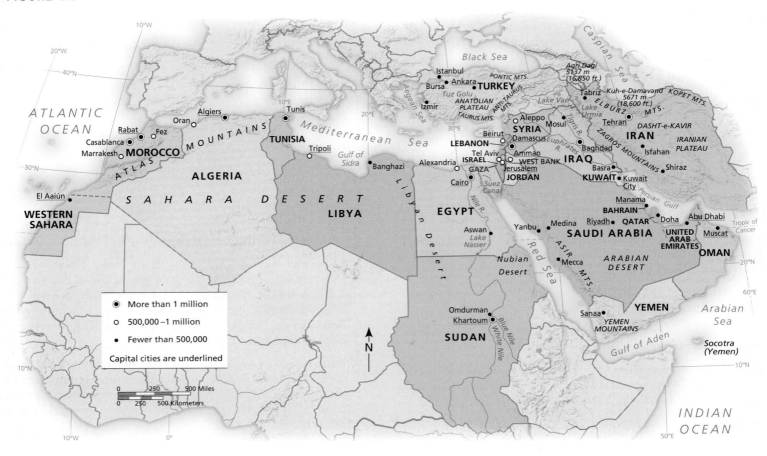

## MIDDLE EAST AND NORTH AFRICA KEY FACTS

- Major subregions: Arabian Peninsula, Persian Gulf, North Africa

- Total land area: 7.3 million square kilometers/2.8 million square miles

- Major physiogeographic features: Sahara and Nubian deserts; Atlas, Taurus, Anti-Taurus, and High Yemen Mountains; Iranian and Anatolian plateaus; Caspian, Arabian, Red, and Black seas. Climate varies throughout the region, the single unifying element being its aridity.

- Population (2010): 484 million, including 78 million in Egypt, 71 million from Turkey, and 40 million in Sudan

- Urbanization (2010): 59%

- GDP PPP (2010) current international dollars: 18,309; highest = Qatar, 97,001; lowest = Sudan, 2,424

- Debt service (total, % of GDP) (2005): 7.05%

- Population in poverty (% < $2/day; 1990–2005): 19.8%

- Gender-related development index value: 0.73; highest = Qatar, 0.86; lowest = Yemen, 0.47

- Average life expectancy/fertility: 72 (M70, F73)/3.1

- Population with access to Internet (2008): 27.4%

- Major language families: Afro-Asiatic, Semitic (Arabic and Hebrew), Indo-European, and Altaic

- Major religions: Islam, Christianity, Judaism

- $CO_2$ emission per capita, tonnes (2004): 15.2

- Access to drinking-water sources/sanitation (2006): 90%/69%

# 4

# MIDDLE EAST
# AND NORTH AFRICA

Dubai, one of the seven emirates and the most populous city in the United Arab Emirates, is becoming known in the early 21st century as the sports capital of the Middle East and North Africa. Horse racing, camel racing, falcon racing, snowboarding, cricket, soccer, rugby, golf—all are available in state-of-the art venues. For instance, the richest horse race in the world isn't the Kentucky Derby (with a purse award of a mere $2 million); it's the Dubai World Cup, with a purse of $6 million. Although wealth and fame tend to dominate the sports landscape of Dubai, a more prosaic sport that has come to capture the enthusiasm of locals in the region is camel racing. Once indulged in only by nomadic Bedouins, in the last 30 years camel racing has risen to become a wildly popular regional sport.

When camel racing first gained popularity, small Indian, Bangladeshi, and Pakistani boys—secured to the saddle with Velcro—were used as jockeys to race the camels. But international human rights organizations protested, and the Japanese came up with an alternative: the robot jockey. Dubai manufacturers soon pirated the far more expensive Japanese model by taking a 14.4-watt power drill and replacing the drill bit with a one-and-a-half-foot long riding crop and wireless receiver. At the Golden Camel Sports Equipment Trading Shop on the edge of Dubai, the locally produced, radio-controlled robot jockeys sell for about $500 in a variety of colors, weighing 4.7 kilograms (about 8 pounds). While assistants clap and whistle to get the camels to commence a trot, trainers like Mohamed Hamed Ali El Wahidi—in their long white robes and headdresses—jump into their waiting jeeps and drive furiously alongside the racetrack, remotely controlling their robot jockeys, sounding their horns, and shouting encouragement into their wireless microphones to their now galloping animals. But if horses and camels aren't your thing, you can still find just about any other sport to occupy you in Dubai from cricket, soccer, rugby, ice hockey, field hockey, and tennis to indoor skiing on the world's first indoor black diamond run.

Although once a major oil producer, Dubai' has diversified its economy so that tourism, real estate, and financial services are the main revenue generators. It is embracing a liberal attitude toward foreign-oriented real estate development, investment, and modernization. Dubai has increasingly chosen to cater to the rich and super rich through lavish hotels, malls, and residential developments. The underside of all the bling and flashiness of Dubai is manifested in its strong connection to South Asia and the some 250,000 foreign workers from there who work for less than $10 a day on its massive construction projects and live and labor under extremely poor—some argue inhuman—conditions. And, although prostitution is illegal in Dubai, this trade in human flesh is also connected to other parts of the world through conspicuously present sex workers from Central Asia, Eastern and Western Europe, Ethiopia, and several other African countries, as well as India. These women are provided visas through an agent, and the hotels extend free accommodation for 3 months while they are expected to service hotel guests. ■

## MAIN POINTS

■ Physically, this region forms something of an east-to-west arc. Iran and Turkey compose the northeastern tier; southward are the Arab states of Lebanon, Jordan, Syria, and Iraq and the Jewish state of Israel. The southernmost boundary includes the states of the Saudi Arabian peninsula. Moving westward are the states of the North African coast from Egypt to Western Sahara, as well as the state of Sudan.

■ The region is largely arid, but is diverse, including not only deserts but also huge mountains, grassy plains, forest ecosystems, wide beaches, and major river systems, as well as subtropical landscapes that include grassy plains and coastal areas. Human-environment interactions in the region have been largely predicated on adaptation, a practice that resulted in the first agricultural revolution: plant domestication.

■ Known as the "cradle of civilization," MENA (as it is called by policy makers and regional experts) has been part of many early empires—including Roman and Byzantine—and has generated important innovations from the invention of writing to the founding of the three world-spanning religions: Judaism, Christianity, and Islam.

■ The mandate system, which was imposed by Europe following World War I and dismantled by the mid-20th century, left a legacy of problematic boundaries and conflicted groupings in MENA that continue to create significant political difficulties today.

■ The most economically important natural resource in the region is oil. Wealth from oil exported across the globe has enabled many parts of the region to achieve modernization in a few short decades.

■ The biggest challenge for the MENA as it moves firmly into the 21st century may be the aftermath of contemporary political conflict in Iraq, Israel, and Palestine, continuing political instability in Lebanon, and the nuclear ambitions of Iran.

# ENVIRONMENT AND SOCIETY

The Middle Eastern and North Africa (MENA) region is environmentally diverse. This diversity is driven by the region's relative location to wider global climatic patterns, which help to create intense dry regions (e.g., the arid desert landscapes of the Arabian peninsula) as well as forested regions (e.g., the pine forests of the Atlas Mountains in Morocco, where snow can fall in the winter). Global plate tectonics have also dramatically affected this region, with uplift creating important highland environments as well as sources of water for some of the world's most historically significant river systems, such as the Tigris and Euphrates. These patterns can be seen from space, as is suggested by **Figure 4.2**. Adaptation to this relatively harsh environment, in fact, led to some of the earliest sedentary societies in the world, and management of these environments are striking for how they have changed the physical landscape of the region. This section, then, examines this region's dynamic physical geography as well as the adaptations that have facilitated the growth of large-scale societies throughout this region.

Perhaps the most popular image is fostered largely by the U.S. commercial film industry, showing vast, blazing hot deserts dotted with lush, but far-flung, oases. The deserts of the region are certainly impressive. The Sahara, the largest desert in the world, has an average annual rainfall of less than 25 millimeters (1 inch). Incorporated within the larger framework of the Sahara are the Libyan and Nubian deserts of Egypt and northern Sudan. The other important desert of the region is known by two names: the Eastern Desert and the Arabian Desert.

Although deserts and oases like those portrayed in Western films do exist, in reality they make up only a small percentage of the total land area of MENA. The predominant landscape of the region is vast grass plains, which receive substantially more precipitation than the deserts. Towering mountain ranges and extensive, treeless plateaus are also common. Trapped among these more predominant landscapes are the isolated deserts, some of which are vast seas of sand (**Figure 4.3**).

The single most important climatic variable unifying the region is **aridity**, in that the climate lacks sufficient moisture to support trees or woody plants. Although MENA is generally characterized by this climatic variable, a wide range of remarkable landscapes exists shaped by some small and some significant differences in the amount of surface water and annual precipitation. The absence or presence of water has strongly influenced the history of the interaction of peoples and environments of MENA.

## Climate, Adaptation, and Global Change

As in any region, global climatic systems interact with the variables of temperature, humidity, and rainfall as they interact with topography, and large water bodies are central to comprehending climate patterns and characteristics. In MENA, temperatures vary dramatically by season and location, and in the desert areas, temperatures exhibit extreme variation between night and day. Humidity and rainfall are also variable across the region. As in other arid lands throughout the world, summers in the lowland areas of MENA are extremely hot and dry, with daily high temperatures often at 38°C (100°F). The highland areas, such as the Atlas Mountains and the Iranian and Anatolian plateaus, and coastal areas of the Atlantic Ocean and the Mediterranean, Caspian, Arabian, Red, and Black seas experience more moderate daily summer temperatures and a predictable influx of visitors escaping the searing heat elsewhere. Winter temperatures, as would be expected, are more moderate in the lowlands and colder in the highlands. The moderating effect of large bodies of water on the coastal areas, as previously mentioned, also adds to climatic variation in the region, producing milder year-round climates with wet winters in those places (**Figure 4.4**).

**Arid Lands**  Except in the coastal mountain areas, precipitation in MENA is low and highly variable. Nearly three-quarters of the region experiences average annual rainfall of less than 250 millimeters (10 inches), which means that agriculture, where it occurs, must be irrigated. Scarce rainfall also means the soils in the region tend to be thin and deficient in a wide range of nutrients. In contrast, agriculture along the coastal plains and lowlands of Turkey and the floodplains of the Nile is able to take advantage of fertile soils. Other exceptional loca-

◀ **FIGURE 4.2 Middle East and North Africa from space** This satellite photo highlights the extreme aridity of MENA. The deserts dominate the southern part of the map. Along the coasts and in the mountains, high plateaus, and steppes, greater moisture availability means that more plants (and humans) can survive and thrive. Few large rivers and lakes exist in the region, but the ones that do are crucial to human, plant, and animal life.

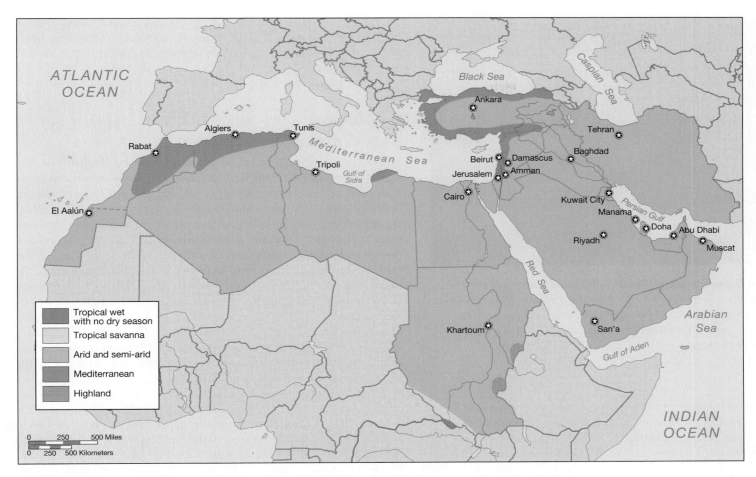

▲ **FIGURE 4.3 Climate of the Middle East and North Africa**  Though clearly dominated at midsection by a continuous swath of dry lands from Western Sahara to northeastern Sudan—MENA is latitudinally broad, stretching from the Mediterranean climes of northern Turkey to the wet, tropical climes of southern Sudan. Generally, however, the region is characterized by aridity: the dry climate of tropical and midlatitude deserts. *Aridity* is a relative term: Some areas are hyperarid and others mildly so.

tions with respect to moisture include southern Sudan, which because of its tropical location in sub-Saharan Africa experiences the wet spring and summer effects of the moisture-bearing storms of the intertropical convergence zone (see Chapter 1, p. 4), and the Central Highlands of Yemen, which experiences abundant rainfall in the summer because of the Indian Ocean monsoon system.

Most of the rain that does fall in the region is affected in some way by numerous mountain ranges. For instance, the winter and spring storms that bring rainfall to the coastal areas of the Mediterranean Sea are orographic in nature (see Chapter 1, p. 5). Moisture-bearing air masses move in from the Atlantic in the west, rise over a coastal range—such as the Atlas Mountains in Morocco or the Judaean hill country of the Levant (the eastern Mediterranean, including Syria, Lebanon, and Israel and the Palestinian territories), where the air is cooled and the moisture is condensed—and then drop their moisture as rain as the air masses rise over the mountains. In some of these mountain ranges, rainfall is quite plentiful. For example, in the Zagros Mountains in Iran, annual rainfall can range between 600 and 2000 millimeters (23 and 80 inches). Snowcapped peaks are found in Turkey, Iran, and Lebanon, where spring and summer snowmelt provides water for lowland human, animal, and plant populations (**Figure 4.5**).

▲ **FIGURE 4.4 Alexandria, Egypt, on the Mediterranean Sea**  Although our image of the region is one of vast, sandy deserts, the reality is that most of the population lives along the coasts and rivers, where the availability of fresh water as well as the moderating effects of the seas on temperature make habitation possible.

▲ **FIGURE 4.5 Mount Ararat, Turkey** High peaks like the one pictured here as well as others in the High Atlas Mountains in northwest Africa, the Elburz and Zagros Mountains in Iran, and the High Yemen Mountains provide water through annual snowmelt. Global climate change, however, is decreasing the annual snowfall in the region and therefore overall water availability.

Most of the coastal areas of the region experience between 375 and 1000 millimeters (15 and 40 inches) of rain a year. Although most rain falls in the winter and early spring, some areas, such as the Black Sea slope of the Pontic Mountains in Turkey, experience summer rains adequate for dry farming—that is, farming techniques that allow the cultivation of crops without the use of irrigation. In some mountainous areas where the peaks are especially high, such as the central Anatolian Plain in Turkey or the Syrian plateau, a rain shadow effect occurs: The mountains cause most of the moisture contained in the air masses passing over them to condense and fall as rain before it can reach the parched interior deserts of the region. This occurs, for instance, in southeastern Saudi Arabia, western Oman, and Dasht-e-Kavir (the Great Salt Desert) of Iran, where there is a complete absence of vegetation and often successive years have only spotty rainfall or none at all.

**Adaptation to Aridity**   People have adapted to the aridity and high temperatures characteristic of the region through architecture, patterns of daily and seasonal activity, and dress. The typical regional architecture features high ceilings, thick walls, deep-set windows, and arched roofs that enable warm air to rise away from human activity. The practice of locating living quarters around a shady courtyard enables residents to move many activities to cooler outdoor spaces that are also highly private. The clothing worn by Middle Eastern and North African people is also an adaptation to the heat and dryness. Head coverings and long, flowing robes made from fabrics of light color lower body temperatures by reflecting sunlight. They also function to inhibit perspiration and thus diminish moisture loss.

Some populations, such as the Berber of North Africa, migrate to mountainous areas in the summer and warmer lowlands in the winter to avoid the extremes of temperature. Plants and animals also have adaptive strategies to deal with the intense heat and aridity. For example, native plant species are typically able to store water for long periods of time or survive on very small amounts of water by keeping their leaves

and stems very small or developing an extensive root system. Animals adapt by lowering their body temperature through sweating or by being active only at night. Irrigation is an important adaptation discussed in the next section.

## Geological Resources, Risk, and Water Management

MENA contains a wide variety of physical landforms (**Figure 4.6**). The region is interspersed with seas, and an ocean hems its western flank. There are substantial mountain ranges, and two major river systems drain through its center. And although the region possesses the gamut of landscapes from tranquil beaches to towering mountains, the landscapes most heavily occupied by humans are the highland plateaus and the coastal lowlands, as well as the floodplains of the major rivers, where rain and surface water are most dependable. As **Figure 4.7** shows, the region is located at the conjunction of three continental landmasses, where active plates make the area highly prone to earthquakes.

**Mountain and Coastal Environments**   The Arabian Peninsula, which is a tilted plateau that rises at its western flank on the Red Sea and slopes gradually to the Persian Gulf on its eastern flank, is part of the Arabian tectonic plate. It is not hard to recognize that the Arabian Peninsula was once part of the African Plate tucked against the coastal areas of present-day Egypt, Sudan, and Eritrea. The separation of the Arabian Plate as it moved eastward from the African Plate millions of years ago resulted in the creation of the Red Sea, which initially formed as a rift valley and later filled with water (see also discussion on Rift Valley in Chapter 5). Both the African and the Arabian plates rub against the Eurasian Plate along the Mediterranean Sea and at the mountains that separate the Arabian Peninsula from the Anatolian and Iranian plateaus. The mountain ranges of Turkey, Iran, and the Transcaucasus radiate out from this feature. Crustal plate contact also means that the subregion surrounding the contact zone is prone to severe earthquakes like the one that shook Turkey in August 1999, killing close to 15,000 people and causing tens of billions of dollars of property damage.

Three mountain ranges dominate the region. Although impressive in terms of their beauty and ruggedness, these ranges, more importantly, generate rainfall and are the source of rivers and runoff for the arid region. The first set of ranges, which contains the most extensive and highest mountains, is the result of contact between the African, Arabian, and Eurasian plates at the center of the region (see Figure 4.7). These include the Taurus and Anti-Taurus mountains in Turkey and the Elburz and Zagros mountains in Iran. These ranges are higher than any in the continental United States and include Kuh-e-Damavand in Iran, which soars to 5,671 meters (18,600 feet), and Agri Dagi, also known as Mount Ararat, which is only slightly less impressive at 5,137 meters (16,850 feet). The second mountain range in the region is the Atlas Mountains of northwest Africa, which stretch along the southern edge of the Mediterranean from Morocco to Tunisia. They are a continuation of the European alpine range and are composed of a complex series of folded ridges separated by wide interior plateaus. The third set of mountains is those that border both sides of the Red Sea, known as the Central Highlands or the High Yemen Mountains.

In addition to being sources of precious water, the mountains provide homes for many people in the region. Though mountain environments

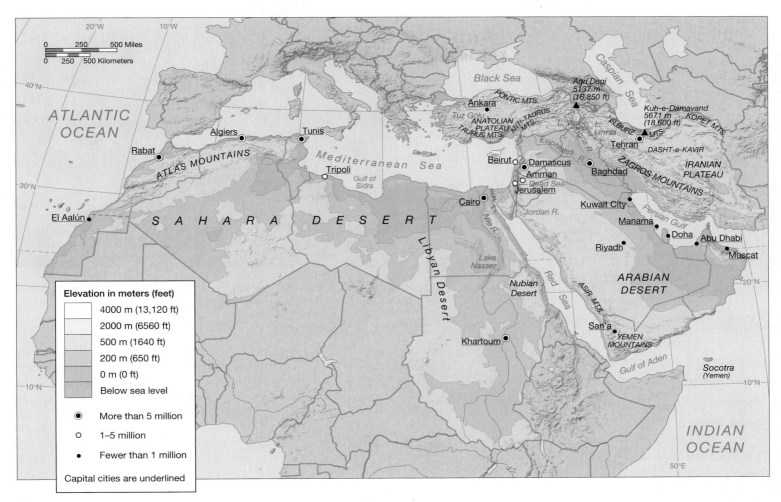

▲ **FIGURE 4.6 Physiographic Features of MENA** Perhaps the most consistent feature of the physiographic map of the region is the way that land and water features seem to alternate in a somewhat regular pattern. Also significant is the scattering of plateaus and mountain ranges that punctuate the vast lowland areas of desert and coastal plains.

can present substantial challenges to human habitation, the availability of moisture means that these environments can support agriculture over a somewhat shortened growing season. Historically, the mountains have also often provided safe havens for minority populations fleeing persecution and discrimination. The Druze in Syria and the Zayidis in Yemen are two such groups of people who have sought mountain refuge from their oppressors.

Whereas the highland areas are home to a small portion of the people of MENA, the coastal areas, the floodplains, and the plateaus are the most densely populated landscapes of the region. The clustering of populations in these landscapes is hardly surprising, given that they are the ones where water is most abundant and the environments the least harsh. The Iranian and Anatolian plateaus are the most obvious examples of these landforms. But the highland plateau of Yemen also contains a sizable population. The coastal areas and floodplains of the region, excluding the coastal areas of the Persian Gulf, are equally attractive to human habitation and constitute some of the most remarkable, highly engineered, and scenic of the region's landscapes.

**Riverine Landscapes**   There are only two major river systems in MENA—the Nile River system and the integrated Tigris and Euphrates rivers system. These systems are essential to the continued growth of the countries through which they flow and, not surprising, are also the

source of conflict—largely over access—because of the precious resource they deliver. The Nile is the world's longest river. Its source is in the mountains of Ethiopia and East Africa. From the Ethiopian Plateau, the Blue Nile flows northward across the Sahara Desert, where it joins the White Nile at Khartoum, Sudan. It proceeds northward as the Nile River, finally emptying into the eastern Mediterranean north of Cairo, as it flows through some of the driest terrestrial conditions on the planet, where no additional moisture is added and high evaporation occurs.

The Tigris and Euphrates system supported the development of a hearth area for the first agricultural revolution. The Tigris River originates in the Anatolian Plateau of Turkey and flows through Iraq. It is joined by the Euphrates River, which flows through Syria, in lower Iraq, where it eventually empties into the Persian Gulf in Iran (**Figure 4.8**).

These two river systems are the lifeblood for millions of the region's inhabitants. Although they are the major source of water for a large proportion of the region's population, there are other sources, though many are highly undependable. In the Sahara Desert, for example, runoff from the Atlas Mountains collects underground in porous rock layers deep below the desert surface. In some places, known as **oases**, land erosion and a high water table have enabled some of the underground water to percolate to the surface. Oases exist in sharp contrast to the dry, largely uninhabitable desert that surrounds them. Oasis soils are usually quite

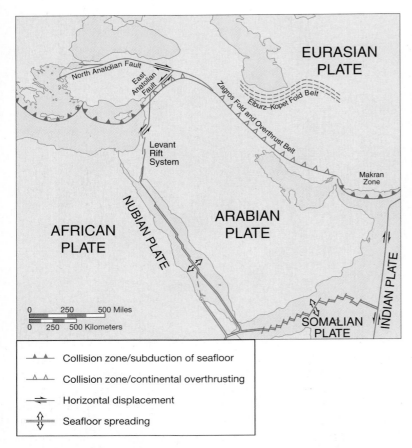

▲ **FIGURE 4.7 Generalized Tectonics of the Middle East** This generalized map shows the concentration of tectonic activity in eastern Turkey. The African and Arabian plates are made of ancient rock and are stable; little tectonic activity occurs there. Along the Eurasian plate are folded and faulted mountains extending from western to eastern Anatolia and then south across Iran and eastward again into the Himalayas. These mountains are the result of active plates colliding.

fertile, and animal and plant life are abundant. Agriculture is frequently undertaken here; dates are a highly successful cash crop in part because of their tolerance of saline soils in oases. Oases also play an economic role in the region when they serve as stopping points for caravans carrying commercial goods across the vast deserts.

Other sources of water in the region include natural springs, perennial streams, and wells drilled largely for irrigation, though some water holes have been drilled to create artificial oases. One of the most ingenious methods for mining water is a system of low-gradient tunnels that collects groundwater and brings it to the surface through gravity flow. The gravity system of water mining is known as *qanat* in Iran, *flaj* on the Arabian Peninsula, and *foggara* in North Africa (**Figure 4.9**). The system is restricted to piedmont—meaning, literally, the foot of the mountains—areas where runoff from orographic rain and snowmelt begins to percolate below ground level. Under ideal circumstances, where the qanat system and its regional relatives are carefully practiced, they provide a highly dependable source of water enabling year-round irrigation.

Because of its scarcity and the difficulties that must be met to secure adequate supplies of it, water is a highly politically charged resource in

▲ **FIGURE 4.8 The Euphrates River** is a critical source of fresh water for Iraq and Syria. Shown here is the river as it flows through Anbar Province in Iraq.

the region, and access rights to water are the source of much conflict. Egypt has threatened military intervention in the Sudan should that country attempt to limit downstream flow of the Nile. Another conflict involves Israel and Jordan, both of which inhabit the Jordan River valley and have access to its waters. A shallow, slow-moving river, the Jordan is nonetheless highly engineered through dams and diversions and is affected by intense groundwater pumping in the region, and at times its flow is little more than a trickle.

**Climate Change** Climate change, caused by the increasing use of fossil fuel globally, is threatening the populous coastal environments of MENA through sea level rise. For example, the Nile delta area is only a few meters/feet above sea level. A small rise in sea level would flood much of the delta, placing important coastline cities such as Alexandria, Damietta, and Port Said at risk. A rise of only a few additional meters/feet would threaten cities further upstream, where industry and shipping facilities are located.

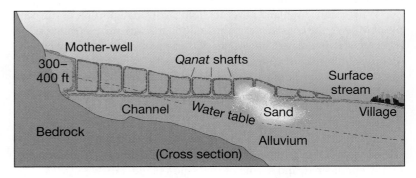

▲ **FIGURE 4.9 Simplified Diagram of Qanat Irrigation** Through a series of low-gradient tunnels, the qanat collects groundwater, bringing it to the surface by way of gravity. The diagram shows the shafts drilled into the loose soil and gravel at the foot of the mountain to reach the water table below. The water is drawn up through these shafts and directed out for use through pipes.

Recognizing the inevitability of environmental problems such as this, the region has become a participant in the UN Framework Convention on Climate Change (UNFCCC) and the Kyoto Protocol, as it is likely to be particularly vulnerable to the effects of climate change, including increasing surface temperature, decreasing precipitation, and sea level rise and drought. With the ominous future predicted by climate change as well as a generally poor record on environmental protection, the state of the environment in MENA is seriously challenged. Structural problems, such as rapid urbanization and a burgeoning population of very poor people, make solutions extremely difficult to implement. On the positive side, the growing global carbon market is highly attractive to countries in this region. The carbon market allows developing countries that have ratified the Kyoto Protocol agreement to receive payments as incentives for investments in climate-friendly projects that reduce greenhouse gas emissions, thus reducing pollution, increasing energy efficiency, and increasing participation in global efforts to halt climate change. Climate-friendly, internationally supported projects were recently launched in Egypt, Tunisia, Jordan, Algeria, Morocco, Iran, and Saudi Arabia.

## Ecology, Land, and Environmental Management

Perhaps the most pivotal event in the environmental history of the region was the domestication of plants that occurred between 9000 and 7000 B.C.E. From the agricultural practices they have developed to the cities, towns and villages they have built, the inhabitants of the region have been ingenious in adapting to environmental constraints.

**Plant Domestication** As mentioned in Chapter 1, the emergence of seed agriculture through the domestication of crops, such as wheat and barley, and animals, such as sheep and goats, replaced hunting and gathering as a way of living and sustaining human life. As cultural geographer Carl Sauer pointed out in his book, *Agricultural Origins and Dispersals* (1952), these agricultural breakthroughs could take place only in certain geographic settings. In these settings natural food supplies were plentiful; the terrain was diversified (thus offering a variety of habitats and a variety of species); soils were rich and relatively easy to till; and there was no need for large-scale irrigation or drainage. The emergence of seed agriculture occurred during roughly the same period (between 9000 and 7000 B.C.E.) in several regions around the world. In the Middle East, Mesopotamia was the source area.

Before the first agricultural revolution, in prehistoric times, hunter-and-gatherer systems were finely tuned to local physical environments and highly vulnerable to environmental change. Because they did not have the ability (or the need) to sustain an extensive physical infrastructure, they were also limited in geographic scale. The domestication of plants and animals represented a transition from hunting and gathering to agricultural-based society. The transition was premised on a series of technological preconditions: the use of fire to process food, the use of grindstones to mill grains, and the development of improved tools to prepare and store food.

Irrigation was the key to the success of the many agricultural-based systems that emerged in MENA when farmers there minimized their dependence on rainfall. Archaeologists and other scholars of the region argue that settled agricultural people were able to control irrigated farming across large areas and to exert dominance over weaker systems. By using food as a weapon of control, some systems were able to thrive, whereas others failed or were incorporated into stronger ones. An example is Babylon, which began in Mesopotamia but was able to turn itself into a 4,000-year empire. Babylon's dominance was achieved by systematically increasing control over regional agricultural production, through a buildup of its military strength (including a walled and fortified city center), the elaboration of a long-distance trade network (through extensive port facilities), and organizing extensive religious and symbolic political control (**Figure 4.10**).

**Human-Environment Interactions** After thousands of years of human occupation and the increasing pressures that population growth has created, the key term for describing human–environment interaction in the region today is *overexploitation*. In the long transition from adaptation to overexploitation, the landscape of the region has been dramatically transformed, in some cases so much so that entire species have been eliminated. The environmental history of MENA, therefore, is a long one that has changed as new technologies have been introduced and more human beings have placed greater pressures on the existing resource base.

It is not surprising that most of the plant and animal life is found where most of the people are: where the climate offers sufficient moisture. In fact, dense and extensive forests used to exist throughout Turkey, Syria, Lebanon, and Iran. Today, after several thousand years of woodcutting and overgrazing, the forests of Turkey and Syria are nearly entirely denuded, with only a few small remnant areas remaining. Although large areas of Lebanon have also been deforested, the Horsh Ehden Forest Nature Reserve, in the northern mountainous part of the country, is one place that has not, and it is home to

▲ **FIGURE 4.10 The Ruins of Babylon** The urban form and function of Babylon was key to the dominance of its extensive empire. Military strength was one of its important features, clearly shown here by fortress walls that are still standing in the right, middle ground of the image.

# THE MAGHREB

The Maghreb is the region of northwest Africa that contains the coastlands and Atlas Mountains of Morocco, Algeria, and Tunisia and the mostly desert state of Libya. Its people, history, and geography make it distinctive. Part of the Maghreb includes a coastal plain along the Mediterranean Sea. Within this narrow band, the region enjoys a moderate, Mediterranean climate, with cool, wet winters and hot, dry summers. Agriculture and tourism thrive here. Within the Maghreb are located the famous cities of Tripoli, Casablanca, Tangier, Marrakech, and the Barbary Coast—places of heroics, legend, and Hollywood-style glamour that conjure in the imagination a landscape of mythic and romantic appeal.

The Maghreb's history differentiates it not only from the rest of Africa but also from the Middle East. In ancient times, the region was influenced by the Phoenicians, Carthaginians, Romans, Christians, Vandals, and Byzantines and then finally by the Arabs, who in the late ninth century C.E. converted the populace to Islam. Although much of the material remains of that period have been destroyed or built over, remnants have survived. For instance, in Tunisia, Phoenician merchants founded a number of trading posts several thousand years ago. The most important one was Carthage, founded in 814 B.C.E. When the Romans defeated the Phoenicians, this area was incorporated into the Roman Empire, providing wheat and other commodities to the population in Rome (**Figure 1**). The ruins of Carthage lie in a suburb of present-day Tunis, the capital of Tunisia.

During the most recent period of imperialism, from the mid-19th to the mid-20th century, millions of Europeans, primarily French but also Spanish and Italian, flocked to the Maghreb and influenced its government, architecture, and language, especially in the city of Algiers, which is where most of the colonial Europeans lived. Significant numbers of Europeans also inhabited Casablanca and Tunis and formed a professional class that introduced many of the local elites to European cultural practices. In the early to mid-20th century, thousands of young people from the Maghreb went to the continent for university educations. They returned with the seeds of nationalist ideology planted in their hearts and minds. Many of these individuals played roles in the independence movements that occurred throughout the Maghreb.

In addition to possessing material remnants of the ancient world, the region is home to the Berbers, a people of ancient origin who preceded the Carthaginians and the Romans (see Figure 4.2). The Berbers appear to have been indigenous to the Maghreb region, though in more recent times, those who are attempting to maintain their traditional ways of life have tended to live in the mountains and deserts, away from the increasingly populated coastal area.

The Maghreb region has a relatively strong economy based largely on oil and mineral exploitation, agriculture, and tourism. Algeria's oil industry provides nearly 90% of its export revenues. Libya, too, has substantial, high-quality

▲ **FIGURE 1 Roman Ruins in Volubilis, Morocco** Volubilis was the administrative center of Roman Africa. The fertile lands of the province produced many commodities such as grain and olive oil, which were exported to Rome, contributing to the province's wealth and prosperity. Pictured here is the ruins of an olive merchant's home.

several species of rare orchids and other flowering plants. Sadly, Lebanon's famous cedars—important to various regional civilizations from the Babylonians to the Assyrians for building and religious purposes—continue to grow only in a few high-mountain areas (**Figure 4.11**). Iran retains a substantial expanse of its deciduous forests, particularly in the Elburz Mountain region. The forests of the Atlas Mountains of Morocco and Algeria continue to be harvested for commercial purposes (see Signature Region in Global Context: The Maghreb, above).

Animal life has been greatly affected by the millennia of human occupation as well as by more recent increases in human population. At one time, and because of the region's location at the crossroads of three continents, a wide variety of large mammals inhabited the region's forests, including leopard, cheetah, oryx, striped hyena, and caracal; crocodiles thrived in the Nile; and lions roamed the highlands of Persia (present-day Iran). Nearly all these species are now extinct or near extinction, with domesticated camel, donkey, and buffalo being the most ubiquitous mammals today. The highland areas of Turkey and Iran still contain a fairly wide variety of mammals similar to those found in parts of Europe, including bear, deer, jackal, lynx, wild boar, and wolf. Birds are also plentiful in the region, with different ecosystems supporting a wide variety of species.

**Environmental Problems** At the same time that the elimination of plant and animal species has been widespread throughout the region, different parts of it are experiencing pressing environmental problems.

▲ **FIGURE 2 Tuaregs Crossing the Sahara** A Tuareg man and his children, dressed in traditional blue robes, cross the Sahara's Erg Chebbi area with camels.

landscapes, and inexpensive and sumptuous food.

Algeria and Morocco are two of the fastest-growing economies in North Africa, with Libya and Tunisia making substantial strides as well. Links across the Mediterranean with the European Union through the Euro-Mediterranean Partnership (recently relaunched as the Union for the Mediterranean) are likely to boost all sectors of the Maghreb economy (as well as other countries of the region, including Syria, Palestine, Israel, Jordan, and Lebanon), from resources to tourism. It should also be pointed out that massive numbers of Maghreb people have migrated to Europe in search of jobs and a better life, so the connections between this formerly colonized group of states and their former colonizers continue to be quite strong.

In the next 10 years or so, it is certainly possible that despite its strong Islamic history and resultant ties to the Middle East, the Maghreb may once again become especially close to Europe. This transformation is something of a tall order, however, as it requires the different states of the Maghreb to overcome their anxieties and animosities as former colonies or occupied territories and learn to wield power alongside their former colonizers and occupiers: Spain, Italy, and France. Despite this potential obstacle, the possibility for profitable cooperation seems more promising than ever, as the Euro-Med network has begun to tackle the depollution of the Mediterranean and develop a regional solar energy plan. ■

petroleum reserves. Both Tunisia and Morocco are significant globally for their phosphate industries. All the Maghreb countries are also agricultural producers, though none is self-sufficient. The most important agricultural products of the region include wheat, barley, olives, dates, citrus fruits, almonds, peanuts, beef and poultry, and vegetables.

Hugging the southern coastline of the Mediterranean with rugged mountains rising up from the coastal plains and then trailing off to the desert, the Maghreb is a spectacularly beautiful setting for tourists (**Figure 2**). The region offers a range of tourist experiences, in both luxury and economy style—from lying on the beautiful beaches to trekking into the Atlas Mountains or the Sahara Desert to visiting ancient archaeological ruins. Europeans are frequent visitors to the Maghreb because a short flight brings them to warm temperatures, exotic

Some of these problems are particularly severe because national governments lack the resources to either prevent or mitigate them. All the states in the region are experiencing serious problems related to water quality, accessibility, and the impact of hydrotechnology on surrounding areas. In many states, excessive drawing of water from oasis wells has been occurring for such a long period that oases are actually dying (**Figure 4.12**). All the states, but especially those in the Persian Gulf, are faced with severe shortages of fresh water and the effects of oil and industrial pollution on air, land, and water. Water control schemes, such as those in Iraq, are destroying the natural habitat of wildlife. In Egypt, Nile irrigation is increasing the amount of salt in the soil, thus decreasing soil fertility. Coastlines along the Persian Gulf are also experiencing erosion and degradation of marine habitats caused by oil spills and the

discharge of ships and industry. And in every state of the region except Turkey, **desertification**—the process by which arid and semiarid lands become degraded and less productive, leading to more desertlike conditions—is an especially troubling problem as precious forests are lost because of overcutting and arable land is lost because of overfarming (see also Chapter 5). Tourism, a source of income for many of the states of the region, is worsening many of these environmental problems. For example, increased coastal development in Lebanon is causing the loss of precious wetlands, and heavy shipping traffic is damaging the coral reefs in the Red Sea.

There is no question that MENA faces severe environmental problems. However, even though the region as a whole has little surplus capital to invest in environmental protection or preservation, some

▲ **FIGURE 4.11 Cedars of Lebanon** For millennia, this coniferous tree has been significant to the region for trade, medicine, religion, and habitation. The national symbol of Lebanon—and shown on its flag—overexploitation has almost completely eliminated the once-abundant cedar forests. Reforestation programs are under way, however, in both Lebanon and Turkey. Pictured here is the last remaining cedar forest in Lebanon.

▲ **FIGURE 4.12 Sahara Desert Oasis** Geological evidence suggests that between 50,000 to 100,000 years ago, the Sahara possessed a system of shallow lakes that sustained extensive areas of vegetation. Though most of these lakes had disappeared by the time the Romans arrived in the region, a few survive in the form of oases. Pictured here is a grid fence that has been erected to slow the advance of migrating sand dunes into an ancient oasis.

important efforts have been made. Egypt and Jordan in the 1990s appointed environmental ministries, and efforts have been made to establish protected areas. Reforestation and **afforestation**—converting previously unforested land to forest by planting seeds or trees—programs are under way in several Middle East states. Oman and the United Arab Emirates have begun to take a deliberate stand against desertification, and greenbelts are being planted. Israel has introduced active breeding programs to encourage the regeneration of endangered animal species.

# HISTORY, ECONOMY, AND DEMOGRAPHIC CHANGE

MENA has long influenced growth and change in the rest of the world and is also deeply influenced by other regions, particularly with respect to its economy, politics, and culture. As mentioned previously, MENA is widely known as a "cradle of civilization." This title derives from the fact that the first known humans to settle there were ultimately responsible for domesticating plants such as wheat, producing some of the earliest integrated civilizations, establishing large cities and networks of villages and towns, and organizing complex religious–political systems. The geographical center of all these accomplishments was the Fertile Crescent, a region arching across the northern part of the Syrian Desert and extending from the Nile Valley to the Mesopotamian Basin in the depression between the Tigris and Euphrates rivers (**Figure 4.13**). Ideas and technologies generated in the Fertile Crescent diffused outward first to similar nearby environments and then beyond the area to reshape landscapes in Europe, the Americas, and Asia. It is also the case that colonization, war, and the thirst for oil in the global economy has also had an enormous impact on cultural change and politics in the region.

## Historical Legacies and Landscapes

Perhaps what is most impressive about the Middle Eastern and North African region is the number of culturally rich and intellectually sophisticated empires that have emerged and flourished in key areas over the last 4,000 years. The most famous empires include the Babylonian, Phoenician, Persian, Byzantine, Roman, and Ottoman. Mesopotamia, Asia Minor (most of present day Turkey), the Nile Valley, and the Iranian Plateau all functioned as centers for the major empires, such that at any one time, each major part of the region was able to extend its control to the rest of the region and, in turn, has been controlled by some other part of the region.

The interaction of dominance and subordination as the region extended its reach to other parts of the world meant that the exchange of ideas, goods, people, and belief and value systems helped tie the region together and promote a more or less unified regional identity. It also meant that a significant amount of social stimulation for developing new ideas and practices existed and enabled technological innovations and cultural revolutions to take place, especially with respect to religion and culture, but also related to trade and the growth of cities.

▲ **FIGURE 4.13 Empires of the Middle East from 1000 B.C.E. to the Rise of Islam** Various empires rose and fell in MENA and encouraged the interaction of people and ideas from across Asia, Africa, and Europe. The four maps show the extent of the most powerful regional empires. An interesting aspect of the many empires that have existed in the region is that they fostered a significant amount of interaction between the region and areas beyond.

▲ **FIGURE 4.14 Influence of MENA on Venetian architecture** Architecture and art diffused out of MENA to shape the landscapes and arts of Europe for centuries but most widely during the Crusades. La Serenissima, or the Serene Republic of Venice, was one Mediterranean location clearly affected by Islamic aesthetics. The Basilica of San Marco distinctly reflects the influence of the Byzantine Empire, the eastern portion of the Roman Empire that flourished in the Middle East from C.E. 330, when Emperor Constantine I rebuilt Byzantium and made it his capital.

**Source of World Religions** Christianity, Islam, and Judaism all developed among the Semitic-speaking people of the deserts of the Middle East. Like Hinduism and Buddhism, originating from the Indo-Gangetic plains (see Chapter 9, pp. 326–27), these three **world religions**—belief systems that have adherents worldwide—are related. Judaism originated about 3,500 years ago, Christianity about 2,000 years ago, and Islam about 1,300 years ago. Judaism developed from the cultures and beliefs of Bronze Age peoples and was the first monotheistic religion (a religion that believes in one God). Although Judaism is the oldest monotheistic religion and one that spread widely and rapidly, it is numerically small because it does not seek new converts. Christianity developed in Jerusalem among the disciples of Jesus, who proclaimed that he was the Messiah expected by the Jews. As Christianity moved east and south from its hearth area, its diffusion was helped by missionizing and by imperial sponsorship.

These three religions have greatly shaped the rest of the world. For instance, zealous Christianity was responsible for the Crusades, military expeditions undertaken through papal sanction by European Christians in the 11th, 12th, and 13th centuries to recover the Holy Land from the Muslims. The Crusades brought Europeans to the Middle East, where they had a modest impact. The reverse impact, however, was momentous: The Crusades were a major stimulus for the European Renaissance, because crusaders returned to Europe with new ideas and practices related to architecture, art, literature, and the sciences (**Figure 4.14**). In addition, the spread of Christianity to Europe is seen by many, including the eminent sociologist Max Weber, as one of the primary foundations for the spread of capitalism following the Renaissance. Capitalism and Christianity were linked in the motivations for conquest by the European colonizers. Spanish colonialism in North and Latin America, for example, was undertaken in the name of the Christian god as missionaries converted the indigenous peoples with whom they came into contact, and trade networks were opened for European markets.

The impact of Islam has been equally substantial as it spread through both empire and, perhaps more important, trade. At the opening of the 21st century, Islam is second only to Christianity in the number of adherents worldwide—about 1.3 billion. **Figure 4.15** illustrates the extent of the contemporary Islamic world. As the map makes clear, the Islamic world includes very different societies and regions from Southeast Asia to Africa.

**Cities and Trade**    In MENA, key urban centers, like Babylon, located at crucial points along natural and well-traveled human routes, were the organizational anchors of the region, shaping distinctive land-use patterns and serving as crucibles for significant cultural developments. In the Nile Valley, for instance, the urban-based Egyptian empire emerged, influenced by the cultures of Mesopotamia at the same time that it developed its own unique way of life. Like their neighbors, Egyptians were active city builders, and they constructed monumental tombs, temples, and palaces, the engineering of which still baffles architectural historians (**Figure 4.16**). The culture was also premised on elaborate rituals and sophisticated body adornments that required significant quantities of gold, cedar, ebony, and turquoise. As a result, the Egyptians entered into active trade with settlements in the northern Red Sea, the upper Nile of sub-Saharan Africa, and the eastern Mediterranean. These trade relations enabled the transfer of ideas and

improvement of technologies that enriched all the cultures of MENA, as well as Greece and Rome.

Considerable evidence suggests that the oldest cities on Earth were constructed along the valleys of the Tigris and Euphrates rivers as well as the Nile River (and possibly in the Indus River valley; see Chapter 9, pp. 312–13) sometime during the fourth millennium B.C.E. The availability of the rivers for transportation and irrigation and the use of wheeled vehicles probably allowed the concentration of a surplus at a few regional centers. The need to protect inhabitants from flooding and to channel river water for irrigation suggests that populations concentrated to take advantage of these opportunities. Walled towns began to appear in Mesopotamia at least as early as 4500 B.C.E. These early cities probably contained between 7,000 and 25,000 inhabitants, and the major producers were fishers and farmers who supported a nonproducing class of priests, administrators, and traders. Artisans also were clearly city dwellers. Although some of the earliest cities were clearly planned, others were more randomly organized. What seems to be consistent across all these early cities are three main elements: city walls, a commercial district, and suburbs, including houses, fields, groves, pastures, and cattle folds. As the historical evidence of urban commercial districts throughout the region suggests, trade was an essential part of life in these early cities.

Many other powerful early civilizations constructed cities and facilitated trade. Most recently, Islamic rule has had the most visible impact on contemporary urban patterns in the region. At its greatest extent, Islamic rule and influence under different dynasties reached westward as far as Tours in France; eastward beyond Turkey and the Iranian Plateau into Afghanistan, Pakistan, and India; and southward into North

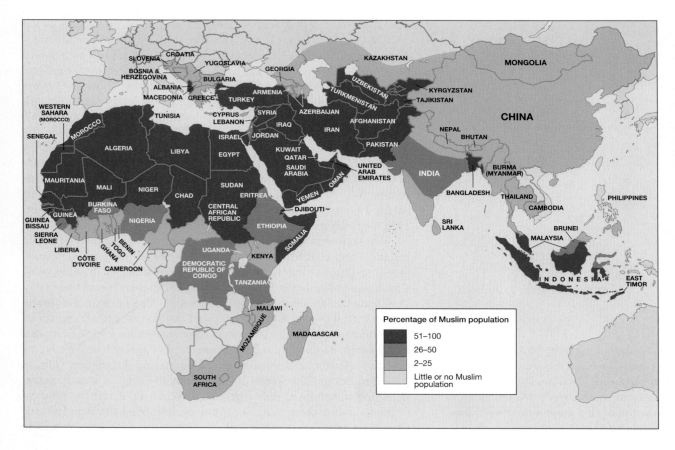

◀ FIGURE 4.15 The Islamic world The distribution of Islam in Africa, Southeast Asia, and South Asia that we see today testifies to the broad reach of Muslim cultural, colonial, and trade activities that carried Islam throughout these regions.

▲ **FIGURE 4.16 Pyramids of Giza, Egypt** The ancient Egyptians' belief in the continuity and stability of the cosmos was supported by a range of cultural activities from the preservation of wood, cloth, people, and animals to the construction of large-scale monuments. The pyramids are one example of this monumental architecture, which represents Egyptians' view that life continues after death.

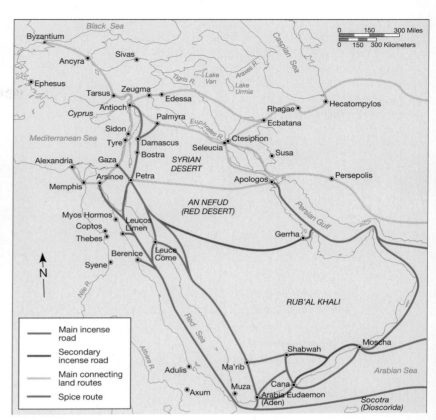

▲ **FIGURE 4.17 Incense and Spice Trade Routes in the Greco-Roman World** For several hundred years before and after the birth of Jesus, the southern rim of the Arabian Peninsula, within current-day Yemen and Oman, was a transshipment area for MENA. Goods that arrived from India, China, Ethiopia, and elsewhere were packed off by camel caravans to Egypt, Persia, Syria, and even Rome. Incense and spices were two of the most valuable commodities traded.

and West Africa, throughout which local variations of the Islamic city can be found today. From the 7th to the 15th centuries C.E., Islamic trade networks were so vast that they linked Mediterranean Europe to parts of the Transcaucasus, Pakistan, and China (**Figure 4.17**). Such extensive trade networks are evidence that parts of the world were highly integrated—politically, economically, and culturally—long before contemporary globalization occurred.

**Ottoman Empire** Based on the Anatolian Plateau in Turkey and with its capital in Istanbul, the Ottoman Empire was a successor to the Byzantine Empire. In power for more than 600 years, its influence began to decline in MENA at the end of the 19th century. The Ottomans were Turkish Muslims who had replaced the Christian Greeks as the political power of the region after 1100 C.E. At its height, the Ottoman Empire (named after the founder of the Ottoman dynasty, Osman) extended from the Danube River in southeastern Europe (including present-day Hungary, Albania, Bosnia, and Kosovo) to North Africa and to the Arab lands of the eastern end of the Mediterranean. Within the region, only the Persian Empire, based on the Iranian Plateau, the central Arabian Peninsula, and Morocco, had been able to resist direct Ottoman control.

By the mid-19th century, Ottoman rule was under siege from Europe through the legacy of the French political revolution and the British Industrial Revolution. By the early 20th century, the edges of the empire were being nibbled away. Egypt was occupied by Britain, and Algeria and Tunisia by France. By the eve of World War I, the Balkans and the remaining European possessions were lost, and the empire had been carved up by various European powers, as well as Russia (**Figure 4.18**).

European occupation exposed MENA to continental ideas about democracy, and as a result **nationalist movements** erupted—groups of people, sharing common elements of culture such as language, religion, or history, who wished to determine their own political affairs. The nation-

alist movements were particularly problematic for the polyglot Ottoman Empire, which had previously held itself together through an elaborate imperial legal and administrative structure that tended to allow for cultural differences. Already weakened by internal conflicts and external challenges, the Ottoman military was defeated by the Allied powers during World War I, resulting in the radical restructuring of the Ottoman Empire.

**The Mandate System** As part of the spoils of war, the Arab provinces of the Ottoman Empire were carved up and were neither colonized nor allowed to be entirely independent. Instead they became **mandates**—areas generally administered by a European power, with a promise and preparation for self-government and future independence. Syria and Lebanon were mandates of the French; the British took Iraq and Palestine and turned part of the latter into Transjordan. The rationale for this political arrangement—neither colony nor independent state—was heavily influenced by U.S. President Woodrow Wilson, who advocated self-determination and freedom over unmitigated colonization. The result was that a new form of external political control was created that legitimized French and British government dominance over their Middle Eastern and North African possessions. The mandate differs from outright colonial status because it requires

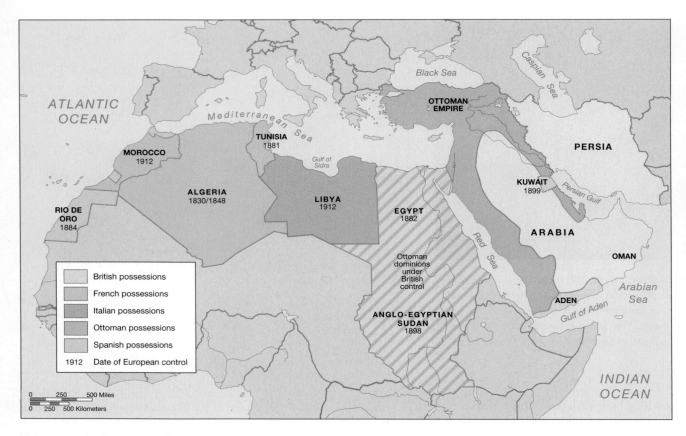

◄ **FIGURE 4.18
Europe in MENA in
1914** The colonial
presence of Europe in
MENA was short-lived—
only about 50 years—but
significant. This map shows
the European possession
of the region before the
Ottoman Empire was split
up in 1920. The breakup
of the Empire lead to the
League of Nations' creation
of the British Mandates of
Palestine and Iraq and the
French Mandates of Syria
and Lebanon.

the mandate holders to submit to internationally sanctioned guidelines. These guidelines require that constitutional governments be established as the first step in preparing the new states for eventual independence.

One of the most consequential mandates was the one determining the future of Palestine. As the mandate holder, Britain was obliged by treaty in 1917 to implement the provisions of the **Balfour Declaration,** named after British foreign secretary Arthur James Balfour, which committed Britain to the establishment of a Jewish national homeland in Palestine (**Figure 4.19**).

The political order that was imposed on MENA following World War I was seriously challenged throughout the region. Egypt, Iraq, Syria, and Palestine, for instance, all revolted violently against the European presence in the region, though none was formally colonized. And although Turkey and Iran were able eventually to establish independent republics, by the mid-1920s, Britain and France exercised control—often somewhat tenuous—over the rest of MENA.

Because of the mandate system of external control, strictly speaking, there were only five true colonies—Aden (British); Libya (Italian); and Algeria, Tunisia, and Morocco (French)—in MENA in the 20th century. But a pattern of control did emerge such that the new states were heavily influenced and, in many cases, overly dominated by their mandate holder.

The negative impact of the mandate system helped foment increasing regional dissatisfaction with outside dominance, and by the mid-20th century, aided by the crushing blows that World War II dealt to Europe, all the states of MENA had gained their independence. However, winning independence from colonizers who have effectively lost inter-

est in their colony is not the same as gaining the allegiance of a diverse collection of new citizens. Many of the challenges that the new states of MENA faced at mid-20th century continue to plague them in the 21st.

## Economy, Accumulation, and the Production of Inequality

The contemporary integration of the MENA region into the global capitalist economy has brought increased wealth for some but also increased poverty or reduced living standards for many others, even in the wealthy oil-producing states. It is important to understand the changes that occurred in the region in the twentieth century as they provide insight into how the region became more centrally integrated into the global economy.

**Independence and Economic Challenges**   In the 1930s, entrepreneurial states such as Turkey adopted policies of **import substitution,** whereby domestic producers provide goods and services that were formerly bought from foreign producers. After World War II, more comprehensive and aggressive approaches to state-led development were also undertaken, often in response to nationalist movements and anti-imperialist sentiments. Iran, Turkey, Egypt, Syria, Iraq, Tunisia, and Algeria were foremost among MENA states that adopted **nationalization** of economic development, which involves the conversion of key industries from private operation to governmental operation and control.

Eventually, states in the Middle East and North Africa began to turn away from the nationalization of industries as their economies began to stagnate, standards of living declined, and national debt sky-

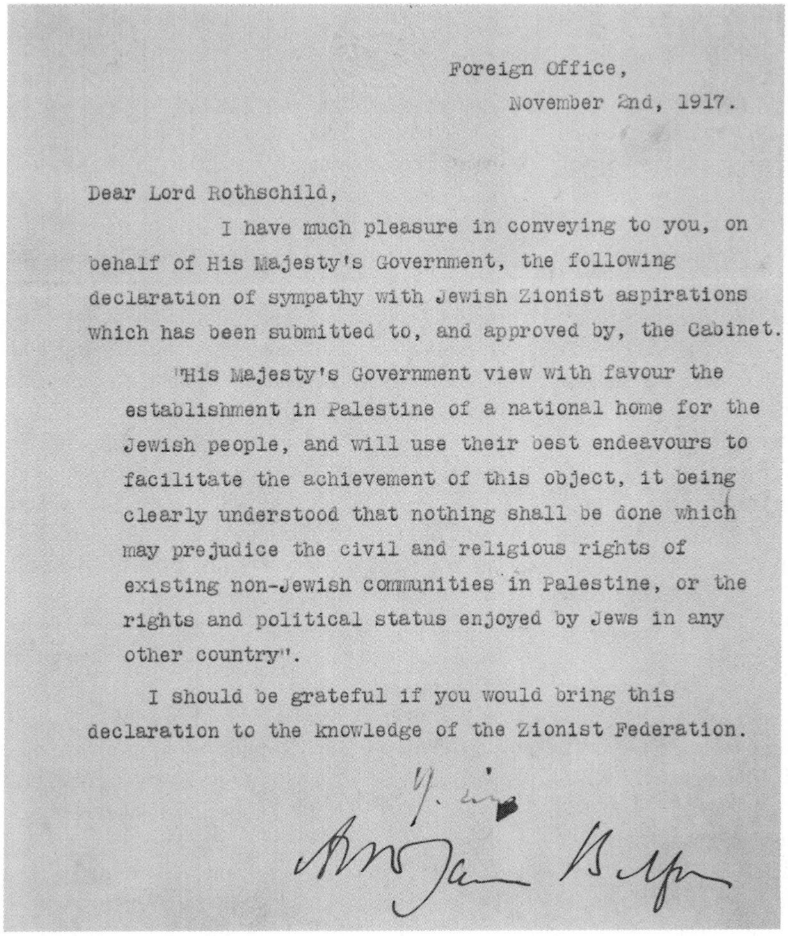

Foreign Office,
November 2nd, 1917.

Dear Lord Rothschild,

I have much pleasure in conveying to you, on behalf of His Majesty's Government, the following declaration of sympathy with Jewish Zionist aspirations which has been submitted to, and approved by, the Cabinet.

"His Majesty's Government view with favour the establishment in Palestine of a national home for the Jewish people, and will use their best endeavours to facilitate the achievement of this object, it being clearly understood that nothing shall be done which may prejudice the civil and religious rights of existing non-Jewish communities in Palestine, or the rights and political status enjoyed by Jews in any other country".

I should be grateful if you would bring this declaration to the knowledge of the Zionist Federation.

▲ **FIGURE 4.19 The Balfour Declaration** The Balfour Declaration was issued in the form of a letter from Lord Arthur James Balfour to Lord Walter Rothchild, head of the British-Zionist organization. Historians believe that behind the simplicity of the letter is a British-Zionist trade: if the Zionists assisted in bringing the United States into World War I (when Britain was sorely in need of support), the British promised them Palestine in return.

lowered the living standards of both urban workers and peasants. This outcome is especially troubling; populations in the region are becoming increasingly urbanized as rural people move to the cities to find employment. Unfortunately, when they arrive, they are confronted with decreased public services, not only in terms of schools and health care but also in terms of the most basic necessities, such as adequate housing and clean water. As a result, many people are forced to live in squatter settlements without sanitation. They are also often forced to eke out a living in the **informal economy**—that is, economic activities that take place beyond official record and are not subject to formalized systems of regulation or remuneration, such as unregulated taxi driving and street vending (**Figure 4.20**).

**The Oil States** Those countries within the region that are recognized as important players in the global petroleum economy and have derived much of their wealth from oil include Bahrain, Iran, Iraq, Kuwait, Oman, Qatar, Saudi Arabia, and the United Arab Emirates. Although Yemen, Algeria, and Libya also produce oil, their situations are substantially different. Yemen's oil reserves are only newly discovered, and the country has not yet exploited them to the level that its Arabian Peninsula neighbors have. Algeria has substantial oil and gas reserves but a far more mixed economy than the states along the Persian Gulf. Its history as a French colony also sharply differentiates it from the others. Libya, like Algeria, has a fairly mixed economy, and its colonial and modern political history make it an exceptional case. Whereas the once active agricultural economies of the oil states have been largely eroded by an emphasis on petroleum production, Algeria and Libya continue to possess productive agricultural sectors in addition to profiting from their oil reserves.

A nearly exclusive dependence on one economic sector, and only one product within that sector, leaves the economy of a state highly

rocketed. Pressured by the International Monetary Fund (IMF), the World Bank, and the U.S. Agency for International Development, states of the region were forced to initiate stabilization and structural-adjustment programs to qualify for new loans and to reschedule old debts. These programs, also known as neoliberal policies (see Chapter 1, p. 23), often raised the cost of food and other necessities, cut government spending on social programs, and generally reduced investments in the public sector. The impact of these programs was felt most directly and significantly by urban workers, government bureaucrats, and people on fixed incomes.

Although the rural peasantry was supposed to benefit most from these neoliberal policies as consumer subsidies were dismantled and markets were privatized (allowing peasants market-based prices for goods and the opportunity to market crops more freely), capitalist farmers have been the main beneficiaries of neoliberalism. Thus, the impact of neoliberal policies has been to put into motion a whole new set of forces in MENA that have improved the lives of some but have mostly

▲ **FIGURE 4.20 Street Vending in Beirut, Lebanon** Informal economic activity takes place throughout the world from babysitting and lawn-mowing jobs among teenage Americans to more sustained practices that constitute livelihoods. Pictured here is informal street vending in the shadow of the more formal corporate economic activity of a McDonald's fast food restaurant.

# PETROLEUM

Petroleum, more commonly known as *oil,* is a naturally occurring liquid composed of various organic chemicals. *Petroleum* usually refers just to crude oil, but the term can also apply to natural gas, tar sands, and shale oil. Petroleum is most widely used as an energy source for industry, commerce, government/military, and residences. Without petroleum, life as we know it in the core of the world-system would collapse, and life in the semiperiphery and periphery would be deeply disabled. In fact, semiperipheral countries base their development goals on petroleum availability.

While contemporary society is utterly dependent on petroleum, ancient peoples also knew about petroleum and made use of surface deposits for waterproofing and as fuel for torches. Later, they learned to distill it and came to use oil as a lubricant and for medicinal purposes. In the 19th century, kerosene was extracted from petroleum and used as a cheap fuel for lamps and lanterns. Recognizing that the crude oil itself might have value, in the mid-19th century drilling began in the U.S. state of Pennsylvania, marking the beginning of the modern petroleum industry. The invention of the internal combustion engine and World War I helped establish the petroleum industry as a foundation of industrial society.

The organization of the petroleum industry took several decades to consolidate. It was more or less complete by the early decades of the 20th century. The U.S. petroleum industry, through its five leading companies, along with two companies in Britain, dominated the petroleum industry worldwide throughout much of the 20th century.

In its early stages, the world petroleum market was organized by these seven companies, which produced and distributed abundant cheap oil first drilled around 1907 in Iran, by the 1920s in Iraq, and not until the 1930s in Bahrain and Saudi Arabia. In the United States, Standard Oil was the most prominent oil company, which by mid-century controlled 95% of the U.S. petroleum industry. In Britain, Shell Oil dominated.

In the middle decades of the 20th century, the Middle East first experienced rumblings of resistance to foreign companies dominating their oil supply. Iran was the first to resist, and by 1960, enraged by the unilateral cuts in oil prices made by the seven big oil companies, the major oil-exporting countries formed the Organization of Petroleum Exporting Countries (OPEC, which is discussed in greater detail later in the chapter) with the goal of controlling the price of oil. The "oil crisis" of 1973 dealt a major shock to the global economy when panic over supply led to wild speculation, and OPEC cut back on production and thus raised the barrel price to cash in on the panic. A worldwide recession resulted, and the oil market suffered greatly because of decreased demand. Numerous conflicts in the region, especially Arab–Israeli ones and the Iranian Revolution, have resulted in the disruption of oil production and increase in prices affecting countries around the world.

The conflicts and economic problems that emerged around the global oil supply in the 1970s made it very clear to the core that oil was a political issue that had to be monitored closely and carefully. The Persian Gulf War in 1990–91 is an illustration of the strategic importance of oil to international relations and its central role in foreign policy. Critics of the 2003 U.S. war in Iraq have also argued that the main reason for the attack by the United States was to gain control over the region's oil production. And although oil prices surged in response to the war, more recently the global financial and economic crisis has caused prices to drop dramatically (**Figure 1**).

Every day a globally coordinated system moves tens of millions of barrels of oil from producers to consumers. Although petroleum allows the core to enjoy a very high standard of living, the widespread and increasing use of petroleum worldwide has serious implications. The most critical problem is environmental pollution of both the air and the water (**Figure 2**). Air pollution generated from industries and vehicles burning petroleum or any of its products is significant but varies from region to region, according to factors such as the amount of petroleum-based energy consumed or the pollution-control system in place. The impact of oil pollution on Earth's oceans and rivers is substantial. The amount of petroleum products ending up in the ocean is estimated at 25% of world oil production. About 6 million tons per year are discharged by tankers, passenger ships, and freighters. But the greatest volume of petroleum products dumped into the ocean is carried there by rivers. Through the discharge of industry, storage installations, refineries, and local gasoline stations, more than three times the quantity coming from all tankers and other ships travels to the ocean by way of the world's rivers.

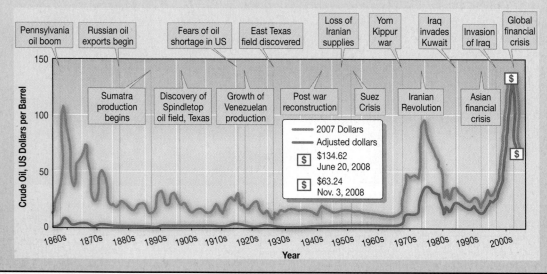

◀ **FIGURE 1 Price of Crude Oil 1860s–2009**
As this graph shows, the barrel price of crude oil was relatively stable until 2003. Production, consumption, and political factors all contributed to its recent escalation and decline.

▶ **FIGURE 2 Oil Landscape of Kuwait.** As seen from the air, the photograph shows the impact of oil production activities on the landscape. The wet-looking patches throughout the image are oil not water.

The collapse of the global economy in 2008 has directly and profoundly struck this region. Worldwide demand for oil has dropped sharply, with 2009 seeing a contraction of 2.56m barrels a day, the biggest annual fall since 1981. With the price of oil hovering at $70 per barrel in October 2009, cities all over the region are facing a tough time and as building construction is curtailed or halted entirely, foreign workers are leaving in large numbers. In 2009, Middle East and North African oil-exporting countries are expected to suffer a $30 billion deficit.

The future of the region as an oil empire is also being seriously called into question by "peak oil," the scientifically informed prediction that the world is at the half-way point of depleting its finite reserves of crude oil. Taking into account production and consumption, two well-respected scientific networks, the Association for the Study of Peak Oil and Gas (ASPO) and Energy Watch Group (EWG), believe that peak oil will be reached in 2010 (ASPO) or that it was already reached in 2006 (EWG) (**Figure 3**). Whether or not it has already been reached or soon will be, the point these organizations aim to make is that oil cannot sustain the world economy forever and that alternative energy sources must be developed and used.

Presently, the total available global oil reserves add up to about 700 billion barrels, more than half of which are in MENA. The peak oil scientists, as well as many energy policy experts, believe that the use of oil as a major source of energy will end up being a very brief affair, lasting little more than a century. Present estimates indicate that the supply of crude oil will probably be entirely exhausted sometime toward the middle of the 21st century, although new discoveries of oil may continue to extend this deadline. Ultimately, as the supply of oil diminishes, its cost will increase sharply, making it too expensive for all but the wealthiest consumers. ■

▶ **FIGURE 3 Predictions of Future Oil Production** This figure illustrates Energy Watch Group's forecast for global oil supply through 2030. The prediction is vastly different from the International Energy Agency's (IES) "World Energy Outlook 2006" forecasts, which are plotted on the graph in red. This divergence reflects not only differences in methodology but also contradictory attitudes toward future ambiguity, with IEA opting for a far more optimistic scenario.

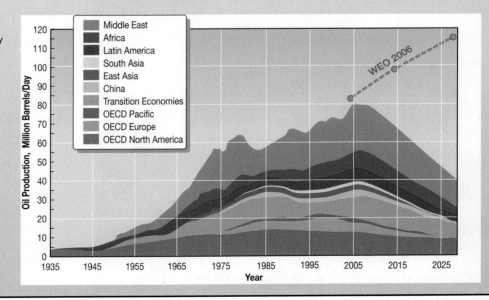

vulnerable to fluctuations in the demand for oil. Recognizing this, all of the oil states on the Arabian Peninsula have begun to diversify their economies such as through tourism or trade (**Figure 4.21**). Some are attempting to introduce new industries, such as textile production and food processing plants. Others are developing port facilities. Still others are resuscitating or introducing agriculture, though the scarcity of water makes irrigation an enormous technological challenge for all of these countries except for Iraq and Iran. Yet, although large-scale, irrigated agriculture is a costly undertaking for any of the Arabian Peninsula Oil States, the prevailing sentiment among them is that food—like oil—is an important security issue.

▼ **FIGURE 4.21 Tourism Development in Dubai** Pictured in the foreground is Madinat Jumeirah Resort and in the background the Burj al Arab Hotel. Tourism is rapidly gaining popularity as a way of diversifying the economies of oil-rich countries. Besides the UAE, Qatar is also involved in the tourism boom, and more than 100 buildings and towers are being erected in Doha, its capital.

The oil states are pivotal in the global economy for at least two reasons. First, much of the rest of the world is highly dependent on them for oil. Without petroleum from the oil states, the ability of a large portion of the globe to maintain productive economies would be severely hampered. Second, the impact of **petrodollars**—revenues generated by the sale of oil—is especially significant for the core economies of the world-system where they are spent, invested, and banked. The oil states of the Middle East are likely to continue to occupy a central role in the affairs of the region as well as the world, where guaranteeing a secure supply of oil is absolutely central to the smooth functioning of the global economy (see Geographies of Indulgence, Desire, and Addiction: Petroleum, p. 126).

**The Eastern Mediterranean Crescent** The Eastern Mediterranean Crescent, made up of Egypt, Turkey, Lebanon, and Israel, has the potential to be a regional economic success story. The greatest differences exist between Israel and the other three states. For example, Egypt, Turkey, and Lebanon have had strong agricultural bases for hundreds, if not thousands, of years (**Figure 4.22**). Israel's commitment to agriculture is more recent. Furthermore, although the former three use agriculture as a major source of export revenues, for Israel, agricultural production is more about achieving national food security, though foodstuffs are also exported. And, whereas Egypt, Turkey, and Lebanon are just beginning to encourage more industrial development—some more successfully than others—Israel already possesses a strong industrial base that is fairly diverse but receives a large share of its income from high-technology production. Finally, and most important politically, Israel is a Jewish state in a predominantly Muslim region (though neither Turkey nor Iran is Arabic). Its economy generates the second highest GDP per capita in the region after the United Arab Emirates.

There are similarities among the four states. For instance, all four possess more of a European orientation than many of their neighbors, certainly more so than the oil states. All four have a sizable middle class that has been important to their political stability. And all four have the potential, because of their resource endowments, to continue to build a diverse economic base. Although Israel coupled with any of the other three states would make strange bedfellows, and strong political and cultural differences have tended to prevent all four of these countries from acting in concert, there is much to suggest that cooperation would be mutually beneficial. The main reason for clustering these four countries is that they appear to possess the necessary ingredients to participate in the world economy because of their histories, their economies, and their roles in regional politics.

Turkey and Egypt once controlled long-lasting, influential, and extensive world empires. Although neither was colonized by Europe, both labored under the conditions of a foreign bureaucracy—Egypt longer than Turkey. Presently, both states have similarly sized populations—Egypt with around 82 million and Turkey with around 72 million—and both are burdened with the problems that large national populations present to economic development. Whereas Turkey possesses a fairly diversified economy with a strong agricultural sector and substantial mineral wealth, national agriculture in Egypt is built on a fairly narrow base, largely because of the environmental constraints of the desert. Both states have significant manufacturing capacity across a range of products from food processing to heavy machinery.

▲ **FIGURE 4.22 Harvesting Cotton in Southeast Turkey** Much of Turkey's climate is conducive to agriculture, and the country contains numerous farming regions. Cotton is a major export crop; Turkey is also the world's largest exporter of sultana raisins and hazelnuts. Other crops are tobacco, wheat, sunflower seeds, sesame and linseed oils, and cotton-oil seeds. Opium was once a major crop, but its exportation was banned by the government in 1972. The ban was lifted two years later as poppy farmers were unable to adapt their land to other crops. The government now controls the production and sale of opium.

Lebanon possesses a strong agricultural base and, for many years before its civil war (1975–90), was a banking and financial center for MENA, connecting it to the core of Europe and North America. It was not until the early 1990s, however, that Lebanon began to recover from the serious political problems that civil war unleashed. The 2006 war with Israel was derailing for Lebanese political and economic stability. With the global economic collapse occurring just 2 years after the cease-fire in Lebanon, the economy remains weak and politics deeply fragmented.

**New Economies and Social Inequalities** MENA is a region of extreme contrasts of wealth and poverty. For example, in 2007 the UN Human Development Report listed the United Arab Emirates as having the highest GDP per capita PPP in the region at $49,116 (5th in

the world), whereas Yemen had the lowest at $2,262 (131st in the world). Not surprising, however, national statistics like per capita income hide all sorts of variation—for instance, between the city and the rural areas and within the same area, where dramatic variation can occur between one urban neighborhood and the next. Most of the extreme wealth of the region comes from oil-based revenues to the states of Saudi Arabia, Kuwait, Iran, Iraq, Oman, Qatar, Libya, and the United Arab Emirates (**Table 4.1**). States with the largest populations tend to have the lowest per capita wealth. Despite the phenomenal wealth generated from oil production for some parts of the region, most of MENA remains poor and highly dependent on an increasingly marginalized agricultural sector.

Neoliberalism has increased the levels of inequality in the region, and other forces limit the life chances and standards of living of the region's population (see Chapter 1). For example, refugees and many migrant workers face very difficult economic circumstances. Since the first refugees left Palestine in 1948, as many as 2 million Palestinians have been displaced to refugee camps in Lebanon, Syria, Jordan, the West Bank, and the Gaza Strip. Many persons are born and grow up in these camps, originally intended as temporary settlements, where basic provisions are poor. Other large refugee populations live in camps in Iran, which shelters over 900,000 Afghan refugees and nearly 600,000 Iraqis. In Sudan, over a half million displaced Sudanese people live outside their home territories, and 1.25 million are internally displaced. Turkey and Iran are host to approximately 1 million displaced Kurds. And in Iraq as of 2007, nearly 5 million people have been displaced or are international refugees because of the war and its aftermath.

Sudan, Turkey, Iran, and Iraq illustrate the recent emergence of a new refugee status category, that of **internally displaced persons** (IDPs), individuals who are uprooted within their own countries because of civil conflict or human rights violations, sometimes by their own governments. The plight of IDPs is actually worse than refugees because their own governments are either unable or unwilling to provide them with the protection or assistance they have a right to expect. Inequality has also been a problem for migrant workers imported into oil states as guest workers because although they receive much higher wages than they would in their home countries, they have substantially lower standards of living than the resident Arab populations.

**The Regional Impact of the Global Economic Crisis** The global financial and economic crisis that unfolded in the fall of 2008 is having dramatic impacts on MENA. The region has been hit differentially but substantially from the once-overheated banking sector to oil and natural gas production to migrant workers who travel to the Middle East to take up jobs that local people are either unqualified or unwilling to accept. According to the World Bank, oil-producing states in the region are expected to lose $300 billion in revenue as crude prices drop and production is curtailed. As a result, nearly all the states have cut spending on public projects such as infrastructure, health care, education, and housing at the same time that many of them have used public monies to shore up the regional banking structure.

In Dubai, which we discussed in the chapter opener, half of the new gargantuan construction projects have been halted or cancelled, leaving a landscape littered with half-built towers on the outskirts of the city stretching out into the desert. Newspapers report that in their rush to

TABLE 4.1   Gross Domestic Product and Inflation Annual Percent Change (2010)

| Country | Gross Domestic Product, Current Prices (Billions $US) | Inflation, Average Consumer Prices, Annual Percent Change |
| --- | --- | --- |
| Algeria | 145.72 | 3.40 |
| Bahrain | 19.91 | 2.50 |
| Egypt | 203.85 | 8.63 |
| Iran, Islamic Republic of | 374.17 | 15.00 |
| Iraq | 80.52 | 8.00 |
| Israel | 204.56 | 0.80 |
| Jordan | 23.33 | 3.58 |
| Kuwait | 121.20 | 4.84 |
| Lebanon | 33.21 | 2.08 |
| Libya | 74.04 | 4.50 |
| Morocco | 90.61 | 2.80 |
| Oman | 49.02 | 6.00 |
| Qatar | 133.30 | 8.43 |
| Saudi Arabia | 423.84 | 4.48 |
| Sudan | 60.66 | 8.00 |
| Syrian Arab Republic | 58.47 | 6.00 |
| Tunisia | 39.27 | 3.40 |
| Turkey | 535.32 | 6.84 |
| United Arab Emirates | 235.52 | 3.12 |
| Yemen, Republic of | 34.98 | 13.30 |

*Source:* International Monetary Fund, World Economic Outlook Database (Accessed August 2009, from http://www.imf.org/external/pubs/ft/weo/2009/01/weodata/index.aspx

leave, many of the once-wealthy foreigners who had either invested in Dubai or were employed in its overheated economy left the emirate in late 2008, simply abandoning their cars at the airport.

## Demography, Migration, and Settlement

For thousands of years in MENA, populations have moved around within the region, at times as refugees, at other times voluntarily. They have also moved out of the region, settling all over the world and shaping the landscapes of cities and regions everywhere. Generally speaking, migration into and out of the region since the end of the colonial period has largely been related to several factors that have pulled immigrants to the region and forced many others to leave. Internal regional and national migration has also been significant and is almost always related to the draw of urban economic opportunity as rural areas experience population increase or economic decline.

The distinctive pattern of population distribution in MENA reflects the influences of environment, history, and culture (**Figure 4.23**). Environment is clearly a key factor in that populations concentrate near rivers, streams, and oases or in areas of dependable precipitation. As a result, the population is heavily concentrated in coastal areas; the floodplains of the Tigris, Euphrates, and Nile rivers and smaller streams; and highland settings such as the Atlas Mountains. Other population clustering occurs around the region's cities, which have been well established for centuries but have grown especially rapidly since the independence period of the 1950s. Even though MENA is more urbanized than is popularly assumed, many of the people of the region still live in rural villages (**Figure 4.24**).

The total population of the 21 countries that make up MENA is well over 450 million, but given the inaccuracy, infrequency, and inconsistency of national censuses in the region, this number is only an approximation. Although most of the region's states have recently conducted

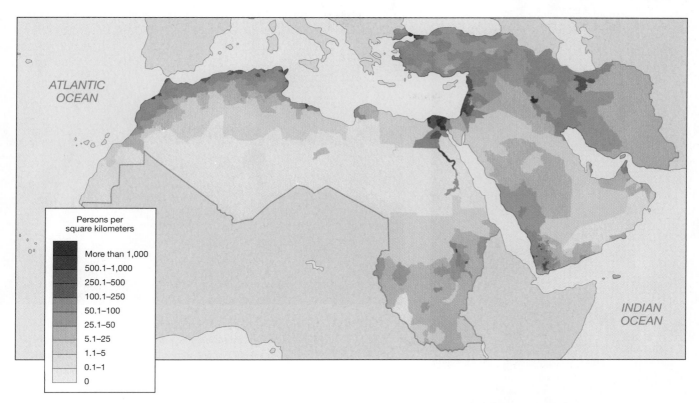

▲ **FIGURE 4.23 Population Distribution in MENA, 2005** Population distribution is heavily influenced by the availability of water. The intensity and unusual linear pattern of population concentration along the Nile River valley is a perfect illustration of this point. Other concentrations, such as along the eastern end of the Mediterranean, the northern edge of Algeria, and the northern parts of Iran, Iraq, and Turkey, also reflect higher availability of fresh water.

censuses that are considered by population experts to be accurate and reliable, others, such as Turkey, tend to underestimate their minority populations (the Kurds in particular) for political reasons.

Recent UN data indicates that the decades-long population boom in the region appears to be coming to an end. And although this means that fertility rates have fallen and are now stabilizing, challenges must still be faced in terms of providing for the health, education, and welfare of both children and the elderly as well as providing jobs for those in between. For instance, several of the most populated countries in the region, including Egypt, Algeria, and Turkey, a large part of the population is younger than 15 years of age, meaning that the populations of these countries will continue to grow as these individuals reach their reproductive ages.

**Pull Factors**  The strongest force drawing migrants to MENA in the last 50 years has been the oil economy. Oil is not the only attractive force, however; other factors have also attracted migrants. For instance, the founding of the state of Israel at midcentury drew large numbers of European Jews to the region, particularly during and after World War II. More recently, Ethiopian and Russian Jews have also migrated, in the former case because of civil war and in the latter because of the end of the Cold War. Non-oil-related economic growth has fostered migration to places like Beirut in Lebanon, Cairo in Egypt, and Istanbul in Turkey, all of which have increased their importance as core cities in the regional economy following the spread of political independence throughout the region.

Still, the most prominent attractive force has been the job opportunities made possible through the wealth generated by the continued development of the petroleum industry. In the states of the Arabian Peninsula, several factors—small populations, lack of skill, and a lack of interest in or possibly cultural resistance to the kinds of jobs made available through the oil economy—have meant that workers had to be imported. Up until the recent economic collapse, nearly three-quarters of the labor force in the oil states were **guest workers**, brought in to work in all aspects of oil production, from exploration and well development to drilling, refining, and shipping. And because oil revenues have increasingly been reinvested in economic development projects in the region, even more jobs came to be created outside the petroleum industry, ranging from service-sector positions to jobs in the building and construction industry (**Figure 4.25**).

To lessen the potentially dislocating impact of foreign workers on local social and cultural systems, immigration policy among the oil-producing states of the Arabian Peninsula has favored Muslim applicants. Within the region, large numbers of guest workers from Syria and Egypt, as well as Palestinian refugees, have been participants in the Arabian Peninsula oil economy, filling both skilled and unskilled positions (**Figure 4.26**). Overall guest workers constitute a rich global mix of nationalities. In the last decade or so, significant numbers have come from outside the region, especially from India, Indonesia, the Philippines, and Pakistan. Most of the workers who have come from other Middle Eastern and North African countries have been male; labor migrants from Indonesia and the Philippines have been female (see also

▲ **FIGURE 4.24 Dar-al-Karn, Yemen** Ancient villages like this one in south-western Yemen are still widely populated in the MENA region, though increasingly young people are migrated to larger settlements for work and education.

▲ **FIGURE 4.25 Labor Migrants** Many immigrant workers live in the oil-rich countries of MENA. In some countries, these workers make up 80% to 90% of the workforce, largely because there are so few local people to fill the jobs. These foreign construction workers are at the site of an apartment complex of 40 skyscrapers on the outskirts of Dubai, UAE.

Chapter 10). Moreover, many sub-Saharan Africans migrate from their countries of origin using North Africa as a point of transit into Europe, where several tens of thousands cross the Mediterranean linking the three regions in complex ways. The migrants are often relatively well educated and from moderate socioeconomic backgrounds. They move because of a general lack of opportunities, fear of persecution and violence, or a combination of both. Those who aren't able to cross join growing immigrant communities in North Africa in Algeria, Libya, Egypt, Tunisia, and Morocco.

**Push Factors**   The most consistent forces pushing migrants out of MENA have been war, civil unrest, and the lack of economic opportunity. Often the latter two have either been fostered or exacerbated by the imposition of the core's political system of territorially bounded nation–states on populations previously organized around very different sociopolitical systems. The case of Lebanon is illustrative. At the beginning of the period of European imperialism in the region, Lebanon became a French mandate under the League of Nations. Instead of promoting national unity among the many ethnic and religious groups of the Greater Lebanese mandate, France created a political administrative system of divide-and-rule that promoted fragmentation and increased the probabilities of sectarian conflict.

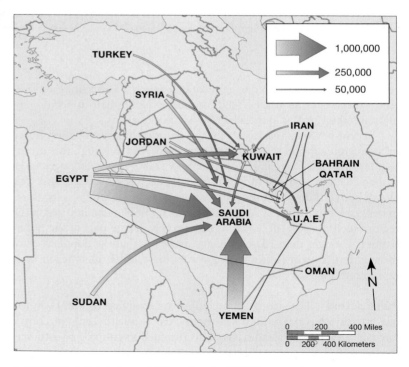

▲ **FIGURE 4.26 Internal Migration in MENA** Internal migration on a massive scale is a fairly recent phenomenon in the region. Until the decline of the Ottoman Empire, most residents of the region lived and died close to where they were born. European colonialism and imperialism in the early 20th century resulted in some migration. But the most significant impetus for mass internal migration really began with the expansion of oil production in the oil states in the 1950s.

Predictably, after independence Lebanon was beset by sectarian divisions, including rebellions, external attacks by Israel, and civil war between factions of Christian and Muslim militias and even within different Christian and Muslim groups. This civil instability compelled tens of thousands of Lebanese people—both Christians and Muslims—to flee the country to escape violence (**Figure 4.27**).

British and U.S. political involvement in Iran also caused many Iranians to flee the country. With the beginning of the Cold War between Western countries and the Soviet Union following World War II, Iran's oil reserves were considered to be of great strategic importance to Britain and the United States. Fearing that a nationalized industry would cut off Britain, which had controlled Iran's oil through the Anglo-Iranian Oil Company, Britain appealed to the United States to help oust Prime Minister Mossadegh from office. The virulent anti-Western fervor that has circulated in Iran since the coup that ousted Mossadegh in 1953, and the revolution in 1979 that eventually deposed Shah Mohammad Reza Pahlavi, are often viewed as the result of the interference of Britain and the United States in Iranian politics. Both events, but especially the fall of the shah, resulted in the exodus of hundreds of thousands of pro-

Western Iranians to the United States, Britain, and Europe, as well as to several Middle Eastern countries such as Egypt.

Other significant instances of migration created in the postcolonial Middle East and North Africa are the Algerians, Tunisians, and Moroccans, who have migrated mostly to Europe; Turks, who have followed a historical migration route to the Balkans and more recently to Germany; and Egyptians, who have migrated to other Arab countries in the region as well as elsewhere in the world. The migration of better-educated as well as low-skilled Turks and Egyptians has both pull and push dimensions to it. The push factor has occurred as educated and skilled workers have had to leave because economic growth has not been able to keep up with population growth by providing high-paying jobs. The pull factors include European policies that enable temporary workers to come and take up low-paying, low-skill jobs that are not economically attractive to Europeans.

The most dramatic instance of massive emigration in MENA is that of the Palestinians, which began to occur when Israel became a state in 1948. Today Palestinians form the most widely scattered diasporic population in the world, with millions living outside the region,

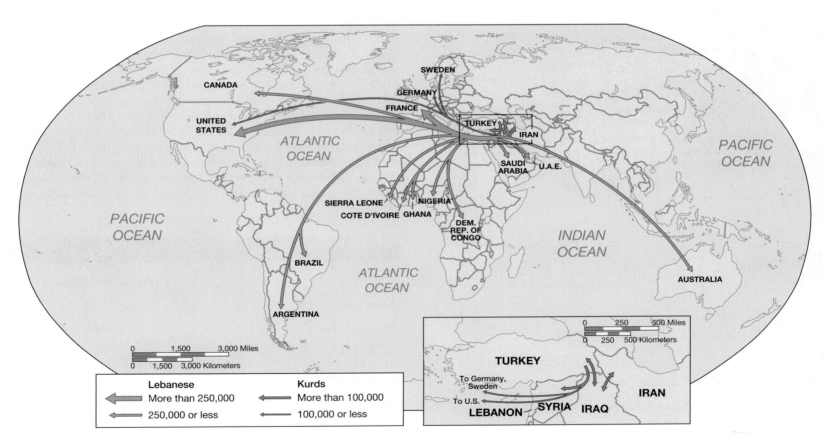

▲ **FIGURE 4.27 Kurdish and Lebanese Diaspora** The most significant diasporic populations of MENA during modern times are the Palestinians, the Lebanese, and the Kurdish people. (Of course, Jews are also a diasporic population, but recently they have returned to—rather than left—the Middle East in very large numbers.) This map shows the scattering of the Lebanese and the Kurds. (The dispersal of the Palestinians within the region is shown in Figure 4.28.) Over the last century, Lebanon has experienced various waves of diasporic migration largely because of war and the tensions within this multiethnic, multireligious society. Many Kurds moved to different parts of the region or left it altogether during the 20th century because of military aggression, persecution, and the repeated failure to establish a Kurdish state.

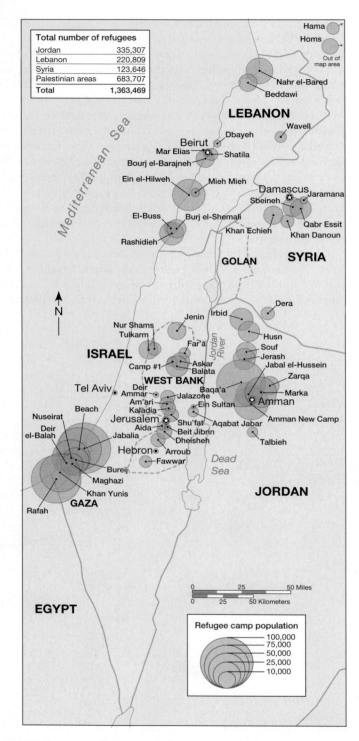

**Total number of refugees**

| | |
|---|---|
| Jordan | 335,307 |
| Lebanon | 220,809 |
| Syria | 123,646 |
| Palestinian areas | 683,707 |
| Total | 1,363,469 |

Refugee camp population

100,000
75,000
50,000
25,000
10,000

▲ **FIGURE 4.28 Palestinian Refugees in the Middle East** This map shows the dispersion of Palestinian refugees—in camps and elsewhere—in the states around Israel and in the West Bank, Gaza, and the Occupied Territories. One of the biggest obstacles in the Israeli–Palestinian peace talks has been the question of refugee return and where Palestinians will be allowed to settle.

whereas others live in various parts of the region, often in refugee camps (**Figure 4.28**).

**Cities and Human Settlement** The predominant pattern of settlement in MENA is a relatively small number of very large cities, a substantial number of medium-sized cities, and a very great number of small rural settlements. Variation on this broad generalization among countries in the region is dramatic. Israel is mostly an urban country, with 90% of its population living in cities. Sudan is largely a rural country, with only about 30% of the population living in cities. Cities in the region are growing dramatically each year, as more and more migrants come to live in them. Only about 50 years ago, most people in the region lived in small scattered rural settlements, but since then political independence and the development of the oil economy have been underlying factors in the increasing urbanization of the population.

MENA has a long and distinguished urban history. Beginning with the period of early empires, cities here have been centers of religious authority, have played pivotal roles in trade networks that extended into Europe and Asia, and have reflected the complex culture that created them. In the early 21st century, cities in the region continue to play central administrative roles—though today their political significance is as strong, if not stronger, than their religious significance. Even though trade continues to be crucial to city building—especially oil-related trade—other economic sectors stimulate urbanization, including processing and manufacturing and services. What has been most remarkable about contemporary urbanization in the region is that its rapid pace has led to the emergence of one or two very large cities in each country that contain a large proportion of the country's population and disproportionately wield political and economic influence. The brisk pace and extreme degree of urbanization in the region can be traced to the migration of rural people in search of economic opportunity in the cities as well as to natural increase among resident urban dwellers. **Table 4.2** lists the major cities of the Middle Eastern and North African region. Cities like Cairo, with nearly 12 million inhabitants, Istanbul, with

**TABLE 4.2   Ten Largest Cities in MENA, 2007**

| City | Population (Millions) |
|---|---|
| Cairo, Egypt | 11.9 |
| Istanbul, Turkey | 10.0 |
| Tehran, Iran | 7.9 |
| Baghdad, Iraq | 5.1 |
| Khartoum, Sudan | 4.8 |
| Riyadh, Saudi Arabia | 4.5 |
| Alexandria, Egypt | 4.2 |
| Ankara, Turkey | 3.7 |
| Algiers, Algeria | 3.4 |
| Casablanca, Morocco | 3.2 |

*Source:* UNPD, World Urbanization Prospects: The 2007 Revision Population Database, http://esa.un.org/unup/ (accessed May 23, 2009).

10 million, and Tehran, with nearly 8 million, are among the world's most populous cities. All are critical nodes in the urban system of the world economy, enabling capital, ideas, and people to come together.

One of the most widespread problems of rapid urbanization in the periphery is the inability of governments to meet the service and housing needs of growing urban populations. As a result, there are inade-

▼ **FIGURE 4.29 Squatter Settlements in Istanbul, Turkey** In Istanbul, squatter settlements are knows as *gecekondu*, a Turkish word meaning that the settlements were built after dusk and before dawn. Geographer Paul Kaldjian's research in Turkey has shown that many residents of these settlements actually own land in the countryside, where they grow some of the food that sustains their lives in the city. With no employment opportunities there, these individuals have been forced to migrate to the city in search of work. Relatives of the urban migrants often stay behind and maintain the family gardens during the growing season. At harvest time, the urban residents return to the countryside to help gather and divide up the crops.

quate and often poor-quality water supplies, electricity, sewer systems, clinics, and schools as well as air pollution and severe traffic congestion. Most critically, governments seem unable to provide housing for all who need it, and squatter settlements have been assembled on unclaimed or unoccupied urban land (**Figure 4.29**). Unfortunately, the very largest cities in the region continue to attract even more migrants, who see the most well-known places as possessing the best opportunities for a better life.

On the other hand, rapid urban growth—directly and indirectly related to the growing oil industry—in a few of the very wealthy oil-producing countries such as Jubail (Saudi Arabia) and Doha (Qatar) has resulted in impressively modern cities with few of the urban problems of their neighbors. Their enormous wealth coupled with their very low populations has made the growth of some of their urban places relatively uncomplicated. Many other cities of the oil-producing region, including Jeddah (Saudi Arabia) and Basra (Iraq), however, have not escaped the erection of shantytowns and the difficult social problems that accompany this type of urban change.

The highly diverse peoples of MENA occupy an ancient region with a complex history and environment. As mentioned earlier, water is a critical variable in shaping where people live. Religion is a central force in shaping social interactions; kinship, tribe, and gender also play roles. Many important cities and several very large ones connect the region to the world-system. Just under half the population still lives in rural settlements, but even with this large proportion of rural dwellers, the Middle East and North Africa is more urbanized than some other world regions, including South Asia and Sub-Saharan Africa. With a large population that derives its livelihood mostly from agriculture, the region, generally speaking, is most heavily involved in international trade around primary products such as minerals and agricultural goods (especially cereals and grains, cotton, and fruits and nuts). The most important of these products is oil, as we shall see in the following section. MENA possesses and trades more oil on the world market than any other region of the world.

## CULTURE AND POLITICS

In the Middle East and North Africa, culture and politics are often seen as combined in deeply passionate ways, particularly around religion and its connection to daily life as well as to formal political structures. In many of the countries of the region, religion and politics are conjoined in states that are constitutionally premised on Shari'a (or Islamic law) as the ideological foundation of their political institutions. These include Saudi Arabia, Iran, Oman, and Yemen. Further, all the other countries of the region except for Syria, Turkey, and Sudan (which are either undeclared or secular states) identify Islam as their state religion. It would be a mistake, however, to assume that the predominance of Islam in this region translates into homogenous culture and politics. For example, although Saudi Arabia is an Islamic state, it is also an absolute monarchy in contrast to Yemen, which is a presidential republic with a two-chambered legislature. Importantly, the culture and politics of the region are the result of ancient histories as well as more contemporary ones that link the region to other parts of the globe through past and present trade routes and patterns (from the ancient Silk Road to the container port facilities in the Persian Gulf), the Muslim Crusades (they preceded the Christians' ones by several centuries), European colonialism, and other sources of external contact (from the invention of writing to the domestication of wheat). The label—the "Middle East"—reflects Europe's more recent connection to the region through its colonial ambitions and a recognition of the region's strategic significance at the intersection of three continents.

## Tradition, Religion, and Language

MENA is infused with especially active religious practices and belief systems. Unlike many parts of the globe, where societies have become increasingly secular, religion in this complicated region is a central feature of everyday life for the vast majority of the inhabitants. Yet, the geography of religion is not uniform, and significant differences in religious practice are likely to occur from the regional level to the neighborhood level. For instance, although Iran is widely recognized as a state that is passionately committed to Islam, so much so that there is little separation between the state and the Islamic religion, in Tehran, the capital city of more than 7 million inhabitants, many upper- and middle-class households are likely to be secular and more aligned with Western values than with the teachings of Islam.

About 96% of the population of MENA is Islamic. The religion with the second-largest number of followers is Christianity (about 3%), with Judaism ranking third (less than 1%). There are also additional religions with much smaller numbers of adherents.

**Islam** Islam is an Arabic term that means "submission to God's will." A **Muslim** is a member of the community of believers whose duty is obedience and submission to the will of God. As a revealed religion (a religion founded primarily on the revelations of God to humankind), Islam recognizes the prophets of the Hebrew and Christian Testaments of the Bible, but Muhammad is considered the last prophet and God's messenger on Earth. The Qur'an, the principal holy book of the Muslims, is considered to be the word of God as revealed to Muhammad by the Angel Gabriel beginning in about 610 C.E.

There are two fundamental sources of Islamic doctrine and practice: the Qur'an and the Sunna (also spelled Sunnah). Muslims regard the Qur'an as directly spoken by God to Muhammad. The Sunna is not a written document but a set of practical guidelines for behavior. It is effectively the body of traditions derived from the words and actions of the prophet Muhammad. Islam holds that God has four fundamental functions—creation, sustenance, guidance, and judgment—and the purpose of people is to serve God by worshiping him alone and adhering to an ethical social order. The actions of the individual, moreover, should be to the ultimate benefit of humanity, not the immediate pleasures or ambitions of the self. A Muslim must fulfill five primary obligations, known as the five pillars of Islam: repeating the profession of the faith ("There is no god but God; Muhammad is the messenger of God."); praying five times a day facing Mecca; giving alms, or charitable giving; fasting from sunup until sundown during the holy month of Ramadan; and making at least one pilgrimage, or **hajj**, to Mecca if financially and physically able (**Figure 4.30**).

The emergence and spread of Islam is linked to the commercial history of MENA. The geographical origin of Islam is Mecca, in present-day Saudi Arabia. When Islam first emerged, Mecca, where Muhammad was born in 570 C.E., was a node in the trade routes that at first connected Yemen and Syria and eventually linked the region to Europe and all of Asia. Today Mecca is the most important sacred city in the Islamic world. It also continues to be a commercial center. Eventually, Medina also

▲ **FIGURE 4.30 Mecca, Saudi Arabia** Pictured here is an Indonesian pilgrim who has come to Mecca to fulfill her religious obligation. Every year, during the last month of the Islamic calendar, more than 1 million Muslims make a pilgrimage, or hajj, to Mecca. In addition to the required pilgrimage, Islamic traditions require Muslims around the world to face Mecca during their daily prayers.

became a sacred city because it was the place to which Muhammad fled when he was driven out of Mecca by angry merchants, who felt his religious beliefs were a threat to their commercial practices.

Disagreement over the line of succession from the prophet Muhammad occurred shortly after his death in 632 C.E. and resulted in the split of Islam into two main sects, the Sunni and the Shi'a. The central difference between them revolves around the question of who should hold the *political* leadership of the Islamic community and what the *religious* dimensions of the leadership should be. The Shi'a contend that political leadership must be divine and therefore must derive from descendants of the Prophet. The Sunni faction, which argued that the clergy (with no divine power) should succeed Muhammad, gained the upper hand and became dominant. In specific countries, however, the pattern varies. It is important to keep in mind that Islam is practiced differently in many different locales throughout MENA and that Muslims who have migrated out of the region—to Europe and the United States, for instance—are shaped by and shape the practice of Islam in MENA.

Perhaps one of the most widespread cultural counterforces to globalization has been the rise of **Islamism**, which is more popularly, although incorrectly, known as Islamic fundamentalism. Whereas *fundamentalism* is a general term that describes the desire to return to strict adherence to the fundamentals of a religious system, Islamism is an anti-Western, anti-imperial, and overall anticore political movement. In Muslim countries, Islamists resist the core, especially Western, forces of globalization—namely modernization and secularization. Not all Muslims are Islamists; Islamism is the most militant movement within Islam today.

The basic intent of Islamism is to create a model of society that protects the purity and centrality of Islamic precepts through the return to a universal Islamic state—a state that would be religiously and politically unified by including principles from the sacred law of Islam into state constitutions. Most Islamists object to secularization because they believe that the corrupting influences of the core place the rights of the individual over the common good. They view the popularity of Western ideas as a move away from religion to a more secular (nonreligious) society.

Another aspect of the Islamist movement is the concept of **jihad**, a complex term derived from the Arabic root meaning "to strive." Current use of the term connotes both an inward spiritual struggle to attain perfect faith as well as an outward material struggle to promote justice and the Islamic social system. *Qital* (fighting or warfare) is one form of jihad and, according to the Qur'an, means a war of conquest or conversion against all nonbelievers. When a war directed against the enemies of Islam occurs, it can be interpreted as a holy war. But jihad can also be a more peaceful struggle to establish Islam as a universal religion through the conversion of nonbelievers. One example of jihad today is the struggle of Shi'ite Muslims for social, political, and economic rights within Sunni-dominated Islamic states.

### Christianity, Judaism, and Other Middle Eastern and North African Religions
Although Islam is the most widely practiced religion in MENA, it is by no means the only religion of political, cultural, or social significance. There are more than a dozen Christian sects—among them Coptic Christians in Egypt, Maronites, the Chaldean Catholic Church, and various orthodox affiliations, including Armenian, Greek, Ethiopian, and even some Protestant faiths. Faiths not associated with any of the three world religions are largely concentrated in Iran and include Bahaism and Zoroastrianism. Generally, Jews in MENA are secular, observing some Jewish traditions. A small percentage are Orthodox or Hasidic, and an offshoot of Judaism, known as Samaritanism, also exists, the adherents of which are largely concentrated in the West Bank.

The three regionally predominant religions have helped shape the people and the landscape of the region. The most obvious and enduring landscape influences have been places of worship and sacred spaces more generally. Nowhere is the enduring interrelationship of the three religions more apparent than in the ancient city of Jerusalem. The centrality of Jerusalem as an ancient religious space, as well as its contemporary significance as a place of pilgrimage for Jews and Christians, is very much tied up with Arab–Israeli conflicts. Modern constructions of the state that link territory with nationality are ill equipped to address the religious significance of Jerusalem to Christians, Jews, and Muslims.

### Regional Languages
Three major language families constitute most of the foundation for the multiplicity of languages spoken in the Middle East. These include Semitic (including Arabic, Hebrew, and Aramaic); Indo-European (Kurdish, Persian, Armenian); and Turkic (Turkish, Azeri). This language diversity is the result of successive migrations of different people into the region. Interestingly, these different languages have also influenced each other. For example Persian is written in Arabic script, whereas Turkish incorporates vocabulary words from Persian and Arabic. And all three are spoken in regional dialectics that are not always mutually understood. In addition to the prominent languages, some ethnic and religious communities have preserved "native" languages for religious use, such as Coptic and Greek. Berber, an Afro-Asiatic language, is also spoken in the region from Morocco to Egypt. Nubian, a Nilo-Saharan language, is spoken by Egyptians and Sudanese. And, there are nearly 1 million Zazaki (an Indo-European language) speakers in southern Turkey. The most important foreign language in the Middle East is English. In North Africa, the colonial languages of French

(Morocco, Algeria, and Tunisia), Spanish (Western Sahara), and Italian (Libya) are also spoken, as is English.

## Cultural Practices, Social Differences, and Identity

The social difference, practice, and identity issues in MENA are as complex as that of any other region of the globe, with the social categories of gender, tribe, nationality, kinship, and family figuring prominently. Global media technologies such as satellite television and the Internet are increasingly penetrating the region, however, with the potential for new social and cultural forms to emerge and old ones to be reconfigured.

**Cultural Practice**   The impact of Middle Eastern and North African culture on the world has transformed the landscapes—in terms of buildings, tastes, sounds, and smells. For example, the cuisine of the eastern Mediterranean, especially that of Lebanon and Syria, as well as of Turkey, Egypt, Iran, Morocco, Tunisia, and Algeria, is available in many large cities in most of the world's regions. In the United States and Europe, it is often available in smaller urban places as well where many young MENA men and women have gone to study at universities and where cafés offering strongly brewed coffee and regional cuisine have sprung up to serve them.

The types of food available from this region include what are known as *meze dishes,* which are predominantly subtly spiced appetizers or small dishes. The range of mezes broadly reflects the tastes and ingredients of a particular country or subregion within that country. Some of the most popular meze dishes include *baba ganoush,* a puree of toasted eggplant, sesame seeds, and garlic; *falafel,* a mixture of spicy chickpeas rolled into balls and deep fried; *fuul,* brown broad beans seasoned with olive oil, lemon juice, and garlic; and *tabbouleh,* a salad of bulgur wheat, parsley, mint, tomato, and onion. The region also specializes in grilled meats, especially lamb and chicken, often served with rice.

A second cultural contribution of the MENA region is music and dance. The most widespread of the region's dances is the traditional belly dance, a women's erotic solo dance done for entertainment. The dance is characterized by undulating movements of the abdomen and hips and by graceful arm movements. Belly dancing is believed to have originated in medieval Islamic culture, though some theories link it to prehistoric religious fertility rites. Middle Eastern and North African music has also become a staple of the contemporary world music scene. One musical superstar is Cheb Khaled, an Algerian and multi-instrumentalist, who sings rai (traditional Algerian music that derives from Arabic poetry and Bedouin folk music) with influences from flamenco to Elvis. His following includes fans in Europe, Asia, and Latin America. Also from Algeria is Souad Massi, a Muslim woman who is a romantic pop vocalist influenced by Western rock, folk, and country music, as well as the chaabi and classical Andalusian music. The target of fundamentalist militias in her native Algeria, she was eventually able to travel to Paris, where she was catapulted into international stardom (**Figure 4.31**).

There are also hugely popular hip-hop artists throughout the region like Ceza in Turkey and Subliminal in Israel and the first Palestinian hip-hop group, DAM, whose music focuses on the Israeli/Palestinian conflict with their signature tune "Meen Irhabi" ("Who's the Terrorist"). From the diaspora, Canadian-Israeli rapper SHI 360 (Supreme Hebrew Intellect), the son of Jewish refugees from Morocco, raps in Hebrew, Arabic, English, and French. In the United States is TIMZ, an Iraqi-

▲ **FIGURE 4.31 Souad Massi** at a music award ceremony in Paris, where she was nominated for the best world music album of the year in 2006.

American rap artist whose award-winning Middle Eastern-tinged music and Tupac Shakur-like lyrics are aimed at shattering stereotypes about both Iraqis and Americans that have emerged since 9/11 and the war in Iraq. Inspired by U.S. hip-hop artists Mos Def and Talib Kweli, TIMZ's first music video premiered on YouTube in early 2007 with the politically controversial song "Iraq."

Perhaps the most substantial if more silent evidence of the globalization of Middle Eastern and North African culture is the appearance of mosques throughout the world. Mosques serve as the main place of worship for Muslims, but they also serve many social and political needs as forums for many public functions. Mosques also function as law courts, schools, and assembly halls. Adjoining chambers often house libraries, hospitals, or treasuries.

**Kinship, Family, and Social Order**   To understand Middle Eastern and North African society, it is important to understand ideas of kinship, family, and other personal relationships. **Kinship** is normally thought of as a relationship based on blood, marriage, or adoption. However, this definition needs to be expanded to include a shared notion of a relationship among members of a group. Not all kinship relations are understood by social groups to be exclusively based on biological or marriage ties. Although in MENA, biological ties, usually determined through the father, are important, they are not the only ties that link individuals and families. In fact, though kinship is often expressed as a "blood" tie among social groups throughout MENA, it is often the case that neighbors, friends, and even individuals with common economic or political interests are considered kin. Kinship is such a valued relationship for expressing solidarity and connection that it is often used to assert a feeling of group closeness and as a basis for identity even where no "natural" or "blood" ties are pre-

sent. It is also important in shaping the spatial relationships of gender determining who can interact with whom under what circumstances.

The idea of the tribe is central to understanding the sociopolitical organization of MENA. Moreover, the term *tribe* is a highly contested concept and one that should be treated carefully. For instance, it is often seen as a negative label applied by colonizers to suggest primitiveness in social organization. Where it is adopted in MENA, however, *tribe* is not seen as a primitive form of social organization but rather a valuable element in sustaining modern national identity.

Generally speaking, a **tribe** is a form of social identity created by groups who share a common set of ideas about collective loyalty and political action. Tribes are grounded in any combination or single expression of social, political, and cultural identities created by those who share them. The result of shared tribal identity is the formation of collective loyalties that result in a primary allegiance to the tribe. External groups may recognize the existence of these self-defined tribal groups and may seek to undermine or encourage their persistence.

One Middle Eastern and North African group that is frequently and proudly tribal is pastoralists. Pastoralism is a subsistence activity that involves breeding and herding animals to satisfy the human need for food, shelter, and clothing. Usually practiced in marginal areas where subsistence agriculture cannot be practiced, pastoralism can be either sedentary or nomadic. Sedentary pastoralists live in settlements and herd animals in nearby pastures, whereas nomadic pastoralists travel with their herds over long distances, never settling in any one place for very long (**Figure 4.32**). Most nomadic pastoralists practice

▲ **FIGURE 4.32 Berber Shepherd Campsite** The Berbers have lived in North Africa since ancient times. *Berber* is the name applied to the language and people belonging to many of the tribes who currently inhabit large sections of North Africa. Pictured here is a Berber shepherd campsite in the High Atlas Mountains, where great herds of sheep are tended. Increasing numbers of Berbers are raising crops, a practice that signals the erosion of their nomadic practices.

**transhumance**, the movement of herds according to seasonal rhythms. Flocks are kept in warmer, lowland areas in the winter and in cooler, highland areas in the summer.

**Gender**   Although gender differences play a part in shaping social life for men and women in MENA, as elsewhere around the globe, there is no single Islamic, Christian, or Jewish notion of gender that operates exclusively in the region. Many people have formed stereotypes about the restricted lives of Middle Eastern and North African women because of the operation of rigid Islamic traditions. These stereotypes do not capture the great variety in gender relations that exist in MENA across lines of class, religion, generation, level of education, and geography (urban versus rural origins, for instance). What pervades MENA, as well as many other societies throughout the world, is an ideological assumption that women should be subordinate to men. This view is held by both men and women in many Middle Eastern and North African societies. Interestingly, it seems that many men regard women's subordination as something natural, something that is effectively determined by biology. In contrast, most women in MENA tend to regard their subordination as something that is the product of the society in which they live and operate, and therefore something that can be negotiated and manipulated. The gender systems that operate in a wide variety of contexts in MENA are derived in large part from some of the same notions about men and women that inform gender systems in Western societies. Although this view is being increasingly critiqued and has begun to be dismantled in the West, it is still powerful there as well as in MENA. Control over women in MENA is frequently exercised by restricting their access to public space and secluding them within private spaces (**Figure 4.33**).

Sexuality and gender roles affect how men and women see themselves and represent themselves publicly in MENA. Some societies, such as in Yemen and Bahrain, exercise very strict control over women's public movements, and women are expected to cover themselves with veils and long, dark clothing when out on the streets. In some more generally secular of the region's societies, such as Turkey and Egypt, women's public movements are much less strictly regulated. The veil—from the all-encompassing full body garment, known as a **chador**, to a simpler head covering—has become the means by which women are able to effectively operate in public and yet remain in their personal space.

National policies and practices vary with respect to women's access to, and behavior in, public space (**Figure 4.34**). Subnational and local variations do exist, particularly differences in urban versus rural practices and even according to class differences within urban areas. Generally, urban women's public movements tend to be more restricted than those of rural women. This is largely because rural villages are usually composed of kin, and women can operate relatively freely among them, whereas urban women must move about in a world of both kin and strangers. In some cities middle- and upper-class women tend to have more constraints on their public social behaviors than poorer women. Again, the strictures placed on women's movements vary throughout the region and even within particular countries and subregions within countries.

The most important aspect of gender systems in MENA is that social reality is not fixed, and that cultural assumptions and practices around gender are subject to negotiation and change. Although the predominant gender theme in MENA is that women are subordinate to men, women can and do exercise significant household as well as political influence and independence across a range of societies in this region.

▲ **FIGURE 4.33 Gendered Architecture**  Islamic architecture reflects gender differences within the culture, and in different places these differences can be either strictly or more loosely observed. A classic aspect of Islamic architecture is the screen placed across windows in the women's parts of the houses and in the interiors of some public buildings. In 1799, the Hawa Mahal in Jaipur, India, was designed to allow the women of the court to watch the activities in the street from behind stone-carved screens without being seen. Although not part of MENA, common practices among Hindus and Muslims as well as the historically significant Muslim population in Jaipur is the reason for this architectural feature.

## Territory and Politics

During the 20th century, the countries of the Middle East and North Africa emerged from their colonial and dependent status with a range of economic and political problems. Many MENA experts believe that most of the political and economic problems of the region are a direct result of artificial political boundaries that united peoples who were

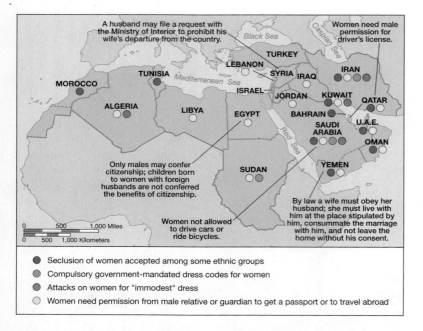

A husband may file a request with the Ministry of Interior to prohibit his wife's departure from the country.

Women need male permission for driver's license.

Only males may confer citizenship; children born to women with foreign husbands are not conferred the benefits of citizenship.

By law a wife must obey her husband; she must live with him at the place stipulated by him, consummate the marriage with him, and not leave the home without his consent.

Women not allowed to drive cars or ride bicycles.

● Seclusion of women accepted among some ethnic groups
◐ Compulsory government-mandated dress codes for women
◑ Attacks on women for "immodest" dress
○ Women need permission from male relative or guardian to get a passport or to travel abroad

▲ **FIGURE 4.34 Women's Mobility** Although Islamic law imposes restrictions on women's movements and dress in public space, there is much variation across the region with respect to adherence to these legal strictures.

previously antagonistic or divided peoples who were once unified. In fact, although the region has experienced wars and conflict for hundreds, if not thousands, of years, and certainly well before the Europeans arrived, it is generally agreed that most of its present conflicts stem directly from either of two things. The first involves the boundaries and borders created by the colonial powers; the second is the strategic importance of MENA to the political and economic interests of the core countries of the world economy.

Remarkably, at the same time that the region has been the site for bitter and, in some instances, seemingly irresolvable conflicts, it has also been the site for much broad and sustained cooperation. The most significant unifying forces have been the religion of Islam and the Arabic language that have helped the vast majority of people appreciate and nurture their common cultural heritage. Moreover, these unifying forces have helped many of the peoples of the region recognize that they have common political and economic goals.

The postcolonial political geography of this region is seen as one of the most serious challenges to stability and peace there. Extreme stereotypes suggest that the inhabitants of the region are naturally bellicose people. Although terrorist organizations do exist in the region (as they do in most regions of the world) and armed conflict has been a sad reality of life in many parts of the region, such characterizations are fundamentally false at the same time that they grossly simplify the region's political, economic, and social history. Both conservative and more radical observers of the region agree that the boundaries drawn by Britain, France, Spain, and Italy in MENA have been the single major source of contemporary conflicts in the region, many of which are decades old. Our treatment of these conflicts in this text is meant to expose their structural sources so that they might be better understood not as stemming from the personal characteristics of the people who inhabit the region but from the difficult political situations they have inherited.

**Tensions and Conflict: Iran, Iraq, and Kuwait**   The tensions that exist and the conflicts that have erupted between Iran and Iraq over the last 30-plus years are the result of several factors. One factor is the cultural differences between Persians (Iranians) and Arabs (Iraqis). Though the majority of both Persians and Arabs are Muslim, their ethnic origins, languages, geographies, and histories are distinctly different. Furthermore, the Persians were unceremoniously conquered by the Arabs in the 7th century and converted to Islam beginning in the 9th century. Since then, both countries have had a very long history of animosity that has more recently been complicated by their different dependent relationships with Britain and, later, the United States. The conflicts between Iran and Iraq remain unresolved today and are a source of continuing concern for the international community because both countries are important in global petroleum production.

Other points of tension have occurred around Iran's support of the Kurdish guerillas, who have been fighting the Iraqi and Turkish governments for autonomous control over their mountainous region. The Kurds are a non-Arabic people, mostly Sunni Muslims, who have been subjugated by their neighbors for most of their history. About 20 million Kurds live in the mountainous areas along the borders of Iran, Iraq, Syria, Turkey, and a small area in Armenia, called Kurdistan by Kurds (**Figure 4.35**). About 8 million Kurds live in southeastern Turkey. Throughout the 20th century and into the present one, the Kurds as an ethnic minority have faced repression and discrimination in the countries in which they live. Many Kurds have agitated peacefully as well as in armed rebellion for outright independence or autonomy. This has prompted recent Turkish incursions into northern Iraq to stop Kurdish groups from agitating for regional autonomy within Turkey. A large number of Kurds have left the region entirely and now live in Western Europe, comprising an important source of financial support for the resistance movement.

In 2008, Iran caught the alarmed attention of the West by declaring its inalienable right to pursue uranium enrichment. Despite ensuing sanctions, as of October 2009 Iran was still enriching uranium—although at a slower rate than previously thought. In 2008 the United States, Russia, China, the United Kingdom, France, and Germany put forward a renewed and enhanced "offer" to deter Iran from continuing its nuclear program, but President Ahmadinejad refused. This refusal has caused tensions to escalate with Israel, with both countries displaying their military capabilities, including Iran's advanced missile and satellite technologies. After decades of delays, Iran's first nuclear power plant at Bushehr is set to come online during 2011, though further delays are still possible.

More newsworthy than Iran's nuclear program is the massive protest that occurred in late spring 2009 in response to the "stolen" reelection of conservative President Mahmoud Ahmadinejad, where a staggering 39 million of the eligible 46.2 million voters (85%) went to the polls. Reformist candidate Mir-Hossein Moussavi and his supporters, protesting election fraud, took to the streets in Tehran as well as in several other cities in numbers that have not been seen since the 1979 revolution. Among the hundreds of thousands of protesters were many students and other young people, who were angry about the lack of freedoms and rights available to them and the mockery of democracy that the fraudulent election revealed.

Since the preliminary results for the Iranian election were announced, a steady stream of updates accumulated on Twitter, YouTube, Facebook, and other social networking sites. Although Ahmadinejad had e-mail services blocked during and after the election, social media

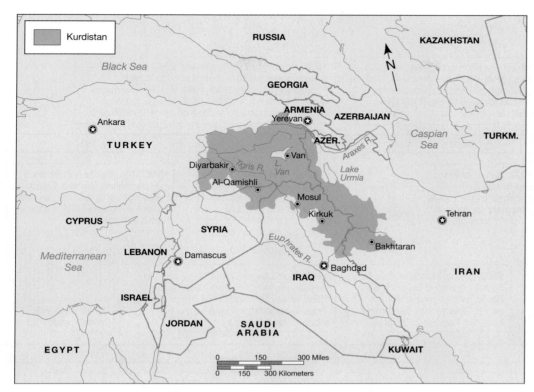

▲ **FIGURE 4.35 Kurdistan , "Land of the Kurds."** This map shows both the area where the Kurdish population is currently dominant as well as its traditional home. Although not an independent state, the map does provide a sense of the boundaries that (from the Kurdish perspective) it might ideally encompass.

like Twitter allowed Mousavi and the protestors to mobilize the marchers as well as to send out critical information to the foreign press.

**The Israeli–Palestinian and Israeli–Lebanon Conflicts** The history of the Israeli–Palestinian conflict is complex and the situation is highly volatile, despite persistent local and international efforts to bring peace to the region. As with the Iran/Iraq/Kuwait case, the chief factors that have inflamed this seemingly intractable political problem were exacerbated by British partitioning of the region.

The official Jewish state of Israel is a mid-20th-century construction that has its roots in the emergence of **Zionism**, a late 19th-century movement in Europe. Zionism's chief objective was the establishment of a legally recognized home in Palestine for the Jewish people. Thousands of European Jews, inspired by the early Zionist movement, began migrating to Palestine at the turn of the 19th century. When the Ottoman Empire was defeated in 1917, the British gained control over Palestine and the Transjordan area and issued the Balfour Declaration, mentioned earlier in this chapter. The Balfour Declaration was highly problematic, however, because a people, the Palestinians, already occupied the area. They viewed the arrival of increasing numbers of Jews and European sympathy for the establishment of a Jewish homeland·as an incursion into the sacred lands of Islam. In response to increasing Arab–Jewish tensions in the area, the British decided to limit Jewish immigration to Palestine in the late 1930s through the end of World War II. In 1947, with conflict continuing between the two groups, Britain announced that it despaired of ever resolving the problems and would withdraw

from Palestine in 1948, turning it over to the United Nations at that time. The United Nations, under heavy pressure from the United States, responded by voting to partition Palestine into Arab and Jewish states and designated Jerusalem as an international city, preventing either group from having exclusive control. Of the mandate of Palestine, the Jewish state was to have 56% and an Arab state was to have 43%; Jerusalem, a city sacred to Jews, Muslims, and Christians, was to be administered by the United Nations. The proposed UN plan was accepted by the Jews and angrily rejected by the Arabs, who argued that a mandate territory could not legally be taken from an indigenous population.

When Britain withdrew in 1948, war broke out. In an attempt to aid the militarily weaker Palestinians, combined forces from Egypt, Jordan, and Lebanon, as well as smaller units from Syria, Iraq, and Saudi Arabia, confronted the Israelis. Their goal was not only to prevent the Israeli forces from gaining control over additional Palestinian territory but also to wipe out the newly formed Jewish state altogether. This war, which came to be known as the first Arab–Israeli War, resulted in the defeat of the Arab forces in 1949, and later armistice agreements enabled Israel to expand beyond the UN plan by gaining the western sector of Jerusalem, including the Old City. In 1950 Israel declared Jerusalem its national capital, though very few countries have recognized this. **Figure 4.36** provides a timeline of key events, and Figure 4.37 depicts the changing political geography that has resulted from the conflict.

The territorial expansion of Israel has meant that hundreds of thousands of Palestinians have been driven from their homeland, and the landscape of Palestine has been dramatically transformed. Today Palestinians live as refugees either in other Arab countries in the region, abroad, or under Israeli occupation in the West Bank, the Golan Heights, and the Gaza Strip (also known as the "Occupied Territories"). By the late 1980s, Palestinians who had remained in their homeland had become so angered by Israeli territorial aggression that they rose up in rebellion. This rebellion, known as the **intifada** (the violent uprising of Palestinians against the rule of Israel in the Occupied Territories), has involved frequent clashes between fully armed Israeli soldiers and rock-throwing Palestinian young men. The intifada is mostly a reaction against 50 years of Israeli occupation of the Palestinian homeland and increasing Israeli settlement, particularly in the West Bank and the Gaza Strip.

In summer 2005, Israel began to cede territory back to Palestine. As Israeli settlers moved from homes they had inhabited, in most cases for decades, critics argued that the return of land was a hollow gesture as Israel continued the construction of physical barriers between Israelis and Palestinians. The Gaza Strip barrier consists of 52 kilometers (30 miles) of mainly wire fence with posts, sensors, and buffer zones. Israel argues that the barrier is essential to protect the security of its citizens from Palestinian terrorism. Palestinians and other opponents of the barrier contend that its purpose is geographical containment of the Palestinians to pave the way for an expansion of Israeli sovereignty and

## Israeli-Palestinian Conflict Timeline

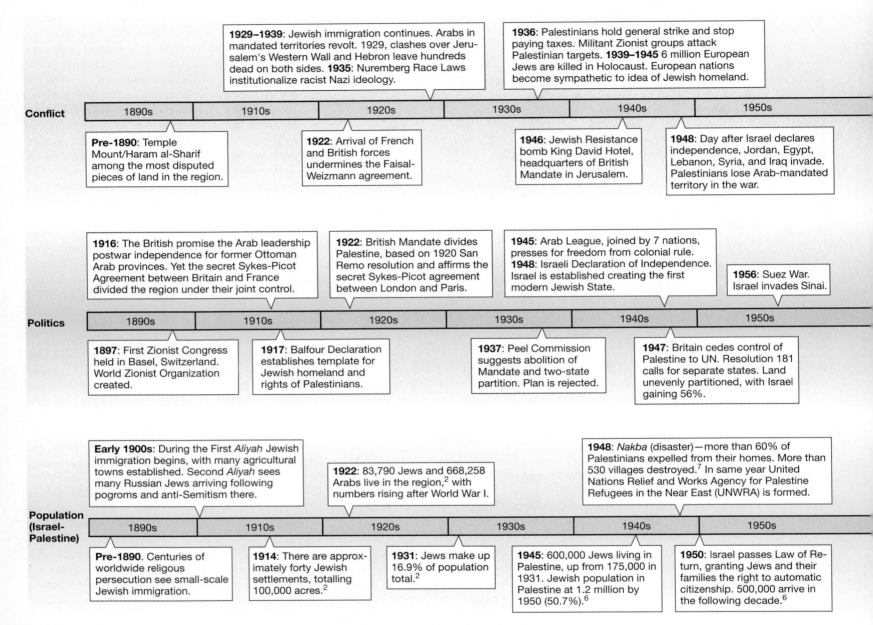

**▲ FIGURE 4.36 Timeline of Israeli–Palestinian Conflict and Change Since 1890** The history of modern Israel/Palestine is one of enduring conflict as Israel increasingly occupies Palestinian territory, and Palestinians continue to fight to keep their homelands. This timeline shows the different conflicts as well as the larger political and population issues around which these conflicts have been played.

to preclude any negotiated border agreements in the future. But Israel argues that the fence is purely a security obstacle, not a part of a future border. With the Gaza Strip security wall complete, in 2002 the West Bank wall was begun. When completed, that wall will seal off another portion of the Palestinian territories from Israel (**Figure 4.38**). These barriers continue to uproot and destroy Palestinian settlements and separate them from their livelihoods. In October 2003, the UN General Assembly voted 144–4 that the West Bank wall was "in contradiction to international law" and therefore illegal. Israel called the resolution a "farce."

With the election of U.S. President Barack Obama, there is renewed hope that through Senator George Mitchell—Special Envoy for Middle East Peace—a resolution to the conflict that involves statehood for Palestine can be achieved. During the first week of his 2009 appointment Mitchell visited Israel, the West Bank, Egypt, Jordan, Turkey, and Saudi Arabia, gathering opinions and information to launch a regional effort

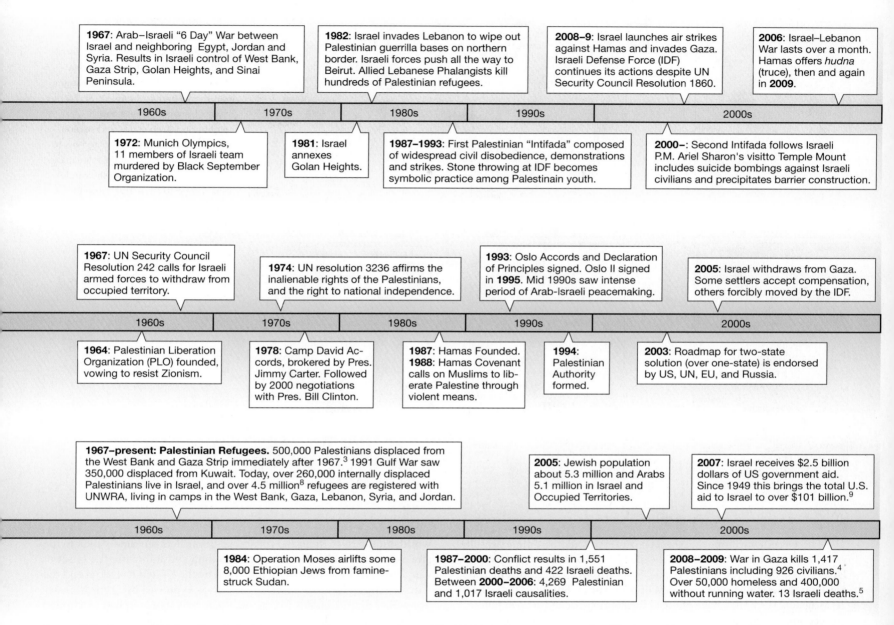

**1967**: Arab–Israeli "6 Day" War between Israel and neighboring Egypt, Jordan and Syria. Results in Israeli control of West Bank, Gaza Strip, Golan Heights, and Sinai Peninsula.

**1982**: Israel invades Lebanon to wipe out Palestinian guerrilla bases on northern border. Israeli forces push all the way to Beirut. Allied Lebanese Phalangists kill hundreds of Palestinian refugees.

**2008–9**: Israel launches air strikes against Hamas and invades Gaza. Israeli Defense Force (IDF) continues its actions despite UN Security Council Resolution 1860.

**2006**: Israel–Lebanon War lasts over a month. Hamas offers *hudna* (truce), then and again in **2009**.

1960s   1970s   1980s   1990s   2000s

**1972**: Munich Olympics, 11 members of Israeli team murdered by Black September Organization.

**1981**: Israel annexes Golan Heights.

**1987–1993**: First Palestinian "Intifada" composed of widespread civil disobedience, demonstrations and strikes. Stone throwing at IDF becomes symbolic practice among Palestinain youth.

**2000–**: Second Intifada follows Israeli P.M. Ariel Sharon's visitto Temple Mount includes suicide bombings against Israeli civilians and precipitates barrier construction.

**1967**: UN Security Council Resolution 242 calls for Israeli armed forces to withdraw from occupied territory.

**1974**: UN resolution 3236 affirms the inalienable rights of the Palestinians, and the right to national independence.

**1993**: Oslo Accords and Declaration of Principles signed. Oslo II signed in **1995**. Mid 1990s saw intense period of Arab-Israeli peacemaking.

**2005**: Israel withdraws from Gaza. Some settlers accept compensation, others forcibly moved by the IDF.

1960s   1970s   1980s   1990s   2000s

**1964**: Palestinian Liberation Organization (PLO) founded, vowing to resist Zionism.

**1978**: Camp David Accords, brokered by Pres. Jimmy Carter. Followed by 2000 negotiations with Pres. Bill Clinton.

**1987**: Hamas Founded. **1988**: Hamas Covenant calls on Muslims to liberate Palestine through violent means.

**1994**: Palestinian Authority formed.

**2003**: Roadmap for two-state solution (over one-state) is endorsed by US, UN, EU, and Russia.

**1967–present: Palestinian Refugees.** 500,000 Palestinians displaced from the West Bank and Gaza Strip immediately after 1967.[3] 1991 Gulf War saw 350,000 displaced from Kuwait. Today, over 260,000 internally displaced Palestinians live in Israel, and over 4.5 million[8] refugees are registered with UNWRA, living in camps in the West Bank, Gaza, Lebanon, Syria, and Jordan.

**2005**: Jewish population about 5.3 million and Arabs 5.1 million in Israel and Occupied Territories.

**2007**: Israel receives $2.5 billion dollars of US government aid. Since 1949 this brings the total U.S. aid to Israel to over $101 billion.[9]

1960s   1970s   1980s   1990s   2000s

**1984**: Operation Moses airlifts some 8,000 Ethiopian Jews from famine-struck Sudan.

**1987–2000**: Conflict results in 1,551 Palestinian deaths and 422 Israeli deaths. Between **2000–2006**: 4,269 Palestinian and 1,017 Israeli causalities.

**2008–2009**: War in Gaza kills 1,417 Palestinians including 926 civilians.[4] Over 50,000 homeless and 400,000 without running water. 13 Israeli deaths.[5]

*Sources:* [1] ProCong.org 2008, http://israelipalestinian.procon.org/viewresource.asp?resourceID=000639. ProCon.org is a non-profit public charity with no government affiliation. It contains an amalgam of population and statistical data on deaths for both Israelis and Palestinians, including multiple sources (e.g., U.N., Israeli Ministry of Foreign Affairs and so on). [2] ProCon.org, http://israelipalestinian.procon.org/viewresource.asp?resourceID=00063. [3] BBC News, http://news.bbc.co.uk/1/shared/sp1/hi/middle_east/03v3_ip_timeline/html/19678.stm. [4] Palestinian Centre for Human Rights 2009, http://www.pchrgaza.org/files/PressR/English/2008/36-2009.html. [5] BBC News 2009, http://news.bbc.co.uk/1/hi/world/middle_east/7838618.stm. [6] Council on Foreign Relations 2009, http://www.cfr.org/publication/15268/. [7] The Electronic Intifada 2007, http://electronicintifada.net/bytopic/197.shtml. [8] UNWRA 2008, http://www.un.org/unrwa/publications/pdf/rr_countryandarea.pdf. [9] Congressional Research Service 2008, http://www.un.org/unrwa/publications/pdf/rr_countryandarea.pdf. All accessed June 29 2009.

for achieving a lasting peace that would affect not only Israel and Palestine but all the states of the region.

For example, the political stability of Lebanon is very much tied to the situation in Israel and Palestine, where Hezbollah (a Shi'a Islamist political and paramilitary organization) militias have stated as one of their main goals the destruction of the state of Israel, a position that is in part instigated by Israel's previous occupation of Lebanon. In late July 2006, in response to the kidnapping of two Israeli soldiers by Hezbollah militias, the Israeli army began sea and air strikes and ground incursions into southern Lebanon. The context for this most recent conflict is complex, but it can be traced in part to the end of the 1967 Arab–Israeli War, when Palestinians began to use Lebanon as a base from which to launch attacks on Israel. Over the rest of the 20th century, despite periods of relative calm, Lebanon—especially the south as well as the cities of Beirut and Biqa—has been a target for Israeli attacks as Hezbollah and Palestinian guerillas continue to make their bases there.

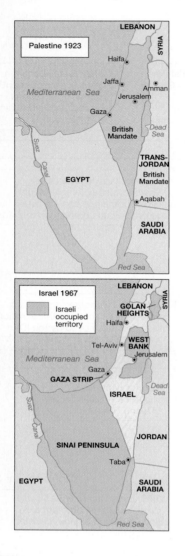

◀ **FIGURE 4.37 Changing Geography of Israel and Palestine, 1923–2009** Since the creation of Israel from part of the former Palestine in 1947, the regional political geography has undergone significant modifications. The changing map shown here is the result of a series of wars between Israelis and neighboring Arab states and several of political decisions regarding how to cope with both resident Palestinians and large numbers of Jewish people immigrating to Israel from around the world.

There was some hope that the 21st century would be a more peaceful and prosperous one for Lebanon. The bloody civil war had ended in 1990, Israeli troops withdrew in 2000, and Syria withdrew its troops in 2005, marking the end of almost 30 years of occupation. But Israel's attacks in 2006 against Hezbollah have plunged the country back into open political and economic turmoil as widespread civilian casualties, the massive destruction of key infrastructure and thousands of homes, and the displacement of approximately one million people have derailed the fragile peace.

**The U.S. War in Iraq** The United States responded to the terrorist attacks of September 11, 2001, by declaring a global war against terrorism and identifying the greatest threats to U.S. security as first Afghanistan and then Iraq. Although evidence of involvement in the 9/11 attacks by Iraq and its leader, Saddam Hussein, was highly questionable, on March 19, 2003, after amassing more than 200,000 U.S. troops in the Persian Gulf region, U.S. President George W. Bush ordered the bombing of the city of Baghdad. The declaration of war and invasion occurred without the explicit authorization of the UN Security Council, and some legal authorities take the view that the action violated the UN Charter. Some of the United States' staunchest allies (Germany, France, and Canada) as well as Russia opposed the attack. Moreover, throughout the world, hundreds of thousands of antiwar protesters repeatedly took to the streets for the weeks and months preceding and following the onset of war, launched by a coalition of forces led by the United Kingdom and the United States. The motivation for the war, as expressed by Prime Minister Tony Blair and President George W. Bush, was that Iraq had stockpiled "weapons of mass destruction"—chemical and biological weapons capable of massive human destruction. In the days leading up to the war, the UN weapons inspector, Hans Blix, and his team were unable to locate any weapons despite an intensive search of the country. President Bush, however, proceeded to justify a dramatically stepped-up "war on terrorism" (following the war in Afghanistan) on the grounds that "neutralizing" Iraq's leader, Saddam Hussein, was necessary to global security.

On May 1, 2003, after landing in a Lockheed S-3 Viking fighter plane on the aircraft carrier USS *Abraham Lincoln,* President Bush announced the end of major combat operations in the Iraq war. The fact that "major combat" ended, however, does not mean that peace has returned to Iraq, as violent conflict continues between U.S. and Iraqi soldiers and Iraqi insurgents as well as through sectarian violence between Sunni and Shi'a.

There have been over 1, 225,000 Iraqi deaths—both civilian and military—from all causes as of summer 2009, according to a well-

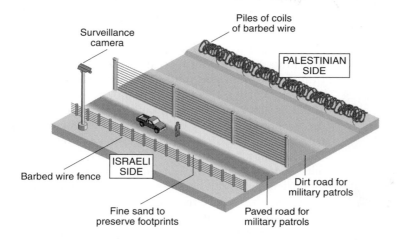

▲ **FIGURE 4.38 Israeli Security Fence, 2005** Shown are the planned and completed portions of the security fence, called "the wall" or the "apartheid wall" by Palestinians and other opponents. It is a physical barrier consisting of a network of fences, walls, and trenches. Israel's stated purpose in constructing it is to create a zone of security between itself and the West Bank.

paid for by the U.S. government and cooperating with U.S. troops that are leading to an overall decrease in sectarian violence. Unfortunately, sectarian violence is still occurring on a daily basis throughout the country, especially in Baghdad. Moreover, the civilian humanitarian situation is among the most critical in the world as millions of Iraqis have insufficient access to clean water, sanitation, and health care.

In spring 2009 President Obama set an August 2010 deadline for withdrawal of U.S. combat forces from Iraq, but the costs of the war are still mounting. U.S. intelligence agencies have suggested that the global jihadist movement has been fueled by the coalition forces' occupation of Iraq to carry out a cycle of revenge attacks known as "blowback" that could threaten the United States and other coalition countries for decades. The U.S. Treasury estimates the cost of the war at $845 billion directly. Nobel-prize-winning economist Joseph Stiglitz and coauthor Linda Bilmes argued that the true cost of the war is closer to $3 trillion when the indirect costs are also assessed, including interest on the debt raised to fund the war, the rising cost of oil, health care costs for returning veterans, and replacing the destroyed military hardware and degraded operational capacity caused by the war.

**Genocide in Darfur**  The conflict that began in early 2003 and continues to unfold in western Sudan has its roots in persistent social, political, and economic inequality between the country's core, centered around the Nile Valley, and the periphery represented by the western province of Darfur (an area about the size of Texas). With more than 450,000 dead and more than 2.5 million displaced either living in refugee camps or having fled the country altogether, the United Nations has called the events in Darfur a humanitarian crisis. Although the Bush Administration called it genocide, it took no protective action.

The brutal violence that has been occurring in Darfur began when native Africans in the province looking for a measure of freedom revolted against Sudan's authoritarian Islamic government. Media reports indicate that the government aimed to end the rebellion by wiping out all the tribal Africans in the area so that Arabs could take over the land. The government provided support for a militia of African Arabs, who call themselves the Janjaweed, to undertake one of the most brutal campaigns of ethnic cleansing that Africa has ever seen.

In May 2006 a peace agreement was brokered after intervention by the United States and Great Britain. Yet, as of summer 2009, reports indicate that the government-sponsored violence in Darfur continues, and the possibility for providing security for those who remain is extremely difficult. It also appears that the government continues to obstruct the delivery of humanitarian aid, creating famine conditions for the vulnerable population.

**The Western Sahara**  Western Sahara is a former Spanish colony that did not gain independence, as did most of the other former colonies or mandate territories of the region in the mid-20th century, but instead was handed over to Morocco by Spain. It was incorporated as a province by the Moroccan state, partially in 1976 and completely in 1987. The population—known as Saharawis—of this coastal area is just 200,000. The landscape is a windswept, flat, and monotonous desert where temperatures routinely reach 45°C (120°F) in the summer. Before the 20th century, Western Sahara was outside the control of any central political or military authority and was considered a geographical backwater. Except for some phosphate deposits and offshore fishing rights, Western Sahara would seem to be a generally resource-poor area and not obvi-

respected British polling agency (**Figure 4.39**). No reliable figures exist for the number of Iraqis injured since the 2003 onset of the war. United Nations High Commission for Refugees has indicated that a total of 4.7 million Iraqis have fled the country since the war broke out and have been displaced internally and to surrounding countries such as Syria and Jordan.

In 2007, President Bush ordered, over Congressional opposition, a surge of 140,000 troops to Iraq with the intention of providing security to the city of Baghdad and the Al Anbar Province. The surge has met increasingly with success, particularly in Al Anbar, where Sunni tribes are cooperating with the U.S. Army to fight al-Qaeda terrorism. And Sunni tribes around the country have formed "Awakening Councils"

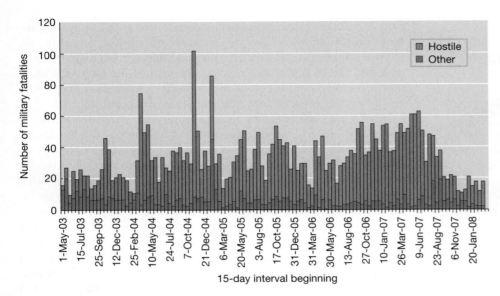

▲ **FIGURE 4.39 United States and United Kingdom Fatalities in Iraq, 2008** This graph shows the fatalities of U.S. and U.K. service people killed in Iraq as of January, 2008. The death toll for Iraqi civilians (not including military personnel) is a controversial figure. Iraq Body Count, a well-respected source on the war (http://www.iraqbodycount.org/database/), has estimated it at between 90,000 and over 100,000.

ously desirable as a territorial acquisition. Yet the Moroccans are aggressively determined to keep possession of Western Sahara, just as the Saharawis are determined to become independent (**Figure 4.40**).

**Regional Alliances**   MENA is a region with more than its share of conflict, yet it is also one where a great deal of cooperation, coordination, and joint action exists. Many political, economic, and cultural cooperative organizations are operating in the region.

The Arab League is a voluntary association of Arab states whose peoples speak mainly Arabic. Formally known as the League of Arab States, it is the most unifying of all the Middle Eastern and North African regional organizations. The stated purposes of the Arab League are to strengthen ties among member states, coordinate their policies, and promote their common interests. The league is involved in various economic, cultural, and social programs, including literacy campaigns and programs addressing labor issues. It is also a high-profile political organization that acts as a sounding board on conflicts in the region, such as the 1991 Persian Gulf War and the Arab–Israeli conflict.

The league was founded in 1945 by Egypt, Iraq, Lebanon, Saudi Arabia, Syria, Transjordan (Jordan, as of 1948), and Yemen. Other countries of MENA that later joined the Arab League are Algeria (1962), Bahrain (1971), Kuwait (1961), Morocco (1958), Oman (1971), Qatar (1971), Sudan (1956), Tunisia (1958), and the United Arab Emirates (1971). The PLO was admitted in 1976. In 1979 Egypt's membership was suspended after it signed a peace treaty with Israel, but it was readmitted 10 years later.

Another central and widely known regional organization, this one based on economic interests, is the Organization of the Petroleum Exporting Countries (OPEC). Whereas organizations like the Council of Arab Economic Unity consider every aspect of economic development and change, OPEC, as its name suggests, is a specialist economic organization. OPEC's central purpose is to coordinate the crude-oil poli-

cies of its member states (see also the Geographies of Indulgence, Desire and Addiction Box, p. 126). Founded in 1960, OPEC has 12 members—Algeria, Angola, Ecuador, Iran, Iraq, Kuwait, Libya, Nigeria, Qatar, Saudi Arabia, United Arab Emirates, and Venezuela—four of which (Angola, Ecuador, Nigeria, Venezuela) are not part of the Middle Eastern and North African region. As is clear from the list, however, Middle Eastern Arab states dominate the membership. OPEC originally was formed in response to the dropping price of oil in the 1950s, when supply greatly outstripped demand. In the 1970s, as oil supplies in non-OPEC countries were reduced, the organization lowered production, which had the effect of raising the price of oil. OPEC also sets production ceilings that specify how much oil may be produced by each member state. This practice ensures that the price per barrel does not fluctuate dramatically because of market gluts or scarcity.

Complementing as well as contrasting with the goals and objectives of OPEC is the Gulf Cooperation Council (GCC). The GCC coordinates political, economic, and cultural issues of concern to its six member states—Saudi Arabia, Kuwait, Bahrain, Qatar, the United Arab Emirates, and Oman. The members of the GCC have come together to coordinate the management of their substantial income from their oil reserves, problems of economic development, and social problems, trade, and security issues. All six of the states in the GCC are politically conservative monarchies wary of the revolutionary republican urges that have swept the region and transformed the previous monarchies of Egypt, Iran, and Iraq, for instance. The GCC has made very large sums of money available to all Arab countries for economic development as well as military protection during political crises.

Many other regional and international organizations have been established in MENA. New regional alliances are being proposed. The Arab Common Market is one such proposed alliance that has strong support in the region. Additionally, some individual states have begun to attempt to make connections with organizations beyond the region, tying MENA more securely to the rest of the globe. For instance, Turkey, already a member of NATO, has applied for full membership in the European Union (an organization of European states dedicated to increasing economic integration and cooperation in Europe—see Chapter 1, p. 36), despite significant political barriers. Morocco, Tunisia, Jordan, and Israel have also signed agreements, so-called Euro-Med agreements, with the European Union that are leading to increased transnational integration beyond the region.

# FUTURE GEOGRAPHIES

MENA came to be known as a coherent region following the 19th- and early 20th-century colonial impulses of Europe. Since the decline of de facto colonialism in the region in the twentieth century, two related concerns have continued to attract worldwide economic and political investment there: oil and regional conflict. A third significant concern is internal: water. With oil as the foundation of the current global economic system and MENA today holding 66% of the world's oil reserves,

▲ **FIGURE 4.40 Saharawi Land Mine Victims** Areas of Western Sahara were littered with land mines during the occupations by Morocco and Mauritania. Many of these mines are still in place, unexploded, a situation that takes its toll on the Saharawi people, who must live, work, or travel across these dangerous areas.

the region will continue to be an important economic factor in the global economy for many decades.

**Oil and Conflict** Critics of the war in Iraq have argued that oil, not terrorism, was the basis for the Coalition of the Willing's (a group of 22 countries led by the United States and the United Kingdom) invasion of Iraq. Whether this assertion is true or not, the fact remains that control over and widespread access to this resource, so critical to the functioning of the global economy, will continue to be a central feature of future geopolitical concerns. And because tensions among different cultural and religious groups in nearly all countries of the region, except perhaps those on the Arabian Peninsula, can complicate global access to oil, MENA will continue to draw the attention (and possibly the military forces) of the world's largest and most powerful states, especially the United States and Russia, as well as China and India. What is not so clear is how climate change and diminishing oil supplies will affect the region. One thing seems clear: OPEC's role as a self-regulating player in the global economy is likely to be challenged. For instance, climate scientists have proposed that OPEC establish a production quota system in response to global warming so that $CO_2$ emissions could be reduced at the initiation of the supply chain. So far, OPEC has resisted this type of external pressure on limiting production, but that does not mean this proposal might not become appealing in the future.

**Water Futures** The third key issue that will shape the future of the region is water. In a region where water is scarce, increasing resources are being invested in desalinization of seawater. Although seawater is abundant, the cost of removing the salt and minerals from it are exorbitant, requiring massive amounts of power. Once a region with a population that had adapted to the extremes of the environment, MENA is increasingly a modern, urban society that has sought to overcome environmental limitations through extreme exploitation of resources. Experiments like Madr City in Abu Dhabai indicate that there is a growing understanding of the need for more sustainable practices. Optimistically, these practices have the potential to reshape the region and provide innovations that can be transferred elsewhere in the world. Whether these sustainable practices will help to alleviate environmental and social problems such as inadequate urban infrastructure, air and water pollution, and unequal access to education and jobs, particularly for migrants, will be the biggest challenge the region faces over the next few decades.

A great deal of hope is being invested in the impact that U.S. President Obama's Middle East policies may have on the region. In late spring 2009, Obama visited Egypt, where he was very enthusiastically welcomed by the Egyptian people and government. Although the popularity of the U.S. president in any Arab country is a good sign, it is also clear from surveys conducted throughout the region that Arab people around the world are deeply distrustful of the United States. How long it might take to thaw that distrust is unclear. But what does seem encouraging is that the Obama administration seems rigorously committed to opening sustained dialogues with countries throughout the region that could certainly lead to more cooperation and productive relations.

# Main Points Revisited

■ Although largely an arid region, the physical geography of the MENA is diverse, including not only deserts but also huge mountains, grassy plains, forest ecosystems, wide beaches, and major river systems, as well as subtropical landscapes in southern Sudan.

In general, aridity makes the region particularly vulnerable to the effects of climate change, including increasing surface temperature, decreasing precipitation, and sea level rise and drought.

■ Human–environment interactions in the region have been largely predicated on adaptation, a practice that resulted in the first agricultural revolution: plant domestication.

The environment of the region has been overexploited through millennia of human occupation. Sustainable practices are increasingly being formulated to address this problem.

■ Known as the "cradle of civilization," MENA has been part of many early empires—including Roman and Byzantine—and has generated important innovations from the invention of writing to the founding of the three world-spanning religions: Judaism, Christianity, and Islam.

■ Religion continues to play a key role for people in the region, connecting them to Muslims and others around the world through a wide range of related cultural practices from language to art.

■ The mandate system, which was imposed by Europe following World War I and dismantled by the mid-20th century, left a legacy of problematic boundaries and conflicted groupings in MENA that continue to create significant political difficulties today.

The most prominent of these is the conflict between Israel and Palestine as it generates both global and regional political antipathies and alignments.

■ The most economically important natural resource in the region is oil. Wealth from oil exported across the globe has enabled many parts of the region to achieve modernization in a few short decades.

Second to oil and of increasing importance is water, particularly as the population of the region has grown dramatically, and water is a very scarce resource.

■ The biggest challenge for the region as it moves firmly into the 21st century may be the aftermath of contemporary political conflict in Iraq and Israel/Palestine, continuing political instability in Lebanon, and the nuclear ambitions of Iran.

The hope of many people throughout the world is that a new U.S. presidential administration may be successful in helping to move the region forward to peacefully resolve these conflicts.

# Key Terms

afforestation (p. 122)

aridity (p. 114)

Balfour Declaration (p. 126)

chador (p. 141)

desertification (p. 121)

guest worker (p. 133)

hajj (p. 138)

import substitution (p. 126)

informal economy (p. 127)

internally displaced person (p. 131)

intifada (p. 143)

Islam (p. 138)

Islamism (p. 138)

jihad (p. 139)

kinship (p. 140)

mandate (p. 125)

Muslim (p. 138)

nationalist movement (p. 125)

nationalization (p. 126)

oasis (p. 117)

petrodollar (p. 130)

transhumance (p. 141)

tribe (p. 140)

world religion (p. 123)

Zionism (p. 143)

# Thinking Geographically

1. Using the Jordan River as an example, how well do countries in this region manage and share their water resources?

2. How is kinship viewed in this region? How do these views affect local government and society?

3. How do views about gender affect the use of public and private space? Are rules concerning the veiling of women uniform throughout the region? If not, how do they vary geographically?

4. Guest workers migrate both into and out of this region. Why do so many guest workers in the oil states come from predominantly Muslim countries worldwide? Why do guest workers from Turkey and North Africa primarily seek work in Europe?

5. Describe the geographic distribution of Palestinians within the Middle East. How have colonial and postcolonial policies in the Middle East left Palestinians with no state of their own?

6. MENA is becoming highly urbanized. What factors are driving urban growth in this region? How are cities with considerable oil wealth handling their rapid urbanization?

7. Why is the Maghreb a geographically distinct area within this region? What sort of tensions exist between indigenous populations who are trying to maintain their traditions and European-influenced populations who are trying to establish their national independence? Why is the Western Sahara a flashpoint for these tensions?

Log in to www.mygeoscienceplace.com for Videos, Interactive Maps; RSS feeds; Further Readings; Suggestions for Films, Music and Popular Literature; and Self-Study Quizzes to enhance your study of Middle East and North Africa.

## ▼ FIGURE 5.1

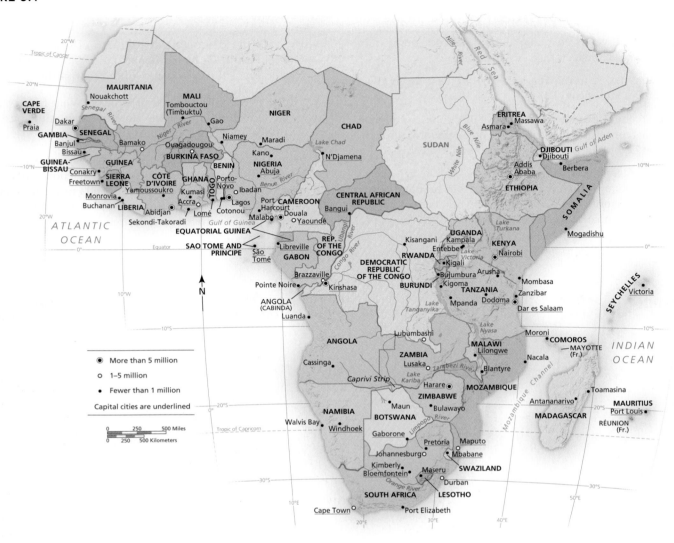

# SUB-SAHARAN AFRICA KEY FACTS

- Major subregions: West Africa, East Africa, Southern Africa, Equatorial Africa, North Africa

- Total land area: 22 million square kilometers/ 8.5 million square miles

- Major physiogeographic features: Sahara Desert, Cameroon Mountains, Fouta Diallon Highlands, the Victoria Falls, Lake Victoria, Congo River, and the African Rift Valley. Tropical climate is dominated by intertropical convergence zone, and the subtropical high leads to frequent drought in the region.

- Population (2010): 778 million, including 156 million in Nigeria, 83 million in Ethiopia, and 67 million in the Democratic Republic of Congo

- Urbanization (2010): 37%

- GDP PPP (2010) current international dollars: 3,526; highest = Seychelles, 20,072; lowest = Democratic Republic of Congo, 340

- Debt service (total, % of GDP; 2007): 5%

- Population in Poverty (%< $2/day; 1990–2005): 72%

- Gender-related development index value: 0.49; highest = Mauritius, 0.8; Cape Verde, 0.73; lowest = Sierra Leone, 0.32

- Average life expectancy/fertility: 50 (M49, F51)/5.4

- Population with access to Internet (2008): 5.6% (Africa)

- Major language families: Afro-Asiatic, Niger-Congo, Nilo-Saharan

- Major religions: Aside from Christianity and Islam, sub-Saharan Africa is dominated by a range of traditional religions, particularly in southern and eastern regions.

- $CO_2$ emission per capita, tonnes (2004): 1.1

- Access to improved drinking-water sources/sanitation (2006): 66%/51%

# 5
# SUB-SAHARAN AFRICA

In the West African country of Togo the successes of women selling wax print textiles in local and international markets became legendary. Some, such as Patience Sanvee, did so well that they were able to buy luxury cars such as Mercedes Benz, which they loaned to the penniless government of Togo to transport visiting heads of state. The popular nickname for these women was "Nana Benz."

The textile trade focuses on the coastal city of Lome, where hundreds of female merchants sell cloth printed with colorful designs and slogans, some based on popular TV programs and others on proverbs. Wearing fashionable prints signifies financial security for women, and the wax cloth often constitutes ceremonial dress for men. The women entrepreneurs of Togo are part of a West African tradition of powerful market and businesswomen, including those of Ghana, Benin, and Nigeria, who dominate trade in food and textiles and often import European goods. These women have considerable autonomy, traveling widely and easily obtaining loans and credit. Business acumen is passed from women to daughters and savings invested in education and real estate. International donors, using these women as examples of African potential in economic development, promote small loans and training targeted at women.

The textiles of West Africa—kente cloth from Ghana, batiklike wax print fabrics from Togo—have become an important part of an African identity, sold to African Americans and others as authentic indigenous clothing imbued with pride and meaning.

The export of these "African" textiles to North America and Europe has historical and contemporary patterns that reveal more complex origins. For example, the wax print textiles not only look like Indonesian batiks, but they also have their origins in colonial trading routes that brought ships of the Dutch East Indian Company to West African ports en route from Java to Europe. Sailors traded Asian batiks, and West African consumers attracted by the designs were then drawn into trade with European companies—such as Unilever and Vlisco—who produced batiklike wax cloth for sale to Africa with designs targeted to particular ethnic groups and distributed through networks of women traders.

In an ironic twist of global competition, West African traders are now being driven out of business by competition from cheap Chinese reproductions of wax prints. The Chinese bypassed the African businesswomen who felt the Chinese imports were of low quality and wished to maintain their links with their Dutch suppliers. Dede Rose Crepy, president of the association of printed fabric sellers, reports that membership has dropped from 1,000 to 22 women and revenue from more than $20 million in the 1980s to less than $2 million now. Another factor was the devaluation of the West African currency in 1994, when it lost half its value, increasing the cost of imported cloth from Europe and undercutting profits. The Chinese have copied the Dutch labels and designs; because there is no trademark protection, the women were unsuccessful at petitioning the government to block Chinese imports. This example of global links to African entrepreneurs illustrates the important role of women in African development, the cultural reach of African symbols, and the vulnerability of economies to shifts in international trade.■

## MAIN POINTS

■ Humans originated in the sub-Saharan African region, where a warm tropical environment is associated with vast deserts and diseases such as malaria, as well as some of the world's richest biodiversity and forests.

■ The legacies of European colonialism include trade patterns oriented to the export of primary products, unequal land distribution, and political instability associated with contested boundaries and Cold War alignments. Overseas countries and firms are especially interested in Africa's mineral and oil resources.

■ Sub-Saharan Africa has shaped other parts of the world through the tragic diaspora of slavery, but interactions continue with labor migration to Europe, refugees fleeing civil wars, and the diffusion of African culture and style. Population growth in Africa is high, but fertility is declining, especially as women's status improves. The region was especially affected by HIV/AIDS.

■ Africa has a large number of poor people and economies where many people are subsistence farmers and countries carried high burdens of debt. Although poverty is still widespread, programs of debt relief, democratic elections, new export opportunities, and service-sector expansion have brought some optimism regarding economic and political futures.

■ There is a great range of landscapes and livelihoods in the 48 countries in sub-Saharan Africa, and it is important to recognize differences in environment, wealth, culture, economy, and political dynamics.

# ENVIRONMENT AND SOCIETY

Africa is a large, complex, and often misunderstood continent. Perceptions range from a fertile tropical forest rife with exotic diseases to an idyllic game reserve, or from a harsh landscape devastated by war and drought to a place where rich cultural traditions reach back to the dawn of humanity. The continental landmass called Africa straddles the equator, stretching from the Mediterranean Sea in the north to the southern tip in South Africa at about 35 degrees south latitude and from Senegal on the Atlantic coast to Somalia on the Indian Ocean (**Figure 5.1**). The total area is about 30.4 million square kilometers (11.7 million square miles)—three times larger than the United States or Brazil. The physical geography has shaped ecosystems and the use of land and the distribution of key mineral and water resources and has, in turn, been transformed by human activities within and beyond the region. The satellite image (**Figure 5.2**) shows many of the key physical features of sub-Saharan Africa.

Sub-Saharan Africa has been defined and divided from North Africa based on historical, physical, and social characteristics that include a legacy of European colonialism and slavery, a mostly tropical climate, and the darker skin of many inhabitants. The race-based definition of sub-Saharan Africa is very controversial but has been used by both Africans and non-Africans to identify the region as "Black Africa."

In this text, we discuss North Africa with the Middle East because of shared characteristics (including similar physical environments of dry climates and human geographies that reflect a dominant Arabic language and ethnicity and Islamic religion). However, the physical and human links across the continent of Africa are such that several sections of this chapter, including the section on humans and the environment, discuss broader patterns across the whole continent and refer to Africa as a whole rather than to sub-Saharan Africa specifically.

The geographical, racial, ethnic, and religious bases for dividing Africa between two world regions is artificial, oversimplifying both the cultural and historical distinctiveness of the two regions, the overlaps between them, and the great variety they contain. For example, the Sahara Desert is a large area, rather than a clear dividing line, and includes territory from both North and sub-Saharan African countries. The Nile River links the North African countries of Egypt and Sudan through a long fertile corridor to the sub-Saharan countries of Ethiopia, Uganda, Kenya, Tanzania, Rwanda, Burundi, the Central African Republic, and the Democratic Republic of the Congo.

## Climate, Adaptation, and Global Change

Most of sub-Saharan Africa has a tropical climate with warm temperatures (higher than 20°C, 70°F) and little frost except in highland areas (**Figure 5.3**). The climate is dominated by two major features of the atmospheric circulation—the intertropical convergence zone (ITCZ) and the subtropical high (see Chapter 1, p. 4). The *ITCZ* is a low-pressure region where air flows together and rises vertically as a result of intense solar heating at the equator, producing heavy rainfall of more than 1,500 millimeters (60 inches) a year over the Congo basin. The subtropical high is a zone of descending air, resulting in dry, stable air that causes desert conditions over the Sahara and Kalahari. The places between, such as the West African coast and East Africa, experience seasonal rainfall as these global circulations shift northward in April

▲ **FIGURE 5.2 Africa from Space** This satellite image clearly shows the major landform regions of the Sahara Desert (in lighter color), the Congo rain forests (darker green), the highlands of Ethiopia, and the line of lakes along the East African Rift Valley, including Lake Victoria. The large island of Madagascar is also clearly visible.

through September and southward in October through March. Where seasonal rainfall is modest, semiarid conditions predominate, and slight variations in circulation can bring variable rainfall, frequent droughts, and great challenges to agriculture and water resources management. During December hot dry winds called the **harmattan** blow out of inland Africa, carrying large amounts of dust and creating stress for humans and animals. In July, the northward shift in circulation allows the southwestern trade winds to blow onto the coast of West Africa, bringing seasonal *monsoon* like rains to inland countries such as Mali.

This general pattern is modified by the regional effects of mountains, lakes, and ocean currents. The cold Benguela current creates cool, dry conditions along the coasts of Angola and Namibia and exacerbates the desert conditions of the Kalahari. Cold water and stable winds that blow along, rather than across, the coast of the Horn of Africa promote dry conditions in countries such as Somalia and Eritrea. High-altitude regions, such as the East African highlands, have higher rainfall and more moderate temperatures, which are favorable to agriculture and human settlement. South Africa is so far south that it is located in more temperate latitudes and experiences a mild Mediterranean climate with dry conditions from October to March and rainfall from the westerly wind belt from April to August.

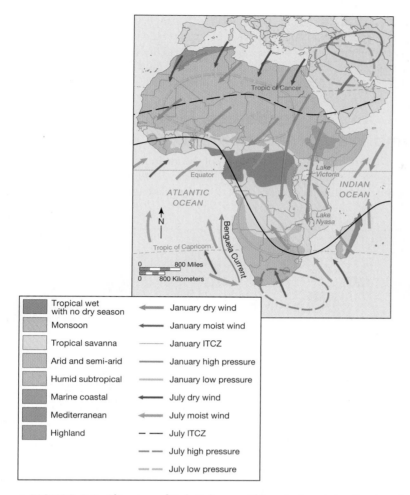

▲ **FIGURE 5.3  Climates of Sub-Saharan Africa** A climate classification of Africa based on the Koppen system shows several major climate zones spanning the equator, including tropical wet and monsoon regions with higher rainfall, arid and semiarid desert and savanna with low seasonal rains. In the temperate latitudes of southern Africa, the climate is milder with a Mediterranean climate on the west coast and marine on the east coast. Highland climates are cooler and often wetter because of altitude. Climate and vegetation zones overlap closely in Africa. They depend mostly on the amount and seasonality of precipitation. Large areas of Africa are covered with savanna vegetation and by deserts.

**Desertification and Famine**  The people of sub-Saharan Africa have made many adaptations to the climate of the continent, including floodplain farming along seasonally variable rivers, traditional irrigation and rainwater harvesting schemes, and herders moving their livestock to follow the rains. However, climate variability—combined with political change, the growth of human and animal populations, and changes in access and use of land—has meant that drought often threatens food security and can sometimes trigger famine. When combined with land degradation, persistent drought can lead to a process called **desertification** (**Figure 5.4a**). Desertification has a variety of meanings but is most generally viewed as the process by which arid and semiarid lands become degraded and less productive, leading to more desertlike conditions through drying, erosion, compaction,

buildup of salts, and loss of fertility and vegetation. There are also disagreements about the relative role of different factors in causing desertification and its permanence. The main culprits are seen as climate change, overgrazing, overcultivation (including of peanuts for exports), deforestation for wood fuel, and unskilled irrigation that results in the buildup of salt in the soil (salinization).

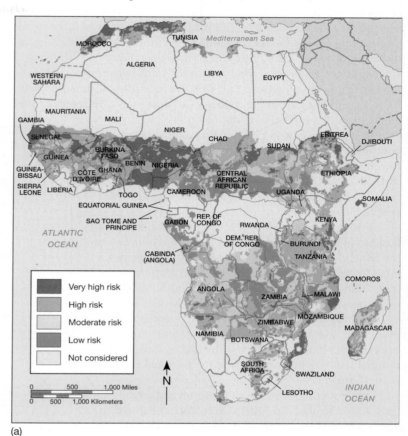

(a)

(b)

▲ **FIGURE 5.4  Desertification in Africa** (a) This map shows physical vulnerability to desertification where high vulnerability is associated with higher climate variability and soil limitations (shallow, poor quality). Vulnerability is very high along the southern borders of the Sahel and in several regions of eastern and southern Africa. (b) Overgrazing goats feeding on acacia in Sahelian desert landscape.

The Sahel region of West Africa borders the Sahara Desert and has highly variable rainfall and a human population dependent on pastoralism. A limited network of meteorological stations provides data that show that rainfall is quite variable across the Sahel, but that wetter conditions seem to have occurred in the 1950s and early 1960s. Rather than adjust their herds to average conditions, Sahelian pastoralists tend to be opportunistic, building up their herds in good years because their livestock are the best way of accumulating wealth and investing capital. Overgrazing can lead to soil erosion through the reduction of a vegetative cover as well as trampling and compaction (**Figure 5.4b**).

Beginning in 1968, it appears that the rains failed in most parts of the Sahel for up to 7 years. As herds began to die off, images of the drought and starving refugees began to appear in the international media, resulting in a relief effort and anguished debates among researchers and policymakers about what had gone wrong and what could be done to avoid future tragedy. Between 1968 and 1973, as many as 3.5 million cattle died, and 15 million farmers lost more than half their harvests. A quarter of a million people died from famine before food relief could reach them.

Some researchers blamed climate and the irrational buildup of herds in the face of regular drought cycles in the Sahel. However, others argued that the roots of the crisis lay in changes in Sahelian political economy stemming from colonial structures and continued overreliance on export cropping. Geographer Michael Watts, for example, showed how in northern Nigeria people were unable to use traditional drought coping strategies because of colonial policies and the loss of land and traditional self-help institutions, and the marginalization of poorer farmers.

Others argued that decades of peanut and cotton production, particularly in Senegal and Mali, exhausted the soil and left it unproductive. Climatologists suggested that deforestation and overgrazing increased the reflectivity of the land surface, which meant less warm air rising to form clouds, and that increased atmospheric dust was reducing the uplift of air. Both processes reduce rainfall. Unfortunately some of the international efforts to respond to the 1970s drought backfired. Drilling deep wells for cattle herds resulted in so many thirsty cattle gathered around the wells that all possible forage vegetation was consumed, and the herds starved. When food aid arrived in communities where some farmers still had crops to sell, food prices dropped, and farmers could not make a living.

**Vulnerability to Climate Change** The latest report of the Intergovernmental Panel on Climate Change (**IPCC**) identifies Africa as one of the most vulnerable regions to climate change and suggests that changes may already be observed with the disappearance of the snow on Mount Kilimanjaro and shifts in ecosystems in southern Africa. In terms of future climate, the IPCC noted considerable uncertainty in the rainfall projections of climate models but indicated that temperatures will warm almost everywhere, and it is likely that southern Africa will have more severe droughts. The concern is the high vulnerability of many regions of Africa to any change in climate because so many Africans are dependent on rain-fed agriculture, live-stock, and already scarce water supplies for their livelihoods and health (**Figure 5.5**). IPCC forecasts that crop yields could fall by as much as half, the population at risk of water problems could increase by 250 million by the year 2020, and the warming temperature could increase vector-borne diseases such as malaria, especially in highland areas.

## Geological Resources, Risk, and Water Management

The continent of Africa is the heart of the ancient supercontinent called Pangaea, the southern part of which broke off to form Gondwanaland about 200 million years ago. The theory of plate tectonics explains that

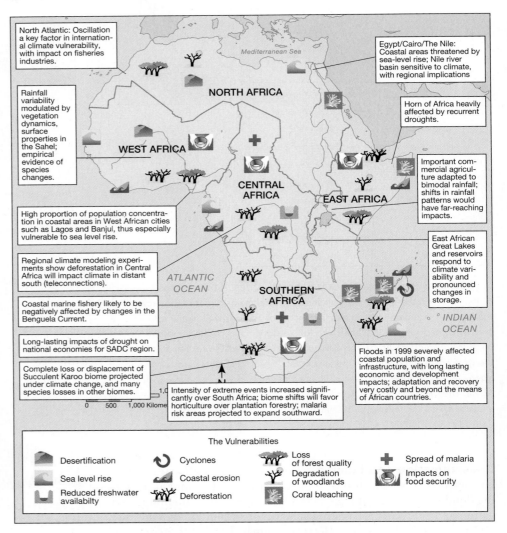

**▲ FIGURE 5.5 Climate Change Vulnerability in Africa** Multiple stresses make most of Africa highly vulnerable to environmental changes, and climate change is likely to increase this vulnerability. This graphic shows which of the regions of Africa (North Africa, West Africa, Central Africa, East Africa, Southern Africa, and the Western Indian Ocean Islands) are most vulnerable to specific impacts of climate change. These impacts include desertification, sea level rise, reduced freshwater availability, cyclones, coastal erosion, deforestation, loss of forest quality, woodland degradation, coral bleaching, the spread of malaria, and impacts on food security.

when the regions we now call Latin America and Asia broke away from Gondwanaland, the high plateau that remained became the continent of Africa. Where the continental plates tore away from Africa during the breakup of Gondwanaland, they left steep slopes (called *escarpments*) that fell from the high plateau to the new oceans. Geologic tensions created trenches and volcanic activity. Most of the rivers that had previously drained into the inland lakes of the supercontinent eventually found outlets to the sea.

**Geomorphology**    Africa is still mainly a plateau continent, with elevations ranging from about 300 meters (approximately 1,000 feet) in the west, tilting up to more than 1,500 meters (approximately 5,000 feet) in the eastern part of the continent (**Figure 5.6**). Some significant mountain ranges are found in western Africa, including the Cameroon Mountains and Fouta Diallon Highlands, with rivers flowing from the uplands. Steep slopes, especially on the western edge of the plateau, drop to narrow coastal plains.

The higher areas of the plateau, whose cooler temperatures and higher rainfall are hospitable to humans, include the High Veld of southern Africa, the highlands of Kenya and Ethiopia, and the Jos plateau of West Africa. Volcanic peaks such as Kilimanjaro (5,895 meters, 19,340 feet), Kenya/Kirinyaga (5,200 meters, 17,058 feet), and the Virungas (4,507 meters, 14,787 feet) rise from the eastern plateau, which is split by a deep trough, where tectonic processes continue to pull the eastern edge of Africa away from the rest of the continent (**Figure 5.7a**).

This trough, called the African Rift Valley, runs more than 9,600 kilometers (6,000 miles) from Jordan and the Red Sea in the north to Mozambique in the south and is called a **rift valley**; it ranges from 50 to 100 kilometers (30 to 60 miles) wide. The African Rift Valley has two major branches and is filled with deep elongated lakes, including Lake Tanganyika at 1,473 meters deep (4,832 feet). Lake Victoria, the third largest lake in the world, lies between the two branches of the rift valley. The age, size, and depth of these lakes make them diverse freshwater ecosystems with important fisheries (**Figure 5.7b**).

**Soils**    African soils tend to be of low fertility because of the great age of the underlying geology and because of high rainfall that leaches (washes out) nutrients from exposed soils. Soil fertility tends to be higher in regions of recent volcanic activity, such as the East African highlands, and in wider river valleys, where sediments settle and create alluvial (river) soils. The tropical soils of wetter zones, such as central Africa, lose their fertility rapidly once the forest is cleared and the soil is exposed to the elements. Between the dry and wet zones, such as between the coastal and Sahel regions of West Africa, soils have more organic material and support crops and pasture. Desert regions can have saline or alkaline soils that are toxic to crops. High iron and aluminum content are also poisonous to plants and crops in some regions.

**Minerals**    Half the continent is composed of very old crystalline rocks of volcanic origin that hold the key to Africa's mineral wealth. Ancient tropical swamps formed sedimentary rocks containing oil and other fossil fuels (Figure 5.6). These include coal in southern Africa and oil and gas in West Africa, particularly Nigeria and Gabon. Iron and manganese are found in western and southern Africa, and most of the world's known chromium is found in southern Africa, especially in Zimbabwe and South Africa. Vast copper reserves are located in the southern Congo and in the copper belt of Zambia, where cobalt is also found; bauxite, which is used in making aluminum, is found in a belt across West

▲ **FIGURE 5.6 Geological and hydrological resources of sub-Saharan Africa** Africa is a plateau continent surrounded by steep escarpments, with rivers that often flow through inland deltas on the plateau or drop over waterfalls at the edge of the escarpment. This creates opportunities for hydrological projects, including dams for irrigation and electricity generation. One of the most significant features is the East African Rift Valley, filled with elongated lakes and several active volcanoes. This map also shows the location of the most important regions of mineral development in Africa, including oil, gold, and diamonds. South Africa, Zambia, and Sierra Leone are particularly rich in minerals, and Nigeria is a major oil producer.

(a)

Africa, and uranium is found in Niger. These minerals are critical to industrial production elsewhere in the world.

Gold is found in several regions of Africa, including Ghana and Zimbabwe, and in South Africa, where as much as half the world's gold reserves lie in the region around Johannesburg. South Africa is also famous for diamonds, which are also found in Botswana and Namibia in southern Africa, at the edges of the Congo basin, and in Sierra Leone in West Africa. Although these resources bring billions of dollars into Africa, they also make national and regional economies vulnerable to fluctuations in world market prices, especially where minerals dominate exports. In Africa as a whole, exports of mining products in merchandise trade were valued at $95 billion in 2003, about 55% of total exports. In South Africa, the world's largest producer of platinum, gold, and chromium, 495,000 people were employed by the mining industry in 2007, contributing 18% to the country's $300 billion GDP.

Mineral resources have played important roles in African history. Salt was a key commodity in trans-Saharan trade from the 10th century to the present day. Gold was valued in West Africa from early times and was worn and traded by kings and leaders; Mansu Musa, the emperor of Mali, carried and traded so much gold on a pilgrimage to Mecca in 1324 that his actions depressed gold prices worldwide. Gold and diamonds spurred European colonial grabs for Africa and conflicts between the core colonial powers. They also created conflicts with

◀ **FIGURE 5.7 Physical landscapes** (a) Desert landscape in Namibia. Desert landscapes are found in southern Africa, where the descending air of the subtropical high and the presence of the cold Benguela current offshore creates dry conditions. The spectacular Sossusvlei Sand Dunes in Namib-Naukluft National Park are some of the highest in the world. (b) Elongated lakes line the bottom of the Great Rift Valley. Lake Bogoria, Kenya, shown here, has several geothermal hot springs as a result of the tectonic activity and a large population of flamingoes colored pink from eating the organisms that live in the lake.

(b)

indigenous groups after the discovery of diamonds in 1867 at Kimberly and gold in 1886 on the Rand, a range of hills to the west of Johannesburg, in South Africa. Gold and diamonds, together with oil, continue to incite conflict within Africa and to amplify interest in African economies on the part of other states and multinational corporations (see Geographies of Indulgence, Desire, and Addiction: Diamonds, p. 160).

The distribution of mineral wealth is uneven between and within countries, with South Africa (gold and diamonds) and Nigeria (oil) accounting for more than half of total value. Sub-Saharan countries with mineral exports accounting for more than one-half of total earnings in 2005 include Angola (oil), Botswana (diamonds), Congo (diamonds, oil, and copper), Gabon (oil), Nigeria (oil), Sierra Leone (diamonds), and Zambia (copper).

**Water Resources**   The routes of Africa's major rivers reflect the legacies of inland drainage on the supercontinent because many of them flow away from the coast and into inland wetlands and deltas before shifting back toward the ocean. For example, the immense Congo River, second only to the Amazon in terms of overall discharge, flows north before turning west toward the rapids that bring it down to the Atlantic. The Niger River flows north toward the Sahara into a large inland delta, before turning south toward its exit to the Atlantic in Nigeria. The Nile, discussed in more detail in Chapter 4, flows into the vast wetland known as the *Sudd*. Several river systems still drain to inland basins, including the Okavango River of southern Africa and the Chari-Logone river system, which drains into Lake Chad in the Sahel. These inland deltas create some of the richest ecosystems in the region, providing habitat for wildlife, fisheries, and grazing and irrigated land for human activities.

Where rivers descend from the high plateaus to the coast, they often cut deep valleys back into the plateaus and drop over rapids and waterfalls, such as Victoria Falls on the Zambezi River in southern Africa, posing a serious problem for navigation by boat into the continent.

**Dams and Development**   However, the rivers of Africa, especially where they descend over the coastal escarpment, provide considerable potential for hydroelectric development. Several large dam projects were initiated around the time of transition to independence in the 1950s to harness the energy of the rivers, to provide electricity to industry and cities, and to irrigate agricultural fields. The successes and failures of these projects are an important illustration of the need to understand both physical and social factors in water resources and other development projects.

For example, the Federation of Rhodesia and Nyasaland (now the countries of Zimbabwe and Zambia) completed the Kariba Dam on the Zambezi at Kariba gorge in 1959. This dam, which produces inexpensive electric power for this region of southern Africa, created Lake Kariba. This required the resettlement of 57,000 people and the evacuation of thousands of wild animals, isolated as the waters rose behind the dam, through "Operation Noah." The areas to which people relocated were infested with the tsetse fly, and many were exposed to the sleeping sickness it carries. Because of poor planning, some people ended up in places resembling refugee camps. The hygiene in these camps was very poor, and epidemics flourished. On the positive side, however, some of the unanticipated benefits include development of a

tourist industry, lush animal habitat around the new lake, and a productive fishery. The electricity produced by the dam supports the copper-mining industry in Zambia.

The Akosombo Dam, completed on the Volta River in 1965, was funded by the governments of Ghana, the United Kingdom, and the United States and by the World Bank. The electricity was targeted for a large aluminum smelter on the coast at Tema, which was supposed to use West African bauxite. This operation is now U.S.-owned and consumes about 80% of the electricity the dam generates; the smelter pays very low taxes, has imported cheap bauxite from the Caribbean, and is thereby able to keep aluminum prices low. The remainder of the electric power either goes to urban domestic consumption or is exported to Togo and Benin. The construction of the Akosombo Dam had a number of social and environmental effects. The huge reservoir behind the dam, Lake Volta, reaches 400 kilometers (250 miles) northward and is the largest human-made lake in Africa. The area now under water included 15,000 homes and more than 700 villages, and 78,000 people had to be resettled to make way for the lake. Many were unhappy with the quality of their new houses and land and have returned to live near the lake.

Prior to the construction of the dam, floodplains downstream had benefited from the annual renewal of sediment, and cattle had grazed on the lush grasses along the river. When the river flow declined and releases from the dam became more sporadic, agriculture and livestock production declined in the area below the dam. Sediment is now building up behind the dam, reducing storage, electrical potential, and the potential lifetime of the dam.

Both Kariba and Akosombo have been affected by drought in recent years and have cut back on the amount of electricity they supply as a result. The slower flow of the river and the resulting stagnant water behind the dams have increased the incidence of several diseases, including schistosomiasis and malaria.

## Ecology, Land, and Environmental Management

**Forests and Savannas**   African ecosystems, as in the rest of the world, are closely tied to climate conditions but also reflect a complex evolutionary history and physical geography that has produced great diversity, unique plants, and perhaps the world's most charismatic community of animal species. The Congo basin hosts Earth's second-largest area of rain forest (after the Amazon), covering almost 2.2 million square kilometers (780,000 square miles). Other forests are found along the West African coasts, along the coast of Kenya, and on the island of Madagascar. Forests make up about 20% of the African land area.

These forests have great biodiversity, including monkeys and apes such as chimpanzees and gorillas and tropical hardwoods of significant economic value, such as mahogany. The forests are threatened by demands for timber and firewood, by poaching and conflict, and by conversion of natural habitats to cropland. One of the ironies of peace agreements in central Africa is that it may open up the region to forest exploitation (**Figure 5.8**).

Madagascar's forests have been a focus of international concern. Less than 15% of the original forests remain. The island has one of the world's richest ecologies, with 25% of all the flowering plants in Africa,

# DIAMONDS

For many consumers around the world, diamonds are associated with love and luxury, the symbol of engagement to marry and of wealth and sophistication. Larger diamonds, after grading, cutting, polishing, and setting in gold or other metals, are sold for high prices in jewelry stores, with about half of all purchases made in the United States. These glittering stones also have considerable industrial value because of their hardness, forming a strong, sharp cutting edge.

An estimated 65% of the world's diamonds originate from African countries, with a total value of roughly $8.5 billion dollars. The diamond industry employs 10 million people directly and indirectly across the globe. About two-thirds of the world diamond trade is controlled by a South African conglomerate, De Beers. The virtual monopoly held by De Beers permits careful control of diamond markets to ensure that prices remain high and the supply stable, and the company has worked hard through advertising to maintain the romantic image of diamonds. Eighty percent of all diamonds are traded through the Diamond Center in Antwerp, Belgium.

Most diamond production takes place in South Africa, Botswana, and Namibia and contributes significantly to export revenue and local employment (see Figure 5.6). In Angola, the Democratic Republic of the Congo, and Sierra Leone, diamonds are also mined or smuggled across borders and have become associated with corruption, violence, and warfare. Easy to transport, diamonds are increasingly used to purchase weapons, and some analysts suggest that these **blood diamonds** may have accounted for 10% to 15% of all global trade in diamonds during their peak 1990s trade. Far from being eliminated, as recently as 2006 a United Nations report concluded that $23 million of illegally sourced diamonds were entering Ghana from rebel-held areas in northern Côte d'Ivoire.

Diamonds are mined in large commercial mines such as those in South Africa, but they are also dug from mud and streams by hundreds of individuals who dream of finding a large gem that will make their fortune (**Figure 1**). Mines and miners in Angola and the Democratic Republic of the Congo are often under the protection of armed guards or military forces. For example, in Angola, the Catoca mine, which produced 1.5 million carats of diamonds in 2001, paid $500,000 a month to the Angolan army for security against the rebel group UNITA (National Union for the Total Independence of Angola), which

▲ **FIGURE 1 Diamond Diggers** Hundreds of people have come to work in the diamond regions of the Democratic Republic of the Congo, where they work deep in the mud of streambeds to sift sediment with the dream of finding especially valuable diamonds. However, they receive only a small portion of the eventual value of the diamonds on the world market and are sometimes harassed by the military or others who use violence to gain access to the diamonds.

including the rosy periwinkle, used to treat the disease leukemia, and unique fauna that include 33 species of lemurs, 800 species of butterflies, and numerous chameleons, cacti, and corals. Deforestation in Madagascar has a long history and is associated with clearing land for rice production, sugar plantations, and cattle ranches and cutting trees to export tropical hardwoods. About 80% of the population is subsistence farmers who use a slash-and-burn technique called *tavy* to clear forests. Although protected areas have been established to protect forests and foster ecotourism, with the support of international conservation organizations, poverty and mining place remaining forests at risk.

In East Africa the loss of forests has made life more difficult for women who have limited access to electricity, gas, or petroleum fuels for heating and cooking and rely on wood or charcoal as their major energy source. The increasing wood and charcoal demands of Nairobi, the capital of Kenya, have had a tremendous impact on forests in the region, with serious deforestation tied to the city's energy needs occurring as far as 200 kilometers (124 miles) away. The difficulties of finding firewood and the increases in prices have disproportionately affected women, who traditionally collect the wood and are responsible for cooking and heating the homes (**Figure 5.9**). Projects to reduce energy demands by

using scrap metal to make more efficient stoves have complemented the efforts of female-led nongovernmental organizations to protect trees in and around Nairobi. The best-known social movement is the Green Belt Movement, which counts 50,000 women as members. Led by Nobel Prize–winning environmental and political activist Wangari Maathai, Green Belt has planted thousands of trees around Nairobi and has been the model for similar groups elsewhere in Africa and the world.

Drier regions have mixed woodlands and grasslands (covering about two-fifths of Africa) with open stands of trees interspersed with shrubs and grasses. The baobab tree is a symbol of this landscape, which is found in West Africa centered on Guinea and in southern Africa near the Zambezi. The **savanna** grassland vegetation is found in tropical climates with a pronounced dry season and periodic fires. Savannas provide extended grazing areas for both wildlife and livestock. Grassy plains such as the Serengeti of Tanzania have some of the densest concentrations of wild, hoofed, grazing mammals (called *ungulates*) in the world, together with their predators such as the big cats—lion, leopard, and cheetah. The larger herbivores include elephants, giraffes, zebras, and rhinoceroses (**Figure 5.10**).

Desert regions (another two-fifths) have very sparse and seasonal vegetation for the most part, with drought-resistant vegetation such as used to control the region. Money associated with diamond mines has funded an increase in number and magnitude of arms on every side of the conflict in Angola, including the purchase of land mines that have maimed thousands of civilians. UNITA, for instance, has used its control of diamond mines in the Lunda provinces to finance a guerrilla struggle against the MPLA government (representing the popular movement for the liberation of Angola).

Angola is covered with 10 to 20 million land mines in a 20-year war against the MPLA over control of Angola's diamonds. In 1994, the Lusaka Peace Protocol called for a cease-fire, so that today an uneasy peace remains in anticipation of another civil war because of UNITA's unwillingness to relinquish its diamond mines to the Angolan MPLA government.

The Democratic Republic of the Congo has also become a pawn in the struggle for diamonds, with Angola, Namibia, and Zimbabwe sending troops to protect the government, and Burundi, Rwanda, and Uganda assisting rebels. The area around the diamond zone near Kisangani in the eastern Democratic Republic of the Congo has been abandoned to fighting, and Zimbabwe and Rwanda have struggled to obtain access to diamond deposits in the southern Democratic Republic of the Congo. Diamonds also funded a brutal civil war in Sierra Leone, where rebels chopped off people's limbs with machetes to intimidate residents into leaving the eastern diamond zones. "Conflict diamonds" from Liberia were being smuggled into neighboring countries for export, and diamonds from the Ivory Coast are finding their way to the British and European markets. Pressure from human rights groups has led De Beers to refrain from purchasing diamonds that originate in conflict zones. De Beers has also agreed to support research and development of a system to fingerprint diamonds based on chemical signatures that can identify legitimate areas of origin.

In contrast, Botswana is producing diamonds under peaceful conditions and with the guidance of traditional leaders. The mines employ more than 25% of the population and are responsible for 33% of the gross domestic product and a standard of living that is much higher than the average for sub-Saharan Africa. Botswana holds the richest diamond mine in the world, named "Jwaneng," meaning "a place of small stones." The open pit mine is owned by Debswana, a partnership between De Beers and the government of Botswana. Production varies between 12.5 and 15 million carats per year, and in 2007 Jwaneng produced 13.5 million carats from 10.3 million tonnes of ore.

On January 1, 2003, the Kimberley Process came into effect, with the chief aim of eliminating the use of diamonds to finance armed conflict. In 2000 the diamond-producing countries of southern Africa had established the Kimberley Process Certification Scheme (KPCS) to protect their legitimate diamond industry; however, the main target of the KPCS was to highlight the illicit trade of rough diamonds, which has fueled much conflict. Canada, a major diamond-producing country, was made responsible for chairing the KPCS at the end of 2003. The mandate is for the global trade of rough diamonds in an open and transparent manner and through legitimate markets. It is recognized as a unique partnership between governments, the diamond industry, and civil society. In November 2007, 74 governments agreed on measures to further strengthen the Kimberley Process, such as intensifying control of rough diamonds across the Ivory Coast. Although the battle against conflict diamonds is far from over, the Kimberley Process is an international attempt at contributing to and promoting international peace and security. Today, the industry reports that less than 1% of diamonds is illegally sourced. *Source:* Partly adapted from B. Harden, "Diamond Wars: A Special Report. Africa's Gems: Warfare's Best Friend," *New York Times,* April 6, 2000.

---

acacias and woody scrub. The Mediterranean climates of South Africa have produced a unique ecosystem dominated by *fynbos* shrubland, the vegetation of which is characterized by waxy or needlelike leaves and long roots that help plants survive long dry periods. Finally, the highland, or montane, vegetation is found on mountain ranges such as the volcanoes of East Africa or the Drakensberg highlands of southern Africa.

As in other world regions, African ecosystems such as those on Mount Kenya show clear altitude zonation, with a vertical change in environment and land use, according to elevation based mainly on changes in climate and vegetation from lower (warmer) to higher (cooler) elevations. The popular tourist hikes up the slopes of Mount Kenya or Kilimanjaro reveal how vegetation and human land use change along these environmental gradients. The base of the mountains is nested in grasslands of the savanna, whereas rocky peaks are covered with snow and ice (now disappearing as a result of climate change). In between, a hiker would pass through zones of dry forest, bamboo forest, heathland, and alpine moorland.

**Diseases and Insect Pests**    Africa's ecologies are notable for several pests and diseases that can have a devastating impact on human populations. Several of these diseases have reservoirs in certain wild species and are then transferred to humans or their domesticated animals by vector (transmission) organisms such as mosquitoes, flies, and snails. Malaria, a disease transmitted to humans by mosquitoes, causes fever, anemia, and often fatal complications. It affects about 400 million people in sub-Saharan Africa each year, killing more than 1.5 million, many of them children (**Figure 5.11a**). Early European explorers and settlers were highly vulnerable to malaria and suffered as much as 75% mortality in some regions. The discovery in 1820 that quinine—an extract from the cinchona tree, thought to have been brought to Europe from Peru by Jesuit priests—could partly control malaria facilitated colonialism and also allowed treatment of some local residents. But it did not cure the disease, and after World War II several synthetic drugs, such as chloroquine, became popular cures. Unfortunately, several strains of the malaria parasite have developed resistance to these drugs, and there is still no certain and cheap cure for the disease, although cheap mosquito nets can help.

Schistosomiasis (also called *bilharziasis*) is associated with a parasite that causes gastrointestinal diseases and liver damage. It is passed to humans who are exposed to a snail vector by working or bathing in

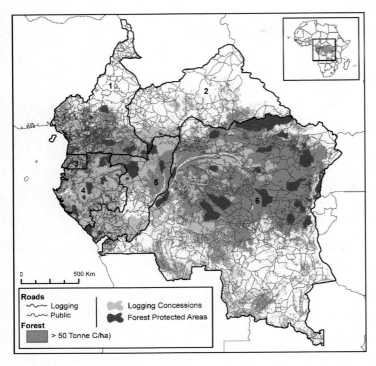

▲ **FIGURE 5.8 Deforestation in Central Africa** Satellite data was used to map logging activity between 1976 and 2003 in six central African countries: Cameroon, Central African Republic, Equatorial Guinea, Gabon, Republic of the Congo, and the Democratic Republic of the Congo (DRC). The results show that 30% of the forests are now owned by private logging companies, and new logging roads (red) go deep into approved logging areas (gray) but also cut into the 12% of forests set aside for conservation (dark green). The most rapid change is in the northern region of the Republic of the Congo.

slow-moving water, such as that in marshes, reservoirs, or irrigation canals. The disease is not fatal but reduces general health and energy levels. The United Nations World Health Organization (WHO) estimates that 200 million people are infected in sub-Saharan Africa. River blindness (onchocerciasis) is transmitted by the bite of the black fly, which

▲ **FIGURE 5.9 Women and forests** Women from Kenya's Greenbelt movement discuss tree planting.

passes on small worms whose larvae disintegrate in the human eye and cause blindness. The eradication of river blindness has been relatively successful in West Africa by controlling fly populations with pesticides and treating victims with drugs. WHO estimates that 18 million people are infected, and 250,000 are blind as a result of this disease in sub-Saharan Africa. Other serious diseases include yellow fever, illnesses caused by rotaviruses where millions of children suffer from diarrhea from contaminated food and water, and sleeping sickness associated with the tsetse fly (**Figure 5.11b**).

Many of these debilitating diseases are associated with the tropical climate and diverse ecologies of Africa. Their spread may have been facilitated by the expansion of human populations and the transformation of natural environments through deforestation and irrigation. Reducing the human toll from these diseases is a major challenge for scientific research, African governments, charitable organizations, and the WHO, which has targeted Africa for extra funds and programs. With an endowment of $27.5 billion in 2008 the Bill and Melinda Gates Foundation has committed over one billion U.S. dollars to fighting malaria, including the development of a malaria vaccine (PATH Malaria Vaccine Initiative) and distribution of mosquito nets, and has spurred a multibillion-dollar global health initiative aimed at achieving major reductions in disease in poor countries, especially in Africa.

**Land Use and Agriculture** The United Nations Food and Agricultural Organization (FAO) has estimated that less than 30% of the soils in sub-Saharan Africa is suitable for agriculture. In addition, agriculture is hindered by an unsuitable climate and an environment that is prone to pests and diseases. However, Africa, as the birthplace of the human species, is also the region where humans first adapted to the constraints of the physical environment, finding sustenance through hunting wild animals, fishing, gathering plants, and domesticating a number of crop and livestock species.

There is some disagreement about whether cattle were domesticated in Africa or introduced from the Middle East and Asia about 8,000 years ago. Archaeological sites from this period have provided evidence of livestock living along with humans and also of domesticated and cultivated grains, including sorghum. The highlands of Ethiopia are considered one of the centers of **domestication**, producing coffee, millet, and an important local cereal called *teff*. Other crops domesticated in Africa include yams, oil palm, cow pea, and African rice. The introduction of maize from Latin America into Africa transformed land use and diets. Other important crops today include millet, sorghum, rice, and yams.

Traditional people developed several strategies for adapting to low soil fertility, including **shifting cultivation**, which involves moving crops from one plot to another to preserve soil fertility. As in other regions of the tropical world, one form of shifting cultivation is **slash-and-burn** agriculture, used to clear patches of forest, shrubs, or grassland through burning and then take advantage of the ash to fertilize crops. When, after a few years, the nutrients are exhausted, farmers move on to a new area and leave the previous plot to return to forest or other vegetation. After a long fallow (rest) period, they return to clear and burn the land again. A modification of shifting cultivation is **bush fallow**, by which crops are planted around a village, and plots are left fallow for shorter periods than in the slash-and-burn system. Soil fertility is often maintained through fallow periods or by applying household waste to the fields. Where household compost is used to grow crops within the village, the technique is called "compound farming" and is popular in forest environments as well as in some urban areas.

(a)

(b)

(c)

▲ **FIGURE 5.10 Flora and fauna** (a) Elephants and Baobab trees in Kruger National Park, South Africa. Kruger gained protection as a game reserve in 1898 and became a national park in 1927. There are more than 10,000 elephants in the park, requiring some population control to avoid land degradation. Baobabs store water inside a trunk that swells during the wet season, and the leaves and fruit are important local foods. (b) Vast herds of wildebeest and zebra migrate across the savannas of Tanzania, where the wildlife attracts millions of tourists to East Africa. (c) The Cape region of South Africa has a cooler, wetter climate that supports the highly biodiverse Fynbos biome. Climate change threatens its existence as climate warms and dries.

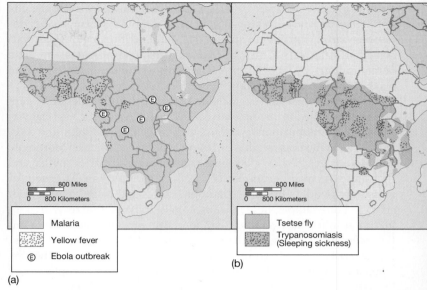

(a)

(b)

▲ **FIGURE 5.11 Maps of tropical infectious diseases and pests** These maps show the distribution of some of the more serious tropical diseases in Africa. (a) Malaria and yellow fever. Yellow fever has a mosquito vector and a reservoir in monkey populations. Mortality in Africa has been reduced through immunization, but many who are poor or live in remote regions still do not have access to vaccines, and as many as 20,000 people died in an outbreak of yellow fever in Senegal in the 1960s. (b) The tsetse fly, which lives in African woodland and scrub regions, is a vector for a virus with a reservoir in wild animals. It is associated with both human and livestock diseases. In humans, the fly's bite causes sleeping sickness, or trypanosomiasis, with fever and infection of the brain that causes extreme lethargy and may end with death of the victim. Half a million people are infected in sub-Saharan Africa. Sleeping sickness can be treated in early stages and can be prevented by a variety of pest-control measures, including burning brush, spraying with pesticides, and removing wild animals (including rodents) that serve as reservoirs for the disease. In domestic animals such as cattle and horses, the tsetse fly causes a disease called nagana, which is similar to sleeping sickness and causes fever and paralysis. This disease prevented the introduction of livestock into many parts of Africa and as a result preserved habitats for wild species. Areas with an elevation above 150 meters (480 feet), those with a long dry season, and those with sparse or no woodland are free from tsetse flies.

Intercropping—planting several crops together—is a technique for keeping the soil covered to reduce erosion, evaporation, and the leaching of nutrients. Where one of the crops, such as beans, can capture nutrients such as nitrogen, intercropping also improves soil fertility. Floodplain farming is used in regions such as the inland delta of the Niger River and the Sudd wetlands along the Nile River in the Sudan.

As noted earlier pastoralism—a way of life that relies on raising livestock—is the human activity best adapted to drier regions of Africa. Nomads, such as the Bedouin, migrate with their animals in search of pastures in the arid landscapes of the Sahel and North Africa. Other groups, such as the Fulani of West Africa, practice a system of seasonal herd movements called *transhumance*. They move their herds to wells and rivers in the dry season and drive them northward to take advantage of new pastures in the wet season. In some regions, farmers let pastoralists graze their herds on harvested fields in the dry season, thereby fertilizing the land with animal manure in a mutually beneficial (symbiotic) relationship with the pastoralists, and in other regions pastoralists are also farmers. Cattle are traded at regional markets and are a family investment for the future in regions where there are few secure ways of accumulating capital (**Figure 5.12**).

Explanations of land degradation in Africa, including desertification (see p. 155), often blame Africans for overusing resources, especially as a result of overpopulation, and for poor management of forests and soils. However, detailed case studies by researchers such as Michael Mortimore and Melissa Leach show that these constructions are often myths, and that there are many examples of careful community management of forests and wildlife based on traditional rules and deep knowledge of ecology. For example, in Machakos, Kenya, a fivefold increase in population resulted in less rather than more soil erosion when communities, especially women, began to terrace steep slopes, manage livestock, and diversify crops. In Guinea, patches of forest were found to be new woodlands planted and maintained sustainably by local people, rather than the remnants of deforestation as perceived by outsiders.

**Conservation and Africa's Wildlife** The rich biodiversity of Africa is valued by local residents, tourists, and international environmental groups alike, but differing views about its protection have resulted in many controversies about conservation. Traditional African societies hunted and gathered wild species for food and also incorporated wildlife into spiritual beliefs. While human populations were low and hunting technologies were less effective, wildlife populations ranged naturally where climate, vegetation, and terrain were most suitable. As population, technology, and land use changed, especially after colonialism, human activity began to modify habitat, and wildlife populations shifted. Europeans contributed to the decimation of African wildlife through indiscriminant hunting expeditions, the elimination of animals along railroads and near farms, and resettlement of local people into regions where they came into conflict with wildlife that encroached on their herds and fields.

Currently about 100 million hectares (5%) of sub-Saharan Africa are under some sort of protected status, and there are more than 1,000 protected areas, more than half in southern Africa (**Figure 5.13**). The major parks in East Africa and southern Africa, such as Serengeti in Tanzania and Kruger in South Africa, have become high-profile international tourist destinations, bringing in millions of dollars to national economies and employing many local people. The parks are not without problems or criticism. Parks have been criticized for providing inadequate benefits to local people who may have been displaced or who lost traditional grazing and hunting rights or whose crops are destroyed by marauding wildlife. In some parks, too much tourism and high animal densities have destroyed fragile habitat, and poaching has pushed some species close to extinction.

Elephants have been a focus of controversy over conservation. They draw attention because of their size, their intelligence, and the value of ivory and meat. Hunting elephants for their ivory tusks and trade in this precious commodity—known as "white gold"—has been carried on for centuries. Pressure on herds in East and Southern Africa grew with colonial demand for ivory décor in Victorian England, and by the

▲ **FIGURE 5.12 African pastoralism** The Masai of Kenya and northern Tanzania are herders—using cattle for meat, milk, and blood and moving their animals as the rains shift. In this photo young men perform a traditional warrior jumping dance next to their cattle.

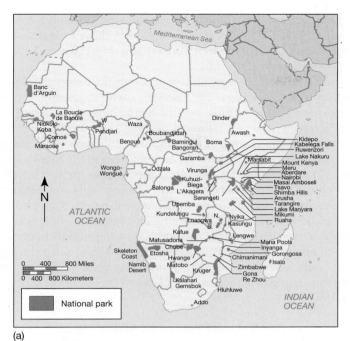

(a)

(b)

▲ **FIGURE 5.13  African conservation challenges**
(a) Africa has large areas of forest and wildlife habitat under protection, although in some regions areas have been invaded or degraded by mining and lumbering and by people fleeing conflict or seeking to survive through farming. National parks are concentrated in southern and eastern Africa, where millions of tourists are attracted by wildlife-viewing opportunities. (b) Many rhinos now have their own bodyguards, such as this group in Kenya. Rhino horns may be removed by wildlife experts and replaced by bright plastic horns to reduce their attractiveness to poachers.

end of the 19th century Africa's elephant populations were significantly diminished.

A second surge in demand occurred in the 1970s, especially in Burundi, Congo, Kenya, and Zaire, when prices for ivory soared with international financial instability and growing demand in Asia. Export demand rose from 220 tons in the 1950s to nearly 1100 tons in the 1980s. A precipitous decline in most African elephant populations from 2.5 million in 1970 to fewer than 500,000 in 1995 was a direct conse-

quence of illegal killing and poaching, fueled by the ivory trade. The situation was aggravated by competition between people and elephants for land and by war and civil unrest, especially as firearms became available to unpaid soldiers and desperate refugees in regions where herds lacked strong protection.

In 1997 elephants were listed under the Convention on International Trade in Endangered Species (CITES) as a Category 2 species, in need of protection, and by 1990 mounting international pressure against the perceived slaughter of elephants resulted in moving elephants to Category 1, the most serious danger of extinction, and in a worldwide ban on the sale of ivory. The ban was opposed by countries in southern Africa, which had seen a less-serious decline in elephant populations and who were funding parks and conservation from the money earned from ivory and hide sales. In southern Africa, park managers had to cull elephants to protect habitat for other species but could not sell the ivory. Eventually, in 2002, the UN granted permission to South Africa, Botswana, and Namibia to sell 60 tons of ivory, even though there has been a general ban on international sales of African ivory since 1989. China was given permission by the UN to purchase ivory in 2008 and is now the world's center of legal and illegal trading of ivory.

The rhino is under much greater threats, especially from poachers who hope to sell rhino horn for dagger handles in the Middle East, especially in Yemen, and for highly valued medicinal powders in Asia. Protecting the rhino from poachers who can make thousands of dollars from selling a horn is a full-time and costly enterprise (**Figure 5.13b**).

# HISTORY, ECONOMY, AND DEMOGRAPHIC CHANGE

Sub-Saharan Africa's role in the world begins with evidence of human origins on the continent more than 2 million years ago and continues with the development of major trading societies about 5,000 years ago and the incorporation into a European-dominated colonial system about 500 years ago. Colonialism included the worldwide trade in African slaves, resulting in a diaspora of African people that has continued to influence the culture and societies of other world regions. It also resulted in political boundaries that split ethnic groups across territories or clustered enemies within one territory.

Peoples from other world regions, including Europe and Asia, came to Africa under colonial rule and created hierarchies of power and politics. Most of sub-Saharan Africa was under European colonial domination by 1900 and did not become independent until after 1950. Independence also coincided with the height of Cold War tensions between the United States and the former Soviet Union and the consequent interventions of the superpowers in African political struggles and civil wars.

At the end of the 20th century, much of sub-Saharan Africa was still struggling with the transition to independent and democratic government and with economies that rely on a narrow set of exports to other world regions. Populations are still growing relatively fast, with many people moving to Africa's cities. Although parts of Africa are becoming highly connected to other regions through migration, telecommunication, and trade, other places are still isolated and rely on subsistence agriculture.

# Historical Legacies and Landscapes

Contemporary African geographies—landscapes, livelihoods, culture and politics—bear the imprint of prior periods of African history, such as colonialism, and of the shifting connections to different regions of the world. These connections in turn have shaped other regions of the world.

**Human Origins and Early African History**   Africa is often called the "cradle of humankind" because archaeologists have shown that the earliest evidence of the human species (*Homo sapiens*) is found in Africa. Fossilized footprints of an earlier ancestor, the hominid (humanlike) *Australopithecus*, were found by archaeologist Mary Leakey at Laetoli in Tanzania and dated to 3.7 million years ago. Two-million-year-old stone tools have been found at several sites in Ethiopia and East Africa, including the famous site at Olduvai Gorge in Tanzania (**Figure 5.14**). Anatomically modern humans, who walked upright and had larger brains, have been dated to at least 100,000 years ago from sites in southern Africa and along the Rift Valley, and many scholars now believe that these humans are the genetic ancestors of all modern humans and thus the most basic link between Africa and the world.

Around 5,000 years ago, written accounts document the development of complex societies in the Nile Valley, with engineered irrigation systems, hieroglyphic writing, and the hierarchical social organization of the Egyptians under their king or pharaoh (see Chapter 4).

From this time onward, explorations, military campaigns, and European trading expanded into Africa from bases in the Nile Valley and North African coast. By C.E. 500 some Indonesians had settled on the island now known as Madagascar, introducing yams and bananas to mainland Africa, and a strong kingdom had emerged at Aksum in Ethiopia and had adopted Christianity. The Bantu people spread from West Africa to central and southern Africa bringing with them technologies such as iron smelting. West African empires from about the eighth century were associated with cities that were great centers of trade, scholarship, and power such as Timbuktu and Djenné in the Sahel (**Figure 5.15**).

**The Colonial Period in Africa**   With the development of faster and larger ships in the 15th century, contacts with Spain, Portugal, and China were added to the regular interaction between the Middle East and Africa. The Portuguese traded for gold from coastal settlements in West Africa, and in 1497 the Portuguese explorer Vasco da Gama rounded the Cape of Good Hope at the southern tip of the African continent, initiating trading with the southeast coasts of Africa en route to India. In return for salt, horses, cloth, and glass, sub-Saharan Africa provided gold, ivory, and slaves to the world through Portuguese and Arab traders. For several centuries, African slaves had been in demand among the Arabs, who used the slaves as servants, soldiers, courtiers, and concubines.

European colonialism took some time to establish control in Africa, and for many years only the coastal ports and trading posts were under European command. The European names for coastal regions along the west coast of Africa clearly indicate the commodities that they provided, from the Ivory Coast in the west to the Gold Coast (now Ghana) and Slave Coast (Nigeria and Benin) to the east. One of the main reasons for European reluctance to move inland was the reputation of Africa as the "White Man's Grave" because so many Europeans were rapidly killed by malaria, yellow fever, and sleeping sickness, diseases against which they had no natural immunity. In addition, African armies attacked ports and fiercely resisted European attempts to move inland.

**Slavery**   Even in the face of native resistance and the ravages of disease, the coastal regions generated enormous profits for European traders mainly through slavery. The Portuguese started to take slaves for their own use on new sugarcane plantations on the Atlantic islands of Madeira and Cape Verde, and in 1530 the first slaves were shipped to the Americas to work on plantations in

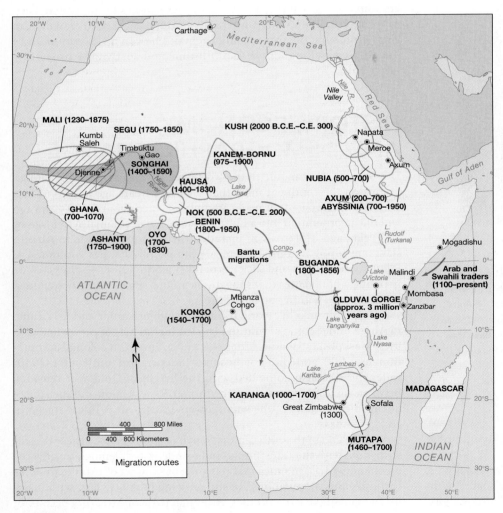

▲ **FIGURE 5.14 Map of African history** Sub-Saharan Africa had a rich historical heritage prior to European arrival in the region. The oldest human remains were found at Olduvai Gorge in present-day Tanzania. This map shows some of the locations and dates of the great kingdoms as well as the early migrations of Bantu people and Arab traders.

(a)

(b)

▲ **FIGURE 5.15 Early centers of commerce and learning** (a) The Sankoré madrassah (school) includes three mosques that are ancient centers of learning and are now the University of Timbuktu. The city was a center for learning from at least C.E. 1000, and by the 1300s Sankoré had 2,5000 students and more than half a million books and manuscripts. It was also one of the key trading centers for West Africa. (b) The island of Zanzibar, just off the coast of Tanzania, was a key port on the Indian Ocean trading with the Middle East and Africa. By 1000 B.C.E., Zanzibar and the islands off the coast of East Africa were familiar to the Egyptians, Phoenicians, Greeks, and Romans. As these Mediterranean empires extended their trade routes to the south and east, Zanzibar became one of several major commercial ports along the East African coast. Around the third century C.E., the trade in goods attracted the attention of merchants from southwestern Arabia, who also began trading with the island residents, bringing weapons, wine, and wheat to barter for ivory and other luxury goods. The Arabs brought the religious traditions of Islam; the local language, Swahili, has many Arabic words and became the trade language for eastern Africa.

Brazil. By 1700, 50,000 slaves were being shipped each year to the Americas to provide labor on colonized lands and new plantations because their potential indigenous labor supply had been decimated by European diseases (see Chapter 6, p. 209). Slavery was an important income source for some African coastal kingdoms, such as Dahomey

and Benin, who captured their enemies or residents of inland villages and sold them to the slave traders. It is estimated that more than 9 million slaves were shipped to the Americas from Africa between 1600 and 1870, with at least 1.5 million slaves dying during the journey (**Figure 5.16**). Most slaves were male, and the conditions of capture and transport were inhuman. Hundreds of slaves were packed into the holds of ships with little food and water and were brutally abused by traders. On arrival in the Americas many of them introduced their own traditions, including ways of cultivating rice, religion, and music that endure today.

Beginning at the end of the 18th century, in Britain and the United States, members of the Quaker community led movements to abolish slavery, with slavery banned in Britain in 1772. The British then abolished the slave trade with their colonies in 1807 and emancipated (freed) slaves in the Caribbean in 1834. Slavery in the Americas was abolished in most countries by the 1850s, and the slavery issue was key in the U.S. Civil War. Slavery was abolished in the United States in 1863 by the Emancipation Proclamation (see Chapter 6, p. 210). Some liberated slaves returned to Africa, especially to Freetown (now the capital of Sierra Leone) and from North America to Liberia.

**European Settlement in Southern Africa**    European settlement was encouraged in southern Africa by the more temperate climate and the strategic significance of the trading routes around the southern tip of the continent. In 1652, the Dutch established a community at Cape Town, which became surrounded with small farms growing wheat and raising cattle for supplying ships and the "Cape" communities, as the region around the Cape of Good Hope was called. As their language, the settlers evolved Afrikaans, a modified version of Dutch; they belonged to the strict puritan Christian Calvinist religion, saw themselves as superior to black Africans, and became known as the *Boers* (Dutch for "farmer"). As their military and trading power grew in the 1800s, the British took control of the Cape trading route, and British immigrants were encouraged to settle in the Cape region from about 1820, mainly in Cape Town and Durban. When the British imposed laws on the Boers, including banning slavery in 1834, the Boers moved north of the Orange River in a great trek, settling on the high pastures called the *veld* in what is now known as the Free State. Some Boers also migrated eastward into the Natal region, where they came into conflict with the powerful Zulus. As we will see, the geography of this colonial settlement framed the 20th-century politics of South Africa (see Signature Region in Global Context: South Africa).

**European Exploration and the Scramble for Africa**    International interest in Africa increased dramatically after 1850, with growing competition among core European powers for colonial control and the discovery that quinine could suppress malaria. Explorers, traders, and missionaries moved to the interior of the continent seeking territory, the source of the Nile, commodities, and souls to convert. Some of the most famous explorers were associated with the British Royal Geographical Society (RGS), which was founded in 1830 for the "advancement of geographical science." The RGS supported and awarded their medal of honor to many explorers of Africa, including David Livingstone, Henry Stanley, Richard Burton, and John Speke (**Figure 5.17**).

These Victorian explorers added greatly to geographic knowledge of Africa, and their reports fueled colonial interest in the continent's resources and people. Their lectures at the Royal Geographical Society and elsewhere increased interest in the discipline of geography and its role in Britain's colonial enterprise. However, their books and those of

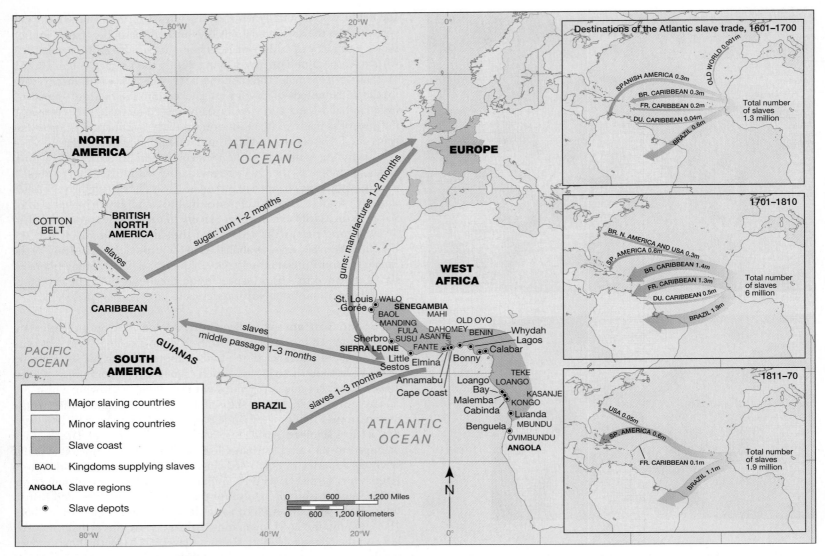

▲ **FIGURE 5.16 The slave trade** Millions of slaves were exported from Africa between about 1600 and 1870, mainly from the West African coast. Some local leaders acted as suppliers in return for guns and manufactured goods. Slaves were sent to work in plantations in the Americas, which then sent sugar, rum, and other products back to Europe in a triangular trade.

other explorers contained many Victorian prejudices and paternalistic attitudes that fostered the popular image of Africa as a barbarous and exotic continent in need of civilization and colonial supervision.

By 1880, new knowledge of African resources—including gold and diamonds—competition among European powers to dominate global empires and markets, and reduced risk of African diseases further increased interest in the continent, especially from the British (in Egypt), French (in West Africa), and Portuguese in Mozambique and Angola (**Figure 5.18**). Inspired by the reports of explorer Henry Stanley, the personal crusade of King Leopold II of Belgium to establish colonies in Africa focused attention on the Congo basin; Germany sought colonies where it had missionaries in what are now Togo, Cameroon, Namibia, and Tanzania. Pressure from commercial companies and even missionaries drove these imperial ambitions and incorporated Africa into the emerging global capitalist economy.

The role of private companies in the colonization of Africa was important because European governments granted exclusive concessions for trade and resource exploitation in Africa. These companies, which were often given the right to police, to conscript, and to tax local populations, included the Royal Niger Company and British South Africa Company, both of which received royal charters in the late 1800s.

This hasty "scramble for Africa" culminated in the **Berlin Conference** of 1884–85, a meeting convened by German chancellor Bismarck to divide Africa among European colonial powers. The 13 countries represented at this conference did not include a single African representative from any state in sub-Saharan Africa, even though more than 80% of Africa was at that time under African rule. The Berlin Conference allocated African territory among the colonial powers, according to prior claims, and laid down a set of arbitrary boundaries

(a)

(b)

▲ **FIGURE 5.17 Colonial explorers** (a) David Livingstone is best known for his explorations of the Zambezi and his encounter with the magnificent waterfalls that he named after Queen Victoria. (b) Henry Stanley, a journalist sent to find Livingstone, poses as the conqueror of a hostile continent.

Nigeria itself comprised several competing groups, including the Yoruba in the southwest, the Ibo in the southeast, and the Hausa in the north.

Mining activities were expanded in many regions, especially in southern and central Africa, with large amounts of gold, diamonds, and copper extracted and exported by European companies. New roads and railways were constructed from inland to the coasts to speed the export of crops and minerals, but few efforts were made to link regions within Africa. The resulting infrastructure facilitated trade beyond but not within Africa.

Colonists established plantations to produce crops, such as rubber, and used a variety of means, including taxation and intimidation, to persuade peasant farmers to produce peanuts, coffee, cocoa, or cotton for global markets. In the temperate climates of the East African highlands and southern Africa, areas that were more attractive to European immigrants, the best land was taken by white settlers for tea and tobacco plantations, livestock ranches, and other farming activities. By 1950, the geography of African agriculture illustrated this export orientation, with vast rubber plantations owned by the Firestone Corporation in Liberia, cocoa dominating the cropland of Ghana and the Ivory Coast, cotton in Sudan, peanuts in French West Africa, and tea and coffee in East Africa (**Figure 5.19**). Traditional African land-tenure systems of communal land and flexible boundaries were often forced into privately owned and bounded plots, and traditional decision-making and legal systems were often replaced with European managers and courts.

that paid little respect to existing cultural, ethnic, political, religious, or linguistic regions.

The next 20 years saw some rearrangement and consolidation of the European colonies. The British created protectorates in what are now Botswana, Zambia, Zimbabwe, and Malawi and expanded their control over the Sudan. The French took control of many regions along the Niger, and the Italians unsuccessfully invaded Abyssinia (now Ethiopia). In southern Africa, British entrepreneurs, including the ambitious Cecil Rhodes, responded to the discovery of gold and diamonds between 1867 and 1886 by acquiring the mines at Kimberly and the Rand and sparking a gold and diamond rush. Growing tensions between the British and Afrikaners resulted in the Boer War (1899–1902), which gave control of much of southern Africa to the British.

By 1914 almost all of Africa was under European colonial control except for Abyssinia, Liberia, and some interior regions of the Sahara Desert. A number of battles were fought in Africa during World War I, and Germany's eventual loss redistributed the German colonies to Britain, France, and Belgium.

**The Impact and Legacy of Colonialism**  All this reshuffling of African territory among European states can overshadow the considerable and everyday impact of colonial rule on African landscapes and peoples. The most general and enduring effects of colonialism include establishment of political boundaries; reorientation of economies, transport routes, and land use toward the export of commodities; improved medical care; and introduction of European languages, land tenure systems, taxation, education, and governance. As noted earlier, many of the new colonial boundaries divided indigenous cultural groups and in some cases placed traditional enemies within the same country. For example, the Yoruba were divided between Nigeria and Benin, and

The effects and process of colonial rule varied among European powers. The British chose a paternalistic indirect rule for most of their African colonies, making preexisting power structures and leaders responsible to the British Crown and colonial administrators in a decentralized and flexible administrative structure. For example, local leaders were required to collect taxes—sometimes a hut tax based on the number of dwellings in a community, sometimes a poll tax based on the number of residents. To obtain money to pay taxes, people had to produce crops for sale to the Europeans, an indirect way of transforming economies and land use to commodity production. Foreign ownership of land was prohibited in some cases, and traditional legal systems were used to resolve local conflicts. The British, preceded by missionaries, also introduced some European-style schools, and by the 1940s a select group of Africans were attending overseas universities and given posts in government administration.

The French colonial policy was one of assimilation, encouraging elites to evolve into French provincial citizens with allegiance to France, but with agriculture and mining under close supervision from the French capital in Paris. By 1946 there were about 20 Africans, elected from West Africa, in the French parliament. The Belgian and Portuguese modes of colonialism are described as much harsher, with direct rule and often ruthless control of land and labor. In the Congo, local people were forced to gather rubber, kill elephants for ivory, and build public works under threat of death or severe punishment. These authoritarian forms of control—with little political participation, dominating official ideology, and frequent use of armed force—provided an unfortunate model for leadership in independent Africa.

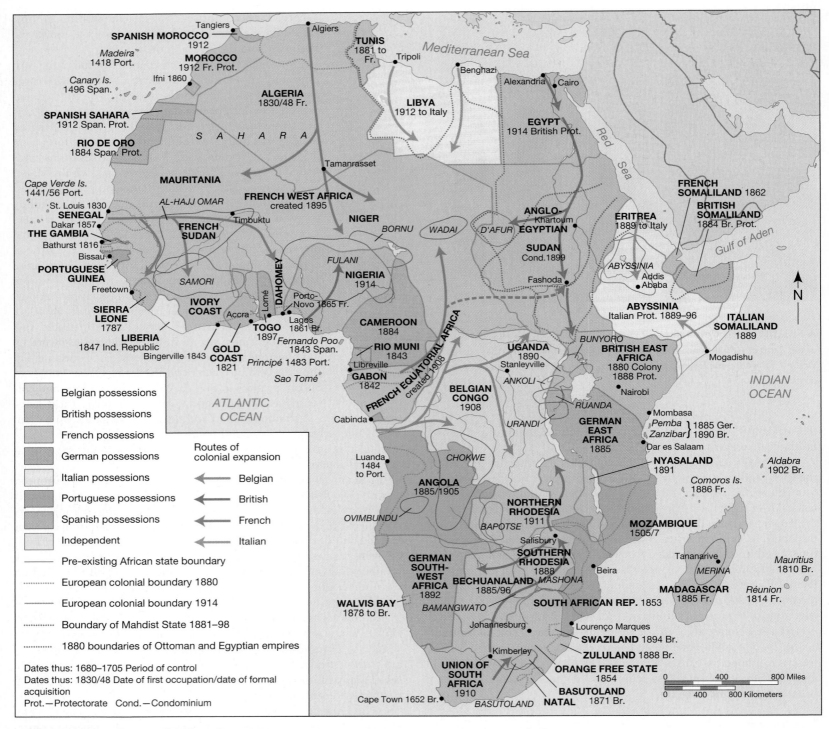

**▲ FIGURE 5.18 The scramble for Africa** Between about 1880 and 1914, European powers, especially the British, French, and Belgian governments, aggressively moved to colonize Africa. This map shows the routes and dates of the takeover of Africa. The British claimed what are now known as Gambia, Ghana, Nigeria, and Sierra Leone in West Africa; Kenya, Uganda, Sudan, and part of Somalia in East Africa; and southern Africa, except for German South West Africa (Namibia), Portuguese Mozambique, the Cape Verde Islands, Angola, and the independent Boer region of South Africa. Germany claimed German East Africa (Tanzania) and Cameroon; Portugal claimed the Cape Verde Islands; and France and Belgium split the Congo. Italy took Somalia, Djibouti, and Eritrea and coastal regions of Libya, with ambitions for Abyssinia (now Ethiopia). The Spanish obtained a small coastal region of northwest Africa and Equatorial Guinea. Most of the remaining territory of West and North Africa was allocated to or taken by the French.

Given the dramatic impact of the colonial period on contemporary Africa, it is significant that in most of Africa, formal colonialism only lasted 80 years, from about 1880 to 1960.

**Independence** Decolonization in sub-Saharan Africa was rapid and ranged from relatively peaceful handovers to well-prepared African leadership to more violent transitions of power to divided or unprepared local elites and militaries (**Figure 5.20**). South Africa was consolidated as an early independent state—the Union of South Africa—in 1910. Other African hopes for independence were encouraged by the British decision to grant India and Pakistan independence in 1947 and demanded by the almost half million Africans who fought with the allies in World War II. The independence movement was led by several foreign-educated activists, such as Kwame Nkrumah of Ghana and Jomo Kenyatta of Kenya, and fostered by organized nationalist groups or African Unions within such key African countries as Tanganyika, Zimbabwe, and Kenya. A *Pan-African* movement, led by black activists in the United States—including W. E. B. DuBois and Marcus Garvey—and others in the West Indies also promoted independence. In 1957, Ghana became the first country in sub-Saharan Africa to have power handed over to local populations; three years later, Nigeria gained its independence.

Although many British handovers were relatively peaceful, countries with significant white settler populations endured more violent transitions. In Kenya, about 3,000 white settlers controlled more than 2.6 million hectares (6.4 million acres) of the best land, especially in the highlands, adjacent to overpopulated indigenous Kikuyu farms. Whites also dominated the government and set policy in the interests of the 60,000 white residents. The *Mau-Mau* rebellion between 1952 and 1956 against white rule resulted in the deaths of 100 whites and more than 10,000 black

▲ **FIGURE 5.19 Export Crops in Africa** (a) A tea plantation below Mount Kenya. Tea was introduced into the Kenya highlands from India in 1903 and grown as a plantation crop on colonial landholdings. There are still 70,000 hectares under production, providing about 10% of the world total after China and India. (b) A child carries a peanut plant near Djiffer, Senegal. Peanuts were introduced as a cash crop in French West Africa during the colonial period. (c) The British promoted massive expansion of cocoa production in Ghana to meet a growing demand for chocolate from around 1880. Currently 800,000 farmers produce about one-fifth of world production (second only to Côte D'Ivoire). The photo shows a woman farmer of the Kuapa Kokoo cooperative that was created in 1993 when cocoa farmers in Ghana united to negotiate better prices for their cocoa and empower small farmers. Kuapa Kokoo now represents 45,000 cocoa farmers and has a stake in the first farmer-owned chocolate company in the world.

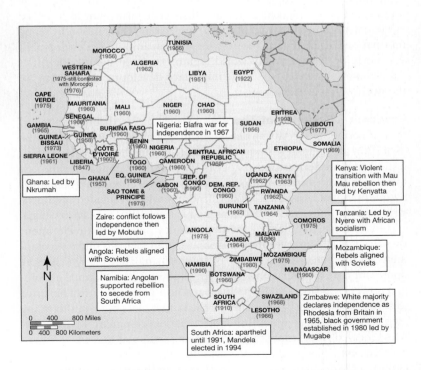

▲ **FIGURE 5.20 Independence** This map shows dates of independence and the characteristics of the transition in selected countries .

Africans. Kenya became independent in 1963. In Southern Rhodesia (now Zimbabwe) the population of about 250,000 white settlers, led by Ian Smith, made a Unilateral Declaration of Independence (UDI) from the UK in 1965 rather than consider the possibility of rule by the 6 million black African population. Only after 15 years of conflict and international trade embargoes did an independent Zimbabwe finally emerge in 1980, with a mostly black government and a legacy of resentment against white residents.

In West and Equatorial Africa the transition occurred dramatically in 1960, with France suddenly recognizing all the independent countries of Mauritania, Mali, Niger, Senegal, Upper Volta (now Burkina Faso), Ivory Coast (now Côte d'Ivoire), Togo, Dahomey, Chad, the Central African Republic, Cameroon, Gabon, and the Congo. In most cases, strong economic and cultural ties were maintained with France, the franc remained the currency, and French troops were stationed in most countries. When Belgium abandoned its African colonies in the early 1960s, the countries of Zaire, Rwanda, and Burundi were left with internal ethnic and political conflicts that produced later problems. Portugal hung onto its colonies of Angola and Mozambique until 1974, despite rebel independence movements sponsored by the Soviet Union and Cuba, who supported violent rebel movements as part of the Cold War.

## Economy, Accumulation, and the Production of Inequality

Sub-Saharan Africa has a very low level of economic development compared with other regions. The region is projected to have an average per capita GNP of $1,823 compared to world average per capita GNP of $11,342 in 2010 (about 1.7% of the total global economy). Many economies are dependent on just a few low-priced exports, and large numbers of people make a living as subsistence farmers or lack formal employment in urban areas. Africa is singled out for attention by many international agencies and receives the highest amount of development assistance per capita of any world region. The World Bank identifies sub-Saharan Africa as "the most important development challenge of the twenty-first century."

**Economic Development Theories** Development theorists agonized over prescriptions for African development in the late 20th century. In the 1950s African underdevelopment was explained by the lack of industrialization, and **modernization theorists** proposed solutions of technology transfer, training, and large projects for power generation, often supported by international assistance. Modernization projects included road construction, the Volta and Aswan dams, and harbors such as Tema in Ghana.

In contrast, scholars of the **dependency school of development theory** such as Walter Rodney and Samir Amin, blamed underdevelopment on the practice and legacy of colonialism. They argued that the dominant core capitalist powers of England and other European colonial powers transformed the political and economic structures of Africa to serve their interests in obtaining cheap raw materials, and in doing so undermined local agriculture and social development. Many countries emerged from colonialism with their economies and trade dependent on just a few products such as minerals or cocoa—for which the export value declined over time in relation to the price of imports.

These deteriorating **terms of trade** often affected small farmers. In 1990, for example, a farmer had to produce twice as much coffee as in 1960 to earn the money needed to purchase a bag of fertilizer (**Figure 5.21**).

For dependency theorists, the remedy was to reduce the dependency on export revenue and avoid the high costs of imports. This could be done by creating local capacity to produce goods that would replace those previously imported from other countries, especially manufactured industrial products (a policy called *import substitution*). Several African nations adopted strict import substitution policies that included subsidization of local industry and protection against foreign imports through tariffs and other mechanisms. As in Latin America (Chapter 7), import substitution was a mixed success and faced greater challenges because of Africa's generally low level of infrastructure, skills, domestic markets, and investment capital. Although it fostered the development of some industries, particularly small-scale manufacturing in African capitals, it also led to inefficiency and poor quality.

Another perspective explains Africa's underdevelopment in terms of its physical geography and the constraints of climate, disease, soils, and distance from markets. Economist Jeffrey Sachs has argued, based on statistical correlations, that many African countries have significantly lower economic growth rates partly because they are landlocked and have tropical climates. This analysis has echoes of **environmental determinism** in blaming nature for economic difficulties.

**Debt and African Economies** The development programs of both modernization and import substitution required capital funds that were not easily available in Africa, and many countries looked outside the region to borrow money. Because many African countries had poor credit ratings with commercial banks, most loans were made from other governments such as the United States or through international agencies such as the World Bank instead of on the private market. Although some funds were invested in infrastructural, industrial, and agricultural development, there is evidence that considerable sums were used to purchase arms or were diverted by ruling elites, increasing their own personal fortunes as overall debt increased.

The total debt of sub-Saharan Africa stood at $224 billion in 2008, only one-third of that of Latin America and Asia, but it is huge as a percentage of GNP in most African countries. In Burundi, Gambia, Guinea-

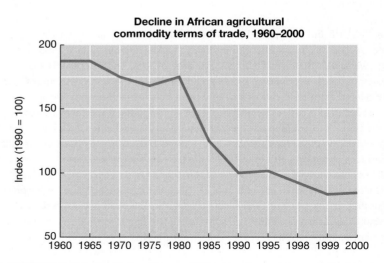

**▲ FIGURE 5.21 Declining Terms of Trade** African exports declined in value in relation to the value of imports between 1960 and 2000. This declining terms of trade has been especially difficult for those countries who do not export petroleum.

Bissau, Malawi, and São Tomé and Príncipe, the total external debt exceeded GNP in 2007 according to the IMF. Several countries were paying more than five times the value of their exports to service their debt each year.

**Structural Adjustment and Debt Relief** As in other regions, the multilateral agencies such as the International Monetary Fund (IMF) and World Bank responded to the debt crisis first by controlling the spiraling size of the loans through extending payment periods and adjusting interest rates and then by demanding structural adjustment policies. These economic policies, mostly associated with the IMF, required governments to cut budgets and liberalize trade in return for debt relief. In many parts of Africa this required devaluation of currencies (to make exports cheaper and thus more attractive to foreign purchases), liberalization of trade (by removing tariffs), reduced public spending, and the privatization of government-held companies.

The first country to accept structural adjustment was Ghana in 1983. By 1990, 30 African countries had implemented programs with the insistence of the IMF. In many of these countries, the impact of structural adjustment was severe, sending food prices spiraling and increasing unemployment as governments cut public-sector jobs. Kenya and Nigeria both fired more than 150,000 government employees in response to IMF policies. Some critics suggested that IMF should be renamed "Imposing Misery and Famine."

The structural adjustment programs fit within the broader program of neoliberalism, promoted particularly by the United States, that includes reducing government subsidies for social and agricultural programs, removing barriers to trade, and privatizing publicly owned land and corporations (see Chapter 7 for a discussion of similar policies in Latin America). But neoliberal policies of free trade clashed with the special concessions that had been granted to Africa by the European Economic Community (EEC). The Lomé Convention, named after the capital of Togo, where it was first signed in 1975, was an agreement between the EEC and 70 countries, 43 of which were in sub-Saharan Africa. In it, the EEC provides economic assistance and trade concessions to promote exports of certain key commodities such as sugar. The convention included access to European markets, stabilization of export earnings on selected commodities, industrial technology transfer, project financing, and development aid. However, such preferential treatment was opposed by the United States and nations in other regions who seek to compete for European markets, and in 2003 the Lomé Convention was replaced the Cotonou agreement, a compromise system of trade and cooperation with conditions on assistance including a requirement for "good governance" and strong participation of the private sector.

The destitution created in Africa by economic crises, structural adjustment, and war prompted international agencies and others to try to cushion the impact of restructuring in Africa by providing programs for alleviating poverty. This assistance is built on several decades of humanitarian and economic assistance to those regions of Africa suffering from disasters and war. Africa is the largest recipient of what is called foreign aid, receiving one-third of the global $105 billion total, equivalent to about 4.5% of the region's total GNP and averaging about $44 per person.

As the 20th century drew to a close, an international campaign was organized to pressure the G8 and major international finance institutions for debt relief, especially for the poorest countries in Africa. Official recognition that many of the countries had debt burdens that would permanently cripple development led to several debt-relief programs. In 1996 the World Bank and IMF introduced the Heavily Indebted Poor Countries (HIPC) initiative to restructure and forgive part of the debt of poor countries that, over a five-year period, showed a willingness to pursue neoliberal economic policies of reduced government spending and free trade. Mozambique was one of the first of about 20 African countries that qualified for this form of debt relief. The HIPC program was severely criticized because it was very specific about changes in policy and told countries how to run their economies and because the debt reductions were too small. An international protest movement, called the Jubilee initiative, petitioned to cancel the majority of debts owed by poorer countries by the year 2000. In 2000 several European countries and the United States did move to cancel the debts owed to them by many countries in Africa, and the IMF increased the amount of debt relief under HIPC.

The "Make Poverty History" movement, supported by popular musicians such as Bono of the band U2, called for action and organized public concerts such as "Live 8" to do more for Africa and debt relief. Pressure to relieve debt focused on the leaders of the wealthy "**G8**" countries (Canada, France, Germany, Italy, Japan, Russia, the United Kingdom, and the United States), who control much of the debt. In 2005 representatives of the countries who assembled at the G8 summit held in Gleneagles, Scotland, agreed to double annual aid to the developing world by $50 billion and to write off the debts of the world's 18 poorest countries, 12 of them in Africa. But Zambian Deputy Finance Minister Felix Mutati and members of the Make Poverty History movement, pointed to the additional need for fair trade so that African farmers can better compete with the West in a global market. The African Growth and Opportunity Act (AGOA) provides 35 sub-Saharan countries with preferential access to U.S. markets for a variety of agricultural, apparel, and steel products (**Figure 5.22**).

▲ **FIGURE 5.22 Make Poverty History** A meeting of the G8 leaders in Edinburgh, Scotland in 2005 sparked protests against the global poverty, especially the crippling debts of poorer countries in Africa.

**Contemporary African Economies** In terms of value, services, minerals, and petroleum dominate the overall economy of sub-Saharan Africa, but in terms of employment, 65% of the adult population works in agriculture. The majority of farmers have small plots and grow just enough food to feed their families (subsistence). The cash and export crop sector is large in some regions such as West Africa and Kenya, where export patterns for cocoa (Ghana, Côte D'Ivoire), coffee (Burundi), cotton (Benin, Burkina Faso), nuts (Gambia), rubber (Liberia), tea (Kenya), and coffee (Ethiopia, Uganda) were established in the colonial period.

Per capita agricultural production has fallen in the last 20 years because growth in agricultural production has not kept up with the growth in population and increased demand from urban populations. Of even greater concern is the fact that the benefits of agricultural progress have not raised the incomes or improved the nutrition of many of Africa's residents. Many regions have become dependent on food aid, and exports have declined relative to imports.

Blame for agricultural problems in sub-Saharan Africa has been attributed to environmental degradation, lack of infrastructure, government policy, and international market structures. Regionwide challenges, identified by geographer Godson Obia, include improving infrastructure for roads and storage and providing adequate incentives and rewards to local producers. Difficulties of getting crops to the market on Africa's dirt roads and tracks, especially in the rainy season, mean that farmers risk having to store grain and other products in granaries that are vulnerable to pests and molds.

Governments have controlled food prices to keep down wages and unrest in urban areas, and this has reduced the prices paid to farmers. The marketing boards established by colonial powers have smoothed out price fluctuations, but they have also kept the profits when world prices are good. International market volatility and a lack of information have made it difficult for farmers to move into new types of crops, and the general decline in the price of agricultural products in comparison to needed imports has also made it difficult to make a living in agriculture.

One often-overlooked success story is urban agriculture in Africa. In cities such as Nairobi (Kenya), Lusaka (Zambia), Kano (Nigeria), and Kinshasa (Zaire), more than half the residents cultivate gardens, either at their homes or on unused land in the city. Kinshasa has been described as a giant garden plot, with crops growing at every roadside, on traffic islands, and on the airport perimeter. Crops from these urban plots contribute to urban food security and incomes.

In Kenya the most rapidly growing export sector is fresh vegetables and cut flowers for export to Europe. Relying on refrigerated air transport out of Nairobi airport, Kenya now provides 40% of the European Union imports of fresh vegetables, sending more than $100 million worth of vegetables and fruit to Britain in 2002. Although these new industries provide employment and higher wages than some other sectors, the strict quality standards, perishability, and need for air transport mean that small producers find it difficult to compete. There are also concerns about pesticide risks to workers in the growing and packing sectors. This has led to a growing Fair Trade movement in which producers are paid a reasonable wage, and crops are produced more sustainably, including flowers, coffee, and crafts (**Figure 5.23**).

The region is less industrialized than others, although South Africa has a substantial manufacturing sector including iron and steel, automobiles, chemicals, and food processing (see Signature Region in Global Context: South Africa). A number of countries have important textile

▲ **FIGURE 5.23 Kenya Flower Industry** Women picking roses at a flower farm near Naivasha, Kenya. The flower industry, centered on Lake Naivasha, sends more than 1 billion cut blooms to Europe each year, including carnations and roses, and is now the fourth-largest flower producer in the world, after the Netherlands, Colombia, and Israel. Some of the farms are enormous, with more than 10,000 workers. Accusations of poor labor conditions and environmental problems of water scarcity and pesticides have resulted in improvements on farms seeking Fair Trade certification.

and clothing industries (see the chapter introduction). One of the major challenges is to keep more of the processing and finishing of goods in Africa rather than exporting raw materials, including the manufacturing of products from fibers such as cotton and foodstuffs. Small businesses, including those owned by women, are benefiting from **microfinance programs**, which provide credit and savings to the self-employed poor, including those in the informal sector, who cannot borrow money from commercial banks. Based on the demonstrated success of the Grameen Bank program in Bangladesh (see Chapter 9, p. 322), which provided small loans to thousands, African microfinance projects offer loans and secure savings opportunities to people who want to start or expand their businesses. Examples include loans to purchase sewing machines, food-processing equipment, agricultural supplies, and shop inventories.

The service sector is large in most countries, with many people working as civil servants in government as well as in the tourist sector (especially in southern and eastern Africa). The telecommunications sector has grown explosively with the rapid adoption of mobile phone technology in Africa, and several countries, especially those with high literacy and English language capability, have companies that perform "back-office services" such as processing insurance claims. Rwanda is establishing itself as a node for Internet communication (**Figure 5.24**).

There are two "hidden" contributions to economic development in most of Africa. The first is the *informal economy,* where millions of people work as street vendors and maids. In countries such as Zimbabwe, Tanzania, and Zambia, as much as half the GNP is associated with the informal economy, and more than three-fourths of the jobs are. The other significant contribution is from Africans working abroad who sent remittances of more than $19 billion home in 2007.

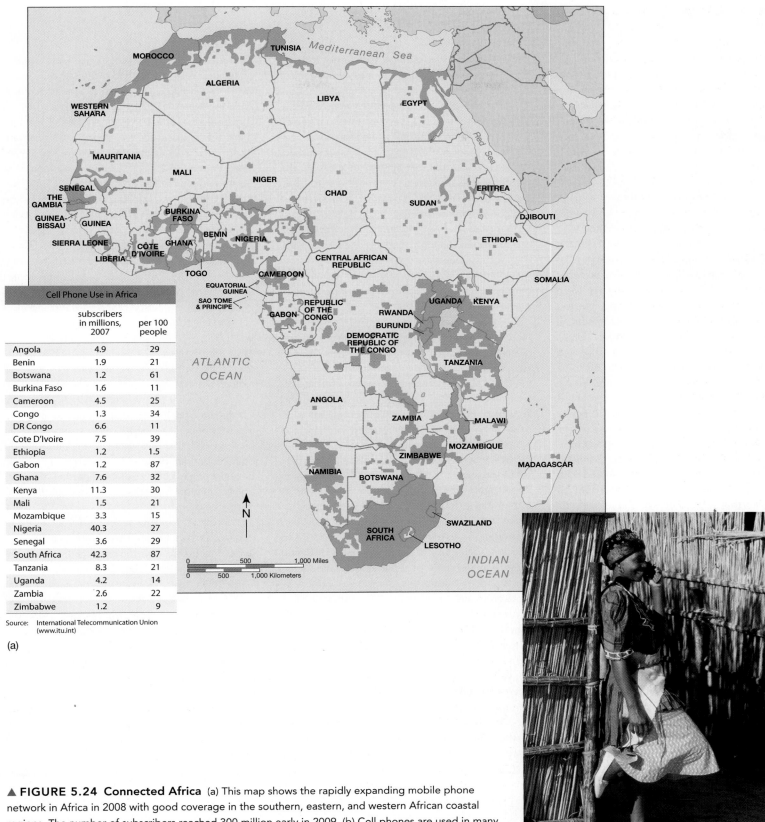

| Cell Phone Use in Africa | | |
|---|---|---|
| | subscribers in millions, 2007 | per 100 people |
| Angola | 4.9 | 29 |
| Benin | 1.9 | 21 |
| Botswana | 1.2 | 61 |
| Burkina Faso | 1.6 | 11 |
| Cameroon | 4.5 | 25 |
| Congo | 1.3 | 34 |
| DR Congo | 6.6 | 11 |
| Cote D'Ivoire | 7.5 | 39 |
| Ethiopia | 1.2 | 1.5 |
| Gabon | 1.2 | 87 |
| Ghana | 7.6 | 32 |
| Kenya | 11.3 | 30 |
| Mali | 1.5 | 21 |
| Mozambique | 3.3 | 15 |
| Nigeria | 40.3 | 27 |
| Senegal | 3.6 | 29 |
| South Africa | 42.3 | 87 |
| Tanzania | 8.3 | 21 |
| Uganda | 4.2 | 14 |
| Zambia | 2.6 | 22 |
| Zimbabwe | 1.2 | 9 |

Source:    International Telecommunication Union
           (www.itu.int)

(a)

(b)

▲ **FIGURE 5.24  Connected Africa**  (a) This map shows the rapidly expanding mobile phone network in Africa in 2008 with good coverage in the southern, eastern, and western African coastal regions. The number of subscribers reached 300 million early in 2009. (b) Cell phones are used in many areas without land line phones, charged by car batteries and solar energy and often shared in a community. They are used to get information about crop prices, call for medical help, access banking services, and keep in touch with friends and family.

# SOUTH AFRICA

South Africa has a distinctive history and particular role in contemporary sub-Saharan Africa. At the southern tip of the continent, with a more moderate climate and controlling a strategic transoceanic transport route, South Africa's Cape was colonized by the Europeans (see p. 168) and gained independence in 1961. Landscapes range from dramatic seacoasts to high mountains and from fertile agricultural valleys to grasslands grazed by cattle and wildlife. Although South Africa is now seen as an international economic and political leader within and beyond the region, there was a period where it was seen as an international outcast because of the policies of **apartheid,** a policy of racial separation. Prior to 1994 space and society were structured to keep black, white, and colored (mixed race) populations apart through control of the movement, employment, and residences of blacks. The goal of apartheid was separate development of the races within South Africa.

The history of racial segregation in South Africa is long, dating back to the establishment of a supply station by the Dutch East India Company in Cape Town in 1652. The Dutch, whose settlement developed slowly at first, were segregationists and attempted to prevent contact between whites and native people, although they did hold Africans as slaves. In 1806 Britain seized political control over the Cape to control the route to its empire in India, and the Dutch settlers migrated north to the Natal, Orange Free State, and Transvaal regions, became known as the Boers, and spoke Afrikaans.

The Boer policies of strict racial segregation between blacks and Afrikaners included the establishment of native reserves and the mandate that blacks needed permission to enter or live in white areas, restrictions known as the *pass laws.* Native people were incorporated into the economy as servants, squatter tenants, or semi-feudal serfs. Ultimately, the "Fundamental Law," established in 1852, legally enshrined the inequality of blacks and whites. Following the eventual British victory in 1902 against the Boers, the Union of South Africa was established under British dominion, including the Cape and the Boer territories to the north, but retained many of the Boer attitudes toward the black population. The first half of the 20th century witnessed the strengthening and extension of the Boer principles of racial segregation through territorial segregation. Black ownership of land was restricted, as was black settlement activity. In addition, the permanent residence of blacks in white urban areas was prohibited. The Natives (Urban Areas) Act codified this latter restriction, defining blacks as temporary urban residents who were to be repatriated to the tribal reserves if not employed. The act also required that blacks, while within urban areas, were to be physically, socially, and economically separated from the white population (**Figure 1**).

Following independence in 1931, Afrikaner-led governments imposed strict racial separation policies, transforming apartheid from practice to rule. Laws included the Group Areas Act of 1950, which established residential and business sections in urban areas for each race, and the Land Acts of 1954 and 1955, which effectively set aside more than 80% of South Africa's land for the white minority. In addition, the pass laws that required nonwhites to carry permits when in white areas were reinforced. The 1950 Population Registration Act classified all South Africans as either Bantu (black), colored (mixed race), or white. "Asian" was later added as a category. Segregation was enforced through regulations to prevent social contact and marriage between races, establishment of separate education standards and job categories, and provision for separate entrances to public facilities such as stations and hotels. Large-scale segregation was established in 1959 through the creation of ten **homelands,** a new version of tribal reserves. The homelands were areas set aside as tribal territories where black residents were given limited self-government but no vote and limited rights in the general politics of South Africa.

For nearly 40 years, apartheid was the method by which a white minority controlled a black majority through processes that were fundamentally geographical. Protests against apartheid were quickly and ruthlessly repressed, with African National Congress leaders such as Nelson Mandela jailed and activists such as Steven Biko killed. Enforcement of the requirement that black students use the Afrikaans language led to riots in Soweto in 1976. International objections to apartheid meant that South Africa was forced to withdraw from the British Commonwealth, and economic and trade sanctions as well as voluntary investment bans were instituted by some major international corporations. A number of white South Africans were also vocal in their opposition to the system.

The 1990s, after years of both domestic and international protest, saw the end of apartheid in South Africa—Nelson Mandela was freed from jail, and President F. W. de Klerk agreed to share political power between blacks and whites. In 1994 South Africa held the first election in its history in which blacks were allowed to vote, and Nelson Mandela was elected the first black president as leader of the ANC (African National Congress). The 1997 postapartheid constitution includes one of the world's most comprehensive bills of rights and prohibits discrimination based on race, gender, pregnancy, marital status, ethnic or social origin, color, sexual orientation, age, disability, religion, conscience, belief, culture, language, and birth.

South Africa now accounts for 27% of the continent's manufactured goods and has the highest overall gross domestic product (GDP at $249 billion. The major cities of Johannesburg and Cape Town host offices of international companies, and there is a well-developed scientific and finance sector (**Figure 2**). But for many poor and black South Africans, the end of apartheid raised unachievable expectations about access to jobs, housing, and land. Although some white residents have left, 10% of the population is still classified as white and another 11% as colored or Asian. Postapartheid governments have struggled to develop the economy and redistribute wealth, crime is common, AIDS is widespread, and social protests are frequent.■

▲ **FIGURE 1 Apartheid in South Africa** Whites only beach sign at Port Elizabeth South Africa in the 1980s in English, Afrikaans, and Zulu/Xhosa. Under apartheid white, colored and black populations were separated in public transport, entertainment facilities and schools.

▲ **FIGURE 2 Johannesburg** High-rise buildings host headquarters for major mining corporations such as Rand Mines and the Johannesburg Stock Exchange. The city is surrounded by townships such as Soweto. Many townships do not have adequate electricity and water supplies, and residents must use wood or coal for heat during the chilly winter season, casting a pall of polluting smoke over the townships that contributes to health problems and high infant mortality rates.

From about 2000, sub-Saharan Africa as a whole had strong economic growth, above the world average at 5% a year and with parallel growth in foreign direct investment from about $10 billion to $50 billion. New investment streams included Chinese interest in minerals and petroleum and a focus on expansion of the IT sector. Higher commodity prices contributed to growth for both mineral and agricultural products, with South Africa and Botswana especially competitive. Even after the global economic downturn in 2008, some experts saw continuing opportunity for Africa, although the falls in commodity prices will hit some countries hard, and foreign investment may decline.

**Social and Economic Inequality**   Economic and social conditions show great geographical and social variation within African countries. People in the core locations—urban areas, the formal sector, and producers of cash crops—generally have longer life expectancies, better service access, and higher incomes than those in the periphery, which includes rural areas and the informal and subsistence agricultural sectors.

Income concentration is high in many parts of sub-Saharan Africa, with the richest 20% of the population receiving more than 60% of overall income in most of southern Africa, the Central African Republic, and Sierra Leone. In Namibia, a staggering 79% of national income is held by the top 20% of society. Throughout sub-Saharan Africa in 2007, 1 billion or 41% of people lived on less than a dollar a day and 72% on less than two dollars.

Sub-Saharan Africa ranks low on many measures compared to other world regions. Life expectancy averages 18 years below the world average of 68 years and is lower than in any other world region. GDP per capita is about $1,823 compared to an $11,342 world average. The United Nations estimates that 40% of sub-Saharan Africans are poor as measured by income and 40% according to the more general measure of human poverty, which combines life expectancy, literacy, and access to basic services such as clean water. Conditions in Africa are also difficult for children, who have an infant mortality rate of 88 deaths per 1,000 children born (almost double the world average) and low levels of nutrition and immunization.

Within Africa, these generally gloomy average statistics do hide some regions with much better conditions. For example, as well as the Seychelles and the Mauritius, the United Nations ranks Cape Verde, Gabon, South Africa, Sao Tome and Principe, and Botswana higher on the Human Development Index (life expectancy, literacy, education, and GDP per capita) than other countries of Africa. Life expectancies in these countries are generally above 50, although many, including Botswana, Zimbabwe and South Africa, have dropped as a result of life expectancies shortened by HIV/AIDS. Levels of literacy, education, and income are also higher than the regional averages. These countries tend to have better provision of basic services. For example, in South Africa, Gabon, and Botswana, 80% or more of the population has access to safe drinking water. In almost all countries life expectancy, incomes, and services are better in urban areas than in rural ones.

Generally, conditions have improved over the last 25 years. For example, from 1970 to 2008, life expectancy has increased from 45 to 50, and infant mortality has dropped from 138 to 88 deaths per 1,000 children born. Literacy has shown dramatic changes, increasing from 38% in 1980 to 61% in 2005.

Improvements in conditions in specific countries are reflected in life expectancy and infant mortality changes. In Ghana, for example, life expectancy increased from 45 to 59 years, and infant mortality decreased from 131 to 71 deaths per 1,000 children. But war and AIDS have also affected Africa. Life expectancy in Botswana and Zambia started to drop in the 1990s as a result of AIDS and in Rwanda as a result of war and genocide. In 1998 AIDS was also mostly responsible for infant mortality increases in Botswana, Kenya, Zambia, and Zimbabwe. In 2007 an estimated 1.9 million people became infected with HIV in sub-Saharan Africa.

The lowest 22 countries on the Human Development Index are all in Africa. Sierra Leone, for example, has a life expectancy of only 42 years and an annual GDP per capita averaging $332, mainly as a result of the loss of life and economic collapse associated with civil war. Low levels of life expectancy reflect some of the deficiencies in service provision in Africa, where an average of 41% of the population lacks access to safe drinking water, and 67% lacks sanitation. Two-thirds of the population lacks safe drinking water in Ethiopia, Angola, Democratic Republic of the Congo, and Sierra Leone.

**The Millennium Development Goals and Africa**   African development is increasingly driven by a new set of eight targets established by the United Nations called the **Millennium Development Goals (MDGs)**, which aim to eradicate extreme poverty and hunger; achieve universal primary education; promote gender equality and empower women; reduce child mortality; improve maternal health; combat HIV/AIDS, malaria, and other diseases; ensure environmental sustainability; and develop a global partnership for development (**Table 5.1**). Sub-Saharan Africa ranked low on many of the indicators with continuing food insecurity, a rise of extreme poverty, stunningly high child and maternal mortality, large numbers of people living in slums, and a widespread shortfall for most of the MDGs. But some of the MDG indicators have shown progress since 1990. Millennium Villages have been established to end extreme poverty by working with the poor, village by village throughout Africa, providing affordable and science-based solutions to help people out of extreme poverty.

# Demography, Migration, and Settlement

The total population of sub-Saharan Africa is estimated to reach 778 million by 2010. The average birthrate, at more than 40 births per 1,000 population per year, is higher than any other continent, and there is a large proportion of young people. The overall population growth rate was about 2% a year, with a doubling time of about 29 years. The Population Reference Bureau has projected the 2025 population at about 1.16 billion.

Nigeria is the most populous country in Africa, with a 2010 population of 156 million, followed by Ethiopia with 74 million, the Democratic Republic of the Congo with 67 million, South Africa with 83 million, Tanzania with 41 million, and Kenya with 37 million. Some populations, such as Nigeria, have not been reliably censused for many years, and the estimates are approximate. Overall population density is relatively low, at 36 people per square mile in 2010, according to the United Nations, compared to the global average of 51. The only countries in the region with population densities higher than the global average are the Indian Ocean islands and the Central African countries of Rwanda and Burundi (**Figure 5.25**).

The density of population per unit of arable land is sometimes considered a better indicator of population pressure on land resources because it measures the ability of the land to support its population, or its carrying capacity. By this measure sub-Saharan Africa appears

**TABLE 5.1  Millennium Development Goals in Sub-Saharan Africa**

| Millennium Goals | Target | Indicator | Sub-Saharan Africa 1990 | Sub-Saharan Africa 2007 |
|---|---|---|---|---|
| 1. Eradicate extreme poverty and hunger | Halve, between 1990 and 2015, the proportion of people whose income is less than $1 a day | Proportion of people living on less than $1 a day, % | 44.6 | 41 (2007) |
| | Halve, between 1990 and 2015, the proportion of people who suffer from hunger | Proportion of people living with insufficient food, % | 36 | 29 (2005) |
| 2. Achieve universal primary education | Ensure that, by 2015, children everywhere, boys and girls alike, will be able to complete a full course of primary schooling | Net enrollment ratio in primary education | 53 | 70 (2006) |
| | | Literacy rate of 15–24-year-olds, % | 67.4 | 72 (2007) |
| 3. Promote gender equality and empower women | Eliminate gender disparity in primary and secondary education, preferably by 2005, and in all levels of education no later than 2015 | Girls' primary school enrollment ratios in relation to boys' (Girls per 100 boys) | 76 | 89 (2006) |
| | | Women in wage employment in the non-agricultural sector | *32.4* | *31 (2006)* |
| 4. Reduce child mortality | Reduce by two thirds, between 1990 and 2015, the under-five mortality rate | Under-five mortality rate per 1,000 live births, % | 185 | 146 (2007) |
| | | Infant mortality rate, % | 111 | 89 (2007) |
| 5. Improve maternal health | Reduce by three quarters, between 1990 and 2015, the maternal mortality ratio | Maternal mortality ratios per 100,000 live births 2000 | 910 | 900 (2005) |
| | | Proportion of births attended by skilled health personnel | 45 (2006) | |
| 6. Combat HIV/AIDS, malaria, and other diseases | Have halted by 2015 and begun to reverse the spread of HIV/AIDS | HIV prevalence in adults aged 15–49, % | *3* | *5 (2007)* |
| | Have halted by 2015 and begun to reverse the incidence of malaria and other major diseases | Number of new tuberculosis cases per 100,000 population (excluding people who are HIV-positive) | *148* | *369 (2007)* |
| 7. Ensure environmental sustainability | Integrate the principles of sustainable development into country policies and programs and reverse the loss of environmental resources | Proportion of land area covered by forests, % | *29.2* | *26 (2005)* |
| | | Halve, by 2015, the proportion of people without sustainable access to safe drinking water and basic sanitation | | |
| | By 2020, to have achieved a significant improvement in the lives of at least 100 million slum dwellers | Proportion of population using improved sources of drinking water, % | 49 | 58 (2006) |
| | | Number of urban dwellers living in slums (millions) | *101* | *166* |
| | | Improved sanitation facilities (% of population with access) | 26 | 31 (2006) |
| 8. Develop a global partnership for development | Address the special needs of the least developed countries, landlocked countries, and small island developing states | Debt service as a percentage of exports of goods and services | 11.5 | 5 (2007) |
| | Develop further an open, rule-based, predictable, non-discriminatory trading and financial system | Internet users per 100 people | 1 (2000) | 3 (2006) |
| | Deal comprehensively with developing countries' debt | Telephone lines and cellular subscribers per 100 people | 1 | 2 telephone lines (2007) and 23 cellular subscribers (2007) |
| | In cooperation with the private sector, make available the benefits of new technologies, especially information and communications | | | |

Source: This table lists the goals, targets, and selected indicators for the MDGs. There have been improvements in food availability and significant improvements in children's schooling, literacy, infant and child mortality, clean drinking water, national debt, and access to technology. But indicators have become worse (data in italics) for infectious diseases, forest cover, and slum dwellers, and almost all indicators for sub-Saharan Africa are a long way from the 2015 targets.

much more densely populated, with an average of 31 people per square kilometer (80 people per square mile) of arable land, which is just about equal to the global average. The total fertility rate (the average number of children born to a woman during her lifetime) is high in most of sub-Saharan Africa, at 5.4 in 2008 and reaching more than 6 children per woman in many regions. As a result, 43% of the African population is under age 15. This has major implications for future population growth and its effects, as this group starts to have children and makes demands on education and employment systems. Fertility rates are lower in southern Africa.

**Explaining Population and Fertility**  What are the reasons for high fertility and birthrates in Africa, and what are the prospects for slowing population growth? Population geographers and other researchers have focused on the study of African demography in response to these questions. They have found that although religious prohibitions of contra-

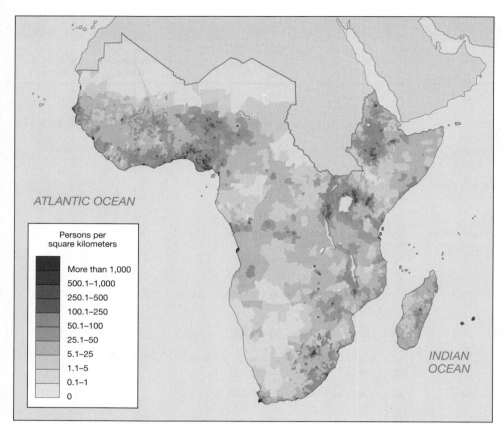

ATLANTIC OCEAN

Persons per
square kilometers

More than 1,000
500.1–1,000
250.1–500
100.1–250
50.1–100
25.1–50
5.1–25
1.1–5
0.1–1
0

INDIAN
OCEAN

▲ **FIGURE 5.25 Africa Population Density** Africa's population is mostly a scattered rural population, with the greatest density along the West African coast, the southeast coast of South Africa, the highlands of East Africa, and along major rivers. Population concentrations tend to be associated with better soils and climate, with colonial centers for mining and export crops, and with coastal ports. For a view of worldwide population density and a guide to reading population-density maps, see Figure 1.8.

ception and lack of access to contraceptive devices may play a small role, other factors are much more important.

Children are valued in Africa for many logical reasons, including their ability to work in agricultural fields and as herders, to help with household work, and to care for younger siblings. Children are also a possible source of financial or other gain when they marry. Children are the main source of security for elderly people in countries where there are few pensions or public services for the aged, and it is traditional for younger generations to respect and care for their elders. Large families are also often perceived as prestigious, a spiritual approach to linking past and present, and a way of ensuring family lineage. In regions of ethnic strife, children represent a way of securing votes, warriors, or political power. Even though infant mortality rates have improved with better nutrition and health care in much of Africa, many African families have internalized the need to have many children to ensure that some survive to adulthood.

Many studies have also shown that conditions for women have a strong influence on fertility rates, with lower age of marriage, minimal female education and literacy, and low rates of female employment all contributing to higher fertility rates. Fertility rates tend to be lower in urban areas with high rates of female education, employment, and later ages for marriage. In Kenya, for example, women with secondary or higher education have a total fertility rate of 4.9 compared to a rate

of more than 7.0 for women with little or no schooling, and in Nigeria the fertility rate is 4.2 for better-educated women compared to 7.2 for women with little education (**Figure 5.26**).

**HIV/AIDS**    Population projections for several African countries have been revised downward because of the high mortality and infection rates associated with the AIDS epidemic. In Zimbabwe and Botswana, where infection rates are more than 23%, estimates of average life expectancy have been revised down by 20 years, and population growth rates have been reduced or even reversed. Overall population projections for Africa for 2025 have been adjusted down by 200 million people.

Sub-Saharan Africa is more severely affected by HIV, the human immunodeficiency virus that causes fatal acquired immunodeficiency syndrome, or AIDS, than any other part of the world. African adult rates of HIV in 2007 were highest in Botswana, where 24% of the population was infected. The United Nations reports more than 22 million people in sub-Saharan Africa infected with HIV in 2007—66% of the worldwide total. The infection rate is estimated at 5% of all adults compared to the 0.8% world rate, and more than 14 million Africans have lost their lives to AIDS since it was identified in 1981. It has become the main cause of early death in Africa, killing more people than malaria and warfare (**Figure 5.27**).

The geography of AIDS in Africa varies by country, by regions within countries, and by social groups. Unlike in other regions, more women than men have AIDS in sub-Saharan Africa, and mothers often transmit the disease to their children. Frequently, married couples are both infected with AIDS and die from it—a situation that has created as many as 11.6 million *AIDS orphans* living in 2007.

Poverty exacerbates the AIDS problem in sub-Saharan Africa because most Africans cannot afford prevention (for example, through the use of condoms), testing, or medicines that prolong the lives of those with HIV/AIDS. Governments have low health-care budgets and may not admit to the severity of the AIDS epidemic. Few people have health insurance, diagnosis and treatment are often delayed, and interaction with other diseases such as tuberculosis increases mortality rates.

Some countries have had success in combating AIDS. Uganda and Senegal have promoted aggressive and successful AIDS education and prevention campaigns and have cut infection rates in half. Unprecedented international agreements with drug companies in combination with new assistance programs from the World Bank, charities, and donor countries are helping lower the cost of drugs. But many are still unable to obtain the drug therapies that will prolong their lives.

There is great concern about the potential impact of some other emerging viruses in central Africa, specifically Ebola fever, which causes severe bleeding and kills more than 50% of its victims. So far, outbreaks such as the ones in the Congo in 1995 and in Uganda in 2000 have been contained, but only after killing more than 200 people in each case.

**Urbanization**    Although Africa is the most rural of world regions, it has been urbanizing rapidly over the last 40 years. In 1960 the urban population of sub-Saharan Africa was only 17 million people, about

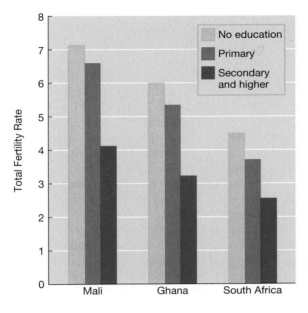

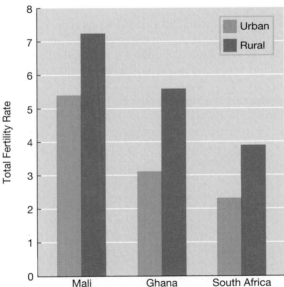

▲ **FIGURE 5.26  Status of Women and Fertility in Africa** Women with higher levels of education have less children, as shown by this 2003 data for three countries. South Africa, with the lowest fertility rates, also has a higher per capita GNP. The second graph shows that urbanization is also associated with falls in fertility rates, mainly because women marry later, take up formal employment, and health care and education are generally better.

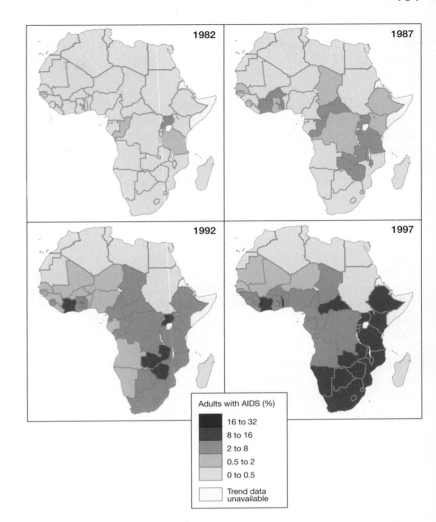

▲ **FIGURE 5.27  AIDS in Sub-Saharan Africa** These maps of the progression of the HIV epidemic in Africa from 1982 to 1997 show the increasing percentage of adults infected. The highest rates of infection were in eastern Africa in the early 1980s but have now shifted to southern Africa, especially Botswana, Zambia, and Zimbabwe, where more than 30% of adults are infected. Urban dwellers who have multiple sex partners, including young office workers and migrant workers, have a higher infection rate, as do women who work in the commercial sex trade and the wives and children of migrant workers. The incidence is lower in rural areas except along major truck routes and in areas where there are a lot of soldiers. The road from Malawi to Durban, where 92% of truck drivers were infected, has been called "the highway of death." Migrant workers have taken the disease back to their homes. The death of skilled farm laborers has resulted in a decline in agricultural output, and many young professionals crucial to the region's future have left their jobs because they have AIDS. Major industries and companies in southern Africa, such as diamond mines and banks, estimate that absence and loss of employees to AIDS is costing them more than 5% of their profits. AIDS is having a serious impact on life expectancy and population size and composition in Africa. Countries such as South Africa and Zimbabwe have adjusted projections of population and economic growth down to take account of the serious negative impact of AIDS on mortality and economic productivity.

20% of the overall regional population. By 2010, the urban population will reach 323 million people, 37% of the total, and was growing at about 3.6% per year. The level of urbanization varies greatly by country, from South Africa with 61% of its people in urban areas, to Ethiopia at 18% and Uganda at 13%. East Africa has a lower level of urbanization (24%) than does southern (59%) or West Africa (45%).

Although urban growth is partly driven by overall population, cities have been growing twice as fast as overall populations. Fertility rates tend to be lower in urban areas. As in many other regions, the major driver of urban growth is migration from rural areas to the cities, and the factors pushing people from rural areas and pulling them to the cities are somewhat similar. People are leaving rural areas because of poverty, lack of services or support for agriculture, scarcity of land, natural disasters, and civil wars. Urban areas are more attractive because they offer jobs, higher wages, better services (including education and electricity), and entertainment. Urban areas have benefited from the **urban bias** of both colonial and independent governments in Africa, which tended to invest disproportionately in capital cities that housed centralized administrative functions. In addition, food prices were kept down in the cities to reduce wage demands and to decrease the risk of civil unrest.

About 30% of sub-Saharan Africa's population lives in the largest city in their country. Lagos (Nigeria), Kinshasa (DR Congo), and Johannesburg (South Africa) are the largest cities with populations of about 10.5 million, 8 million, and 3.6 million, respectively, and there are about 40 cities with more than 1 million inhabitants. Geographer David Simon has used several indicators to identify African cities of greatest regional importance and links to the global system, including the presence of stock markets, large numbers of embassies, air traffic, and headquarters of international or regional organizations and corporations. According to these criteria, Johannesburg, South Africa; Nairobi, Kenya; and Lagos, Nigeria, lead with more than 70 embassies and 40 regional or international headquarters in each city.

Life in African cities has great contrasts of wealth and poverty, with many of the inhabitants living in the informal settlements that surround major cities, with inadequate social services, sanitation, water supply, energy, waste disposal, and shelter. Over time, some of the most notorious urban slums such as Soweto outside Johannesburg in South Africa or Kibera in Nairobi, Kenya, have become important to overall functioning of cities, housing many of the lower-paid workers, slowly improving services, and becoming centers of social movements that make political claims for the poor.

For example, Lagos—the colonial capital of Nigeria until the capital was changed to Abuja in the interior in 1992—has a metropolitan population of more than 13 million people and many serious problems (**Figure 5.28**). Sited on a natural harbor, Lagos was developed by the British as a rail terminus beginning about 1880. It also became a leading cargo port, industrial area, and center for production of consumer goods. About 80% of Nigeria's trade goes through Lagos, although some is shifting to the oil regions to the east. The city region includes 53% of Nigeria's manufacturing, 62% of the gross industrial output, and 22 industrial estates. When structural adjustment caused food price increases and unemployment after 1989, city residents expressed their frustrations by rioting. Lagos is also a cultural center, with many well-known writers and musicians, and has become the home of the West African film industry (see p. 186).

Lagos is infamous for its traffic and crime problems. The average commute to work is more than 90 minutes in polluted air and tangled traffic jams, made worse by inadequate bridges between islands. Electricity and other services are also insufficient; as a result of interruptions and lack of service, the National Electric Power Administration (NEPA) has been given the nickname of "Not Expecting Power Anytime."

One of the ways in which urban Africans reduce food expenses is through urban agriculture and in cities such as Kampala, Uganda, any unused land—along roads and railways and around airport and factories are covered with small gardens or grazed by domestic animals.

**The Sub-Saharan African Diaspora**   Migration into Africa from other regions is overwhelmed by the immense African diaspora and emigration from Africa to other regions. Millions of black Africans were captured and sent as slaves, initially to the Middle East, but

▼ **FIGURE 5.28 Lagos** Pedestrians avoid polluted water in a poor neighborhood of Lagos. The city is built on wetlands on the coast and often floods in the rainy season.

more significantly to the Americas, where their descendants represent a high percentage of the populations in Brazil, the Caribbean, and the United States. A second wave of emigration was associated with the aftermath of colonialism, a time when many Africans retained British Commonwealth passports or French citizenship and moved to Britain and France (or other Commonwealth countries such as Canada) in search of work. This included many people living in the Caribbean, often the descendants of slaves, who then were part of a secondary migration back to England or to Canada and the United States.

Other recent emigrations from Africa include movements of white populations from South Africa and other countries to Europe, North America, and Australia and a "brain drain" of 20,000 African students and professionals, especially doctors, per year to universities and companies in the core regions of Europe and the Americas. There are also African refugee populations in several world regions.

**Migration within Africa** Contemporary migration within Africa is mainly associated with movements in search of work and with refugees fleeing famine, floods, and violent conflict. Labor migrations emerged during the colonial period when loss of traditional land to colonists and high taxes forced people to look for other work, and employment became available in mines and on plantations. For example, Sahelian residents migrated to work in peanut-, cotton-, and cocoa-producing areas in Senegal, Côte D'Ivoire, Nigeria, and Ghana; in East Africa, Hutus and Kikuyus migrated to work on Kenyan and Ugandan coffee and tea plantations and European farms. The most significant labor migration of the last 100 years is from southern Africa, especially Botswana and Zimbabwe, into South Africa to work in the mining industry. In 1960 more than 350,000 foreign workers were employed in South Africa. These migrations have disrupted family life, but the remittances that are sent back by workers have become an important contribution to local and, in the case of countries such as Botswana and Lesotho, national economies. Labor migration also continues from inland West Africa to coastal cities such as Abidjan.

**Refugees** The United Nations High Commission on Refugees (UNHCR) estimated that there were more than 2.7 million refugees in sub-Saharan Africa in 2007 (down from 4.5 million in 2002), more than 20% of the world total and the largest number of any world region. UNHCR reported that the "total population of concern," which takes into account refugees, stateless persons, asylum seekers, and internally displaced persons, was 10.5 million throughout the region. Armed conflict and human rights violations in the Central African Republic, Chad, the Democratic Republic of the Congo, Somalia, and Sudan led to outflows of refugees numbering close to 120,000 people, mainly to Kenya (25,000 arrivals), Cameroon (25,000), Sudan (22,500), and Uganda (9,400). Other refugees include those from conflicts in Rwanda and Burundi, although many of these have been recently repatriated, contributing to an overall decline in the refugee population.

Sub-Saharan Africa continued to host over one-fifth of the world's refugee population at the end of 2007. There are great regional variations within Africa. For example, West Africa contained the region's lowest number of refugees at 174,000, whereas the number was significantly higher in the East and Horn of Africa at 815,000 and Central Africa and Great Lakes at 1.1 million. The Republic of Tanzania was the world's sixth-largest asylum country in 2007, taking in 436,000 refugees. The largest individual refugee camp is also located in Tanzania, with some 91,000 inhabitants.

The refugee populations place serious burdens on neighboring countries that lack the resources to feed and resettle the impoverished and starving arrivals. Guinea, for example, absorbed almost a half million refugees from Liberia and Sierra Leone, and Tanzania took in a similar number from Burundi. Most international refugees are housed in camps that are supported by international organizations and charities (**Figure 5.29**). Disease spreads rapidly in the crowded conditions of the camps, and food supplies are sometimes interrupted or diverted by military groups and governments. Refugees are often accused of spreading HIV and other diseases, but excluded from HIV/AIDS programs. Long-standing conflicts and loss of livelihoods at home mean that many refugees spend long periods in the

▲ **FIGURE 5.29 Refugees** Millions of Rwandans fled the country in the 1990s to camps such as this in Katale, Democratic Republic of Congo. Many have now returned with peace in Rwanda.

camps with little hope of return. However, more peaceful conditions and carefully monitored repatriation have resulted in the return of refugee populations to countries such as Rwanda and Mozambique.

People forced to move within their own countries because they do not fall under international definitions or assistance for refugees are some of the most desperate migrants. For example, UNHCR estimates that there are one million internally displaced people in the Democratic Republic of the Congo, most of them inaccessible to relief organizations.

# CULTURE AND POLITICS

Sub-Saharan Africa is a large and extremely diverse region that should not be overgeneralized, with hundreds of different ethnic groups with traditional culture, religions, and languages, and a wide range of political systems and social organizations. Although many African countries have suffered serious conflicts in recent decades, including those in the aftermath of independence and with interventions by non-African superpowers, many African countries have moved toward democracy.

## Tradition, Religion, and Language

It is hard to draw cultural generalities from a continent as large and diverse as Africa. Those who do generalize highlight the importance of the extended family, ties to the land, oral tradition, village life, and music in traditional African culture.

The importance of the extended family is linked to the supremacy of *kinship* ties in social relations and obligations and to a widespread respect for elders as sources of wisdom. Kinship ties going back multiple generations in the same region may define "clan" allegiances and may drive primary loyalties, as in contemporary Somalia, where inter-clan conflict has dominated recent political events. Different extended families are often linked through intermarriage, with a transfer of wealth, sometimes in the form of cattle, from the husband to the wife's family as a mark of respect and value of the woman's labor and companionship.

The tie to the land is connected to traditional forms of land tenure in Africa, where in many regions land was viewed as given by the spirits or held in trust for ancestors and future generations. The Elesi of Odogbolu (a traditional leader) in Nigeria expressed this view in these words: "The land belongs to a vast family of which many are dead, few are living, and countless members are still unborn." This gives land a communal nature such that it could not be bought or sold by individuals. In some cases, land rights are held by the extended family or the community, rather than by the individual, and in other societies the chief or king controls land.

Traditions of reciprocity, where a gift is given to obtain a favor, and of helping family members are a major source of cultural confusion, according to African historian Ali Mazrui. He suggests that these traditions provide an explanation for the way in which some leaders have favored family members with jobs in their administrations and for the use of bribes when making requests from government officials. He also notes that under colonialism, local residents viewed stealing from the government as a legitimate form of resistance because they felt that foreigners were robbing Africa of resources and funds through taxation.

**Religion**   Traditional African religions have been described as animist (worship of nature and spirits), but this overgeneralizes the wide variety of local religious beliefs in Africa. Although natural symbols, sacred groves of trees, and landforms may have religious significance, many African religions also feature a belief in a supreme being, several secondary gods or guardians, good and evil spirits, and ancestor worship. Ancestors, priests, or witch doctors mediate and interpret the wishes of the gods and spirits, and rituals ensure the stability of society and relations with the natural world. More than 70 million people (about 10% of the total population) are reported to practice traditional religions in the continent of Africa.

Christianity spread into sub-Saharan Africa via North Africa and Ethiopia from about C.E. 300, but the pace of conversion accelerated rapidly under European colonial rule and European missionaries. Dutch Calvinism in southern Africa; Catholicism in French, Spanish, and Portuguese colonies; and Anglican beliefs in the British colonies all had strong influences. Of the 360 million estimated Christians in all of Africa, there are about 125 million Roman Catholics and 114 million Protestants. The vibrancy of Christianity in many countries draws more young people to a career in the church than in many other world regions, and Africa-trained pastors are now taking leadership positions in European and North American churches.

Islam is another major religion in Africa as a whole, with 308 million adherents, and it is predominant in the Sahel, North Africa, and some East African coastal communities. It was spread by traders and drew some fierce adherents among West African groups, such as the Fulani, who went to war to eradicate animistic beliefs. As in Latin America, traditional religion has blended with Christianity and Islam to create forms in which local traditional rituals are incorporated into religious services. Another parallel to Latin America is the recent rapid spread of evangelical Christianity in many regions of Africa.

Religious differences have fueled political conflict in some regions of Africa, most notably where Muslims and Christians were forced into the colonial national boundaries. In West African countries bordering the Sahel, such as Nigeria, tensions exist between northern Islamic groups, such as the Hausa, and southern Christians such as the Igbo. Sudan is another region of religious conflict (see Chapter 4).

**Language and Ethnicity**   The geography of languages in Africa is incredibly complex, with more than 800 living languages, 40 of them spoken by more than 1 million people (**Figure 5.30**). The dominant indigenous languages, spoken by 10 million or more, are Hausa (the Sahel), Yoruba and Ibo (Nigeria), Swahili (East Africa), and Zulu (southern Africa). Hausa and Swahili are known as trade languages (or **lingua franca**), spoken as second languages by many groups to facilitate trade. English, French, Portuguese, and Afrikaans are also spoken in regions of recent colonial control and education systems or white settlement, and Arabic is common toward northern Africa. Arabic has strongly influenced Swahili along the east coast of Africa. Because most countries have no dominant indigenous African language, they have often chosen a European language for official business and school systems. The countries with the most coherent overlap between their territory and a dominant African language are Somalia (Somali), Botswana (Tswana), and Ethiopia (Amharic).

The multiplicity of languages and dialects reflects the large number of distinct cultural or ethnic groups in Africa. Some writers use the term *tribe* to define these groupings and describe Africa as a *tribalist* society.

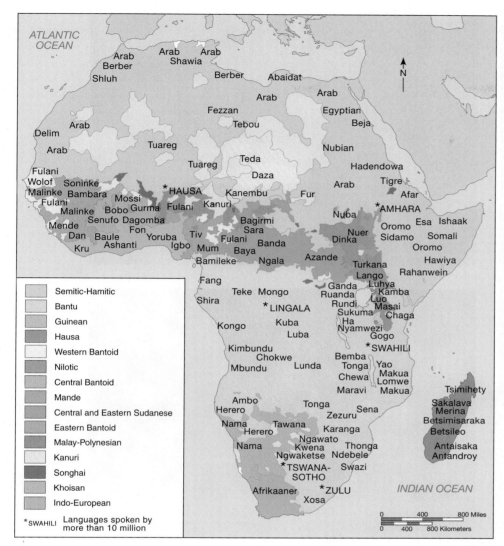

ATLANTIC OCEAN

Semitic-Hamitic
Bantu
Guinean
Hausa
Western Bantoid
Nilotic
Central Bantoid
Mande
Central and Eastern Sudanese
Eastern Bantoid
Malay-Polynesian
Kanuri
Songhai
Khoisan
Indo-European

*SWAHILI Languages spoken by more than 10 million

INDIAN OCEAN

▲ **FIGURE 5.30 African Languages** The cultural complexity of Africa is clearly demonstrated in the variety of indigenous languages shown on this map. The indigenous languages of Africa can be grouped into larger language families, including the Afro-Asiatic languages of North Africa such as Somali, Amharic, and Tuareg; the Nilo-Saharan languages such as Dinka, Turkana, and Nuer in East Africa; and the largest Niger-Congo group, which includes Hausa, Yoruba, Zulu, Swahili, and Kikuyu. A small family is the group of Khoisan languages spoken by the bushmen of southern Africa, which have a distinctive "click" vocalization. Swahili and Hausa are spoken by millions as the trade languages of East and West Africa, respectively.

The term *tribe* describes a form of social identity created by groups who share a common set of ideas about collective loyalty and political action, with group affiliation based on shared kinship, language, and territory. Although it is used by many groups to identify themselves, other groups see the term as negative (related to colonial perceptions of savagery) and now prefer to use the term *ethnic group*.

The largest ethnic groups in Africa are associated with the dominance of certain languages, such as Hausa, Yoruba, and Zulu, but almost all groups were either split geographically by colonial national boundaries or grouped together with their neighbors, enemies, or others with whom they shared no affinity. Attempts to consolidate ethnic groups across boundaries and struggles for power between groups within countries are a major cause of conflict in contemporary Africa.

For example, tensions between the Ibo and Yoruba in Nigeria led to civil war when the Ibo declared the independence of eastern Nigeria as Biafra in 1967. The conflict, which drew international attention and intervention because of starvation in Biafra and the presence of oil in the region, resulted in as many as 1 million deaths (mainly from hunger and disease) and lingering ethnic resentments after Nigeria was reunited. Ethnic and tribal tensions are also high in the Horn of Africa, especially in Somalia, where tribal warfare frustrates attempts at peace.

## Cultural Practices, Social Differences, and Identity

Africa is associated with a rich tradition of music and the arts that has increasingly influenced other regions of the world. Slavery was one way in which African musical, artistic, and food customs spread around the world, especially to the Americas. The arts are dynamic in much of Africa, often reflecting international influences and new youth identities.

**Music** Traditional music of the West African Sahel may have influenced the development of American blues, and West African coastal traditions influenced Afro-Caribbean music styles such as the Cuban rumba.

Africa is often associated with music from percussion instruments, especially drums. Other instruments with metal keys that are plucked or tapped, as well as flutes and harplike stringed instruments, are also commonly used in traditional and popular music. Such musical traditions vary widely across the continent and include the complex rhythms of women drummers in Tanzania and of xylophone players in Uganda and the chanting of the Zulu of southern Africa. In West Africa, oral traditions are associated with the singers and storytellers, some of whom receive the respected name of *griots*.

African popular music mixes indigenous influences with those of the West, especially those of the United States and the Caribbean. For example, the highlife music of West Africa is derived from Caribbean calypso and military brass bands, adding stronger percussion, soul influences, and exchange between lead and background singers to emerge as the now internationally popular *juju* or Afrobeat sounds of Nigeria's King Sunny Adé and Fela Kuti. In French West Africa, singers such as Youssou N'Dour blend traditional African beats with powerful vocals, and in South Africa, singers such as Miriam Makeba received international recognition in the 1950s, presaging the popularity of the a cappella style of South African black musicians such as Ladysmith Black Mambazo. Popular musicians have often expressed political opinions against apartheid or corruption, and Senegalese musician Youssou N'Dour campaigns to combat malaria (**Figure 5.31a**). The Festival in the Desert, held in Mali each year, brings together musicians from across West Africa with local pastoralists, tourists and international media.

▲ **FIGURE 5.31 International Fame for Africans** (a) Senegalese musician Youssou N'Dour sings at the Africa Live concert to fight malaria. (b) Footballer Didier Drogba from the Ivory Coast has gained international fame as a player for the English team Chelsea. He began his career in France, was named African footballer of the year in 2006, and is known for charity work, including as a UN Goodwill ambassador. (c) Haile Gebrselassie from Ethiopia and Paul Tergat from Kenya at the finish line of the men's 10,000-meters of the 2000 Olympic Games in Sydney.

**Art** African art is also incredibly varied and includes painting, metalwork, and sculpture. In traditional Africa, artists were valued specialists, often under the patronage of kings and producing works of spiritual value. The masks and wood sculptures of the Dogon and Bambara people of West Africa are now collected around the world while maintaining cultural significance within the region. Kente cloth designs from northern Ghana have become meaningful in African-American identity and clothing. As interest in travel and world culture has grown, artists and others have started to produce items for sale to tourists and to international distributors, including some organizations that try to ensure fair trading principles of returning as much value as possible to local people. But not all African art is traditional, and galleries in major cities and Web sites such as Novica promote young modern artists to a global market.

**Film** Nigeria hosts a booming film industry that has become known as "Nollywood," producing hundreds of videos for the home video market each year in an industry now third only to the United States and India. These films are often shot on location for very little money but are incredibly popular, often dealing with moral questions and social issues such as AIDS, corruption, and witchcraft. The films are shown throughout Africa and also in diaspora communities in the United Kingdom and United States.

**Sport** For many young Africans their heroes are sports stars, especially those who make it internationally in football or athletics. Even in remote villages boys will be playing with improvised footballs, hoping to be spotted and eventually recruited to play for a European team.

Superstars such as Didier Drogba from the Ivory Coast, who plays for the English football club Chelsea (**Figure 5.31b**), or Samuel Eto'o from Cameroon, who plays for Spanish club Barcelona, become very rich and enormously popular. Another group of African sports stars are the long-distance runners of East Africa, where Ethiopia and Kenya have produced numerous Olympic gold medalists in both men's and women's events (**Figure 5.31c**).

**The Changing Roles of Women**   Women in Africa tend to have less education and lower incomes than men, but in most countries they live slightly longer. The average female annual income (PPP) in sub-Saharan Africa is about $2000 less than for men, and literacy rates are 54% for women compared to 69% for men. Gender differences in Africa demonstrate the process of the **feminization of poverty**, whereby more than two-thirds of all people who join the ranks of the poor are women. African women are more likely to be poor, malnourished, and otherwise disadvantaged because of inequalities within the household, the community, and the country. Women are less likely to receive an education, and overall pay rates are lower in the workplace. Patriarchy and cultural traditions mean that women may be required to eat less than men and only after the men in the family have eaten, and women bear disproportionate responsibility for heavy household work such as collecting fuel wood and water (**Figure 5.32**). The tradition of female circumcision has become a controversial struggle between those who see it as a human rights abuse involving brutal genital mutilation and health risks and others who see it as an important religious and symbolic experience. More than 100 million women in Africa are estimated to have undergone the ritual.

Women are also disadvantaged by many traditional and modern institutions that define property rights. Land may be passed on only to male children, and new land titles are often granted to male household heads. Development policies for agricultural training and technology to make work easier have been directed at men, and new projects for tree planting or cash crops often focus on men. Although some African governments and development agencies have recognized these disparities and established programs meant to improve conditions for women, poverty reduction among women has been patchy and partial.

There are places in Africa where significant numbers of women are seen as powerful, especially in urban areas, where female entrepreneurs have been successful (see the chapter introduction). An early call for more attention to the role of women in Africa was that of economist Esther Boserup, whose 1970 book, *Women's Role in Economic Development*, documented the importance of women as farmers and resource managers in Africa. Geographers, among others, subsequently documented the work of African women in meeting the basic needs of families through preparing food and collecting wood and water, as income generators in crafts and community work, and as agricultural producers in both subsistence and commercial sectors. They also showed distinctly gendered spaces in the African landscape, with parts of the home, the market, and certain trees and crops reserved primarily for women's lives and work. A 1996 study by the United Nations Food and Agriculture Organization (FAO) found that women's contribution to the production of food crops ranges from 30% in Sudan to 80% in the Republic of the Congo and that women in sub-Saharan Africa are responsible for 70% of overall food production, 100% of food processing, 50% of animal husbandry, and 60% of agricultural marketing.

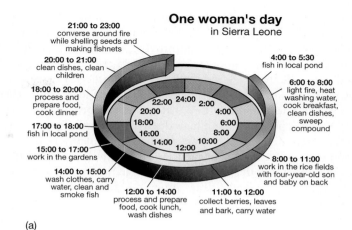

(a)

(b)

(c)

▲ **FIGURE 5.32 Women's Work in Africa** (a) This graphic from the United Nations Food and Agricultural Organization shows the typically large burden of work carried by many African women. (b) Ethiopian women carry heavy loads of firewood. (c) Women drawing water from a well in Senegal.

Those who understood the role of women's work in African communities and economies criticized development policies that ignored, undervalued, or displaced women. They also showed how women were often disproportionately affected by environmental degradation, as deforestation and drought made more difficult the work of collecting wood, water, and food.

Development policies gradually began to incorporate these ideas. The "Women in Development" (WID) approach focused on women's productive roles with projects that provided technology, credit, and training to women. This approach was in turn criticized for ignoring women's reproductive roles and the larger social processes such as discriminatory land-tenure policies that shape women's lives. In the 1980s the **gender and development (GAD)** approach was promoted as better linking women's productive and reproductive roles and trying to understand the gender-related differences and barriers to better lives of both men and women. Development agencies such as the World Bank incorporated elements of both approaches into programs that supported education, credit, and land-titling programs for women, women's organizations, and recognition of women's work.

## Territory and Politics

The search for peace and for democratic politics in Africa has been frustrated by the legacies of colonial frontiers and the Cold War, ethnic rivalries, and the special interests of powerful individuals and sectors. Although some countries were able to create (or re-create) a sense of national identity following independence, others are still coping with internal struggles, contested nationalisms, and claims on land beyond their current borders. The cost of wars in sub-Saharan Africa has been a hindrance to investments in development. The Stockholm International Peace Research Institute reports that military expenditures by governments in sub-Saharan Africa ranged between $6.6 and $9.5 billion in the 1990s and that many governments were spending more on arms and the military than on education or health.

**The Cold War and Africa**  Independence movements and transitions in Africa coincided with the global tensions associated with the Cold War between the United States and the Soviet Union. Several African independence leaders had been introduced to socialist ideas during education overseas, and many other Africans found communist and socialist ideas of equity and state ownership appealing after the repression, foreign domination, and inequality of colonial rule. In Tanzania, President Julius Nyerere developed the concept of an African socialism based on the traditional values of communal ownership and kinship ties to extended family expressed as *ujamaa* (familyhood). He believed that a socialist system of cooperative production would be more compatible with African traditions than would individualistic capitalism.

In Angola and Mozambique, where the Portuguese fiercely repressed independence movements, revolutionary movements espousing leftist ideals attracted the interest of the Soviet Union, China, and Cuba, which provided military and economic assistance and trained young Africans in their universities. Angolan troops supported rebels in Namibia (at that time under South African control). The United States and South Africa both sought to support pro-Western movements that came into conflict with Marxist–Leninist governments and movements.

By the 1980s, 50,000 Cuban advisors and troops and millions of dollars' worth of Soviet military aid were flowing into Angola and Namibia and only when South Africa agreed to grant Namibia independence did Angola agree to reduce Cuban and Soviet presence and to seek peace with the pro-Western rebel groups.

The Cold War had a different manifestation in the Horn of Africa, where independent countries came into conflict and sought arms and assistance from the superpowers. Somalia had sought Soviet aid as early as 1962 because the United States supported Ethiopian and Kenyan regimes that were resisting Somali expansion into adjacent territories with large Somali populations (a process called *irredentism;* see Chapter 1, p. 36). In Ethiopia, Emperor Haile Selassie had ruled for more than 40 years over an economy, organized on semifeudal lines, in which the concentration of land and wealth contributed to periodic famines. Growing demands for reform grew during the 1960s, culminating in 1974 in a takeover by a council of junior military officers (the "Derg"), who promoted an Ethiopian socialism of self-reliance and widespread land reform in support of peasants and workers.

Africa provided fertile soil for Cold War rivalry as newly independent nations searched for political ideals, dealt with civil wars and incursions from their neighbors, and sought assistance to develop their economies. In many countries, millions of dollars were expended on arms and other military assistance, thousands were killed, and rural areas were abandoned because of land mines. Many countries are still recovering from these conflicts.

Geographers Samuel Aryeetey-Attoh and Ian Yeboah have identified multiple causes for continuing political instability in Africa, including ethnic conflict, poor leadership, outside interference, and the legacies of recent independence struggles and racist government. They noted the frequency of military coups, with Ghana, Nigeria, and Uganda all experiencing at least five coups since independence about 50 years ago. They are also concerned about the number of elected leaders who eventually drifted toward one-party states and dictatorships, with accompanying repression and restrictions on freedom of speech. They do see signs of optimism emerging in the political geography of Africa, as many countries moved toward democratic elections, and as political and ethnic tensions were reduced with the end of the Cold War and of apartheid in South Africa.

**Recent Conflicts**  There are a number of unresolved and serious political conflicts and crises in sub-Saharan Africa, although others have reached peaceful resolution in the last few years (**Figure 5.33**).

**Somalia**  Some of the most difficult problems are associated with the *failed state* of Somalia in the northeastern Horn of Africa. The postwar boundaries of the state of Somalia left out some traditional Somali cultures and territories that were granted to Kenya, Djibouti, and Ethiopia. Military governments have failed to regain these areas and have lost internal legitimacy. Within Somalia, rivalries between different clans have undermined stability, and there has been a state of civil war for much of the time since 1990, with clans in the northeast and southwest declaring autonomy and being ruled by warlords. The combination of natural disasters (drought, tsunami, and floods) and conflict means that thousands of Somalis are internal or international refugees. Civil unrest, collapse of fisheries, unemployment, and a location on the Gulf of Aden have provided opportunities for young Somali men to engage in piracy. They seize ships and demand ransoms in the millions of dollars, rejuvenating some local economies with their spending. The UN Security Council has introduced sanctions against Somalia for its failure to control piracy, and other countries have offered antipiracy assistance.

**Zimbabwe**  The economy of Zimbabwe has spiraled downward over the last decade. GDP has declined, inflation has soared into the thousands of percent, by 2008 unemployment was over 90%, and 80% of the population is below the poverty line at risk of hunger, cholera, and other disease outbreaks. The problems in Zimbabwe partly stem from a difficult transition from British colonial rule, with a white apartheid

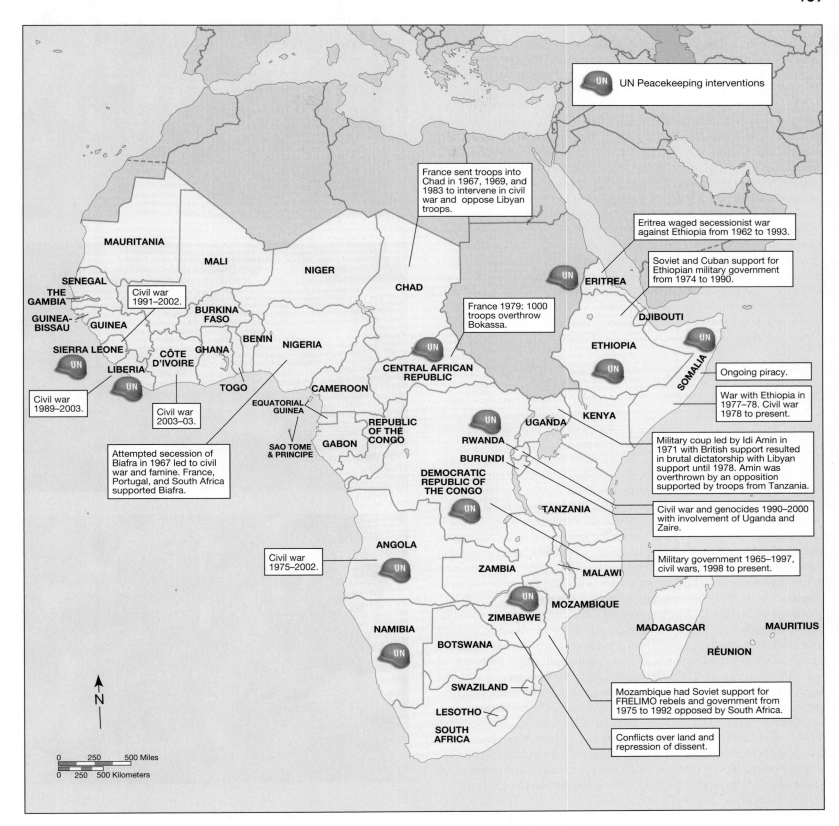

▲ **FIGURE 5.33  Map of Conflict in Africa**  Africa has experienced many wars and conflicts since independence, some of them fueled by foreign intervention, including Cold War politics involving the Soviet Union, Cuba, and the United States. France and Britain have continued to provide some military assistance in former colonies, and countries such as South Africa, Liberia, Libya, Uganda, and Angola have provided military support for groups in neighboring countries.

government declaring independence in 1965, controlling most of the good land and opposed by black guerilla movements. In 1980, white rule was replaced by elections, and President Robert Mugabe of Zimbabwe was elected and promised to redistribute land. Although title to the lands was given to white settlers by the colonial governments, and the farms were well run, many black Africans believe that these lands were seized unfairly and should be returned to the indigenous owners. In 2000, claiming that the 1% of the population that was white owned 70% of the productive agricultural land, Mugabe began a compulsory land distribution to blacks with sometimes violent invasions of squatters onto white-owned farms. Critics of Mugabe argue that previous land redistribution failed because peasant farmers did not have the knowledge or resources to succeed in commercial and export-oriented farming and that many farm workers are at risk of losing their jobs. Zimbabwe's problems have escalated beyond land conflicts, with rigged elections, human rights abuses, little health care, suppression of the media, and corruption. Three million refugees have fled to South Africa. Although elections in 2008 were apparently won by the opposition, Mugabe has retained considerable power. As of 2009, a power-sharing agreement with some opposition ministers has been facilitated by South Africa, but the economy is still in chaos (**Figure 5.34a**).

**Nigeria**  With extensive oil reserves and a population of over 150 million, Nigeria is one of the most powerful countries in Africa, formed when the British brought together several ethnic groups, including Yoruba, Igbo, and Fulani, into a large colony, which became independent in 1960. Since independence government has been dominated by military dictatorships who often brutally suppressed opposition and did little to resolve highly unequal distributions of wealth. In the 1960s the Niger delta became the core of an oil-producing region that triggered the secessionist Biafra war (the Ibo seceded from the mostly Yoruba Nigerian government) and promised an economic boom for Nigeria. As Nigeria has become even more dependent on oil exports, which are more than 90% of all exports by value, political unrest has continued in the oil region. This is partly associated with ethnic conflicts and from resentment that oil wealth is not benefiting the residents of the oil region; tensions also relate to environmental pollution of the lands and waters of the delta (**Figure 5.34b**). The Nigerian government's hanging of Ken Saro-Iwa—a novelist and activist who protested oil spills in the area occupied by 500,000 Ogoni people—for his alleged role in political unrest in 1995 provoked international outrage and sanctions against the government. The conflicts in the oil region have had global implications for major oil companies such as Shell, who have been accused of complicity with the government and subject to consumer boycotts but who claim that they are working to protect the environment and human rights in the region.

Civilian rule and democratic elections have reduced tensions since 1999, and the economy has been growing fast with an expanding manufacturing and service sector, including banks and telecommunications, car manufacturing, film production, and textiles. Crime and corruption is still a serious problem, including international Internet and phone scams.

**Peacekeeping**  Some of the most searing images of conflict in sub-Saharan Africa came from Rwanda and Sierra Leone during the 1990s, where genocide and violence carried out by child soldiers seemed to present irreconcilable legacies. But over the last decade peace has broken out in several previous conflict situations, and recoveries have been remarkable. In some cases peace settlements have been brokered and monitored by the United Nations and regional security forces. The United Nations

(a)

(b)

(c)

▲ **FIGURE 5.34 Continuing Conflict and Hope for Peace in Africa** (a) People collect water from a factory in Harare, Zimbabwe, in 2008, where cholera has caused hundreds of deaths and hyperinflation is causing food insecurity. (b) In the Niger Delta, local Urohobo people bake tapioca in the heat of a gas flare from a Shell pipeline. Pollutants from the flare cause serious health problems. (c) Liberian president Ellen Johnson-Sirleaf holds a white dove in her hands soon after her election in 2006.

Peacekeeping Forces operate under the authority of the United Nations Security Council to help establish and maintain peace in areas of armed conflict with the permission of disputing parties. In Africa, UN forces, with their distinctive pale blue helmets, have been deployed in Angola, the Democratic Republic of the Congo, Sudan, Ethiopia, Eritrea, Rwanda, Namibia, Somalia, Sierra Leone, Liberia, and Mozambique.

After initial success in monitoring transitions to peace in Mozambique and Namibia, the success of these missions has been mixed. UN forces failed to prevent massacres in Rwanda and Sierra Leone, were initially unable to establish peace in Somalia and the Democratic Republic of the Congo, and had inadequate human or financial resources to sustain several operations. Early in 2006, the Security Council adopted Resolution 1653, which called on the Governments of Uganda, Rwanda, Burundi, and the Democratic Republic of the Congo to disarm and demobilize militias and armed groups. African leadership—especially from South Africa and Nigeria—in promoting peace within the region is of growing importance, including negotiations led by former South African president Nelson Mandela and a West African peacekeeping force and monitoring group, ECOMOG, led by Nigeria under the auspices of the Economic Community of West African States (ECOWAS). The recent head of the United Nations, Kofi Annan, was from Ghana and was particularly concerned with improving conditions in Africa.

**Rwanda** The scars of conflict and genocide are still evident in Rwanda, where tensions between the majority Hutu and powerful Tutsi people erupted into civil war in 1994. The ethnic divisions between these two groups were created or exacerbated by Belgian colonists, who gave the Tutsis control over the Hutus, mostly peasant farmers. The Tutsis received education, training, and other benefits, whereas the Hutus were taxed heavily and given few privileges. The rapid withdrawal of Belgium resulted in a Hutu majority government in Rwanda and a population that harbored resentment against the Tutsi minority (about 20% of the population of the two countries). In neighboring Burundi, the Tutsis maintained power until the late 1980s, but with several internal military coups and severe repression of Hutu uprisings.

Burundi finally moved toward multiparty and multiethnic government in 1993, but the new president, Melchior Ndadaye, was killed in a coup, and subsequent ethnic violence killed more than 200,000 and sent 800,000 refugees into neighboring countries. When a 1994 plane crash killed the presidents of Burundi and Rwanda (Cyprien Ntaryamira and Juvenal Habyarimana), some Hutus blamed Tutsis and moderate Hutus. They initiated a massacre in which more than 500,000 died in Rwanda, and many Tutsis fled to neighboring countries, where rebel forces were organized. When these Tutsi rebels won control of Rwanda and Burundi, thousands of Hutus fled to avoid retribution. Two million refugees ended up in Zaire (now the Democratic Republic of the Congo) and thousands in Uganda, Kenya, and Tanzania.

The challenge of reconciliation in these two countries is enormous because the memories of violence are so fresh and the divisions so deep. Since 2000 considerable energy has been focused on reconciliation and peace, thanks to the mediation of respected leaders such as Nelson Mandela of South Africa and Julius Nyerere of Tanzania. In Rwanda there has been an effort to bring to justice those most responsible for the genocide, to encourage forgiveness, and to de-emphasize ethnicity in politics. There are also serious efforts to rebuild the economy, tourism, and agriculture, including plans for nationally available high-speed broadband. In 2008, Rwanda elected a legislature in which the majority of politicians were women.

**Liberia and Sierra Leone** In Liberia and Sierra Leone, civil wars and authoritarian leadership produced conditions of anarchy in the late 1990s. The economies of these two countries depend on a very narrow range of exports with volatile prices—rubber in Liberia and cocoa in Sierra Leone. Resettlement of freed slaves from other regions, who saw themselves as elites, has also been a long-term cause of conflict with the indigenous residents, and there is considerable resentment of urban wealth by rural residents. In the 1990s, struggles for power between opposition power groups within the countries were fueled by arms and capital obtained through the sale of diamonds, with many ordinary people fleeing as refugees. In Sierra Leone civil war between rival paramilitaries killed more than 200,000 people. Peace was established in Sierra Leone in 2002 as UN forces disarmed rebels and militias, and in 2003 hopes for peace in Liberia were encouraged by new leadership that sought to disarm rebels and control military violence. In 2006 U.S.-educated economist Ellen Johnson-Sirleaf won the Liberian presidential election by vowing to sustain democracy and rebuild the economy. She became Africa's first elected woman head of state (**Figure 5.35c**).

**Peacekeeping Regional Organizations** ECOWAS is one example of programs for economic integration and political cooperation in sub-Saharan Africa, established in 1975 to promote trade and cooperation with West Africa. ECOWAS includes the countries of Benin, Burkina Faso, Cape Verde, Côte d'Ivoire, Gambia, Ghana, Guinea, Guinea-Bissau, Liberia, Mali, Mauritania, Niger, Nigeria, Senegal, Sierra Leone, and Togo. African integration was the dream of several independence leaders, most notably Kwame Nkrumah of Ghana, who called for a Union of African States in his famous 1961 speech, "I Speak of Freedom." Nkrumah believed that only by joining together could independent Africa reach its full potential. Unable to convince others that complete unity was desirable, Nkrumah was able to lead the establishment of the Organization of African Unity (OAU) in 1963. The OAU, now called the African Union and based in Addis Ababa, Ethiopia, promotes solidarity among African states, the elimination of colonialism, and cooperative development efforts. It was successful in mediating boundary disputes between Ethiopia and Somalia in the 1960s and in pressing for the end of the apartheid regime in South Africa.

In southern Africa, economic integration has been promoted since 1979 through the Southern African Development Community (SADC), which promotes trade and development coordination, especially improvement of transport links. Members include Angola, Botswana, Lesotho, Malawi, Mauritius, Mozambique, Namibia, Swaziland, Tanzania, Zambia, and Zimbabwe. South Africa finally joined in 1994 with the advent of black majority rule. Similar regional programs have included the East African Cooperation (EAC) and East African Economic Union between Kenya, Uganda, and Tanzania, and the larger Common Market for East and Southern Africa (COMESA), which replaced the Preferential Trade Area (PTA) in 1993.

## Future Geographies

The future of sub-Saharan Africa is difficult to forecast because there are both positive and negative signs throughout the region. The region is still adjusting to the enduring legacies of colonialism that include economies dependent on mineral or agricultural exports, unequal land distribution, and boundaries that divide or cluster groups with little attention to cultural values or political expediency. Many countries are still struggling with the transition to independence and the creation of representative governments and continue to rely on foreign assistance. Although some countries have improved living standards, established democratic governments, and increased food and economic production, others are still mired in conflict over resources and political futures or face forbidding health challenges of malaria and AIDS. Social inequalities persist in many regions and contribute to migration, unrest, and famines. Although some Africans maintain rural subsistence lives, disconnected from the world economy, others are working in transnational corporations or are producing new exports for the global market. Sub-Saharan Africa has many connections to other world regions and to global systems through trade, migration, and the diffusion of rich and varied cultures, yet it is often overlooked by the media and the core economies and portrayed only through negative images of poverty, war, and disease or as a vast nature reserve.

On many indicators the region is doing much better than when the first edition of this book was published in the 1990s Cell phones have rapidly connected many Africans to each other and the wider world; economies have been growing relatively rapidly at 5% averaged across the region since 2000, and most countries, except Zimbabwe, have brought down inflation.

Looking to the future there are several key challenges for Sub-Saharan Africa that would provide rather different alternative futures. Although these include the Millennium Development Goals, generally targeted at the poorest people, there are also exciting opportunities for sub-Saharan Africa within vibrant national and international flows of people, goods, and ideas. Some of these challenges include the following.

- Maintaining peace and seeking reconciliation in regions of recent conflict
- Reducing rates of infection and finding cures for HIV/AIDS, malaria, and other diseases that are particularly prevalent in Africa
- Increasing food security through creation of regional food systems, improving yields, and providing more attractive prices for farmers, including participation in a fairer trading system worldwide where African exports, especially those with value added by processing, can generate good incomes for producers
- Expanding the opportunities created by the rapid adoption of cell phone technology and the Internet to encourage literacy, cultural exchange, and entrepreneurship
- Taking advantage of new economic and cultural links with other regions such as Asia and Latin America

Whatever the future may hold for sub-Saharan Africa, it is a region that the rest of the world can ill afford to ignore.

## Main Points Revisited

■ **Humans originated in the sub-Saharan African region, where a warm tropical environment is associated with vast deserts and diseases such as malaria, as well as some of the world's richest biodiversity and forests.**

Challenges for the region include ensuring food and water security in the face of drought, reducing vulnerability to tropical diseases and pests, and conserving wildlife. Climate change poses major risks to many people and landscapes in Africa.

■ **The legacies of European colonialism include trade patterns oriented to the export of primary products, unequal land distribution, and political instability associated with contested boundaries and Cold War alignments. Overseas countries and firms are especially interested in Africa's mineral and oil resources**

Many countries in the region are now more politically stable and are trying to develop new trade links that allow African citizens to benefit from exporting not only from its primary resources such as minerals and oil but also by selling manufactured goods and services to each other and other regions

■ **Sub-Saharan Africa has shaped other parts of the world through the tragic diaspora of slavery, but interactions continue with labor migration to Europe, refugees fleeing civil wars, and the diffusion of African culture and style. Population growth in Africa is high, but fertility is declining, especially as women's status improves. The region was especially affected by HIV/AIDS.**

Labor migrants provide an important source of income through remittances and African culture has in turn been transformed by global trends, especially as mobile phones and TV spread to remote regions. Populations are starting to urbanize with large squatter settlements around growing cities.

■ **Africa has a large number of poor people and economies where many people are subsistence farmers and countries carried high burdens of debt. Although poverty is still widespread, programs of debt relief, democratic elections, new export opportunities, and service-sector expansion have brought some optimism regarding economic and political futures.**

Although a number of countries show improved conditions on many of the Millennium Development Goals, others still have many people without basic needs such as clean water and adequate food and housing.

■ **There is a great range of landscapes and livelihoods in the 48 countries in sub-Saharan Africa, and it is important to recognize differences in environment, wealth, culture, economy, and political dynamics**

It is important to avoid simple stereotypes that portray the region as poor, diseased, disaster prone, economically underdeveloped, and conflict ridden and to understand the variations between and within countries.

# Key Terms

apartheid (p. 176)

Berlin Conference (p. 168)

blood diamonds (p. 160)

bush fallow (p. 162)

dependency school of
development theory (p. 172)

desertification (p. 155)

domestication (p. 162)

environmental determinism
(p. 172)

feminization of poverty
(p. 187)

G8 (p. 173)

gender and development
(GAD) (p. 188)

harmattan (p. 154)

homelands (p. 176)

IPCC (p. 156)

*lingua franca* (p. 184)

microfinance programs
(p. 174)

Millennium Development
Goals (MDGs) (p. 178)

modernization theory (p. 172)

rift valley (p. 157)

savanna (p. 160)

shifting cultivation (p. 162)

slash-and-burn (p. 162)

terms of trade (p. 172)

urban bias (p. 182)

# Thinking Geographically

1. How has the physical geography (climate, soils, geology) posed challenges for the people of sub-Saharan Africa, and how have people adapted to these challenges and taken advantage of these resources?

2. What are the enduring legacies of colonialism in different regions of sub-Saharan Africa?

3. How might the following improve living conditions and economies in sub-Saharan Africa: debt relief, women and development programs, microfinance programs, climate adaptation, and regional peacekeeping?

4. Why has wildlife conservation become a priority in some regions of sub-Saharan Africa, and what successful strategies have been used to protect wildlife?

5. What are the arguments for and against treating sub-Saharan Africa as a major world region?

6. How do the following affect the geographical distribution of population in sub-Saharan Africa and/or population growth rates: high infant mortality, high HIV/AIDS rates, education levels of women, coastal ports, highland areas, mineral deposits?

7. What are the current major patterns of migration between and within African countries and beyond, and how do they relate to employment opportunities, natural disasters, and conflict?

8. How is sub-Saharan Africa progressing in achieving some of the targets set out in the UN's Millennium Development Goals? Why are some countries making more progress than others?

Log in to www.mygeoscienceplace.com for Videos, Interactive Maps; RSS feeds; Further Readings; Suggestions for Films, Music and Popular Literature; and Self-Study Quizzes to enhance your study of Sub-Saharan Africa.

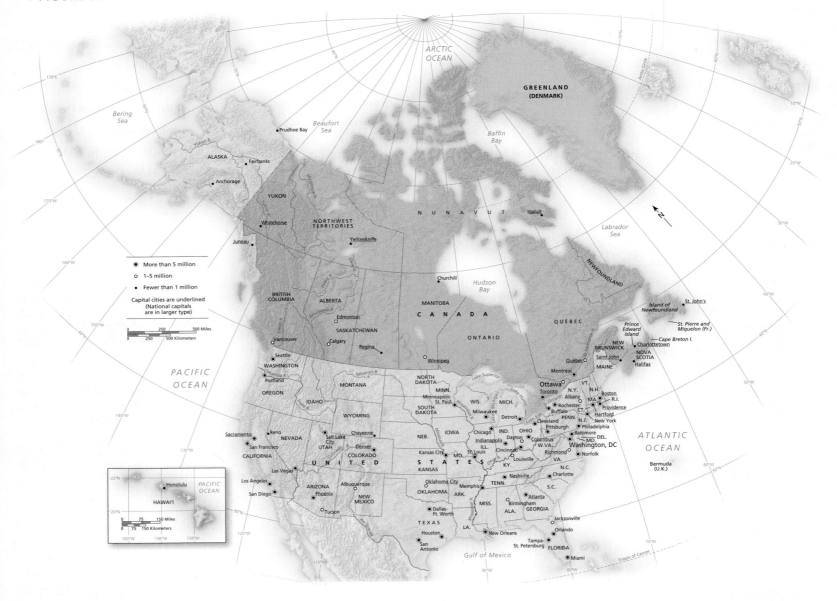

## THE UNITED STATES AND CANADA KEY FACTS

- Major subregions: United States of America, Canada

- Total land area: 19.6 million square kilometers/7.6 million square miles

- Major physiographic features: Great Lakes, Rocky Mountains, Great Plains, Appalachian Mountains, Sierra Nevada, Grand Canyon, Southwestern deserts.

- Climate and temperature is variable across the continent from cold at the Arctic Circle to warm and moist in the southeast and warm and dry in the southwest.

- Population (2010): 354 million, including 310 million in the United States and 44 million in Canada.

- Urbanization (2010): 82%

- GDP PPP (2010) Current international dollars: 41,866; GDP PPP per capita 45,254 US; 38,478 Canada

- Debt service (total, % of GDP) (2005): 0%

- Population in poverty (%< $2/day) (1990–2005): 0%

- Gender-related development index value: 0.95

- Average life expectancy/fertility: 78(M76, F81)/2.1

- Population with access to Internet (2008): 74.4%

- Major language families: English, Spanish, and French—all of which descend from the Indo-European family.

- Major religions: Predominantly Christian (Protestant and Roman Catholic)

- $CO_2$ emission per capita, tonnes (2004): 20.3

- Access to drinking-water sources/sanitation (2006): 99%/100%

# 6

# THE UNITED STATES AND CANADA

In Chino Hills, a suburb of Los Angeles, 30 miles (50 km) east of its downtown, household income is twice that of the national average. In Vista del Sol, a quiet neighborhood there, Greg Garland, a local narcotics cop, found more than 800 marijuana plants in one of its latté-colored stucco houses, a house that had sold for $600,000 the year before. In 2008, Garland raided more than 40 houses like this Vista del Sol McMansion that turned out to be marijuana factories. Between 2007 and 2008, the State of California's "Campaign against Marijuana Planting" destroyed over three million plants worth nearly $10 billion from suburban grow houses, back gardens, timber forests, and state lands. Marijuana has surpassed grapes as California's most valuable agricultural crop.

Outdoor plantations are nourished with fertilizers and extensive irrigation systems and are tended by workers specifically imported to the United States to work them. Although there certainly are counter culture people who make their livelihood by operating marijuana plantations throughout the United States and Canada, California narcotics agents estimate that four-fifths of outdoor marijuana plantations are run by Mexican criminal gangs.

Indoor factories are mostly operated by East Asians. In an effort to keep landlords off their property, these "potrepreneurs" buy big, expensive houses in good neighborhoods. The pot plants propagated in these high-end grow houses are enabled by state-of-the-art systems of heat and artificial light, as well as extravagant ventilation systems that filter the strong smell of growing plants. Interestingly, the increase in marijuana propagation in California is not because of an increase in demand. There, only about 11% of the population is thought to be users, which is less than the estimated rate of marijuana use in every state in New England. The Federal Drug Enforcement Administration credits the booming industry to the steady tightening of U.S. borders after the terrorist attacks of September 11, 2001. Before 9/11, California was far more likely to import high-grade marijuana from Canada and low-grade product from Mexico (see Figure 7.34b). With both borders now being far more risky to cross, producers have moved to where the consumers are. Asian gangs, who dominate the hydroponic marijuana trade in British Columbia, Canada, have moved operations south, and Mexican distributors, who are also likely to handle cocaine and methamphetamine as well as marijuana, have moved north to become local producers. ■

## MAIN POINTS

■ The United States and Canada constitute the predominantly English-speaking countries of the North American continent (**Figure 6.1**). The United States includes 48 contiguous states and two noncontiguous ones: Alaska (northwest of Canada) and the volcanic islands of Hawai'i, southwest of the North American continent in the Pacific Ocean. Canada includes 10 provinces and 3 territories and is the world's second-largest country after the Russian Federation.

■ The high economic prosperity of the region has resulted in dramatically altered land and waterscapes as populations have adapted to and modified the world around them. These transformations continue to shape the relationship between people and their environment in the United States and Canada.

■ Migration has been responsible for the population of this region, beginning with the indigenous people who crossed the Bering Strait 30,000 years ago to the arrival of European colonists in the 15th century and to the more recent waves of immigrants, who have again come from across the globe to live there.

■ The United States and Canada are key forces in global geography, shaping not only economic and political events near and far, but also cultural trends and social practices. Few people in the world have been unaffected by "the American way of life."

■ Canada and the United States control vast natural, human, and financial resources such that events in this region, both historically and today, reverberate throughout the globe from the colonial wars that changed the course of European history to the role of fast food and the Internet in daily life.

# ENVIRONMENT AND SOCIETY

The United States and Canada occupy the land mass of North America, whose position with respect to climate patterns, plate tectonics, and ocean currents and human adaptation to them make it key to the unfolding of global environmental processes. The people of Canada and the United States have occupied the region for millennia, though the most profound global environmental impacts and innovations emanating from the region have come about only since the last century. The region is a highly prosperous one, and the advanced standard of living enjoyed by people in the United States and Canada has had dramatic effects on the region's landscapes from water environments to land uses to **airsheds**, a unit for air quality measurement. This section discusses the complex biophysical processes that characterize the region, as well as how they shape environment and society relations and practices domestically and across the globe.

## Climate, Adaptation, and Global Change

**Climate Patterns and Agricultural Practices** Because temperature, precipitation, and terrain patterns combine to influence vegetation and to some extent soil, it is important to understand the role that climate plays in this region. Temperatures range from those in the Canadian Arctic and Alaska in the north, where it is very cold for most of the year, to those in the southern United States, where it is warm for most of the year because the Tropic of Cancer is just a few hundred miles south. In short, the United States and Canada contain nearly every type of climate condition possible, with temperatures varying quite dramatically on any one day of the year from north to south (**Figure 6.2**). Of course, within-region differentials also occur because of the moderating influence of the oceans on three sides, the Gulf of Mexico, and the interior Great Lakes, as well as the presence of mountains and plateaus. Central United States and Canada experience the most dramatic temperature ranges, whereas the coasts have a far narrower range because of the moderating influence of the oceans.

Variations in amounts of precipitation also have a profound effect on climate. The east and west coastal areas tend to be mild and moderately wet, and the interior is largely arid because the north–south mountain chains prevent moisture-bearing clouds from moving inland to drop their moisture. Because of this, a moisture gradient exists that declines slowly but continuously from east to west as far as the three significant mountain ranges—the Rockies, Sierra Nevadas, and Cascades. Once beyond these mountains, the moisture gradient rises dramatically toward the Pacific.

In the southeastern part of the United States, where no significant coastal ranges exist, moisture-bearing clouds are able to more readily condense into rain, and the warm Gulf of Mexico is an important source of moisture and storms, including hurricanes. In the Arctic north, annual precipitation approximates desert conditions because of the dominance of very stable air masses with low moisture content and the dryness of cold air. The jet stream (see Chapter 1, p. 4) brings precipitation to most of the continent in the winter months. The warmer parts of the region—in the southern United States and Hawai'i—experience this precipitation as rain, but the colder, more northerly parts experience snow. In areas around the Great Lakes, the warming effects of these large bodies of water add even more moisture to the mix, bringing especially heavy snowstorms (called "lake-effect" snow)

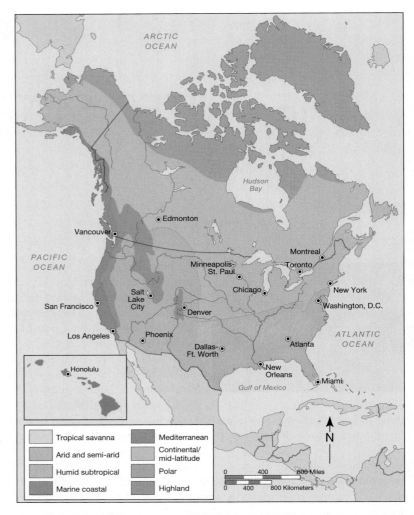

▲ **FIGURE 6.2 Climate of the United States and Canada** This is the only region in the world that contains the full range of climate types that occur globally, stretching from Alaska and the Canadian Arctic to the tropical Hawai'ian Islands.

to places like Buffalo, New York, on the northeastern tip of Lake Erie, and Sault St. Marie, Ontario, on the channel between Lake Superior and Lake Huron.

**Environmental Modification and Impacts** The United States and Canada possess a physical geography that contains a bounty of resources, from the whole range of minerals to vast forests; fertile, highly productive land; extensive fisheries; varied and abundant wildlife; and magnificent physical beauty (**Figure 6.3**). Accompanying this extensive physical wealth has been the technological capability and the drive to exploit these resources to an extent achieved by few other regions on Earth. One result of this combination of physical resources and human ingenuity is a region that experiences an extraordinarily high quality of life. A second result is a high level of material consumption that results in a region with elevated levels of air and water pollution, numerous sites—often on the fringes of major urban settlements—of soil contamination, solid waste, an ongoing problem of nuclear waste disposal, acid rain, extinct and endangered species, and increasingly frequent reports of insect and animal genetic mutations.

► **FIGURE 6.3 The United States and Canada from space** This image shows a region surrounded by oceans, gulfs, and bays with vast interior plains and high mountains on both the east and west coasts. The region also contains vast mineral wealth, extensive forests and supplies of clean water, and high agricultural productivity.

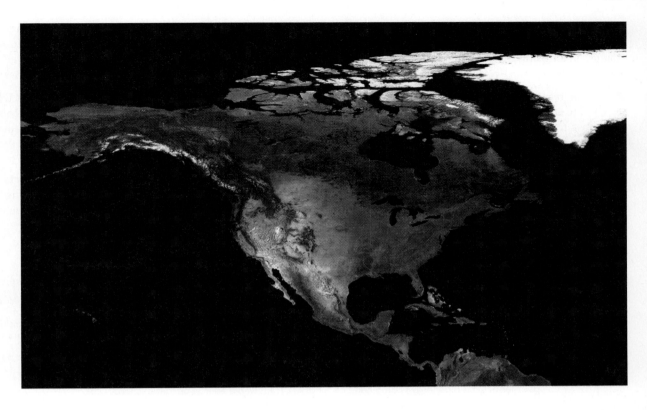

Certainly the most significant and pressing environmental problem in the United States and Canada today (as well as for many other places around the globe) is the impact of global climate change. The United States and Canada play a major role in causing the problem through high total and per capita greenhouse gas emissions resulting from the production and consumption patterns associated with its high standard of living. These emissions are based on the region's heavy reliance on fossil fuels, including high gasoline use and coal-fired electricity, as well as agricultural and industrial production that generates not only massive amounts of carbon dioxide but also other greenhouse gases (GHGs) like methane and nitrous oxide. Although climate change, as discussed in Chapter 1, is indisputably a global problem with serious projected impacts, its effects are already being felt regionally and locally. In the United States and Canada, changes have already been detected in terrestrial and marine biological systems, such as through upward and northward shifts in the ranges of plant and animal species responding to increases in temperature and the melting of glaciers and snow in arctic and mountain regions.

Knowing the **carbon footprint**—the total set of GHG emissions caused directly and indirectly by an individual, organization, event, product, or place—is a way of understanding how GHGs are contributing to climate change and where their impacts might be reduced. **Figure 6.4** is a map of the carbon footprint of the continental United States. It shows that the older, more heavily populated and urbanized eastern half of the country and the West Coast have the highest carbon emissions.

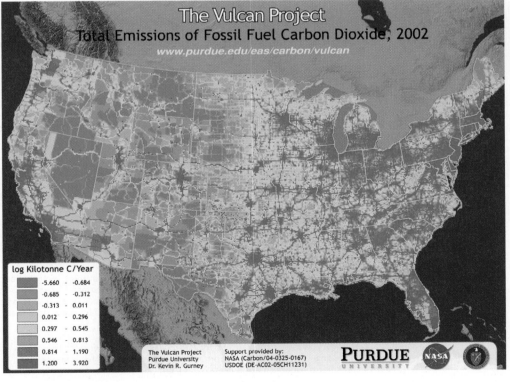

The Vulcan Project
Total Emissions of Fossil Fuel Carbon Dioxide, 2002
www.purdue.edu/eas/carbon/vulcan

log Kilotonne C/Year

| | |
|---|---|
| -5.660 | -0.684 |
| -0.685 | -0.312 |
| -0.313 | 0.011 |
| 0.012 | 0.296 |
| 0.297 | 0.545 |
| 0.546 | 0.813 |
| 0.814 | 1.190 |
| 1.200 | 3.920 |

The Vulcan Project
Purdue University
Dr. Kevin R. Gurney

Support provided by:
NASA (Carbon/04-0325-0167)
USDOE (DE-AC02-05CH11231)

PURDUE
UNIVERSITY

NASA

▲ **FIGURE 6.4 Carbon footprint of the United States** Shown here is the carbon dioxide emissions inventory of the United States plotted down to 100-square-kilometer chunks.

The likely impacts of climate change on the United States and Canada have been projected by scientists using complex models that estimate changes in temperature, precipitation, and sea levels at the regional scale. The most significant effects of climate change in North America include:

■ An increase in sea levels along much of the coast that would exacerbate storm-surge flooding and shoreline erosion, placing at risk those low-lying cities such as Miami and other coastal communities where uncontrolled growth has made homes vulnerable and increased insurance costs.

■ Temperature increases that could cause adverse health impacts from heat-related mortality and mosquito-borne diseases, and stress agricultural crops and livestock. In more northerly regions, warming might increase agricultural productivity.

■ More severe droughts, especially in the western United States, which would increase competition among agricultural, municipal, industrial, and ecosystems over the region's overallocated water resources and could increase risks of wildfire.

■ Changes in natural ecosystems that could alter the current configuration of landscapes and valuable species such that they would no longer be protected by the current pattern of parks and conservation areas.

**Adaptation and Mitigation Efforts**    At the same time that the problems of climate change are becoming more clearly understood, a wide range of mitigation measures are being identified and developed that can begin to address the impacts. **Table 6.1**, adapted from the

### TABLE 6.1   Corporate and Government Responses to Climate Change

| Sector | Key Mitigation Technologies and Practices Currently Commercially Available | Policies, Measures and Instruments Shown to Be Environmentally Effective | Key Constraints or Opportunities (Constraints above; Opportunities below) |
|---|---|---|---|
| Energy Supply | Improved supply and distribution efficiency; fuel switching from coal to gas; nuclear power; renewable heat and power (hydropower, solar, wind, geothermal and bioenergy); combined heat and power; early applications of carbon dioxide capture and storage (CCS) (e.g. storage of removed $CO_2$ from natural gas) | Reduction of fossil fuel subsidies; taxes or carbon charges on fossil fuels; feed-in tariffs for renewable energy technologies; renewable energy obligations; producer subsidies | Resistance by vested interests may make them difficult to implement  <br><br> May be appropriate to create markets for low emissions technologies |
| Transport | More fuel-efficient vehicles; hybrid vehicles; cleaner diesel vehicles; biofuels; modal shifts from road transport to rail and public transport systems; non-motorised transport (cycling, walking); land-use and transport planning | Mandatory fuel economy; biofuel blending and $CO_2$ standards for road transport Taxes on vehicle purchase, registration, use and motor fuels; road and parking pricing | Effectiveness may drop with higher incomes  <br><br> Particularly appropriate for countries that are building up their transportation systems |
| Buildings | Efficient lighting and daylighting; more efficient electrical appliances and heating and cooling devices; improved cook stoves, improved insulation; passive and active solar design for heating and cooling; alternative refrigeration fluids; recovery and recycling of fluorinated gases | Appliance standards and labelling; building codes and certification; public sector leadership programmes; including procurement | Periodic revision of standards needed, Enforcement can be difficult  <br><br> Government purchasing can expand demand for energy-efficient products |
| Industry | More efficient end-use electrical equipment; heat and power recovery; material recycling and substitution; control of non-$CO_2$ gas emissions; and a wide array of process-specific technologies | Provision of benchmark information; performance standards; subsidies; tax credits, tradable permits, voluntary agreements | Predictable allocation mechanisms and stable price signals important for investments  <br><br> May be appropriate to stimulate technology uptake |
| Agriculture | Improved crop and grazing land management to increase soil carbon storage; restoration of cultivated peaty soils and degraded lands; improved rice cultivation techniques and livestock and manure management to reduce $CH_4$ emissions; improved nitrogen fertiliser application techniques to reduce $N_2O$ emissions; dedicated energy crops to replace fossil fuel use; improved energy efficiency | Financial incentives and regulations for improved land management; maintaining soil carbon content; efficient use of fertilisers and irrigation | May encourage synergy with sustainable development and with reducing vulnerability to climate change, thereby overcoming barriers to implementation |
| Forestry/forests | Afforestation; reforestation; forest management; reduced deforestation; harvested wood product management; use of forestry products for bioenergy to replace fossil fuel use | Financial incentives (national and international) to increase forest area, to reduce deforestation and to maintain and manage forests; land-use regulation and enforcement | Constraints include lack of investment capital and land tenure issues  <br><br> Can help poverty alleviation |
| Waste | Landfill $CH_4$ recovery; waste incineration with energy recovery; composting of organic waste; controlled wastewater treatment; recycling and waste minimisation | Financial incentives for improved waste and wastewater management | Most effectively applied at national level with enforcement strategies  <br><br> May stimulate technology diffusion |

Source: IPCC, Climate Change 2007: Synthesis Report, p. 60. http://www.opcc.ch/pdf/assess,emt-report/ar4/syr/ar4_syr.pdf., accessed 6 July 2009.

Intergovernmental Panel on Climate Change 2007 report, provides a wide range of adaptation and mitigation measures to help governments and corporations address climate change. Many greenhouse gas reductions, especially those associated with energy efficiency such as home insulation and fuel efficient transport, can be made at little cost and may even save people money. Of course, individuals can also attend to the effects of climate change in numerous of ways: from buying local produce (which reduces the energy expended in transportation and refrigeration) to saving energy in the home, to using reusable water bottles instead of disposable ones (which minimizes waste and the costs of disposing it). In the absence of serious federal effort to reduce emissions many cities and businesses across the United States and Canada made their own commitments to mitigate climate change.

The U.S. Conference of Mayors formed a Climate Protection Agreement under the leadership of Seattle's mayor, Greg Nickels. The aim of those who have signed the agreement is to reduce carbon emissions at the sites of their production by achieving the following goals:

1. Strive to meet or beat the Kyoto Protocol targets in their own communities, through actions ranging from antisprawl land-use policies to urban forest restoration projects to public information campaigns.

2. Urge state and federal governments to enact policies and programs to meet or beat the GHG emission reduction target negotiated for the United States in the Kyoto Protocol (7% reduction from 1990 levels by 2012).

3. Urge the U.S. Congress to pass the bipartisan GHG reduction legislation, which would establish a national emission trading system (which may pass both houses sometime in late 2009).

**Figure 6.5** is a map produced for the U.S. Western Governors' Association that identifies areas of high potential for renewable resources with low environmental impact and GHG emissions in the western provinces of Canada and the western United States. These resources are capable of enabling the states and provinces to achieve goals 1 and 2 listed earlier. The overall renewable energy supply in the region includes wind, solar, hydropower, and geothermal sources.

Several regional alliances have been created to reduce greenhouse gas emissions, including the Regional GHG Initiative (RGGI) among the northeastern states of Connecticut, Delaware, Maine, Maryland, Massachusetts, New Hampshire, New Jersey, New York, Rhode Island, and Vermont and the Western Climate Initiative (Arizona, California, New Mexico, Oregon, Washington, Utah, and Montana), as well as British Columbia, Manitoba, Ontario, and Quebec. These regional alliances are taking on commitments to reduce their emissions, but have made it easier to meet

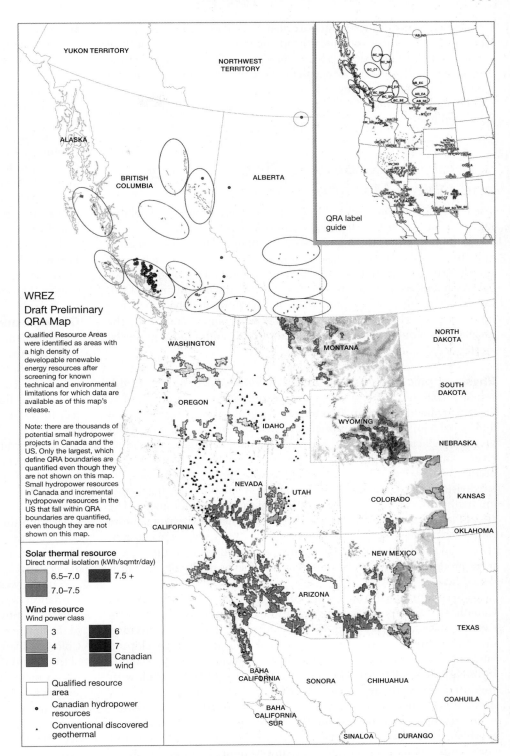

▲ **FIGURE 6.5 Renewable Energy Sources in the West** This map indicates the areas in the western part of the United States and Canada that possess the potential for renewable energy development with low environmental impacts.

these goals through the use of **cap-and-trade programs**. This is a three-part regulatory system in which the "cap" is a government-imposed limit on carbon emissions, emission permits or quotas are then given or sold to states (or firms), and the "trade" occurs when unused permits are sold by those who have been able to reduce their emissions beyond their quota to those who are unable to meet theirs (as discussed in

Chapter 1). Many of these states are also working together to plan on how to adapt to climate changes already occurring or that will occur if we do not reduce emissions. These plans include coastal protection as well as rethinking water management and conservation. Proposals for a national emissions trading system for the United States would use a similar logic to achieve goals that might be as ambitious as an 80% reduction in emissions by 2050.

## Geological Resources, Risk, and Water Management

The United States and Canada are centered on a vast central lowland that includes the Canadian Shield, the Interior Lowlands, and the Great Plains. To the east are the Appalachian Mountains, which descend gradually to the Gulf–Atlantic Coastal Plain, which becomes broader as one travels farther south and southwest. To the west are three distinct topographical regions. Moving from west to east are the mountains and valleys of the Pacific coastal ranges; then an **intermontane**—lying between mountains—set of basins, plateaus, and smaller ranges; and finally the great Rocky Mountain range, which rises steeply and imposingly at the western edge of the central lowlands.

**Physiographic Regions**   Figure 6.6 provides an overview of the different physiographic regions of the United States and Canada. The Canadian Shield, shown in light brown, is a geologically very old region and is particularly rich in minerals. The Interior Lowlands, in light green, is a glaciated landscape with fertile soils and abundant lakes and rivers. This part of the continental land mass is devoted mostly to agriculture, with some industry as well. The third primary physiographic subdivision is the Great Plains—shown in light blue—an area that slopes gradually upward as you move west toward the Rocky Mountains. The Great Plains region is also delimited by the fact that it experiences more rainfall than the surrounding regions. This is an area of extensive gently rolling and flat terrain with excellent soils and some of the world's most productive farms. Beef, pork, corn, soybeans, and wheat are produced here.

The major landform feature of the eastern United States and Canada is the Appalachian mountain chain. Surrounding it on its eastern and southern flanks is a coastal plain that can be divided into two subregions, the Piedmont area of hills and easterly sloping land and the Gulf–Atlantic Coastal Plain, a lowland area that extends from New York to Texas. The coastal plains are generally level, with soils that are sandy and relatively infertile. Where agriculture does occur, it is intensive and scattered. For instance, the lower Mississippi is an area of good soils and intensive agriculture. The Piedmont is the historic region of plantation agriculture—cotton, tobacco, and corn—but much of the soil in this region is depleted or eroded from overfarming.

**Riverine Landscapes**   The Mississippi–Missouri river system is at the heart of U.S. economic geography. As the most extensive and navigable river system in the world, it, along with the five Great Lakes (Superior, Michigan, Huron, Erie, and Ontario), undeniably enabled the early and formidable growth of the interior of the United States in ways that no railway system could have. Importantly, the engineering that has been applied to the Mississippi–Missouri system (including its major tributaries) over a century or more has also irretrievably altered

its ecosystem and exposed vast numbers of human settlements to extreme danger from flooding.

The physical geography of the United States/Canada world region is the context for, and an inescapable reminder of, the enormous benefits and the sometimes devastating costs of the "American way of life," supported as it is by a rich bounty of environmental resources. An example of this is the impact of Hurricane Katrina on New Orleans, Louisiana, in late August 2005—a disturbing demonstration of the disastrous potential of riverine engineering. Built on the delta marshland of the Mississippi River, New Orleans has been a city of levees, flood walls, and drainage canals since the 19th century, and nuisance flooding has been a perennial problem in the city. Hurricane Katrina was different, however, and far more destructive; high winds and rising water, combined with flawed levee design and lack of critical maintenance, caused widespread breaching or failure, subjecting 80% of the population to flooding. There were more than 1,800 deaths related to the flooding. More than a million people were evacuated, and tens of thousands were rendered homeless.

**Geologic and Other Hazards**   The western coastal formation sits along the fault line of two active crustal plates, the Pacific Plate and the Juan de Fuca Plate (see Figure 1.8, p. 8). Both plates are moving northward, rubbing against the more stationary North American Plate on which the continental landmass sits. As a result of this friction, the coastal area from San Diego through British Columbia to Alaska is subject to frequent tremors. Extreme and devastating earthquakes have occurred in the past and are likely to occur again in the near future. The extraordinary views of the Pacific Ocean provided by the mountainous topography of the Pacific coastal region have attracted extensive home development on mountain slopes, and past earthquakes have destroyed many of these homes. The extensive rainfall in this area has also wreaked havoc, despite extraordinary feats of engineering that have placed multimillion-dollar homes on steep mountainsides. When the soil becomes saturated with rainwater, liquefaction occurs, and the soil literally moves in one massive slump, carrying very large structures along with it.

In the western United States, the intermontane basin and plateau formation between the Pacific coastal range and the Rockies includes four major physiographic provinces (**Figure 6.7**): the Columbia Plateau to the north, which begins at the headlands for the Columbia River; the Colorado Plateau in the south, which includes the erosional landscape of the Grand Canyon; in between, the Great Basin in Nevada and Utah, which includes extinct lakes as well as the Great Salt Lake; and the southwestern deserts (Mojave, Sonoran, and Chihuahuan). Its landscape has many contrasts, with spectacular scenery that includes deep canyons, majestic mountains, and unique deserts. It occupies an area of dry climates, thin vegetation, and a general absence of many perennial surface streams. Wildfires, droughts, floods, and landslides are a continuing threat in the region.

**Mineral, Energy, and Water Resources**   The United States and Canada possess great energy and mineral resources. They are among the top 20 producers of natural gas in the world. Canada is second only to Saudi Arabia in its proven oil reserves, and although China is the largest producer of coal in the world, the United States possesses the largest proven coal reserves globally. Canada's oil is mostly located in the province of Alberta in sand deposits that pose difficulties for extraction. The United States' oil deposits are mostly located in Texas,

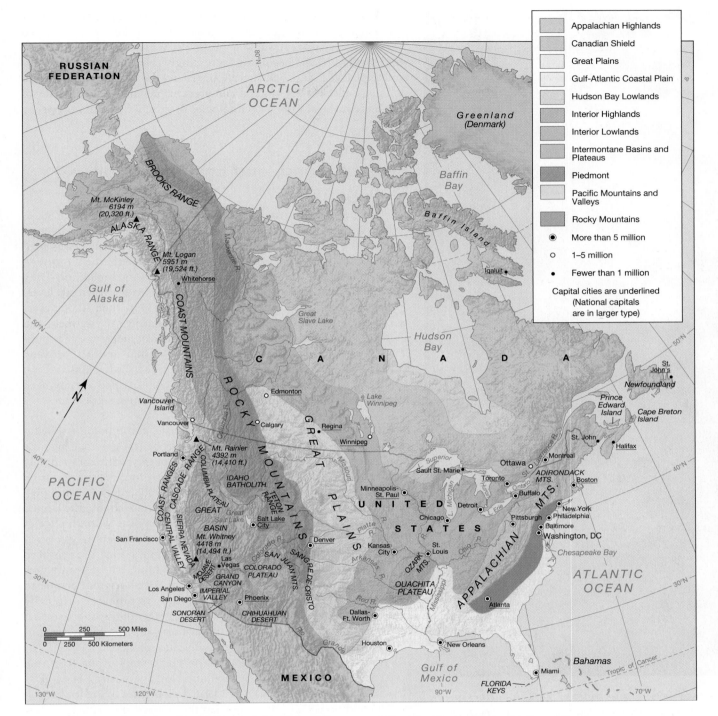

**▲ FIGURE 6.6 Physiographic Regions of the United States and Canada** The Interior Lowlands and Great Plains are central to agricultural production. The mountain ranges in the west are significantly higher and steeper because of their more recent emergence, whereas the Appalachian Mountains in the east are lower and more sloping because of their greater age and the effects of erosion. The eastern coastal region slopes toward the sea, but the West Coast is characterized by steep cliffs and fairly narrow beaches. Waves are also more dramatic in the west, and the western seashore is more dangerous and prone to riptides and deadly undertows.

Whereas gas, coal, and oil, as well as uranium to fuel nuclear power plants, are all nonrenewable sources, in the western United States and Canada renewable hydropower is available, and solar and wind power are also becoming increasingly prominent on the energy landscape (see Figure 6.5). Both Canada and the United States are also rich in a wide array of mineral and precious metal resources, many of which are integral to new technologies of production from copper to silicon.

## Ecology, Land, and Environmental Management

The Europeans who arrived on the Atlantic Coast of the United States and Canada beginning in the late 15th century encountered an environmentally diverse landscape thinly populated by indigenous peoples. These were the peoples who first "discovered" the continent, having traveled as hunters and gatherers across the broad, low-lying belt of tundra known as the Bering Strait land bridge when sea levels dropped and exposed land. It is estimated that more than 30,000 years ago, these peoples began the process of populating the continent and altering (and permanently changing) the environments they encountered. Despite what written accounts of initial contact between European explorers and native people recorded, Europeans did not discover a "pristine" or "virgin" land, but one that had already been transformed by tens of thousands of years of human settlement.

**Indigenous Land Use**   Most archaeologists, anthropologists, historians, and geographers believe that during the last great ice age, the huge ice sheet that had once covered much of the continent gradually retreated. This enabled the first group of Arctic hunters to cross from Siberia into present-day Alaska, between 20,000 and 35,000 and possibly up to 60,000 years ago (**Figure 6.8**). They moved into the United States and Canada by traveling southward along the western edge of the continental icecap. Because Canada was probably entirely covered by ice during this period, it is likely that people settled the southern part of the continent. It is also likely that the Inuit, who currently live in Canada's Arctic region, were the last of the aboriginal people to arrive. These Neolithic, or Stone Age, hunters probably originally came from northern China and Siberia. Descendants of these first hunters gradually moved farther and farther southward into the continental landmass, advancing eventually into Mexico.

Abundant evidence suggests that the ancestors of the Neolithic hunters eventually spread throughout North America (as well as Central and South America) and adapted their ways of life to the particular conditions they encountered as they moved and settled. As they began to settle, different native groups introduced agriculture in different places. With game, fish, and wild and cultivated foodstuffs available, an economic system based on subsistence production and trade took form. The cultivation of maize spread to wherever it could be grown, and new wild foods, like potatoes and tomatoes, were eventually domesticated.

In New England, prior to European contact, hunting, gathering, and some shifting cultivation existed among the indigenous peoples. Hunter–gatherers were mobile, moving with the seasons to obtain fish, migrating birds, deer, and wild berries and plants. Shifting agriculture was organized around planting and harvesting corn, squash, beans, and tobacco (see **Geographies of Indulgence, Desire, and Addiction: Tobacco**). For

▲ **FIGURE 6.7 Athabasca Falls, Jasper National Park, Alberta, Canada**
The Canadian Rockies, of which Jasper National Park is a part, are distinct from their U.S. counterparts in that they are older granite formations and are primarily the result of overthrusting. The U.S. Rockies were formed by uplifting and are mostly sedimentary rock.

the Gulf of Mexico, Louisiana, Alaska, and California. Importantly, in addition to being a leading energy producer, the United States is the leading consumer of energy resources (though China is fast catching up), relying mostly on Canadian imports of natural gas, coal, and oil to supplement this thirst for energy.

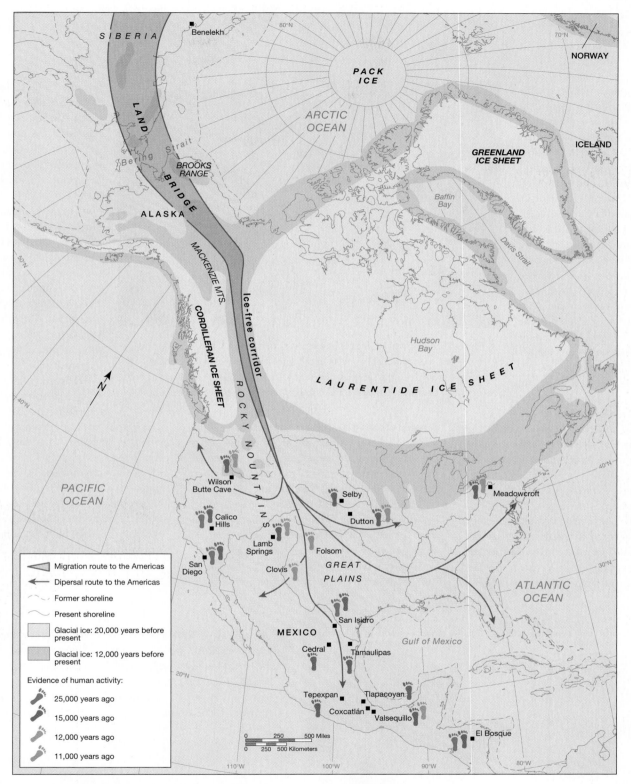

▲ **FIGURE 6.8 Migration of Neolithic Hunters into the Americas** It is believed that the first humans to enter the North American continent by way of the Bering Strait land bridge, which had been exposed through the gradual retreat of ice sheets, were hunters pursuing large mammals and mastodons. When these animals grew scarce, the small bands of hunters moved on, looking for new prey. They also fed themselves with roots, plants, and berries and learned to fashion clothing, weave baskets, and construct fishing nets. Because of the limits of the environment, the bands tended to remain small, and as the population grew, small groups would break off and move on to increase their chances of survival in more plentiful surroundings as yet unoccupied or hunted.

# TOBACCO

The global spread of tobacco cultivation, manufacturing, and consumption over the past five centuries has made tobacco one of the world's most widespread agricultural products. Tobacco's development as a global commodity presents an excellent example of the diffusion and diversification of social practices and values.

Based on evidence such as clay pipes found at several prehistoric sites, archaeologists believe indigenous people domesticated the wild tobacco plant more than 5,000 years ago. Native cultivators grew two main types of tobacco, *Nicotiana rusticum* in the northeastern United States and Canada and *Nicotiana tabacum* in Central and South America. Practically all the tobacco produced today is of the *tabacum* species. Native populations ascribed to tobacco a number of economic, social, and cultural purposes, and by the time of sustained European contact in the 15th century, tobacco use was ubiquitous in the Americas.

Native people also developed all of the principal means of tobacco consumption—smoking, inhaling, and chewing—before European contact. Sailors and merchants carried these customs back to Europe during the 16th century, where tobacco found favor in the ports and royal courts of Spain and Portugal. From the ports of these early colonial

▲ **FIGURE 1 Tying Tobacco Leaves on a Farm in China** The world's three largest multinational cigarette companies grow tobacco in scores of countries throughout the world. China is the world's largest producer and one of the many countries that has a favorable climate for tobacco production.

powers, tobacco use spread along trade routes to Africa, Asia, and the rest of Europe. By the end of the 17th century, the tobacco trade formed one of the most important parts of the colonial economy connecting the Old and New Worlds.

With the constant spread of tobacco consumption came the extension of tobacco cultivation. Tobacco is a highly adaptable plant that can grow in a wide range of climatic and soil conditions. It has been successfully cultivated as far north as Sweden and as far south as New Zealand

(**Figure 1**). Despite its adaptability, tobacco is also quite sensitive to climatic and soil conditions. The result has been the development of hundreds of different tobacco types, each with its own regional complex of production and specific commodity uses based on the leaf's subjective qualities, such as taste and aroma. Tobacco types have changed, evolved, and even disappeared over the centuries of cultivation and use, with changes in consumer preference and market structure. The history of one particular tobacco commodity, cigarettes, illustrates the changing social and economic geography of tobacco production and consumption (**Figure 2**).

The modern cigarette did not enjoy widespread popularity until the middle of the 20th century. Spanish and French consumers were the first to take up cigarette smoking in the early 19th century, and British soldiers brought Turkish cigarettes (probably developed from Spanish and French varieties) back with them from the Crimean War in the late 1850s. From the fashionable officers' clubs of London, cigarette smoking spread to New York's social elite in the 1860s. Manufacturers sprang up across Britain and the United States to feed and expand the growing demand for cigarettes, but several obstacles stood in the way of large-scale

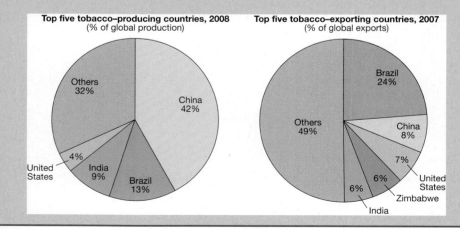

◄ **FIGURE 2 Tobacco-producing and tobacco-exporting countries, 2007** (a) Global tobacco production serves 1.1 billion smokers. In China, which is the largest producer, 63% of males between the ages of 15 and 69 smoke. Only 3.8% of Chinese women smoke. (b) Despite the fact that cigarettes are a multibillion-dollar industry worldwide, in only four countries—Kyrgyzstan, Macedonia, Malawi, and Zimbabwe—do tobacco exports amount to more than 5% of total export earnings.

▶ **FIGURE 3 Increases in Smoking by Region Between 1998 and 2008** As this figure shows, nearly all the increase in smoking during this 10-year period occurred in the semiperiphery and periphery. The difference between core and periphery suggests an inverse relationship between smoking and level of development.

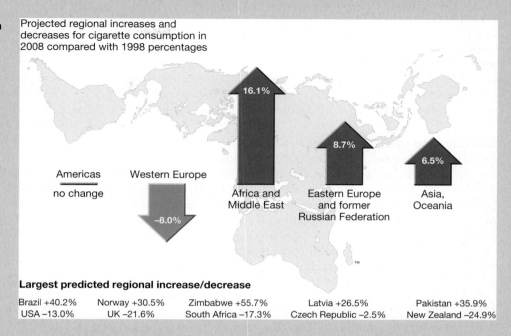

Projected regional increases and decreases for cigarette consumption in 2008 compared with 1998 percentages

Americas — no change

Western Europe −8.0%

Africa and Middle East 16.1%

Eastern Europe and former Russian Federation 8.7%

Asia, Oceania 6.5%

**Largest predicted regional increase/decrease**

| | | | | |
|---|---|---|---|---|
| Brazil +40.2% | Norway +30.5% | Zimbabwe +55.7% | Latvia +26.5% | Pakistan +35.9% |
| USA −13.0% | UK −21.6% | South Africa −17.3% | Czech Republic −2.5% | New Zealand −24.9% |

production and consumption. These early cigarettes required imported Turkish tobacco and expensive skilled labor to roll and package the finished product. In general, cigarettes were an upper-class urban luxury item, too expensive for the common consumer. In the United States, many considered cigarettes a passing fad, too expensive, European, and effeminate to be a viable long-term commodity.

The mechanization of cigarette production in the United States during the 1870s and 1880s made cigarettes available and affordable to most tobacco users. Large-scale mechanized production reduced labor and material costs and changed the structure of the entire tobacco industry, encouraging makers to expand production and create demand through intensive advertising campaigns, brand-name recognition (brand names often referred to British aristocracy or Middle Eastern luxury), and worldwide sales distribution. By the beginning of the 20th century, American and British producers had pushed into markets from Shanghai to Cairo. Cigarettes became an increasingly globalized commodity.

By 1910, the tobacco industry had developed into a "big business" dominated by the American Tobacco Company and its subsidiaries

and partners around the world. The U.S. Supreme Court dissolved this monopoly in 1911 and divided the tobacco industry among a handful of successor companies carved from American Tobacco. The dissolution sparked a new round of market competition and brand innovation, the most important of which was the development of the blended American cigarette. This new product added sweetened Kentucky burley tobacco to the Turkish tobacco and Virginia and North Carolina bright tobacco that had originally filled mass-produced cigarettes. The contours of the modern industry developed at this time, with several well-known manufacturers and their associated brands appearing in the aftermath of the Supreme Court's decision.

Through the 20th century, the cigarette became the most popular and widespread tobacco product, constituting more than half of British tobacco sales by 1920 and more than half of U.S. sales by 1941. Some of this market expansion came at the expense of other tobacco commodities but much also came during the 1920s, when gendered constructions of cigarette smoking changed, and large numbers of urban women took up the habit. Manufacturers diversified their brands and advertising strategies accordingly, a process they have

repeated to tap growing markets in countries in the global South. In 2007, an estimated 43.4 million U.S. residents were cigarette smokers, whereas worldwide estimates place the number of adult smokers at more than 1 billion (**Figure 3**).

Even as cigarette makers have opened new markets and realized astounding profits, they have had to answer to concerns about their products' social costs, particularly in relation to the impact on public health. Several state- and federal-level legal battles in the past decade have forced "Big Tobacco" to reassess its product and marketing strategies and pay out billions of dollars to cancer victims. These court decisions have set off further political debates about corporate responsibility, public health, and the allocation of settlement money and prepared the ground for a June 2009 law signed by President Barack Obama that gave the U.S. Food and Drug Administration strong new regulatory authority over tobacco. The trend of litigation against tobacco companies and tougher regulation in the United States and elsewhere in the global North may set precedents for makers in other countries as well and has forced a reassessment of tobacco's social value.■

*Jamey S. Essex, Ph.D., University of Windsor*

both hunters and gatherers and shifting cultivators, a wide range of resources was identified and used. The economy was based on need. Need was met by planting or foraging or through barter (for example, trading corn for fish). Moreover, the prevailing practice was to take only what was needed to survive. In addition to native people having no concept of private property or land ownership, there is also no evidence that a profit motive existed before contact with Europeans. Land and resources were shared. Still, although New England native people did appear to live in something of a balance with the natural resources they exploited, substantial vegetation change occurred as a result of their settlement and hunting activities (often using fire), which resulted in some species depletion.

**Colonial Land Use**  Geographer and environmental historian William Cronon has shown that Europeans saw the natural world they encountered in the United States and Canada much differently from how native peoples saw it. Most important, they viewed resources as commodities to be accumulated, not necessarily for personal use but to be sold for profit or export. The arrival of the Europeans meant that pressures on natural resources were hugely accelerated, especially for wood, furs, and minerals. In the Atlantic region, where European settlement first occurred, there was extreme exploitation of white pine, hemlock, yellow birch, beaver, and whales that led ultimately to deforestation and extinctions.

The arrival of Europeans in the region also meant a dramatic change in prevailing social understandings of the nature of land. Native perspectives about the communality and flexibility of land were replaced by European views of land as private and as having fixed boundaries. European settlers wanted to own and fence a plot of land, which led eventually to the concentration of land in large private farms, plantations, and haciendas. Increasingly, the native people of the United States and Canada were forced onto less-productive land or reservations and were prohibited from hunting or gathering on private lands or from moving with the seasons as they had before.

**Contemporary Agricultural Landscapes**  Overall conditions are very good for agriculture as one moves from east to west until the precipitation gradient drops to a very low level just beyond the Interior Lowland and Great Plains regions in the United States. Beyond that point, soil fertility is low, and rainfall is limited and infrequent. Conditions then become favorable again in the valleys along the Pacific Coast. Most of the agricultural productivity of the United States and Canada is concentrated in the Interior Lowland and Great Plains regions. Although natural conditions favor these areas for the highest agricultural productivity, other parts of the region are also important agriculturally, largely because farmers there have overcome the natural barriers to production through fertilizers, irrigation, pesticides, and other technological applications. This is the case in the Pacific valley areas, for instance, where irrigation water drawn from the Colorado River enables agriculture to flourish, and in the Southwest, where intensive irrigation supports significant cotton and citrus production. For places like Arizona and New Mexico, northward-flowing air masses bring summer monsoonal precipitation patterns that support the region's unique desert vegetation. The summer and winter rains there allowed the ancestors of contemporary Native Americans to cultivate beans, squash, and corn using sophisticated irrigation systems in a landscape that would otherwise seem inhospitable to subsistence agriculture. The absence of mountainous barriers to the north means that polar air sweeps down regularly in the winter months all the way into the midsection of the Interior Lowlands.

The agriculture of the Great Plains states and Prairie provinces is large scale and machine intensive, dominated by a few crops, the most

important of which is wheat (**Figure 6.9**). Winter wheat is planted in the fall and harvested in late May and June. Winter wheat is grown across much of the United States, but it predominates in the southern Plains from northern Texas to southern Nebraska. Spring wheat is grown primarily from central South Dakota northward into Canada and is planted in early spring and harvested in late summer or fall. It is suited to areas where winters are so severe that germinating winter wheat would be killed. Most of the wheat grown in these former grassland ecologies uses dry farming techniques, meaning the crop is not irrigated.

In addition to wheat, sorghum is also an important agricultural commodity for southern Plains farmers. A grain that was domesticated in Africa, more sorghum is grown in Texas and Nebraska than wheat, where it is destined for stock feed. On the northern Plains, barley (first domesticated in the Himalayas) and oats (first domesticated in the Fertile Crescent in the Middle East) are major secondary crops, with most of the continent's barley crop coming from the Lake Agassiz Basin of North Dakota and Minnesota. Nearly all flaxseed (first domesticated in the Fertile Crescent) produced in North America also is grown in the northern Plains. Sunflowers (native to the Americas), a source of the vegetable oil canola and important ingredients in many livestock feeds, are rapidly increasing in importance in the Red River valley of Minnesota and North Dakota.

The Gulf–Atlantic Coastal Plain is a lowland area that extends from New York to Texas. The coastal plains are generally level, with soils that are sandy and relatively infertile. Where agriculture does occur, it is intensive and scattered. For instance, the lower Mississippi is an area of good soils and intensive agriculture. The Piedmont is the historic region of plantation agriculture—cotton, tobacco, and corn—but much of the soil in this region is depleted or eroded from overfarming.

The interior valleys along the western coastal areas of the United States and Canada contain agricultural landscapes of fruit and vegetable production. These agricultural landscapes are irrigated ones, meaning that rainfall is inadequate to sustain the crops. More than half of the fruits, vegetables, and nuts consumed in the United States are produced in California. Washington State accounts for 50% of the U.S. apple sup-

▲ **FIGURE 6.9  Winter Wheat, Manitoba, Canada**  Pictured here are both fields and storage silos on a large farm on the Canadian prairie. Canada is one of the world's largest wheat producers.

ply. Interestingly, California is also the place where many new food movements have begun and disseminated across the region. These movements include a move away from conventionally produced food to organic production, slow food, and local food, all of which are alternatives to industrial food production.

**Alternative Food Movements**   Organic farming describes any farming or animal husbandry that occurs without commercial fertilizers, synthetic pesticides, or growth hormones. This type of farming can be contrasted with conventional farming, which uses chemicals in the form of plant protectants and fertilizers, or intensive hormone-based practices for breeding and raising animals. Organic production can be seen as a rejection of industrial farming practice and the increasing use of genetically modified organisms, or GMOs, on U.S. and Canadian conventional farms. A GMO is any organism that has had its DNA modified in a laboratory rather than through cross-pollination or other forms of evolution (**Figure 6.10**). Examples of GMOs are a bell pepper with DNA from a fish added to make it more drought tolerant, a potato that releases its own pesticide, a soybean that has been engineered to resist fungus, and rice that has been modified to include more nutrients. Although conventional framing practices continue to dominate the region's agricultural sector, organic practices are a growing force not only among small farmers but also with larger corporate entities such that Wal-Mart has introduced organic foods into its superstores.

Local food is usually organically grown, but its designation as local means that it is produced within a fairly limited distance from where it is consumed. Most understandings of local food set a 100-mile radius as the border of what is truly local. Thus, if one is a "locavore," the food one consumes should be produced on farms that are no further than 100 miles from the point of distribution. Local food movements have resulted in the proliferation of communities of individuals who have joined together to support the growth of new farms within the local "food shed" (the area within the 100-mile radius). Slow food, as the name suggests, is an attempt to resist fast food by preserving the cultural cuisine and the associated food and farming of an ecoregion.

▲ **FIGURE 6.10 Protests against GMO food in Mexico**  Greenpeace activists in Mexico City are shown protesting against the importation of GMO rice from the United States.

Although slow food, local food, and organic and ecologically sustainable agricultural practices and the small farming landscapes they produce are proliferating across Canada and the United States, they in no way challenge the dominance of more conventionally produced, distributed, marketed, and consumed food. Moreover, as several critics of the movements have pointed out, these alternative practices are largely organized and promoted by white, middle-class people and exclude, often simply through cost and associated accessibility, poor people of color everywhere. In the United States and Canada, the result is that more than any other economic class, poor people turn to cheap, easily accessible food, also known as fast food.

Fast food was born in the United States as a product of the post–World War II economic boom and the social, political, and cultural transformations that occurred in its wake. The concept of fast food, edibles that can be prepared and served very quickly, sold in a restaurant, and served to customers in packaged form, actually preceded the word for it. What made fast food *fast* was the adoption of industrial organizational principles applied to food preparation in the form of the Speedee Service System, first introduced in 1948. Today, fast food is so ubiquitous in the United States—and is becoming increasingly so throughout the world—that it is taken for granted by many consumers. But health practitioners, alternative food activists, journalists, and others have begun to turn a critical eye on fast food to expose some of its dietary, labor-related, and ecological shortcomings. The concerns about the effects of fast food on the nation's health have risen as our consumption of (or, as some health experts would argue, addiction to) fast food increases and our levels of routine exercise decrease. There is also concern about the impact of fast-food ingredients on the landscape. The Beyond Beef Campaign, an international coalition of environment, food safety, and health activist organizations, lists the following statistics about beef production and its environmental impacts. Beef raised for fast-food hamburgers is one of the largest components of the fast-food sector.

- One quarter-pound of hamburger beef imported from Latin America requires the clearing of 6 square yards of rainforest and the destruction of 165 pounds of living matter, including 20 to 30 different plant species, 100 insect species, and dozens of bird, mammal, and reptile species.

- Cattle degrade the land by stripping vegetation and compacting the earth. Each pound of feedlot steak costs about 35 pounds of eroded topsoil.

- Nearly half the total amount of water used annually in the United States goes to grow feed and provide drinking water for cattle and other livestock. Producing a pound of grain-fed steak requires almost 1,200 gallons of water.

- Cattle produce nearly 1 billion tons of organic waste each year. The average feedlot steer produces more than 47 pounds of manure every 24 hours.

- Much of the carbon dioxide released into the atmosphere is directly attributable to beef production: burning forests to make way for cattle pasture and burning massive tracts of agricultural waste from cattle feed crops.

- U.S. cattle production has caused a significant loss of biodiversity on both public and private lands. More plant species in the United States have been eliminated or threatened by livestock grazing than by any other cause.

**Environmental Challenges**   Decades of federal environmental protection legislation have forced U.S. industries to curtail much of their polluting processes or move to manufacturing sites outside the United States. Nonetheless, various regions of the country still face serious and persistent environmental challenges. Acid rain, generated by industrial processes and automobile emissions, continues to pose a challenge around the Rust Belt on both sides of the U.S.–Canada border. Along the U.S.–Mexico border, air and water pollution are also a problem.

The legacy of past pollution-generating industrial practices also remains a problem. The U.S. government currently is overseeing the cleanup of hundreds of **superfund sites,** locations officially deemed by the federal government as extremely polluted and requiring extensive, supervised, and subsidized cleanup.

The most serious environmental challenge Canadians and Americans face today is their seemingly insatiable appetite for resources, especially energy resources. The impact of the high level of energy consumption not only seriously challenges Earth's supply of renewable energy resources, but it is also leading to global climate change as discussed earlier. In an attempt to begin to address this challenge and to reduce energy consumption in the region, a growing consumer interest in sustainability has created a large and growing market for environmentally friendly—also known as "green"—goods and services. In fact, leading corporations such as the large automobile manufacturers and energy companies like General Electric and Shell Oil are making substantial investments in energy efficiency and energy technologies, and consumers are responding positively. A good example is the largely unexpected growth in the purchase of hybrid cars. Local governments are also converting their fleets to hybrids along with large corporations like Federal Express (FedEx). Green technology is not only reshaping automobile supply and demand, but it is also affecting architecture, furniture, and clothing and other consumer items, as well as urban and regional planning. The Obama administration has actively encouraged the public adoption of more green technologies through legislative initiatives and tax breaks. One such initiative was the 2009 "cash for clunkers" program, wherein car owners received a voucher worth $3,500 if they traded in a vehicle getting 18 miles per gallon or less for one getting at least 22 mpg. The voucher was worth $4,500 if the new car's mileage was 10 mpg higher than the old vehicle. Canada provides tax rebates to consumers who purchase hybrid electric vehicles before 2010. The Obama administration is also providing generous tax rebates for consumers who purchase residential solar energy systems.

# HISTORY, ECONOMY, AND DEMOGRAPHIC CHANGE

Today the United States and Canada are established democracies modeled on European political traditions. But both have populations of native peoples that predated European colonization. Both consolidated their leadership roles in the global economy early in the 20th century through an effective strategy of economic development, and global geopolitical maneuvering. Thus, although much of the recent history of these two countries is the result of European colonization, the longer-term historical geographies of this region are tied into the broader historical movements and people from without and within the region. Because of this and their comparable economic status, academics as well as policymakers and government agencies treat the two countries as a coherent region, although there

is certainly quite a bit of difference across these two national spaces. That said, the continuities provide a framework for thinking about this space as a world region, one with a set of parallel historical experiences. There is no doubt that the United States and Canada have had and continue to have an important affect on global relations. But, the growing interconnectivities of the recent period suggest that this region is also subject to dynamic change as it evolves in relation to the changing geopolitical and sociocultural realties of our globalized world.

## Historical Legacies and Landscapes

In the 15th century, European views of the world did not include the existence of the North American landmass. And although 16th-century Spanish missionaries and explorers identified the southernmost section of present-day United States and Mexico as of interest to their exploration and missionary efforts, most of the rest of North America was considered of little consequence because it presented none of the appearances of grandeur and resource potential that the Aztec, Maya, and Inca empires of Latin America did (see Chapter 7 and **Figure 6.11**).

**Indigenous Histories**   The distribution and subsistence practices of the indigenous people of the United States and Canada around 1600 reflected the great diversity of cultural groups that occupied the continent (**Figure 6.12**). When Christopher Columbus arrived in the Caribbean, he assumed he had reached the Far East, or the Indies, and he called the aboriginal people he encountered *los indios,* or indians. And just as thousands of native languages existed at the time of European contact, there was no single word of self-description com-

▲ **FIGURE 6.11 Norse Settlement in L'Anse Aux Meadows, Newfoundland** The earliest contact between the Old World and the New is likely to have been in the 10th century, when seaborne Norse adventurers reached Greenland and ventured to the coast of North America, establishing themselves at L'Anse aux Meadows on the northern tip of the island of Newfoundland. This photo shows the remains of what are believed to be as many as three Norse settlements. According to available evidence, the Norse settlers and the Inuit at first fought each other, but then established a regular trading relationship. The Norse settlements were soon abandoned, probably as the Norse withdrew from Greenland.

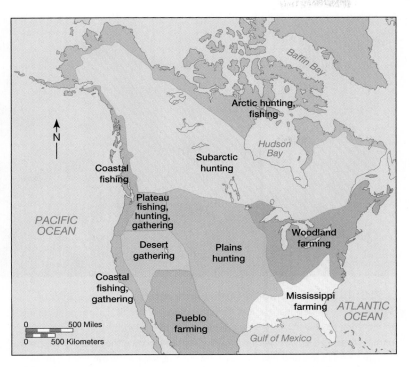

▲ **FIGURE 6.12 Subsistence practices of U.S. and Canadian native peoples around 1600** This map, which shows the subsistence practices of indigenous peoples of the region at the point of European contact, testifies to the extensive spread of native population on the continent before the Europeans began to colonize. The geographic distribution of different tribes indicates their relationship to cultures and subsistence practices developed by the ancient people who preceded them.

mon to the diverse people who occupied the continent. As a result, the word *Indian* has endured as an extremely misleading and sometimes derogatory term for describing a wide range of regional cultural groups.

The estimates of the native population of the region at the point of European contact range widely, between 1 million and 18 million individuals. More than anything, the differing estimates indicate how little is known about the people who were thriving in the United States and Canada before the Europeans arrived. What is well known is that there was no common culture, particularly no common language, among the many who first populated the region. Native American scholar Jay Miller estimated that in 1492 the indigenous inhabitants of the United States and Canada spoke 2,200 different languages, with many regional variations as well. Tribal culture and local environmental conditions were the frameworks within which daily life was governed and lived.

Europeans originally made contact with various individuals and tribes for help with extracting resources, including animal furs; naval stores, such as tar and turpentine for shipbuilding; fish; and other primary-sector products that would be exported back to the continent. The experiences of the Dutch, French, and English in North America also differed substantially from that of the Spanish in Latin America (Chapter 7). The Spanish conquistadors vanquished sophisticated civilizations to plunder their gold and silver treasuries and make themselves rich in the process. The Dutch, French, and English were also interested in improving their financial situations, but they encountered a very different set of cultures with no centralized system of social control that they could exploit as the Spanish had

done with the Aztecs in Mexico. Instead, the eastern tribes that Europeans first encountered in the United States and Canada were small, autonomous groups, possessing a lively sense of rivalry and competition with their neighbors. The Spanish were interested in massive occupation and exploitation, whereas the Dutch, French, and English considered exploration and colonization to be commercial ventures.

**Colonization and Independence**   No other landmass already occupied by a diverse range of complex and widely distributed societies, with the possible exception of Australia, has undergone such a dramatic transformation in such a short period of time. Moreover, lumping the settlers into one category of Europeans vastly simplifies the very complicated process of settlement that different people from different parts of Europe brought with them to the United States and Canada. So, although the occupation by missionaries and settlers who came to the region in the 16th and 17th centuries is widely known as the period of **Europeanization**, the process was actually highly selective. Only a few Western European countries—France, Spain, the Netherlands, and Great Britain—dominated colonization in the United States. In Canada, the European colonizers were predominantly Britain and France. In both Canada and the United States, Great Britain was by far the most influential of the four, though others did have substantial impacts, particularly the French in Canada and the Spanish in the United States.

Even within the four countries that dominated colonization, different groups from different regions settled in different parts of the United States and Canada. The unusual individuals and groups who assumed great risks in coming to the region represented only a small sample of their national cultures. As a result, the Europeanization process, as it unfolded along the Atlantic seaboard of the United States and Canada, was accomplished through mixing a very wide range of native and imported traditions that eventually created distinct colonial cultures and societies in different places.

The process of European settlement routinely resulted in the overwhelming exploitation and abuse of the native people by colonists and later citizens. The history of colonial settlements, although at first peaceful, over time erupted into disputes over land claims that ended in violence and often outright massacres on both sides. After a time, moreover, not only were there conflicts between tribal people and colonists, but also direct conflict emerged among the various European groups that vied for control over land. In addition, colonists fanned the flames of rivalry between opposing tribal groups by providing them with arms, thus elevating the level of technology and intensifying the degree of violence in armed conflict. Finally, as has been widely documented, exposure to the Old World diseases that the colonists brought with them had a devastating impact on native populations (see Chapter 7, p. 242). As native populations were decimated, defeated, or demoralized and pushed farther into the interior of the continent, the various European groups increasingly came into direct conflict, leading eventually to the Seven Years' War (1756–63), the U.S. phase of which is known as the French and Indian War (1754–63). This war left the British more or less triumphant over the whole of the European-inhabited territory of the United States and Canada.

In the following two decades, residents of the original 13 colonies of the United States became disillusioned with their administrators in Britain—who were taxing them to help recoup the high cost of the Seven Years' War—and launched their own war, leading to the creation of a new, independent nation in the late 18th century. In Canada, a bloodless separation from Great Britain would not occur until well into the 19th century.

Even before the American Revolution, however, a process of **Americanization** had begun, as a generation of individuals of European parentage born in the U.S. colonies felt less loyalty and fewer cultural ties to the mother country. As a result, a new ethos of liberalism, individualism, capitalism, and Protestantism emerged, gained currency, and ultimately came to define a U.S. national character. The successful outcome of the Revolutionary War with Britain (1775–83) left the continent with a robust new nation dominated by Anglo-American institutions and with the addition of slavery. Canada remained a colony, under British control, composed of both French- and English-speaking settlers.

**The Legacy of Slavery in the United States**    Although the impact of European colonization in the 16th and 17th centuries was felt all along the Atlantic seaboard of the United States and Canada, development of the U.S. South following the end of the revolutionary period differed dramatically from that of its northern neighbors. Before the arrival of the Europeans, the area that now forms the southeastern United States was inhabited by a wide range of native tribes, among them the Cherokee, Choctaw, Chickasaw, Creek, and Seminole people. Early on, the region was occupied by military personnel living in scattered outposts like Jamestown in Virginia. However, by the mid-17th century the military outposts had given way to tobacco farms (see Geographies of Indulgence, Desire, and Addiction: Tobacco, pp. 204–5). At first, **indentured servants**—individuals bound by contract to the service of another for a specific term—from Britain were the primary source of labor on the tobacco and later indigo and cotton plantations. Increasingly, however, servants earned their freedom and were replaced by slaves from Africa, at the same time that disease, armed conflict, and demoralization reduced the native populations that would have been another source of laborers.

African slaves had been a well-established commercial staple of the Mediterranean well before the Spanish and Portuguese introduced them to their newly captured territories in Latin America and the Caribbean, thereby establishing the Atlantic slave trade. By the early 15th century, Dutch and English raiders attacking Spanish and Portuguese ships were able to take control of the slave trade. By the early 17th century, England became the dominant slaving nation. As a result, slaves were a part of the social and economic system of the American colonies beginning, practically simultaneously, with their founding. As an institution of formal social and economic organization, slavery endured in the South for more than 250 years, ending officially in 1870, following the end of the U.S. Civil War (1861–65). Its legacy, however, continues to shape the landscape and identity of the region.

**European Settlement**    With the creation of new nations and the transformation of colonies into states, the relentless European settlement of the North American continent accelerated, more so in the United States than in Canada. By the middle of the 19th century, settlement in the United States had pushed beyond the Appalachian Mountains into the Interior Lowlands, including the upper Ohio and Tennessee river valleys and the interior South. France lost control of Canada to Britain in 1763, which had little impact on new settlement there. By the end of the century, however, southern Ontario, in and around present-day Toronto, became attractive to settlers (**Figure 6.13**).

Historians have argued that frontier settlement involved a continual process of national and personal reappraisal, as well as of increasing geographical divergence, as new settlers encountered new landscapes. It was also a process of sustained mobility, so much so that mobility has come to be seen as characteristic of the region's inhabi-

▲ **FIGURE 6.13 Canadian settlement patterns** In Québec the first settlers laid out long, narrow lots from the shores of the St. Lawrence River into the interior, as shown. As settlement moved farther inland, roads were built parallel to the waterways with narrow lots extended on either side. The pattern is duplicated in the Red River valley of Manitoba, where the early settlers were also French. In Ontario and the eastern townships of Québec, land subdivision was made according to preconceived plans. Although the townships were more or less square, the grid became irregular because it was started from a number of different points, each of which used a differently oriented base. In the prairies, the grid is much more regular, partly as a result of the topography and partly because a plan for the subdivision of the whole region was laid out in advance of settlement.

tants, especially in the United States. There, the movement of the frontier was continuous *and* mostly contiguous, at least until settlers reached the Great Plains in the middle of the 19th century and confronted significant mountain ranges at its western edge. In Canada, westward expansion was interrupted early on by the vast, generally infertile, though heavily forested, Canadian Shield, which separated Ontario from the prairies. Many Canadian settlers leapfrogged across to the northern midsection of the country to acquire suitable farmland.

In the United States, the pace of westward expansion was accelerated by the federal government's decision in the late 1780s to sell public lands cheaply to citizens. By 1850, the development of the railroads reoriented the pace and direction of continental settlement—eastward from the Pacific Coast to the interior west rather than from the East Coast westward—at the same time that it diminished the previous isolation of pioneer settlements. By the close of the 19th century, the frontier process had resulted in a set of rural and agrarian regions and subregions that stretched across the continent. Each region was defined by its own experiences of the history and particular conditions of settlement and by its distinctive regional economic development.

The regional economy during this period was oriented to agro-mercantile activities. This means that trading agricultural crops and primary resources, such as fish, timber, and minerals, provided an economic base for the expanding population. Yet in the United States by the mid-19th century, a new economy based on manufacturing was rapidly gaining momentum along the northern Atlantic seaboard, especially in and around southern New England and New York. At the same time, the rest of the United States and Canada was being settled by European immigrants or their descendants, making their livelihoods largely by farming.

By the early 20th century, however, with the continent occupied from east to west mostly by Europeans and Euro-Americans, the industrialization of the U.S. economy was well on its way to transforming the landscape from one of rural agricultural settlement to one of urbanization and industrialization. The 1920 census documented for the first time in U.S. history that there were as many people living in cities as there were in rural areas. From that point onward, the United States and, soon after, Canada became increasingly urbanized. By the late 20th century, 75% of the U.S. and Canadian population lived in cities, up from 25% in the mid-19th century.

**Urbanization, Industrialization, and Conflict**   Long before explorers and colonists arrived, native people built cities in the regions that would become the United States and Canada. However, a far more extensive and intensive urbanization process began in earnest during the colonial period. The Europeans who colonized the United States and Canada were usually part of commercial urban systems in their home countries. As they colonized the new lands, they responded to the need for central places for organizing commerce, defense, communication, and, later, administration and worship by building cities. In Florida and in the southwestern United States, the Spanish founded cities, which became symbols of political and military authority (**Figure 6.14**). French explorers came not to settle but to reap commercial rewards, and they established urban centers to facilitate the exchange of goods. The Dutch also established urban settlements for trading centers, including for furs and slaves.

With few exceptions, however, the British played the largest role in shaping U.S. urbanization and urban life. Sustained by trade based on an agrarian economy, the U.S. Atlantic coastal cities established by the English colonists were also oriented toward a kind of corporate communalism tempered by notions of social and religious harmony. As geographer Alan Pred has shown, in addition to being administrative centers, Boston, Providence, Baltimore, Philadelphia, Charleston, and Savannah were also ports and key nodes in a globally expanding mercantile system. As such, they enabled the transfer of resources, goods, and people, not only from the interior hinterland into the cities but also outward to Europe. At the same time, these cities received goods from England for U.S. consumer markets. Colonists saw their burgeoning cities not only as commercial centers but also as places where new ideas could be hatched and nurtured and as hearths of "civilization," where Old World cultural practices confronted those of the New World, creating in the process uniquely North American urban places.

As sites of innovation and cradles of culture, by the early 19th century cities were also largely the places where a new economy based on manufacturing was born and flourished. While industrialization was fueling the economies of the northern and midwestern United States, by the mid-19th century the differences between the northern and southern regions of the United States had become increasingly pronounced. The commerce and industry fueling the national economy made the cities of the North the most important ones economically. Although Charleston, Savannah, and Norfolk in the South were all ports, only Charleston was a deepwater port. More important, the southern ports were less crucial as transfer points than were those in the North because the South had many navigable inland rivers. Ships could simply pick up agricultural products and raw materials or drop off their own cargoes by way of the rivers, thereby eliminating the need to stop at the coastal ports.

Whereas the North's economy was more diversified—based on commerce, agriculture, and industry—the South's was more simply tied to staple crop agriculture. Because the southern plantations produced much of the food and clothing for the region, and because most of the capital was invested in slave labor (which was unpaid and thus had no income to stimulate commercial exchange), the economy, as well as society and culture, made the South very different from the North and Canada. All these differences—especially slavery and the economic issues that surrounded it—divided the southern states from the northern ones in highly volatile ways. By 1861 the division was so deep that the South formed a new government, the Confederate States of America, and attempted to secede from the United States to protect slavery and their agricultural economy. Civil war ensued.

With the South in defeat following the end of the Civil War in 1865, territories in the West joined the Union as free states, contributing people, resources, and capital to the burgeoning U.S. economy. Fearing that the United States, emboldened by its Civil War victory, might launch an invasion of Canada and responding to agitation among the French-speaking minority, the British Parliament passed the North America Act of 1867. The result was the creation of the Dominion of Canada, dissolving its colonial status and effectively establishing it as an autonomous state within the world-system, with its own constitution and parliament. All the existing Canadian colonies joined the new confederation except British Columbia, which waited until 1871; Prince Edward Island, which joined the dominion in 1873; and Newfoundland, which remained independent until 1949.

## Economy, Accumulation, and the Production of Inequality

By the end of the 19th century, Canada and the United States were fast becoming key players in the global economy. The United States, politically independent just before the onset of the Industrial Revolution, was able to

▼ **FIGURE 6.14  Fort Matanzas, St. Augustine, Florida**  Military defense sites were an important part of the early settlement of North America by Europeans as they fought each other for control of the territory. *Matanzas* means slaughter in Spanish and refers to the massacre of nearly 250 French Huguenots at the hands of the Spanish at the site, 175 years before the fort was constructed.

become economically competitive because of several favorable conditions. Vast natural resources of land and minerals provided the raw materials for a wide range of industries that could grow and organize without being hemmed in and fragmented by political boundaries. Populations, which were growing quickly through immigration, provided a large and expanding market and a cheap and industrious labor force. Cultural and trading links with Europe provided business contacts, technological know-how, and access to markets and capital (especially British capital) for investment in a basic infrastructure of canals, railways, docks, warehouses, and factories.

**The Two Economies**    Canada's path to global economic competitiveness was distinctive. Although most of its population enjoys a high quality of life and high levels of economic productivity, Canada is certainly an atypical economy largely because it had, until very recently, never been highly industrialized. The primary sector (see Chapter 1, p. 23) was the major pillar of its economic prowess. Though that sector has declined in

centrality, it continues to play an important role in its economic structure. When industry did begin to grow and flourish in Canada after World War I, most of it occurred in the midsection of the country along a swath of land at the U.S.–Canadian border. A substantial proportion of the industries built there were branch plants of U.S. manufacturers. Furthermore, Canada's major trading partner is the United States, which imports more than half of all Canadian exports. What has been most remarkable about Canada's place in the world economy is that it became so successful as a **staples economy**, one based on natural resources that are unprocessed or only minimally processed before they are exported to other areas, where they are manufactured into end products.

Over the last several decades of the 20th century, Canada's economy shifted so that today its largest and most dynamic sectors are first, by a large margin, real estate, finance, and insurance, and second manufacturing. Although an important energy-trading partner to the United States, the mining and energy sector constitutes only 4.5% of the Canadian GDP. As **Figure 6.15a** and **b** show, by comparison, the U.S.

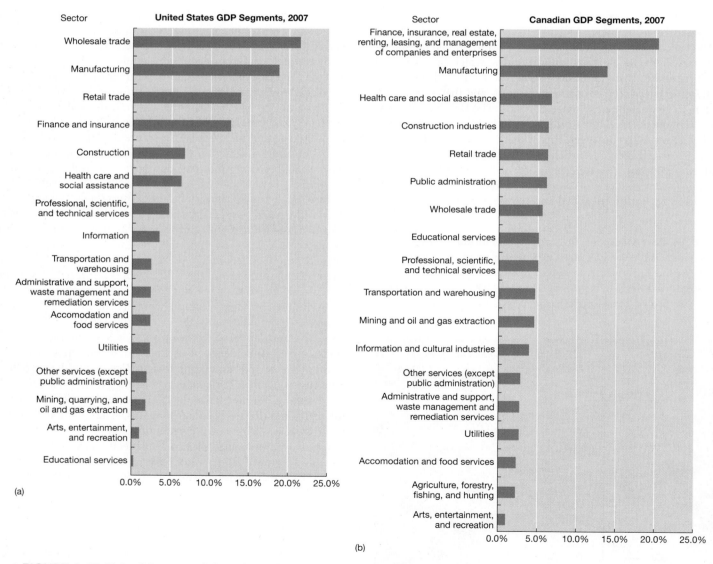

▲ **FIGURE 6.15  United States and Canadian GDP segments, 2007** (a) Trade and manufacturing dominate the U.S. economy, whereas manufacturing and services dominate the Canadian one (b). Despite those differences, the economies of the two countries have much in common.

economy is more concentrated in trade and manufacturing. What the graphs do not show is that while the Canadian economy has increased its share in manufacturing, the United States has decreased its share. Both countries possess very large and powerful economies, with the U.S. economy being dominant.

Today the United States and Canada together produce more than one-quarter of the world's GNP. The United States has the world's largest economy and Canada the ninth largest. Their resources are extensive and varied, and their ability to exploit them is high. As part of the recent restructuring of the global economy discussed in Chapter 1, the various regions of both the United States and Canada have experienced significant transformations in their economies, societies, political institutions, and even their physical environments.

**Transforming Economies**    Political, economic, and social geographers would agree that the most important regional transformation of the last 25 years has been the rise of the service economy based on finance, real estate, and insurance. In the United States, the most recent wave of internal migration, the movement of U.S. residents from the Rust Belt (called this because of the aging factories) to the Sun Belt (called this because of the warm weather), was part of this shift. During the 1960s and the 1970s, the historic core of North American industrialization, the Manufacturing Belt, began to experience economic problems in the form of high labor costs and aging infrastructure, mostly manifested in outdated technology systems. Once-peripheral regions of the country, the South and Southwest, began to attract investors. The military had invested in this region during World War II, establishing bases and holding training exercises. Following the war, the government continued to invest in this region as it built up its military capacity during the Cold War.

As the computer age dawned, numerous places in the South and the Southwest were ripe for civilian investment opportunities, possessing abundant land and labor forces that were highly educated. Labor was also not used to unions and high wage rates. The result was a shift, which has since been rebalanced, as the profitability of old, established industries in the Manufacturing Belt—or the Rust Belt, as it came to be called—declined compared to the profitability of new industries in the fast-growing new industrial districts of the Sun Belt. Once the profitability differentials between the two places became significant, disinvestment began to occur in the Rust Belt. Manufacturers there began to reduce their wage bill by cutting back on production, to reduce their fixed costs by closing down and selling off some of their factory space and equipment, and to reduce their spending on research and development for new products. This disinvestment, in turn, led to deindustrialization in the formerly prosperous industrial core regions of the Midwest and Northeast.

Deindustrialization involves a relative decline (and in extreme cases, an absolute decline) in industrial employment in core regions as firms scale back their activities in response to lower levels of profitability (**Figure 6.16**). In effect, technological innovations in computerized production systems facilitated new industrial applications, and investors and manufacturers began to look around for new places in which to invest and build. Innovations in transport and communications technology, combined with these production innovations, created windows of locational opportunity. The result was the movement of capital investment in manufacturing away from the old industrial districts of the Manufacturing Belt and into small towns and cities in the Sun Belt, to suburban fringe areas near some of the old industrial districts, and offshore to countries that had lower-cost workforces.

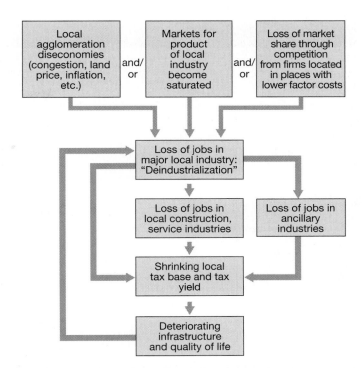

▲ **FIGURE 6.16 Spiral of Deindustrialization** When the locational advantages of manufacturing regions are undermined for one reason or another, profitability declines, and manufacturing employment falls. This can lead to a downward spiral of economic decline, as experienced by the traditional manufacturing regions of North America during the 1970s and 1980s.

Meanwhile, the capital made available from disinvestment in the Rust Belt became available for investment in new ventures based on innovative products and production technologies. Old industries and a large proportion of an established industrial region were "dismantled" to help fund the creation of new centers of profitability and employment. This process is often referred to as **creative destruction**, something that is inherent to the dynamics of capitalism. Creative destruction provides us with a powerful image to understand the need to withdraw investments from activities (and regions) yielding low rates of profit and to reinvest in new activities (and, often, in new places).

**Reorganization of Regional Landscapes**    The dramatic and often very painful changes that occurred around the decline of the Rust Belt and the rise of the Sun Belt have resulted in a major reorganization of the U.S. regional landscape. The new technology systems—using robotics, telematics, biotechnology, and other knowledge-based systems that have emerged and been refined and improved over the last 25 years—have helped encourage the growth of new regions with very different economic bases than they possessed only 25 years ago. Although the Rust Belt experienced a crippling decline in the 1970s and early 1980s, by the mid-1990s it was again booming, having reorganized its economy and political institutions around service-related employment. The Sun Belt has maintained its strong economy and has continued to experience phenomenal population growth into the 21st century. Its economic base is a rich mix of sectors from resource extraction to knowledge-based industries.

The new regional geographies of the United States have resulted in a massive population redistribution. More than 50% of U.S. residents

| TABLE 6.2 Income Disparity in the United States | | | | | | | | |

| Household Groups | Share of All Income | | | Average After-Tax Income (Estimated) | | | Change | |
|---|---|---|---|---|---|---|---|---|
| | 1977 | 1999 | 2007 | 1977 | 1999 | 2007 | 1977–99 | 1977–2007 |
| One-fifth with lowest income | 5.7% | 4.2% | 3.4% | $10,000 | $8,800 | $11,551 | −12.0% | 15.5% |
| Next lowest one-fifth | 11.5 | 9.7 | 8.7 | 22,100 | 20,000 | 29,442 | −9.5 | 33.2 |
| Middle one-fifth | 16.4 | 14.7 | 14.8 | 32,400 | 31,400 | 49,968 | −3.1 | 54.2 |
| Next highest one-fifth | 22.8 | 21.3 | 23.4 | 42,600 | 45,10 | 79,111 | 5.9 | 85.7 |
| One-fifth with highest income | 44.2 | 50.4 | 49.7 | 74,000 | 102,300 | 167,971 | 38.2 | 127.0 |
| 1% with highest income | 7.3 | 12.9 | 21.2 (2005) | 234,700 | 515,600 | 868,000 (2004) | 119.7 | 269.8 (1977–2004) |

Source: U.S. Census Bureau, 2000 and 2004 Population Survey, Annual Demographic Supplements. + Current Population Survey, 1968 to 2008 Annual Social and Economic Supplements
http://www.census.gov/hhes/www/income/histinc/p60no231_tablea3.pdf

now live west of the Mississippi River, whereas 100 years ago most people lived in the East. In the last 15 years, as a result of the dramatic changes brought about by the emergence of a fifth technology system based on solar energy and other green technologies, robotics, microelectronics, biotechnology, advanced materials, and information technology, a "new economy" has emerged in the United States and Canada. The new economy, though caused by the practical implementation of the fifth technology system, is also about the dramatic transformation of markets, work, and the labor force.

**Wealth and Inequality** Although globalization and the new economy have helped improve the employment opportunities and level of wealth of many in the United States, it has also seriously set back many others (**Table 6.2**). In 2007, in the United States, 12.5% of the population lived in families that did not earn enough to rise above the official poverty threshold. The poverty line for a family of two adults and two children was set at $21,027. The poverty rate among children was 18%. Statistics such as these underscore the likelihood that up to 40% of U.S. children will experience poverty at one time or another, because many families move in and out of poverty over time. According to a 2007 UNICEF report, the United States ranked 20th out of 21 industrialized nations for child material well-being. Released in October 2008, the most recent data brief from the U.S. Centers for Disease Control and Prevention's National Center for Health Statistics ranked the United States 29th globally in infant mortality in 2004, the latest year such data were available for all countries. The U.S. ranking, which has declined from 12th in 1960 to 23rd in 1990, currently ties the United States with Poland and Slovakia. Child advocates argue that one of the reasons that the United States does so poorly as compared with other industrial countries is that it also has the lowest government benefits to families with poor children. Another reason is the stagnation of wages at the lower end of the wage spectrum.

In Canada, the poverty rate for an equivalent family of four is 8.2%. In 2008, more than 5 million Canadians lived in poverty. And poverty is increasing for youth, workers, young families, and immigrant and visible minority groups. Poverty among aboriginal groups is higher than among the rest of the Canadian population both on and off reserve. In fact, if the statistics for Canadian aboriginal people were viewed separately from those of the rest of the country, Canada's aboriginal people would slip to 78th on the UN Human Development Index—the ranking currently held by Kazakhstan. Over most of the 20th century, Canada provided a strong safety net for poor families, but in the last two to three decades, the poverty rate among children there has grown to the second highest (after the United States) among all developed countries.

The gap between rich and poor in the United States continues to widen. Between the mid-1980s and 1990s, average household income (in real terms) rose by less than 1%. But the average income of the poorest one-fifth of households in the United States increased by only 0.1%, while the average income of the top one-fifth of households jumped by 20%. In the late 1990s, 5% of the population held 21% of the wealth. In 1999 the bottom 80% of the population claimed only 50% of the wealth. One reason is that nine-tenths of the growth in wealth went to the richest 1% of households, whose annual income in 1999 averaged $515,600 after taxes. In 2005, the richest 1% of Americans possessed wealth worth $16.8 trillion, nearly $2 trillion more than the bottom 90%. One worker making $10 an hour would have to labor for more than 10,000 years to earn what 1 of the 400 richest Americans pocketed in 2005. **Figure 6.17** illustrates the enormous wealth disparity that characterizes U.S. society.

**The New Economy** In the last decade and a half, a new economy has emerged, especially in the United States and Canada, that has fundamentally transformed industries and jobs through information technologies (IT). These changes have been facilitated by a high degree of entrepreneurialism and competition, transforming the United States and Canada, as well as many other regions around the globe. But the new economy was born in the United States, sired by the technological changes that emerged from Silicon Valley, California, nearly 50 years ago.

It is generally agreed that the previous economic order, the "old economy," lasted from 1938 to about 1974. The year 1974 was a critical year in economic history: Oil prices were skyrocketing, and the corporate rate of profit was falling in the core. That economy's foundation was manufacturing geared toward standardized, mass-market production and run by stable, hierarchically organized firms focused on the

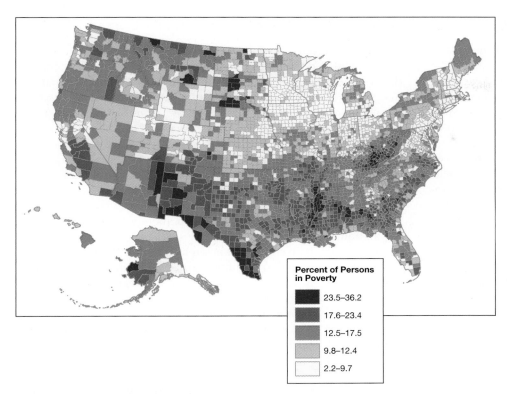

▲ **FIGURE 6.17 Poverty in the United States, 2008** One of the most startling aspects of this map is the geographical proximity of astronomical wealth and debilitating poverty. Some of the richest individuals in the country live side by side with some of its poorest. It is perhaps one of the most challenging paradoxes that in a "land of plenty," the plenty is concentrated among so few.

**Percent of Persons in Poverty**

- 23.5–36.2
- 17.6–23.4
- 12.5–17.5
- 9.8–12.4
- 2.2–9.7

in the value of the Canadian dollar has harmed exports. The impact of the crisis on the United States—which takes 80% of Canada's exports—has a direct impact on the Canadian economy. For example, the U.S. housing market collapse hurt Canada because much of the wood in new U.S. houses comes from Canada.

### The United States and the Global Economic Crisis

The failure of several major private financial institutions in October 2008 prompted panic among international financial markets. Suddenly, the economy was facing the prospect of recession, and millions of households in affluent countries were facing the very real prospect of the widespread loss of jobs, savings, and pension funds. The government of the United States, along with those of other leading economies, intervened with hundreds of billions of dollars of support for U.S. private financial institutions in an attempt to prop up the international financial system and, with it, their national, regional, and local economies. In the United States, this involved the partial nationalization of financial institutions: a dramatic reversal of the neoliberal philosophy that had come to dominate the political economy of most countries over the previous 30 years.

How could this have happened? Part of the explanation lies in the steady increase in debt that had been fuelling every aspect of the global economy and, in particular America's, since the late 1970s. Consumer spending had been financed increasingly by credit card debt. A housing boom had been financed by an expanded and aggressive mortgage market, and wars in Iraq and Afghanistan had been financed by U.S. government borrowing from overseas. By mid-2008, private debt in America had reached $41 trillion, almost three times the country's annual gross domestic product; the external debts of the United States had meanwhile reached $13.7 trillion. All sorts of financial instruments had emerged in the speculative free-market climate engendered by neoliberalism: securitization, derivatives, hedge funds, collateralized debt obligations, mortgage-backed securities, and so on. Everyone, it seemed, was borrowing from everyone else in an international financial system that had become extremely complex, increasingly leveraged, and decreasingly regulated.

The first signs of the crisis came in 2007, when credit losses associated with "subprime" mortgages—ones given to a borrower with poor credit—led to several large investment firms getting into difficulty. Taking advantage of the relaxed controls on U.S. financial institutions resulting from neoliberal policies, mortgage lenders had been selling their mortgages on bond markets and to investment banks to fund the soaring demand for housing. But this also led to abuses, as mortgage lenders no longer had an incentive to check carefully on borrowers. Many lenders began to offer mortgages that included "subprime" lending to borrowers with poor credit histories and weak documentation of income. But it appears that few in the financial services industries fully understood the complexities of the booming mortgage market, while the various risk-assessment agencies were seriously at fault in underestimating the risks associated with these loans. Eventually, when interest rates increased, many households began to default on their monthly payments, and as the bad loans added up,

U.S. market. Massive political and economic restructuring rocked core regions, the United States and Canada among them. Many regard 1975–90 as the transitional period, from the old manufacturing-based economy to the new IT economy.

The new economy, however, is about more than just new technology. It is also about the application of new technologies to the organization of work—from the impact of biotechnology on farming to the impact of IT on organizing management hierarchies in the insurance industry. In short, the new economy has applied IT to transform the organizational practices of firms and industries. Dynamism, innovation, and a high degree of risk are at the center of the new economy.

Canada has also been an active participant in the new economy, and part of the shift it has experienced into services and high-tech manufacturing is evidence of this. A particularly dynamic aspect of this shift has been in science-based industries, as well as research, development, and manufacturing in communications and transportation technologies. Quaternary sector transformations have also been important. It should be pointed out as well that the global financial crisis has been less pronounced in Canada, where the banks are more tightly regulated, more liquid, and less highly leveraged. Instead of being high-flying investment banks, they tend to operate in a more traditional manner, with large numbers of loyal depositors and a more solid base of capital. But Canada's economy has not been entirely trouble-free. The Toronto Stock Exchange has also fallen significantly since late 2008, though in late summer 2009 it began to show signs of recovery. And the increase mortgage lenders found themselves, in turn, in financial trouble. The impact of the mortgage market/housing crisis has been especially difficult for the states of Arizona, California, Nevada, and Florida because of speculation and overbuilding above actual demand.

Because the entire world economy has become so interdependent, the problem quickly spread. The internationalization of finance meant that the banks in other countries were caught up in the complex web of loans, and by the fall of 2008 the banking industry had been thrown into crisis worldwide. As credit markets seized up, manufacturers and other businesses found it difficult to get the credit they needed to keep going. Understandably, investors big and small were shaken; stock markets collapsed, and car manufacturers went bankrupt. Consumer confidence also plummeted, prompting retailers to cut back on orders. The sophisticated flexible production system of global commodity chains (see Chapter 7) meant that the impact across the globe was almost instantaneous.

Since the collapse in October 2008, some of those banks that survived have begun to recover, with Bank of America, for example, arranging to pay off its government bailout. The housing market is showing signs of bottoming out, and consumer spending is somewhat improved. In late 2009, unemployment is still high as are residential foreclosures, and the stock market continues to remain shaky. Although President Obama has poured billions of dollars into the banks and financial institutions and is aiming to stimulate new economic growth through tax incentives on green technology adoption for consumers, it is still too soon to declare that the crisis is over.

## Demography, Migration, and Settlement

The European settlers who colonized the United States and Canada largely displaced the indigenous groups who were already occupying the continent. Through colonization and later westward expansion, indigenous groups were reduced in number and pushed into new and usually more marginal areas of occupation. In the United States, native people were mostly settled onto reservations through treaties with the U.S. government; living on reservations granted native people a form of limited political sovereignty (**Figure 6.18**). As a result, Whites constitute nearly three-quarters of the total U.S. population; African Americans, about 13%; Asians and Pacific Islanders, about 4%; and Native Americans, about 1%. Hispanics, who may also be counted among other groups, make up about 14% of the total U.S. population.

In Canada, native people, known as the First Nations or aboriginals, were also settled onto more than 2,250 separate pieces of land known as *reserves*. The interactions between immigrants and the native people of Canada were little better than they were in the United States. Encounters between aboriginals and Europeans in Canada began to increase in the 16th century through trade deals, the exchange of goods, intermarriage, and friendships. Contact between the Old World and the New resulted in high rates of mortality for the First Nations, as diseases (typhoid, influenza, diphtheria, plague, measles, tuberculosis, and scarlet fever) ravaged the vulnerable populations. It is estimated that after nearly 200 years of European contact, which began in the

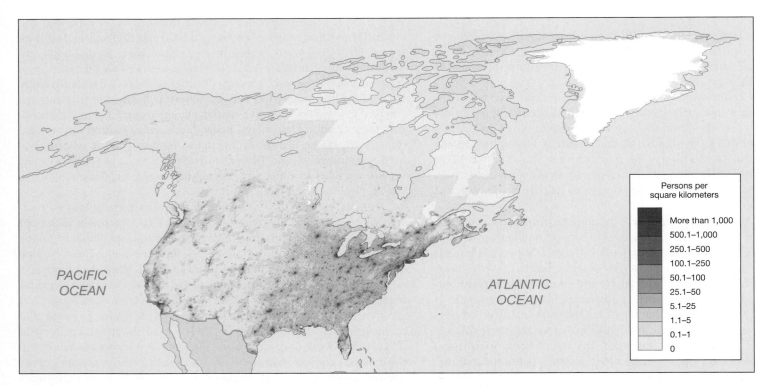

Persons per
square kilometers

More than 1,000
500.1–1,000
250.1–500
100.1–250
50.1–100
25.1–50
5.1–25
1.1–5
0.1–1
0

▲ **FIGURE 6.18 Population Map of the United States and Canada** The U.S. population is distributed thinly across the country, with the heaviest concentrations across the East and West coasts; the Canadian population has coastal clusters but only a thin band of settlement from east to west, hugging the U.S.–Canada border.

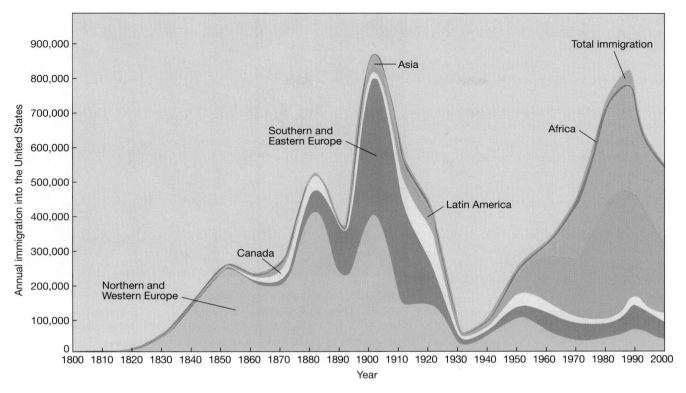

▲ **FIGURE 6.19 U.S. Immigration, 1800–2000** Immigration often is conceived as a series of waves, as the graph shows very dramatic peaks and troughs over this 200-year period. However, numerous groups object to the term *waves*, which they feel reduces people to the flotsam and jetsam of oceanic movements. In their view, immigration can be undertaken for any number of reasons, but none of them is as impersonal or random as a debris-saturated wave.

17th century, the aboriginal population of Canada was reduced by as much as 95%.

**Immigration**   As societies founded on European colonization and settlement, both the United States and Canada have varied and extensive immigration histories. The history of U.S. immigration is frequently discussed in terms of waves, because the numbers and types of immigrants ebbed and flowed over time (**Figure 6.19**). It is generally agreed that there have been three major waves of immigration into the United States, with the first occurring between 1820 and 1870, the second between 1870 and 1920, and the third beginning in 1970 and continuing through the present. In between these waves is an extended period of declining immigration that lasted from the mid-1930s to the 1970s. At the beginning of the 19th century, the population of the new nation was largely dominated by English colonists and African slaves. There were small numbers of Irish, Dutch, French, and Germans.

The first large wave (1820–70), in which overall immigration rose sharply to 2.8 million individuals, involved large numbers of Irish and German immigrants. The number of English immigrants declined. The newly arriving Irish were mostly peasants who fled the potato famine that had devastated the Irish economy and daily life. The Germans who came were mostly skilled craft workers who were seeking new opportunities in a burgeoning country.

In the second wave (1870–1920), in addition to the continuing stream of "old" immigrants from northern and western Europe, "new" immigrants from southern and eastern Europe joined the flow into the United States.

Between the 1870s and the 1880s, the absolute number of immigrants rose dramatically from 2.8 million to 5.2 million. Widespread economic depression in Europe and North America in the 1890s led to a decline in absolute numbers of immigrants (3.6 million). The numbers rose again in the first decade of the 20th century to an all-time high of 8.8 million. This wave carried peasants, skilled workers, and successful merchants.

The second wave of immigrants caused widespread backlash from first-wave immigrants. By the early 20th century, anti-Catholicism flared up against immigrants from Italy, Sicily, Poland, and Ireland. The new immigrants were blamed for everything from causing economic depression to destroying the character and moral fiber of the U.S. worker. Viewing immigrants as dangerous to U.S. values, Congress passed the Johnson-Reed Immigration Act in 1924 and imposed quotas on the numbers of new immigrants who would be allowed to enter the United States. The intention of the quotas was explicitly racial, and the impact was to reduce by half the foreign-born population. The "national origins" quota system remained in force until 1965, when the Hart-Cellar Act restructured the entire immigration system.

The third immigration wave (1970–present) is substantially different from the other two in that large numbers of migrants have been coming from Asia and Latin America. Although Asians have been part of U.S. immigration history since the mid-19th century, Latin American migration to the United States is a 20th-century phenomenon that has increased since the changes brought about by the Hart-Cellar Act and the Immigration and Reform and Control Act of 1980. By 1990, Mexico had become the largest source of immigrants to the

United States and continues to be so. Most recent Mexican immigration into the United States has been to California, Texas, and Arizona, although there is a sizable Mexican population in Chicago and in many smaller cities because Mexican workers have been hired to harvest fruits and vegetables and to work in the meatpacking industry of the Midwest.

The most recent wave of immigration is largely the result of the dislocating effects of contemporary globalization and the attraction of the United States as a mythical place to live out a dream; where the rewards of hard work and self-improvement are a better life for the current generation and an even better one for the next. Although over time the "new" immigrants of the second wave have largely been assimilated into mainstream U.S. life, experiencing increasing prosperity and social mobility in later generations, third-wave immigrants continue to confront racism and bigotry. They have been frequent victims of **hate crimes**, acts of violence committed because of prejudice against women; ethnic, racial, and religious minorities; and homosexuals.

**Immigrant Canada**  The immigration history of Canada is very similar to that of the United States, with one significant difference. The French dominated the settler stream into Canada well into the 18th century, but by around 1750, other immigrant groups from Britain and Ireland joined the stream as they had in the United States. In addition, Canada received some immigration from the United States at least until 1810, when restrictive British policies made it difficult for Americans to immigrate. By the beginning of the 20th century, Canada and the United States had very similar experiences of immigration, including the restrictions that curtailed inflows of new migrants until the late 20th century. Today, Canada is a primary destination for Asian migrants, who make up nearly 50% of the immigrant stream.

As with all European settler societies, including those in Australia and South Africa, the range of ethnic groups that have migrated—whether voluntarily or not—creates a complex and diverse culture. Both Canada and the United States have experienced migration streams from all over the world. For example, in the populations of Los Angeles, New York, Toronto, and Vancouver—all popular destinations for aspiring immigrants—the number of nationalities represented is impressive. In Garden Grove, California, a city that has one of the highest immigrant populations in Los Angeles County, the school system must cope with 65 different foreign languages among its student population. Although the variety and range of national groups can potentially contribute to rich and interesting local and national expressions of culture, culture difference can also be seen as threatening to those who wish to protect a particular view of what it means to be American or Canadian (**Figure 6.20**).

**Internal Migration**  In addition to foreign immigration, **internal migration**—the movement of populations within a national territory—has also played a role in both countries. In the United States three overlapping waves of internal migration over the past two centuries have altered the population geography of the country. These three major migrations were tied to broad-based political, economic, and social changes.

The first wave of internal migration began in the mid-19th century and increased steadily through the 20th century. This wave has two aspects: (1) a massive rural-to-urban migration associated with industrialization and (2) a large movement of people from the settled eastern seaboard and Europe into the interior of the country. Westward expan-

▲ **FIGURE 6.20  Protesting immigration reform**  In spring of 2006, hundreds of thousands of people took to the streets in cities across the United States to protest a federal immigration reform bill—H.R.-4437—aimed at tightening the US borders through more security infrastructure and personnel. In response, proimmigration rallys took place throughout the country, this one in Chicago in March.

sion took off in the early 19th century, when an official settlement policy was created (**Figure 6.21**).

Despite this emphasis on western expansion and rural settlement, between 1860 and 1920 the United States was transformed from a rural to an urban society. Industrialization created new jobs, and unneeded agricultural workers (along with foreign immigrants) moved to urban areas to work in the manufacturing sector. Since the 1920s, this pattern of urbanization has continued, and in 2000, 80% of Americans lived in metropolitan areas.

The process has been quite similar for Canada, making urbanization the dominant settlement process. In addition, there was large-scale internal migration westward in Canada, mostly during the early decades of the 20th century. But important differences in the immigration history of the United States and Canada also exist. During the colonial period, most of the immigrants arriving in Canada were British and French, whereas the immigrant stream to the United States was more broadly based. Moreover, whereas immigrants from all parts of Europe migrated to Canada during the late 19th and early 20th centuries, the number of people immigrating to Canada was smaller than the number immigrating to the United States. Finally, there has long been migration interaction between the two countries, with many Americans immigrating to Canada during the American Revolutionary War. In the 20th century, Canadian immigration to the United States exceeded American immigration to Canada, as Canadians sought to secure a higher standard of living south of the international border.

The second wave of internal U.S. migration, which began early in the 20th century and continued through the 1950s, was the massive and very rapid movement of mostly African Americans out of the rural South, where they had made livelihoods picking cotton, to cities in the South, North, and West (**Figure 6.22**). Although African Americans already formed considerable populations in cities such as Chicago and New York, large numbers of blacks moved out of the rural areas when

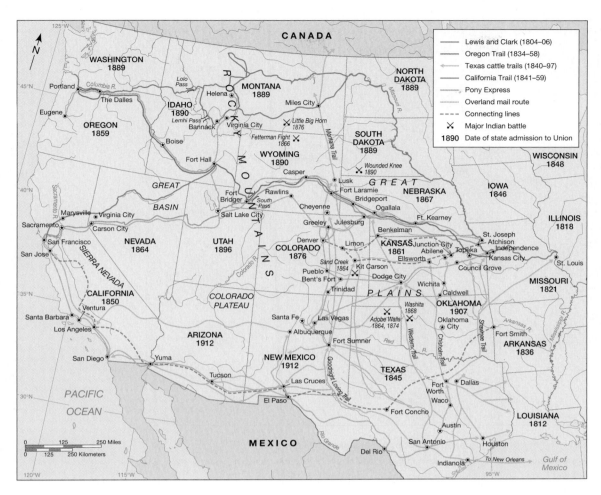

**◄ FIGURE 6.21 U.S. Frontier Trails, 1834–97** The frontier was part of the history and mythology of the United States (and Canada), and the paths individuals traveled to move westward suggest the many ways in which they experienced the passage. The topography of the land west of the Mississippi was intimidating, with vast grasslands, arid plains, searing deserts, and unimaginable insects and animals, like the vast herds of buffalo that the pioneers encountered. As more and more European and American migrants moved westward, the American Indians, the native inhabitants of these regions, saw their ways of life being threatened by these newcomers and often attacked the newcomers' wagon trains. The westward travelers who successfully navigated all the dangers and hardships of the trail then had to confront the seemingly insurmountable obstacle of the Rocky Mountains.

mechanization of cotton picking reduced the number of jobs available. At the same time, pull factors attracted African Americans to the large cities. In the early 1940s, for example, large numbers of new jobs in the defense-oriented manufacturing sector became available because many urban workers left their jobs and entered the military when the United States entered World War II. This second wave of migration can be seen as part of a wider pattern of rural-to-urban migration among agricultural workers as industrialization spread globally. After the war, a more important catalyst drove this migration: the increasing emphasis on high levels of mass consumption, which reoriented industry toward production of consumer goods and in turn stimulated large increases in the demand for unskilled and semiskilled labor. The impact on the geography of racial distribution in the United States was profound.

The third wave of internal migration began shortly after World War II ended in 1945 and continued into the 1990s. Between the end of the war and the early 1980s—and directly related to the impact of governmental defense policies and activities on the country's politics and economy—the region of the United States now commonly called the Sun Belt, and including most of the states of the U.S. South and Southwest, experienced a 97.9% increase in population. During the same period, the Midwest and Northeast, known as the Rust Belt, together grew by only 33.3%. **Figure 6.23** shows suburban development Santa Fe, New Mexico. Though it has not grown as dramatically as Houston or Phoenix or Las Vegas, it still manifests the low-density, low vertical profile, and geographically extensive form expressed in most Sun Belt cities.

**Urban to Suburban Migration** The first evidence of **suburbanization**—the growth of population along the fringes of large metropolitan areas—can be traced back to the late 18th and early 19th centuries, when real estate developers looked beyond the city for investment opportunities, and wealthy city-dwellers began seeking more scenic residential locations. Later, residents fled to the suburbs to get away from the new immigrants and their increasing hold over urban machine politics.

The process was rapidly accelerated, however, with the introduction of new transportation technologies—first horse-drawn streetcars, then commuter rail services, and finally, automobiles. Each innovation in transportation allowed people to travel longer distances to and from work within the same or shorter time period. North Americans chose to move to the suburbs in massive numbers, not in the least because the suburbs were, arguably, considered by many to be more healthful places to raise a family. Suburbanization continues today in both the United States and Canada with a new wrinkle—a slight reversal of migration from urban to rural areas, as retirees especially search out the good life on the far fringes of the metropolitan core in small towns like Bisbee, Arizona, and the Okanagan Valley in British Columbia (**Figure 6.24**).

The most compelling explanation for the large-scale population shift characteristic of the third migration wave is the pull of economic opportunity. Rather than reinvesting in upgrading the aged and obsolescent urban industrial areas of the Rust Belt, venture capital was invested in Sun Belt locations, where cheaper land and lower labor costs made it more profitable to introduce manufacturing and service-sector activ-

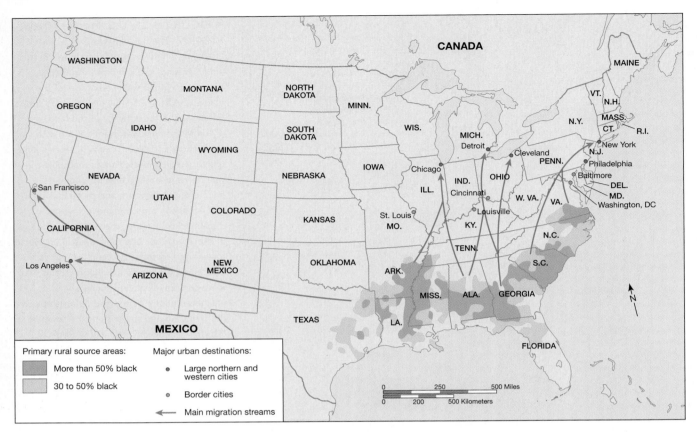

▲ **FIGURE 6.22 African-American migration** The northward migration of African Americans in the early part of the 20th century was nothing new in the United States. Since the Civil War, blacks had been leaving the South in a slow but steady stream. By the turn of the new century, "Jim Crow" laws that legalized segregation and discrimination against blacks had been enacted; the boll weevil had devastated the cotton crop; and the cotton gin (which mechanized cotton processing and put tens of thousands of pickers out of work) had been introduced. An increased flow of new migrants joined already established black communities in most major northern cities.

ity. The 2000 census shows a decrease in the rate of in-migration to the Sun Belt, but the changes in the geography of population at the beginning of the 21st century are dramatically different from the patterns of 150 years ago. This new population distribution illustrates the way political and economic transformations play an especially significant role in shaping individual choice and decision making.

## CULTURE AND POLITICS

The United States and Canada constitute a region that is culturally and politically diverse and whose cultural and political influences extend far beyond the region to even remote corners of the globe. From jazz to hip-hop, horror movies to hamburgers, as well as with respect to poli-

◀ **FIGURE 6.23 Sun Belt city: Santa Fe, New Mexico** Although the residential architecture and design varies from place to place in the Sun Belt, the overall layout of residences set up side by side in numbing repetition is a classic manifestation of urban growth there. This mass production approach enables builders to keep construction costs low as materials can be ordered en masse and the building process broken down into separate tasks.

▲ FIGURE 6.24 Okanagan Valley, British Columbia  Pictured here is the town of Kelowna on Lake Okanagan. This town and the valley surrounding it have become an attractive site for retirement living for Canadians because of its sunny climate and the wide range of outdoor activities available, including hiking and fishing. The region is also a cultural attraction because of its First Nations history.

tics, language, literature, fashion, and activism, the influence of American culture on the rest of the world has been and continues to be dramatic. Importantly, the influence of the rest of the world on the region should not be ignored. Perhaps what the region is most adept at is "sampling"—that is, taking lots of what the world has to offer and translating it into something hybrid, popular, and desirable.

This section first discusses modern-day religion and language geographies of the United States and Canada. It then looks at contemporary cultural practices and phenomena as they have been produced in the region and disseminated widely. It ends with a treatment of regional social movements as well as national politics and the way they have shaped and continue to have an impact on the rest of the globe.

## Tradition, Religion, and Language

The main language in the United States is English, with Spanish also widely spoken. Although the United States is popularly considered a Protestant country (including large numbers of Baptists, Methodists, Presbyterians, Lutherans, Pentecostals, and Episcopalians), Roman Catholics form the largest single religious group. The largest non-Christian religion is Judaism, and other non-Christian religions, such as Islam, Buddhism, and Hinduism, also have substantial followings.

In Canada, where there are two official languages (English and French), the situation for immigrants has always been somewhat different from **assimilation**. Canadian popular opinion and public policy have advocated something more akin to **multiculturalism**, the right of all

ethnic groups to enjoy and protect their cultural heritage. Multiculturalism in practice includes protection and support of the right to function in one's own language, both in the home as well as in official or public realms. The ideal form of multiculturalism is premised on the belief that immigrants should not have to give up any of their original cultural attributes or practices. Under multiculturalism, the emergence of a single unified national identity, to which all Canadians could relate, is not technically possible. Instead, multiculturalism leads to ethnic coexistence, in which diverse groups share the same national space but not the same cultural systems.

A significant aspect of the geography of U.S. religion is its regional variation. For example, the Bible Belt, which stretches from Texas to Missouri, is dominated by Protestant denominations, many of them fundamentalist and evangelist. Mormons, or members of the Church of Jesus Christ of Latter-day Saints, are concentrated in Utah, where more than 75% of the population are adherents. Large numbers of Mormons also live in Nevada and Idaho. Large Catholic communities exist throughout the Southwest and in many of the large cities of the Northeast and upstate New York. Sizable concentrations of Jews occur in large U.S. urban centers like New York, Los Angeles, and Miami.

Canada has a population of roughly 33 million people. Canadians are predominantly of British origin (about 33%). The second-largest majority is people of French origin, who make up about 25% of the population. There is also a large population of mixed British and French origin. The remaining population groups are small, with blacks making

# THE NORTHERN FRONTIER

The Northern Frontier sits at the apex of the North American continent. It is a vast area, including the relatively new territory of Nunavut (created in 1999 from the eastern portion of the Northwest Territories and turned over to the control of the region's First Nations). Although it occupies about one-third of Canada (an area roughly the size of India), it is home to only 60,000 people, nearly half of whom live in or around the settlement of Yellowknife. The distinctiveness of the Northern Frontier landscape is reflected in a few key facts:

- Wildlife vastly outnumber people (and in the most northerly parts, human habitation is impossible).

- Much of the region is beyond the tree line, whereas mountain chains in the east and west rise majestically from the slope and sedimentary plain of the interior of the region.

- Short growing seasons make it impossible to sustain agriculture.

- Ice, snow, or permafrost are the norm.

The Northern Frontier is a place of incredible, near-unimaginable beauty with relatively little impact from humans across its vast extent.

Nunavut (which means "our land" in Inuktitut) is the largest territorial unit within Canada and is home to about 31,000 people, most of whom are Inuit. Its spectacular beauty and developing economy makes it a key element in the Northern Frontier. In response to claims by native people for land and political power, the Canadian government began to negotiate settlements beginning in the 1970s, and over the last 25 years has ceded to native people millions of acres of land and control over it. Nunavut stands out as a monumental concession and an indication of Canada's ability to make substantial reparations for its imperial past.

Until its transfer to native people, Nunavut was the central and eastern part of the Northwest Territories. Covering about 2 million square kilometers (about 772,000 square miles), Nunavut includes Baffin and Ellesmere islands and the surrounding region, stretching almost to the North Pole. For the most part, Nunavut is a flat tundra where average temperatures range from –32°C (–25°F) in January to 5°C (41°F) in July. Abundant wildlife includes white fox, caribou, and seals (**Figure 1**). Geological surveys have shown that Nunavut is rich in diamonds, gold, uranium, copper, lead, silver, zinc, and iron. The severity of the climate makes any kind of large-scale mining a distinct challenge. At present, however, there are numerous mining operations in Nunavut, including the first diamond mine, which opened in 2006, with another four to be in operation by 2010. The settlement pattern of the Inuit is along the coast of Hudson Bay and the Labrador Sea.

Although the northern portion of the Northern Frontier is sufficiently inhospitable to deter human settlement, the southern part is

▲ **FIGURE 1  Musk Ox Bull, Ellesmere Island** In addition to about 60,000 people, Nunavut is also home to a wide variety of wildlife, including caribou, polar bears, Arctic foxes, whales, and seals. The fur and skin of these animals is used for clothing and their flesh for food. Tourists also come to fish, hunt, and camp in this land of austere beauty.

up less than 2% and indigenous people making up nearly 4%. Whereas the U.S. population is distributed widely, if somewhat thinly, across the country from east to west and north to south, with heavy concentrations along the coastal areas and in parts of the Midwest, Canada's population distribution reflects difficult environmental constraints and crucial U.S. economic connections. Nearly 75% of the population inhabits a narrow belt along the U.S. border.

Canada's largest religious denomination is Roman Catholic, with nearly half of that population living in Québec as a result of early French influence. The largest Protestant denomination is the United Church of Canada, followed by the Anglican Church of Canada. There is also a substantial Jewish population in Montréal and Toronto. Ukrainian Orthodox communities are clustered throughout the provinces of Alberta, Saskatchewan, and Manitoba, and Buddhists and Hindus are concentrated mostly in the Canadian cities where large numbers of Asian immigrants have settled, such as Vancouver and Toronto.

During the second half of the 20th century, many members of the Canadian First Nations organized to protect their cultural rights and reclaim land that was taken from them. New treaties were signed, and vast areas of territory were returned, as hunting and fishing rights were also restored (see Signature Region in Global Context: The Northern Frontier, above). Success at reclaiming lands by native peoples in the United States has been far more limited. Over the last several decades, as in the United States, a free-enterprise economy has grown among inhabited and is a destination for adventure tourists. For instance, the area in and around the settlement of Yellowknife, the largest city in the Northern Frontier, is developed with transportation linkages and other aspects of urban infrastructure. Communication linkages are particularly innovative such that Nunavut has become a regional leader in wireless communications technology, which is critical in a place where there are no roads connecting the widely dispersed communities.

The impact of climate change on the fragile ecosystem of the Northern Territories and Nunavut in particular is already being felt through warmer year-round temperatures, unpredictable weather, a shift in prevailing wind directions, less snow and rain, and changing snow and ice conditions. Changes such as these are having a wide range of effects on everything from limiting people's travel options to decreasing the fertility of female polar bears. Projections for future impacts include rising sea levels, a reduction in the extent and thickness of sea ice, and more extreme weather events, all of which are sure to increase erosion and flooding of coastal communities—where most of the population of Nunavut live (**Figure 2**). But these are only some of the possible impacts, many of which would affect marine and terrestrial fauna and flora and dramatically change the landscape itself. Although life for the Inuit has for millennia been about adaptation to this extreme environment, these new changes are likely to occur over decades rather than centuries and will create very different kinds of challenges. Pan-Arctic alliances of indigenous peoples are calling for emissions reductions and considering lawsuits for compensation for damages. ■

▲ **FIGURE 2 Retreating Ice Shelves on Ellesmere Island, Nunavut** An image from the Terra satellite captured in summer 2008 shows the ice shelves on Ellesmere Island. Prior to this date, only five ice shelves remained in the Canadian Arctic. With an estimated age of 4,500 years, these ice shelves were the remnants of a once-massive "glacial fringe" that explorer Robert Peary described in his trek along the Ellesmere Island coast in 1907. In July 2008, these shelves began disintegrating rapidly. By late August, Ellesmere ice shelves had lost a total of 214 square kilometers (83 square miles).

Canada's aboriginal groups. Many are now involved in various sectors of the economy—from oil production to tourism.

In the province of Québec, the Québeçois have organized to protect their cultural heritage. Their political movement, which seeks to separate Québec from the rest of Canada, is premised on a deep desire to preserve and enrich Québeçois culture. Since 1980, the province has attempted twice to secede from Canada; both efforts have failed. Lately the talk is less of secession and more about extending Québec's ties to Europe through the North American Free Trade Agreement. In short, globalization seems to be opening up new opportunities for Québec to maintain its distinct identity, prosper economically, and control more of its own destiny without having to secede from Canada.

## Cultural Practices, Social Differences, and Identity

Music, art, literature, dance, architecture, film, photography, sports, fashion, journalism, and cuisine, not to mention science, medicine, and technology, have all been shaped by the generations of immigrants who have come to occupy the United States and Canada. The influence of immigrants on music has been particularly impressive. Country, bluegrass, jazz, the blues, and rap all originated in the United States but have deep roots in the Old World. From jazz to rap, African Americans have been responsible for musical innovations that have been widely accepted, applauded, and imitated throughout the world.

**Arts, Music, and Sports** The early 20th-century origins and particularly U.S. expressions of jazz have been influential worldwide at the same time that they reveal a complex but clear lineage back to the African musical roots left behind in the wake of their terrible Atlantic passages. West African folk music forms one of the central foundations of jazz. But jazz was also influenced by European popular and light classical music of the 18th and 19th centuries. The earliest documented jazz style was Dixieland jazz, which emerged from New Orleans and was played by white musicians who recorded the new music form on phonograph records. The spread of these recordings helped jazz become a sensation in the United States and Europe.

Soon African-American jazz groups—the originators of the jazz style that was expressed through the related styles of ragtime, marches, hymns, spirituals, and the blues—were able to capitalize on the popularity of white Dixieland largely through the improvisational style of trumpeter Louis Armstrong. Armstrong migrated to Chicago in the 1920s, influencing local musicians and stimulating the evolution of the Chicago style.

About the same time that jazz caught on in Chicago, Harlem in New York was emerging as a center for jazz, organized around a highly technical, hard-driving piano style. Regional variations on the original Dixieland style emerged in the urban areas, where significant populations of African Americans had settled. Jazz continued to flourish from the 1930s through the 1950s. In the 1960s jazz began to lose popularity, as audiences embraced mainstream rock and roll, which had itself been influenced by jazz and the blues. In the 1980s, jazz experienced a revival as a serious form of music, which it continues to enjoy today. Other distinctly U.S. musical and performance styles include rap, bluegrass, and musical theater, the last having roots in European opera. Native populations in both the United States and Canada have also made significant cultural contributions to music, handicrafts (especially basketry, rugs, jewelry, and pottery), and contemporary literature. People from all over the world travel to visit First Nations and Native American sites to view and collect their distinctive commercial products such as Navajo rugs from the U.S. Southwest and the wood carvings of the Haida people of Pacific Canada.

The game of baseball is another U.S. innovation, and it too has enjoyed widespread popularity beyond North America, especially in Caribbean countries like the Dominican Republic, Venezuela, and Cuba, but also in South Korea and Japan, among other places. The composition of many U.S. and Canadian major league baseball teams (one—the Toronto Blue Jays—is Canadian) demonstrates just how popular this sport has become worldwide. As a high-stakes commercial enterprise, baseball has traveled well. Consider the following: At the opening of the U.S. baseball season in 2009, nearly a third of all the players on the 30 team rosters were from countries and territories other than the United States. These include Aruba, Australia, Canada, Colombia, Cuba, Curaçao, Dominican Republic, Japan, South Korea, Mexico, Nicaragua, Panama, Puerto Rico, Taiwan, and Venezuela (**Figure 6.25**).

**U.S. Cultural Imperialism** Many scholars argue that "globalization" is really just a euphemism for "Americanization," and it is difficult to argue against this perspective. It is also important to recognize, however, that just as U.S. culture is circulating intensively beyond its national borders, other cultures have come to influence U.S. culture in numerous and distinctive ways. The following examples help illustrate the fact that the globalization of culture, though largely dominated by the United States, is not exclusively so. For decades, other core countries have been exerting important cultural influences on the United States, especially Japan. After the rage of Japanese *Pokémon* trading cards in the late 1990s, for example, *Yu-Gi-Oh!* trading cards became wildly popular among U.S. children. Gwen Stefani toured with *harajuku* girls in

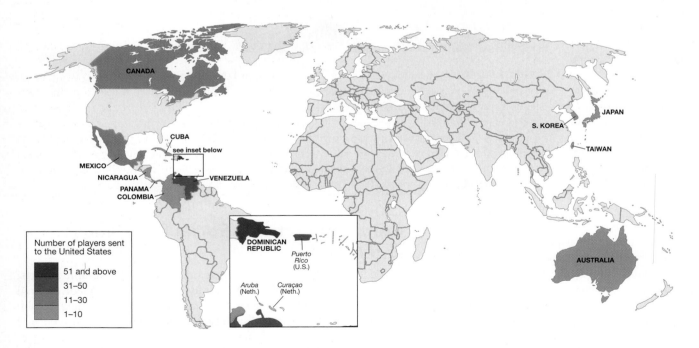

▲ **FIGURE 6.25 The globalization of U.S./Canadian major league baseball, 2009** The old saying that something is as "American as baseball or apple pie" may need revision, given the dramatic transformations that have occurred over the last 10 to 15 years in the demographics of players on major league baseball teams and the growth of interest and participation in baseball worldwide.

2004, the name given to teenage girls who dress up in trendy clothes and hang out in the area around Tokyo's Harajuku Station. But it's not just Japan that is influencing U.S. culture; other countries are as well. Britain continues to deliver on the Harry Potter children's fiction series in both film and literary form, and although it is clearly a children's genre, it captivates adults and children alike. The *No. 1 Ladies Detective Agency,* now an HBO television series, is about a quirky female detective in Botswana; books on which the series is based have been bestsellers in the United States for several years. They are written by Zambian-born Alexander McCall Smith, who resides in Edinburgh, Scotland. And, also very popular in the United States are the Hong Kong action movies, including *House of Flying Daggers* (2004) and *Kung Fu Hustle* (2004).

Peripheral countries influence U.S. culture as well. For instance, the film industry in India, Bollywood, is increasingly coming to shape U.S. film in both its content and look. Depending on one's viewpoint, the very popular film *Moulin Rouge* (2001) either imitated or stole Bollywood formats and dance sequences. Clearly, globalization has made it possible not only for U.S. and Canadian culture to circulate widely but also for cultural products from other parts of the world to penetrate the region.

**Canadian Cultural Nationalism**   Canadian culture tends mostly to represent a mix of immigrant and British settler influences. One of the most significant aspects of Canadian culture is the fact that it has had to battle the tremendous influence of commercialized U.S. culture, which because of its geographical proximity has been enormously difficult to resist. Canada has been very aggressive in its attempt to ward off the invasion of U.S. cultural products and has developed an extensive and very public policy of cultural protection against the onslaught of music, television, magazines, films, and other art and media forms.

Government bodies, such as the National Film Board of Canada and the Canadian Radio and Television Commission, actively monitor the media for the incursion of U.S. culture. For example, 30% of the music on Canadian radio must be Canadian. Nashville-based Country Music TV was discontinued from Canada's cable system in the early 1990s and replaced with a Canadian-owned country music channel. Besides regulating how much and what type of culture can travel north across the border, the Canadian government also sponsors a sort of "affirmative action" grant program for its own culture industries, including music, radio, and the print media (**Figure 6.26**).

## Territory and Politics

In the last 25 years, the different clusters of states and provinces within the United States and Canada have shifted. Even areas that have been less pivotal to the U.S. and Canadian economies have reorganized their economies to some extent so they can participate more actively in the revolutionary changes that are underfoot. Three territorial clusters are the clear pacesetters in innovation and dynamism: the Pacific Rim (western California); Cascadia (the northwestern United States, Alaska, and parts of Canada); and the U.S.–Canadian Core (the old Manufacturing Belt consisting primarily of the region around New York, Chicago, Toronto, and Montréal and including the corridor between Boston and Washington, DC). In these regions the population has concentrated in large cities, manufacturing products and providing services for the areas surrounding them (**Figure 6.27**).

Beyond the U.S.–Canadian key territorial clusters and principal metropolitan areas lie numerous other landscapes and regions whose relationship to the environment, history, economic contribution, or political

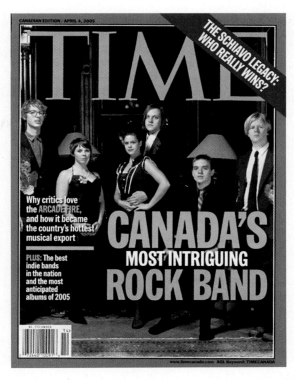

▲ **FIGURE 6.26  The Arcade Fire**  Popular Canadian musical exports include a wide range of vocalists, such as Avril Lavigne, Kevin Parent, and Diana Krall, as well as rock groups like Barenaked Ladies and The Arcade Fire, pictured here on the cover of *Time* magazine, April 4, 2005.

background makes them distinctive. The Prairie provinces of Canada are critical to the world wheat market and to Canada's strong economy. Most important, this part of Canada produces more agricultural output than any other place of comparable size on Earth. Whereas the U.S. portion of the Plains and Prairies tends also to include large-scale industry, the Canadian side is involved in mineral and oil extraction.

The New West, or Intermountain West, an area known historically for ranching and other primary-sector activities, has expanded its economy to include more service-based activities by mixing tourism and recreational activities with product-support service activities, as well as second-home and retirement residential developments. Some high-technology development is also part of the mix, especially in and around the places where universities are located, such as the Denver, Colorado, and Albuquerque, New Mexico, metropolitan regions.

New England and the Maritime provinces are not only visually charming, as they reflect some of the earliest artifacts of colonial architecture, but they are also economically and politically significant for both historical and contemporary reasons. New England contains important high-technology research and development centers and the universities needed to sustain them. The Maritime provinces have developed their economies around call centers and related service-sector employment.

The U.S. South is also distinctive in terms of environment, history, politics, and culture. Even within the region there is significant variation, particularly with respect to coastal and inland areas as well as the Deep South (the states of Alabama, Georgia, Louisiana, Mississippi, and South Carolina) and the other Southern states (Kentucky, Florida, North Carolina, Tennessee, Virginia, West Virginia, and sometimes Texas).

▲ **FIGURE 6.27 U.S. and Canadian territorial clusters** It is a fairly consistent phenomenon of U.S. journalism that every five or so years, another popular book on the territorial reorganization of the United States is published. This phenomenon of serial publication reflects the fact that capital and people are increasingly mobile, thereby creating new regional clusters as they participate in contemporary economic transformations; it also reflects the public's hunger for new ways to think about itself.

Whereas the Deep South is often called the Bible Belt because of the attachment of much of the population to Protestant fundamentalism, states like Florida, for instance, hardly qualify as part of the southern region, as their populations are ethnically diverse and their orientation is more cosmopolitan than rural. Hawai'i, disconnected from the mainland United States far off in the Pacific, is also unique. A thick gloss of Americanization overlies Polynesian cultural roots, and a strong connection to East Asia, especially Japan, also exists.

**The Pacific Rim and the New Economy**   The Pacific Rim encompasses the western portion of the state of California, oriented around the three key coastal cities of San Francisco, Los Angeles, and San Diego and the inland state capital of Sacramento (**Figure 6.28**). In addition, the Pacific Rim includes all the nations bordering the Pacific Ocean as well as the island countries situated in it. As a global region, the Pacific Rim has grown dramatically, both economically and politically, over the last 25 years. In an attempt to improve trading relations among the many countries that are

▲ **FIGURE 6.28 Los Angeles, California** Los Angeles is a key city in the Pacific Rim. Its port connects the United States to Asia, and its economy includes a strong manufacturing base, extensive service-sector employment, and tourism. Because of its burgeoning economy, Los Angeles is the destination for hundreds of thousands of immigrants.

part of the Pacific Rim, the Asia–Pacific Economic Cooperation was established in 1989. Both the United States and Canada are members.

This region is the birthplace of the new economy conceived in Silicon Valley and now widely dispersed throughout many parts of the region. Through flows of trade, information, capital, and people, the global Pacific Rim region connects the U.S. economy to the diverse and growing economies of East, Southeast, and South Asia and the many Pacific island nations thousands of kilometers across the Pacific Ocean.

**Cascadia and Internationalism** Linking the U.S.–Canada north of the Pacific Rim is Cascadia, which includes Alaska, parts of the Canadian Yukon and Northwest Territories, British Columbia, and Alberta, as well as parts of the states of Washington, Oregon, and California. The Cascadia region's name comes from the Cascade Mountains range, which runs from northern California through Oregon and Washington and into southern British Columbia. At its core are the key cities that anchor the populous corridor—Vancouver, Seattle, and Portland, Cascadia is a relatively new and dynamic region oriented around high technology. An important additional hallmark of Cascadia is its internationalism, which is manifested in ignoring, if not actually erasing, the cultural, political, and economic borders of three northwestern states, two Canadian provinces, and two territories (see Figure 6.26).

Cascadia is unique in that it is an unusually unified international region that has a global outlook. The Washington–British Columbia border is the busiest of any along the U.S.–Canada border, and the region contains four exceptional ports (Portland, Tacoma, Seattle, and Vancouver; **Figure 6.29**). Seattle is the leading West Coast container port; Vancouver is the busiest in terms of volume shipped. Most important for its global positioning, the region is the closest place in North America to Japan, Hong Kong, and China. As geographer Matthew Sparke has written, despite all this international cooperation and phenomenal economic growth, there are fears that there may be too much growth in Cascadia.

**The U.S.–Canadian Core** Moving east across the middle of the continent, a third key territorial cluster is the old Manufacturing Belt encompassing the key cities of New York and Chicago, as well as Montréal and Toronto. The U.S.–Canadian Core also encompasses the most populous corridor in the United States that runs from Washington, DC, to Boston, known as **Megalopolis,** and a similar one that runs from Windsor to Québec City in Canada, known as "**Main Street**" (**Figure 6.30**). Although dubbed *the Manufacturing Belt* in the late 19th and early 20th centuries because it constituted the industrial powerhouse of the two countries, the label at the beginning of the 21st century is a misnomer. Manufacturing employment has dropped dramatically since the 1960s, and quaternary economic employment now dominates the region. We call this the core of the core: the U.S.–Canadian Core.

▲ **FIGURE 6.29 Peace Arch at the United States/Canada border** The Arch flies the flags of the United States and Canada mounted on its crown and two inscriptions on both sides of its frieze. The inscription on the U.S. side of the Peace Arch reads, "Children of a common mother," and the words on the Canadian side read, "Brethren dwelling together in unity." Within the arch, each side has an iron gate hinged on either side of the border with an inscription above reading "May these gates never be closed."

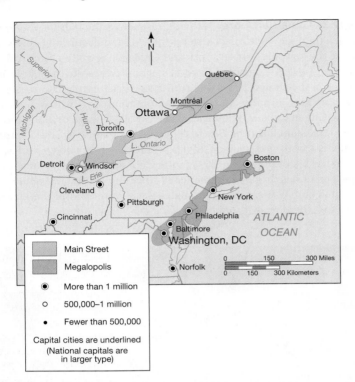

▲ **FIGURE 6.30 Megalopolis and Main Street** The term *megalopolis* was originally coined by the Greeks to refer to places where people lived, rather than just to abstract space. French geographer Jean Gottman resuscitated the term to describe the dense, extended, and connected urban settlements, a "super-metropolitan" conurbation, along the coast of the eastern United States; hence, the proper name "Megalopolis." Main Street is Canada's answer to Megalopolis. Also a heavily urbanized corridor, Main Street connects some of Canada's major cities. The two terms are also interesting in their contrast, with Main Street suggesting a homey, familiar space, and Megalopolis implying something of extraordinary scale.

▲ **FIGURE 6.31 Minneapolis, Minnesota** Minneapolis is the largest city in Minnesota and one of the key points along the Mississippi River. The city's economy developed largely around the processing of grain from the Great Plains, where companies like General Mills and Pillsbury became giant corporations. Today Minneapolis is more noted for its medical facilities and financial institutions.

The history of this territorial cluster is instructive because it embodies the damage as well as the new growth that cycles of capital investment and disinvestment can visit on people and place. The cities of the old Manufacturing Belt (New York, Chicago, Toronto, Boston, Montréal, Philadelphia, Baltimore, Chicago, Detroit, Windsor, Hamilton, Québec City, Cincinnati, Minneapolis, Milwaukee, St. Louis, and Cleveland, to name the most populous ones), already thriving industrial centers in the late 19th century, were connected through the early railroad network. Individual cities specialized in particular industrial products (for example, grain milling in Minneapolis, agricultural tools and equipment in Chicago, brewing in Milwaukee and St. Louis, coach building and furniture in Cincinnati), producing in volume for the national market rather than local ones (**Figure 6.31**).

By the mid-20th century, the requirements of a new technology system—using robotics, telematics, biotechnology, and other knowledge-based systems—overwhelmed the ability of this manufacturing powerhouse to respond, and disinvestment in the form of deindustrialization occurred (see Figure 6.23). For much of the 1970s and 1980s, the old Manufacturing Belt suffered population loss and capital flight as workers, entrepreneurs, and investors went to other parts of the region and the world to take advantage of more attractive employment and investment opportunities. But by the early 1990s, the old Manufacturing Belt was showing signs of recovery, and by the middle of the decade it had regained its dominance and political clout.

**States and Government** The United States and Canada are **federal states**, part of a form of government in which power is allocated to units of local government (province/state, county, and city/town government) within the country. Federalism leaves many political decisions to the local governments. Canadian federalism, however, allows more power to provinces than states have under U.S. federalism. The extremely close vote in Quebec in 1995—which would have enabled the province to become fully independent from Canada—is an illustration of the political power of provinces. Such an election or outcome would not be constitutionally possible in the United States.

During the first 100 years of the U.S. republic, the federal government spent most of its time regulating commerce. But beginning in the late 19th century, urged by constituents across the country, the federal government began to take an increasingly active and direct role in regulating and supporting all aspects of U.S. social and economic life, particularly with respect to providing for social welfare; developing infrastructure, such as dams and highways; and transferring large amounts of tax dollars to contractors for the buildup of U.S. defense systems, especially during the Cold War.

Because the state was so heavily invested in all aspects of U.S. society, but especially in the economy, when the global economy experienced shock waves during the 1970s, the government was hit very hard. And as corporations and business in the United States and elsewhere searched for remedies to their economic problems, the government did likewise and imposed dramatic restructuring on its own operations and programs.

The view that government's primary responsibility was as a guarantor of social welfare had dominated popular understanding since the 1930s. By the 1970s and 1980s, as local governments in the Rust Belt were declaring bankruptcy and the federal government was accumulating massive debt, popular opinion changed, and the role of government was reconfigured. Since the late 1980s, it has become routine for local governments to act more as entrepreneurs than as managers of the social welfare. As deindustrialization accelerated in the Rust Belt, government agencies in the Sun Belt helped lure investment to the region by offering tax breaks, creating needed infrastructure, and providing subsidies for private investment.

As a way to reduce its mounting debt, the federal government began to shed its responsibilities for social welfare, passing these responsibilities on to state governments. The federal government also began to shut down military bases throughout the country as the fall of the Berlin Wall signaled the end of the Cold War. With decreased responsibilities for social welfare and lower military spending until the recent military campaigns in Iraq and Afghanistan, the federal government has oriented its role toward more actively facilitating the free flow of trade and the operations of transnational corporations abroad.

Since independence, Canada has fostered a government that has been far more inclined to guarantee social welfare than has the United States, though Canada also has a tradition of entrepreneurialism in government. Many scholars of Canadian history believe that federation of the former Canadian colonies into the Dominion of Canada was driven by capitalists interested in supporting the burgeoning industrialization of the country. More recently, in addition to continuing its tradition of providing social welfare, the state in Canada has accelerated its entrepreneurialism by directing support to expanding its tertiary sector (activities involving the sale and exchange of goods and services) and quaternary sector (activities involving the handling and processing of knowledge and information), particularly with respect to high-technology development.

**U.S. Political and Military Influence**  In addition to its economic strength, the United States achieved its place as the world's most powerful country through its political and military strength. The first sign of the United States flexing its political and military muscle came with the Monroe Doctrine, which, although issued by President James Monroe in 1823, eventually became the foundation of U.S. foreign policy in Latin America. Monroe contended that European powers could no longer colonize the American continents and should not interfere with the newly independent Spanish-American republics. So long as Europe stayed out of the Americas, Monroe promised that the United States would not interfere with existing European colonies or with Europe itself.

Although desirous of new territories for commercial potential, and often urged to war or other acts of aggrandizement by business leaders, U.S. militarism did not extend to an eagerness to get involved in European affairs, and the country only reluctantly entered World War I in 1917. Stunned by the nearly five million war casualties and the horror of the first highly technological engagement in history and eager to protect its growing economy, the United States subsequently entered a period of relative isolationism, rallying around the slogan of "America First." Although President Franklin Roosevelt declared U.S. neutrality with respect to the European war in 1939, by 1942 the country had entered World War II. The end of the war in 1945 marked a turning point in U.S. political and economic prowess, as U.S. loans helped rebuild war-torn Europe and Japan. Allied victory and U.S. participation in the war effectively solidified the country's status as a world leader and ensured the dominant participation of the United States in several of subsequent wars, including those fought in the Persian Gulf, Korea, and Vietnam.

All these wars were fought in the name of the Cold War, which pitted the capitalist United States and western Europe against the communist Soviet Union for the hearts, minds, and territories of peoples throughout the globe. The Cold War came to an end with the disinte-

gration of the Soviet Union in 1991. A new era of global cooperation dawned, with the U.S. government and U.S. transnational corporations still leading the way, though with markedly less militaristic fervor. The recent wars on terrorism in Afghanistan and Iraq have halted the brief hiatus in significant U.S. military involvement in global affairs, however (**Figure 6.32**).

**Social Movements**  Two of the most enduring, widespread and effective social movements in the United States and Canada have unarguably been the women's movement and the environmental movement. Inspired by their sisters in the United Kingdom, the origin of the women's movement in the United States can be traced to the 1848 women's rights convention held in Seneca Falls, New York. From that point, women continued to agitate for the right to vote, to have the same civil rights as men, to have access to birth control, to improve working conditions for themselves and enjoy equal pay for equal work and fight discrimination in the workforce, and to fight for disarmament and peace. A whole host of other issues from access to safe abortions to prosecuting rape have all followed from that 1848 activism. Campaigns today for the civil rights of blacks and other minorities, as well as rights for the disabled, gay rights, and children's rights, have derived inspiration as well as tactics and rhetoric from the women's movement. The vice-presidential campaigns of Geraldine Ferraro (1984) and Sarah Palin (2008) and Hilary Clinton's (2008) presidential campaign are all testimony to the effectiveness of the women's movement in the United States. Women's rights movements in Canada followed a similar trajectory with women receiving the right to vote in 1918, two years before U.S. women.

▲ **FIGURE 6.32 U.S. War in Iraq**  Military operations to remove Saddam Hussein from power began in mid-March 2003. Major combat operations were terminated by mid-April, although hostilities between the coalition forces and Iraqis continue. The war, prompted by the Bush administration's assertion that Iraq possessed weapons of mass destruction that could be used for terrorist ends, is an illustration of the new belligerence displayed by the United States on the world stage. As of late 2009, over 4,200 American troops had died and over 30,000 had been wounded in the war. Pictured here is an American mother viewing the photograph of her son at the "Faces of the Fallen" exhibit at Arlington National Cemetery.

The environmental movement in the United States can also be traced to the 19th century and the pioneering works of men like Henry David Thoreau and George Perkins Marsh, who in the 1850s and 1860s began to lecture and write about the destructive impact of humans on the natural world. The Canadian environmental movement emerged at roughly the same time and has continued to play an important role in Canadian politics and the economy since then. The creation of national parks in both countries can be traced to the influence of these early environmental pioneers, and contemporary environmental protection policy and movements around sustainability and green technology echo the sentiments and commitments of these earlier movements. Today, citizens in both Canada and the United States are deeply engaged in a wide range of movements and practices that reflect concern with the environmental, as well as the cultural, social, and economic, impacts of global warming

Other movements of prominence today in both countries include the **antiglobalization movement** (groups that stand in opposition to the unregulated political power of large, multinational corporations and to the powers exercised through trade agreements and deregulated financial markets; **Figure 6.33**). Movements that operate in more covert ways because of their far-right, antigovernment positions include survivalist organizations like the Christian identified Michigan Militia.

# FUTURE GEOGRAPHIES

The United States is currently the most powerful nation on Earth, and Canada is one of its staunchest allies. Both countries possess broad resource bases; a large, well-trained, and very sophisticated workforce; and a high level of technological sophistication. The United States has a domestic market that has greater purchasing power than any other single country, as well as the most powerful and technologically sophisticated military apparatus, and it has the dominant voice and the last word in international economic and political affairs. It also has a distinctive message as a global leader: free markets, personal liberty, private property, electoral democracy, and mass consumption.

But, as discussed in Chapter 1, this is a period of dramatic and rapid change as the world's political borders are being dismantled or rearranged around new economic relationships wrought by globalization. It is impossible to know what these new relationships will mean for the future of today's world regions. For example, although the regionalization framework for this book locates Mexico with Latin America, despite the fact that it too occupies the North American continental landmass, it is not inconceivable that in 20 years or so, a new regionalization will combine Mexico (and possibly other countries in Central America) with the United States and Canada because of immigration, internal demographic changes, or the impact of economic alliances such as the North American Free Trade Agreement (NAFTA).

The future for the United States, at the moment, is unclear. Its economic dominance is no longer unquestioned in the way it was in the 1950s, 1960s, and 1970s. The 2008 collapse of the U.S. financial and housing industry has left it reeling, and there is every prospect that its economic prowess will soon be overtaken by China. On some measures of economic development, the European Union has already overtaken the United States. More important, the globalization of the economy has severely constrained the ability of the United States to translate its economic might into the firm control of international financial markets

▲ **FIGURE 6.33 World Trade Center** The financial heart of New York City, the twin towers of the World Trade Center were proud symbols of the heart of global capitalism. On September 11, 2001, a terrorist attack completely destroyed the towers, other parts of the complex, and surrounding buildings, killing thousands of people, paralyzing the city and the region, and profoundly disrupting the U.S. economy. The attacks caused changes in U.S. politics at home and abroad, leading to a war on terrorism in Iraq as well as one at home. The war on terrorism in the United States has affected immigration, air travel, and a wide range of everyday practices. It is also responsible for the creation of a globally transformative federal initiative, called "homeland security," that reaches from Washington, DC into the neighborhoods of cities and towns across the country and the world.

that it used to enjoy. In contrast, Canada has weathered the impacts of the global economic crisis rather well, with its banking institutions still solid and its housing sector relatively unaffected.

The terrorist Al-Qaeda attacks of 2001 gave a new focus for U.S. geopolitical strategy. But the invasion of Iraq in the absence of evidence of weapons of mass destruction, together with U.S. refusal to participate in high-profile international agreements such as the Kyoto Protocol and its lack of cooperation with the International Criminal Court and the United Nations, resulted in a significant weakening of the political, cultural, and moral leadership of the United States in global affairs.

The bright spot has been the presidential election in 2008 of Barack Obama, whose administration has already begun to turn around many of these crises in geopolitics and the environment, raising hope around the globe that the United States will resume some of its global leadership roles, particularly with respect to issues like global climate change, and a more stable global capitalist economic system.

# Main Points Revisited

■   The United States and Canada constitute two predominantly English-speaking countries of the North American continent. The United States includes 48 contiguous states and two noncontiguous states: Alaska (northwest of Canada) and the volcanic islands of Hawai'i, southwest of the North American continent in the Pacific Ocean. Canada includes 10 provinces and 3 territories and is the world's second-largest country after the Russian Federation.

■   The high economic prosperity of the region has resulted in dramatically altered land and waterscapes as populations have adapted to and modified the world around them. These transformations continue to shape the relationship between people and their environment in the United States and Canada.

New global agreements like the Kyoto Protocol and the IPCC report on global change are likely to result in more sustainable human–environment interactions.

■   Migration has been responsible for the population of this region, beginning with the indigenous people who crossed the Bering Strait 30,000 years ago to the arrival of European colonists in the 15th century to the more recent waves of immigrants who have come from across the globe to live here.

Nineteenth- and 20th-century migration has changed the composition of the region very dramatically and is likely to shift the population away from its white, Protestant roots.

■   The United States and Canada are key forces in global geography, shaping not only economic and political events near and far, but also cultural trends and social practices. Few people in the world have been unaffected by "the American way of life."

The impact of the failure of the banking system can be traced directly to the U.S. mortgage market, making it clear that global economic recovery will depend on the ability of the United States to rebuild some of its key economic foundations.

■   Canada and the United States control vast natural, human, and financial resources such that events in this region, both historically and today, reverberate throughout the globe from the colonial wars that changed the course of European history to the role of the Internet and fast food in daily life.

The technologies of the new economy are taken up differently in different places both in this region and beyond, highly influenced by local culture and practices.

# Key Terms

airshed (p. 196)

Americanization (p. 210)

antiglobalization movement (p. 230)

assimilation (p. 221)

cap-and-trade programs (p. 199)

carbon footprint (p. 197)

creative destruction (p. 213)

Europeanization (p. 209)

fast food (p. 207)

federal state (p. 228)

GMO (p. 207)

hate crime (p. 218)

indentured servant (p. 210)

intermontane (p. 200)

internal migration (p. 218)

local food (p. 207)

Main Street (p. 227)

Megalopolis (p. 227)

multiculturalism (p. 221)

organic farming (p. 207)

slow food (p. 207)

staples economy (p. 212)

suburbanization (p. 219)

superfund site (p. 208)

# Thinking Geographically

1. Which areas of the United States and Canada suffer most from acid rain, deforestation, chemical or nuclear toxic waste, and flooding? What can be done to reduce or remedy ecological damage?

2. Describe the factors that make Québec unique. Why do so many Québecois want to secede from Canada? How might increasing globalization affect Québec?

3. Discuss the three main waves of immigration into the United States. When did they occur, and where did the immigrants come from? Discuss the three main waves of internal migration within the United

States. How were African Americans, many of whose ancestors were brought to the United States long before other immigrant groups arrived, generally affected by immigration and internal migration?

4. Which areas have been primarily Hispanic since their incorporation into the United States? Which areas have seen a large increase in Hispanic population over the past 30 years, and why?

5. Contrast and compare the changing geography (since 1980) of the Pacific Rim, Cascadia, and the North American Core. What factors make Silicon Valley, Vancouver, and New York City distinctive?

Log in to www.mygeoscienceplace.com for Videos, Interactive Maps; RSS feeds; Further Readings; Suggestions for Films, Music and Popular Literature; and Self-Study Quizzes to enhance your study of The United States and Canada.

▼ FIGURE 7.1

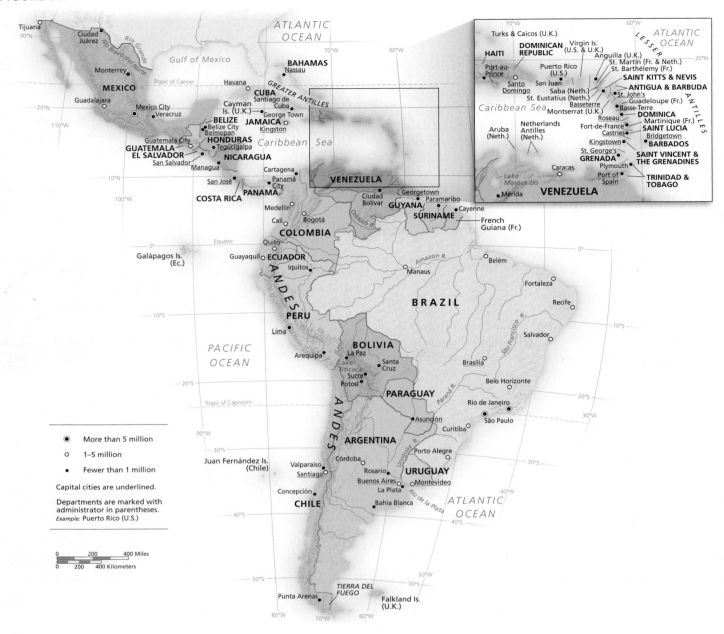

# LATIN AMERICA AND THE CARIBBEAN KEY FACTS

- Major subregions: Central America, the Southern Cone, Caribbean Region, Gulf of Mexico

- Total land area: 20 million square kilometers/ 7.7 million square miles

- Major physiogeographic features: Andes Mountains, Amazon basin, rainforest and river, Mexican plateau, Lake Maracaibo. Influenced by global atmospheric circulation, the climate is varied across the region and includes tropical, midlatitude, and Mediterranean systems.

- Population (2010): 574 million, including 196 million in Brazil, 108 million in Mexico, and 50 million in Columbia

- Urbanization (2010): 79%

- GDP PPP (2010) current international dollars: 10,345; highest = The Bahamas, 35,799; lowest = Haiti, 1,325

- Debt service (total, % of GDP) (2005): 6.4%

- Population in Poverty (%< $2/day) (1990–2005): 29.8%

- Gender-related development index value: 0.78; highest = Barbados, 0.89; lowest = Guatemala, 0.68

- Average life expectancy/fertility: 73 (M69, F76)/2.4

- Population with access to Internet (2008): 34%

- Major language families: Spanish and Portuguese, both of which descend from the Romance branch of Indo-European language; also indigenous languages such as Maya and Quechua

- Major religions: Mainly Christian (mostly Roman Catholic), yet indigenous religions are widely practiced

- $CO_2$ emission per capita, tonnes (2004): 3.2

- Access to improved drinking-water sources/sanitation (2006): 91%/ 80%

# 7
# LATIN AMERICA AND THE CARIBBEAN

The Sierra Norte of Oaxaca is a region of forested mountains where mist and rain supply streams that replenish rivers and groundwater in the valleys below, and where thousands of different plants and animals live in one of Mexico's most biodiverse regions. Ancient cities such as Monte Alban were constructed more than 2,000 years ago by indigenous Zapotec people; some of their descendants still live in mountain communities, where they cultivate corn as a foundation of traditional diets and coffee and charcoal as a source of commercial incomes. Older residents such as Don Miguel talk about traditional uses of local ecosystems, including the use of plants for medicine, and of the social bonds whereby ceremonial and labor responsibilities are rotated between different members of the community, and decisions about the use of land are made in community assemblies.

The forests have been an important source of income, especially for making charcoal and for wood used to make furniture, household implements (such as the small wooden whisks used to whip hot chocolate drinks), and the popular painted wooden animals sold to tourists in the streets and markets of the city of Oaxaca (pronounced wa-ha-ca). Forests have also been leased to paper companies and other commercial loggers, who build roads into remote mountain locations but provide some local employment. In many families, several generations of men have worked in the forest, now using modern chainsaws and lathes to cut and shape wood.

Many Zapotec and other Oaxacans spend part of their lives as migrant workers in the United States, where in Los Angeles and Chicago you can hear people speaking indigenous languages and women sell traditional foods from their kitchens and street corners in the neighborhoods where migrants stay while working in restaurants, in construction, and as maids in homes and hotels. Wages are often saved to invest in homes and small businesses back in Oaxaca, sent back through companies that specialize in the transmission of these remittances from U.S.-based migrants to their homes in Latin America. For some young people, it is difficult to return to the remote villages, and they may stay in the United States or settle in Mexico City. Remittances are also used to purchase satellite dishes and trucks that sustain links to wider culture and markets.

Other global connections are emerging as business opportunities reach the Sierra Norte in the form of new markets. Shiitake mushrooms are cultivated for export to Japan, and wood and coffee can be certified as sustainably produced for sale to "green" consumers. But the process of getting organic or fair trade certification can be complex and expensive for a small amount of added value for the products and requires a level of outside monitoring that is unacceptable to some community members.

Other communities are developing ecotourism projects to take advantage of the thousands of tourists that visit the city of Oaxaca in the valley below the mountains, and leaders are also investigating the option of demanding a payment for the environmental service of maintaining the forest watersheds to provide water to the valley, which suffers frequent droughts and where water demand has increased with urban growth and tourism. These ecotourism projects and other ways of making money from environmental protection are often supported by international nongovernmental organizations (such as the World Wildlife Fund) that wish to protect the forests and biodiversity through supporting sustainable livelihoods.[1] ■

## MAIN POINTS

■ The region includes 30 independent countries, where colonial global connections resulted in many countries now using Spanish as the official language. European languages and institutions are also evident in Portuguese-speaking Brazil and in Caribbean territories of the United States, the United Kingdom, France, and the Netherlands. Although a region of conflict and authoritarian rule for much of the 20th century, with several interventions by the United States, Latin American countries are now mostly governed democratically.

■ The physical geography of this region is situated within a wider set of climatic and geological processes that link this region to the global patterns of atmospheric circulation and tectonics. These include the atmospheric systems of the tropical convergence zone over the Amazon and the wind systems that cross the Atlantic and Pacific Oceans as well as the complex geology of uplift that creates high mountain ranges from Alaska to the Andes and the arc of volcanic Caribbean islands.

■ Key historical periods include the early complex societies of the Aztecs, Incas, and Mayas, who made large-scale modifications to their environment and European colonialism which left a legacy of demographic collapse and of export crops such as sugar, coffee, and bananas and metals such as silver and copper dominating the economies of many countries.

■ The 20th-century economic development policies parallel those in many other developing regions, and have included import substitution and then neoliberalism in which many countries expanded trade, privatized resources and institutions, and cut government spending.

■ Following worldwide patterns, fertility has dropped significantly, and many people have moved to rapidly growing urban areas that have serious environmental problems and large underserviced housing areas.

---

[1]This story is based on information from the Mountain Voices project of the Panos Institute.

# ENVIRONMENT AND SOCIETY

Latin America is the southern part of the large landmass of the Americas that lies between the Pacific and Atlantic oceans (**Figure 7.1**). Traditionally, the Americas are divided into the two continents of North America (Canada, the United States, and Mexico) and South America (all the countries south of Mexico). Latin America includes all the countries south of the United States from Mexico to the southern tip of South America in Chile and Argentina. It forms a world region of considerable physical and social coherence covering more than 20 million square kilometers (7.7 million square miles). This chapter also includes the Caribbean region—the arc of islands that sweep from the U.S. state of Florida to Colombia.

Latin American landscapes have long been transformed by humans. Research has shown that many of the seemingly pristine forests were cleared centuries ago and have grown back, the grasslands selectively burned or grazed, the mountains carved into terraces (**Figure 7.2**), and the waters of the deserts stored or diverted. Similarly, studies show that contemporary human settlements and agriculture are extremely vulnerable to natural disasters and epidemics and, therefore, that the human geography of Latin America continues to be influenced by environmental conditions, climate change, and climate events. An understanding of the physical geography and the ways in which people have adapted to and modified their environmental surroundings is integral to appreci-ating the historical and contemporary human geography of the Latin American region and the challenges to its sustainable development.

## Climate, Adaptation, and Global Change

The overall climates of Latin America are determined by global atmospheric circulation, including the positions of the equatorial high- and tropical low-pressure zones and the major global wind belts (see Figure 1.3). Climatic patterns provide good general examples of how global circulation affects regional climate, vegetation, and human activity (**Figure 7.3**).

As we discussed in Chapter 1 (p. 3), the general circulation of the atmosphere is driven by the differential heating and rotation of Earth, with warm air rising at the equator at the intertropical convergence zone, then cooling as it rises, thereby producing high rainfall, then flowing poleward, and finally sinking over regions around the tropics of Capricorn and Cancer. The equatorial zones of high temperatures and rainfall provide conditions for the rapid growth of vegetation in the form of the rainforests of the Amazon, where annual rainfall ranges from 1.5 to 2 meters (60 to 80 inches). Where the air sinks over the tropics, it becomes warmer and drier, holding so little moisture by the time it reaches ground level that these regions are characterized by the very

▲ **FIGURE 7.2 Andean terraces and alpaca at Machu Picchu, Peru** The Inca constructed terraces, such as these around the Inca ruins at Machu Picchu, not only to reduce soil erosion and provide a flat area for planting, but also to decrease frost risks by breaking up downhill flows of cold air and allow for irrigation canals to flow across the slopes in efficient ways. Constructing and maintaining these terraces and irrigation systems required large-scale social organization in which the Incan empire excelled. People were organized into small groups called ayullu, and labor for agriculture and mines was commanded through the mita labor system, by which communities were required to provide a certain number of days of work to the central authorities. In the 16th century, many terraces were abandoned as the native labor force was reduced because of the ravages of newly introduced European diseases and the need to shift laborers to the Spanish mines. Alpaca graze on terraces below Machu Picchu, now a major tourist attraction.

Central America, and in the southern hemisphere bring rain to the east coast of Brazil. Easterly winds moving across the warm Caribbean absorb a lot of moisture, especially during the fall, when the sea surface is warmest. When storms start to circulate, the warm sea fuels both the moisture and energy of the storms, producing the hurricanes that regularly cross the Atlantic coast of Latin America.

The regions on the margins of the trades and at the edges of the equatorial rainfall zone have highly seasonal climates with a distinct rainy season, as all zones shift north in the June to August period and south in November to February.

The westerlies bring heavy rains to the southern South American nation of Chile. When the global circulation shifts southward in December, storms spinning out of the Northern Hemisphere westerlies also bring rain to northern Mexico. Latin America's extensive grasslands occur where seasonal shifts in wind and pressure belts result in a distinct rainy season, especially on the margins where the rains are fairly moderate.

The coastal mountain chains of Latin America, especially the Andes, clearly illustrate the role of topography in regional climate, as ocean winds that encounter coastal mountains are forced to rise even higher, cooling to the point that they release most of their moisture in the form of rain and snow. The high precipitation over the Andes feeds the rivers that water the lowlands east and west of the mountains, most notably the Amazon. However, mountains also create a *rainshadow* effect because winds passing over mountains from the coast to the interior lose their moisture over the higher altitudes and then become warmer and drier as they descend to the interior, creating arid conditions to the leeward of mountain ranges (see Figure 1.4). Examples include the dry region of southern Argentina, known as Patagonia, to the east of the Andes and the drier regions of Chihuahua in northern Mexico in the lee of the Mexican Sierra Madre.

Higher altitudes are also cooler, so despite the intensity of the sunlight, extensive highland regions of the Latin American tropics have cooler temperatures more conducive to human activities and agricultural crops, such as wheat, apples, and potatoes, generally found in colder climates.

**Altitudinal zonation** is a vertical classification of environment and land use according to elevation based mainly on changes in climate and vegetation from lower (warmer) to higher (cooler) elevations. In Latin America these changes are defined in a simple classification of Latin American mountain environments into the tierra caliente (Warm), tierra templada (Moderate), and tierra fría (Cold). The very high altitudes are called the tierra helada (Icy). Each of these zones is associated with characteristic vegetation types and with agricultural activities (**Figure 7.4**). Communities locate fields at different elevations to adapt to and take advantage of different climatic and soil conditions. At higher altitudes, potatoes grow and animals graze; lower down, grains such as wheat and corn grow; and finally, vegetables and fruit are found at lower levels with more tropical climates.

Latin America also provides a classic case of how the temperatures of the ocean can influence the climate of adjacent landmasses. Winds flowing across the very cold ocean current that normally flows northward off the coast of Peru and Chile pick up very little moisture and exacerbate the already dry conditions promoted by descending air over the tropics forming the Atacama Desert, which is one of the driest spots on Earth.

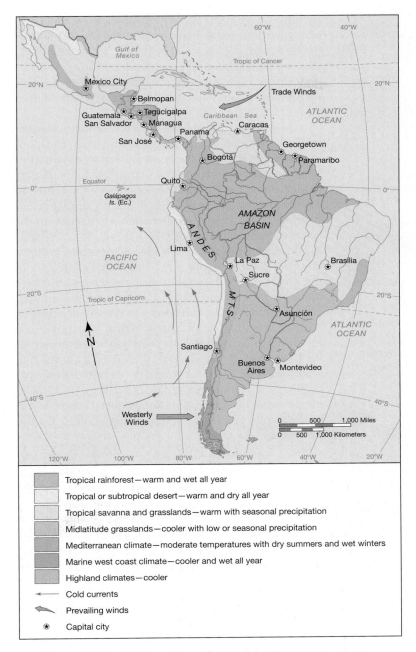

▲ **FIGURE 7.3  Climate regions of Latin America** Latin America's climate is influenced by major wind and pressure belts and the configuration of land and oceans. The average pattern shown in this map tends to shift northward in June and southward in December, bringing seasonal changes to many regions.

low rainfall, sparse vegetation, and dry conditions of deserts that in Latin America include the Sonoran and Chihuahuan deserts of Mexico and the Atacama Desert of Chile.

Air flowing from the tropics to the equator is dragged by the spinning Earth into an east–west flow called the *trade winds,* and air flowing poleward from the Tropics is dragged into a west–east flow called the *westerlies.* The trade winds flowing across the Atlantic north of the equator frequently produce rain on the Caribbean islands and east coasts of

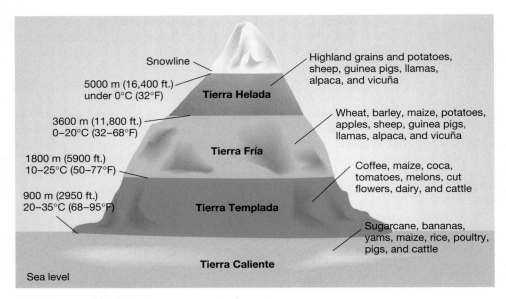

Snowline

Highland grains and potatoes, sheep, guinea pigs, llamas, alpaca, and vicuña

5000 m (16,400 ft.)
under 0°C (32°F)

**Tierra Helada**

3600 m (11,800 ft.)
0–20°C (32–68°F)

Wheat, barley, maize, potatoes, apples, sheep, guinea pigs, llamas, alpaca, and vicuña

**Tierra Fría**

1800 m (5900 ft.)
10–25°C (50–77°F)

Coffee, maize, coca, tomatoes, melons, cut flowers, dairy, and cattle

900 m (2950 ft.)
20–35°C (68–95°F)

**Tierra Templada**

Sugarcane, bananas, yams, maize, rice, poultry, pigs, and cattle

**Tierra Caliente**

Sea level

▲ **FIGURE 7.4 Altitudinal zonation** The altitudinal zonation of climate and vegetation in mountainous regions such as the Andes creates vertical bands of ecosystems and provides a range of microenvironments for agricultural production. The tierra caliente runs up to 900 meters (2,950 feet), the tierra templada ranges from 900 to 1,800 meters (2,950 to 5,900 feet), the tierra fría from 1,800 to 3,600 meters (5,900 to 11,800 feet), and the very high altitudes (higher than 3,600 meters, or 11,800 feet) are called the tierra helada. These zones are associated with characteristic vegetation types such as rainforest in the tierra caliente and grasslands in the tierra fría, and with agricultural activities such as growing sugar and bananas in the tierra caliente, coffee and cattle in the tierra templada, and potatoes and barley in the tierra fría. Global warming is shifting these zones upwards in many parts of the tropics.

**Climatic Hazards** In October 1998 Hurricane Mitch dumped a year's worth of rain (about 1.2 meters, or nearly 4 feet) on Central America in 48 hours (**Figure 7.5a**). Flash floods and mudslides on deforested slopes left nearly 10,000 people dead, almost 20,000 missing, and more than 2.5 million temporarily dependent on emergency aid. Honduras, the second-poorest nation in the Western Hemisphere, was the hardest hit. Of the 6 million people living in Honduras, nearly 2 million were affected by the storm, 1 million lost their homes, and 70% of the country's productive infrastructure was damaged or destroyed. This disaster created immediate food shortages and decimated the vital export crops of bananas, coffee, and shrimp, which are responsible for half the country's annual export revenue of $3 billion. Ten years later some regions of the country and the poor still have not recovered fully.

What turned Mitch from a natural hazard into a human disaster was a chain reaction of social vulnerabilities created by long-term climate change, environmental degradation, poverty, social inequality, gender discrimination, racism, population pressure, rapid urbanization, and international debt. Deforestation increased the magnitude of the floods and the risk of landslides. When the forest is cleared, the ground is exposed; when rains come, the result is an increase in surface runoff and soil erosion. Nature creates hazards, but it is humans who create *vulnerability*, through social inequality and unequal access to resources. Some 72% of the farmers of Honduras have access to only 12% of its arable land, whereas 1.5% control 40%. Large-scale beef ranching and banana plantations have displaced poor peasants over decades, forcing them to live in isolated valleys, where they have carved

out subsistence farms on surrounding hillsides and riverbanks, further destabilizing these areas. Almost 80% of the Honduran population already lives below the poverty line, with widespread malnutrition and illiteracy. The poor find it difficult to recover from disasters because they lack the money and other resources to rebuild their homes and livelihoods.

Latin America's climate does not remain constant from year to year. One of the most important causes of this climate variability is the phenomenon known as **El Niño**—a periodic warming of sea surface temperatures in the tropical Pacific off the coast of Peru that results in worldwide changes in climate, including droughts and floods. El Niño brings warmer and wetter winds to the coasts of Peru and Ecuador with high rainfall and flooding. But the sensitivity of the global atmospheric circulation is such that the links between Pacific Ocean temperatures and conditions elsewhere produce droughts in northeast Brazil, floods in southern Brazil and northern Mexico, and fewer Caribbean hurricanes as well as droughts in southern Africa, Australia, and Indonesia (**Figure 7.5b**). In some years, the ocean off Peru gets colder than normal, producing a contrasting global pattern called **La Niña** (the periodic abnormal cooling of sea surface temperatures in the tropical Pacific off the coast of Peru), with floods in northeast Brazil, drought in northern Mexico, and more intense Pacific hurricanes.

**Climate Change** Latin America is already experiencing the effects of global warming, especially in the Andes, where glaciers and snowfields are shrinking (**Figure 7.6**). This is creating serious problems for communities that rely on runoff from ice and snow for irrigation and drinking water and for larger-scale hydroelectric schemes, where dry conditions mean electricity cutoffs in major cities. Climate change may also cause more intense hurricanes. The Amazon forest is an important "sink" for greenhouse gases because the vegetation takes up carbon dioxide and a critical rainfall source from evaporated moisture. Unfortunately, there are some indications that higher levels of carbon dioxide, and droughts associated with global warming, combined with deforestation, are drying the Amazon and reducing its ability to absorb carbon. If this continues, the Amazon could become a globally significant "tipping point," where drying, fires, and associated carbon emissions flip the vegetation from forest to grasslands and make the region a source rather than sink for carbon. Although there are many traditional adaptations to climate variability in Latin America—including complex irrigation systems developed by the Aztecs, Maya, and Incas (discussed later)—global warming may test the limits of these systems if it brings drier and warmer conditions.

## Geological Resources, Risk, and Water Management

The physical landscape of Latin America varies widely and includes striking mountain ranges, high plateaus, and enormous river networks (**Figure 7.7**). The two largest-scale physical features in Latin America are

(a)

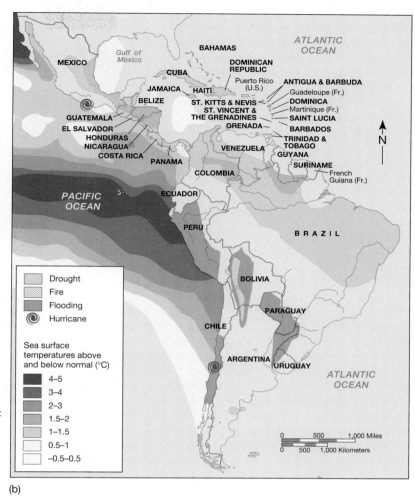

▲ **FIGURE 7.5  Climate hazards** (a) Hurricane Mitch was one of the strongest hurricanes on record when it hit Central America after moving across the Caribbean in October 1998. With winds of more than 180 miles per hour, Mitch produced heavy rains that intensified over Central America's coastal mountains, causing severe flooding, especially below deforested hillslopes. Hurricane Mitch inundated vast areas of Honduras, dropping about 1.2 meters (4 feet) of rainfall over a 48-hour period and caused tremendous damage along the Nicaraguan coast, shown here. (b) During an El Niño, the ocean warms off the coast of Peru and causes heavy rain and flooding along the usually dry Peruvian coast. The Pacific warming is linked to other changes in atmospheric circulation that produce drought in northeast Brazil, the Caribbean, and the altiplano; floods on the Parana River; and droughts and fires in Mexico and Central America.

(b)

▼ **FIGURE 7.6  Andean glaciers at risk**  Climate changes caused by increasing levels of greenhouse gases from fossil fuels and deforestation have already produced higher average temperatures in mountain regions. Andean glaciers and snow cover such as on the volcanoes Paronacota and Pomerata in Bolivia are shrinking, and the rivers that depend on snow and ice melt are drying up, creating problems for irrigated agriculture, drinking water, and hydroelectric generation in the valleys below.

▲ **FIGURE 7.7 Physical regions and landforms of Latin America** The Amazon basin and the Andes mountain range are the two largest physical features in Latin America. Major South American rivers such as the Plata and the Orinoco provide transport routes into the interior as well as water resources for agriculture and hydroelectricity generation. The two major deserts—the Sonoran and Atacama—are located along the Pacific coasts of Mexico and Chile, and important grasslands are located in the pampas region of Argentina and the llanos of northern South America. Past and current human populations settled on highland plateaus, such as the altiplano of the Andes and the Mesa Central of Mexico. The region has many active volcano and earthquake zones associated with the movements of tectonic plates. Older geological formations are associated with important mineral deposits, such as the silver mines of the Andes and central Mexico, copper in central and northern Chile and in northern Mexico, and gold and iron along the edges of the Brazilian and Guiana highlands. World-class oil deposits occur in the Gulf of Mexico and in Venezuela, especially Lake Maracaibo, and in the western Amazon basin.

the Andes Mountains and the Amazon basin, both easily seen from space (**Figure 7.8**). The Andes are an 8,000-kilometer-long (5,000-mile-long) chain of high-altitude peaks and valleys that for the most part run parallel to the west coast of South America, with the highest peak, Aconcagua, at 6,960 meters (22,830 feet).

Other important physical features include the mountainous spines of Mexico and Central America and the high-altitude flatter areas, or plateaus that lie between or next to the mountain ridges such as in the Andes at more than 3,000 meters (9,500 feet), where the plateau region is called the **altiplano** (Figure 7.6). This region, as well as the Mexican plateau, or Mesa Central, is an important area of human occupation because it provides flatter, cooler, and wetter environments for agriculture and settlement than do the adjacent steep-sloped mountains, dry lowland deserts, and humid lowlands.

The Caribbean basin has large areas of limestone geology, where water tends to flow underground and create large cave systems, such as those in the Yucatán Peninsula of Mexico and in Puerto Rico. Coral reefs, a key feature of the Caribbean landscape, are created when living coral organisms build colonies in warm, shallow oceans. These reefs, hosts to myriad other marine animals, are fragile ecosystems that are easily damaged by boats, divers, pollution, and environmental change.

**Tectonic Hazards**    Many of the mountain areas and the location of island chains are the result of a long history of tectonic activity in Latin America at the intersection of several major tectonic plates (see Chapter 1, p. 6, and Figure 1.8). Central America is one of the most tectonically active regions in the world. The region south of Panama sits on the South American plate, the slow westward drift of which causes it to collide with the adjacent Nazca plate. This activity results in folding and uplifting of the western edge of the South American plate to form the Andes Mountains, thus forcing the Nazca plate downward and under South America in the process of subduction, and producing active volcanoes and severe earthquakes. Similarly, Mexico sits on the North American continental plate, also drifting westward and causing the sub-

▶ **FIGURE 7.8  Satellite image of Latin America**  Certain physical features of Central and South America are clearly visible in this satellite image, including the verdant green of the Amazon basin and the mountain ridges of the Andes and Central America, as well as Lake Maracaibo on the northern coast of Venezuela.

duction of the Cocos plate and the uplift and folding of the Sierra Madre Mountains. The Caribbean plate, in contrast, is moving eastward, pulling away from the Cocos, moving under the South and North American plates, and causing geological tensions and cracks that produce earthquakes and volcanic activity in Central America and produced several volcanic islands in the Caribbean.

Volcanic activity poses threats to human activity when eruptions and ash destroy crops and lives (**Figure 7.9a**). The tensions associated with shifting tectonic plates have also produced devastating earthquakes that have ravaged the capital cities of some Latin American countries—for example, in Mexico City, Mexico; Managua, Nicaragua; Guatemala City, Guatemala; and Santiago, Chile (**Figure 7.9b**). Such natural disasters cannot be blamed solely on geophysical conditions but also on vulnerability. The greatest damages occur when people are forced to live in unsafe houses or on unstable slopes because they lack the money or power to live in safer places, cannot afford insurance, or are unable to obtain warnings of impending natural disasters such as volcanic eruptions and hurricanes.

**Mineral Resources**  The mineral wealth of Latin America is typically found on the old crystalline Guiana shield (or highlands) and where crustal folding brings older rocks near the surface (see Figure 7.7). Major precious-metal mining districts include the Peruvian and Bolivian Andes, where mountains of silver were excavated in the Spanish colonial period at Cerro de Pasco and Potosí, and lead, zinc, and tin are still important; the silver region of the Mexican Mesa Central; and the gold and iron mines at Carajas on the edge of the Brazilian plateau. World-class iron deposits are found on the southern edge of the Brazilian shield at Itabira, on the northern edge of the Guiana highlands at Cerro Bolivar, and in northern Mexico. Copper is the geological treasure of the southern Andes, especially northern Chile, and is also important in northern Mexico. The

shores of the Caribbean, including the Guianas and Jamaica, have deposits of bauxite (a mineral used in the aluminum industry). These minerals, especially gold and silver, were foundations of the European colonial economies and now dominate the export economies of countries such as Chile and Bolivia. They are a focus of foreign interference and ownership and have often transformed local labor and environmental conditions.

The other critical resources associated with Latin America's geology are oil, gas, and coal. Coal is found in northern Mexico, Colombia, Brazil, and Venezuela. The earliest oil booms and later gas developments occurred in Venezuela around Lake Maracaibo and on Mexico's Gulf Coast. Oil was discovered in Venezuela in 1917, and the country became one of the founding members of the Organization of Petroleum Exporting Countries (OPEC) in 1960. Oil became an important foundation of the national economy, and the oil industry was nationalized in 1976, but the economic benefits of oil production reached only about 20% of the population. Lake Maracaibo, the site of about 4% of world oil reserves, is now crowded with thousands of oil derricks that produce about two-thirds of Venezuela's oil output (**Figure 7.10**).

The Mexican oil deposits were first commercially exploited in the 1890s and were nationalized in the 1930s under the government oil company PEMEX (Petróleos Mexicanos). In all of Latin America's oil regions, environmental pollution has been a serious problem, leading to waterways contaminated with waste oil, widespread ecosystem damage, and serious health problems among local residents.

The most recent oil developments are in the Amazon, where oil and gas were discovered in 1967. The Amazonian oil deposits are found mostly in remote forest areas, where land rights of indigenous people are not secure, and as a result, conflicts have erupted between Peru and Ecuador and among governments, corporations, and indigenous groups. Mining in the Amazon has also created controversy in regions where

(a)

(b)

▲ **FIGURE 7.9 Natural disasters** (a) The people of the island of Montserrat in the Caribbean have lived with the threat of volcanic eruptions for centuries. Eruptions in the 1990s in the Soufriere Hills forced the evacuation of more than two-thirds of the residents. Most left their homes, farms, beach houses, and hotels, abandoning their entire lives at the foot of the volcano. The capital city, Plymouth, was burned and half-buried by waves of gas and ash. Assistance was provided by the British government, which controls Montserrat as an overseas territory. (b) The 1985 earthquake in Mexico City killed as many as 10,000 people and devastated downtown buildings. The city is especially vulnerable because it is built on a former lake bed with sediments that act almost as liquids when shaken by earthquakes.

Paraguay, and Uruguay rivers) originate in the Andes and the Brazilian highlands. The Plata system has become a major source of energy through large hydroelectric dams such as the Itaipu dam on the Paraná, and there are controversial plans to dam and divert the water resources of the vast wetlands of the Pantanal. The Orinoco drains the *llanos* grasslands of Colombia and Venezuela. The Amazon tributaries flow downward and eastward from the Andes into an enormous river network that covers a basin of more than 6 million square kilometers (2.3 million square miles). The Amazon basin includes the river itself and the surrounding landscape, about two-thirds in Brazil and including parts of Peru, Ecuador, Bolivia, Colombia, Venezuela, Guyana, Suriname, and French Guiana. The Amazon River and its tributaries carry 20% of the world's fresh water and provide transport, sediment, and fish that support the agriculture, diets, and mobility of the people of the basin. This watershed also nourishes the Amazon rainforest, which is home to more than 100,000 species.

Latin America has several large freshwater lakes, among them Lake Nicaragua in Nicaragua and Lake Titicaca at the border of Bolivia and Peru. Major waterfalls such as Iguaçu Falls—where Brazil, Argentina, and Paraguay meet (**Figure 7.11**)—and Angel Falls, Venezuela—the tallest waterfall in the world at a height of 985 meters (3,230 feet), falling off the flat-topped mountain of Auyantepui—have become increasingly popular tourist destinations.

## Ecology, Land, and Environmental Management

The diversity of Latin America's physical environments has produced a large number of different species, or **biodiversity**. Latin America's biodiversity is substantial because of the size of the continent, the range in climates from north to south, altitudinal variations within short distances, and a comparatively long history of fairly stable climates and isolation from other world regions. Many tourists are attracted to the col-

▲ **FIGURE 7.10 Oil wells in Lake Maracaibo, Venezuela** Venezuela is a major petroleum producer with associated problems of pollution, such as in Lake Maracaibo, shown here. The Venezuelan economy is vulnerable to changes in world oil prices, and the distribution of profits is critical to the political popularity of leaders such as President Hugo Chavez, who used oil in negotiations with other countries.

migrants have gathered to work in the gold mines. The mining process uses mercury, a hazardous chemical element that is now polluting ecosystems and causing health problems among the miners.

**Water Resources** The three largest river basins in Latin America are the Amazon, the Plata, and the Orinoco, all flowing to the Atlantic Ocean (see Figure 7.1). The rivers in the Plata basin (including the Paraná,

▼ **FIGURE 7.11 Iguaçu Falls** Iguaçu Falls, on the Paraná River, where Brazil, Argentina, and Paraguay meet, has become a major tourist destination.

orful birds and verdant plants associated with the ecological region between the tropics of the Americas, also called the **neotropics**.

Desert ecosystems, such as in the Atacama Desert of northern Chile and Peru, are associated with drier climates, where species have developed many interesting adaptations to water scarcity (**Figure 7.12a**). Between the moist forests and dry deserts lie ecosystems where alternating wet and dry seasons produce vegetation ranging from scattered woodlands to dry grasslands. Grasslands are also found at higher altitudes, where there is not enough precipitation or temperatures are too low to support highland forests. In Argentina the *pampas* grasslands cover more than 750,000 square kilometers (300,000 square miles) and have become important to the cattle economy (**Figure 7.12b**). Other large grassland ecosystems include the *llanos* of Colombia and Venezuela and the *cerrados* of Brazil. The high grasslands of the Andean altiplano provide habitat for grazing animals such as the llama, wild guanaco, and vicuña. Natural ecosystems vary with elevation and include the *páramo* of the northern Andes, with such unusual cold-adapted plants as the *frailejón*. The long coasts of Latin America and the islands of the Caribbean include about 50,000 square kilometers (about 19,000 square miles) of mangrove ecosystems, or about 25% of the world's total (**Figure 7.12c**).

The wetter climates of Latin America are associated with magnificent forest ecosystems, including the tropical rain forests of the Amazon, Central America, and southern Mexico and the temperate rain forests of southern Chile. The Amazon forest ecosystems are notable for the sheer number of species found within small areas of forest. A constraint on human development, particularly in the warm and wet climates of much of Latin America, is the large diversity and prevalence of pests and dis-

eases that weaken and kill plants, animals, and humans in the tropics around the world. For example, malaria is endemic in much of the Amazon basin and lowland Central America. Humans have massively transformed the ecology of Latin America through the domestication of plants and animals and the conversion of forests and grasslands to agriculture.

**Domestication**   The most dramatic transformation of nature by early people in Latin America was domestication. Starting more than 10,000 years ago, wild plants and animals were domesticated into cultivated or tamed forms through selective breeding for preferred characteristics. Many of the world's major food crops were domesticated by indigenous Latin Americans, including the staples of maize (corn), manioc, and potatoes, as well as vegetables and fruits such as tomatoes, peppers, squash, avocados, and pineapples (**Figure 7.13**). Tobacco, cacao (chocolate), vanilla, peanuts, and coca (cocaine) were also domesticated in Latin America. In dry areas, people tried to ensure water supplies by building small dams and channels to bring water to these crops. Latin America has very few indigenous domesticated animals. The camelids (llama and alpaca) were tamed and bred for wool, meat, or transport, and dogs and guinea pigs were also bred for pets and meat.

As in other regions of the world, the increased yields from domesticated crops created a surplus that permitted the specialization of tasks, the growth of settlements, and ultimately the development of highly complex societies and cultures. In Latin America, the complex societies included the great Mayan, Incan, and Aztec empires (**Figure 7.14**). These groups all modified their environments to increase agricultural

(a)

(c)

(b)

◄ **FIGURE 7.12 Latin American ecosystems**  Latin American ecosystems range from forests and grasslands to deserts and coastal mangroves. (a) The Atacama Desert of Peru and northern Chile is one of the driest locations on Earth. However, the area is economically important because of the copper and other minerals that lie beneath the surface and that support one of the world's most significant mining areas. (b) The fertile soils of the extensive grasslands of pampas of southern South America traditionally supported a livestock economy associated with the gaucho cowboy. It was the wheat and beef of the pampas that made Argentina one of the richest countries in the world a century ago. (c) Shrimp farms in the Gulf of Fonseca, Honduras. Mangrove ecosystems protect coasts from storms and provide breeding areas for fish and other marine animals. In countries such as Honduras and Mexico, mangrove ecosystems are being destroyed to build seafood farms (called mariculture) for export. The farms provide some local employment, yet often pollute the water and are vulnerable to disease.

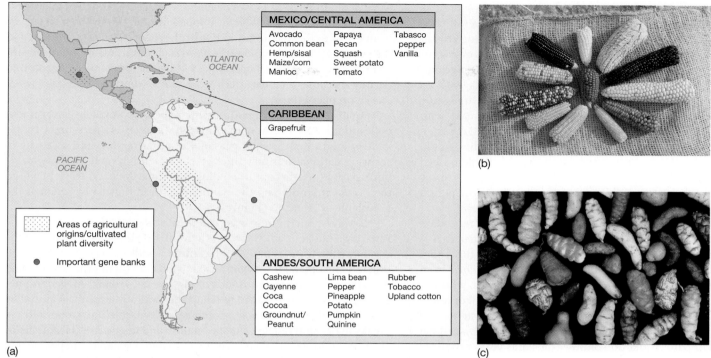

(a)

**MEXICO/CENTRAL AMERICA**

| | | |
|---|---|---|
| Avocado | Papaya | Tabasco |
| Common bean | Pecan | pepper |
| Hemp/sisal | Squash | Vanilla |
| Maize/corn | Sweet potato | |
| Manioc | Tomato | |

**CARIBBEAN**

Grapefruit

**ANDES/SOUTH AMERICA**

| | | |
|---|---|---|
| Cashew | Lima bean | Rubber |
| Cayenne | Pepper | Tobacco |
| Coca | Pineapple | Upland cotton |
| Cocoa | Potato | |
| Groundnut/ | Pumpkin | |
| Peanut | Quinine | |

Areas of agricultural origins/cultivated plant diversity

● Important gene banks

(b)

(c)

▲ **FIGURE 7.13  Domestication of food crops in Latin America**  (a) Latin America has two important centers of domestication in Mexico and the Andes. The red dots on the map denote research centers, where attempts are being made to preserve crops' genetic diversity in "gene banks." (b) Maize (corn) was domesticated in Mexico from a wild grain called teosinte. Bred for a variety of microenvironments and tastes, traditional maize has many shapes and colors. (c) The many colors and shapes of potatoes grown in the Andes are indications of the diversity of varieties domesticated in these regions.

production and to exploit water, wood, and minerals to support their cities, metal production, and trade.

**Mayan, Incan, and Aztec Adaptations to the Environment**  These environmental transformations were widespread, and in some cases—most notoriously the Mayan civilization—people placed so much pressure on regional landscapes that environmental degradation may have precipitated social collapse. The Mayas occupied the Yucatán Peninsula as well as a considerable portion of Guatemala and Honduras (Figure 7.14), with a period of expansion beginning about 3000 B.C.E. and reaching a peak of control and social development from about 600 B.C.E. to 800 C.E. Faced with rapid declines in the fertility of soils after clearing the rainforest, the Mayas adapted by burning the forest to capture the nutrients in the trees through the ash and then by moving on to clear another patch of forest once the declining fertility of the previously cleared area resulted in reduced yields. There is evidence that the Mayas cleared vast areas of forest during this period. It took up to 30 years for the forest to regrow on a plot that had been cleared, farmed, and then abandoned. This adaptation to rainforest environments mirrors those in other parts of the world and has been termed *slash-and-burn* (see p. 160 in Chapter 5), or *swidden,* agriculture. It is the agricultural system often used in tropical forests that involves cutting trees and brush and burning them so that crops can benefit from cleared ground and nutrients in the ash. The Mayas also developed methods for growing crops in wetland areas by building *raised fields* that lifted plants above flooding, but took advantage of the rich soils and reliable moisture of wetland environments.

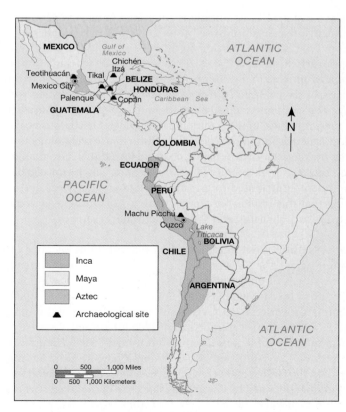

| | |
|---|---|
| ▨ | Inca |
| ☐ | Maya |
| ▨ | Aztec |
| ▲ | Archaeological site |

▲ **FIGURE 7.14  Map of the Maya, Inca, and Aztec Empires**

Between 500 C.E. and 1000 C.E., the great Mayan cities such as Copán, Palenque, Tulum, and Tikal were abandoned, and overall population declined dramatically. Many scholars believe that one reason for the Maya collapse was their overuse of the soils. Population growth and tribute demands by the Maya elites required increases in agricultural production, often involving the clearing of sloped lands, with no time to allow plots to recover before planting them again. Large-scale forest clearing has also been linked to regional changes in climate, with increases in temperature and decreases in rainfall that would threaten agriculture. Soil erosion, droughts, and declining soil fertility would have contributed to a decline in the amount of food available to feed the large population and would cause some of the nutritional stresses that archaeologists have detected in human skeletons from the period of collapse.

The Incas also responded to the difficulties of living in a mountain environment in a variety of ways, including through the construction of many miles of agricultural terraces on the steep hillslopes of the Andes during the height of their empire, which in 1400 C.E. stretched from northern Ecuador to central Chile (Figure 7.14). The Aztecs, who settled in central Mexico in the 1300s, were experts in the control of water. They constructed an extensive network of dams, irrigation systems, and drainage canals in the basin of Mexico to cope with the highly seasonal and variable rainfall pattern that in some seasons and years produced droughts and in other seasons and years produced large lakes and wetlands. They also developed the *chinampa* agricultural system, which permits agriculture in lake and wetland environments by building an island of soil and vegetation. Evidence suggests that as the Aztecs cleared forests in the basin of Mexico and this may have contributed to a drop in the water table and to a resulting water crisis that led to the abandonment of some settlements.

These widespread modifications of the environment of Latin America are evidence that has been used to debunk what geographer William Denevan has called the **pristine myth,** the erroneous belief that prior to European arrival in 1492 the Americas were mostly wild and untouched by humans and that indigenous peoples lived in harmony with nature. In fact, large areas had been cultivated and deforested by indigenous populations. The environmental adaptations and impacts of the Mayas, Incas, and Aztecs still echo in the traditional technologies used in some regions of Latin America and in the continual efforts to use technology to benefit from the physical environment and avoid its hazards. Geographers are among those who have pointed to the ways in which overuse of their environment contributed to the collapse of Mayan society and to the warning that this implies for current patterns of widespread deforestation, overuse of the land, and depletion of water resources.

**Deforestation**   Covering more than a half-billion hectares (about 1.2 billion acres), the Amazon basin contains water, forest, mineral, and other resources of great value, yet has had relatively low population density until recent years (**Figure 7.15a**). The colonial image of the Amazon basin varied from a vision of a tropical Eden with untapped resources to an impenetrable, disease-ridden jungle hell of savage tribes. The region was of botanical interest, but it held little economic attraction until the late 19th century, when development of the automobile industry in the United States and Europe exploded the demand for rubber, a product obtained by tapping the latex sap of scattered rubber trees in the forest. Local rubber tappers, or *seringuieros,* sold the rubber to middlemen. They in turn traded with the "rubber barons," who constructed enormous mansions and a magnificent opera house in the

Amazonian port of Manaus. The end of the rubber boom is said to have occurred when Henry Wickham shipped thousands of rubber tree seeds from Brazil to Kew Gardens, in England, where they were cultured and the seedlings exported to Southeast Asia. The success of the more efficient Asian plantations, especially in Malaysia, drove the Amazon into decline, because Brazilian trees were too susceptible to disease when grown on plantations.

Amazonian development became a focus of Brazilian government policy in the 1970s because the Amazon was seen as a pressure valve for landless and impoverished peasants in other regions and as a way of securing national territory through settlement. Several highways were built across the Amazon, including the Trans-Amazon from Recife to the Peruvian border and the Polonoreste from Brasilia to Belém at the mouth of the Amazon (**Figure 7.15b**).

Government policy was specifically designed to colonize the Amazon. Political leaders saw it as a frontier region similar to that of the western United States in the 19th century. Landless peasants were given title to plots of land if they promised to develop them productively, and peasants migrated in thousands along the new roads. As geographer Susannah Hecht has shown, much of the land was actually acquired by large landholders who took advantage of favorable incentives and tax breaks to develop ranches for speculation and tax havens. Hecht's fieldwork also found that when both small holders and large ranches cleared the land of forest, often by burning, soil fertility declined rapidly, leading to further deforestation as farms and pastures were abandoned.

In satellite images of the region, the process of Amazonian deforestation—the networks of new roads and associated forest clearance—can be clearly seen (**Figure 7.15c**). Satellites also show the thousands of fires that are set each year to clear land. These fires produce a dense layer of smoke that closes airports and chokes local residents. But the photos also show that the pattern of development and deforestation varies spatially, with some remote areas still relatively untouched and others along roads and around cities almost totally transformed to agriculture. At the southern margins of the Amazon basin, global and national markets for soybeans are driving deforestation and land use changes over large areas.

Estimates of the rate of Amazon basin deforestation do not always agree because of differences in the way in which forests are defined and satellite images are analyzed and because clouds and smoke prevent accurate assessments in some regions. But the general consensus is that perhaps 15% of the Amazon forest has been cleared and that the current rate in Brazil varies between 10,000 and 30,000 square kilometers (4,000 to 11,000 square miles) a year. The fate of the Amazon has attracted global attention, led by scientists and environmental organizations concerned about the impacts of such large-scale forest loss on biodiversity and climate. The Brazilian government has responded by removing some of the tax breaks for development, by intensifying monitoring and control of deforestation, and by establishing parks and reserves. One of the best-known reserves is named for Chico Mendes, a rubber tapper, who organized resistance to deforestation by large ranchers and was murdered in 1988. He pushed for the establishment of areas that were protected for appropriate extractive uses called *extractive reserves.* Other parks and reserves, such as Manu in Peru and Cuyabeno in Ecuador, are becoming tourist destinations, where international tourists stay in jungle lodges and are able to observe the rich bird and animal life of the forest.

The coastal zones of central and southern Chile are also important producers of timber for world markets. The forest industry has started

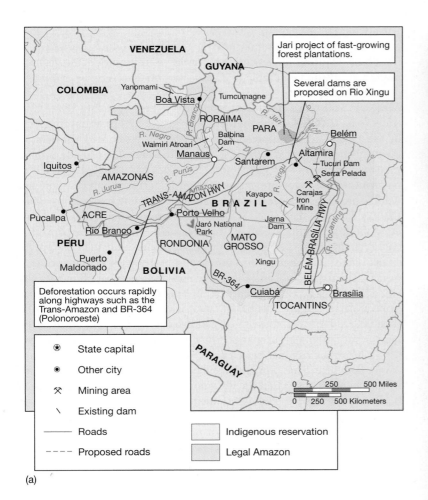

(b)

(c)

▲ **FIGURE 7.15 The Amazon basin** (a) Map of development in the Amazon. This map shows some development projects that are causing deforestation and other environmental problems in the Amazon basin, including major roads such as the Trans-Amazon highway, dams such as Tucuri, and agricultural development around cities such as Acre. The basin also has many indigenous groups, some of which have gained reserve status for their lands. (b) Wide highways cut into the Amazon forest. (c) Satellite image of deforestation. This satellite image from the southern region of the Brazilian Amazon in the state of Rondonia shows how the forest is cleared as roads and people move into the Amazon area. The cleared areas are shown in lighter brown and the forest in green. A distinctive grid pattern is clear, as people farm rectangular plots along roads.

to encroach on the groves of towering old-growth alerce trees that are similar to the redwoods of the western United States, because both species rely on the heavy seasonal rains from westerly storms. Conservationists have been able to obtain some protection through environmental legislation and purchase of remaining forest land. One of the more controversial conservation efforts is that of U.S. millionaire Douglas Tompkins, who purchased a swath of 700,000 acres of forest land that cuts across almost the entire width of Chile, upsetting Chileans, who saw this as a threat to their national security. Tompkins established a national park there, called Pumalin Park, which the government of Chile declared a nature sanctuary on August 19, 2005.

**The Green Revolution** An alternative to deforestation and a solution to food insecurity is to increase yields on existing land rather than clear new fields. For the first part of the 20th century, the yields of most agricultural crops in Latin America were very low (less than 1 ton per hectare), and farmers with small plots of land could not produce enough to feed themselves, let alone sell in the market. As population and urban-consumption demands increased, countries such as Mexico and Brazil had to import basic food crops such as wheat and corn. A solution to low productivity and poverty in rural areas was the **Green Revolution**—the process of agricultural modernization that used a technological package of higher-yielding seeds, especially wheat, rice, and corn, that in combination with irrigation, fertilizers, pesticides, and farm machinery increased crop yields in several world regions from about 1950 to 1980. Mexico was a global center for Green Revolution technology, hosting the International Center for Improvement of Maize and Wheat (CIMMYT) near Mexico City. Other Latin American countries such as Argentina and Brazil also promoted Green Revolution agricultural modernization, including key crops such as rice and soybeans.

Although the Green Revolution increased crop production in many parts of Latin America, it was not an unqualified success because of its role in increasing inequality and in environmental degradation. The Green Revolution has been criticized because it increased dependence on imports of chemicals and machines from foreign companies and thus contributed to the debt problem. The benefits tended to accrue to wealthy farmers, who could afford the new inputs and to irrigated regions, while poorer farmers on land watered only by rainfall fell behind or sold their land. In some cases, such as in southeastern Brazil, machines replaced workers, thus leading to unemployment, and Green Revolution technology and training also tended to exclude women, who play important roles in food production.

The new agricultural chemicals, especially pesticides, contributed to ecosystem pollution and worker poisonings, and the more intensive use of irrigation created problems of water scarcity and of salinization (buildup of salt deposits in soil because of evaporation of water that includes salts drawn up from the subsoil). The increased use of imported pesticides on export crops is associated with damage to ecosystems and workers' health. Using imported pesticides on crops in developing countries, then exporting the contaminated crops to the regions where the pesticides were manufactured, is termed the **circle of poison**. The most serious criticism of the Green Revolution was that it contributed to the worldwide loss of genetic diversity by replacing a wide range of local crops and varieties with a narrow range of high-yielding varieties of a few crops. Planting single varieties over large areas (monocultures) also made agriculture vulnerable to disease and pests.

**Market Solutions for Conservation** Led by Costa Rica, Latin America has seen a boom in tourism, often geared to the natural attractions of the coasts and rainforests. Environmentally-oriented tourism, or **ecotourism**, is designed to protect the environment and provide employment opportunities for local people. Costa Rica has won high praise from environmentalists for protecting 30% of its territory in biosphere and wildlife preserves (**Figure 7.16a**). Costa Rica has more bird species (850+) than are found in the United States and Canada combined and more varieties of butterflies than in all of Africa. It has 12 distinct ecosystems that contain more than 6,000 kinds of flowering plants, more than 200 species of mammals, 200 species of reptiles, and more than 35,000 species of insects. The payoff for Costa Rica is the escalating number of tourists who come to visit its active volcanoes, palm-lined beaches, cloud forests, and tropical parks (**Figure 7.16b**). In 1995, when Costa Rica received more than 800,000 tourists, tourism exceeded banana exports as the country's main source of foreign exchange. Since 1995, the number of tourists visiting Costa Rica has increased by 10% each year (reaching 2 million in 2008). Ecotourism has brought mixed benefits to rural areas of Costa Rica. The benefits are not shared equally among residents, and some regions are becoming so crowded that environmental degradation is occurring.

The ecological diversity of Latin America also supports biological prospecting, or **bioprospecting**, for new medicines and products with commercial uses. For example, Costa Rica has signed agreements with multinational pharmaceutical companies that give the companies rights

(a)

(b)

▲ **FIGURE 7.16 Conservation in Costa Rica** (a) Costa Rica protects a high percentage of its land in parks. (b) Ecotourism has become a source of income and employment, attracting thousands of tourists each year to forest and coastal ecosystems. This photo shows tourists on walkways high in the rainforest in the Arenal National Park, Costa Rica.

to prospect and develop in return for turning over a share of profits to the national government and to local people.

Other new approaches to conservation include Payments for Environmental Services (PES), a program that pays local residents to protect the environment because of its value to others. For example, in Costa Rica, upstream communities are paid to protect their forests because of the role they play in sustaining the flow of rivers for downstream water users such as plantations and hydroelectric schemes. Another market has been created as a result of the Kyoto agreement to reduce greenhouse gas emissions: Many Latin American communities are now receiving payments under Kyoto's Clean Development Mechanism (CDM) for energy efficiency, renewable energy, and sequestering carbon dioxide through reforestation schemes, thus capturing and reducing greenhouse gases to create carbon credits for northern countries to meet their commitments.

# HISTORY, ECONOMY, AND DEMOGRAPHIC CHANGE

Much of Latin America can be characterized by a common experience of colonialism that included the dominance of the Spanish and Portuguese languages, religion (Roman Catholicism), legal and political institutions, demographic collapse of indigenous populations, European migration and settlement, and European control of resource extraction, trade links, and other economic activity. Most of the region became independent in the 19th century and was drawn into global trade relations, especially with Britain and the United States. In the 20th century, the Latin American region experienced rapid integration into global markets and the transition from revolutionary and military governments to democratic ones. Contemporary Latin America has urbanized rapidly with several megacities and industrial regions acting as international centers of commerce.

## Historical Legacies and Landscapes

**Early Latin America** Archaeological evidence from sites in Chile, Mexico, Peru, and Ecuador suggests a human presence in Latin America from at least 10,000 years ago following migrations from Siberia into North America across the Bering Straight. The descendants of these migrants constitute the indigenous people of Latin America, many of whom lived by hunting, fishing, and gathering but also eventually domesticated some of the world's most important agricultural crops such as corn (see p. 242). For example, most of the Amazon region was occupied by up to 2 million seminomadic people in several thousand groups and tribes, often living along rivers, and speaking different but related languages. An earlier section of this chapter also described the environmental impacts of the complex societies that emerged in the Andes (the Inca from about 1200 C.E.), Mexico and Central America (Maya from about 2000 B.C.E. and Aztec from about 1200 C.E.). The Maya constructed monumental cities and had a written language, mathematics, and astronomical observatories. People of the Caribbean islands included the Carib and Taino/Arawak, who traded with cities on the Meso-American mainland growing corn and cassava. The Aztec and Inca empires were expansionist, often coming into conflict with neighboring groups such as the Mapuche (who

resisted Inca domination) and the Tlaxcalans (who eventually allied with the Spanish to conquer the Aztecs).

**The Colonial Experience in Latin America** The integration of Latin America into a global system of political, economic, ecological, and social relationships began more than 500 years ago with the arrival of Spanish and Portuguese explorers at the end of the 15th century and the onset of *colonialism* (see Chapter 1) that inexorably linked a Latin American periphery to a European core. The 15th and 16th centuries were a period of innovation in Europe, with changes in manufacturing technology and the development of an economic policy in which government controlled industry and trade (a system known as *mercantilism*; see Chapter 1, p. 16). Improvements in shipbuilding and navigation allowed Europe to explore—and then expand trade with—other regions of the world, including Asia to the east and Africa to the south. Europeans also sailed west in search of new routes to Asia.

The most famous of these European explorers was Christopher Columbus (known in Latin America as Cristóbal Colón), an Italian from Genoa, who lived in Lisbon, Portugal. Under the sponsorship of Queen Isabella of Spain, Columbus was commissioned to search for new territory and trading opportunities on a western route to the Indies (as Asia was then known). Having set sail from southern Spain on August 3, 1492, with three small sailing ships—the *Santa María*, the *Pinta*, and the *Niña*—Columbus, commanding the *Santa María*, arrived in the Caribbean in October and landed on Watling Island, a small island in the Bahamas, to which he gave the name San Salvador (**Figure 7.17**). On his first voyage, Columbus also visited Cuba and another island that he called Hispaniola (now Haiti and the Dominican Republic). When the *Santa María* was wrecked on the north coast of Hispaniola, Columbus left behind 21 volunteers to found a colony and returned to Spain on the *Niña*, bringing with him six local people, several parrots, and some gold ornaments.

Columbus's second voyage in 1493 was much larger, with 17 ships and 1,500 men because he intended to establish permanent settlements, but he was frustrated by divisions within his team and by hostility from local residents. On his third and fourth voyages, he explored the island of Trinidad and coasts of Venezuela and Central America. With the promise of new lands in the Western Hemisphere, the Spanish wanted a ruling that would assign the new lands to Spain rather than to its competitor, Portugal. The resulting **Treaty of Tordesillas** was an agreement made by Pope Alexander VI in 1494 to divide the world between Spain and Portugal along a north–south line 370 leagues (about 1800 kilometers, or 1100 miles) west of the Cape Verde Islands. Portugal received the area east of the line, including much of Brazil and parts of Africa, and Spain received the area to the west.

Columbus was followed in subsequent decades by others seeking gold, territory, and other resources in Latin America. The most notable explorers, or *conquistadors,* included Hernán Cortés, who landed in Veracruz, Mexico, in April 1519 and went on to conquer the Aztec empire and its capital of Tenochtitlán in the basin of Mexico; and Francisco Pizarro, who seized control of the Incan empire centered in Cuzco, Peru, in 1533. The Portuguese began their colonization of Brazil with the landing of Pedro Alvares Cabral in April 1500 at Porto Seguro in southeast Brazil.

The Spanish expanded and administered the new Latin American colonies through the **viceroyalty** system, the largest scale of Spanish colonial administration. In this system, Mexico City and Lima emerged as the headquarters of the viceroyalties of New Spain (Mexico and

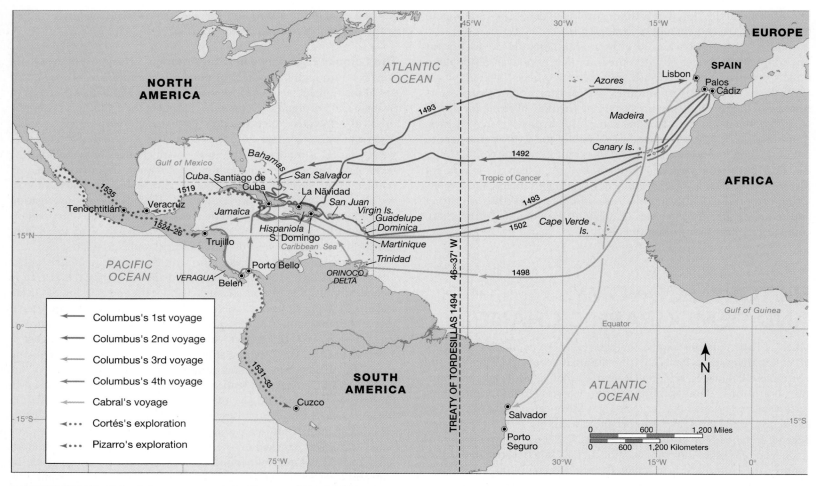

▲ **FIGURE 7.17 Colonial voyages and the Treaty of Tordesillas** The map shows the major voyages and missions of Columbus, Pizarro, Cabral, and Cortés and the division of Latin America between Spain and Portugal under the Treaty of Tordesillas in 1494. The initial line set by the pope in 1493 was contested by Portugal and shifted farther west.

Central America) and Peru (Andean and southern South America), respectively. It is important to recognize that the colonial effort in Latin America was a process that took place over at least two centuries, with some places incorporated earlier than others and some regions never really coming under complete colonial control because of their remoteness (the Amazon) and local resistance (parts of the Andes). The Spanish charged the new administrators with obtaining gold and silver for the Spanish crown, converting the indigenous people to the Catholic religion, and making the colonies as self-sufficient as possible through the use of local land and labor. The Spanish crown demanded 20% of all mine profits, the so-called *Quinto Real,* or royal fifth.

**The Demographic Collapse and the Columbian Exchange** The search for local labor to work in the mines and fields of the Spanish colonizers was frustrated by one of the most immediate and significant impacts of the European arrival in Latin America, the **demographic collapse** after about 1500, with the rapid die-off of the indigenous populations of the Americas as a result of diseases introduced by the Europeans to which residents of the Americas had no immunity. Because of the long isolation of the Americas from other continents, indigenous people lacked resistance and immunity to European dis-

eases such as smallpox, influenza, and measles. When they caught these diseases from Europeans and then from each other, mortality rates were very high. Researchers have estimated that up to 75% of the population of Latin America died in epidemics in the century or so after contact. This massive mortality demoralized local people, led to the abandonment of their settlements and fields, and meant that there was a scarcity of labor to work in the mines, missions, and agricultural activities with which the Spanish, for example, hoped to support their colonial enterprise. The introduction of European diseases into the Americas is just one example of the interaction between the ecologies of the two continents that historian Alfred Crosby has called the **Columbian Exchange**—the interchange of crops, animals, people, and diseases between the Old World of Europe and Africa and the New World of the Americas beginning with the voyages of Christopher Columbus in 1492 (**Figure 7.18**).

When the Spanish and other colonial powers arrived in new lands, they brought with them favorite plants and animals that they planned to introduce into the new colonies, but also, unintentionally, diseases, weeds, and pests such as rats that were stowaways on their ships. In return, the explorers and colonists collected species that they hoped could be sold or traded back to Europe and elsewhere.

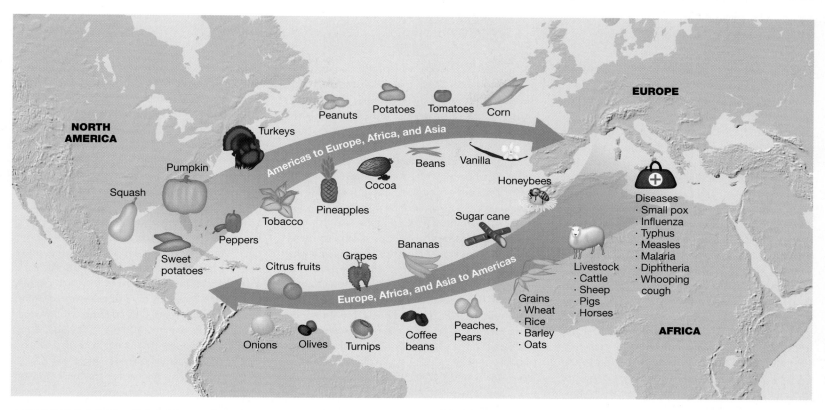

▲ **FIGURE 7.18 The Columbian exchange** The arrival of Europeans in the Americas initiated the extensive exchange of crops, animals, and diseases between the continents, including the introduction of European diseases to the Americas, thus killing millions in a demographic collapse of indigenous populations. The introduction of wheat, sugar, and livestock transformed the landscapes of Latin America, and European and African diets (and health) were altered forever with the arrival of sugar, corn, potatoes, and tobacco.

In Latin America, the Spanish introduced the crops and domesticated animals of their homeland—especially wheat and cattle, but also fruit and olive trees, horses, sheep, and pigs. Sugar, rice, citrus, coffee, cotton, and bananas, which had originally been brought to Spain from North Africa and the Middle East after the Moorish invasions of the 6th century, were also transported to the Americas. The Spanish colonizers took back to Europe corn, potatoes, tomatoes, tobacco, and possibly the human disease syphilis.

Over longer periods these exchanges had other effects. The clearing of land for European crops such as wheat and sugar and the overgrazing by cattle and sheep contributed to soil erosion and deforestation in Latin America and the Caribbean. Newly introduced rats, pigs, and cats ate the food that traditionally supported local species or consumed the local fauna, especially ground-dwelling birds. One of many species of Andean potatoes carried back to Europe became the foundation of the Irish diet, and one cause of a famine and migration to the Americas when disease destroyed potato harvests in Ireland in the mid-19th century. Corn and manioc were introduced into Africa and became new staples, whereas peanuts and cacao were the basis of new African export economies. Cotton was introduced from Latin America to India and grown for British textile mills; pineapple was distributed to the Pacific, including Hawaii; and tobacco became an addictive habit, eventually throughout the world.

## Colonial Landholding, Labor Relations, and Exports
To wrest profits and products from their new lands, the Spanish introduced several new forms of land tenure and labor relations into Latin America

that still influence contemporary landscapes. Where the colonizers wished to directly control the land, they granted land rights over large areas to Spanish colonists, often military leaders, and to the Catholic Church, ignoring traditional local uses and establishing fixed property boundaries. These *latifundia* (large rural landholdings or agricultural estates) typically occupied the best land, forcing other farmers onto small plots of land, or *minifundia*. **Haciendas** were established to grow crops such as olives and wheat, mainly for local consumption in mines, missions, and cities, rather than for export.

**Plantations** were large agricultural estates (usually tropical or semitropical and commercial- or export-oriented). Most were established in the colonial period, growing single crops (monocultures) such as sugar or tobacco for export, mainly in the wetter coastal areas. Labor for the haciendas, plantations, and mines was obtained initially through the institution of **encomienda**, a system by which groups of indigenous people were "entrusted" to Spanish colonists, who could demand tribute in the form of labor, crops, or goods. The colonists in turn were responsible for the indigenous groups' conversion to the Catholic faith and for teaching them Spanish. These forms of labor control did not produce a large enough workforce, especially where the Europeans wanted to establish export plantations with high labor requirements, and in the tropics where the demographic collapse had devastated local populations that may have been small to begin with. In this case, colonial trading routes were used to import slaves, mainly from Africa, to the Caribbean, Central America, and Brazilian plantations to work in the

# SUGAR

No other food has had the historical and geographical impact of sugar, a sweet substance prevalent in most contemporary diets, yet for much of human history unknown outside some South Pacific islands and Asia. Millions of people now consume sugar in almost addictive quantities—in soft and alcoholic drinks and in candy and as an additive to most processed foods. Too much sugar can cause health problems such as diabetes and tooth decay, and sugar is especially appealing to children because it provides a quick boost of energy when it is absorbed into the bloodstream. About 60% of sugar is produced from a tropical grass known as *sugarcane*, and the remainder comes from sugar beets, which are grown in many of the world's temperate regions.

As demand for sugar, known as "white gold" because of its high value, grew in the 17th century, millions of Africans were transported to the Caribbean and Latin America and forced to work as slaves on sugarcane plantations. Sugarcane production is labor intensive, especially the arduous process of burning the cane fields and then cutting the cane with machetes. The cane is then ground into pulp and boiled at the sugar mill to make sugar. The initial expansion of sugar production to the New World occurred in the 1500s, when the Portuguese established plantations along the coast of northeast Brazil. Today the descendants of the African slaves who worked on those plantations make up a large segment of Brazil's large Afro-Brazilian population and dynamic culture. Geographer Jock Galloway described how, by 1800, the Caribbean, under British and Dutch colonialism, had become the world's most important sugar production region, providing 80% of Europe's supply. It was also the primary destination of slave arrivals from Africa (**Figure 1a**).

Sugar provided enormous profits to those capitalists who controlled its production and who often gained considerable political power in the core countries. George Washington and Thomas Jefferson owned sugar plantations on the British Caribbean island of Barbados, for example. Anthropologist Sidney Mintz, in his book *Sweetness and Power*, argued that there is a clear but complex link between slavery, sugar plantations in the New World, and the rise of industrial capitalism in Europe. It is sugar, and the greed that sought slave labor to produce it, that is mostly responsible for the suffering of millions of slaves in the Americas and for the high proportion of people of African heritage in countries such as Brazil, Cuba, Jamaica, and Haiti. After slavery ended, sugar prompted another diaspora of indentured contract workers from Asia, who were lured to Caribbean islands such as Trinidad and Jamaica. Sugar, in the form of rum, was the drug that propelled the British and French navies in their battles and colonizing expeditions into Asia and the Pacific. Sugar producers in Britain encouraged the consumption of sweetened tea and thus drove the development of tea plantations in Africa, Australia, and Southeast Asia, as well as on Pacific islands such as Hawaii and Fiji, and caused the shift from sugar as a luxury item to a virtual necessity in the diet of working-class Britain.

Slavery persisted longest in Cuba under Spanish control, and the sugar barons started to shift their operations there in the 19th century. The Cuban cane was of high quality, and before long the large island was covered with fields of sugarcane. By the mid-20th century, sugar dominated the economy, and many fields had foreign, and especially U.S., owners. After the Cuban Revolution in 1959, the sugar industry was nationalized by Fidel Castro's government, which found ready markets in the Soviet Union.

Brazil is now the world's major sugarcane producer, followed by India, China, and Thailand. Sugar consumption is very high within Latin America, especially in Brazil and Cuba. Sugar has had profound economic and social consequences

production of sugar. Slave imports from Africa to Latin America and the Caribbean eventually totaled more than 5 million people, including 3.5 million to Brazil and 750,000 to Cuba (see Geographies of Indulgence, Desire, and Addiction: Sugar, above). In many parts of Latin America, the colonial legacy of land grabs, labor exploitation, and racism still frames contemporary attitudes toward indigenous people, who are still trying to regain land and dignity lost during the colonial period.

The most important export commodities in Spanish colonial America were silver, produced mainly from mines in Mexico and Bolivia; sugar, grown on plantations in Cuba and southern Mexico; tobacco from Cuba; gold from Colombia; cacao (for chocolate) from Venezuela and Guatemala; and indigo, a deep blue dye, from Central America. In the first phase of the developed colonial economy (1540–1620), Spain derived enormous wealth from the bonanza of the silver mines at Potosí (now within Bolivia) that produced half of the world's silver in the 16th century. This rapid influx of money led to inflation in Europe, and Spanish industry suffered as upper classes in both Spain and the colonies chose to purchase luxuries from other parts of Europe.

By 1620, Spain was embroiled in expensive wars with England, France, and Germany, partly over control of trade with the Americas. The demographic collapse and exhaustion of surface silver deposits was resulting in lower revenues from the colonies. Merchants and landowners in the colonies were also starting to resent the strict control of trade and taxation by Spain and were using positions of power and smuggling to keep revenue for themselves, thereby contributing to the overall weakening of Spanish power and economy. This local resentment built over the next century, eventually producing calls for independence.

**Independence** Independence movements surged in 1808 when France, led by Napoleon conquered Spain and threatened to tighten trade controls. Revolutionaries, drawn from both Spanish-American and indigenous leaders, set out to liberate Latin America from Spain, partly inspired by the French and American revolutions. Between September 16, 1810, when priest and peasant leader Miguel Hidalgo called for Mexican independence in the famous *Grito* ("cry"), and 1824, when Simon Bolívar finally led northern South America to independence, a series of regional revolts led to the formation of independent republics in Mexico, Argentina, Peru, Colombia, Chile, and Brazil. In 1848 Mexico was defeated in its war with the United States and was forced to cede large portions of its territory to the State of Texas and to what would become the states of Arizona, California, Colorado, Nevada, and New Mexico, leaving in return an enduring Hispanic cultural legacy in these regions.

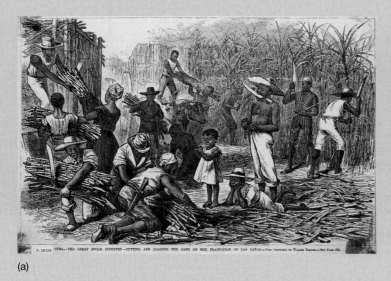

(a)

(b)

▲ **FIGURE 1 Sugar production** (a) This print shows workers on a sugarcane plantation in the 19th century in Cuba. Labor needs for the sugar plantations drove the importation of millions of Africans as slaves to Brazil and the Caribbean until the mid-19th century. The African diaspora to Latin America and the Caribbean is evident in the racial composition of contemporary populations of countries such as Jamaica, Cuba, Haiti, and Brazil. (b) A contemporary photo of sugarcane cutters from Marie Galante in the French West Indies shows that conditions remain very difficult.

on those parts of the world in which it became a major crop. It has created highly divided societies, with a wide gap between owners and workers, and has produced multiethnic societies with internal disparities and conflicts (**Figure 1b**).

Where sugar remains a main export crop, it preserves a legacy of dependency and vulnerability to world markets, exacerbated by surpluses and new diet trends that have reduced the demand for sugar and depressed the price of sugar

relative to other crops. However, the search for alternatives to fossil fuels such as oil is stimulating a new interest in fuels made from biomass, including ethanol from sugarcane, especially in Brazil. ■

In the Caribbean, after several rebellions against France (inspired by the French Revolution), former slaves declared Haitian independence in 1804 and occupied the rest of the island of Hispaniola until the Dominican Republic gained independence in 1844.

When slavery was abolished by the British in 1807, many emancipated slaves eventually became small farmers. But a decline in sugar prices produced poverty and unemployment. These provoked rebellions in Jamaica beginning in 1865 that eventually resulted in a more representative government—but one still controlled by Britain as one of its crown colonies. Independence came late to most of the Caribbean, especially to countries like Trinidad and Tobago (in 1962), which were under British control, and the Caribbean retains distinctive links to Britain, France, and the Netherlands. Other British island colonies such as Antigua and Barbados, St. Lucia, and St. Kitts and Nevis did not become fully independent until the beginning of the 1980s.

Cuba remained under Spanish control until after the Spanish-American War in 1898, when limited independence was granted under U.S. influence and frequent intervention. Puerto Rico also shifted from Spanish control to a U.S. territory after 1899 and is still currently part of the United States, with some residents desiring independence and others full status as a U.S. state. Six islands remain as Dutch protectorates with full autonomy for internal affairs (Aruba, Bonaire, Curaçao,

St. Martin, Saba, and St. Eustatious), and Martinique and Guadeloupe are overseas departments of France. The British Virgin Islands, Turks and Caicos, Montserrat, Anguilla, and the Cayman Islands are still colonies of Britain, and the U.S. Virgin Islands are part of the United States.

**The Export Boom** In the first half of the 19th century, the loss of colonial trade routes and protections against competition, civil wars led by regional strongmen, foreign reluctance to invest capital in the new and unstable republics, and a brain drain of skilled Spaniards back to Europe all combined to produce an economic decline in Latin America. But around 1850, as the political situation stabilized and industrialization in Europe and North America created investment profits and new demands and consumers. Capital became available for the Latin American economies in which liberal political thought supported free trade and foreign investment.

Foreign capital helped develop export economies for nitrate (used to make fertilizer) and copper in Chile; livestock in Argentina; coffee in Brazil, Colombia, and Central America; bananas in Central America and Ecuador; tin in Bolivia; and silver and henequen (a fiber used in making sacks and matting) in Mexico. Foreign-owned companies, which were mostly British in the 19th century, ran many of the new export activities and developed railroads and banks. The basic mineral and agricultural exports did not command high prices in relation to the cost of manufactured imports (what is called poor *terms of trade*), and exports were very vulnerable to changes in world prices. Countries that relied on these exports developed economies highly dependent on, and closely linked to, a volatile world market—a condition still evident today in the vulnerabilities of Chile (49% of total export value in 2008 from copper), Panama and Ecuador (10%, bananas in 2006), and Venezuela (90%, crude petroleum).

## Economy, Accumulation, and the Production of Inequality

The 1929 stock market crash and ensuing world depression demonstrated the extent to which Latin America had become integrated into the global economy: Throughout the region there were declines in exports, restrictions on investment, and a general economic crisis. This, together with a general awareness that foreign ownership and poor terms of trade for unprocessed exports made Latin American economies vulnerable to world conditions, led to the development of the new economic strategy of *import substitution*, a process by which domestic producers provide goods or services that were formerly bought from foreign producers.

Known as Import Substitution Industrialization (ISI) in Latin America, import substitution derived from critical views of global integration associated with the **dependency school of development theory** that argues that the developing world should reduce their links to the core developed economies.

Mexico, Brazil, and Argentina moved aggressively to implement import substitution policies from the 1930s to the 1960s, including protection of domestic industries through tariffs and import quotas. Government nationalization (the process of converting key industries from private to governmental organization and control) and investment in new manufacturing industries fostered production of chemicals, steel, automobiles, and electrical goods. Import substitution policies temporarily slowed Latin America's integration into global markets and stimulated the growth of domestic industry and workforce in regions such as northeastern Mexico (steel), Mexico's Gulf Coast (petrochemicals, **Figure 7.19**), and São Paulo, Brazil (automobiles). Growing criticisms of import substitution highlighted an oversized government bureaucracy and the high costs of subsidizing industries that were inefficient and produced goods of poor quality because of a lack of competition and government protectionism.

**The Debt Crisis**    A new infusion of capital into the world economy, associated with the increased oil profits that followed the formation of the Organization of Petroleum Exporting Countries (OPEC), brought banks to Latin America seeking to invest in what they viewed as stable and rapidly growing economies. Mexico, with the promise of its own oil bonanza and industrial expansion, as well as the more industrialized countries of Brazil and Chile, were offered the largest loans, but almost all Latin American governments took advantage of the initially low-interest loans to support development and other projects. When interest rates rose and debt payments soared in the early 1980s, Latin American governments were unwilling to cut back on popular subsidies and programs and instead borrowed more money, ran budget deficits, and overvalued their currencies. The resulting runaway inflation and debt reached unprecedented levels. By 1989 Brazil owed $111 billion; Mexico, $104 billion; and Venezuela, $33 billion—with annual payments reaching more than half of the annual gross national

▲ **FIGURE 7.19 Industrialization for import substitution in Latin America**  Between 1930 and 1970, many Latin American countries, especially Argentina, Brazil, and Mexico, implemented import substitution policies to develop national manufacturing capacity. Governments invested heavily in steel, automobile, and chemical plants and protected them against competition from imports through tariffs. This petrochemical plant in Mexico processes petroleum from the oil fields in the Gulf of Mexico.

product (GNP). The 1980s have been called Latin America's "lost decade" because of the slowdown in growth and deterioration in living standards that occurred during that decade.

The resulting decline in purchasing power and living standards, and the likelihood that suspension of debt payments and default would destabilize the international financial system, prompted international financial institutions and the U.S. government to seek a solution to Latin America's debt crisis. The United States extended the repayment period for debts and lent more money, while the International Monetary Fund moved to restructure loans on condition that governments initiate stabilization and structural adjustment policies. Mexico got a $48 billion bailout, paid mainly to its banking sector.

**Structural Adjustment and Neoliberalism**    Stabilization policies set out to curb inflation by cutting public spending on government jobs and services, increasing interest rates, controlling wages, and devaluing currencies to increase exports. **Structural adjustment policies** required the removal of subsidies and trade barriers, the privatization of government-owned enterprises such as telephone and oil companies, reductions in the power of unions to demand higher wages, and an overall focus on export expansion. These policies, while reducing inflation and debt in many countries, had very negative effects on some people and sectors. Increased food prices and reduced health and education services because of the withdrawal of subsidies, as well as rising unemployment as government jobs were cut, hit the poor particularly hard with increases in malnutrition and destitution. As a result of structural adjustment in Peru in 1990, gasoline prices went from 10 cents to $2 per gallon.

Free-trade policies were introduced in many countries as political power shifted to those with a belief in *neoliberalism,* echoing the views of the 19th-century liberals who believed in free trade and reduced government. Neoliberalism promoted free trade and a reduction in the role and budget of government, including reduced subsidies and the privatization of formerly publicly owned and operated concerns such as utilities. Neoliberal governments were open to the possibility of expanding free trade through regional agreements that would take down barriers among trading partners.

The most dramatic step was taken by Mexico, which in 1994 joined the **North American Free Trade Agreement** (NAFTA) with the United States and Canada. NAFTA set out to reduce barriers to trade among the three countries, through, for example, reducing customs tariffs, quotas, and other trade protections. Other initiatives include MERCOSUR (Spanish acronym for "southern common market"), initiated in 1991 and linking Chile, Argentina, Brazil, Paraguay, and Uruguay in a trade agreement, and CARICOM, formed in 1973 to create a trade zone in the Caribbean. The Andean Pact originally linked Peru, Ecuador, Colombia, Venezuela, Bolivia, and Chile in a 1969 agreement, but Chile and Venezuela have withdrawn, leaving four countries in what is now called the Andean Community of Nations.

**NAFTA** Advocates of NAFTA argued that free trade would create thousands of jobs in Mexico with higher wages and that these opportunities would reduce migration to the United States. Mexican agriculture would shift to growing high-value fruit and vegetables, where it had a comparative advantage during the winter, and Mexico would be able to reduce food prices by importing low-cost grain from the United States and Canada. Free trade was also linked to financial stabilization and to promises of more democratic government in Mexico.

The coalition to oppose NAFTA brought together environmentalists and labor activists in all three countries working within trinational coalitions that successfully petitioned for two side-agreements to be signed as part of NAFTA. The environmental side-agreement established the Commission for Environmental Cooperation (CEC) and led to the creation of two related institutions, the Border Environment Cooperation Commission (BECC) and the North American Development Bank (NADB). The CEC would monitor environmental impacts and enforce regulations, the BECC would certify as sustainable new water and sewage projects along the U.S.–Mexico border, and the NADB would fund environmental infrastructure improvements certified by the BECC. The labor side-agreement included commitments regarding minimum wages, child labor, and rights to unionize, and was matched by a U.S. program to compensate U.S. workers who could prove they had lost their jobs as a result of NAFTA.

Studies suggest that NAFTA has led to the creation of thousands of new jobs in Mexico and to increased wages in some industries. However, many of the hoped-for benefits of NAFTA were frustrated by the economic crises that followed currency devaluation in 1994 and by continuing inequality in both urban and rural areas, which means the benefits have not reached the poor. The problems of pollution and waste associated with urban and industrial development along the border between the United States and Mexico have led a variety of nongovernmental organizations (NGOs) and community groups to demand improved environmental protection. In Nogales, Mexico, women in informal *colonia* settlements have organized to demand safe drinking water and the cleanup of wastes from factories, and they have also created a recycling and tree planting program. The new environmental agreements have created an important space for Mexicans to protest lack of environmental enforcement and to seek funding for water and sanitation projects, but this has not yet resulted in an overall improvement in environmental

conditions in Mexico. Criticisms of NAFTA have fueled subsequent protests against neoliberalism and globalization in Mexico (**Figure 7.20**).

**Contemporary Economic Conditions** The 1994 Mexican peso devaluation followed a period of high spending, overvaluation, and the Chiapas rebellion and triggered what was called the Tequila Effect, where investors withdrew funds across much of Latin America. In the late 1990s several Latin American countries were economically unstable, including Argentina, where a recession from 1999 to 2002 included freezing of bank deposits, default on international debt payments, and increased poverty and unemployment. But many economies have stabilized and grew considerably from 2000 to 2008, including Brazil with a 3.7% annual growth in GDP from a diverse economy that included automobiles, aircraft, steel, oil, minerals, petrochemicals, biofuels, soybeans, and consumer durables such as TVs and appliances.

The World Bank estimated the overall GNI (gross national income) of the Latin America and Caribbean region in 2007 at $3.2 trillion, about 5% of the world's total, higher than that of Africa, South and Southeast Asia, the Middle East, East Asia, and the Pacific. Per capita GNI at $5,801 was also higher than that of any other low- or middle-income region (all world regions except Europe, Australia, New Zealand, Japan, the United States, and Canada). Brazil is ranked the world's ninth-largest economy, and Mexico is ranked eleventh.

Total exports in 2007 were valued at about $900 billion and were almost balanced by imports (each about 5% of the global total). The

▼ **FIGURE 7.20 NAFTA and neoliberalism** A protester in Monterrey, Mexico, opposes U.S. imperialism in front of a banner that says, "No to neoliberalism."

Latin American region also received more foreign investment in private capital than any other low- or middle-income region but also had the largest total debt—more than $825 billion in 2007. The strongest export economy was that of Mexico ($265 billion in 2007).

The agricultural economy of many parts of Latin America has changed considerably in recent years. Many governments have shifted from giving top priority to self-sufficiency in basic grains to encouraging crops that are apparently more competitive in international trade, such as fruit, vegetables, and flowers. These **nontraditional agricultural exports** (NTAEs) have become increasingly significant in areas of Mexico, Central America, Colombia, and Chile, replacing grain production and traditional exports such as coffee and cotton (**Figure 7.21a**). Rather than grow these and other crops on large company landholdings, the current strategy is *contract farming* for multinational corporations such as Del Monte. Farmers sign contracts with companies to produce crops to certain production and quality standards in return for a guaranteed price. Central Chile is one of the most productive agricultural export zones in Latin America and is increasingly compared to the U.S. State of California, with which it shares a moderate Mediterranean climate of warm, wet winters, and moderate summer temperatures (**Figure 7.21b**). Flowers from Colombia and Ecuador are exported worldwide. Some farmers, especially of coffee and bananas, have turned to organic production and Fair Trade labels as a way to save on inputs and access new consumer markets in North America and Europe (**Figure 7.21c**).

Fisheries are another critical component of Latin American and Caribbean food and export systems, and activities range from subsistence fisheries in small coastal villages to large-scale commercial exploitation of offshore fisheries. The overall catch in the region was more than 10 million metric tons in 1994, contributing on average about 10% of overall food supply and making a significant contribution to exports in Chile, Ecuador, and Costa Rica. But offshore catch has halved in Chile as a result of overfishing and climate variability. Aquaculture and mariculture (the cultivation of fish and shellfish under controlled conditions) in coastal lagoons has become an important export sector in countries such as Chile (salmon), Ecuador, and Honduras (see Figure 7.12c). The manufacturing and service sectors are strong in many countries with both heavy manufacturing from the import substitution period and newer industries focused on exports such as clothing and computers. Southeast Brazil and central and northern Mexico have important industrial zones, with iron and steel, chemicals and aerospace around cities such as Sao Paulo in Brazil, iron and steel in the northern Mexico city of Monterrey, light industrial production along the border with the United States, and car manufacturing and food processing plants near Mexico City.

Low wages in Central America have attracted labor-intensive manufacturing, such as garment industries, to urban areas in El Salvador and Honduras. Costa Rica, with a better-educated workforce and more stable economy, has lured high-technology companies such as Microsoft, General Electric, and Intel to build factories near San Jose. Central American governments are now seeking development through proposals for a Central American free trade agreement (CAFTA) and through infrastructure projects such as Plan Puebla Panama, which would link countries of the region through highways, hydroelectric developments, and tourist development.

The mining industry is still important in many parts of the Andes and includes copper in Chile and Peru, tin in Bolivia, emeralds in Colombia, and silver and gold in several countries. Despite some unionization and attempts to improve technology and working conditions, many miners still face very difficult circumstances, and they endure high levels of respiratory diseases and accidents. The benefits of mining have also tended to concentrate in multinational corporations or in a few families that control the major companies. Prices have been very volatile, recently falling to low levels on world markets.

When the economic crisis hit North America and Europe in 2008, some commentators believed that increasingly diverse trade links, reduced debt, and government regulation of the Latin American financial sector would buffer the region against the economic downturn. But the credit crisis reduced foreign investment, and export opportunities declined, bringing economic difficulties to many firms. The flow of remittances from people working overseas also slowed as unemployment increased across the industrial economies.

The Latin American region is highly linked into the global economic system, with considerable flows of capital and goods to other world regions, especially North America. Although its GNP makes this region rank higher than many others on economic indicators, as we will see, these indicators hide tremendous variations in economic conditions within the region and living conditions that do not always reflect seemingly favorable economic statistics.

**Inequality**   Indicators of inequality show some Latin American and Caribbean countries as having greater concentrations of wealth and land in the hands of an elite few than in other parts of the world, with a majority of the population remaining poor and landless. However, recent data suggests that the gap is narrowing in many countries (**Figure 7.22**).

The highest average national incomes in Trinidad and Tobago and Venezuela are about 50% of those in the developed world, whereas in the poorest countries such as Haiti, there is a low-income economy of less than $1,000 per capita. Recent reports from the Inter-American Development Bank suggest that 25% of all income in Latin America is received by only 5% of the population, compared to 16% in Southeast Asia and 13% in developed countries. At the other end of the scale, the poorest 30% receive only 7.5% of total income, compared to more than 10% in the rest of the world. Although income distribution became more equal from 1960 to 1982, conditions became more unequal during the late 1980s. These high levels of inequality are associated with high levels of poverty, with more than 172 million Latin American and Caribbean people earning less than a subsistence income of $2 U.S. per day, and around 30% of the population earning less than $1 per day in countries such as El Salvador, Venezuela, and Honduras.

The vast gap between the richest 10% and the poorest 30% is reflected in other social measures. For example, the richest heads of households average 11.3 years of education and the poorest 4.3 years, with even larger gaps (of 9 years) in Mexico and Brazil. The income ratio of the richest 10% of Bolivians compared to the poorest 10% is a staggering 168%, and with Gini indexes of 59, both Colombia and Haiti possess the highest levels of inequality in the region—and indeed the world.

The informal sector, or **informal economy**, includes economic activities that take place beyond official record and are not subject to formalized systems of regulation or remuneration. In Latin America this comprises a variety of income-generating activities of the self-employed that do not appear in standard economic accounts, including street sell-

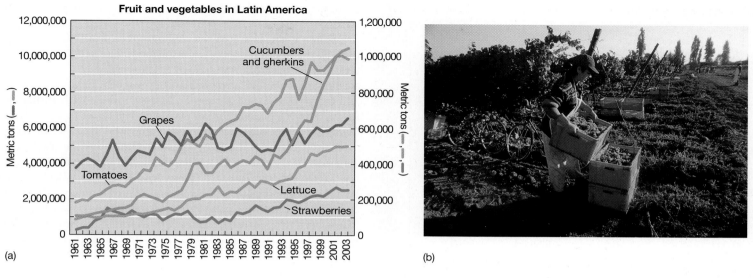

**Fruit and vegetables in Latin America**

(a)

(b)

(c)

▲ **FIGURE 7.21 Nontraditional agricultural exports** (a) Production of fruit and vegetables has increased in Latin America, partly as a result of export of nontraditional crops. Tomato, cucumber, and lettuce production has grown rapidly in response to demand from North America and Europe. However, consumer demand for unblemished fruit and vegetables requires heavy applications of pesticides and water to meet export quality standards and distribution schedules. The crops are vulnerable to climatic variation and to the vagaries of the international market, including changing tastes for foods and health scares about pesticide or biological contamination. In countries such as Chile, annual pesticide use doubled between 1984 and 1998, from just under 6,000 metric tons to more than 14,500 metric tons. (b) Spanish land grants distributed the fertile land of the Central Valley of Chile into large haciendas that produced wheat and raised cattle in the colonial period, especially to supply the growing cities of Santiago, Valparaíso, and Concepción. Wheat became an important export after independence, but the major boost to exports came in the 20th century with the development of refrigerated transport and shipment by air. This allowed Chile to take advantage of the hemispheric contrast in seasons, selling fruit and vegetables grown in the Chilean summer (November to March) to North American winter markets. Production and export of fruit and vegetables—especially fresh grapes, apples, peaches, and berries—grew dramatically, together with fruit packing and processing industries that employed thousands of people. Agricultural expansion has been aided by the Green Revolution package of technologies and by some land reforms that provided plots of land to those who would farm it intensively. Chile has also developed a wine industry that exported more than $800 million worth of wine in 2004. (c) Fair Trade and organic certification provides new markets for sustainable production from Latin America.

ing, shoe shining, garbage picking, street entertainment, prostitution, crime, begging, and guarding or cleaning cars.

Social and health conditions are often considered a better measure of overall inequality within and between countries than economic measures. National improvements in life expectancy, infant mortality, and literacy, for example, tend to reflect improvements at the lower end of the

scale rather than for the better-off segments of the population. Latin America tends to compare more favorably to the rest of the world on social and health indicators than on measures of income and income inequality. For example, life expectancy averages 73 years, higher than any other region in the developing world and compared to a world average of 69 years. Adult literacy rates are also relatively high, averaging

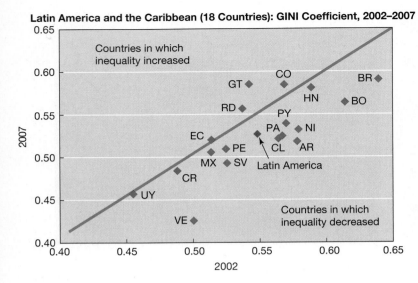

**Latin America and the Caribbean (18 Countries): GINI Coefficient, 2002–2007**

▲ **FIGURE 7.22 Inequality in Latin America and the Caribbean** This graph of the GINI index of inequality (where higher values show greater inequality in incomes) shows that inequality declined in all countries except Guatemala (GT), Dominican Republic (RD), and Colombia (CO) between 2002 and 2007. Conditions became considerably more equal in Argentina (AR), Bolivia (BO), Brazil (BR), Chile (CL), Nicaragua (NI), and Venezuela (VE), although countries such as Brazil and Bolivia still had very unequal income distributions.

91%, compared to 62% in sub-Saharan Africa, 62% in South Asia, and 81% worldwide. According to the United Nations International Children's Fund (UNICEF), only 8% of children under age 5 are defined as underweight in Latin America, compared to 27% worldwide.

However, there are wide gaps in social and health conditions within Latin America. Haiti, Central America, and the Andes tend to have much worse conditions than do Argentina, Chile, Uruguay, northern South America, Costa Rica, Mexico, and the English-speaking Caribbean. For example, in 2007, the average Haitian lived only 60 years, and the literacy rate was only 61%, compared to a life expectancy of 79 years in Costa Rica with literacy reaching 96%.

These national indicators also hide large variations in economic and social conditions within Latin American and Caribbean countries. In Mexico, the southern regions of the country have lower incomes and life expectancies than do the northern and central areas of the country. In Brazil, the northeastern and Amazon zones have higher infant mortality, lower life expectancy, and lower average monthly incomes than the southern parts of the country. Each Latin American and Caribbean country has its own geography of inequality, with the more rural regions generally having lower social and economic conditions.

## Demography, Migration, and Settlement

Prior to the arrival of the Europeans around 1500, Latin America is estimated to have had a population of approximately 50 million people, including large concentrations within the empires of the Aztecs and Incas and many smaller groups of hunters, gatherers, and agricultural communities. The demographic collapse dramatically reduced indige-nous populations, but significant Indian populations remained in Mexico, northern Central America, and the Andes.

Colonialism changed the demographic profile of Latin America through the intermixing of European and Indian peoples and the importation of slaves from Africa to the Americas. Few European women accompanied the early Spanish and Portuguese explorers and settlers, and many of the newcomers fathered children with Indian women through force, cohabitation, or marriage. The resulting mixed-race populations were classified according to their racial mix. The most common category was that of **mestizo**, a person of mixed white (European) and American Indian ancestry. Others included *mulatto* (Spanish/African) and *zambo* (African/Indian). These racial categories reflected racist perceptions that permeated society and correlated strongly with social class and culture. Even the Spanish divided themselves between *peninsulares* (those born in Spain) and *criollos* (those born in the Americas), with the elite sending their pregnant wives to Spain so that their children would be born there and thus have the highest social status.

**The African Diaspora to Latin America** Slave imports to Latin America from Africa totaled more than 5 million people during the colonial period, including 3.5 million to Brazil and 750,000 to Cuba. Many of the Caribbean islands, including Haiti and Trinidad, with very small indigenous and European populations, had a large number of African slaves working on plantations, and African populations also settled along the plantation coasts of Mexico, Central America, northern South America, and Ecuador. Although slavery was not abolished until the mid-1800s (1888 in Brazil), escaped and freed slaves formed communities as early as 1605, most famously the African community of Palmares, which was an autonomous republic from 1630 to 1694 in the Brazilian interior. These settlements, also called *maroon communities,* were created by escaped and liberated slaves in other regions such as Jamaica. Racial mixing occurred among European, Indian, and African populations, especially in Brazil.

**Other Diasporas** There is a legacy of other diasporas in contemporary Latin American populations. Asian immigration to the region began during the colonial period and picked up after the end of slavery, with Chinese, Indian, and Japanese workers brought to work on plantations and in construction. The workers had to pay off the cost of their travel and sustenance (indentured workers). Europeans other than the Spanish and Portuguese settled in the more temperate climates, especially in Argentina, where many families have Italian, German, or British names. Six million Italians and Spanish migrated to Argentina. Some regions, such as Patagonia, are associated with Welsh immigration and culture.

**Population Growth** The overall population of contemporary Latin America will total about 570 million people in 2010. The distribution is clustered around the historical highland settlements of Central America and the Andes and in the coastal colonial ports and cities (**Figure 7.23**).

Population has grown rapidly since 1900, when the regional total was 100 million, mainly as a result of high birthrates and improvements in health care. Brazil (196 million) and Mexico (108 million) have the largest populations today. Fertility rates—the average number of children a woman in a particular population group is projected to have during the childbearing years, ages 15 to 49—reach four children per woman in Haiti. Many countries are still growing at more than 2% per year, and population-doubling times are at less than 35 years,

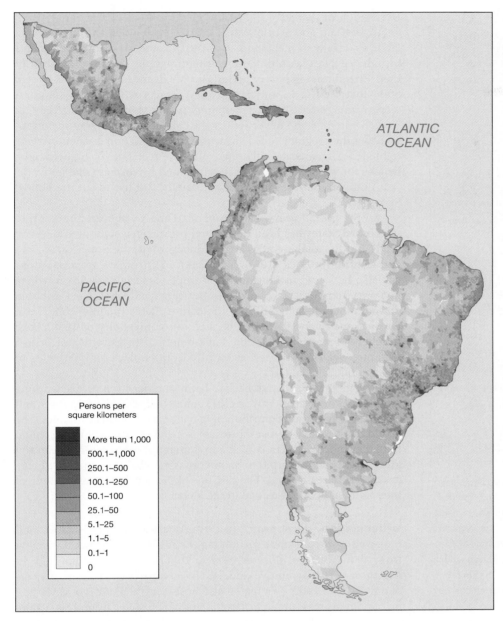

▲ **FIGURE 7.23  Population distribution of Latin America, 1995** The general distribution of population in Latin America includes sparsely settled interiors and high population densities around the historical highland regions of the Andes, Mesoamerica, and the coastal regions, especially former colonial ports and cities and their hinterlands.

tion. Fertility rates have tended to drop as people move into the cities, as health care improves, and as more women work and are formally educated. Mexico illustrates this pattern with lower fertility rates in urban, industrial, and higher-income states near Mexico City and the U.S. border and higher fertility rates in the poorer, more rural southern states. Attitudes toward family size in Latin America are also affected by the Catholic Church's position against contraception and the culture of machismo, which views high male fertility as a measure of status.

**Migration**   More than 150 million people are estimated to have moved from rural areas to cities in Latin America in the 20th century. The reasons for this massive rural–urban migration include factors that tend to push people out of the countryside and others that pull people to the cities (**Figure 7.24**). People leave rural areas because wages are low; because services such as safe drinking water, health care, and education are absent or limited; or because they do not have access to land to produce food for home consumption or for sale. Unemployment as a result of agricultural mechanization, price increases for agricultural inputs, and the loss of crop and food subsidies have also driven people from rural areas to the cities. Other push factors have been environmental degradation and natural disasters, such as Hurricane Mitch in Honduras, as well as long-running civil wars or military repression of rural people as in Guatemala.

Cities pull migrants because they are perceived to offer high wages and more employment opportunities, as well as access to education, health, housing, and a wider range of consumer goods. Governments often have an *urban bias* in providing services and investment to cities that are seen as the engines of growth and the locus of social unrest. Social factors that encourage migration to the cities include the promotion of urban lifestyles and consumption habits through television and other media, and long-standing social networks of friends and families that link rural communities with people in cities who can provide housing, contacts, and information to new migrants.

Although most people have migrated to cities within their own countries, there are several other important migration flows within the Latin American region. Several

placing pressure on food, water, housing, and infrastructure. Fertility rates have declined throughout much of the region. However, because a large percentage of the population is under age 15, especially in Central America, populations are likely to continue to grow as this cohort enters its reproductive years. For example, Guatemala's population is predicted to almost triple to 32 million people by 2050. The highest population densities (more than 200 people per square kilometer, or about 500 people per square mile) are found on the Caribbean islands and in El Salvador.

High fertility rates are characteristic of poorer, rural regions, where infant mortality is high, children can contribute labor in the fields, and women do not have access to education, employment, or contracep-

countries have encouraged the colonization of remote frontier regions by providing cheap land and other incentives to migrants. For example, the building of roads and availability of land in the Amazon created a stream of migrants from coastal regions of Brazil to the interior, and the development of irrigation in Mexico and Chile attracted migrants to desert regions. People have moved among countries in Latin America in search of work or fleeing from war and repression, with major population movements out of the Andes to work in mining, agriculture, and oil in Argentina and Venezuela, and out of Central America to Mexico either as refugees or workers seeking higher wages. Some of the smaller migrant streams have included better-off sectors of society—for example, many intellectuals left Chile, Argentina, and Brazil for Mexico,

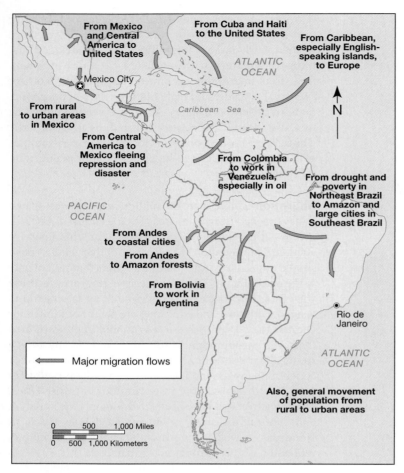

▲ **FIGURE 7.24 Major migration streams in Latin America** There are major migration streams within countries and between countries in Latin America and to the rest of the world from Latin America. The most significant overall trend is rural-to-urban migration, but poor people are also moving into frontier regions in the Amazon and southern Mexico, from the Andes to work in Argentina and Venezuela, and away from political unrest and natural disasters. The Caribbean has flows to Europe, especially from the English-speaking islands, and to the United States, especially from Cuba, Haiti, the Dominican Republic, and Puerto Rico.

Venezuela, and Costa Rica during times of repression of leftists and students by military governments.

**The Latin American and Caribbean Diaspora**   Latin American and Caribbean people have also left the overall region in considerable numbers, creating a global Latin American and Caribbean diaspora. The United States hosts the largest number of people who define themselves as being of Latin American or Hispanic heritage. Many Mexican families became part of the United States when the land they lived on became U.S. territory following the U.S.–Mexican War in 1848. They use the phrase "the border crossed us, we didn't cross the border" to emphasize that they are not migrants but long-standing residents. Between 1900 and 1930, 1.5 million Mexicans (10% of the total population) migrated to the United States to escape the chaos of the Mexican Revolution and partly to fill labor shortages created by World War I. Although 400,000 Mexicans (some of them U.S. citizens) were deported during the Great

Depression in the early 1930s, the growth of the U.S. economy from about 1940 on and World War II created such a demand for low-cost labor, especially in agriculture, that the U.S. and Mexican governments introduced a formal guest farm worker program. This program distributed 4.6 million temporary permits for Mexicans to work in the United States between 1942 and 1964. Many **braceros** (defined as a guest worker from Mexico given a temporary permit to work as a farm laborer in the United States) never returned to Mexico, and migration continued after the program ended, even as U.S. immigration restrictions were tightened. Migrants are still drawn to the United States by higher wages, by jobs for women in the service sector, and by strong social networks that link communities in Mexico to family and friends in the United States.

In the last 50 years, Latin American and Caribbean migration to the United States has been dominated by Mexicans (about 40% of the total), but the total includes large numbers of people from Cuba (15%) and Central America (10%). Significant Latin American populations can also be found in Canada and Europe (especially in Spain, where migrants from Andean countries work in low-paid jobs). The Caribbean diaspora includes migration to the United States (mainly from Cuba, Jamaica, and Puerto Rico), but because of colonial links to Britain, large numbers of Caribbean people have migrated to Europe and British Commonwealth countries, especially from Jamaica and Barbados to Britain and Canada.

The money that is sent back to Latin America from people working temporarily or permanently in other countries is called **remittances** and can make a significant contribution to national and local economies. The World Bank estimated that more than $60 billion was sent to the region from the United States in 2007. Many communities in the Caribbean and Mexico rely on these funds to build houses, purchase agricultural inputs, or educate their children. They are one of the new but informal flows of international financial capital in the global economy.

**Urbanization**   Most people in Latin America now live in cities, and the levels of urbanization are among the highest in the world, ranging from about 71% in most of Central America to more than 90% in Argentina, Uruguay, and Venezuela. This compares to a regional average of only 10% in 1900. The region also hosts several of the world's largest metropolitan areas (or *megacities*), including Mexico City and São Paulo, second and third in the world with about 18 million people each in the metropolitan areas, Buenos Aires at 12 million, and Rio de Janeiro at 6 million. A major cause of urban growth is migration, although the redefinition of city boundaries (to include metropolitan regions) and internal population growth have also played a role. In many countries, the population of the largest city in an urban system is disproportionately large in relation to the second- and third-largest cities in that system and this so-called *urban primacy* is characteristic of Argentina (Buenos Aires has 32% of the national population), Peru (Lima, 29%), Chile (Santiago, 34%), and Mexico (Mexico City, 17%). This concentration of population and development in one or two cities within a country can create problems when physical and human resources, political power, and pollution are all focused in one major settlement.

The megacities of Mexico City and São Paulo are global centers of commerce but also have many urban social and environmental problems. Mexico City is the economic, cultural, and political center of Mexico producing 40% of the gross national product with headquarters of leading Mexican companies and regional offices of multinational

▲ **FIGURE 7.25 Mexico City** Mexico City is located in a high basin surrounded by mountains, including the volcanic pair of Popocatépetl on the left (5,426 meters, 17,800 feet) and Iztaccihuatl on the right (5,288 meters, 17,340 feet). Air pollution obscures the view of the volcanoes many days of the year.

at risk from the volcano Popocatépetl, which overlooks the southern part of the basin (see Figure 7.25).

Mexico City's most infamous environmental problem is air pollution, which currently reaches levels that are dangerous to human health on more than 100 days a year. Thousands of automobiles, trucks, and buses, many with inadequate emission controls, are responsible for about 75% of the air pollution, with dust, fires, industrial plants, and miscellaneous energy use responsible for the remainder. The location of the city adds to the pollution problem because polluted air is often trapped in the basin by the surrounding mountains and by inversions where warm air traps cold air near the ground. The high altitude of the city at more than 2,000 meters (6,000 feet) means that fuel burns less efficiently and that humans must breathe more air because of the lower oxygen levels. Although the government has implemented air pollution controls, the continuing growth of the city and of car ownership has prevented any significant decline in pollution levels.

Located on a high plateau about 50 kilometers (30 miles) from the Atlantic coast, the city of São Paulo, Brazil, has wide avenues and many skyscrapers around the central business district with surrounding neighborhoods of poorer and slum housing (**Figure 7.26**). It is the major financial center for Brazilian and international banks and has recently developed a large telecommunications and information sector, with more than a million technical and scientific workers. Brazilian geographer Milton Santos reports that São Paulo employs more than 2 million manufacturing workers and produces 30% of Brazil's gross national product, having moved from a commercial center to a manufacturing hub to a service and information core for the global economy.

corporations. On a clear day, the view from the top of the modern high-rise office buildings includes elegant colonial plazas and administrative buildings, modern skyscrapers owned by international corporations, and the gleaming snowy peaks of the volcanoes that ring the basin (**Figure 7.25**).

The city grew very rapidly throughout the 20th century, from a population of 500,000 in 1900 to almost 20 million in 2000, expanding from 27 to 1,000 square kilometers (10.4 to 386 square miles) as the city embraced many satellite communities. Migrants who were pushed out of rural areas and attracted by the opportunities of the city drove most of the growth.

As in other large Latin American cities, many of the new migrants to Mexico City could not afford to rent or purchase homes and settled in irregular settlements, or *barrios*, that surround the city. As much as 50% of Mexico City's housing stock is defined as self-help construction, ranging from cardboard and plastic shanties to sturdier wood and brick structures with aluminum or tile roofs. Many of these settlements occupy steep hillslopes, valley bottoms, and dry lake beds that are vulnerable to flooding, landslides, and dust storms. The *barrio* of Netzahualcóyotl houses more than three million people on the shores of Lake Texcoco.

The location of Mexico City on a former lake bed, with unconsolidated sediment, adds to the risks from earthquakes in this seismically active zone. The earthquake that woke residents in the early hours of October 19, 1985 killed as many as 10,000 people and destroyed more than 100,000 homes and other buildings (see Figure 7.9b). The city is also

▲ **FIGURE 7.26 São Paulo** High-rise office buildings are surrounded by poorer slum housing called favelas.

# THE CARIBBEAN

The Caribbean islands are a diverse mix of cultural traditions, political systems, and environments. They include some of the poorest countries in the region and some of the wealthiest, many linked closely to the international economy. The distinctive physical geographies include extensive coral reefs and mangrove forests, small islands dominated by active volcanic peaks, and vulnerability to the hurricanes that arrive each fall. The Caribbean is divided into several subregions: the *Greater Antilles*, which include the islands of Cuba, Hispaniola (divided into Haiti and the Dominican Republic), Jamaica, and Puerto Rico, and the *Lesser Antilles*, which include the British and U.S. Virgin Islands, the Windward Islands, the Leeward Islands, and the islands of the southern Caribbean Sea north of Venezuela. Political groupings and affiliations include the many islands of the Bahamas, the former British colonies that remain within the Commonwealth, the French protectorates of Guadeloupe and Martinique, and the U.S. territory of Puerto Rico.

Caribbean culture is heavily influenced by the African traditions of the millions of slaves who were brought to the region as labor for the colonial plantations. Many countries, such as Haiti and Jamaica, have a predominantly black population. Although most countries use the colonial languages of English, Dutch, French, and Spanish as official languages, millions of Caribbean residents speak versions of Creole, languages that blend English or French with African or even indigenous words to create distinct languages, such as Haitian patois. Reverberations of Africa are also found in Caribbean foods and music and in spiritual traditions such as Voodoo, Santería, and Rastafarianism. The Caribbean has produced internationally renowned writers that provide fascinating insights into Caribbean life, such as V. S. Naipaul, Jamaica Kincaid, and Derek Walcott. Outmigration to the United States, Canada and the United Kingdom has brought Caribbean influences to these countries.

The economy of the Caribbean still has strong echoes of the colonial past, with many islands specializing in plantation export crops, such as sugar, tobacco, and coffee. Sugarcane long dominated the land and economy of islands such as Cuba, Jamaica, and the Dominican Republic, but sugarcane has declined with competition from sugar beets—and in Cuba's case, the disintegration of Soviet markets—although the production of rum is still extremely important. Coffee is grown in the cooler highlands of countries such as Jamaica, Puerto Rico, and Haiti, with Jamaica's Blue Mountain coffee receiving premium prices on world markets. Tobacco is the basis of Cuba's renowned cigar industry. Bananas are grown on the British Caribbean islands of St. Vincent, St. Lucia, and Jamaica, with preferential access to the European Union under the Lomé Agreements, which are intended to assist former British colonies with trade. Such preferences are likely to disappear with moves toward tariff removal and open markets within the European Union and under the World Trade Organization. In all cases, a country's primary exports are vulnerable to world market fluctuations and to the tastes of global consumers.

Jamaica and Trinidad and Tobago have economies partly reliant on mineral and energy resources. Jamaica exports bauxite for aluminum production; Trinidad and Tobago export oil and gas derivatives such as ammonia fertilizer and methanol.

The Caribbean has both benefited and suffered from its location in "America's Backyard" as the impact of the Monroe Doctrine brought the region within the U.S. sphere of influence. U.S. political goals included protecting the route to the Panama Canal, preventing the spread of communism, maintaining stability for U.S. corporations with assets in the region, and aligning trade in U.S. interests.

Using these and other goals as justification, the United States has intervened repeatedly in the Caribbean since the Spanish-American War of 1898, which freed Cuba from Spain and made Puerto Rico a U.S. colony. The United States sent troops to the Dominican Republic, Cuba, and Haiti several times in the earlier part of the 20th century; invaded Grenada in 1983 to oust a socialist government and rescue U.S. medical students; and sent troops as peacekeepers to Haiti in 1994. The U.S. relationship with Cuba remains complicated with limitations on trade and travel—slowly being lifted as politics shifts in both countries (**Figure 1**).

Export manufacturing activities have developed in many Caribbean countries, especially where formal **free trade zones** have been established within which goods may be manufactured or traded without customs duties. The Caribbean Development Initiative (CDI), designed to promote development and discourage civil unrest and socialism, guaranteed access to U.S. markets and resulted in the growth of export industries such as textiles and clothing.

Although some countries and communities have benefited from export agriculture and new industrial development, there is serious poverty and inequality in many Caribbean countries. Haiti is the poorest country in the entire Latin American and Caribbean region and the only one on the UN list of "least developed countries," with high infant mortality, low life expectancy, and difficult living conditions for most of the population that many blame on the semifeudal concentration of wealth and land under the Duvalier dictatorship from 1957 to 1976. High population densities and a search for land on which to grow a few subsistence crops has driven people to clear forests on steep slopes where soil erosion can be so serious that agriculture is impossible and the land is left desertified and degraded. Foreign aid makes up more than a third of the national budget, and the debt has been reduced under the Highly Indebted Poor Countries initiative (see Chap-

---

The cultural and media center of Brazil is the city of Rio de Janeiro, which has been overshadowed by the economic growth of its rival São Paulo (**Figure 7.27**). Rio was the capital of Brazil from 1822 to 1960. The urban structure includes an older city center with a wealthier residential zone and beaches such as Copacabana toward the south, and a poorer, more industrial zone to the north. Rio and São Paulo followed the same pattern as other Latin American cities, attracting millions of migrants who settled informally around the urban core. The **favelas**, a Brazilian term for informal settlements that grow up around the urban core, lack good housing and services. In São Paulo, 28% of residents have no drinking water, and 50% have no sanitation. The crowding, high land costs, violent crime, poverty, and pollution of the city are starting to cause economic development to shift to smaller neighboring cities.

As the megacities become more polluted, crowded, and expensive, migration is now shifting to smaller cities in many countries, such as Monterrey or Guadalajara in Mexico.

► **FIGURE 1 Havana, Cuba**
Havana, located on the northern shore of the Caribbean's largest island, was one of Spain's most important ports and the gate to the wealth of the Americas. As trade in slaves, sugar, and tobacco grew during the colonial period, imposing buildings rose around the main plaza and along the harbor front. A recent resurgence in tourism, mainly from Europe and with modest investment from the Cuban government, has brought new prosperity to some sectors of Havana. These areas have seen construction of new hotels, the opening of stores accepting U.S. dollars, and renovation of older colonial architecture. Restrictions on vehicle imports mean that many are still driving old U.S. automobiles from the 1950s.

lion of them on cruise ships. Almost half the population of the Bahamas and the Virgin Islands is employed in tourism in hotels, restaurants, bars, casinos, ocean sports such as diving and fishing, and small shops (**Figure 2**). The downside of tourism includes pollution of oceans, reefs, and beaches by ships; competition for fresh water and higher food prices for local residents; cultural and social stresses from interactions between the wealthy visitors and poorer residents; and prostitution and drugs. Critics also point out that many profits flow out of the region because many tourist enterprises are foreign owned. ∎

▼ **FIGURE 2 Tourism in the Caribbean**
Cruise ships tower over the town of Nassau on the island of New Providence in the Bahamas. The area near the cruise terminal is crowded with restaurants, bars, and souvenir shops catering to the tourists who may spend only a few hours on shore. But the Bahamas, like other Caribbean islands, also have large hotel resorts that entice visitors with beaches, warm weather, and exotic food and entertainment.

ter 5). Millions of Haitians have fled conflict and poverty to migrate to the United States, Cuba, Canada, and France, bringing their distinctive culture that mixes French, African, and indigenous traditions to create art, music, and rituals.

Tourism and offshore financial services are the two new foundations of the Caribbean economy. **Offshore financial services** include the provision of banking, investment, and services to foreign nationals and companies who wish to avoid taxes, oversight, or other regulations in their own countries. Such services are an important economic sector in the Bahamas and the Cayman Islands. The latter has thousands of registered companies and, with improved telecommunications, is integrally tied to the global financial sector. The sector does not employ a large number of people and is under suspicion of laundering drug money, but the fees support government programs that benefit local residents.

An image of the Caribbean as a paradise of golden sands, warm turquoise seas, and friendly festive people has been constructed to attract more than 18.5 million tourists in 2008, 17 mil-

# CULTURE AND POLITICS

Latin America is a dynamic world region where economic, political, cultural, and social changes have been rapid in the 20th century and have varied in their nature and impact among and within countries. Latin American culture and language is heavily influenced by colonial legacies, but has gained worldwide influence and still maintains important pockets of indigenous traditions. Latin American countries took divergent political paths that have included socialist and military governments; authoritarian, single-party, and multiparty systems; and highly centralized and localized administrations. The challenges of creating functioning national governments and promoting economic growth dominated the postindependence period in the 19th century. The 20th century saw regional factions, the working class and the poor demanding reform through revolution and populism, and threats to the distribution of wealth and elite power met by military and authoritarian rule.

▲ **FIGURE 7.27  Rio de Janeiro** Rio de Janeiro has a stunning location with a harbor overlooked by Sugar Loaf Mountain and the beaches of Copacabana. The commercial harbor is now a center for shipbuilding and for agricultural exports from the southeast of Brazil, including soybeans and orange juice in specially constructed tankers.

One of the most dramatic shifts was from a continent dominated by military and authoritarian governments in 1970 to almost regionwide democratic systems by 2000.

## Tradition, Religion, and Language

The mixed racial and ethnic composition of Latin America (see p. 256) is echoed in many aspects of cultural heritage and social practices in the region. Indigenous culture, including traditional dress, crafts, ceremonies, and religious beliefs, persists in regions such as highland Guatemala and Peru, partly as a result of colonial policies that kept Indian communities separate while demanding tribute and labor, and partly because of resistance to the adoption of European culture by conservative indigenous religious and political leaders. Cultural traditions are now promoted to tourists and revalued through indigenous social movements seeking political rights and recognition. For example, indigenous Mayan centers such as Quetzaltenango in Guatemala are promoted as tourist destinations, where traditional crafts may be purchased and photos taken (often for a price) of women and children in traditional colorful woven garments (**Figure 7.28**).

**Languages**  Indigenous languages endure in several regions of Latin America (**Figure 7.29**). The most widely spoken languages are *Quechua* in the Andean region (spoken by 13 million people), English Creole and French Creole in the Caribbean (10 million), *Guaraní* in Paraguay (4.6 million), *Aymara* in the Andes (2.2 million), *Mayan* in Guatemala and southern Mexico (1.7 million), and *Nahuatl* in Mexico (1.3 million). Spanish is the dominant language across most of mainland Latin America, except for Brazil (Portuguese), Belize and Guyana (English), Suriname (Dutch), and French Guiana (French). The foods of Latin America blend indigenous crops such as corn or potatoes with European influences, especially from Spain. Although Mexico and Jamaica are associated with spicy dishes that include *chile*, the food is quite mild in the rest of Latin America. In livestock-producing areas, such as Argentina, grilled meat is extremely popular, but in much of Latin America, the poor eat simple meals of rice, corn, potatoes, and—for protein—beans. Modified versions of Mexican cuisine have diffused throughout North America and include many chain restaurants. Foods in the Caribbean reflect the medley of cultures in the region with African, Asian, and European influences combining to create dishes of fish, chicken, pork, a range of vegetables, fruits, and starch crops (rice, potatoes, and yuca).

**Religion**  One of the main objectives of Spanish and Portuguese colonialism was the conversion of indigenous peoples to Catholicism. Although some indigenous people fiercely resisted missionary efforts,

others found ways to blend their own traditions with those of the Catholic Church to create new *syncretic* religions. The process of conversion was facilitated by the reported appearance of the brown-skinned Virgin Mary of Guadalupe to an Indian convert in Mexico on December 9, 1531, and by the efforts of some priests to protect local communities from the Spanish efforts to obtain land, tribute, and labor by force. More than 400 million people in Latin America are followers of Catholicism, Islam and Judaism draw about 2.4 million and 500,000 million, and more than 10 million people have traditional Mayan beliefs.

The slave trade brought African religious traditions to Latin America and the Caribbean, and these often merged with indigenous and Catholic beliefs to construct syncretic contemporary rituals, followed by more than 30 million people such as Candomble and Umbanda in Brazil, Voodoo in Haiti, and Santería in Cuba and other islands. Candomble and Umbanda are both sects of the Macumba religion, with rituals that involve dances, offerings of candles and flowers, sacrifice of animals such as chickens, and mediums and priests who use trances to communicate with spirits that include several Catholic saints. Voodoo (also spelled Voudou) rituals include drumming, prayer, and animal sacrifice to important spirits based on traditional African gods and Catholic saints. These rituals are led by priests who act as healers and protectors against witchcraft. Santería, which is closely connected to the Yoruba religion of West Africa, blends Catholic saints

▲ **FIGURE 7.28 Guatemalan handicrafts** Tourists browse textiles and other crafts sold by indigenous Maya at a market in Guatemala.

Spanish
Portuguese
English
Dutch
French
Mayan
Kuna
Miskito
Zapotec
Mixtec
Garifuna
Aymara
Guarani
Quechua
Mapuche
Yanomama
Other indigenous languages

▲ **FIGURE 7.29 Languages of Latin America** This map shows the major regions of living indigenous languages in Latin America. Most of mainland Latin America uses Spanish as an official language, except for Brazil, which uses Portuguese; French Guiana, French; Suriname, Dutch; and Belize and Guyana, English. The Caribbean has more language variation than the mainland, and the dominant European languages reflect the colonial histories of the islands.

40 million Latin Americans are now members of such churches.

Latin American and Caribbean art and literature have incredible variety and regional specialization. Traditional textiles, pottery, and folk art are sold to tourists and by import stores in North America and Europe. Literary traditions include magical realism (where authors such as Gabriel García Márquez blend imaginary and mystical themes into their fiction), and Latin American and Caribbean authors have won six Nobel prizes for literature. Works of noted Mexican muralists, including Diego Rivera, and the tortured paintings of his companion, Frida Kahlo, are numbered among the masterpieces of 20th-century art (**Figure 7.30**).

In the 1990s several strands of Latin American and Caribbean music became popular. Traditional music, such as the Andean pipes of the groups Inti Illimani and Los Incas, has formed the soundtracks of documentary films and is available in music stores worldwide. The music of Nueva Canción, or New Song movement, with its social conscience, and singers such as Mercedes Sosa, who fled repression in their home countries, has become a part of the global folk music scene. Caribbean global influences include the reggae of Jamaica and steel-drum bands of Trinidad, which resonate with the rhythms of Africa (**Figure 7.31a**). Latin pop and rock music has become extremely popular, often produced from Miami, Latin America's business capital in the United States (**Figure 7.31b**).

## Cultural Practices, Social Differences, and Identity

For the most part, cultural practices in Latin America have been evolving toward global cultural trends promoted by formal education and popular media, especially television. Differences have been partly erased by the common experience of popular television shows—especially the "telenovelas" (soap operas)—and by education systems and media that promote modern urban lifestyles.

**Gender Relations** Certain cultural views of the family and gender roles are characteristic of Latin America. Multiple generations often live and work together, individual interests are subordinated to those of the family, and the traditions of machismo and marianismo define gender roles within the family and the society. Machismo constructs the ideal Latin American man as fathering many children, dominant within the family, proud, and fearless. Marianismo constructs the ideal woman in the image of the Virgin Mary as chaste, submissive, maternal, dependent on men, and closeted within the family. Latin American society is generally patriarchal, with institutions that have prohibited or limited women's right to own land, vote, get a divorce, and secure a decent education.

These stereotypes are, of course, contradicted by individual cases and are breaking down in the face of new geographies and global cultures. Family links are weakened through migration and the isolation of many living spaces, from each other and from those of other family members in

with African spirits associated with nature, using rituals similar to other Latin American and Caribbean religions.

The emergence of a new form of Catholic practice, **liberation theology**, focused on the poor and disadvantaged. Liberation theology is informed by the perceived preference of Jesus for the poor and helpless and by the writings of Karl Marx and other revolutionaries on inequality and oppression. This new orientation to the poor was espoused by the Second Vatican Council, called by Pope John XXIII in 1962. Priests preached grassroots self-help to organized Christian-base communities and often spoke out against repression and authoritarianism. In recent decades evangelical Protestant groups with fundamentalist Christian beliefs have grown and spread rapidly in Latin America. Their message of literacy, education, sobriety, frugality, and personal salvation has become very popular in many rural areas. Estimates suggest that up to population defined as Indian, and Colombia, Chile, El Salvador, Mexico, Nicaragua, Panama, Paraguay, and Venezuela are more than half mestizo.

These numbers hide subtle differences in how different countries record, construct, and perceive race and ethnicity. For example, a tendency to identify with Europe may increase the proportion of those who report themselves as European in Argentina, whereas a national pride in mestizo heritage increases self-identification as being of mixed-race heritage in Mexico. Brazil has promoted an image of Brazilian racial democracy and equality, where skin color is called "coffee," and musical, religious, and dietary traditions merge into a uniquely Brazilian culture.

This myth of racial democracy is contradicted by evidence of continuing racism in Brazil and other Latin American countries. Studies show that race and class correlate strongly, with Afro-Brazilians being on the whole poorer, less healthy, less educated, and more discriminated against in employment and housing. In Mexico, the media have tended to promote lighter skin as more desirable through the choice of more European-looking actors in commercials and other programs, and job advertisements still ask for "good appearance," hinting at a preference for nonindigenous features.

**Sports**   Whereas the English-speaking Caribbean, especially the West Indies, produces some of the world's best cricket players, several Spanish-speaking islands, such as the Dominican Republic, are associated with famous baseball players, such as Sammy Sosa, Pedro Martinez, and Albert Pujols, who gained fame and fortune in the U.S. leagues. Soccer (futbol) is the mostly popular sport in mainland Latin America, with enormous stadiums hosting tens of thousands of fans in countries such as Argentina, Brazil, and Mexico and children in the poorest communities playing with makeshift balls and goals.

## Territory and Politics

Latin American politics in the 20th century became caught up in Cold War politics between the United States and the Soviet Union, with the United States seeing European intervention in the region, and especially

▲ **FIGURE 7.30 Latin American art**  A self-portrait by Mexican artist Frida Kahlo standing on the United States–Mexico border contrasting the indigenous tradition of Mexico with modern industry in the United States.

urban environments. Men's and women's roles are changing as fertility rates decline and women enter the workforce and politics. Latin American and Caribbean feminists have organized to obtain the right to vote; to effect changes in divorce, rape, and property laws; to gain access to education and jobs; and to elect women to political office.

Gender inequality is widespread in Latin America. Female literacy, on average, is 2% to 15% lower than that of male populations. Women tend to earn much less money on average than men. In Latin America and the Caribbean, women earned 49% of the male rate in 2007. In Ecuador, for example, female income (PPP US$) was $5,189 in 2007 compared to $9,075 for men. This inequality has been associated with systematic institutional biases that denied women in many countries the right to vote or the right to marital property until the 1950s, with cultural traditions that discourage more than a few years of education for women, and with employment structures that pay women less than men or pay less for traditionally female work such as domestic service work and food processing.

**Race**   Recent population censuses attempting to record race and ethnicity show some general patterns that correlate with the population history described here. Large proportions of the populations of Brazil (50%), Colombia (29%), Cuba (34%), and Haiti (90%) are of African heritage, and Argentina, Costa Rica, and Uruguay report significant numbers of Europeans. Indian populations are high in Peru (47%), Ecuador (43%), Bolivia (71%), and Guatemala (66%), which have a significant percentage of their

(a)

(b)

▲ **FIGURE 7.31 Latin Music**  (a) Years after his death, Bob Marley of Jamaica is still popular as a reggae artist and representative of the Rastafarian religion. (b) Gloria Estefan, who is originally from Cuba, is one of the superstars of Latin pop and rock based in Miami.

left-wing ideologies, as threats to its own security. This stemmed from 1823, when U.S. President James Monroe issued his **Monroe Doctrine**, stating that European military interference in the Western Hemisphere, including the Caribbean and Latin America, would no longer be acceptable, would be considered a threat to the peace and security of the United States, and would be considered a hostile act. It also stated that in return for European noninvolvement in the Western Hemisphere, the United States would not interfere in European affairs. This doctrine set the stage for subsequent U.S. involvement and intervention in Latin America and the growth of U.S. economic and political dominance in the region.

Latin America was drawn more explicitly into the new U.S. political and economic sphere of influence with U.S. interventions (**Figure 7.32**) to maintain stability and economic access in Cuba (1896–1922), Haiti (1915–34), Nicaragua (1909–33), and Panama (from 1903–1999 to control the canal).

**Panama Canal**   The Panama Canal provides a fascinating case of a strategic investment by the United States that led to long-term political tensions. The Panama Canal joined the world's two great oceans—the Atlantic and the Pacific—across the Isthmus of Panama and shortened the ocean trade route between them by weeks, with considerable reduction in costs (**Figure 7.33**). In November 1903, after the failure of a French attempt to build a canal, with U.S. encouragement and French financial support, Panama proclaimed its independence from Colombia and concluded the Hay/Bunau–Varilla Treaty with the United States. The treaty conceded rights to the United States "as if it were sovereign," in a zone roughly 16 kilometers (10 miles) wide and 80 kilometers (50 miles) long. In that zone, the United States would build a canal, then administer, fortify, and defend it "in perpetuity." In 1914 the United States completed the existing 83-kilometer (50-mile) lock canal, one of the world's greatest engineering triumphs with three sets of locks lifting enormous ships up and across a lake 24 meters (80 feet) above sea level. More than 25,000 people lost their lives trying to build the canal, mainly from tropical diseases such as malaria and yellow fever.

The early 1960s saw the beginning of sustained pressure in Panama for the renegotiation of the Hay/Bunau–Varilla Treaty. During the 1970s, negotiations between the United States and Panama over the status of the Panama Canal concluded with two new canal treaties in 1977, effective in 1979. These treaties handed control of the Canal Zone to Panama, with a joint U.S.–Panamanian Panama Canal Commission administering the canal until the end of 1999, at which time the U.S. would withdraw forces from Panama.

During the 1980s, when Panama's military government, led by Manuel Noriega, was implicated in drug trafficking and refused to recognize civilian elections supporting opposition candidates, the U.S. interests in the region were threatened. After economic sanctions failed, President George H. W. Bush ordered the U.S. military into Panama on December 20, 1989, citing the need to protect U.S. lives and property, to fulfill U.S. treaty responsibilities to operate and defend the canal, to assist the Panamanian people in restoring democracy, and to bring Noriega to justice.

Panama's celebrations on the handover of the canal in 1999 reflected liberation from U.S. domination as well as acquisition of full control of an economic asset. Tolls for the canal bring in more than $1.2 billion a year, and Panama plans to privatize the ports and invite foreign investors to build hotels, industrial parks, ecology-based tourism, ship-repair facilities, and private housing.

**Military Rule**   Many other Latin American countries experienced political upheavals in the 20th century. In the 1960s and 1970s the dual threats of economic instability and communist ideas contributed to a rise in authoritarianism and military governments. Seeking financial order and control of socialist movements, the military took control of governments in Brazil in 1964, Chile and Uruguay in 1973, and Argentina in 1976. Although central authoritarian control certainly provided some degree of economic stability and growth, the military governments aggressively kept social order by repressing dissent, especially among students and workers perceived as having socialist ideals. In Argentina, the military government's so-called Dirty War is alleged to have killed 15,000 people and forced many others to leave the country. In Chile, the military government of General Augusto Pinochet has been accused of similar disappearances and human rights abuses.

**Revolutions**   Increasing the concentration of land and wealth in the hands of the few and expanding foreign ownership and export orientation at the beginning of the 20th century produced growing frustration among the poor, the landless, and opposition or regional factions in several countries. Internal tensions between elites and other groups, especially landless peasants, complicated relationships with the United States, especially as the Cold War between the democratic West and the Soviet Union intensified in the 1950s. A series of 20th-century revolutions in Mexico, Guatemala, Cuba, Bolivia, and Nicaragua reverberated around the hemisphere and the world (see Figure 7.32).

The Cuban Revolution, led by Fidel Castro in 1959, created a socialist state only miles from the U.S. mainland on the largest island in the Caribbean. The United States has taken an aggressive stance against the Cuban government, including the Bay of Pigs invasion in 1961 and a series of embargoes on trade. The United States maintains a large military base at Guantánamo Bay in eastern Cuba, infamous as a terrorist holding center, where high-wire fences separate the base from the rest of communist Cuba.

Most of the opposition to the Cuban government within the United States today has been associated with Cuban-Americans based in Florida, who left Cuba as a result of the revolution. Other U.S. groups, especially agricultural and pharmaceutical companies and states with economies based in these exports, are pressuring the federal government to open trade with Cuba. At one time, sugar and tobacco were the major Cuban exports, especially to the former Soviet Union. They were mostly produced on collective farms. Both scholars and politicians have debated the successes and failures of Cuban socialism, highlighting the positive elements of sustainable agriculture and widespread access to health and education, reflected in favorable social indicators, and the negative aspects of the absence of free elections and inefficiencies of a centrally planned economy.

The spread of socialist ideas about working-class activism and the need for land reform led to the election of socialist governments in Guatemala in 1954 and Chile in 1970. In both cases, redistribution of land and nationalization of key industries threatened the local elite and U.S. interests to the extent that the United States was implicated in assassinations and military coups that overthrew socialist leaders Jacobo Arbenz Guzmán in Guatemala and Salvador Allende in Chile within three years of their elections.

In Nicaragua, concentration of wealth and land under the Somoza dictatorship fostered rebellion that resulted in the establishment of the socialist Sandinista government in 1979. Again, Cold War anticommu-

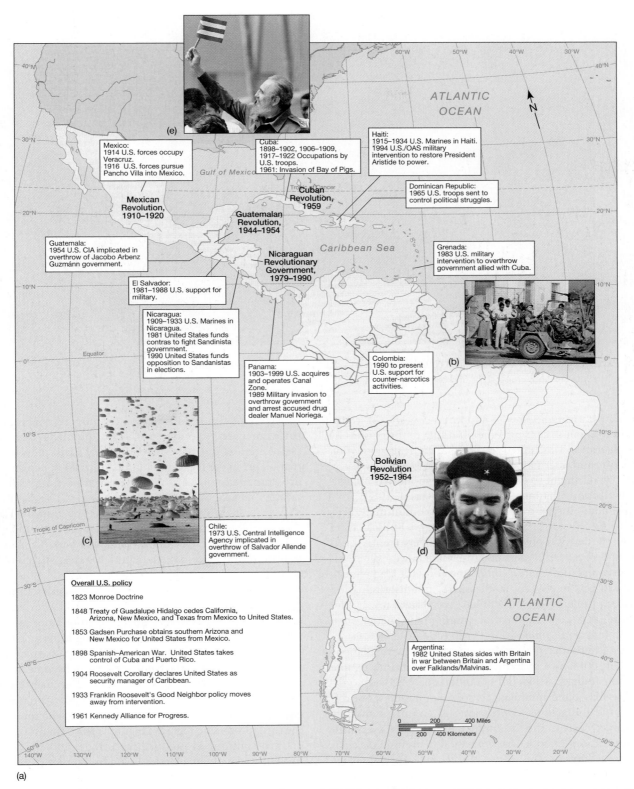

**Mexico:**
1914 U.S. forces occupy Veracruz.
1916  U.S. forces pursue Pancho Villa into Mexico.

**Cuba:**
1898–1902, 1906–1909, 1917–1922 Occupations by U.S. troops.
1961: Invasion of Bay of Pigs.

**Haiti:**
1915–1934 U.S. Marines in Haiti.
1994 U.S./OAS military intervention to restore President Aristide to power.

**Mexican Revolution, 1910–1920**

**Cuban Revolution, 1959**

**Guatemalan Revolution, 1944–1954**

**Dominican Republic:**
1965 U.S. troops sent to control political struggles.

**Guatemala:**
1954 U.S. CIA implicated in overthrow of Jacobo Arbenz Guzmánn government.

**Nicaraguan Revolutionary Government, 1979–1990**

**Grenada:**
1983 U.S. military intervention to overthrow government allied with Cuba.

**El Salvador:**
1981–1988 U.S. support for military.

**Nicaragua:**
1909–1933 U.S. Marines in Nicaragua.
1981 United States funds contras to fight Sandinista government.
1990 United States funds opposition to Sandinistas in elections.

**Panama:**
1903–1999 U.S. acquires and operates Canal Zone.
1989 Military invasion to overthrow government and arrest accused drug dealer Manuel Noriega.

**Colombia:**
1990 to present U.S. support for counter-narcotics activities.

**Bolivian Revolution 1952–1964**

**Chile:**
1973 U.S. Central Intelligence Agency implicated in overthrow of Salvador Allende government.

**Overall U.S. policy**

1823 Monroe Doctrine

1848 Treaty of Guadalupe Hidalgo cedes California, Arizona, New Mexico, and Texas from Mexico to United States.

1853 Gadsen Purchase obtains southern Arizona and New Mexico for United States from Mexico.

1898 Spanish–American War.  United States takes control of Cuba and Puerto Rico.

1904 Roosevelt Corollary declares United States as security manager of Caribbean.

1933 Franklin Roosevelt's Good Neighbor policy moves away from intervention.

1961 Kennedy Alliance for Progress.

**Argentina:**
1982 United States sides with Britain in war between Britain and Argentina over Falklands/Malvinas.

▲ **FIGURE 7.32  U.S. interventions in Latin America and Latin American and Caribbean revolutions**  (a) This map shows the dates and locations of U.S. intervention in Latin America and those areas where there were major Latin American revolutions in the 20th century. In most cases the United States intervened as a result of Cold War politics and in the interests of U.S. national security, desiring to prevent formation of governments perceived as allied with the Soviet Union. Most revolutionary movements have been inspired by calls for land reform and socialist policies. (b) U.S. marines in Grenada in 1983. (c) U.S. paratroopers during the invasion of Panama, December 21, 1989, ordered by U.S. President George H. W. Bush. The military were ordered to seize Manuel Noriega to face charges on drug trafficking in the United States. (d) Guerilla leader Che Guevara, who fought in the Cuban and Bolivian revolutions. (e) Cuban revolutionary leader Fidel Castro.

▲ **FIGURE 7.33 Panama Canal** The Panama Canal cuts through the isthmus that joins North and South America and is a critical transport route for both cruise ships and cargo.

nist sentiments led the United States to support—covertly—a counterrevolutionary movement of *contras*. When funding to the contras by the Reagan Administration was linked to illegal arms deals with Iran, the resulting domestic and international opposition to U.S. covert operations led the U.S. government to change tactics: It supported opposition candidates in elections that ousted the Sandinistas in 1989.

Guerilla movements inspired by socialist and communist ideas, which emerged in El Salvador, Colombia, Peru, and Bolivia, were severely and often violently repressed by ruling governments. In El Salvador, the attempts of powerful oligarchies to retain power were associated with massacres and murders. Despite some efforts at land reform, the colonial legacy lingers in poverty, landlessness, and discrimination against many Andean Indians. Resentment has fueled rural revolt, illegal coca production, and some support for guerilla movements, such as the Shining Path (*Sendero Luminoso*) in Peru.

In Guatemala, authoritarian governments forced Mayan populations to flee into neighboring Mexico or farther north or retreat into remote mountains. There the military, which did not differentiate between ordinary people and guerillas, often annihilated whole communities. The difficulties faced by the indigenous populations of countries such as Guatemala are characteristic of the discrimination that persists against indigenous cultures in Latin America. Discrimination stems from the colonial period. It has been reinforced by a series of elites, who used an image of indigenous people as uncivilized, underdeveloped, and rebellious, to take their lands, undermine their religion and language, and force them into low-paid occupations as farm workers, miners, or servants.

**The Drug Economy** The drug economy has seriously destabilized politics in some regions of Latin America, especially Bolivia, Colombia, Mexico, and Peru.

Latin America produces drugs that are illegal in many countries, including cocaine (from the coca plant), heroin (from poppies), and marijuana. Latin American farmers grow drugs because of their high price compared to other agricultural products. In regions where crop yields are low, where people have only small plots of land, and where market prices for legal agricultural crops do not cover production costs, drug production is an attractive or even necessary survival option. The farmers receive only a fraction of the street value of the drugs when they are sold. Most of the drug exports are controlled by powerful families in Colombia and Mexico, who manage the transport systems from rural Latin America by land, air, and boat into the main distribution and consumption centers in the United States, such as Los Angeles and Miami.

In some areas the drug trade has exacerbated political conflicts, such as in Colombia, where several guerilla movements control large areas, despite opposition by government military units. The groups are alleged to have links with powerful drug lords and to offer protection to farmers involved in drug production. Because the conflicts are fueled by the drug economy, the United States has become involved in the strife through its support for the Colombian government's antidrug activities. The United States sprayed pesticides and provided military training and equipment to the Colombian army groups, who repress rebel groups in the drug-producing regions (**Figure 7.34a**). Analysts contend that farmers will continue to produce crops that can be processed into drugs, until they can obtain a better living from other crops or other means of employment, or until the demand is controlled in the United States. They argue that the United States should be focusing on the control of demand within its own borders or even on limited legalization of consumption, rather than on fighting a "war on drugs" in Latin America.

Drug traffickers control production areas but also influence the police, the army, judges, and political leaders through intimidation or bribery. In Mexico, power struggles between three major cartels of narcotic traffickers have escalated to the point where more than 4,000 people were killed in drug-related violence in 2008 (**Figure 7.34b**). The Mexican army is

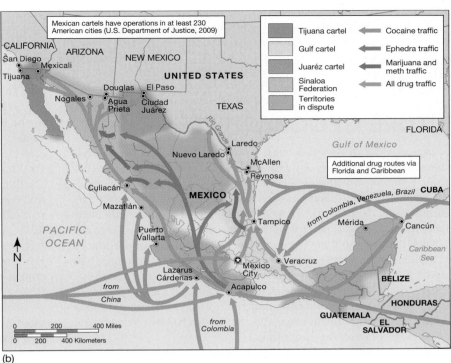

▲ **FIGURE 7.34 Drug production in Latin America** (a) Cocaine derives from the coca plant, the leaves of which have been chewed by Andean residents for centuries to provide energy and alleviate the effects of high altitude. Peru, Colombia, and Bolivia produce about 98% of world cocaine supplies; Jamaica, Colombia, and Mexico are major producers of marijuana (cannabis). The coca-growing regions in Colombia are mostly located south of Bogota and near the border with Ecuador and Peru. In Peru and Bolivia, coca has been grown on the eastern slopes of the Andes. Because of aggressive eradication efforts, including herbicide spraying funded by the United States, the area in coca cultivation in Latin America fell from 210,000 hectares (741,000 acres) in 1990 to 173,000 hectares (445,000 acres) in 2002 and shifted from Peru and Bolivia to Colombia. But production techniques have improved, and the overall production has remained around 300,000 tons; the distribution is increasingly controlled by organized crime throughout the global world-system. (b) Mexico is a major route for drugs into the United States, and the trade is dominated by powerful cartels, who control different regions and engage in violent conflict over disputed territories.

struggling to control the situation, which is fuelled by a flow of guns south across the border from the United States and by corruption of police forces.

**Democratization and Social Movements** Public and foreign outrage at authoritarian repression and human rights violations, the inability of military governments to solve economic problems, the end of the Cold War, and international and internal pressures that linked economic globalization to democratic governance eventually resulted in gradual transitions to democratic governments in most Latin American countries—including Argentina in 1983, Brazil in 1985, and Chile in 1989. In Argentina the departure of the military government was hastened by the loss of a war with Britain when Argentina invaded the Falkland Islands in 1982 (called the Islas Malvinas in Latin America). However, democracy has been fragile, with economic and other crises regularly toppling governments in countries such as Bolivia and Argentina and with heavy-handed tactics and questionable elections maintaining one-party systems in others (including Mexico for most of the 20th century).

Political opposition and activism have often taken the form of organized **social movements** that have also pressured for specific resources and issues, such as housing, water, human rights, or environmental protection. Social movements have also filled a gap in service provision and local administration created by the economic crisis and neoliberal policies that have shrunk government in many Latin American countries. New indigenous social movements have organized to promote language, culture, and land rights. In Mexico, the Zapatista rebellion of 1994 focused on opposition to NAFTA, but unrest has continued in protests against neoliberal policies and in support of indigenous rights. Other popular protests include those in Argentina against privatization and high food prices and in Bolivia against the privatization of water.

In Brazil, a landless movement with an estimated 1.5 million people became the largest social movement in Latin America. From 1985, *Movimento dos Trabalhadores Rurais Sem Terra*, or MST, Brazil's Landless Workers Movement, won land titles for more than 350,000 families in 2,000 settlements. It has reformed land through forced land redistribution and by demanding legal rights and political change with considerable public support.

Some of these social movements translated into changes in electoral politics—the so-called shift to the left—as a backlash against budget cuts, privatization, and inequality brought more left-wing, antiglobalization, and populist leaders to power after 2000, including former MST labor leader President Lula in Brazil, President Chavez in Venezuela, and indigenous leader President Morales in Bolivia. In 2006 Chile elected Michelle Bachelet, who campaigned on a platform of reducing inequality through social programs.

# FUTURE GEOGRAPHIES

The region of Latin America, with its diverse lands and people, has changed dramatically in recent years as the processes of market liberalization, economic integration, democratization, urbanization, and environmental degradation have transformed the region and changed its relationships with other world regions. Each of these processes has interacted with local conditions to produce a new mosaic of distinct regional geographies throughout the region and has produced new opportunities and challenges for people and policymakers. Costa Rica is often seen as a model for the region, where investments in education and social services, a stable government (focused on peace without a standing army), and protection of the natural environment are associated with per capita incomes higher than the average and foreign investment in well-paid industrial sectors such as computing.

In terms of human–environment relations, the future of the region will depend on the ability to manage the resource demands and pollution emissions of new industrial development and farming technologies, the pressures of new consumption habits, and growing populations. The region also faces the risk that climate change will degrade water supplies and ecosystems, increase disaster losses, and encroach on coastlines. New low-carbon energy sources and the protection of forests will reduce the region's contribution to greenhouse gas emissions. Mexico was the first developing country to take on formal targets to reduce carbon emissions.

The economic future may depend as much on markets within the region as beyond, especially if the global economy remains unstable and as the United States and China negotiate their complex and dominant trade relationship. Countries with more diverse economies such as Brazil and Mexico are capable of meeting many of the consumption needs within the region, and even smaller economies such as Nicaragua are starting to produce higher-value processed products such as clothing and food. But Latin America economies will remain sensitive to global markets in key areas such as minerals and basic grains. For example, in 2008, increase in U.S. demand and policy incentives for ethanol made from corn resulted in reduced availability of corn for food in Mexico, sending food prices higher and causing social unrest. Fluctuations in copper and oil prices are tightly coupled to economic performance in Chile, Venezuela, and Mexico.

Politically, even Cuba is likely to move toward an elected government in the near future, and although economic instability or corruption can still trigger military intervention (as in Honduras in 2009), citizens and international allies seem firmly committed to democratic rule in the region. In many countries the challenge is how to build a firm basis of governance at regional and local levels, and to balance the privatization and government cuts under neoliberalism with a stronger safety net for the poor and a more stable tax base.

# Main Points Revisited

■ **The region includes 30 independent countries, where colonial global connections resulted in many countries now using Spanish as the official language. European languages and institutions are also evident in Portuguese-speaking Brazil and in Caribbean territories of the United States, the United Kingdom, France, and the Netherlands. Although a region of conflict and authoritarian rule for much of the 20th century, with several interventions by the United States, Latin American countries are now mostly governed democratically.**

The term Latin America was coined in the 19th century by the French, who sought to discourage British interests in the region and justify their own imperial ambitions there by asserting that the shared Romance languages of Spanish, French, and Portuguese—all of which were derived from Latin—were the defining characteristic of the region and because of a shared legacy of European colonial domination. Although most countries are now democracies, French Guiana and a dozen Caribbean islands are still governed by European countries or the U.S. There are continuing tensions around rights to land, indigenous concerns, and the violence and corruption associated with the drug trade.

■ **The physical geography of this region is situated within a wider set of climatic and geological processes that link this region to the global patterns of atmospheric circulation and tectonics. These include the atmospheric systems of the tropical convergence zone over the** Amazon and the wind systems that cross the Atlantic and Pacific Oceans as well as the complex geology of uplift that creates high mountain ranges from Alaska to the Andes and the arc of volcanic Caribbean islands.

Several globally important crops were domesticated from the natural environment of the region including corn and potatoes. The valuable biodiversity of the region is under threat from deforestation and climate change, which is also creating risks to water resources, especially as Andean glaciers disappear.

■ **Key historical periods include the early complex societies of the Aztecs, Incas, and Mayas, who made large-scale modifications to their environment and European colonialism which left a legacy of demographic collapse and of export crops such as sugar, coffee, and bananas and metals such as silver and copper dominating the economies of many countries.**

Continuing reliance on export crops and minerals makes some countries vulnerable to shifts in global markets. The contemporary composition of the region includes those indigenous people who survived the demographic collapse (especially in the Andes and Central America), many people of African descent brought to the Caribbean and Brazil as slaves and a mixed (mestizo) population who have some European ancestry.

■ The 20th-century economic development policies parallel those in many other developing regions, and have included import substitution and then neoliberalism in which many countries expanded trade, privatized resources and institutions, and cut government spending.

Countries such as Brazil and Mexico have developed diverse economies including manufacturing and services but experienced economic shocks associated with debt and financial crises and have high levels of income inequality. A backlash against neoliberalism in countries such as Brazil and Bolivia has led to the election of left wing and populist leaders who oppose the privatization of water resources and the loss of government safety nets for the poor.

■ Following worldwide patterns, fertility has dropped significantly, and many people have moved to rapidly growing urban areas that have serious environmental problems and large underserviced housing areas.

Latin America's megacities, such as Mexico City and São Paulo, dominate the economies of their countries and are surrounded by informal settlements called barrios or favelas, and have high levels of air pollution and problems with waste disposal. Emigration from the region, especially to the United States, has diffused Latin American culture around the world. Social and health indicators are better than in many other developing regions.

# Key Terms

altiplano (p. 239)

altitudinal zonation (p. 235)

biodiversity (p. 241)

bioprospecting (p. 246)

bracero (p. 258)

circle of poison (p. 246)

Columbian Exchange (p. 248)

demographic collapse (p. 248)

dependency school of development theory (p. 252)

ecotourism (p. 246)

El Niño (p. 236)

encomienda (p. 249)

favela (p. 260)

free trade zone (p. 260)

Green Revolution (p. 245)

hacienda (p. 249)

informal economy (p. 254)

La Niña (p. 236)

liberation theology (p. 264)

mestizo (p. 256)

Monroe Doctrine (p. 266)

neotropics (p. 242)

nontraditional agricultural exports (NTAEs) (p. 254)

North American Free Trade Agreement (p. 253)

offshore financial services (p. 261)

plantation (p. 249)

pristine myth (p. 244)

remittances (p. 258)

social movements (p. 269)

structural adjustment policies (p. 252)

Treaty of Tordesillas (p. 247)

viceroyalty (p. 247)

# Thinking Geographically

1. In what ways did the people of Latin America adapt to and modify nature prior to the arrival of Europeans, and how did colonization then change ecology and environment through the Columbian exchange and other processes?

2. Which regions of Latin America are particularly vulnerable to natural hazards, and why and how might climate change and alterations in land use such as deforestation increase these risks?

3. How did colonialism restructure the economies of Latin America? How did import substitution attempt to change the structure, and why was it then rejected in favor of neoliberalism and free trade?

4. How does the movement of people and drugs affect the relationship between the United States and Mexico, and what are alternative solutions to problems associated with these movements?

5. How have urban-rural contrasts driven the growth of major cities in Latin America, and what are the main environmental and social problems of these urban areas?

6. To what extent did Latin America shift from authoritarianism and revolution to democracy in the late 20th century? How has the rise of social movements influenced politics in the region?

7. Explain how and where nontraditional export crops, aquaculture, tourism, bioprospecting, offshore financial services, and U.S. involvement affect the economies of the Caribbean and Central America. Discuss the criticisms of some of these activities.

Log in to www.mygeoscienceplace.com for Videos, Interactive Maps; RSS feeds; Further Readings; Suggestions for Films, Music and Popular Literature; and Self-Study Quizzes to enhance your study of Latin America and the Caribbean.

## EAST ASIA KEY FACTS

- Major subregions: Tibetan Plateau, Central mountains, Continental margin, Japanese archipelago

- Total land area: 9.6 million square kilometers/ 3.7 million square miles

- Major physiographic features: The Tibetan Plateau holds the Himalayan Mountains (including Mount Everest) and is the source of the great Mekong and Indus Rivers. The plateau is responsible for the low levels of precipitation in the region, as well as the cool summers and cold winters

- Population (2010): 1.56 billion, including 1.354 billion in China, 128 million in Japan, and 49 million in South Korea

- Urbanization (2010): 49%

- GDP PPP (2010) current international dollars: 29,337; highest = Hong Kong, 41,964; lowest = Mongolia, 3,730; China = 6,786

- Debt service (total, % of GDP) (2005) : 1.8%

- Population in poverty (%, < $2/day) (1990–2005): 39.8%

- Gender-related development index value (2005): 0.83; highest = Japan, 0.94; lowest = Mongolia, 0.70

- Average life expectancy/fertility rate (2008): 72 (M71, F77)/1.7

- Population with access to Internet (2008): 45.8%

- Major language families: The dominant Sino-Tibetan family contains the various subbranches of the Chinese language. Japonic and Altaic families are also widely spoken.

- Major religions: Buddhism, Daoism, Confucianism, and variously practiced folk religions

- $CO_2$ emission per capita, tonnes (2004): 6

- Access to improved drinking-water sources/sanitation (2006): 87%/72%

# 8

# EAST ASIA

At undisclosed locations over several years between 2001 and 2007, somewhere on Chongming Island at the mouth of China's Yangtze River, 15 alligator hatchlings were released into the wild. Scientists watched the small reptiles make their way into the mud and vanish into the undergrowth, hoping that this small population would establish itself, reproduce, and expand across its native range.

This Chinese alligator, known locally as *tu long*, or "muddy dragon," is one of only two existing alligator species in world. Its more famous cousin, the American alligator, thrives in the Southeastern United States. Here in the Yangtze, however, the official status of the alligator is now "critically endangered," according to the International Union for Conservation of Nature, the key organization that tracks global biodiversity.

Originally *tu long* could be found throughout the deltas of the Yangtze, which is the world's third-longest river and home to numerous endemic (regionally unique) species. With decades of agricultural and industrial pollution on the Yangtze, however, the ecology of the system has been radically changed. Massive flood control efforts have also transformed the environment. The Three Gorges Dam is in its final stage of construction on the Yangtze. As the world's largest hydroelectric dam, it has fundamentally transformed the flow of the waters in which these alligators are struggling to survive.

The plan to reintroduce Chinese alligators was hatched a decade earlier, in 1999, when as few as 130 individuals remained in the wild. The animal had become near extinct by this time and was only being bred in captivity. Indeed, several of the hatchlings released included individuals that were actually offspring of alligators from the Bronx Zoo in New York, where they have been bred in captivity for many years, long isolated from their home on the Yangtze. Cooperation between scientists from around the world made it possible to collect a genetically diverse set of eggs and begin the slow effort to bring the animals back. Chinese authorities, increasingly aware of the threat to the environment that river use has created, helped to support the program.

Whether or not these 15 little alligators become the source of a healthy new population, the case of the Chinese alligator demonstrates several of things. First, it stresses the enormous impacts that generations of agricultural and industrial development have had on the ecologies of East Asia. As a densely populated and economically vibrant region, the hidden ecological costs of the "economic miracles" in Taiwan, South Korea, and China are only now coming to light. In 2006, for example, a search for the Yangtze River dolphin, or *baiji,* which long thrived in these waters, failed to find a single one.

Second, it shows how the government of China, like other countries in East Asia with their peculiar mix of free market capitalism and state control, has come to reevaluate and consider these hidden costs of development. The future economic and ecological character of the region depends on where future government priorities go.

Finally, the case reminds us that regions become the way they are precisely through their linkages to *other* regions. Consider that this effort to reestablish a peculiarly indigenous species, endemic to the muddy flats of the Yangtze, required exchanges of knowledge, eggs, and animals across the globe. The Bronx, in a very clear way, is an essential site for the conservation and reestablishment of native environments on the Yangtze. ■

## MAIN POINTS

■ East Asia is a region physically distinguished by the vast continental interior of China in the West and center, coupled with the peninsular and island coasts in the east formed at the edge of the Ring of Fire.

■ The landscapes of the region have been massively modified over the centuries, especially for the management of water flow.

■ The historical geography of the region is characterized by the core economic power of China, which was at the center of global trade networks for more than a thousand years prior to colonization and is becoming so again.

■ The development of postcolonial nation states in the late 20th century resulted in the creation of profoundly different economic experiments in China, South and North Korea, and Japan.

■ The cultural landscapes of the region are rooted in many dominant religious and philosophical traditions, including Confucianism, Taoism, and Shinotism, but these have been challenged and reformulated both through internal upheavals and ongoing interactions with cultural groups on the region's periphery.

■ The geopolitics of East Asia revolve around the shifting balance of economic power between Japan and China and the unresolved problem of the Korean Peninsula, where the Korean War has never officially come to an end.

# ENVIRONMENT AND SOCIETY

The most striking and geographically significant environmental feature of East Asia (**Figure 8.1**), with implications for regional climate, resource endowments, and historical flows of people, trade, and culture, is its position on the east end of the vast Eurasian continent, across which its strong connections to Europe, Africa, and South Asia have been secured for millennia. Its formidably high interior plateaus and towering mountain folds drive the engine of river systems outflowing onto the plains, and its peninsular and island coasts create a set of oceanic subregions abutting the continental mass (**Figure 8.2**). These characteristics are themselves the product of a long geotectonic history that links the region to other world regions, insofar as the Tibetan plateau and Himalayan ridge are the product of a collision with the South Asian plate, and the island boundaries (including all of contemporary Japan) are the result of tectonic activity along the western boundary of the **Ring of Fire** (Chapter 1, page 5). Together, these set the stage for the complex adaptation of regional cultures to climate, landscape, and a diverse set of ecosystems.

## Climate, Adaptation, and Global Change

The geological uplift of the Tibetan Plateau between 65 million and 2 million years ago is the key to understanding certain aspects of the climate of East Asia (**Figure 8.3**). The elevation of the Tibetan Plateau cut off the moisture that was formerly brought into the interior of East Asia from the Indian Ocean by monsoonal winds. This contributed to the gradual dessication of the Tibetan Plateau, the numerous lakes of which are now much reduced in size. Protected by mountain barriers and sheer distance from the coast, much of western and northwestern East Asia today averages less than 125 millimeters (5 inches) of rain a year. On the Tibetan Plateau, high elevations make for cool summers and extremely cold winters. Further north, in Xinjiang, Qinghai, Gansu, and Mongolia, summer temperatures may be extremely hot. The Turfan Depression, some 154 meters (505 feet) below sea level, is one of the hottest places in East Asia, with recorded temperatures in excess of 45°C (113°F).

To the east of this vast arid region are two distinctive climatic regimes. The northern regime is subhumid (slightly or moderately humid, with relatively low rainfall). It is bounded to the west by the Lüliang Mountains and the Greater Khingan Mountains and extends southward as far as a latitude of about 35° N, encompassing the Northeast China Plain, the North China Plain, the northern parts of the Korean Peninsula, and the Japanese archipelago. Winters here are cold and very dry. Summers are warm, with moderate amounts of rain from the southeasterly summer monsoon winds. Rainfall, however, is extremely variable, so that both drought and flooding occur frequently. The southern regime is humid and subtropical. It extends west from the plains of the Middle and Lower Chang Jiang valley as far as the Sichuan basin, and south as far as the southernmost coastlands of South China. Winters here are mild and rainy, and summers are hot with heavy monsoonal rains. Overall, annual rainfall is 1,200 millimeters (47 inches) or more higher than it is in the north.

In the arid and subhumid regions of East Asia, drought is a critical natural hazard, causing widespread famine as a result of crop failures in drought years. In addition, the subhumid parts of East Asia tend to be prone to flooding. To some extent, this pattern is a result of the irregular summer monsoon rains, which can produce a sudden deluge.

### Adapting to Semiarid and Subtropical Climates

Each of these distinctive climates has influenced the creative adaptations of regional cultures, especially as they adapt their production to very different rainfall patterns and temperature ranges. The western deserts and grasslands of Xianjiang and Mongolia, for example, have a long and well-developed herding tradition that mixes settled agriculture with herd mobility to utilize infrequent and unevenly distributed rainfall.

The rice cultures of southeastern China and Japan, in contrast, have developed systems to capture and control the flow of plentiful water to maximize production year-round, and sometimes combine rice production with fish production. This rice–fish culture, which dates to at least as early as 300 C.E., allows the production of protein by raising fish in wet rice ponds, where rice plants provide shade and organic material for the fish as the fish eat the pests that might attack the rice plants, while also oxygenating the water and the soil. The system also produces its own fertilizer, often making sup-

▲ **FIGURE 8.2 East Asia From Space** As this image shows, much of East Asia consists of vast areas of upland plateaus and mountain ranges.

▶ **FIGURE 8.3 Climate map of East Asia**
East Asia is divided between four main climates, including the variable highland climate of the Tibetan Plateau, the arid climates of the western interior, the continental climates of the north, and the humid subtropical climates of coastal China and southern Korea and Japan.

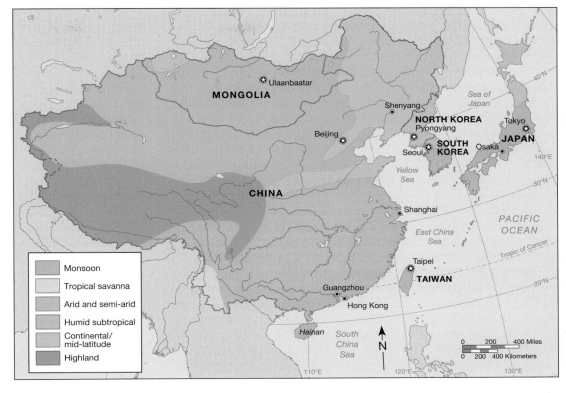

plementary chemical inputs unnecessary. This system operates precisely because of the plentiful water made available by the humid climate and the many flowing rivers, harnessed through complex systems of carefully designed irrigation canals and levees (**Figure 8.4**). This technique likely originated in China but has since diffused across East Asia and is found in use as far away as the island of Java in Southeast Asia and in parts of India. Innovative adaptations to climate change have, in this way, been a cultural connection of East Asia to the rest of the tropical world.

**Climate Change**   With a trend of warming over the last 50 years, which has been especially pronounced during the winters, future climate change presents serious problems and challenges. The impact of this overall warming is unclear, but it portends decreases in catch in the Pacific fisheries on which most countries in the region depend heavily, for example, and may do serious damage to the delta ecosystems along the coasts of China and Japan.

Rainfall regimes are less easy to predict because trends over the last century include drier average conditions in some parts of the region (including northeast China) but wetter trends in others (including arid western China). Intense rainfall appears to be on the increase, with more dramatic single rain events

▲ **FIGURE 8.4 Fish farms and rice paddies**   An aerial view of the fish farms and rice paddies along the border of Shenzhen and Hong Kong. This ancient technique of flooded fields maximizes production of both proteins and grains, by tapping the climatological advantages of subtropical rainfall and the vast river resources of the plains.

causing increased risks of flooding in western and southern China and Japan. Droughts, meanwhile, have become more common in areas to the north and east, with associated dust storms and crop failures in north China and livestock die-offs in Mongolia.

These conditions need not necessarily lead to destitution or starvation, and crisis and may open onto new regional adaptations. Drought-tolerant crop options are available and part of the historical crop diversity of the region. Similarly, traditional livestock breeds—including hardy Mongolian cattle varieties (the Ujumqin and Halhīn Gol varieties) and a range of Tibetan goats—are well adapted for water scarcity and variability. Some of the secrets for adapting to climate change in the future likely lie in the adaptive history of East Asian producers of the past.

Given the critical influence of urban activities on greenhouse gas emissions that drive climate change, urban areas in Japan and China represent an enormous challenge for mitigating climate change. China's $CO_2$ emissions (6.2 billion tons) surpassed those of the United States (5.8 billion tons) in 2006 (Netherlands Environmental Assessment Agency, http://www.mnp.nl/en). This figure continues to rise, largely as a result of ongoing Chinese urbanization and the transportation, construction, and lifestyle demands that go with this transition. Japan is already a major emitter of greenhouse gases (emitting 1.2 billion metric tons in 2003), though its rate per capita far exceeds that of China, due to Japan's much smaller population and its much more energy-demanding lifestyle. Where each person in China emits approximately 3 tons per year, those in Japan emit more than 9, whereas in the United States the rate is more than 24 tons per person. Future rounds of negotiations over the control of climate change will increasingly focus on China, therefore, especially as urban areas continue to grow.

## Geological Resources, Risk, and Water Management

Much of East Asia consists of juxtaposed plateaus, basins, and plains separated by narrow, sharply demarcated mountain chains. These broad physiographic regions contain a great diversity of river systems and elevational gradients, which present the region with water resources as well as hazards. The satellite photograph of East Asia in Figure 8.2 suggests a broad, threefold physical division of the Tibetan Plateau; the central mountains and plateaus of China and Mongolia; and the continental margin of plains, hills, continental shelves, and islands (**Figure 8.5**).

**Tibetan Plateau**    In the southwest, the Tibetan Plateau, an uplifted **massif** (a mountainous block of Earth's crust) of about 2.5 million square kilometers (965,000 square miles), forms a plateau of several mountain ranges—including the Himalaya Mountains—with peaks of 7,000 meters (22,964 feet) or more. The Tibetan Plateau is a unique physical environment, a vast area that has been violently uplifted in relatively recent geological time to produce the youngest, highest plateau in the world. It is surrounded by a series of lofty mountains that tower to heights of between 6,000 and 8,000 meters (19,684 to 26,245 feet): The Himalayas contain the world's highest peak, Mount Qomolangma (Mount Everest), at 8,848 meters (29,027 feet).

**Plains, Hills, Shelves, and Islands**    The bulk of the population of East Asia lives in this continental margin, which includes the great plains

of China: the Northeast China (Manchurian) Plain, the North China Plain, and the plains of the Middle and Lower Chang Jiang (Yangtze River) valley. Most of these plains lie below 200 meters (656 feet) in elevation. South of the Chang Jiang is hill country with elevations of around 500 meters (1,640 feet), and along the coast, including the Korean Peninsula, are uplifted hills and mountains of 750 to 1,250 meters (2,460 to 4,100 feet). The Japanese archipelago of Hokkaido, Honshu, Shikoku, and Kyushu forms the outer arc of East Asia's continental margin. Its backbone of unstable mountains and volcanic ranges that project from the shallow seafloor extends to the island of Taiwan and is part of the Ring of Fire that girdles the Pacific Ocean.

**Earthquakes in a Still-Forming Region**    This broad division is the product of a long and complex geological history. The key to understanding the basic physiography of the region is the plate tectonics that have had a dramatic influence in recent geological times. The entire Tibetan Plateau was uplifted between 65 million and 2 million years ago. The Indian–Australian plate moved northward and pushed up against the Eurasian plate, not only uplifting the Tibetan Plateau, but also causing the mountain-building episode that resulted in the Himalayas. Subsequently, the weathering of the newly created mountains provided huge quantities of clay, silt, and other fluvial deposits that now blanket the plains. To the east, the movement of the Pacific plate toward the Eurasian plate caused the folding and faulting that have resulted in the mountains of the peninsulas and islands of the continental margin, including the Korean Peninsula and the Japanese archipelago.

The geological uplift of the Tibetan Plateau also had consequences for geomorphic processes. It increased the gradient of the rivers that flowed off the borders of the plateau, enabling them to incise (or cut their way into) the plateau and produce a great number of gorges and canyons of considerable length and depth. The rivers are still incising rapidly, and the Tibetan Plateau itself is still not stable: Seismic disturbances continually occur along the Himalayan foothills, often bringing disaster to settlements there.

The physical environments of East Asia are, in fact, relatively hazardous. In addition to regular earthquakes along the Himalayan foothills, much of the North China Plain is subject to seismic activity. In May of 2008, a massive earthquake destroyed schools, houses, and factories across the Sichuan province of central China, taking the lives of approximately 70,000 people. Much of Japan, too, is subject to earthquake hazards. The Japanese archipelago is situated within one of Earth's most geologically active zones, at the junction of three tectonic plates: the Eurasian, Philippine, and Pacific plates. In addition to volcanic activity—at least 60 volcanoes have been active within historic times in Japan—there is almost perpetual earthquake activity. Most of the 1,000 or so earthquake events that occur in Japan each year are minor, but a few are serious enough to cause damage to property, and occasionally they can be devastating. The most recent disaster was in Kobe in 1995, when a severe earthquake killed 6,452 people.

**Flooding, Flood Management, and a History of Massive Hydroengineering**    In addition, the rivers that flow through the plains carry a tremendous amount of silt, which is then deposited in their more sluggish lower reaches, building up the height of the riverbed and making their course unstable. The Huang He (Yellow River) is the largest and most notorious of these rivers, having changed its course several times in the last two centuries. In between these events, the

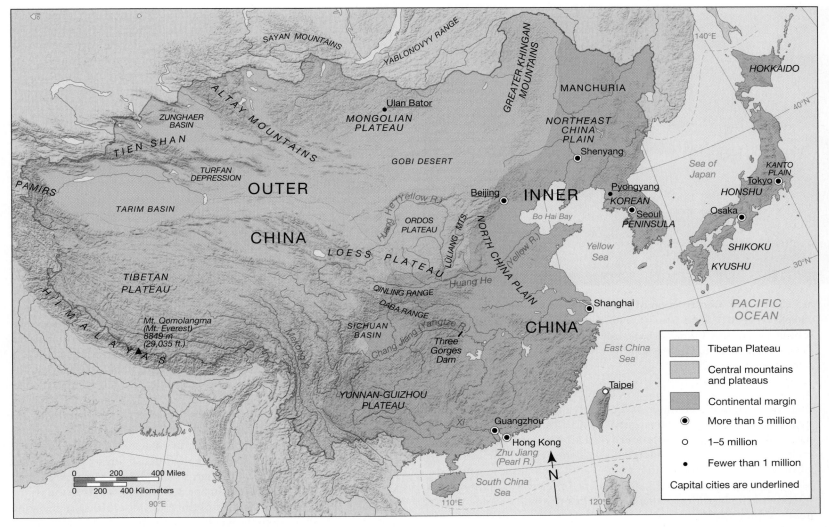

▲ **FIGURE 8.5 East Asia's Physiographic Regions** East Asia can be divided into three broad physiographic units: the Tibetan plateau; central mountains and Outer China; and the continental margins approximating to Inner China, the Japanese archipelago, the Korean Peninsula, and Taiwan.

Huang He regularly bursts its banks, flooding the farms and villages of the densely populated North China Plain.

Flooding events in China bring with them inundation of crops and massive mudslides, which can bury roads, disrupt power transmission, and on occasion, sweep away whole villages. In July of 2009 alone, for example, a single three-day continuous rainstorm affected 10 million people by flooding, caused 300,000 people to leave their homes, and killed 20 people.

Unsurprisingly, the people of East Asia have over the millennia developed sophisticated techniques to modify the landscape and achieve water control. Over the centuries, marshes were drained, irrigation systems constructed, lakes converted to reservoirs, and levees raised to guard against river floods. These traditions of water management are remarkable examples of how coordinated action from strong imperial states have tamed highly irregular and unpredictable systems. This success has caused some researchers to term the regions' historic empires as **hydraulic civilizations**, held together and expanded based on their capacity for controlling water.

The tradition continues into the present. A massive public works program aimed at hydroelectric facilities, irrigation and drainage, and improved waterways has been the cornerstone of the Chinese central government approach to the economic development of the country's interior. The Three Gorges Dam on the Chang Jiang (**Figure 8.6**) has become emblematic of this strategy and of China's ambitious determination to modernize. It has also, unfortunately, become emblematic of the problems of graft, corruption, and disregard for the social and environmental consequences of large-scale infrastructure development. The dam itself will span 2 kilometers (1.25 miles) from bank to bank. The reservoir will extend 650 kilometers (404 miles) upstream, submerging 19 cities, 150 towns, and 4,500 villages and hamlets. On completion, 32 giant generators will provide for almost 10% of China's energy output, bringing electric power to millions of rural households and opening central China to a much broader range of industries. The increased water level in the enormous reservoir will improve river navigation and should help flood control along the Chang Jiang. One significant potential problem is that the dam sits on an active fault

▲ **FIGURE 8.6 Construction of the Three Gorges Dam** Scheduled for completion by 2011, Three Gorges Dam is one of the world's largest-ever engineering projects. The dam is being built in central China at a spot where the Chang Jiang narrows to form the Xiling, Wu, and Qutang gorges.

line. The weight of the water in the reservoir could trigger an earthquake, with devastating consequences. Critics also predict that the dam will threaten the region's ecosystem. The river blockage will endanger several species indigenous to the region, including the Chinese alligator, the white crane, and the Chinese sturgeon. Meanwhile conservationists have no hope of recovering more than a fraction of the historical artifacts and relics that will become permanently submerged.

## Ecology, Land, and Environmental Management

The ecologies of East Asia follow closely on the broad land and climate forms that divide East Asia, but these have been subject to dramatic transformation under successive civilizations that have radically adapted and sculpted the environment over millennia. The principal human impact has been through clearing land for farming. Over the millennia, as the population grew and premodern civilizations flourished, much of humid and subhumid East Asia was cleared of its forest cover. Even the topography was altered to suit the needs of growing and highly organized populations, with hills and mountainsides being sculpted into elaborate terraces to provide more cultivable land.

Meanwhile, several of the world's most important food crops and livestock species were domesticated by the peoples of East Asia, beginning around 6500 B.C.E. Millet, soybeans, peaches, and apricots were domesticated in the more northerly subhumid regions, and rice, mandarin oranges, kumquats, water chestnuts, and tea were domesticated in the more humid regions to the south. Mulberry bushes were domesticated as silkworm fodder, and hemp was domesticated as a source of fiber and oil. Chickens and pigs were among the livestock species domesticated.

Only in the arid western parts of East Asia and the inhospitable Tibetan Plateau—a broad region often referred to as "Outer China"—do contemporary regional landscapes reflect large-scale natural environments that are relatively unmodified by human intervention. Outer China is an outback region that contains barely 4% of the population of East Asia. Large areas are effectively uninhabited, and in the greater part of it, population density does not exceed one person per square kilometer. The scenery is often spectacular, encompassing snow-clad mountains, vast swamps, endless steppes, and fierce deserts.

**Tibetan and Himalayan Highlands** Not surprisingly, the rugged terrain of the Tibetan Plateau is quite sparsely inhabited. Nomadic pastoralism remains the chief occupation within the region; the principal source of livelihood is the yak, along with sheep and goats. Yaks supply milk and wool as well as meat. Yak wool is the fabric used for the darkish, rectangular tent that is a distinctive component of the Tibetan landscape. In recent decades, Chinese reforms have brought an expansion of farming to the more sheltered valleys of the region, with wheat, highland barley, peanuts, and rapeseed as the principal crops. The mixed forests of Gaoshan pine, Manchurian oak, and Himalayan hemlock are increasingly being brought into timber production, accelerating a long history of deforestation in the region and extending the imprint of humankind on this wild region. At the eastern edge of the plateau, overgrazing, combined with a decade of hotter, drier weather, has destroyed the thin topsoil, with the result that this legendary horse country is now gradually being degraded.

**Steppe and Desert** Northwestern China (**Figure 8.7**) and much of Mongolia is a territory of temperate desert—dunes, scrub-covered hillslopes, and basins with plant communities that are tolerant of high levels of mineral salts in the soil—although dense forests of tall spruce clothe the cool, wet flanks of the Tien Shan and Altay mountains. Inner Mongolia is mostly a mixture of steppe and semidesert. As in the steppe lands of Central Asia (to the west of the Altay Mountains), a gradation of landscapes ranges from the wooded steppe through the grassy steppe proper and the desert.

Much of Mongolia is taken up by the Gobi Desert, where bare rock and extensive sand dunes dominate the landscape. Several spectacular natural features are found in the Gobi, including huge basalt columns arranged in clusters resembling groups of pencils. These natural landscapes were impacted significantly from the 1950s to the 1980s by an extensive virgin lands program similar to that in the Soviet Union (see Chapter 3), whereby hundreds of thousands of acres of virgin land were plowed and large-scale irrigation systems were installed to allow for farming—especially wheat, root vegetables, and fodder crops. Meanwhile, Mongolia's extensive mineral resources (including oil, coal, copper, lead, and uranium) began to be exploited, and an industrial sector emerged, stimulating urban growth.

**Inner China, Korea, and the Japanese Archipelago** The remainder of East Asia—"Inner China" plus the Korean Peninsula, the Japanese archipelago, and Taiwan—can be divided broadly into the landscapes of the subhumid regions to the north and those of the humid and subtropical regions to the south. The natural landscapes of the northern part of Inner China, along with the northern part of the Korean Peninsula and the northern half of the Japanese archipelago, are dominated by mixed temperate broad-leaf and needle-leaf forests on the hills and mountains. Higher elevations, and more northerly latitudes, are dominated by spruce, fir, and birch trees; lower elevations and more southerly lati-

▲ FIGURE 8.7 Northwest China Surrounded by mountain ranges, the vast plains and deserts of northwest China stretch for thousands of kilometers.

tudes are dominated by maple, basswood, and oak. The North China Plain has been modified by thousands of years of human occupancy and agriculture (Figure 8.8). The capitals of the Shang and Qin dynasties were established here, and the Mongol Yuan dynasty established the imperial capital in Dadu, which was later renamed, under the Ming dynasty, as Beijing.

In contrast, the winter landscape of the North China Plain is brown and parched, and on windy days the air is thick with the dust that blows easily from the silty soils. The rivers that flow through the region—especially the Huang He and the Huai He—are the agents that have nourished the plains, a region of relatively low rainfall; they are also the agents of disaster, bringing occasional floods of devastating impact. The Huang He carries massive quantities of silt that make for productive farming, but that make its course unstable, with the result that it regularly bursts its banks.

To the south of the Qinling and Daba ranges, including Taiwan, the southern part of the Korean Peninsula, and the southern half of the Japanese archipelago, contemporary landscapes are dominated by low

▼ FIGURE 8.8 Inner China's northern farming landscape The Northeast China Plain, the initial core region of Chinese civilization, has the largest concentration of cultivated land in China. Winter wheat, corn, sorghum, millet, and cotton are the chief crops.

# CENTRAL CHINA

Central China—the middle and lower reaches of the Chang Jiang (**Figure 1**)—is a hearthland area of Han China and of Chinese agricultural civilization. By the late 17th century, the middle reaches of the Chang Jiang had become the largest center of commercial food (mainly rice) production in China, while the Chang Jiang Delta flourished as a center of handicraft industries. Today, Central China accounts for almost a quarter of all cultivated land in China. It is one of the most densely populated regions of East Asia and contains China's largest city, Shanghai (population 13.1 million in 2005), and several other key metropolitan centers, including Wuhan (5.93 million), Nan-jing (2.84 million), Hangzhou (1.96 million), Nanchang (1.99 million), Hefei (1.33 million), and Suzhou (1.36 million).

Central China has developed around waterways. A series of lake basins stretches from the Three Gorges in the middle reaches of the Chang Jiang (see Figure 8.6) eastward to the sea for a length of more than 1,800 kilometers (1,119 miles) and a total area of about 160,000 square kilometers (61,776 square miles, roughly the size of Michigan). The Chang Jiang has long been the west–east "Golden Waterway," whereas its tributaries, together with the Grand Canal, have provided north–south route ways. The Chang Jiang is prone to flooding, however. The flood of 1887 was probably the most disastrous

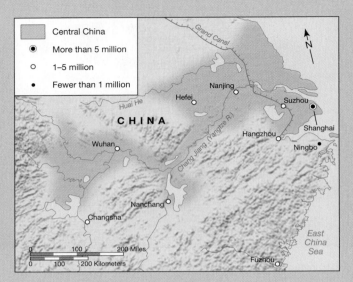

▲ **FIGURE 1  Map of Central China**  Reference map showing principal physical features and major cities.

ever recorded in the world; an estimated 7 million people lost their lives. In 1931 another disastrous flood claimed approximately 200,000 lives. More recently, floods in 1998 led to the loss of more than 3,000 lives, the evacuation of 14 million people, and damage amounting to $24 billion. Throughout the region, levees must be continuously repaired and improved to guard against river floods.

Water dominates the landscapes of Central China. The large lakes that border the Chang Jiang occupy more than 18,000 square kilometers (6,950 square miles), and all along the river itself are smaller lakes and ponds, together with tens of thousands of kilometers of canals, irrigation ditches, and thousands of linear ponds (**Figure 2**). The countryside is characterized by the rectilinear patterns of the drainage channels and the canals that link the smaller settlements of the region (**Figure 3**). This "land of rice and fish" is extremely productive; in many locales the humid subtropical climate allows for a triple-cropping system (two crops of rice, plus one of winter wheat or barley). Since 1949, Central China has been developed as a core industrial region as well as a key food-producing region. The middle Chang Jiang plains contain one of the most important oil fields in China, and the region's energy resources will be significantly increased by hydroelectric power from the Three Gorges Dam, due for completion in 2011 (see p. 277). The industrial belt of the middle Chang Jiang includes two concentrations of heavy industry: steel, engineering, and shipbuilding in Wuhan and automobiles in western Hubei. The Delta region of the Chang Jiang is the most important economic hub, however, with Shanghai standing at the center of an extensive industrial complex dominated by engineering, textiles, chemicals, and electronics. ■

---

hills that are covered with a secondary growth of evergreen monsoon forest, leaving less than 10% of the land for agriculture. Nevertheless, the cultivable land is very productive, and the climate allows for double-cropping of paddy rice and for cash crops such as sugarcane, mulberry trees, and hemp. Water and waterways are of vital importance in this region (see Signature Region in Global Context: Central China).

Around Guilin and in the Zuo River area to the south, near Nanning, erosion has dissolved areas of soft limestone, leaving spectacular pinnacles several hundred feet tall. These improbably shaped hills contrast with intermediate areas of intensely cultivated flatlands and winding rivers, making for some strikingly beautiful landscapes (**Figure 8.9**).

**Impacts of the Green Revolution**  The land management systems of East Asia are thousands of years old, well established, and rooted in large-scale manipulation of both land and water. Technological innovations in production have swept the region since the 1960s, with significant implications for environmental quality and human health. These green revolution strategies involved the introduction of new varieties of crops, which produced far more food, but also required higher levels of industrial inputs, including fertilizer, water, and pesticides. The environmental impacts of these changes are potentially enormous, especially considering the vast areas of East Asia under cultivation. In China, for example, 554 million hectares of land are under agricultural cultivation, almost 13% of the global total. In China, 257 kilograms of fertilizer are used per hectare (more than twice the global average). Released in waterways, fertilizers can radically damage biodiversity and the health of streams and rivers.

More dramatically, the heavy use of herbicides and insecticides, which are a crucial component of Green Revolution technology, have resulted in widespread illness in rural areas. According to Chinese researchers, there were more than 214,000 cases of acute pesticide poisonings in China between 1992 and 1995 alone. This high incidence

◀ **FIGURE 2 Central China** Rice fields with fishponds in the Chang Jiang Delta region.

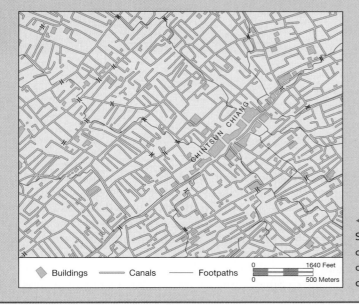

◀ **FIGURE 3 Central China: Rural settlement pattern**
Settlements in the watery landscapes of Central China are dispersed along the rectilinear patterns of drainage channels and canals. Shown here is the settlement pattern of the Fengshien district of the Chang Jiang Delta.

rate has led in recent years to new provisions for rural health care in the country. It has also encouraged the adoption of genetically modified varieties of crops, especially so-called Bt crops, which are modified to produce a protein that causes the crop to require fewer pesticide inputs. The result has been a decline in such deaths overall. Nonetheless, the high incidence of pesticide poisonings suggests that many other environmental effects from intensive agriculture may be occurring in waterways and soils in China that have gone unmeasured and unobserved.

**Conservation Problems**    The dense settlement of East Asia also holds implications for native fauna. Five important mammal species have become extinct in Japan (including the wolf). In China there are more than 385 threatened species. Tigers are all but gone in China, where 90% of their population vanished in the last century. More prominent and iconic among threatened species is the giant panda. These ani-

mals lost half their habitat (upland forest) in the 1970s and 1980s, bringing them to the edge of extinction. They present an especially difficult conservation challenge due to the difficulty of breeding in captivity (**Figure 8.10**).

There may be a location where East Asian biodiversity has thrived over the last few decades, though it remains hard to tell. This area is the so-called DMZ (or demilitarized zone), a no-man's land located between North and South Korea and established at the end of hostilities between the two powers in the 1950s (see later discussion). Since that time, the area has been entirely uninhabited and has been heavily strewn with land mines. As a result, wildlife has come to thrive in the long corridor between the two states. There are even rumors that tigers survive here, though these reports are so far unsubstantiated.

**Environmental Impacts of Urbanization**    Throughout East Asia, cities continue to grow at a rapid rate, stretching outward to devour not

▲ **FIGURE 8.9 The Guangxi Basin** In this region a unique combination of geomorphology and agricultural development has led to one of the world's most spectacular landscapes. From classic old landscape paintings to contemporary tourist posters, the dramatic karst (eroded limestone) landscape of the Guangxi Basin has come to embody a stereotypical traditional Chinese landscape.

only agricultural land, but also natural areas. This problem is most acute in Taiwan, which is an island of merely 36,000 square kilometers, two-thirds of which is a mountainous and hilly region surrounded by burgeoning coastal urban development. Here the continuous increase in urbanized areas moves inevitably inward toward the island's natural core areas (**Figure 8.11**).

Efforts to protect core areas from rapid urban growth take two forms. First, establishment of national parks and protected areas in East Asia is increasing rapidly. China claims 145 areas larger than

▲ **FIGURE 8.10 Giant Pandas Eating Bamboo** Since most of the native forest habitat of giant pandas (Ailuropoda melanoleuca) in China has been destroyed, many pandas live in captivity in China, at places like the Wolong Giant Panda Captive Breeding Center in Sichuan Province, China.

100,000 hectares under protection, Japan claims 7, and South Korea 1. The total land area dedicated to these is inevitably limited, however, and rarely are they situated in the path of urban growth. Although six national parks have been established on Taiwan, they cover only 308,000 hectares, less than 13% of the core nonurbanized area. The second mode of protection is through deliberate planning efforts to manage density and capitalize on high-density residential and industrial development. In other words, many East Asian cities are increasingly working to achieve urban economic growth without paving over adjacent natural areas.

Like much of East Asia, Taiwan has a significant cultural advantage when it comes to preserving open lands in this way. People in the region are more historically tolerant of high-density urban living, which allows more people to use less space. Vertical urban development in major Taiwanese cities (in high-rises and closely packed homes) helps offset the pressure of urban growth on the environment. Whether this culture persists, on the other hand, as wealthier residents become capable of buying and developing larger properties, remains to be seen.

**The Asian Brown Cloud** Emblematic of urban growth problems is the Asian Brown Cloud, a blanket of air pollution 3 kilometers (nearly 2 miles) thick that hovers over most of the tropical Indian Ocean and South, Southeast, and East Asia, stretching from the Arabian Peninsula across India, Southeast Asia, and China almost to Korea. The brown haze was first identified by U.S. Air Force pilots but is now clearly visible in satellite photographs and from the Himalayas (**Figure 8.12**).

The Asian Brown Cloud consists of sulfates, nitrates, organic substances, black carbon, and fly ash, along with several other pollutants. It is an accumulated cocktail of contamination resulting from a dramatic increase in the burning of fossil fuels in vehicles, industries, and

▲ **FIGURE 8.11 City of Taipei** The sprawling city of Taipei is characterized by high-rises and pollution, but it is within close proximity to large natural areas, like Yangmingshan Mountain, from where the picture is taken.

power stations in Asia's megacities, from forest fires used to clear land, and from the emissions of millions of inefficient cookers burning wood or cow dung. A study of the Asian Brown Cloud sponsored by the UN Environment Programme and involving more than 200 scientists suggests that the Asian Brown Cloud not only influences local weather but also may have worldwide consequences.

The smog of the Asian Brown Cloud reduces the amount of solar radiation reaching the Earth's surface by 10% to 15%, with a consequent decline in the productivity of crops. But it can also trap heat, leading to warming of the lower atmosphere. It suppresses rainfall in some areas and increases it in others, while damaging forests and crops because of acid rain. The haze is also believed to be responsible for hundreds of thousands of premature deaths from respiratory diseases.

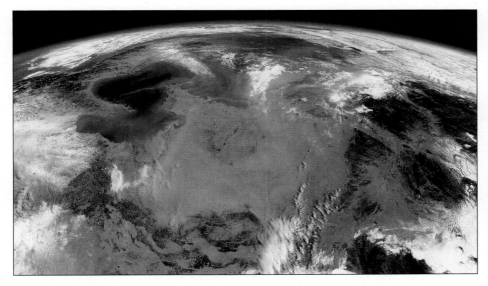

▲ **FIGURE 8.12 The Asian Brown Cloud** This view shows a dense haze over eastern China, looking East across the Yellow Sea toward Korea.

# HISTORY, ECONOMY, AND DEMOGRAPHIC CHANGE

East Asian civilizations are indisputably some of the oldest in the world, and the mark of these cultures is indelible on the contemporary landscape (think of the Great Wall of China, which is 2,500 years old and still visible from space). The four key essential geographic elements of this long history are enormously successful agricultural production established by large, well-settled civilizations; persistent state territories created by political consolidation; the region's role as the historical eastern anchor for a global trade system long before the colonial era; and the postcolonial era leading to highly divergent economic experiments established in differing parts of the region.

## Historical Legacies and Landscapes

**Empires of China**   China has had a continuous agricultural civilization for more than 8,000 years. The first organized territorial state was that of the Xia dynasty, a Bronze Age state that occupied the eastern side of the Loess Plateau (in present-day Shanxi Province) and the western parts of the North China Plain (northwestern Henan Province) between 2206 and 1766 B.C.E. It was succeeded by the Shang dynasty (1766–1126 B.C.E.), during which walled cities appeared. The first unified Chinese empire, though, was that of the Qin dynasty. Emperor Shih Huang-ti (221–209 B.C.E.) established an imperial system that lasted for 2,000 years. He abolished feudalism, replacing the feudal hierarchy with a centralized bureaucratic administration, and had the Great Wall (**Figure 8.13**) built to protect China from "barbarian" nomads. The stability brought about by the Qin dynasty was a precondition for the success of the Silk Road (**Figure 8.14**) as an economic and cultural link among the civilizations of China, Central Asia, India, Rome, and later, Byzantium. This trade route made East Asia the effective center of gravity for global trade for centuries, a role the region is beginning to reestablish in the present era.

The history of the Chinese empire is complex, with constantly shifting territorial boundaries and successive dynasties that tended to move from vigorous beginnings, with power concentrated around a strong center, followed by a slow loss of control as regional centers gained more power, to a final collapse as a forceful new dynasty was established. Sometimes, the empire was fragmented or subdivided for extended periods, but over two millennia the overall trend was for a larger and more consolidated continental empire.

The first major dynasty to run through the cycle was the Han dynasty, from 206 B.C.E. to C.E. 220. Han emperors extended the Great Wall westward, allowing the Chinese to control more trade routes along the Silk Road. Another important economic development took place early in the 7th century, under the Sui dynasty, when the northern and southern regions of Inner China were linked by the first of a series of Grand Canals. Its purpose was to bring the plentiful rice of the south to the Sui capital in the northwest (present-day Xi'an), and to the armies stationed in the northeast.

Over the next several centuries, the canal system was enlarged and modified. By the 15th century, China's canal

▲ **FIGURE 8.13 The Great Wall** The Qin dynasty (316–209 B.C.E.) began construction of the Great Wall. The Wall that visitors see today dates from the Ming period and was built at various times between the late 14th and mid-16th centuries.

rigid system in which a subjugated peasantry sustained a relatively sophisticated but inward-looking civilization. Japan's distinctive civilization was largely a result of the introduction of Buddhism, which arrived from India via China and Korea. Buddhist influence became so pervasive during the 7th and 8th centuries that the ruling elite established a new capital in C.E. 794 to make a clean break from the powerful temples in Nara, the old capital. The new capital, Kyoto, was to be the residence of Japan's imperial family for more than 1,000 years, during which it became the principal center of Japanese culture.

For the last 200 years of this period, just as an industrial system was developing in the Western Hemisphere, the Tokugawa dynasty (1603–1868) strove to sustain traditional Japanese society. To this end, the patriarchal government of the Tokugawa family excluded missionaries, banned Christianity, prohibited the construction of ships weighing more than 50 tons, closed Japanese ports to foreign vessels (Nagasaki was the single exception), and deliberately suppressed commercial enterprise. At the top of the imperial social hierarchy were the nobility (the shogunate), the barons (daimyos), and the warriors (samurai). Farmers and artisans represented the productive base exploited by these ruling classes, and only outcasts and prostitutes ranked lower than merchants.

In terms of spatial organization, the Japanese imperial economy was built around a closed hierarchy of castle towns, each the base of a local shogun. The position of a town within this hierarchy depended on the status of the shogun, which in turn related to the productivity of the agricultural hinterland. As a result, the largest cities emerged among the rich alluvial plains and the reclaimed lakes and bay-heads of southern Honshu. Largest of all was Edo (known today as Tokyo), which the Tokugawa regime had selected as its capital in preference to the traditional imperial capital of Kyoto and which, bloated by military personnel, administrators, and the entourages of the nobility in attendance at the Tokugawa court, reached a population of around 1 million by the early 19th century. Kyoto and Osaka were next largest, with populations of between 300,000 and 500,000, respectively, and they were followed by Nagoya and Kanazawa, both of which stood at around 100,000.

system was more than 1,000 kilometers (621 miles) in length. The imperial grain transportation system along the canals employed up to 150,000 soldiers to man its fleet and required the compulsory labor of many more to dredge and maintain the channels. (Canals of comparable scale only began to be cut in Europe—most notably in France—in the 18th century.)

The infrastructure of China's canal system created a complementary relationship between the north and the south. The economic center of gravity was in the agriculturally prolific south, but the political center was almost always in the north. In 1279 the Mongols, under Kublai Khan, conquered China, establishing imperial rule for the first time over most of both Inner and Outer China. The Mongol dynasty, known as the Yuan dynasty, was expelled after less than 100 years and was succeeded by the Ming (1368–1681) and the Qing (1681–1911) dynasties.

**Japanese Civilizations and Empires** The Chinese concept of the centralization of power in an imperial clan spread to Japan in the 6th century C.E., and as in China, a succession of dynasties maintained a

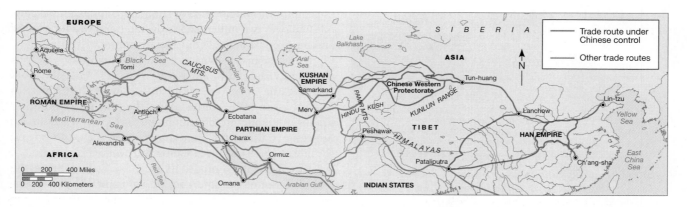

▲ **FIGURE 8.14 The Silk Road** The political and economic stability created by the Qin dynasty was a precondition for success of the Silk Road, the overland system of trade routes that connected China with Mediterranean Europe. The map shows the trade routes of the Silk Road as they existed between 112 B.C.E. and C.E. 100.

**Imperial Decline**   The dynasties of Imperial China and Imperial Japan both eventually succumbed to a combination of internal and external problems. Internally, the administration and defense of growing populations began to drain the attention, energy, and wealth of the imperial regimes. The imperial system itself also meant that cultural and social elites tended to be focused on the arts, humanities, and self-promotion at the imperial court, rather than on economic or social development. As the economy stalled under these constraints, peasants were required to pay increasingly heavy taxes, driving many into grinding poverty and thousands to banditry.

By the early 19th century, both China and Japan had moved into a phase of successive crises—famines and peasant uprisings—presided over by introverted and self-serving leadership. As in feudal Europe (see Chapter 2), the peasantry began to flee the countryside in increasing numbers in response to a combination of rural hardship and the lure of the relative freedom and prosperity of the cities. Finally, the imperial courts of both China and Japan suppressed the spread of knowledge of modern weapons because they feared internal bandits and domestic uprisings. As a result, both empires relied on antiquated military technologies, even as Europeans and Americans were racing ahead with new weapons developed through the new technologies of the Industrial Revolution.

The suppression of weapons technology contributed to the external problems that undermined both imperial regimes. European traders had been a growing presence in East Asia in the 18th century but had been restricted to a few ports. Initially, Europeans bought agricultural produce (mainly tea) in exchange for hard cash (in the form of silver currency). Over time, this proved to be an unacceptable drain on European treasuries, and the British eventually provoked China into a military response by insisting on being able to trade opium (grown in India for export by the East India Company) for tea and other Chinese luxury goods. In 1839 the Chinese, having prohibited the sale of opium several decades before, destroyed thousands of chests of opium aboard a British ship. This was just the excuse the British needed to exercise their superior weaponry. The so-called Opium War (1839–42) ended with defeat for the Chinese and the signing of the Treaty of Nanking, which ceded the island of Hong Kong to the British and allowed European and American traders access to Chinese markets through a series of treaty ports (ports that were opened to foreign trade as a result of pressure from the major powers).

Shortly afterward, in 1853, U.S. Admiral Matthew Perry anchored his flagship in Edo Bay (now Tokyo Bay) to "persuade" the Japanese to open their ports to trade with the United States and other foreign powers. The Japanese quickly complied, and the lesson learned, on both sides, was East Asia's abject weakness in the face of superior Western military technology. In both China and Japan, fear of the Western powers galvanized feelings of nationalism and xenophobia. The same fear precipitated a period of civil war among the feudal warlords that culminated in revolutionary change.

**Japan's Industrial Revolutions**   Japan was first to react to the humiliation of Western assertiveness. Japan's transition from feudalism to industrial capitalism can be pinpointed to a specific year—1868—when the Tokugawa dynasty was toppled by the Meiji imperial clan. In the Meiji clan, a clique of samurai and daimyo were convinced that Japan needed to modernize to maintain her national independence. Under the slogan, "National Wealth and Military Strength," the new elite of ex-warriors set out to industrialize Japan as quickly as possible.

A distinctive feature of the entire process was the very high degree of state involvement. Successive governments intervened to promote industrial development by supporting capitalist monopolies (called **zaibatsu**). The Japanese financed this modernization by harsh taxes on the agricultural sector, but several other factors helped foster rapid industrialization in Japan in the late 19th and early 20th centuries, including a strong feudal cultural order, a well-developed educational system, and a very lucrative sericulture (silk production) system (see Geographies of Indulgence, Desire, and Addiction: Silk, p. 286). Finally, and most important, were the spoils of military aggression (**Figure 8.15**). Naval victories over China (1894–95) and Russia (1904–05) and the annexation of Taiwan (1895), Korea (1910), and Manchuria (1931) provided expanded markets for Japanese goods in Asia. In 1931 the Japanese army advanced into Manchuria to create a puppet state. In 1936 a military faction gained full political power and, declaring a Greater East Asian Co-Prosperity Sphere, began a full-scale war with China the following year. The military leadership overplayed its hand, however, by attacking the United States at Pearl Harbor in December 1941. The resulting sustained U.S. offensive against Japan, ultimately culminating in the dropping of atomic weapons on the cities of Hiroshima and

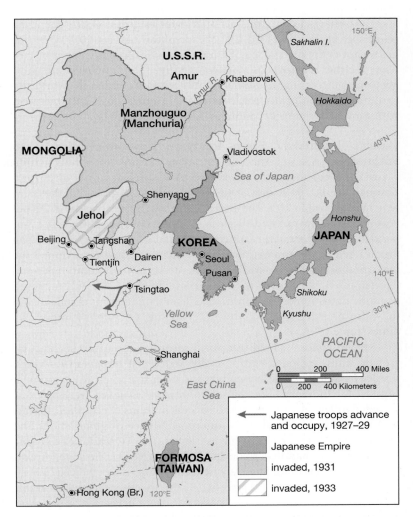

▲ **FIGURE 8.15 Japanese Expansionism**  In the late 1920s and early 1930s, Japan deliberately pursued a policy of military aggression to secure a larger resource base for its growing military-industrial complex.

Nagasaki (resulting in hundreds of thousands of deaths), led to Japan's ultimate defeat in World War II in 1945. Japan's industry laid in ruins. In 1946, output was only 30% of the prewar level.

# Economy, Accumulation, and the Production of Inequality

Resistance to colonialism and the formation of new nation states in East Asia during the postcolonial era brought with it turbulence and violence, but also an inflorescence of economic experiments. The result has been the establishment of several contrasting systems of economic management in the states of the region today, each with its own pattern of accumulation and uneven economic development. Japan, China, and South Korea, most notably, each have large economies, but their road to modern economic development has led down three very different regional paths.

These very different paths to economic development have led to a current economic climate in the region that is marked by interesting similarities. All the states in the region have geographically uneven patterns of development and accumulation. Agriculture and rural areas, most notably, have suffered in both Korea and China. All of them, moreover, are moving toward greater regional integration. Mutual investment and integration of firms between states are hallmarks of East Asian economies at the end of the first decade of the 21st century. What this has meant is the emergence of a new kind of economic region in East Asia, one that is *externally* more competitive internationally and shifting toward increasing accumulation and centrality within the world system, but one that is *internally* increasingly competitive and fractured, with some subregions expanding while others contract.

**Japan's Economy: From Postwar Economic Miracle to Stagnation**  Though the war left Japan in ruins, incredibly, within 5 years the economy had recovered to its prewar levels of output. Throughout the 1950s and 1960s the annual rate of growth of the economy held at around 10%, compared with growth rates of around 2% per year in North America and Western Europe. Having begun the postwar period at the bottom of the ladder of international manufacturing, Japan found itself at the top by 1963. By 1980, Japan had outstripped the major industrial core countries in the production of automobiles and television sets, for example. The Japanese, in short, not only achieved a unique transition directly from feudalism to industrial capitalism, but they also presided over a postwar "economic miracle" of impressive dimensions.

Several factors helped transform reconstruction into spectacular growth. These included exceptionally high levels of personal savings, rapid acquisition of new technology, and extensive levels of government support for industry. The economic miracle achieved by Japan was remarkable not only for its overall success in terms of economic performance, therefore, but also the interdependence of government and industry, characterized by some as "Japan, Inc." Orchestrated by the Ministry of International Trade and Industry (MITI), the state bureaucracy guided and coordinated Japanese corporations—organized in business networks (known as **keiretsu**). These together created favorable trade policies, technology policies, and fiscal policies to help Japanese industry compete successfully in the world economy.

Today, Japan is a globally powerful economy and has developed extensive linkages not only within East Asia but also throughout the

**Pacific Rim,** a loosely defined region of countries that border the Pacific Ocean. Though in the late 1980s the success of the Japanese economy made "Japan, Inc." the envy of global capitalism, the nation's economy has suffered significantly in the intervening period.

Starting in the early 1990s, Japan entered a period of stagnant economic growth, which persists to the present. In that decade, GDP grew at approximately 1.5% annually (while it grew in excess of 4% each year in the prior decade). The previous guarantees of Japanese economic culture (including long-term employment) were lost, suicide rates rose, and the government was unable to pull the economy from its nosedive. During the most recent economic crisis, this stagnation has turned to outright recession (as it has in many places). The strong role of banks in providing capital in the Japanese business model may have been a significant contributor to the crisis, in a period when global liquidity has declined around the world. It is possible, therefore, that the elements of the Japanese economy that allowed its miracle to occur may have contributed to the miracle's end. The most recent global economic recession has only added to the crisis. Dependent on export markets in the United States and elsewhere that have all but halted consumer spending since 2008, Japan's industries are facing their greatest challenge since World War II.

Perhaps more important, Japan has been forced to compete for foreign investment with China. Having developed manufacturing industries that undercut those of the United States in the 1970s and 1980s, Japan faces deindustrialization itself through the inexorable process of "creative destruction." Several Japanese electronics giants, including Toshiba Corp., Sony Corp., Matsushita Electric Industrial Co., and Canon, Inc., have expanded operations in China even as they have shed tens of thousands of workers at home. Olympus manufactures its digital cameras in Shenzhen and Guangzhou. Pioneer has moved its manufacture of DVD recorders to Shanghai and Dongguan. China's gains have often occurred at Japan's expense.

**China's Economy: From Revolutionary Communism to Revolutionary Capitalism**  Revolution came to China in 1911, when the Qing dynasty was overthrown and replaced by a republic under the leadership of Sun Yat-Sen's Nationalist Party (or Kuomintang). In the 1920s an alliance between China's fledgling Communist Party and the Nationalist Party, now led by Chiang Kai-shek, flourished for a while, but it ended abruptly in 1927 when Chiang Kai-shek attempted to quash the Communist Party. The Communists organized a strategic retreat in what became known as the Long March (1934–35), covering 9,600 kilometers (5,965 miles) through the rugged interior, in which more than 100,000 people perished. Mao Zedong emerged as the leader of the party. Mao devised a new strategy, aimed at gaining support for revolution from China's rural peasantry—who made up 85% of China's population. During the war years, Communist forces fought hardest and suffered most against the Japanese, and their experience proved crucial in fighting the Nationalists. By October 1949 the Communist Party had control over almost all China except for the island of Taiwan, where the Nationalist leadership had retreated under U.S. protection.

Mao Zedong now faced the task of reshaping society after two millennia of imperial control. In 1958, in what became known as the **Great Leap Forward,** Mao launched a bold scheme to accelerate the pace of economic growth. Land was merged into huge communes, and an ambitious Five-Year Plan was implemented. The impact of the new Five-Year Plan on the landscape was dramatic. Whereas pre-Communist China had an average farm size of 1.4 hectares (3.5 acres), the new agricultural

# SILK

Silk has long been symbolic of opulence, luxury, and social status. From its origins in China, silk became one of the principal items of trade along the trade routes between Asia and Europe. In Renaissance Europe, sericulture—the production of silk—and silk weaving were regarded as lucrative, leading-edge industries. Today, synthetic fibers have displaced lower-grade silks in most world markets, but good-quality silk fabrics remain highly desirable. Not only does their cost set them off as status symbols, but also their light weight, sensuous feel, strength, and resistance to creasing make them a preferred indulgence of many of the world's more affluent consumers.

Silk is manufactured from the fibers of the cocoons produced by silkworms, which live on the leaves of mulberry bushes. Sericulture was first practiced in North China around 2700 B.C.E. and was well established in that region by the 1st century B.C.E. Silk was used as currency when Qin dynasty and Han dynasty envoys traveled in Central Asia, and was the principal item carried by their merchants along the Silk Road (see Figure 8.14) through Central Asia and on toward India, Persia, and Rome. For centuries, China was the only country in the world capable of producing silk. The secrets of sericulture and silk manufacture were jealously guarded by early emperors: The penalty for attempting to export silkworm eggs or divulging the technique of silk weaving was death. Within China, silk fabrics came to embody the refinement and sophistication of the elite.

Sericulture reached Japan through Korea by the 3rd century, thus facilitating the first development of silk production outside China. Shortly afterward, sericulture was established in India. According to legend, two monks smuggled silkworms from China to Europe in the 6th century C.E.

Promoted by the Byzantine Emperor Justinian I (C.E. 527–65), sericulture spread to Syria, Turkey, and Greece. The more prosperous cities of the Mediterranean, pursuing an import substitution strategy, deliberately invested in silk manufacture, buying the raw silk from Greek and Arab merchants. By the end of the 14th century, silk production had been established in Bologna, Florence, Genoa, Lucca, and Venice. By the end of the

▲ FIGURE 1 Silk A woman waits for customers at a store on Silk Street, Beijing.

15th century, silk production played a vital economic role throughout the Italian Peninsula, from the Alps to Sicily.

Chinese silk fabrics continued to be imported into Europe, however, because of their superior quality. Chinese governments under the Ming dynasty (1368–1681) vigorously promoted sericulture, and great quantities of Chinese silk were exported to Southeast Asia, Japan, and Spanish America as well as to Europe. By the 18th century, Britain led European silk manufacturing because of British innovations in silk-weaving looms, power looms, and roller printing. At the beginning of the

19th century, a Frenchman named Joseph Jacquard developed a machine for figured-silk weaving that gave French manufacturers an edge. Then, with the opening of the treaty ports in the 19th century, exports of silk from China more than doubled. This was partly because of a silkworm disease that devastated sericulture in France and Italy in 1847, and partly because of the expansion of the world economy.

Silk was also important in Japan's entry to the modern world economy. With the Meiji revolution in 1868, the Japanese government organized a domestic silk industry, deliberately developing it with modern technologies as a major export sector. This enabled Japan to earn a significant portion of the foreign exchange it needed to purchase the machinery and raw materials necessary for its industrialization. Between 1870 and 1930, Japan's raw silk trade financed about 40% of its imports of foreign machinery and raw materials.

Japan's success came at the expense of China. Silk continued to be China's major export until the 1930s, but the Chinese silk industry was undercapitalized and unresponsive to changing market conditions. Political instability, followed by Japanese invasions, further weakened the industry's effectiveness, and after 1949 it was relegated to a very low priority in Mao Zedong's communist regime.

China's silk industry bounced back with the overall liberalization of the economy under Deng Xiaoping's leadership in the late 1970s. Since that time, China has once again become the world's main producer of silk (**Figure 1**), and world silk production has almost doubled, despite the fact that synthetic fibers have displaced silk for many uses. Together, China and Japan are responsible for more than half the world's total annual production of silk fabrics.■

communes averaged 19,000 hectares (46,949 acres, or 73 square miles) in size, with between 30,000 and 70,000 workers. Instead of a patchwork of fields, each with a different crop and presenting a rich palette of browns, yellows, and shades of green, vast unbroken vistas now appeared, planted with crops dictated by the central planners.

China's planners, however, were concerned only with increasing overall production. They paid little or no attention to whether a need for the products existed, whether the products actually helped advance modernization, or whether local production targets were suited to the geography of the country. Several years of bad weather, combined with the rigid and misguided objectives of centralized agricultural planning, resulted in famine conditions throughout much of China. It is estimated that between 20 and 30 million people died from starvation and malnutrition-related diseases between 1959 and 1962.

These outcomes in part explain the radical economic changes that followed in the 1980s and onward, but make them no less startling. Following this long era of economic stagnation, the Chinese economy has been growing at double-digit rates for much of the past 20 years. Under the leadership of Deng Xiaoping (1978–97), China embarked on a thorough reorientation of its economy, dismantling central planning in favor of private entrepreneurship and market mechanisms and integrating China into the world economy. Agriculture was decollectivized, state-owned industries were closed or privatized, and centralized state planning was dismantled.

In the 1980s and early 1990s, when the world economy was sluggish, China's manufacturing sector grew by almost 15% each year. Almost all the shoes once made in South Korea or Taiwan are now made in China. More than 60% of the toys in the world, accounting for $9 billion in trade, are made in China. Since 1992, China has extended its trade policy, permitted foreign investment aimed at Chinese domestic markets, normalized trading relationships with the United States and the European Union, and joined the World Trade Organization.

Since the time of Deng Xiaoping, the establishment of **Special Economic Zones** (SEZs) has been an important part of economic planning in China (**Figure 8.16**). The zones were set up as carefully segregated export-processing areas that offered cheap labor and land, along with tax breaks, to transnational corporations. The first four SEZs—Shantou, Shenzhen, Xiamen, and Zhuhai—were all located in South China, deliberately building on the prosperity of Hong Kong, the former British colony that was returned to China on July 1, 1997. The SEZs were experimental, as the Communist leadership in Beijing looked for ways to maintain political control while absorbing foreign capital, technology, and management practices.

Since then, China's openness to capitalist investment and trade has become a signal to which regional investors have responded positively and enthusiastically. Networks of business connections quickly sprang up in the coastal cities and regions adjacent to Hong Kong and Taiwan. Once the business networks were in place, capital flowed in from all over the globe. Meanwhile, a new web of interdependent trade and investment has emerged, linking coastal China with Singapore, Bangkok, Penang, Kuala Lumpur, Jakarta, Los Angeles, Vancouver, New York, and Sydney.

In 2001, China was admitted to the World Trade Organization, allowing China to trade more freely than ever before with the rest of the world. In 2005 the United States bought more than $243.5 billion worth of goods made in China. Wal-Mart was the single largest U.S. importer, according to a study by the Citizens Trade Campaign, procuring at bargain prices from China everything from T-shirts to car stereos.

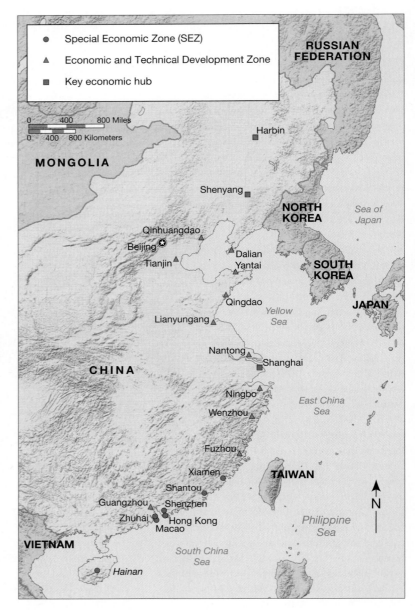

▲ **FIGURE 8.16 China's Special Economic Zones** Since the 1980s, China has established a series of Special Economic Zones and Economic and Technical Development Zones as a way to attract foreign capital, technology, and management practices while maintaining political control.

The low cost of labor underpins the preponderance of "Made in China" apparel labels in U.S. stores. Whereas skilled clothing workers in Toronto's garment district, for example, are paid $350 to $400 a week, plus benefits, for a 33-hour week, their counterparts in Hong Kong are paid $250 for a 60-hour week, with no benefits; and in China skilled garment workers can be hired for $30 for a 60-hour week. Studies of the industry have shown that some of the female workers in Chinese garment factories are as young as 12 years old, girls from rural villages who have been sent to work as sewing machinists in city workshops, sleeping eight to a room, sewing 7 days a week from 8 A.M. to 11 P.M. As a result of these conditions, the retail margin (gross sales profit) on clothing made in workshops in China and sold in Europe and the United States is 200% to 300%, compared to a margin of only 70% or so on domestic-made clothing.

As a result, today most enterprises in China, both private and collective, contract with one another rather than with the central government and interact directly with consumers. After they have fulfilled their contracts and paid their taxes, they are free to pay incentive bonuses and to invest any remaining profits in improved equipment or expanded production facilities. About 80% of China's GDP is now generated by these nonstate enterprises. The growth of towering urban financial centers like Shanghai is the signature geographic expression of these new nodes of economic power in China (**Figure 8.17**).

This pattern of coastal accumulation occurs at a sacrifice, however. Environmental degradation and pollution have been concomitant consequences of the speed of industrialization. Meanwhile, the success of industrial modernization intensified the gap among different regions and between urban and rural areas. The countryside and interior have not gained at the rate the coastal industrial belt has, which has created potential political tensions within China.

The current global economic crisis has suggested several other important things about the current trajectory of Chinese economic development. First, the increasingly globalized economy is dependent on consistent markets for Chinese industrial products (ranging from toys to shower curtains to processed food). This makes China somewhat vulnerable to downturns in consumer spending in places like the United States and Europe. Conversely, China's massive international mobilization of capital has included the acquisition of significant debt from other nations. As of May 2009, China holds $801.5 billion of U.S. debt, making it the largest holder of U.S. debt. This gives the country great political power in negotiations with the United States, but again makes it overwhelmingly concerned about the status of other economies around the world.

▲ **FIGURE 8.17 Shanghai** The Pudong district, shown here, reflects the government's success in reestablishing Shanghai as a major international center.

**Two Koreas: Two Economies**   Meanwhile, the end of World War II brought about an important geopolitical change to Korea. Under the sponsorship of Soviet troops that had moved south from Manchuria into the northern part of Korea, Kim Il Sung, an anti-Japanese Marxist–Leninist nationalist, came to power in 1949. Almost immediately, U.S. troops occupied the southern part of the country. Like East and West Germany, Korea was suddenly divided into two. The 38th parallel was agreed on as the border, and both Soviet and U.S. military forces withdrew.

In June 1950, just a few months after Mao Zedong's victory in China, Kim Il Sung's troops carried out a major attack on South Korea, seizing the capital, Seoul, and quickly extending control over almost the whole country. This led the United Nations to intervene, and U.S. forces rapidly rolled back the North Korean forces all the way to the Chinese border. This, in turn, prompted China to become involved on behalf of the North Koreans, and there followed a devastating war. Casualties were horrendous, with hundreds of thousands killed and much of the country destroyed. In 1953 the 38th parallel was finally restored as the border, but the two Koreas have remained bitter rivals.

The result was two very different economic experiments on the Korean Peninsula. In South Korea, whose economy was crippled by the lengthy period of wars between the 1930s and the 1950s, the government crafted a state-led development effort that carefully controlled imports, managed internal competition between firms, and took out enormous international loans to invest heavily in private industrial development. The initial emphasis was on **import substitution**, developing domestic industries with government protection to produce goods for the domestic market. The second stage, between the mid-1960s and mid-1970s, saw the growth of export-oriented, labor-intensive manufacturing. The South Korean government facilitated the development of these export industries by providing incentives, loans, and tax breaks to firms and by encouraging the growth of giant, interlocking industrial conglomerates called **chaebol**. In the late 1970s, South Korean economic planners decided to diversify the economy with greater emphasis on heavy industry, chemicals, and assembly. By the mid-1980s, manufactured goods accounted for 91% of total exports, and more than half were in the form of ships, steel, and automobiles. The country is now one of the world's leading exporters and is the home of major international companies making products familiar across the world, including electronics (Samsung) and automobiles (Hyundai-Kia).

The radical success of South Korea's economy made it the leading so-called **Asian Tiger** economy during the 1980s and 1990s, and demonstrated how capitalist development might be tied to very strong state policies regulating the economy. In addition to Korea's established strength, East Asia has two other Asian Tigers, newly industrialized territories that have experienced rapid economic growth that has lifted them from the periphery of the world-system to the semiperiphery. These are Hong Kong and Taiwan (the fourth Asian Tiger is Singapore, in Southeast Asia; see Chapter 9).

North Korea, conversely, remained a highly isolated and economically stagnant state over the same period. During this time, central planning was the hallmark of North Korea's economic strategy, which sometimes retards capitalist investment and adaptation to changing market conditions. Beyond this, however, a major economic barrier for the country is the incredibly disproportionate fraction of the economy dedicated to military spending, perhaps as much as 25%. This, coupled with sanctions that periodically isolate the country further from global trade, has meant a North Korean economy that has been largely stagnant, small, and marked by shortages and hunger for citizens.

South Korea was hit particularly hard by the Asian financial crisis of 1998. The preceding boom had led South Korean banks to make loans to their parent companies for investments throughout Asia, and when the bubble burst, many of the investments had to be written off. South Korea's economy had to be propped up with loans and guarantees from the International Monetary Fund, which in turn has required the liberalization of the economy.

The pace of South Korean economic growth and urbanization has also brought heavy pollution, acute housing shortages, and rampant corruption. Socioeconomic inequality is also a characteristic feature of South Korea, both within the cities and between the country's subregions. The government has attempted to improve equality among the subregions through spatial-planning policies that have directed industrial growth away from Seoul toward provincial towns and cities. Currently, the major priorities of domestic policy are keeping up with the need for basic infrastructure improvements to maintain economic efficiency, and keeping up with educational spending to be able to further develop a high-tech sector. Government spending on social and environmental issues remains minimal, however, so that there is little immediate prospect of a reduction in social inequalities or of improvements in the quality of life for the many millions who live in cramped housing and degraded environments.

As in Japan, a similar regional competitive effect can also be seen in Korea. Pusan, the center of the South Korean footwear industry that in 1990 exported $4.3 billion worth of shoes, is now full of deserted factories. South Korean footwear exports are down to less than $700 million, whereas China's footwear exports have increased from $2.1 billion in 1990 to $13 billion in 2002.

At the same time, North Korea has begun a cautious transition of its economic policy during the first years of the 21st century toward a more competitive stance. Specifically, it has begun to experiment with Chinese-style Special Economic Zones to foster some foreign investment, and created nodes of accumulation within an otherwise authoritarian state management of the economy. The impact of this experiment is impossible to determine at this early stage. It certainly represents part of an ongoing trend, however, that makes East Asia more externally competitive as a world region for investment, while also becoming a place of more bare-knuckle internal competition, as subregions and nations compete for scarce investment capital.

## Demography, Migration, and Settlement

The most striking demographic characteristic of East Asia as a whole is the sheer size of its population. With a total population of some 1.55 billion in 2005 and almost 14% of Earth's land surface, East Asia contains about 22% of its population at an overall density of only 11 persons per square kilometer (28 per square mile). As **Figure 8.18** shows, the bulk of the population is distributed along coastal regions and in the more fertile valleys and plains of Inner China. Population densities in Outer China are very low, lower than 1 person per square kilometer (between 2 and 3 per square mile) in most districts—the

same as in the far north of Canada. In contrast, population densities throughout Inner China and in Japan, North and South Korea, and Taiwan are very high: between 200 and 500 per square kilometer (518 to 1295 per square mile) on average.

**Population Density** High densities of population are in part a reflection of the very high levels of agricultural productivity in these regions and in part a reflection of past population growth rates. High densities often mean overcrowding, especially in areas where farmland is valuable and where the topography restricts settlement. The most striking examples of crowding are from Japan, where mountainous regions preclude urban development in much of Honshu, the main island. The great majority of Japan's 127 million people live in crowded conditions in the towns and cities of the Pacific Corridor, where space is so tight and expensive that millions of adults live with their parents because they cannot afford places of their own. The average dwelling in metropolitan Tokyo is about 60 square meters (646 square feet). (By comparison, the median dwelling size in metropolitan Los Angeles is just over 150 square meters, or 1,615 square feet.).

These levels of crowding are reflected not only by crowded streets and tiny homes (**Figure 8.19**), but also by people's behavior and social organization. Japanese daily life is filled with rules, and many of them have evolved because of space shortages. In many Tokyo neighborhoods, for example, trash is picked up four times a week, mainly because people have no place to store it. The trash itself must be sorted carefully into burnable and nonburnable loads and put out in government-approved see-through bags so it is easy to spot violations.

**Population Control** In China, crowded towns and cities are more often a reflection of high rates of the rural-to-urban migration of low-income households than of a sheer lack of space. China certainly does not lack space, and until relatively recently China's population was officially regarded as being too small rather than too large. Mao Zedong wanted a large population with which to fully exploit China's large territory. When Ma Yinchu, an eminent economist and the president of Beijing University, cautioned the leadership in 1957 about the rapid growth of China's population, he was dismissed from his post and became a "nonperson," his existence not recognized by the authorities, for the next 22 years. Between 1961 and 1972, China's population spurted by 210 million people as a demographic transition took effect (see Chapter 1, p. 27). Improved medical care and public health reduced death rates, while birthrates remained high, encouraged by the political leadership. In 1970 the average family size was 5.8 children, and China's population had been growing at 2.6% per year for the previous decade.

By the early 1970s, it had become clear that China's communal mode of production could not sustain such a large population. Increasingly, population growth was seen as a threat to the country's chances of economic development. In response, China's Communist Party instituted an aggressive program of population control. A sustained propaganda campaign (**Figure 8.20**) was reinforced in 1979 by strict birth quotas: one child for urban families, two for rural families, and up to three for families that belonged to ethnic minorities. The policy involved rewards for families giving birth to only one child, including work bonuses and priority in housing. The only child was later to receive preferential treatment in university admission and job assignments (an aspect of the policy that was later abandoned). In contrast, families with more than one child were to be penalized by a 10%

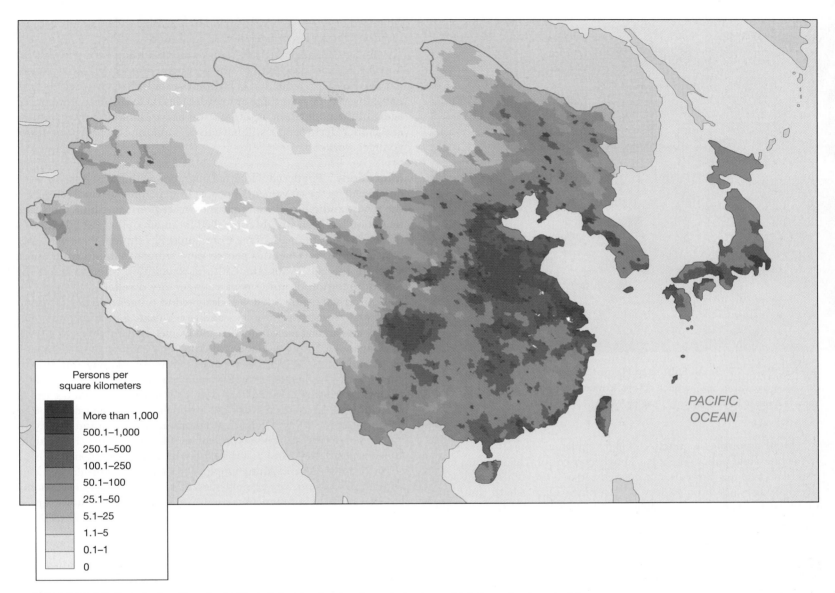

▲ **FIGURE 8.18 Population Density in East Asia** The density of population is very high throughout most of the continental margins of East Asia and in the fertile basins and river valleys of Inner China, but population densities are very low in the Tibetan Plateau, northwestern China, and Mongolia.

decrease in their annual wages, and their children would not be eligible for free education and health-care benefits.

In China's major cities, the **one-child policy** has been rigorously enforced, to the point where it has been almost impossible for a woman to easily plan a second pregnancy. Who is allowed to have a child, as well as when she may give birth, is rigidly controlled by the woman's work unit. Workers are usually organized into groups of 10 to 30 individuals. If any woman in the group gives birth to more than one child, the entire group can lose its annual bonus. Neighborhood committees and "granny police" also watch over the families in their locale, adding social pressure and acting as distributors of birth-control devices. Abortions are freely available for unsanctioned and unwanted pregnancies.

In terms of reducing population growth, the policy has been very successful. The average family size in China in 2005 was 1.8 children,

whereas the overall annual growth rate of the country's population between 1995 and 2005 was 0.9%. China, like Japan, Taiwan, and North and South Korea, has completed its demographic transition and now has low birthrates as well as low death rates. Without such an aggressive population policy, China's population today would have been more than 300 million larger.

Nevertheless, the policy has not been without problems. In addition to the personal and social coercion involved, the one-child policy has led to the problem of spoiled children—"little emperors," who are the center of attention of six anxious adults (the parents plus two sets of grandparents). One aspect of this is an increasing incidence of obesity in Chinese children. Being fat used to be considered a hedge against bad times, but it is now coming to be seen not only as unhealthy but also as symptomatic of the cultural shift that has produced so many spoiled children.

▲ **FIGURE 8.19 Small Suburban Homes, Japan** High population densities have driven up the price of land so much that even suburban homes, such as these in Kyoto, are very small by Western standards.

More serious is the practice of aborting female fetuses in response to the one-child policy. There has long been a cultural bias toward male children in China. Boys are not only considered inherently more worthy than girls but are also seen as insurance against hard times and providers for their parents in old age. In the past, this bias took the form of abandonment of girl children and even infanticide. Today, such practices are much less common, but widespread selective abortion means that currently there are close to 120 boys for each 100 girls. One consequence of this is that within a few years there will be 50 million

▼ **FIGURE 8.20 China's Population Policy** This billboard proclaiming the importance of family planning embodies the ideal of China's policy of limiting urban families to a single child.

Chinese men who will have no prospect of finding a wife, simply because there will not be enough women to go around.

Today, China's population policies are beginning to be relaxed. Not only has the rate of growth of China's population slowed satisfactorily, but also the sharp reduction in birthrates over the past two decades will mean that for the next three or four decades there will be a pronounced aging of the population, creating a top-heavy situation in which more and more elderly people will have to be supported by fewer and fewer younger workers.

**Migrations**    For decades internal migration within East Asia has been dominated by the movement of people from peripheral, rural settings to the towns and cities of more prosperous regions. In China, however, migration has been strictly controlled for much of the past 50 years. In the immediate aftermath of the Communist Revolution, there was a significant local movement as people were reorganized into communes. Then, in the late 1960s, the Cultural Revolution brought about the forced migration of millions of younger city dwellers, a program of "reverse urbanization" that was designed to purge the cities of "decadent" and "revisionist" thinkers and to inculcate a proletarian revolutionary zeal among the young men and women sent out to live and work in villages. According to Maoist thinking, cities tended to harbor materialist and counterrevolutionary values. A household registration program effectively operated as an internal passport system, allowing Mao's planners to restrict urbanization. Even after Mao's death and the end of the Cultural Revolution, Chinese authorities remained opposed to urbanization, seeking to keep rural labor forces in place and to prevent the kind of overurbanization typical of peripheral countries. Regulations aimed at restricting rural-to-urban migration included raising housing costs for migrants and fining employers who hired transient workers without permission. The gradual liberalization of the Chinese economy has inevitably led to increased rural-to-urban migration, however. By 2006, roughly 44% of China's population was urban (compared to 17.3% in 1975).

In Japan, levels of urbanization have increased dramatically since World War II. In 1950 almost half of Japan's population was dispersed throughout the country in farming households in rural areas. During the period of postwar economic recovery and growth, rural-to-urban migration occurred very rapidly as manufacturing industries expanded, mainly along the Pacific Corridor from Tokyo to northern Kyushu. As in the United States, there was a brief phenomenon of **counterurbanization** (the net loss of population from cities to smaller towns and rural areas) during the 1970s, as some businesses sought to escape the congestion and inflated land prices of metropolitan areas and as some people sought out quieter and more traditional settings in which to pursue slower-paced lifestyles. This counterurbanization, however, was selective, affecting only a few places and regions.

By 1990 some 46.8% of Japan's total land area was officially designated as "depopulated" and eligible for special funding. With less than 7% of Japan's population, these rural areas were left with declining economies and aging populations. Japan's cities, in contrast, grew at a terrific pace—partly through migration and partly through natural increase of their younger populations. By 2005 nearly 80% of Japan's population was urbanized, the Pacific Corridor having become a megalopolitan region with several major metropolises.

South Korea has experienced a similar pattern of rural-to-urban migration, with Seoul the focus of a shift that has seen overall levels of urbanization increase from 21.4% in 1950 to 84.2% in 2005. Even North

Korea, with a strictly controlled Communist regime and poor levels of productivity, experienced a steady increase in urbanization: from 31.0% in 1950 to 64.8% in 2005.

**Diasporas**   East Asia is a world region that is distinctive insofar as it has few immigrant populations of any significant size. Distrust of foreigners of all kinds has been a long-standing condition within East Asia. The Communist regimes of China and North Korea contributed to this closure as well by maintaining tightly closed borders for several decades. In contrast, there has been considerable emigration from East Asia. The Chinese diaspora (**Figure 8.21**) dates from the 13th century, after the conquest of China by the Mongols in 1279. Some Chinese took refuge in Japan, Cambodia, and Vietnam, but it was the Yuan (Mongol) dynasty that promoted the basis for a wider Chinese diaspora through its trading colonies in Cambodia, Java, Sumatra, Singapore, and Taiwan. Under the Ming and Qing dynasties, Chinese communities developed around additional trading colonies, including several in Indonesia, the Philippines, and Thailand.

After defeat in the Opium War (1840–42) and the opening of trade with Western powers, many more Chinese emigrated, creating the basis for the modern Chinese diaspora. The Industrial Revolution and the opening of the world economy through imperialism and colonization provided many opportunities. Between 1845 and 1900, 400,000 Chinese were estimated to have emigrated to the United States, Canada, Australia, and New Zealand. Over the same period, an additional 1.5 million emigrated to Southeast Asia (Indonesia, Malaysia, Singapore, Thailand, and Vietnam), working in mines, on road- and railroad-building, and as farm laborers. Another 400,000 emigrated to the West Indies and Latin America, mainly Chile, Cuba (after 1847), and Peru (after 1848).

Chinese immigration to the United States began with the California gold rush of the 1850s. By 1860 there were 35,000 Chinese in the United States, concentrated in San Francisco, Los Angeles, Seattle, and Portland and scattered in the mines, railway construction projects, and ranches of California. Other significant concentrations of Chinese immigrants developed in Boston, New York, and Philadelphia. But intense discrimination against the Chinese—who were, as geographer Susan Craddock has shown, wrongly blamed for everything from outbreaks of cholera to the moral corruption of 19th-century cities—led to a decade of federal restrictions on Chinese immigration, beginning in 1882. It was only after World War II that Chinese immigration to the United States once more became a significant flow. Today, the Chinese diaspora is the largest in the world and one of the most prosperous. The past quarter-century has seen a great increase in Chinese immigration to the United States, Canada, and Western Europe. In 2000 the Chinese population of the United States was 1.6 million.

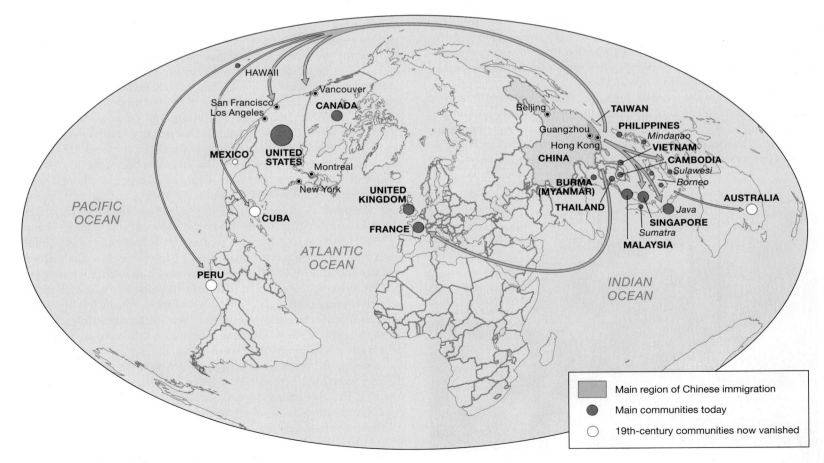

▲ **FIGURE 8.21  The Chinese Diaspora**  Most emigration from China has been from the heavily populated regions of Inner China. Chinese emigration began on a large scale during the 19th-century Industrial Revolution, when Chinese laborers found opportunities for employment in newly colonized lands.

The Korean and Japanese diasporas are much smaller but have become important elements of the contemporary geography of both East Asia and North America. In Japan and elsewhere in Asia, Korean immigrant populations have remained a discriminated minority, even after three generations. More recent Korean immigrations have been to Hawaii and North America, mainly Los Angeles and Vancouver. The strongest component of the Japanese diaspora is in North America. Today, the Japanese-American population numbers almost 900,000, with concentrations in Honolulu, San Francisco, and Los Angeles.

**Great Asian Cities** Japan and Korea have become highly urbanized in the years since World War II, and China is following quickly. The resulting cities have created numerous environmental problems and stresses (see earlier discussion), but they have also radically propelled new forms of architecture, public transportation, and urban planning and design. Tokyo, for example, is a conurbanized conglomeration of 23 wards, each with its own governing body, all linked together through an overarching planning handled through a central metropolitan government. This innovation allows this city, with its 35 million people stretched across the entire metropolitan area, to manage city affairs with flexibility and local initiative, even while coordinating key city functions. For example, 27 million people ride Tokyo's highly organized rapid public transportation system every day across distant parts of the city.

The new cities of China are all the more spectacular for the recentness of their growth. The city of Guangzhou, for example, though long a bustling trading port, recently grew to a population of more than 10 million people across its metropolitan area (**Figure 8.22**). The bursting economic activity of the city is embodied in a construction boom, where innovative structure, curvilinear design, and expressive experimentation are creating whole new skylines.

In a reminder of the regional interconnectedness of rural and urban areas, however, it is critical to recall that the glistening new buildings of Guangzhou are, at least in part, constructed by impoverished migrants from the countryside (more than 150 million rural laborers live in China's cities as temporary, and typically second-class, citizens).

▲ **FIGURE 8.22 Construction Boom in Guangzhou** People walking near a construction site in Guangzhou. Hundreds of new skyscrapers now form the skyline of this bustling city at the mouth of the Pearl River in southern China.

# CULTURE AND POLITICS

Though urbanization is the unquestionable focus of current attention in East Asia, traditional societies in the region revolved around family, kin networks, clan groups, and language groups developed in more largely rural contexts and state systems, with a strong bureaucracy enforcing social order. In the larger state societies of China and Japan, society was historically rigidly hierarchical, and individuals were subsumed within the family unit, the village, and the domain of the local lord. In these environments, important social values were those of humility, understatement, and refined obsequiousness, with particular deference being shown to older persons and those of superior social rank.

To a degree, these traditional values persist as a distinctive dimension of the cultural geography of the region. Not surprising, in rural areas and smaller towns, and among older people in general, traditional cultural values are held most strongly. Traditional rural communities with a great variety of distinctive ways of life still characterize much of the region (**Figure 8.23**). There are, however, marked regional variations in cultural traditions within East Asia. In addition to the variations in religious adherence associated with different ethnic groups, there are some striking regional differences in language, diet, dress, and ways of life. The changing relations of the region to other global cultures, moreover, has fully transformed local ways of living.

## Tradition, Religion, and Language

**Languages** The languages of East Asia, and their variety, emphasize that the region as a whole is rooted in numerous diverse and independent cultural traditions, brought into dialogue. Most exemplary in this regard, "Chinese" is linguistically a term that is somewhat imprecise. The region of East Asia dominated by Chinese people includes not only many Chinese dialects, but also an enormous range of diverse minority languages (like Uigur in the West, Sibo in the Northwest, and Chuang, Yi, Nakhi, and Miao in the South). The dominance of Han people within China is reflected in the geography of language, however. Mandarin, the language of the old imperial bureaucracy, is spoken by Hui and Manchu people (as well as the people of Taiwan), together with almost all Han people, though significant regional variations still remain in the Mandarin dialects that people use. The other 53 ethnic minorities in China have their own languages.

Chinese writing dates from the time of the Shang dynasty (1766–1126 B.C.E.), with tens of thousands of characters, or **ideographs**, each representing a picture or an idea. It is possible for a single dictionary to contain 50,0000 distinct characters. Chinese children may learn a few thousand characters in their first 10 years, but advanced reading or writing may take several times that number. Characters themselves are enormously complex, especially in their studiously artful calligraphic form, whereas a single Chinese character might take as many as 33 brush strokes to complete (**Figure 8.24**).

During the Communist period, the government of China facilitated literacy by reducing the number of characters in use and by simplifying them. It has also adopted a new system, **pinyin**, for spelling Chinese words and names using the Latin alphabet of 26 letters. Previously, Chinese words and names had been translated (usually by Westerners) into a Latin-based version using a Western system, known as the Wade-Giles system, with often misleading results and a proliferation of mispronunciations. Older textbooks, reference books, and atlases, for example, refer to Chongqing (pinyin) as Chunking (Wade-Giles), Mao Zedong as Mao Tse-Tung, and so on.

(b)

(a)

(c)

▲ **FIGURE 8.23 Rural China** A slight majority of China's population lives in rural settings, and rural ways of life remain dominant throughout much of the country. (a) Vegetable market in Shanxi, Xian. (b) Girl herding ducks, Guangdong province. (c) Women with cattle, Yunnan province.

Japanese is an entirely distinct language from Chinese, and indeed from almost all other languages, developing independently. Though the Japanese adopted some Chinese pictographic characters in the 3rd century C.E., the two languages are largely unrelated. Japanese ideographs—symbols representing single ideas or objects (*kanji*)—number more than 10,000, and these are joined by numerous other symbols (*kana*) that represent phonetic syllabus or word fragments. Collectively, the language is wholly unique.

The affiliation of Korean to other languages in the region is uncertain. Although it contains Chinese words, its structure and grammar is far closer to Japanese. Even so, the linkage between those two languages is completely unclear. The Korean alphabet, unlike Japanese and Chinese, is phonetic (where characters represent independent sounds), though the text is written is clustered groups, which are wholly unlike that of either of the other two languages. The fundamental differences between the languages of East Asia suggest the independent development of diverse cultures within the region, each with its own linguistic and symbolic logic. Their clear divergences remind us that the region is a product of interactions, not a pregiven or inevitable coherent whole. Their shared symbols and mutual influences suggest their ongoing development of strong regional ties, however.

**Religion**   Chinese culture found spiritual expression in the philosophy of **Confucianism**. Unlike formal religions, Confucianism has no place for gods or an afterlife. Also distinctive is Confucianism's emphasis on ethics and principles of good governance and on the importance of edu-

cation as well as family and hard work. Confucianism proved to be ideally suited to Imperial China, and it remains the most widely recognized belief system in China, even after several decades of discouragement from the Communist regime. Formalized religions in China include Daoism (throughout much of the country), Tibetan Buddhism (in Xizang), and Islam (in large parts of Inner Mongolia and Xinjiang).

For most Chinese, however, folk religions are far more important than any organized religion. Animism—the belief that nonliving things have spirits that should be respected through worship—continues to be widely practiced, as does ancestor worship—based on the belief that the living can communicate with the dead and that the dead spirits, to whom offerings are ritually made, have the ability to influence people's lives. The costs of ancestor worship can be financially burdensome. The offerings involve burning paper money and hiring shamans and priests to perform rituals that will heal the sick, appease the ancestors, and exorcise ghosts at times of birth, marriage, and death.

Japanese indigenous culture was expressed in **Shinto**, which does not have a distinctive philosophy but, rather, a belief in the nature of sacred powers that can be recognized in every individual existing thing. The traditions of Shinto may be thought of as the traditions of Japan itself. Seasonal festivals elicit widespread participation in present-day Japan, regardless of people's religious affiliation. These traditions usually entail ritual purification, the offering of food to sacred powers, sacred music and dance, solemn worship, and joyous celebration. Buddhism is also important in Japan. It was introduced to Japan in the are common, whereas herds of sheep, goats, cows, horses, camels, and yaks also provide meat dishes. Inner China, by contrast, has little stock-raising and, consequently, few milk-based dishes; its principal meat dishes derive from scavengers such as ducks, poultry, and pigs. Within Inner China, the humid and subtropical south has developed a cuisine based on rice, while in the subhumid north, noodles form the staple diet.

Conversely, the "distinctive" character of East Asian cuisine more generally is heavily influenced by historical linkages with the rest of the globe. Many Chinese dishes, for example, are marked by the creative use of hot chilies. This Central American domestic plant could not have been a part of Asian cuisine before 1500. The meat-filled dumpling, a standard fundamentally "Chinese" staple on Asian menus in Britain, France, and the United States, was likely introduced to China in the 3rd century C.E. from elsewhere in Asia. Similarly, vegetarian traditions in East Asian cuisine thrived under the influence of Buddhist teaching (originally from South Asia—see Chapter 9), but these competed with meat-loving habits of people from different regions and faiths (**Figure 8.25**).

In other words, East Asian foods have become regionally distinct in part because of the peculiar characteristics of their local conditions, including diverse local plant and animal species and cultural habits and traditions. But they have also become distinctive through their unique interaction with other food traditions from around the world. Korean food is Korean as a result of its unique adoption and adaptation of cooking techniques and ingredients from the Americas and Europe.

**Feng Shui** Throughout East Asia there is widespread adherence to **feng shui** (pronounced fung shway)—not a religion, but a belief that the physical attributes of places can be analyzed and manipulated to improve the flow of cosmic energy, or *qi* (pronounced "chi"), that binds all living things. Feng shui involves strategies of siting, landscaping, architectural design, and furniture placement to direct energy flows

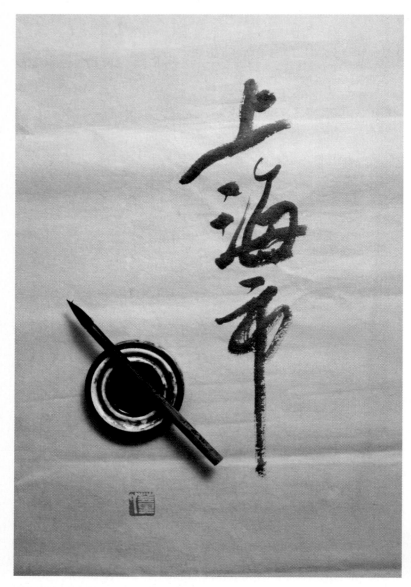

▲ **FIGURE 8.24 Chinese Calligraphy** The words "The City of Shanghai" written in Chinese calligraphy characters; Shanghai, 1992.

6th century C.E. from Korea. During the Nara period (C.E. 710–84), Buddhism was vigorously promoted, leading to the first blossoming of distinctive Japanese art and architecture. Later, between the 12th and 14th centuries, Zen Buddhism was introduced from China, adding new painting styles, new skills in ceramics, and the custom of tea drinking.

**Regional Traditions of Food** For many people outside East Asia, "Chinese food" or "Korean food" may appear as an undifferentiated set of generic dishes (from Kung Pao chicken to Mongolian barbecue). The diversity of food traditions in East Asia, however, emphasizes both the regionally unique character of food practices and their dependence on traditional food sources, as well as the complex influences from other regions that have long affected and transformed East Asian cuisine.

In terms of regionally distinct foods, for example, within China, there are fundamental differences between the cuisines of Inner and Outer China and between those of the north and south within Inner China. In Outer China, milk-based dishes—yogurt, curds, and so on—

▲ **FIGURE 8.25 Chinese Street *Bao zi* (Dumplings) in Shanghai** Bao zi are steamed then fried dough filled with vegetables or meat, common as a snack or a main meal. This is classic Chinese street food and is common all over China. Very cheap, tasty, and filling, bao zi are part of the dumpling family of food, which come in all shapes and sizes and styles; here steamed and fried, others are boiled and fried, some come with soup. Spicy or pungent, sweet or savory, dumplings provide a cornerstone of Chinese food. Here the bao zi are in a large frying pan next to some *jiao zi*, which are folded with a different technique.

and is often known as geomancy. The oldest school of feng shui, known as the Land Form (or simply Form) School, dates back to the Tang dynasty (C.E. 618–907). From its origin in the jagged mountains of southern China, the Form School used hills, mountains, rivers, and other geomorphological features as a basis to evaluate the quality of a location. Subsequent modifications introduced the idea that specific points of the compass exert unique influences on various aspects of life. For example, the south, with its orientation toward the Sun's path and away from cold north winds, was declared a most auspicious direction, conducive to longevity, fame, and fortune, whereas influence over career and business success was attributed to the north.

In East Asia today, feng shui practitioners focus most on the layout and interior design of homes and offices (**Figure 8.26**). The principles and beliefs are complex, but some of the key ideas are as follows: Water and mirrors enhance qi flow. Narrow openings or hallways cause qi to flow too quickly, negating its beneficial effects. Straight lines are a frequent cause of high-velocity qi that can be dangerous to health and well-being. The numbers 8 (representing "prosperity") and 9 (representing the fullness of heaven and earth) are good, but 4 is bad ("death"). The placement of furniture and other objects, as well as the use of certain colors (purple is popular) and motifs (birds are favored), are all believed to enhance qi flow and, therefore, career success, strong family relationships, and good health. Ideally, each room should include some representation of each of the five fundamental elements (fire, metal, water, wood, and earth).

## Cultural Practices, Social Differences, and Identity

**Gender and Inequality**  Just as economic reform in the region has been accompanied by the reemergence of regional inequality, economic and social reforms have also led to changes in the status of women.

▲ **FIGURE 8.26 Feng Shui** The principles of feng shui are as popular as ever throughout much of East Asia.

Until Deng Xiaoping's reforms of the 1990s, gender equality was a central tenet of government policy. One of Mao Zedong's much-quoted slogans was, "Women hold up half the sky." During the Cultural Revolution, society was strictly gender-neutral, images of female beauty and ideas of romance were suppressed, and the participation of women in political and economic activities reached almost 100%.

The effects of economic and social reform were mixed. Competitive markets in education resulted in a fourfold expansion of women college students in the 1990s, and these better-educated women have gone on to make their mark in China's labor markets, earning high salaries and challenging traditional gender boundaries for top jobs. Between 1991 and 2001, the number of marriages in rural areas arranged by parents—a tradition that largely escaped the attention of the Cultural Revolution—fell from 36% to 16%. At the same time, there has been an explosion of femininity and sexuality in the media, not always with positive results for women. Traditional gender stereotypes have resurfaced, with women associated with domestic responsibilities and unskilled work. This has played out in labor markets as older and less well-educated women have borne the brunt of layoffs in the state sector and have been left with lower-paid occupations. As a result, there is a widening income gap between men and women. Urban women who made, on average, 70% of men's wages in 1990 made less than 63% of men's wages in 2005. Gender inequality is most marked in rural settings, where women's wages, on average, are only 40% of men's and where gender roles have regressed farthest toward traditional stereotypes.

**Ethnicity and Ethnic Conflict**  Although East Asia has few immigrant populations of any significant size, there are some important ethnic variations. According to China's census of 2000, its population contains 56 different ethnic groups. The dominant group consists of Han people, who make up more than 91% of the population. The Han people originally occupied the lower reaches of the Huang He and the surrounding North China Plain. They spread gradually inland along river valleys, into present-day Korea, and toward the humid and subtropical south. With the Ming dynasty, Han colonialism and imperialism established them as the dominant group throughout the jungle lands of the south and southeast (as far as present-day Vietnam) and in Taiwan. In the 19th century, they spread north into Manchuria. Today there are more than a billion Han in China. The remaining 55 minority groups add up to fewer than 110 million people. They are mostly residual groups of indigenous peoples, such as the Miao, the Dong, Li, Naxi, and Qiang and are found in the border regions, removed from central authority in Beijing, relatively remote, and economically disadvantaged.

Under these circumstances, it is not surprising to find that tensions exist between several of the larger minority groups and the dominant Han. Perhaps the best-known case is that of the Tibetans, who came under the rule of the Han as recently as 1950, when the Chinese People's Liberation Army marched into Tibet as "liberators." Ethnic communities like the Tibetans, with their own highly distinctive cultural traditions and religions, face explicit state policies of "sinification"—the ostensible "modernization" of their economies and communities in a way that appears unmistakably colonial for minority communities.

The same is true of ethnic Uyghurs, who predominate in westernmost China (and in many post-Soviet central Asian states). This Muslim minority group has in recent years demanded greater autonomy for self-governance and the practice of their own cultural traditions. In July 2009, violence broke out between Uyghurs, ethnic Han people, and state police in the city of Ürümqi in far Western China (**Figure 8.27**). The Chinese state has looked upon Uyghur activism as hostile to the interests of the nation and has branded some Uyghurs as terrorists. When the United States sought to release four Uyghurs held at Guantanamo Bay (who, like many detainees, were never convicted of any crime) as part of their effort to close that facility, the Chinese government demanded their extradition back to China, where they would likely face trial as terrorists.

**Politics of Culture and Culture Change**   Culture in East Asia, as elsewhere, is therefore partly a product of intensely political struggles and debates. The enunciation of subregional cultures—as in the case of the Tibetans and Uyghurs—has historically implied autonomy, independence, or a challenge to the authority of dominant cultures and their associated ethnicities and often the states that these groups governed.

But revolutionary history of the region has at times been conversely hostile, even to the traditional values and practices of dominant cultures. Nowhere was this more clearly seen in East Asia than during the Cultural Revolution. In an attempt to restore revolutionary spirit and to reeducate the privileged and increasingly corrupt Communist Party officials, in 1966 Mao Zedong launched what he called a "Great Proletarian Cultural Revolution." The Cultural Revolution brought a sustained attack on Chinese traditions and cultural practices and a relentless harassment of "revisionist" elites, which was defined broadly as anyone not belonging to the rural peasantry. Millions of people were displaced, tens of thousands lost their lives, and much of urban China was plunged into a terrifying climate of suspicion and recrimination. U.S. President Richard Nixon made a path-breaking visit to China in 1972 in an attempt to reopen China's relations with the Western world, but only with the death

of Mao Zedong and the subsequent arrest of the politically extreme "Gang of Four" (which included Mao's wife) in 1976 did the Cultural Revolution come to an end. China's new leader, Deng Xiaoping, charted a more pragmatic course, gradually achieving stability and economic growth and opening China to Western science, technology, and trade.

In the aftermath of the Cultural Revolution, though, little faith remained in Communist Party doctrine. Traditional values had been severely eroded. The result was that Western values of materialism, democracy, and individualism began to spread into the ideological vacuum along with a substantial revival of traditional Chinese culture. To combat the threat to the established order, China's leadership therefore launched a series of mass campaigns—first a campaign against "spiritual pollution," then a repressive campaign against opponents of the ruling Communist Party (following the violent crackdown against protesters who challenged the legitimacy of the Party in Beijing's Tiananmen Square in 1989), followed by a campaign against corruption, and then another to promote civil and respectful behavior.

**Emerging Cultural Traditions**   Despite the resiliency of traditional cultural practices, even in the face of dramatic efforts to change and reform them, new vibrant cultures are emerging. In cities, and especially among youth, traditional ideas are giving way to new value systems and cultural practices, partly hybridized with imported Western cultural norms and partly original. The rise and commercialization of sports is one example, as evidenced by the popularity of baseball and golf in Japan and of soccer everywhere. Another is the emergence of distinctive youth cultures in the major metropolises of China, Japan, Taiwan, and South Korea. In Tokyo, for example, more than 70% of the single women live at home as "parasite singles." The number of Japanese women in their late twenties who have not married has risen from 30% to about 50% in the last 15 years—and their opinions and lifestyles help define a kind of Tokyo yuppie devoted to leisure and luxury (**Figure 8.28**).

▼ **FIGURE 8.27 Uneasy Calm in Ürümqi China**   A Muslim Uyghur couple observes a unit of Chinese policemen on patrol in an Uyghur neighborhood in Ürümqi, China, July 2009. The unrest between Muslim Uyghurs and Han Chinese left at least 170 people dead.

East Asian cultures are also simultaneously being drawn into the globalization of culture. Chinese art and artifacts were popularized in Europe and North America in the 19th century; Chinese and Japanese cuisines have been introduced to cities throughout the rest of the world; the simplicity of Japanese architecture and interior design has influenced Modernist design; religions such as Tibetan Buddhism and Zen Buddhism have a small but growing following in both Europe and North America; and feng shui has recently found adherents in many Western countries. Some new East Asian cultural products have become popular beyond the region. One example is Hong Kong movies, with their distinctive genre that mixes martial arts, metropolitan youth culture, and traditional values. Another is manga and anime.

**Manga and Anime**   Two stylized artistic entertainment forms have swept the globe in recent years: **manga** and **anime**. Manga refers to print cartoons books (*komikku*), which date at least to the last century. Contemporary manga novels cover topics from fantasy to history, usually with an easily recognized and common style, typically involving characters with large eyes and childlike features (**Figure 8.29**). Anime,

▲ **FIGURE 8.28 Dining Out** Miki Takasu, foreground, and her girlfriends dine at a Tokyo restaurant.

▲ **FIGURE 8.29 Manga** These manga novels are part of the publication of the French publication company "Editions Glenat," whose global reach is testimony to the international success of the manga Japanese comic tradition.

in a general sense, refers to all Japanese animated cartoons. More specifically, however, it is associated with a distinctive style of animated film. Though stylistically anime can vary from highly realistic to more abstract, internationally it is most closely associated with the styles of manga, with chapters, settings, and stories drawing heavily from manga print cartoons and human forms with large eyes and exaggerated expressions. Early popularity of anime film in the United States came in the form of cartoon television series, including *Speed Racer* (originally *Mach GoGoGo*), a cartoon that ran in the late 1960s and made global audiences familiar with the styles, themes, and colorful palette of both Japanese manga and anime. Since then, this art form has exploded internationally. Nintendo's Pokémon cartoons (and their associated game products) are a multi-billion-dollar global business. Elements of anime are also evident in the characterization and style of figures in Nintendo's Wii game system.

### Global Chinese Culture—2008 Olympics

It is also increasingly evident that Chinese culture is entering a period of global ascendance. Chinese music and products are increasingly familiar throughout Asia and internationally. Most obviously, the 2008 Beijing Olympics, which was watched by 4.7 billion viewers worldwide (according to Nielsen Media Research), put "new" China on the map as a global cultural leader, and its cities—Beijing first among them—as destinations on par with New York, Tokyo, Paris, or Sydney. Many in the international community expressed concern about the ability of the Chinese to prepare the venue and gain control over Beijing's runaway pollution. Indeed, for many months of the year, Beijing's air typically measures pollution levels higher than 250 micrograms of particulate matter per cubic meter (far in excess of the World Health Organization's target of 20 micrograms/cubic meter).

The event, however, was largely seen as a tremendous success, involving spectacular architecture (**Figure 8.30**), well-coordinated events, and the hosting of tens of thousands of overseas guests. Although pollution was not wholly tamed, control of traffic and industry brought pollution down to levels in which visitors and athletes could function.

The event was also an enormous source of pride for China's government and its people.

The hidden costs of that success are notable. Stringent police control was instituted throughout the city, free expression was quashed, and to make way for construction (and as part of ongoing urban development of the city), many small traditional neighborhoods (or *Hutongs*) were demolished. Chinese culture, in this case, was projected globally in part through the use of force over which the Communist Party still maintains a monopoly.

## Territory and Politics

**An Unfinished War in Korea**   The partition of North and South Korea in 1953 (described earlier) left North Korea behind the world's most heavily militarized border. The governing of North Korea in the period since has reflected the leadership of Kim Il Sung, the leader, whose vision for North Korea was what he called *juche*: a mixture of Stalinist socialism, self-reliant nationalism, and the cult of the personality. The people soon learned to refer to Kim Il Sung as the "Great Leader" (**Figure 8.31**). With good natural resources, including significant reserves of coal and iron ore and good potential for hydropower, North Korea set out to become a major power within East Asia. For decades, more than one-quarter of the country's GDP has been expended on the military, including the development of nuclear power capability that could lead to nuclear weaponry.

To carry out this strategy of national development, Kim Il Sung imposed an austere regime that resulted in drab cities with bleak housing and a population whose bland and regimented quality of life was poisoned by the paranoia that was generated by a "Big Brother" government. The everyday landscapes of rural and small-town North Korea entered a time warp that has given them the appearance of late-1950s environments, with old buildings, old technologies, and clunky-looking consumer goods.

In 1994 Kim Il Sung died and was succeeded by his son Kim Jong Il, who immediately came to be referred to as the "Dear Leader." Soon afterward, in 1995 and 1996, North Korea experienced unusually bad

▲ **FIGURE 8.30 Beijing National Stadium: "The Bird's Nest"**
Constructed for the 2008 Beijing Summer Olympics, the Bird's Nest cost approximately $423 million and is one of the world's largest steel structures, seating more than 80,000 people.

floods, followed in 1997 by a severe drought. By 1998, agricultural production was down to 50% of its previous levels. Worse, the floods had affected the country's coal mines, so that industrial production was disrupted, along with supplies of coal for winter heating. North Korea's landscape became one of famine and distress. Estimates of the total number of deaths because of North Korea's food shortages between 1995 and 2005 range between 1 and 3 million.

By 2009, after 56 years of sacrifice and austerity, personal incomes in North Korea were only one-tenth those in South Korea, and infant

▼ **FIGURE 8.31 The Great Leader**  A commune group visiting Pyongyang bows to the Grand Monument to Kim Il Sung, the "Great Leader."

mortality rates were five times higher. North Korea remains not only one of the world's most impoverished societies but also one of the most closed and rigid. Listening to foreign radio broadcasts is punishable by death, citizens can be detained arbitrarily, and there are an estimated 150,000 political prisoners. Society is stratified according to loyalty to the Communist Party, and citizens' level of commitment determines their access to housing, education, and health care.

North Korea is also one of the most highly militarized countries in the world. It has the world's fifth-largest standing army (after China, the United States, Russia, and India), even though it has a relatively small population (an estimated 23.1 million in 2005). North Korea's weapons and missiles are sold indiscriminately on world markets to raise badly needed foreign exchange, whereas its own arsenal— including possible nuclear capability—is pointed menacingly at South Korea and Japan. This threat represents a major geopolitical problem for the region, and North Korea has attempted to leverage this threat for economic aid and political concessions from both China and the United States.

Most recently, North Korea has made efforts to expand its missile capacity, testing increasingly long-range missiles that might—in theory—carry nuclear payloads in the future. The country may possess as many as 800 ballistic missiles, the longest-range of which reaches deep into the Pacific Ocean, making it within potential firing range of Japan and, perhaps eventually, Europe and the United States. Abortive North Korean tests of missiles in 2009, however, revealed that this capacity is not yet well developed and is still error prone.

**Unresolved Geopolitics in Taiwan**   When the Communist revolution created the People's Republic of China in mainland China, the ousted Nationalist government established itself in Taiwan (then called Formosa) as the Republic of China. Granted diplomatic recognition by Western governments but not by the People's Republic, Taiwan immediately became a potential geopolitical flashpoint in the Cold War. Large amounts of economic aid from the United States, provided because of Cold War geopolitics, helped prime Taiwan's economy. Strict currency controls, the creation of government corporations in key industries, and strong trade barriers to protect domestic industries from foreign competition were introduced by the Taiwanese government in an effort to establish a self-sufficient economy. An authoritarian regime suppressed opposition to government policies. By the early 1960s, Taiwan's political stability and cheap labor force provided a very attractive environment for export processing industries.

Taiwan lost its full international diplomatic status in 1971, when U.S. President Richard Nixon's rapprochement with the People's Republic led to its entrance into the United Nations. In 1987 Taiwan's government lifted martial law, began a phase of political liberalization, and relaxed its rules about contact with mainland China. Meanwhile, mainland China still claims Taiwan as a province of the People's Republic and has offered to set up a Special Administrative Region for Taiwan, with the sort of economic and democratic privileges that it has given Hong Kong.

Taiwan's geopolitical problems have not prevented it from achieving great economic success. With a land area of 36,000 square kilometers (13,900 square miles), only 25% of which is cultivable, and a population of 22 million, Taiwan would be the second smallest of China's provinces and seventh from the bottom in population. Yet Taiwan's per capita GDP (in PPP "international" dollars) in 2005, $27,572, was almost four times that of mainland China (excluding Hong Kong and Macao). In 2005 Taiwan was the world's 19th-largest exporter, with a total export trade of $189.4 billion. Taiwan's economic growth rate over the past three decades has been phenomenal, averaging more than 8% per year. Sometimes referred to as "Silicon Island," Taiwan has 1.2 million small- and medium-sized enterprises. Most make components, or entire products, according to specifications set by other, often well-known, international firms, whose brand names go on the final product. Because Taiwan's firms tend to be small, they have been able to be flexible in responding to changes in technology. They have also been able to move quickly to take advantage of China's "open door" policy. Between 1990 and 2005, Taiwanese businesspeople invested more than $150 billion abroad, the greater share of it in mainland China. In South China and in Fujian Province, across the Taiwan Strait from Taiwan, Taiwanese enterprises operate what amounts to a parallel economy.

Taiwan's economic success as one of the newly industrialized "Asian Tigers" is not without growing pains. In Taipei and other cities, Taiwan's breathtakingly fast modernization has brought heavy pollution, acute housing shortages, and rampant corruption. The rapid acquisition of automobiles, motor scooters, and air conditioners has made the environment unbearable and transportation a nightmare. Growth has been so rapid that the government has been unable to solve the problems of water supply and waste disposal, and "garbage wars" over the issue of sanitary landfill placement have occasionally led to huge quantities of uncollected garbage. Within Taipei (see Figure 8.11), the capital and the industrial and commercial heart of Taiwan (population 2.95 million in 2005), these problems have been intensified by an influx of young people from other parts of the island, drawn by the educational and economic opportunities and by the exciting sense of rapid change.

### The Contested Peripheries of China

As early as the 7th and 10th centuries C.E., with the expansion and unification of the Chinese empire, the Han Chinese had begun to move into the margins of the Tibetan region. By that time, Tibet had developed a distinctive culture based on Tibetan Buddhism—developed from a fusion of the region's ancient animistic religion (known as Bon) with Tantric Buddhism imported from India. During the Ming (1368–1644) and Qing (1644–1911) dynasties, the entire Tibetan Plateau became part of the Chinese empire. Far away from the center of power, Tibet was divided into numerous small tribes and districts, with feudal chieftains ruling autocratically, generation after generation. As a result, Tibet became a highly repressive theocracy based on serfdom. In 1950 China seized on this fact as justification for its invasion of the strategically important region.

Between 1950 and 1970 the Chinese "liberated" the Tibetans, drove their spiritual leader, the Dalai Lama, into exile, destroyed much of Tibetan cultural heritage, and caused the death of an estimated 1 million Tibetans. In 1959 the Tibetans rose in an attempted revolt, and their spiritual leader, the Dalai Lama, fled into exile. In 1965, Tibet was granted the status of an autonomous region, called Xizang, but by that time large numbers of Han had flooded in, often taking positions of authority, leaving Tibetans as disadvantaged, second-class citizens. There were serious anti-Chinese disturbances in 1988, and since then the volatile situation has continued to simmer. Today in Xizang, all Buddhist monasteries are rigidly controlled by the police and Communist Party officials, and expressions of devotion to the Dalai Lama are banned. About 60% of the area's population is now of Han Chinese descent. These developments have given rise to significant ethnic tensions within the region, while the exiled Dalai Lama has acquired global celebrity as a result of his international tours promoting Buddhism and publicizing the "cultural genocide" committed by China in Tibet. The Dalai Lama does not want national independence for Tibet, but greater autonomy. For their part, the Chinese cannot understand the ingratitude of the Tibetans. As the Chinese see it, they have saved the Tibetans from feudalism and built roads, schools, hospitals, and factories.

As noted earlier, another region of tense interethnic relations is that of Xinjiang, a vast territory that occupies one-sixth of China. Its multiethnic population totals only 15 million, including an estimated 8 million Uyghurs and a sizable number of Han colonists who arrived after 1950. The majority of Xinjiang's population are Muslim, but their agitation for greater autonomy and respect for Islamic values has been ruthlessly suppressed by the Chinese authorities. In 1962 minority factions rebelled against the Chinese, calling for the establishment of an independent East Turkestan Muslim Republic. Tens of thousands of nomadic peoples fled across the mountainous border with the Soviet Union, and for several decades now there has been ongoing feuding in the frontier region. In recent years, separatist sentiments in Xinjiang have been fueled by the collapse of the Soviet Union and the realignment of Central Asia's newly independent Muslim states that share borders with Xinjiang. Protests and bombings increased sharply in the 1990s, and the traffic of weapons, political literature, and insurgents has prompted the Chinese to increase surveillance in Xinjiang and, intermittently, to close its borders with Pakistan and neighboring Central Asian states.

### The Geopolitics of Globalized Chinese Investment

The enormous success of capitalism in China has resulted in two simultaneous trends that have put pressure on global geopolitics. First, the enormous production system of the industrial zones of China have tested the limits of available raw materials and energy resources within the country. Petroleum and minerals essential to sustaining production in Chinese factories are found in insufficient quantities at the current rate of use, even though China is well endowed with resources. This has led to a search for new sources of inputs, taking China abroad into new resource frontiers, especially in Africa.

Simultaneously, the industrial growth of the country has made a handful of business operators (and government officials) enormously wealthy. As a result, there is significant capital available in China for investment abroad. Although some of this capital is sunk in purchasing the debt of other industrial nations (especially the United States), more speculative investment has become attractive for the Chinese, especially in underdeveloped areas, again including Africa (**Figure 8.32**).

The convergence of these two trends has given China a large and quickly growing presence throughout the developed world, with implications for international conflict. In a notable case, Chinese development money has flowed heavily into Sudan, a country under international scrutiny for the high levels of uncontrolled and state-tolerated violence (indeed, genocide) in its Darfur region (see Chapter 5).

From the Chinese point of view, the supply of development capital to countries like Sudan will provide a net benefit to Sudan's citizens. Moreover, the Chinese position remains that the mode of governance a nation chooses is not the business of other members of the world community. So, too, China points to the historical development of poor nations by wealthier ones in the past and stresses that international condemnation of such investment is merely a symptom of jealousy on the part of European nations, who have long colonial histories in places like Africa.

From the point of view of the United States and the European Union, conversely, Chinese support for the Sudan regime subverts any efforts to bring the horrible genocide under control there. The moral ambivalence of Chinese investment therefore represents an impediment to peace and human rights.

Whatever the outcome in the case of Sudan, the rise of China as an international development and investment actor on a global scale will be a key factor in the formation of new regions. Consider, for example, how

▲ **FIGURE 8.32 Chinese investment in Africa** A Chinese engineer supervises work at a construction site in Sudan's capital, Khartoum, February 16, 2009. China's interests in Africa have multiplied in recent years, with trade rising tenfold since 2000 to nearly $107 billion last year, and Beijing has been at pains to reassure African nations that it will not desert them during the economic slowdown.

the strong historic association of China with Africa was a part of the global trade system 1,000 years ago. This was only broken during the colonial era. With the rise of a powerful China, sub-Saharan Africa and East Asia are becoming relinked, with transformative possibilities for each region. Again then, the characteristics of regions emerge from their relations, and changing global relations portend the creation of new regions.

# FUTURE GEOGRAPHIES

The future geographies of East Asia over the next 20 years rest on the trajectory of three ongoing trends: (1) the rise of China as a central geopolitical player, (2) the fate of the Korean Peninsula, and (3) the internal transformation of China.

**Emerging Chinese Hegemony** As we have seen, the growth of the Chinese economy has placed that country in a new political position relative to the rest of the globe. Will the 21st century be the "Chinese Century" or, rather, a century of diverse economic players, including China among many? How central and autonomous will the Chinese economy be as global trade reduces the absolute dominance of historically critical players (like the United States)? An extremely strong political China portends interesting possibilities because that country is in a

unique position to oversee negotiations with North Korea (see below), for example, and to provide a strong counterweight to Russia and other global resource powers. As noted earlier, a strong political China is also in position, however, to support and extend the power of problematic and dangerous states (like Sudan) and therefore subvert growing mandates to ameliorate the dramatically negative effects of resource development or to hold international criminals accountable for their actions. The future of China is the linchpin for the future of East Asia.

**Emerging Korean Conflict or Reconciliation** On the one hand, it is easy to forecast a gloomy scenario on the Korean Peninsula. The behavior of the North Korean state, especially regarding nuclear capacity and long-range missile capabilities, suggests that country wishes to negotiate, if at all, from a position of terrific and dangerous strength. The nontransparent nature of the government and the extreme political power of a single family of rulers also make the regime thoroughly unpredictable.

On the other hand, rapid and precipitous change is also possible. Efforts at détente between the Koreas have been extremely limited to date. But the Korean people continue to hold a tacit interest in unification (in a strange echo of the populations of the two Germanys prior to the fall of the Berlin Wall—see Chapter 2). Also, the current leader of North Korea, Kim Il Jong, is declining rapidly and is possibly in ill health. His passing may signal an opportunity for political change. These trends make it possible to at least imagine a very different outcome, in which a unified Korea emerges in the 21st century, which would hasten regional disarmament and bring East Asia back from the brink of a conflict simmering for 50 years.

**Emerging Internal Chinese Political Change** To date, the Chinese government has been able to reconcile central party rule with capitalist development by providing ongoing growth, a flood of new consumer goods, and better living conditions for a great number of its people. This difficult compromise between an antidemocratic state and an open (if not totally "free") market would be tested, however, if growth were to flag, and the guarantee of prosperity were not met. On the former point, the recent economic downturn showed little serious impact on the Chinese capacity to produce, sell, and compete globally. On the latter point, however, it is apparent that many parts of the country have been left behind during this period of rapid growth. The countryside, most obviously, has been sacrificed for the benefit of the cities, and the resource-rich west has been mined for the benefit of the industry-heavy east. Could these tensions lead to a more transparent and democratic China, and would such a state be more or less amenable to the kinds of economic growth we see today? The central party has tacitly answered this question with a "no." Should a slow reconfiguration of politics within China occur, especially as new generations of leadership emerge, however, it is possible that China might look radically different than the country we see today in only a few decades.

# Main Points Revisited

■ East Asia is a region physically distinguished by the vast continental interior of China in the west and center, coupled with the peninsular and island coasts in the east formed at the edge of the Ring of Fire.

Risks associated with these features especially include flooding from the enormous drainages across China and earthquakes characteristic of the mountain building on the islands of Japan.

■ The landscapes of the region have been massively modified over the centuries, especially for the management of water flow.

The management of waterscapes has historically included large and sophisticated irrigation systems, often overseen by expert centralized authority. This continues to be the case, especially in China, where hydroelectric investment and water diversion are key parts of development strategy.

■ The historical geography of the region is characterized by the core economic power of China, which was at the center of global trade networks for more than a thousand years prior to colonization and is becoming so again.

The colonial era and the upheavals of the Second World War drove China from its historically dominant position in the region, allowing for a reconfiguration of the region dominated by trans-Pacific trade between Japan and the United States.

■ The development of postcolonial nation states in the late 20th century resulted in the creation of profoundly different economic experiments in China, South Korea, and Japan.

Japan used conglomerate-oriented and banking-dependent industrialization. South Korea utilized import substitution to grow its industrial capacity. Chinese development dramatically switched from central planning to state-led capitalism. All of these developments led to spatially uneven patterns of development.

■ The cultural landscapes of the region are rooted in many dominant religious and philosophical traditions, including Confucianism, Taoism, and Shinotism, but these have been challenged and reformulated both through internal upheavals and ongoing interactions with cultural groups on the region's periphery.

The large number of ethnolinguistic minority communities on these regional margins, especially within China, has led to cultural and political conflicts.

■ The geopolitics of East Asia revolve around the shifting balance of economic power between Japan and China and the unresolved problem of the Korean Peninsula, where the Korean War has never officially come to an end.

The possibilities for conflict resolution are complicated by the uncertain capacity and willingness of China to intercede and influence North Korea's behavior.

# Key Terms

anime (p. 298)

Asian Tigers (p. 289)

chaebol (p. 289)

Confucianism (p. 295)

counterurbanization (p. 292)

feng shui (p. 296)

Great Leap Forward (p. 286)

hydraulic civilizations (p. 277)

ideograph (p. 294)

import substitution (p. 289)

keiretsu (p. 286)

manga (p. 298)

massif (p. 276)

one-child policy (p. 291)

Pacific Rim (p. 286)

pinyin (p. 294)

Ring of Fire (p. 274)

Shinto (p. 295)

Special Economic Zones (SEZs) (p. 288)

zaibatsu (p. 285)

# Thinking Geographically

1. What are the three main physiographic provinces of East Asia? How do the Tien Shan and Himalaya mountains affect East Asia physically and culturally? How has human activity modified the landscape in East Asia?

2. How and to what degree was East Asia economically linked to the rest of the world in the precolonial era? What is the Silk Road?

3. What triggered the Korean War? What is the significance of the 38th parallel? Who is Kim Jong Il?

4. Why is clothing made in China produced more profitably than in Hong Kong or Canada?

5. State-assisted capitalism occurs frequently in East Asia. Explain how the central government involves itself in corporate activities in China, Japan, and South Korea.

6. How are private investors in Hong Kong, Taiwan, and elsewhere changing the geography of China, especially in the Special

Economic Zones? How has the People's Republic of China's attempt to create a classless society been affected by capitalist investment in China?

7. In 1970, how many children were born to the average Chinese family? By 2005, what was the average number of children per family? Who helps enforce China's one-child policy? Aside from lower birthrates, what are two main results of the one-child policy?

8. How do Confucianism and Shinto differ from each other, and how have they influenced the cultural geography of East Asia? What is animism?

9. What are some dominant East Asian cultural traditions? How is geomancy (feng shui) used in everyday life? How have East Asian beliefs and traditions affected other world regions?

10. How do food and entertainment traditions reflect not only the regional distinctiveness of East Asia but also its linkages to other global regions?

 Log in to www.mygeoscienceplace.com for Videos, Interactive Maps; RSS feeds; Further Readings; Suggestions for Films, Music and Popular Literature; and Self-Study Quizzes to enhance your study of East Asia.

# ▼ FIGURE 9.1

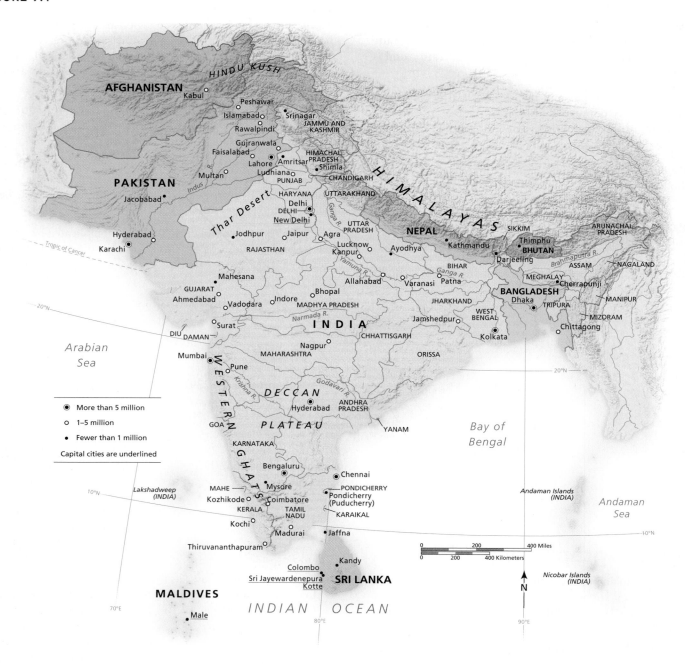

## SOUTH ASIA KEY FACTS

- Major subregions: Deccan Plateau, Mountain Rim, Peninsular Highland, Coastal fringe

- Total land area: 5 million square kilometers/ 1.9 million square miles

- Major physiogeographic features: Bay of Bengal; Himalayas (including K2 Mountain); Khyber Pass; Indus, Ganga, and Brahmaputra rivers; Indus Plains. Dominant aspect of climate is southwesterly summer monsoons—where the Western Ghats can receive up to 4000 mm of rain.

- Population (2010): 1.64 billion, including 1.2 billion in India, 164 million in Bangladesh, 165 million in Pakistan

- Urbanization (2010): 29%

- GDP PPP (2010) current international dollars: 3,153; highest = Bhutan, 5,857; lowest = Afghanistan, 850

- Debt service (total, % of GDP) (2005): 1.2%

- Population in poverty (%, < $2/day) (1990–2005): 69.6%

- Gender-related development index value (2005): 0.61; highest = Sri Lanka, 0.74; lowest = Nepal, 0.52.

- Average life expectancy/fertility rate (2008): 64 (M63, F65)/3.5

- Population with access to Internet (2008): 4.8%

- Major language families: Indo-Aryan (including Hindi, Bengali, and Urdu), Dravidian

- Major religions: Hinduism, Islam, Buddhism

- $CO_2$ emission per capita, tonnes (2004): 0.8

# 9

# SOUTH ASIA

Pranita Joshi, a woman living in the sprawling suburbs of Mumbai, finishes cooking her husband's lunchtime meal in the middle of a busy workday morning. She closes it, still piping hot, into her small metal *tiffin*, a set of stacked containers or lunchboxes for carrying food. From here, the tiffin begins an epic journey of dozens of miles, changes hands five times, and somehow, remarkably, winds up on the desk of her husband, a systems engineer working in a high-rise in the city's downtown. Nor is Mr. Joshi's lunch alone. Tens of thousands of these lunches miraculously travel similar routes every day, from all around the outskirts of this city of 15 million people, traveling from the kitchens of workers to their offices by way of a complex and highly organized labor system.[1]

The tiffin delivery system depends on a huge network of *dabbawallas*—food delivery men—who pick up the food from countless homes, deliver them by bicycle to commuter rail heads, organize them into groups, pick them up at local train stations, carry them on their heads through crowded streets, and finally hand them off for final delivery in offices across the city. A system of complex codes and markings on every tiffin ensures that they arrive, still warm, in the hands of the right office worker. A fully organized union of these delivery men, the "Mumbai Tiffin Box Carriers Association," sees to it that every member of the working chain receives a fair wage. New experimentation with Internet communication has made the system even more efficient. The startling complexity of this network, together with the remarkable simplicity of its core concept, say a lot about the always-innovative households and economies of South Asia.

First, although this system is old, indeed more than a century old, it continues to innovate, precisely as Mumbai has grown more financially successful, more internationally connected, and more central to India's economy. As Mumbai's labor force grows and its middle class thrives, the demand for hot food delivery from home has only increased. The global connections of South Asia to other world regions has both created new economies in high-tech businesses like those of Mr. Joshi, and also expanded traditional ones, like that of the dabbawallas.

Second, although the business thrives on information technology, both in the ancient form of colored markings on each tiffin and the expanding use of the Internet, it also depends heavily on labor. In a country of more than one billion people, high technology and working hands together propel daily life.

Finally, the tiffin system shows the way the urban core still depends on, and is linked to, more rural areas and demonstrates the way home labor, largely performed by women, is critical to the thriving office economy. In South Asia, the emerging global, urban, modern, and high-technology office world is not so much replacing the rural, traditional, and manual systems of the past as much as it is becoming more firmly interlinked with them. ■

## MAIN POINTS

■ South Asia is a region physically distinguished by its ringing arc of mountain ranges and the annual rhythm of its summer monsoon rains.

■ The landscapes of the region are characterized by inventive adaptation to regional climate, especially including highly productive agricultural systems, carefully adapted to the exigencies of drought and flooding.

■ The historical geography of the region is characterized by its presence as a center of global trade, especially in Asian/African trade across the Indian Ocean, which was later exploited, and overridden by, colonial governance.

■ South Asia has one of the largest and densest populations of all world regions, and its growth is ongoing, but growth rates are dropping rapidly, especially where health care, birth control, and education are available, especially to women.

■ The cultural landscape of South Asia is influenced by endemic traditions, including Buddhism and Hinduism, but is also a product of continuous adoption and adaptation of other cultural traditions, including religions, artistic traditions, and languages from around the world.

■ The geopolitics of the region revolve around the relations of India to Pakistan, and the future fate of the region rests on resolution of deep antagonisms between these nuclear states, rooted in the partition of the region after colonialism.

■ Economically, South Asia is at a critical juncture, with considerable potential in its human and natural resource base and emerging new economy. Foreign investment has flowed into the country, helping generate exceptionally high economic growth rates and a huge, well-educated and sophisticated consumer society.

---

[1] Saritha Rai, "In India, Grandma Cooks, They Deliver," *The New York Times*, May 27, 2007.

# ENVIRONMENT AND SOCIETY

Two aspects of South Asia's physical geography have been fundamental to its evolution as a world region, each of which is rooted squarely in the region's connections to others, through the global climate and tectonic systems (**Figure 9.1**). The first striking feature is the system of the **monsoon**, seasonal torrents of rain on which the livelihood of the people of South Asia depends. This system is propelled by the annual heating and cooling of the land mass of Central Asia to the north and the constant feeding of moisture from the Indian Ocean (see Figure 9.3). The signature physical condition of this region is therefore a result of South Asia's position between other world regions.

Second, as the satellite image reveals, South Asia is clearly set apart from the rest of Asia by a forbidding mountain rim (**Figure 9.2**). This arc of mountain ranges has created a porous boundary around the people of South Asia, creating a large-scale natural setting in which distinctive human geographies have flowed back and forth between surrounding regions. Like the monsoon, this distinctive feature is born from the interaction of multiple regions, as the South Asian plate rams into the Asian plate, creating massive uplift and ongoing mounting growth.

## Climate, Adaptation, and Global Change

The climate of South Asia is distinguished by the dramatic effect of the southwesterly summer monsoon, a deluge of rain that sweeps from south to north across the region. The word *monsoon* derives from the

Arabic word *mausim,* meaning "season." Although now widely used to describe any seasonal reversal of wind flows in the lower-middle latitudes, *monsoon* was originally applied to the distinctive seasonal winds in the Indian Ocean on which Arab traders relied to power their sailing ships on their annual voyages to and from the East Indies (the name formerly applied loosely to present-day Malaysia and Indonesia) in quest of spices, ivory, and fine fabrics.

**The Monsoon** In most of South Asia, the seasonal pattern of climate is as follows: a cool and mainly dry winter, a hot and mainly dry season from March into June, and a wet monsoon that "bursts" in June and lasts into September or later. The engine behind this remarkable phenomenon is the heating and cooling of the Asian continent to the north. During the early summer, the interior parts of Asia begin to heat more rapidly than the areas of the ocean to the south, creating low pressure convection as hot air rises all across Asia. By midsummer, the jet stream and the atmospheric circulatory system move north, allowing moist maritime air to invade, drawn onward by low pressure to become the southwest, or "wet" monsoon. Producing strong wet winds that draw ocean air from the South, the monsoon moves inland starting in June and brings drenching rains. These slowly progress from the island of Sri Lanka, across the southern coastal zones of India, across Bangladesh and the Gangetic Plain of India, and finally into the northwestern parts of India and Pakistan (**Figure 9.3**). The northern hills and uplands exert a strong **orographic effect**, causing moist air from the sea to lift and condense, producing heavy rainfall during this period. The arrival of the wet monsoon season is announced by violent storms and torrential rain. This is the "breaking" or "bursting" of the monsoon (**Figure 9.4**).

In winter, conversely, a major branch of the jet stream tends to fend off the low-pressure systems of the circumpolar atmospheric whirl, helping maintain stable high-pressure conditions over the Mountain Rim and the Tibetan Plateau. The prevailing winds are northeasterly, blowing from the interior toward the sea—these are the so-called dry monsoon winds.

One exception to this pattern is in parts of Afghanistan, Pakistan, and northwest India, where shallow low-pressure systems move through from the eastern Mediterranean, bringing light but useful rainfall in late winter. Another is northern and eastern Sri Lanka, where trade winds bring some winter rains. The southern part of Sri Lanka, the Maldives, and the Nicobar Islands are far enough south to be affected by intertropical convergence (see Chapter 1), and so they rarely have a dry month all year. The result is the overall pattern of climate across the region, with tropical climates hugging the coasts and the islands, a broad belt of humid subtropical climate following the Ganges and abutting the Mountain Rim, and arid climates in the desert northwest of India and Pakistan in the central interior of the Indian plateau (**Figure 9.5**).

**Adapting to Rainfall and Drought** For much of South Asia, there is a significant risk of drought and famine as a result of a late or unusually dry monsoon season. In 2000 and again in 2002 the late arrival of the summer monsoon left more than 50 million people in west and central India facing acute water shortages and widespread crop failure. In Afghanistan, seven years of drought culminated in 2004 in disastrous crop failures that drove an average of 500 families into refugee camps each day. Even where the monsoon does not fail completely, it can be highly uneven, so that villages in one region may receive sufficient rainfall to plant and harvest, while nearby settlements do not.

▲ **FIGURE 9.2 South Asia from Space** This image clearly shows how South Asia is naturally bounded by mountains and oceans.

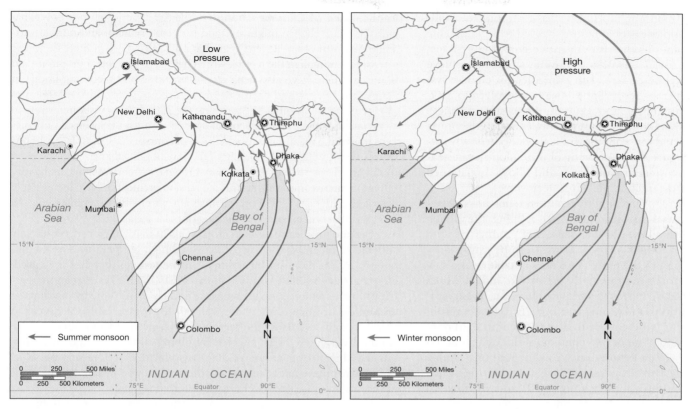

▲ **FIGURE 9.3 Summer and Winter Monsoons** The wet summer monsoon "bursts" in June and lasts into September or later. In winter, the prevailing "dry monsoon" winds are then northeasterly.

▼ **FIGURE 9.5 Climate Map of South Asia** South Asia is marked by great heterogeneity in climate regions, with desert in the northwest and wet tropics in the far south.

▼ **FIGURE 9.4 Monsoon in Dhaka** Commuting to work during the monsoons in Bangladesh often means improvising and depending on human labor. In these extreme 2007 storms, 15 people were killed in floods, but daily urban life continued.

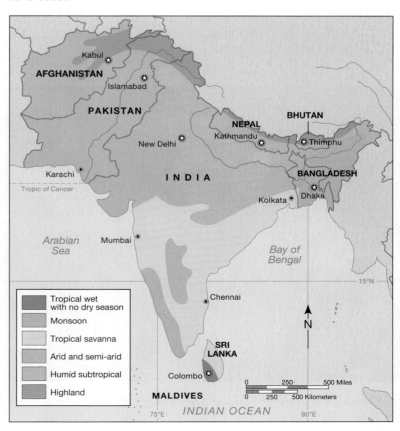

The fierceness of the monsoon rains, as well as their unpredictability, spatial unevenness, and occasional failure, has led to a remarkable range of ingenious adaptations that use inventive cropping systems, agricultural biodiversity, social networks, and water storage to achieve stability and survival, even in the face of an unpredictable climate system. Systems for growing crops across South Asia have evolved to maximize efficiencies in the use of soil and water. In the mountain fringes of the Himalayas and the Karakorums, most notably, enormous labor has gone into stabilizing the land for cropping in the face of potentially catastrophic soil erosion (**Figure 9.6**). **Terracing** of these slopes over generations has created a distinctive landscape and allowed sustained agricultural yields amid torrential rainfall and steep slopes.

In more arid regions, conversely, agricultural systems have developed over thousands of years that maximize intercropping, the mixing of different crop species that have varying degrees of productivity and drought tolerance. In bad rainfall years, a harvest can be salvaged by relying on the heartiest part of the crop. Especially notable in the cornucopia of South Asian foods are the hardy grains like millet, which need only a handful of rainfall events to survive to harvest. The agrodiversity of traditional South Asian farming is extremely high as a result.

During the 1960s, efforts were made to boost agricultural production through the imposition of modern technologies through what has come to be known as the **Green Revolution**. Many international actors, including countries like the United States, global agencies like the United Nations Food and Agriculture Organization (FAO) and multilateral lending agencies like the World Bank, extended new technologies to the region, including high-yielding varieties (HYV) of wheat and corn, which could double or triple production per hectare of land. These crops and cropping technologies required far more inputs than their traditional counterparts, however, including more water, fertilizer, and pesticides. The result has been a massive increase in productivity, but it has come with a devastating increase in the dependency of framers on purchased products, a decrease in native seed diversity, and an exhaustion of soil quality and water supplies.

▲ **FIGURE 9.6 Terraced Fields in River Valley** Overlooking terraced fields of Duikar in the Hunza River valley in the Karakoram Range, Hunza, Pakistan.

**Implications of Climate Change**　It is too soon to tell precisely what global climate change might portend for water scarcity and rainfall variability in South Asia. Recent years have actually seen an increase in the intensity of rainfall in the region, which has led to dramatic rain events and flooding. There have also been record-breaking and recurring floods in Bangladesh, Nepal, and northeast India in recent years; a record 944 mm of rainfall fell in Mumbai, India during the 2005 July monsoon, leading to 1,000 fatalities and a financial cost of over 250 million $US. Uncertainty prevails in making meaningful predictions, of course. For example, typhoons (hurricanes), which are endemic to the Bay of Bengal and a serious environmental hazard in Bangladesh, have grown less frequent in recent history but more intense, meaning storms are hard to predict but have increasingly devastating impacts. These coastal areas are extremely vulnerable to the sea level rise that will accompany melting of polar ice as well.

Despite this heavy rainfall, crop yields in many regions have already suffered in recent years from unexpected high temperatures as the frequency of hot days and heat waves in the last century has increased. Despite heavy rainfall events, water scarcity in cities is also an increasing problem, which climate change will likely exacerbate by making the critical occurrence of monsoon rainfall less predictable from year to year. Drought events in the region are closely tied to El Niño, whose variability is likely to be affected by global climate change. More frequent and recurring failures of the monsoon are therefore very possible, especially for arid and semiarid northern India and Pakistan, where the monsoon often fails to reach, resulting in drought.

One of the most important and already-evident impacts of global climate change on the region is the decline in Himalayan glaciers, which poses serious questions about water security in the region. As the Intergovernmental Panel on Climate Change (IPCC) reports, these glaciers, which cover 17% of the mountain areas of the Himalayas and represent the largest body of ice apart from the polar ice caps, are receding faster than in any other area of the planet. At the current rate of warming, they could disappear altogether by 2035, destroying all but the seasonal (monsoonal) flow of the major rivers of the region: Ganges, Indus, and Brahmaputra. This threatens agriculture as well as the drinking water supplies for more than a half-billion people.

Finally, one of the less well understood, but equally imperative, problems posed by climate change is the fundamental transformation of disease ecology in the region. For example, nearly the entire population of the region is currently vulnerable to various vector-born diseases, especially those spread by mosquitoes, including dengue fever, yellow fever, and malaria, which alone killed 935 people in India in 2008, with 1.5 million debilitating cases overall. With overall global warming, it is reasonable to anticipate that cities with historically cooler winters, especially in the northern part of the subcontinent, will experience increasingly lengthy mosquito breeding seasons and a significant, if not exponential, increase in rates of disease.

## Geological Resources, Risk, and Water Management

In geological terms, South Asia is a recent addition to the continental landmass of Asia. The greater part of what is now South Asia broke away from the coast of Africa about 100 million years ago. It drifted slowly on a separate geological plate for more than 70 million years until it collided with the southern edge of Asia. The slow but relentless impact crumpled the sedimentary rocks on the south coast of Asia into a series of lofty mountain ranges and lifted the Tibetan Plateau more than 5 kilometers (3.1 miles) into the air. The Himalayas, which stand at the center of South Asia's Mountain Rim, are still rising (at a rate of about 25 centimeters—9.8 inches—per century) as a result of this geological event.

Not surprisingly, the principal physiographic regions of South Asia also reflect this major geological event and include: the Peninsular Highlands, the Mountain Rim, the Plains and the Coastal Fringe (**Figure 9.7**).

The Peninsular Highlands of India form a broad plateau flanked by two chains of hills. The highlands rest on an ancient shield of granites and other igneous and metamorphosed sedimentary material, together with some very old sedimentary rocks. This shield has remained a relatively stable landmass for much of the past 30 million years. However, between 65 and 55 million years ago, immense fissure eruptions of lava buried the northwestern part of the peninsula beneath up to 3,000 meters (9,842

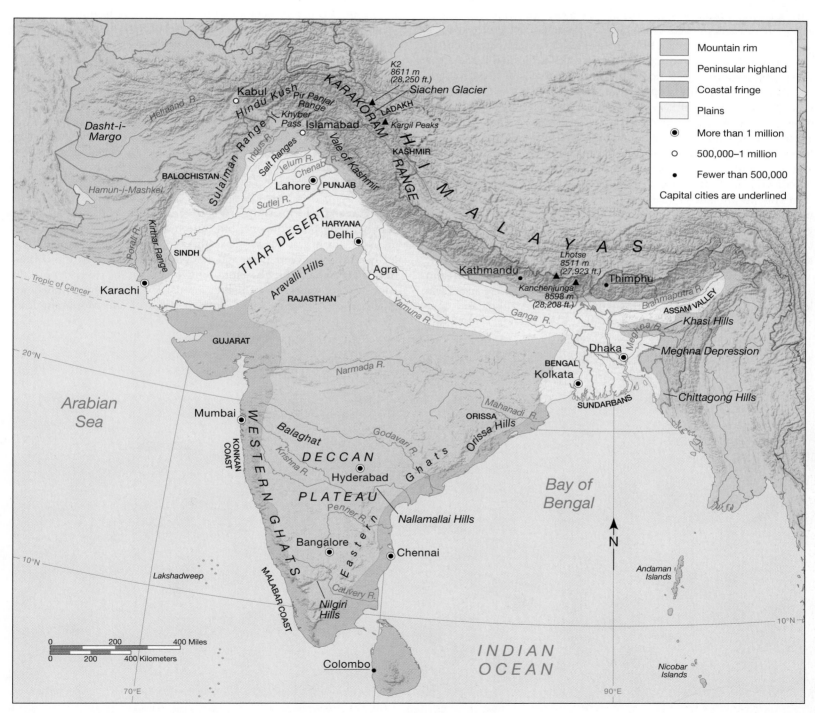

▲ **FIGURE 9.7  South Asia's Physiographic Regions**  The physical geography of South Asia is framed by the highland plateaus of an ancient continental plate and a young mountain rim, separated by broad plains of alluvium washed down from both.

feet) of basalt, a dense volcanic rock. Today, this lava plateau—the Deccan Lava Plateau—covers about one-third of the Peninsular Highlands.

The Mountain Rim is a vast region of spectacular mountain terrain, remote valleys, varied flora and fauna, ancient Buddhist monasteries, and fiercely independent tribal societies. The physical geography of the region is complex, with several mountain ranges together sweeping in a 2,500-kilometer (1,554-mile) arc that contains numerous high peaks including K2 (8,611 meters; 28,250 feet). At the heart of the Mountain Rim are the young but spectacular Himalaya and Karakoram ranges. Interspersed among the high peaks and the foothills are protected gorges and fertile valleys that sustain isolated settlements. The largest of the fertile valleys is the Vale of Kashmir, a verdant basin that is 130 kilometers (81 miles) long and between 30 and 40 kilometers (19 to 25 miles) wide. In this and other valleys, it is possible to grow rice, with corn and wheat at higher elevations and orchards—especially of apricots and walnuts—on the valley slopes.

The broad plains of young sedimentary rocks and alluvium have been created by the deposition of material eroded from both the Peninsular Highlands and the Mountain Rim. Three river systems—the Indus, the Ganga (Ganges), and the Brahmaputra—begin within 1,600 kilometers (994 miles) of one another in the Himalayas but flow in three different directions through the mountains and into the plains. All three river systems provide the Plains region with a steady, if uneven, flow of melting snow. As a result, the Plains region has long been widely irrigated and has supported a high population density (**Figure 9.8**). Cultivation of grains and rice is the predominant activity (see Signature Region in Global Context: The Indus Plains, p. 312).

The Bengal Delta, which covers a large portion of Bangladesh and extends into India, is the product of three major rivers, the Ganga (also known as the Ganges), the Brahmaputra, and the Meghna and their **distributaries** (river branches that flow away from the main stream) that sluice down to the Bay of Bengal, creating a vast web of waterways. Because these rivers lie in deep deposits of sand and clay and carry such enormous quantities of water, especially in flood stage, they are almost impossible to control with engineering works; as a result, flooding is normal. In the monsoon season, about 70% of the delta region is flooded up to 1 to 2 meters (3 to 6 feet) deep.

The narrow Coastal Fringe is the product of marine erosion that has sawed into the edge of the ancient shield of the Peninsular Highlands. Elsewhere it is the product of marine deposits, and in some places the plains consist of alluvial deposits in the form of deltas and mudflats. Many of South Asia's largest and most prosperous cities developed from trading posts that were established along the Coastal Fringe in the 17th century. The largest among them—Chennai (formerly Madras), Colombo, Karachi, Kolkata (formerly Calcutta), and Mumbai (formerly Bombay)—were great centers of commerce under British imperialism. Far out into the Bay of Bengal are India's Andaman Islands and the Nicobar Islands. During the rainy monsoon seasons, the Coastal Fringe is filled with luxuriant growth, especially along the southwest Malabar Coast of India and the southwest coast of Sri Lanka, where rich harvests of rice and fruit support dense rural populations.

**The Hazard of Flooding**   Where the low-pressure systems of the summer monsoon flow over hills and mountains, the monsoon rains are especially heavy. The Western Ghats and adjacent coastal plains typically receive between 2,000 and 4,000 millimeters (79 to 158 inches) of rainfall as the southwest monsoon winds meet the steep slope at the edge of the Peninsular Highlands. Similar levels of annual rainfall are received in the central and eastern parts of the Mountain Rim.

▼ **FIGURE 9.8 The Ganga Plains** Although the Plains have been the hearth of successive empires because of their agricultural productivity, most of the Plains farmers rely on irrigation for their crops, and some areas are very dry.

▲ **FIGURE 9.9 Monsoon Season** Flooding occurs regularly during monsoon season in the Bengal Delta. Though the floods are troublesome, the silt deposited during the floods helps replenish soil nutrients. Serious problems occur when the rains are unusually heavy and when high tides or exceptionally high rainfall are accompanied by strong onshore winds.

▲ **FIGURE 9.10 South Asian Marble and Sandstone in Monumental Architecture** Long a crucial part of monumental construction in India and Pakistan, the sandstones and marbles of the region are reflected in classical architecture, including the Jama Masjid in Old Delhi, the largest and most important mosque in the area.

Cherrapunji, in the Khasi Hills south of the Assam Valley, boasts the world's average annual rainfall record of 11,437 millimeters (450.6 inches). It once recorded as much as 924 millimeters (36.4 inches) in one day at the onset of the monsoon season.

It is in Bangladesh, however, that monsoon rains produce regular and widespread flooding. Swollen by monsoon rains, the distributaries of the Ganga, Brahmaputra, and Meghna river systems regularly spill over into the low-lying delta and plains areas (Figure 9.9). If monsoon rains are unusually heavy, flooding can be disastrous, inundating villages, drowning people and livestock, and ruining crops. In 1988 all three of the major rivers reached flood stage at the same time, with the result that floods drowned more than 2,000 people. In 1970, 1991, and 1999, cyclones hit the delta area during especially heavy flooding and exceptionally high tides, leading to devastating damage and widespread loss of life. The 1970 cyclone killed between 300,000 and 500,000 people. The 1999 cyclone hit the low-lying coast of northeast India, pushing rivers backward and flooding much of the province of Orissa, killing an estimated 20,000 people, and leaving 278,000 families homeless. Some geographers have suggested that annual flooding is becoming more pronounced and pointing to deforestation in India and Nepal as the cause of increased runoff.

**Energy and Mineral Resources** The geological resources of the region are modest by global standards, but there are several important deposits. There are valuable deposits of iron ore, manganese, gold, copper, asbestos, and mica in the Peninsular Highland region, for example, and thick seams of coal are present in the northeastern and east-central parts of the highlands. Coal is found in workable quantities in parts of the Mountain Rim; oil-bearing ranges have been located in the Salt Ranges (also a source of rock salt), the Assam Valley, and Gujarat; and natural gas is present in Bangladesh and the Sindh region. New oil deposits have recently been discovered in the northwesternmost parts of India and in the Barmer region of the state of Rajasthan.

More dramatically, the geology of the region yields large quantities of sandstone and marble, construction materials that have long been a fabric of the region's architecture, including monumental works like the Taj Mahal in Agra and numerous major structures in Delhi (**Figure 9.10**). These materials are quarried in many locations, but are especially found in and around the Aravalli hills. Although these have proven incredibly valuable resources in construction and for export on international markets, the labor and environmental conditions at quarries are highly problematic (**Figure 9.11**). Workers, including young people, work in conditions that expose them to fine particles, leading to silicosis in their lungs. Pay and schooling conditions are equally poor in mining areas. The high levels of disturbance of the earth's surface in these areas also damages watersheds, wildlife habitats, and forest cover.

**Arsenic Contamination** Perhaps the most dramatic case of geology's impact on humans in the region came to light in 2000, when it was discovered that millions of tube wells in Bangladesh were drawing arsenic-contaminated water. Tube wells are water wells that are lined with a durable and stable material, usually cement, that makes it possible to sink wells to

▲ **FIGURE 9.11 Quarrying Marble** Workers extract a marble block at a quarry at Makrana, India.

# THE INDUS PLAINS

The Indus Plains (**Figure 1**) have a long history of agricultural productivity that has supported a succession of empires. Harappan agriculturalists, flourishing between 3000 and 2000 B.C.E., produced enough surplus to sustain a civilization that was centered in the cities of Kot Diji (near present-day Sukkur), Moenjodaro (near Larkana), and Harappa (near Sahiwal). The center of gravity of later empires shifted north as it became more important to command access to the overland routes to Central Asia. For a thousand years, from the 6th century B.C.E. until around C.E. 450, the northern Indus Plains took over as the core area of civilization, where a fusion of Greek, Central Asian, and Indian art and culture developed. Taxila (near present-day Islamabad) was of particular significance as a cultural center.

From these earliest times, the region's productivity depended on irrigation, for the climate is hot, the rainfall irregular, and the soils sandy (**Figure 2**). The plains consist of a great mass of alluvium brought down by the Indus and its five tributaries (from west to east, these are the Jhelum, Chenab, Ravi, Beas, and Sutlej) that flow across the Punjab (*panj ab* means "five rivers"). The river floodplains are naturally fertile, but the interfluves (areas of runoff between river valleys, known in this region as *doabs*), though they have good soils, are semiarid and require irrigation.

Two traditional methods of irrigation established the plains as the granary of successive empires. One was a series of inundation canals—channels constructed to carry the floodwaters of the monsoon season beyond the regular floodplains of the rivers. The other was the tube well, a simple shaft sunk to the level of the

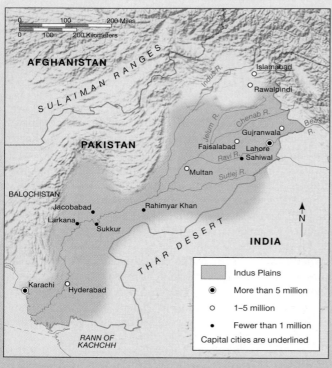

▲ **FIGURE 1 Indus Plains** Reference map showing principal physical features, political boundaries, and major cities.

▲ **FIGURE 2 Indus Plains Agriculture** Centuries of labor-intensive investment in irrigation systems have made productive agriculture possible in the sandy soils and semiarid climate of the Indus Plains.

a greater depth than traditional water wells. They were installed throughout the country as a result of a campaign in the 1970s by UNICEF, the United Nations Children's Fund. The purpose of the wells was to provide drinking water free of the bacterial contamination of the surface water that was killing more than 250,000 children each year in Bangladesh.

Unfortunately, the well water was never tested for arsenic contamination, which occurs naturally in the groundwater, and for many years the well water was believed to be completely safe. By the 1990s, high rates of certain types of cancer throughout much of Bangladesh led researchers to investigate, resulting in the identification of the cause as arsenic-contaminated water from tube wells. Medical statistics indicate that 1 in 10 people who drink such water over a prolonged period will ultimately die of lung, bladder, or skin cancer. In a 2000 report, the World Health Organization described the crisis as the largest mass poisoning of a human population in history. The scale of the environmental disaster far exceeds that of Chernobyl (see Chapter 3, p. 86): As many as 85 million people still draw arsenic-contaminated water from their local wells, and although the technology is available to purify Bangladesh's plentiful supplies of surface water, it will take many years to replace the estimated six million tube wells that are affected.

water table, from which the water is raised by a variety of means, the most common being the "Persian wheel," driven by bullocks or other farm animals (**Figure 3**). British colonial engineers extended the inundation canals in the 19th century, and in the early 20th century they added a series of dams and irrigation schemes that extended irrigation to a greater portion of the doabs in the Punjab and to the lower Indus plains known as the Sindh. The new farmlands created by these irrigation schemes came to be called the Canal Colonies. Today they are distinctive within the landscape of the plains for their severely rectilinear field patterns, in contrast to the small, irregular-shaped field systems of the older-established areas.

With irrigation, farmers on the Indus Plains can grow two sets of crops. The first set of crops, sown to take advantage of the monsoon rains and harvested by early winter, includes rice, millet, corn, and cotton. The second, sown at the start of the cool, dry season and harvested in March or April, includes wheat, barley, rapeseed, mustard, and tobacco. This productivity has traditionally supported a high population density, mostly in large, nucleated villages. Recently, however, the Indus has become so overused that fishermen and farmers are being forced to migrate. A series of grandiose dams and barrages on the river, intended to boost irrigated crop production, have so damaged the water flow that Pakistan is now forced to import an increasing volume of grain.

Craft industries are present throughout the towns and villages of the region. These are dominated by the traditional manufacture of homespun and woven fabrics in cotton, silk, and wool, carpets, footwear, pottery, and metalworking. Modern manufacturing industry is dominated by cotton textiles and woolen knitwear and is concentrated in the larger cities. Lahore (population 6.5 million in 2006) is the cultural, educational, and artistic capital of Pakistan and has a relatively large engineering and electrical goods sector. Multan (metropolitan population 3.8 million) is situated at the center of the country's main cotton-growing region, and its manufacturing sector is dominated by cotton textiles. Gujranwala (population 2 million) is an engineering and metalworking center. Hyderabad is a tobacco-processing and textile-manufacturing center. Like all cities in this region, Karachi (population 11.9 million) has a large informal economic sector and correspondingly extensive slums and squatter neighborhoods. But it also has a thriving commercial center with towering hotels, tourist shops, shopping centers, wide boulevards, and some exclusive residential districts. The region's character has evolved, therefore, as a result of internal relationships, where policies and governments have changed over time, but also as a result of external connections, as new markets emerge for products from the Indus and new technologies allow changes in systems of production and exchange. ■

▶ **FIGURE 3 Irrigation** Raising water from tube wells in the Punjab was traditionally accomplished by "Persian wheels," but is now accomplished using modern power pumps.

# Ecology, Land, and Environmental Management

The ecologies and landscapes of South Asia are remarkably diverse and so support a huge range of life. Most notably, the ecosystems of the region include deep deserts, thick forests, isolated mountains, and a critical coastal zone. The great diversity of habitats has led to a wide-ranging diversity of species, although many of these are at risk. The desert regions are home to several of antelope and deer species and an impressive range of endemic (uniquely native) bird species, including the great Indian bustard. The thick forests of the region represent a valuable trove of biodiversity and are home to a range of monkeys, deer, sloth bear, and panthers. They also serve as the last remaining breeding sites of the Indian wolf, a species that may have broken off from other Asian wolves so long ago as to constitute an entirely separate and uniquely regional species. The mountains of the region, including the Hindu Kush of Pakistan and the Himalayas of Nepal, India, and Bhutan, sustain animals like the one-horned rhinoceros in the foothills and the rare clouded leopard in the high country abutting Tibet.

One of the most notable and endangered ecosystems in the world, the world's largest contiguous area of tidal mangrove forests, which stretch through the delta regions of the Ganges across the coastal edge of West Bengal in India and into Bangladesh. Much of the youngest part of the delta is tidal, and in the southernmost reaches of the delta saltwater penetrates the channels, creating a distinctive ecology of untouched mangrove and tropical swamp forest—the **Sundarbans** ("beautiful forest")—that is home to crocodiles, Chital deer, and one of the largest single populations of Bengal tigers in the world.

These critical species and the landscapes and ecosystems that sustain them, although distinctive to the subcontinent, are in fact part of a larger distributive pattern of flora and fauna, which reflects how the biodiversity, ecology, and land uses of South Asia are deeply influenced by the region's connection and disconnection from other parts of the world. Biozones of the northwestern part of the country, along with their key plant and animal species, are shared with the region's arid and mountainous neighbors stretching west across Iran to Southwest Asia and North Africa, whereas the forests and mountains of the northeast are contiguous with biogeographic zones of Southeast Asia. More dramatically, many of the remarkable winter bird species of the subcontinent migrate from their summer homes as far away as northern Siberia. The Siberian Crane has historically migrated 5,000 kilometers from Russian to northern India every year, crossing Kazakhstan, Uzbekistan, Turkmenistan, Afghanistan, and Pakistan in the process (**Figure 9.12**).

**Land Use Change**    Human influence on the landscape began from a very early period. The first extensive imprint of human occupation in this region dates from at least 4,500 years ago, when the people of the Harappan culture began to irrigate and cultivate large areas of the Indus Valley. Between 3000 and 2000 B.C.E. Harappan agriculturalists produced enough surplus, primarily in cotton and grains, to sustain an urban civilization. Following the disappearance of the Harappans, the arrival of Aryan (after the language that they spoke) civilization (1500 to 500 B.C.E.) into the plains of the Ganges was accompanied by widespread settlement, arable farming, new trade links, and cities, and this began the transformation of the plains of the northern part of the subcontinent from a green wilderness to a heavily agricultural zone, where forests remain, but only in fragments (**Figure 9.13**). Although the plains region remained the most intensively developed, human settlement spread throughout South Asia, and a succession of kingdoms, sultanates, and empires slowly brought more land under cultivation. For the most part, human occupation was sustainable. Subsistence farmers, herders, fisherfolk, and artisans drew on local resources for their food, traditional medicines, housing materials, and fuel, but apart from clearing part of the forest cover, their activities could be sustained from one generation to another without significant harm to the environment.

It was the arrival of European traders, and especially British rule, that accelerated the deforestation of South Asia. In 1750, when the British were beginning their imperial conquests, more than 60% of South Asia was still forested. British imperial rule brought systematic clearing of land for plantations and the methodical exploitation of valuable tropical hardwoods for export to Europe and North America. By 1900, only 40% of South Asia remained forested.

The most rapid period of change has been the past 50 years, as the independent countries of South Asia have sought to modernize and expand their domestic economies. Between 1951 and 1976, for example, about

▲ **FIGURE 9.12  Siberian Cranes, Bharatpur India**  A pair of habituated Siberian cranes, *Grus leucogeranus*, stand together calling in Indian marshland. The species is seriously endangered. The marshland that these cranes seasonally flock to is itself artificial, however, as it was formed through deliberate flooding by the maharaja here centuries ago. The area is now a national sanctuary.

15% of India's land area was converted to cropland. Meanwhile, population growth in rural India has led to more and more wood being taken as fuel. Only 20% of India remains forested today, and less than half of that is intact, natural forest—the rest consists of forest plantations, which have displaced natural ecosystems with monocrops. About one-third of the forest plantations in India consist of eucalyptus, a fast-growing, nonindigenous species that is very demanding of soil moisture.

**Agriculture and Resource Stress**    The rural population of South Asia represents 65% to 75% of the total. Researchers Madhav Gadgil and Ramachandra Guha of the Indian Institute for Science estimate that approximately 400 to 500 million people remain what they call "ecosys-

▲ **FIGURE 9.13 Deforestation Near Ooty, Western Ghats, India**
Much of the original forest cover of South Asia has given way to agricultural land uses. Many productive forest fragments continue to exist, however, in and around dense human settlements.

tem people," living at subsistence levels but in sustainable ways that have protected and preserved the environment. Increasingly, however, these ecosystem people are being pushed onto unproductive soils and arid hillsides as commercial forestry, mining, road and dam construction, and the spread of industry limit their access to the land. As the environmental commons diminishes and populations increase, a destructive cycle is set in motion. Ecosystem people are forced to use their limited resources in increasingly unsustainable ways, depleting sources of fuel wood, exhausting soils, and draining water resources.

The combination of a marginal rural population and the desire on the part of newly independent countries to jump-start industry and agriculture has put pressure on other natural resources, especially water. In parts of Punjab and Haryana, the "breadbasket" of India where almost a third of the country's wheat is grown, the water table has fallen more than 4 meters (13 feet) in the last decade. In the southern Indian state of Tamil Nadu, groundwater levels have fallen more than 25 meters (82 feet) in the last decade as a result of overpumping, leaving Chennai, like many other large cities, dependent on supplementary water supplies hauled in by tanker.

**Environmental Pollution**   In parallel with the acceleration of resource depletion, environmental pollution has accelerated. In India about 200 million people do not have access to safe and clean water; about 690 million lack adequate sanitation, and an estimated 80% of the country's water sources are polluted with untreated industrial and domestic wastes. The Asian Development Bank has estimated that less than one in ten of the industrial plants in South Asia comply with pollution-control guidelines. Only 10% of all sewage in South Asia is treated. India alone generates about 50 million tons of solid waste each year, most of which is disposed in unsafe ways: burned, dumped into lakes or seas, or deposited into leaky landfills.

Air pollution has also become a serious issue. According to the Tata Research Institute in New Delhi, air pollution in India causes an estimated 2.5 million premature deaths each year. Motor vehicle emissions are a major contributor to urban air pollution. During the 1990s, the number of vehicles on Indian roads increased by 300%, and there has been a corresponding rise in rates of respiratory diseases.

The immediate costs of this situation are significant. A 1998 World Bank study estimated that India loses $13.8 billion every year—equivalent to 6.4% of the country's gross domestic product (GDP)—as a result of environmental degradation. The largest share of this cost—$8.3 billion—is associated with the health effects of water pollution. The health impact of urban air pollution and consequent loss of productivity account for an estimated loss of $2.1 billion. Soil degradation and the consequent loss of agricultural output is estimated to cost $2.4 billion a year, and rangeland degradation, resulting in a loss of livestock carrying capacity, costs $417 million each year. Deforestation is estimated to cost $244 million annually.

One of the most horrific environmental disasters of all time took place in Bhopal, India, in 1984, when lethal methyl isocyanate leaked overnight from a Union Carbide plant, killing around 4,000 people in nearby neighborhoods and permanently damaging the health of hundreds of thousands more. The exact causes and responsibility for the event have still not been settled conclusively. However, to many the Bhopal disaster is emblematic of the potentially disastrous effects of lax attitudes toward environmental planning and regulation on the part of plant owners and managers. Most of the time, such laxity does not involve loss of human life. Nevertheless, the results can be calamitous both to communities and to the environment.

**South Asia's Disappearing Megafauna**   In 2000 a World Conservation Union (IUCN) study identified 11,046 plant and animal species from around the globe as being at risk, including 180 species of mammals and 182 species of birds that are critically endangered. Symbolic of this acute problem are the so-called charismatic megafauna: large and exotic species such as the rhinoceros, the elephant, whales, and tigers. In South Asia, some charismatic megafauna have already disappeared. Cheetahs disappeared from the wild in India more than 50 years ago, the last sighting being in 1948, when three young males were shot dead by a hunting party in the jungles of Bastar in Madhya Pradesh, central India. Today, the Bengal tiger (**Figure 9.14**) and the Asian elephant are emblematic of critically endangered species in South Asia.

Experts put remaining numbers of Bengal tigers at somewhere between 3,060 and 3,985, although this could be an overestimate. Tiger populations are notoriously difficult to evaluate, despite recent technological advances that help keep track of their movements. What is clear is that the tiger is being lost. Animal by animal, its footprints are vanishing from South Asia's forests. The causes of the animal's decline are the same ones that are killing off many other species: loss of habitat and remorseless poaching.

In April 2000, the United Nations Convention on International Trade in Endangered Species (CITES) issued a report criticizing how India is caring for its tigers, claiming that the Indian government has displayed a lack of concern and effort. Tiger losses to poachers, concluded CITES, are being covered up by officials, and figures for the remaining animals are deliberately inflated. Poachers hunt tigers in response to the huge demand for tiger body parts in traditional Chinese and Japanese medicine. A single tiger is worth more than $50,000 to poachers. The skin alone fetches an estimated $11,000, whereas a 10-gram tablet containing tiger bone sells for $25, and a bowl of tiger penis soup sells for $53.

▲ **FIGURE 9.14 Bengal Tiger** The most charismatic of all "charismatic megafauna," the Bengal tiger is one of the most seriously endangered species: its habitat is fast diminishing, and it is hunted by poachers, with inadequate protection from government agencies.

# HISTORY, ECONOMY, AND DEMOGRAPHIC CHANGE

South Asia has developed distinctive cultures and generated influential concepts and powerful ideals that have spread around the world. Receptive to sociocultural and economic practices from other regions, the subcontinent has also been the source of ideas that are now fully global. From the hearth areas (the areas of origin) of Harappan and Vedic civilizations in the Plains, sophisticated cultures and powerful political empires spread across vast sections of the region. South Asia's resources and its geographic situation on sea-lanes between Europe and the East Indies made it especially attractive to European imperial powers, and in modern times it has become a pivotal geopolitical region with an emergent industrial and high-tech sector and a large market for core-region products. Since the time of colonialism and into the present, the dense interconnections of South Asian economy to that of the rest of the world has resulted in an explosive diaspora, and South Asian laborers, engineers, administrators, and scientists are increasingly present in regions throughout the world.

## Historical Legacies and Landscapes

**Early Empires**   The Mauryan Empire (320–125 B.C.E.) was the first to establish rule across the greater part of South Asia. By 250 B.C.E., the emperor Ashoka had established control over all but present-day Sri Lanka and the southern tip of India. Securing control had wrought such havoc and destruction, however, that Ashoka renounced armed conquest and adopted a policy of "conquest by *dharma*," that is, through the example of spiritual rectitude and chivalrous obligations. Dharma, a key element of Buddhist teachings, is part of the legacy of Ashoka's reign over Southwest Asia, which also includes Buddhist principles of vege-

tarianism, kindness to animals, nonacquisitiveness, humility, and nonviolence. During this period, the culture of South Asia became closely interlinked with that of other regions, as Ashoka sent Buddhist missionaries throughout Asia to spread the faith. Ironically, this successful evangelization led to the survival of Buddhism throughout Asia, centuries after its disappearance in peninsular South Asia.

After Asoka's death in 232 B.C.E., the Mauryan Empire fell into decline, and northern India soon succumbed to invaders from Central Asia. After more than four centuries of division and political confusion, the Gupta Empire (C.E. 320–480) united northern India and came to control all but the northwestern hill country, the Peninsular Highlands, the southwestern coastlands (modern Kerala), and Sri Lanka. The Gupta period is generally regarded as the classical period of Hindu civilization. It produced the decimal system of notation, the golden age of Sanskrit and Hindu art, and contributions to science, medicine, and trade.

**Mughal India**   Toward the end of the 15th century, a clan of Turks from Persia (now Iran) moved east in an attempt to evade the control of Timur's (Tamerlane's) Tartar Mongol Empire. These Turks were the Mughals. Led by Babur, they conquered Kabul, in what is now Afghanistan, in 1504. By 1605, Babur's grandson, Akbar, had established control over most of the Plains, and in the next century Mughal rule extended to all but Sri Lanka and the southern tip of India from 1526–1707 (**Figure 9.15**).

Akbar's rule was an extraordinary time, his achievements driven by his personal desire to synthesize the best of the many traditions that fell within his domain while maintaining strict control. Traditional kingdoms and princely states were kept intact, but they were integrated in a highly organized administrative structure with an equitable taxation system and a new class of bureaucrats. Persian became the official language, yet Akbar abolished the tax on non-Muslims that had been instituted by his grandfather. Mughal rule did not seek to impose Islam on indigenous populations. However, Mughal commitment to the religious precepts of Islam, together with the equitable system of Mughal governance, gave great stature to Islam. Over time, Islam proved attractive to many, especially in the northwest (the Punjab) and the northeast (Bengal), both areas where Buddhism had previously been dominant. By 1700, mosques, daily calls to prayer, Muslim festivals, and Islamic law had become an integral part of the social fabric of South Asian life.

Mughal rule was also characterized by a luxurious court and by extensive support for creativity in art, architecture, music, and literature. Spectacular architecture became a signature of Mughal rule, with the landscape of the northern regions becoming punctuated not only with lavish mosques and palaces but also with forts and citadels, towers, and gardens, including the iconic Taj Mahal in Agra.

The last of the great Mughal emperors, Aurangzeb (1658–1707), provoked a series of rebellions and uprisings with his anti-Hindu policies and his reinstatement of the tax on non-Muslims. At the same time, he had to respond to raids from the Marathas, an empire from the Konkan coast, and was forced to conduct military campaigns on several fronts. With his death in 1707, the Mughal Empire collapsed, leaving South Asia open to the increasing interest and influence of European traders and colonists.

**British Imperialism**   European traders were a regular presence along the coasts of South Asia long before the dissolution of the Mughal Empire.

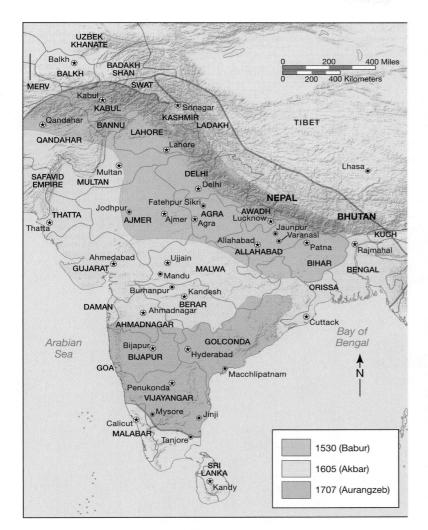

▲ **FIGURE 9.15 Mughal India** Two centuries of Mughal rule began in 1504, with the conquest of Kabul, in present-day Afghanistan; by the time of Aurangzeb's death in 1707, Mughal rule had extended to almost all the subcontinent.

The Portuguese were the first, with the arrival of Vasco da Gama in India in 1498. Early in the 17th century the British East India Company (established in 1600) and the Dutch East India Company (established in 1602) set out to deliberately contest the Portuguese monopoly of the Indonesian spice trade. South Asia, situated as it was on the route between Europe and the East Indies, provided an attractive array of intermediate stops for trade in the calicos, chintzes, taffetas, brocades, batiks, and ginghams of Gujarat, Bengal, Golconda, and the Tamil country.

By the 1690s, European trading companies had established a permanent presence in several ports, though they had no interest in establishing colonies or exerting any kind of political authority, even when South Asia fell into disarray at the end of Mughal rule. The British East India Company was most successful of these, and in 1773 the British government transformed its East India Company into an administrative agency. Soon afterward, during the French Revolution and Napoleonic Wars, the British pushed ahead with aggressive imperialist policies in South Asia, using a mixture of force, bribery, and political intrigue to gain exclusive control over more and more of the region, which by this time was riddled with political and religious disunity.

**The *Raj***  The focus of British imperialism now shifted beyond trade and territorial control to social reform and cultural imperialism. In a famous memo written by East India Company Supreme Council member Thomas Macaulay in 1853, British administrators were urged to create a special class of South Asian people who would be "Indian in blood and colour, but English in taste, in opinions, in morals, and in intellect." One by one, the territories of native rulers were annexed or brought under British protection (**Figure 9.16**).

Under the governor-generalship of Lord Dalhousie (1848–56), there was a push to bring Western institutions and a modern industrial infrastructure to South Asia. Railroads, roads, bridges, and irrigation systems were built, and restrictions were placed on slave trading, *suttee* (a widow's ritual suicide on her husband's funeral pyre), and other traditional practices. Western educational curricula flourished in private colleges; British-style public universities were established. All this provoked an anticolonial reaction, which came to a head in 1857, when an Indian Army unit rebelled purportedly because 85 of its soldiers were jailed for upholding their religious principles, refusing to use ammunition greased with animal fat. The incident quickly spread into a year-long civil uprising—the "Indian Mutiny"—throughout the north-central region. Massacres were carried out on both sides, but the mutiny was eventually quelled, and in 1858 the British Crown assumed direct control over India. In 1876 Queen Victoria was declared Empress of India.

Thus emerged the **Raj**, British rule over South Asia, which by 1890 extended to the entire region with the exception of present-day Afghanistan and Nepal. The British brought plantation agriculture to South Asia, producing food crops for the British domestic population and commodity crops for British industry and British merchant traders. Among plantation crops were coconuts, coffee, cotton, jute, rubber, and tea (see Geographies of Indulgence, Desire, and Addiction: Tea). The Raj also introduced Western industrial development and technology to South Asia, displacing indigenous crafts and industries. It fostered Western political ideas of national territorial sovereignty and the materialism that accompanies free markets in land, labor, and commerce. This was to be an explosive legacy at the conclusion of the Raj in 1947, when Britain partitioned colonial India into separate independent national states.

**Partition**  Grassroots political resistance to British imperial rule had been institutionalized through the Indian National Congress Party, formed in 1887 to promote greater democracy and freedom, not only from imperial rule but also from the traditional and autocratic rule of hereditary maharajas (leaders of the princely states that were under British protection). A leader and the inspirational figure of this movement was Mohandas Gandhi, whose vision of social justice and accountability used methods of nonviolent protest, including boycotts and fasting. Under Gandhi's leadership, the case for national independence became irrefutable. Soon after the conclusion of World War II, the British set about withdrawing from South Asia altogether.

In creating new, independent countries, Britain sought to follow the European model of building national states on the foundations of ethnicity, with particular emphasis on language and religion. As a result, it was decided to establish a separate Islamic country, called Pakistan ("Land of the Pure"). Administrative districts under direct British

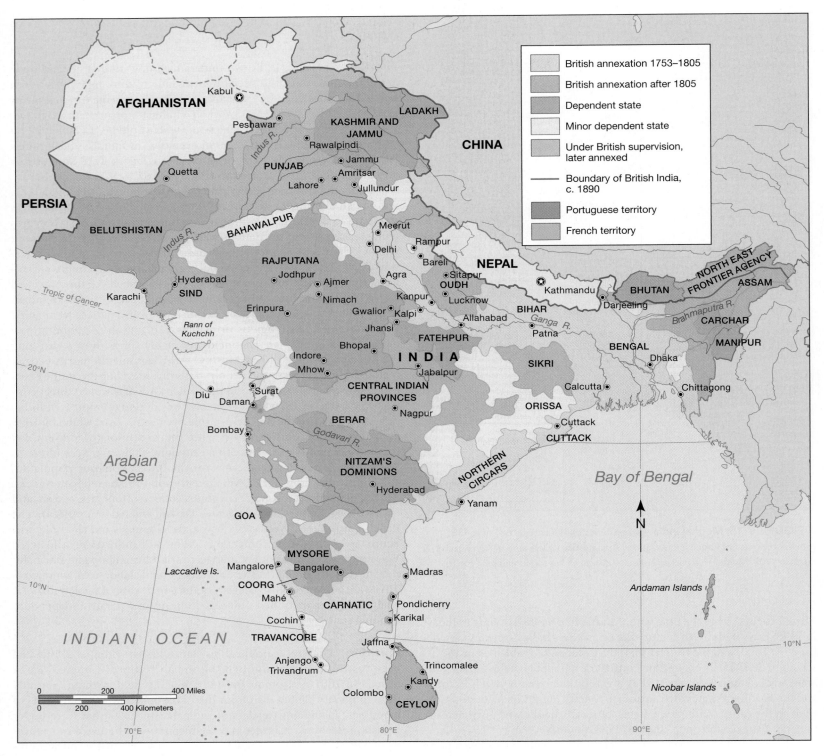

▲ **FIGURE 9.16 The British Conquest of India** British interest in India began as a consequence of merchant trade in the early 18th century, but by 1890 the British had come to control, directly or indirectly, most of South Asia.

control that had a majority Muslim population were assigned to Pakistan, together with those princely states whose ruling maharajas wished to join Pakistan rather than India. The result was that Pakistan was created in two parts, East Pakistan and West Pakistan, one on each shoulder of India, separated by 1,600 kilometers (994 miles) of Indian territory.

In 1947, when Pakistan and India were officially granted independence, millions of Hindus and Sikhs found themselves as minorities in Pakistan, whereas millions of Muslims felt threatened as a minority in India. Communal violence erupted across the region. In desperation, more than 12 million people fled across the new national boundaries—

# TEA

Tea, a mild drug that makes a refreshing drink, was cultivated in China and Japan for centuries before becoming a commodity in international trade. Thereafter, tea became a catalyst of economic, social, and political change. The Opium War (see Chapter 8) was fought in large part over tea; the American Revolution was sparked by riots over a tax on tea; and the social and economic fabric of large parts of South Asia was destroyed by colonial mercantile forces that transformed land into tea plantations.

Tea was first brought to Europe by Portuguese traders and marketed as a medicine. Tea drinking was adopted by the royal court of King Charles II in England (1660–1685), and tea quickly became an indulgence of the European bourgeoisie. By 1700 more than 17 million pounds of tea were being exported to Europe each year, mostly to Holland, Portugal, and England. As East Asian trade came to be dominated over the next century by the British East India Company, the cost of tea fell, and the taste for it spread. It soon became an addiction for the middle and working classes in Britain and a desirable indulgence for many in the American colonies. A significant market for tea had meanwhile also developed in Russia and in many countries of the Middle East and North Africa.

South Asia became a source of tea only in the 19th century, when the British sought to find commercial advantage in their newly acquired territories in Assam (northeast India). In 1848 the Assam Company hired botanist Robert Fortune to travel incognito to China to discover the secrets of cultivating and processing tea. After four expeditions, he brought to Assam not only the knowledge that the company needed but also 12,000 seedlings, the specialized tools used in processing tea, and a skilled Chinese workforce. The British colonial government granted land on inexpensive leases to anyone who had the capital to establish a tea plantation. Inevitably, those with the capital were European settlers, and thousands of would-be planters flocked to Assam. Valuable timber was cut, forests were cleared, and the land not needed for tea plantations was rented to tribal people from Bihar and Orissa, who were contracted to work—for miserably little pay and in dreadful conditions—as laborers in the plantations.

The opening of the Suez Canal in 1869 cut transport costs, made tea even cheaper, and allowed producers a bigger profit margin. Tea plantations in Assam and neighboring Bengal increased sixfold in size between 1870 and 1900, by which time there were half a million plantation workers in the region. In this same period, Ceylon—now Sri Lanka—emerged as a major tea-producing area. In the 1870s, 250,000 acres of coffee plantations in Ceylon were destroyed by blight, leaving coffee planters bankrupted and land extremely cheap. Thomas Lipton, a prosperous grocer from Glasgow, Scotland, arrived in Ceylon in 1871 while on a world cruise, bought dozens of the plantations at bargain prices, and turned them over to the production of tea.

Lipton was a pioneer of "vertical economic integration," where a single company controls all the operations from growing tea to processing, management, transport, blending, packaging, and marketing. Through this system, Lipton was able to cut the cost of tea to European consumers by 35%. Lipton and other tea planters brought in low-caste Tamil Hindus from famine-stricken areas of South India to work as laborers in the Ceylonese plantations. The conditions in which they were forced to work were atrocious, as bad as those of their counterparts in Assam and Bengal. Being low caste, low paid, isolated on estates in the highlands, and with no political voice, the "estate Tamils" were effectively enslaved on the plantations.

After independence in India, Bangladesh, and Sri Lanka, most European planters were forced to sell their tea plantations, either to the respective governments, under nationalization programs, or to indigenous business interests. The conditions for plantation workers have remained relatively unchanged, however. Plantation workers are still an impoverished group, a fact that is often belied by the image of smiling tea-pickers dressed in traditional costume, adorned with jewelry, working in sunshine amid beautiful scenery (Figure 1). ∎

▲ **FIGURE 1 Tea Pickers** At the beginning of the economic chain that links South Asian producer regions with consumers in Europe and North America, tea pickers work long hours in harsh conditions for meager pay.

the largest refugee migration ever recorded in the world. As Hindus and Sikhs moved toward India and Muslims moved toward Pakistan in opposite directions between the two countries, many hundreds of thousands were senselessly killed. In Kashmir, a Hindu maharaja had elected to join India, but Pakistani forces intervened on partition to protect the majority Muslim population, with implications that have never been resolved.

Having withdrawn from the greater part of South Asia, the British granted independence to the island of Ceylon as a Commonwealth dominion in 1948. In 1949 Britain handed to India its formal control over the external affairs of the kingdom of Bhutan, and in 1968 Britain granted independence to the Maldives. The Raj was finally over, but the legacy of partition remains a dimension of the geography of South Asia in today's world. In some ways, the states of South Asia are still adjusting to the 1947 partition of India and Pakistan. In Pakistan, divergent regional interests in East and West Pakistan quickly developed into regionalism, with East Pakistani leaders calling for secession. As a result, the country was split into two independent states in 1971: West Pakistan became Pakistan, and East Pakistan became Bangladesh. Meanwhile, neither India nor Pakistan could agree on the status of Kashmir, and the two countries briefly went to war over the region in 1948, in 1965, and again in 1971.

## Economy, Accumulation, and the Production of Inequality

In the decades since, South Asia—and India, in particular—has come to play a larger role within the economic world-system. India is the world's largest democracy and has maintained stable parliamentary and local government through elections and rule of law since adopting its constitution in 1950. In 1992, after losing its major trading partner with the collapse of the Soviet Union, India embarked on a further series of economic reforms as a condition of a structural adjustment program attached to a loan from the World Bank. Before these reforms, many key institutions—including banks, utilities, airlines, railways, radio, and television—were government owned and operated. High tariffs, restrictions on foreign ownership, high taxes, widespread corruption, and a complex system of permits, licenses, quotas, and permissions all kept foreign investors away and suppressed the energy of Indian entrepreneurs.

**Economic Reforms**  The reforms created a more open and entrepreneurial economy. Key institutions have been privatized, and foreign investment has been flowing into the country, helping generate exceptionally high economic growth rates. The fact that India's middle class conducts business in English gives India a comparative advantage in today's world economy. Over the past 15 years, India has been the second-fastest-growing country in the world—after China—averaging above 6% growth per year. This booming economy, moreover, shows terrific resilience in the face of recent financial turmoil. While most economies around the globe shrank during 2008 and 2009, India's (almost alone) actually posted job growth across industrial and technical sectors.

India now has an affluent middle class estimated at about 200 million—a huge, well-educated, and sophisticated consumer market that has become part of the "fast" world and an agent of globalization. Market reforms have meanwhile triggered an associated cultural change: Flaunting success is no longer frowned on, and so India's expanding middle class is increasingly unabashed about its cars, Palm Pilots, mobile phones, and vacations in Phuket and Singapore.

This rapid growth of India's affluent middle class serves not only to accentuate the contrasts within South Asia between the traditional and the modern but also to highlight the desperate situation of an even larger group: the extremely poor. According to the United Nations Development Fund, approximately 45% of India's population (that is, almost half a billion people) lives on less than a dollar a day—the World Bank's definition of dire poverty. In fact, most of India's poor (390 million of them) somehow exist with an income of a dollar a *week*.

The other countries of South Asia have not experienced the kind of economic boom enjoyed by India, though there have been attempts to foster regional economic integration. In 1985 the South Asian Association for Regional Cooperation (SAARC) was established. Although progress has been slow—mainly because of the friction between India and Pakistan—the member states did sign a South Asian Preferential Trade Agreement in 1996, which established some modest mutual tariff concessions. Impatient with the pace of SAARC, India also set up two subregional cooperation groups: one with Nepal, Bhutan, and Bangladesh, and another with Sri Lanka and the Maldives. India has also signed a free-trade agreement with Sri Lanka and has pursued wider avenues of economic and diplomatic cooperation, becoming a "dialogue partner" in the Association of Southeast Asian Nations (ASEAN) and lobbying for a seat as a permanent member of an expanded United Nations Security Council.

The liberalization of India's economy has shown less-predictable consequences. Lifting export controls has enabled farmers with access to large amounts of capital to reorganize their production toward lucrative overseas markets, with the result that domestic consumers have to pay more for traditional staples. Thus, for example, many farmers are switching from growing grains for local consumption to cash crops like cotton and tobacco, whereas others are turning to the cultivation of flowers and strawberries to be shipped to newly affluent urbanites or to be air-freighted abroad. Now that a global market has become aware of high-quality local specialties, such as the fragrant Basmati rice of the Himalayan foothills and the short-season Alphonso mangoes of Maharashtra, their price within India has put them in the luxury class, out of reach of many of the consumers who have traditionally regarded them as staples.

**From Central Planning to Turbulent Markets**  India's market reforms of the early 1990s triggered a transformation in the country's economic development with implications for regional change and interdependence within South Asia. Breaking with socialist principles of centrally planned development and social and regional equality has unleashed the spatially uneven economic development processes of capitalism. The growth and the wealth have not been evenly distributed throughout India. Certain industries and certain places and regions have grown dramatically, whereas elsewhere there has been disinvestment and recession—the "creative destruction" inherent to capitalism. One of the most dramatic examples of regional growth is that of the software industry in south India. More generally, the growth has been centered in larger metropolitan areas and preexisting industrial centers in the Upper Ganga Plains, Damodar-Hooghlyside, and Mumbai-Pune, again following classic principles of capitalist economic development.

The booming software and service industries of southern India are paradigmatic in this regard. The towns and cities of southern India together account for about 20% of the country's industrial employment. Bangalore (population 6.1 million in 2006), in particular, has become a thriving industrial and business center. Following the location of key defense and telecommunications research establishments by the government in the 1960s, Bangalore became the premier science and technology center of India, attracting investment from a variety of transnational corporations as a result of the quality of its workforce. Industries here include the manufacture of aircraft, telecommunications equipment, watches, radios, and televisions. However, the city has become world famous for its software industry, the affluent employees of which have contributed to the city's progressive and liberal atmosphere and its lively commercial centers featuring fast-food restaurants, yuppie theme bars, and glitzy shopping malls (**Figure 9.17**).

(a)

(b)

▲ **FIGURE 9.17  Bangalore**  (a) A rave party at a Bangalore warehouse, organized by MTV and Kingfisher Beer. (b) Downtown Bangalore at dusk.

First impressions of Bangalore are similar to those of other major cities in India: a sprawl of decaying single-story houses and shops, cramped apartment buildings, crumbling colonial offices, mile after mile of squatter slums, and the pervasive sights and smells of poverty. Yet within Bangalore is a parallel universe of high-tech industry and the Western-style materialism and fast-world lifestyles of its workers. Altogether, Bangalore is now home to more than 500 high-tech companies that employ 100,000 people. Bangalore is also home to the Indian Institute of Science, a world-renowned technical school. After the structural economic reforms of the Indian government in the early 1990s, free-enterprise capitalism was able to flourish, and in Bangalore there was a pool of Indian programmers who had become experts at writing concise, elegant code on their rather limited hardware. When American software companies began to encounter rising costs in Silicon Valley, they found in Bangalore a large pool of highly trained, English-speaking, ambitious, and inexpensive software engineers.

**Inequality and Gender**   As the case of Bangalore shows, places and regions with an initial advantage in terms of factories, skilled labor, specialized business services, and affluent markets can attract more investment, faster, through "cumulative causation," the self-reinforcing spiral of regional growth. The corollary is that places and regions with a weak industrial base, with a weak or obsolescent infrastructure, and with an unskilled or poorly educated workforce tend to experience a downward spiral of recession. In India today, the remoter rural regions are experiencing most acutely the negative consequences of the country's economic reforms.

Against the background of acute and chronic poverty in South Asia, the material wealth and Western lifestyles of the growing middle classes serve to highlight the extreme inequality that also characterizes the region. Official statistics reveal that hundreds of millions in South Asia live not just in poverty, but in ignorance and destitution (**Table 9.1**). If anything, poverty and inequality are increasing.

In rural South Asia, scarcity remains the norm. Illiteracy is common, and even the most basic services and amenities are lacking. Life expectancy is low; hunger and malnutrition are constant facts of life

## TABLE 9.1   Indicators of Poverty in South Asia, 2007/2008

| | Percentage of Total Population not Expected to Survive to Age 40 | Adult Illiteracy Rate | Percentage of Total Population not Using an Improved Water Source | Percentage of Total Population not Using Improved Sanitation | Percentage of Population with an Income of Less than $1 a Day |
|---|---|---|---|---|---|
| Afghanistan | 43.1 | 72.0 | 61 | 66 | No data |
| Bangladesh | 16.4 | 52.5 | 26 | 39 | 41.3 |
| Bhutan | 16.8 | 53.0 | 38 | 30 | No data |
| India | 16.8 | 31.0 | 14 | 67 | 34.3 |
| Maldives | 12.1 | 3.7 | 17 | 41 | No data |
| Nepal | 17.4 | 51.4 | 10 | 65 | 24.1 |
| Pakistan | 15.4 | 50.1 | 9 | 41 | 17.0 |
| Sri Lanka | 07.2 | 9.3 | 21 | 9 | 5.6 |

▲ **FIGURE 9.18 Rural Poverty** Grinding poverty is the norm in rural South Asia. This woman ekes out a living cracking stones on the banks of the Mahananda River in northeastern India.

(**Figure 9.18**). In urban areas, poverty is compounded by crowding and unsanitary conditions. In South Asia's largest cities, one-third or more of the population lives in slums and squatter settlements, and hundreds of thousands are homeless (**Figure 9.19**). Clean drinking water is limited, and most poor households do not have access to a latrine of any kind.

**Poverty** The worst concentrations of poverty are characterized by overcrowding, a lack of adequate sanitation, high levels of ill health and infant mortality, and rampant social pathologies. Consider, for example, the squatter settlement of Chheetpur in the city of Allahabad, India. The settlement's site is subject to flooding in the rainy season, and

▲ **FIGURE 9.19 Urban Poverty** For many of the urban poor, poverty means homelessness. Here in Karachi, Pakistan, men warm themselves over a small fire in the street.

a lack of drainage means living with stagnant pools for much of the year. Two standpipes (outdoor taps) serve the entire population of 500. There is no public provision for sanitation or the removal of household wastes. In this community, most people eat less than the recommended minimum of 1,500 calories a day; 90% of all infants and children under age four have less than the minimum calories needed for a healthful diet. More than half the children and almost half the adults have intestinal worm infections. Infant and child mortality is high—though nobody knows just how high—with malaria, tetanus, diarrhea, dysentery, and cholera the principal causes of death among children under age five.

Much of this poverty results from the lack of employment opportunities in cities that are swamped with people. To survive, people who cannot find regularly paid work must resort to various ways of gleaning a living. Some of these ways are imaginative, some desperate. Examples include street vending, shoe-shining, craftwork, street-corner repairs, and scavenging on garbage dumps (**Figure 9.20**). This informal economic sector consists of a broad range of activities that represent an important coping mechanism. More than a half billion people in South Asia must feed, clothe, and house themselves entirely from informal sector occupations.

**Women and Children** Among South Asia's poor, women bear the greatest burden and suffer the most. South Asian societies are intensely patriarchal, though the form that patriarchy takes varies by region and class. The common denominator among the poor throughout South Asia is that women not only have the constant responsibilities of motherhood and domestic chores but also have to work long hours in informal-sector occupations (**Figure 9.21**). In many poor communities, 90% of all production is in the informal sector, and more than half of it is the result of women's efforts. In addition, women's property rights are curtailed, their public behavior is restricted, and their opportunities for education and participation in the waged labor force are severely limited.

Women's subservience to men is deeply ingrained within South Asian cultures, and it is manifest most clearly in the cultural practices attached to family life, such as the custom of providing a dowry to daughters at marriage. The preference for male children is reflected in the widespread (but illegal) practice of selective abortion. Within marriages, many (but by no means all) poor women are routinely neglected and maltreated.

The picture is not entirely negative, however, and one of the most significant developments has been the emergence of women's self-help movements. Perhaps the best known of these is the **Grameen Bank**, a grassroots organization formed to provide small loans—**microfinance**—to the rural poor in Bangladesh. The Grameen Bank runs completely against the established principles of banking by lending to poor borrowers who have no credit, but now claims to be a financially sustainable, profit-making venture with 17,366 employees. Cumulatively, the bank has loaned more than $5.2 billion, 98% of which has been repaid, according to Grameen, a rate that is far higher than that for any conventional financial institution operating in the country. The average size of a Grameen loan is about $120, typically enough to purchase a cow, a sewing machine, or a silkworm shed. Studies have shown that the bank's operations have resulted in improvements in nutritional status, sanitation, access to food, health care, clean drinking water, and housing, and that more than one-third of all borrowers have risen above the poverty line, with another third close to doing so. The most distinctive feature of the Grameen Bank is that 96% of its borrowers are women (**Figure 9.22**).

▶ **FIGURE 9.20 Informal Economic Activities** Where jobs are scarce, people have to cope through the informal sector of the economy, which includes a broad variety of activities, including agriculture (backyard hens, for example), manufacturing (craftwork), and retailing (street vending). (a) A jewelry repairman works on a street at a pearl market in Hyderabad, India. (b) An Afghan man sells beef at a roadside in Kabul, Afghanistan, where the long history of civil war, Taliban rule, and U.S. invasion has led to food shortages and a lack of formal economic opportunities.

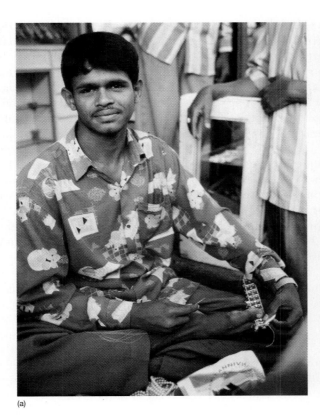
(a)

(b)

In India, the Self-Employed Women's Association (SEWA) has made a major contribution to building self-confidence and self-reliance among poor working women by mobilizing and organizing them. SEWA was formed in 1972 in Ahmedabad in the State of Gujarat. It evolved from a trade union of textile workers, but unlike conventional trade unions, SEWA organizes women workers in the informal sector: vegetable vendors, rag and paper pickers, bamboo workers, cart-pullers, and garment workers. SEWA has given its members a degree of independence from middlemen and, consequently, an invaluable sense of independence.

▼ **FIGURE 9.21 Women's work** In most households, women must not only raise children, prepare food, and do most of the domestic chores, but also work in informal-sector activities or as unskilled laborers. Here, Bhil women work on widening the road to their village, Damodra, in Jaisalmer, Rajasthan, India.

▼ **FIGURE 9.22 Rural enterprise** Microcredit programs such as those pioneered by the Grameen Bank have enabled tens of thousands of rural women to begin small businesses.

Children in impoverished settings are even more vulnerable than women. Throughout South Asia, the informal labor force includes children (**Figure 9.23**). In environments of extreme poverty, every family member must contribute something, and so children are expected to do their share. Industries in the formal sector often take advantage of this situation. Many firms farm out their production under subcontracting schemes that are based not in factories but in home settings that use child workers. In these settings, labor standards are nearly impossible to enforce.

The International Labour Office has documented the extensive use of child labor in South Asia, showing that many children under 10 years of age are involved in a great variety of work—tending animals, carpet-weaving, stitching soccer balls, making bricks, handling chemical dyes, mixing the chemicals for matches and fireworks, sewing, and sorting refuse. Most of them work at least 6 and as many as 12 hours a day. A particularly cruel type of exploitation of child labor is **bonded labor.** This kind of bondage occurs when persons needing a loan but having no security to back up the loan pledge their labor, or that of their children, as security for the loan.

## Demography, Migration, and Settlement

South Asia has the second-largest and the fastest-growing population of all world regions. The total population of South Asia in 2008 stood at 1.59 billion, with India accounting for just over 1 billion.

▼ **FIGURE 9.23 Child Labor** The exigencies of poverty mean that children are required to contribute to the household economy, often from a very early age. Here a boy works in a garbage dump in Lahore, Pakistan.

With overall growth rates in the region of 1.7% per year (compared to 0.5% per year in China), South Asia is headed for a population of 1.64 billion by 2010.

**Figure 9.24** shows the distribution of population within South Asia. The first thing to note about this map is the very high density of population throughout most of the region. The overall density of population in India is 316 persons per square kilometer (819 per square mile), compared to 131 persons per square kilometer (338 per square mile) in China and 29 persons per square kilometer (75 per square mile) in the United States. In detail, patterns of population density reflect patterns of agricultural productivity. The combination of good soils with a humid climate or with extensive irrigation supports densities of more than 500 persons per square kilometer (1,300 per square mile) in a belt extending from the upper Indus Plains and the Ganga Plains through Bengal and the Assam Valley. Similar densities are found along much of the Coastal Fringe.

**Declining Growth**    In many parts of the subcontinent, however, population growth is slowing or halting. Although some rural areas have fertility rates as high as six children per women, a rate consistent with historic averages, others have declined to far lower figures, some as few as two births per family.

Little of this change has resulted from explicit and harsh population policies favored by regional governments in the past. These efforts, which began in the 1950s and continued into the 1970s, were largely misguided. In the mid-1960s, for example, the government of India announced specific demographic targets and opened "camps" around the country for the mass insertion of intrauterine devices (IUDs). The program soon failed, mainly because of negative public reaction to the poor training of health workers and unsanitary conditions in the camps.

Next came vasectomy camps. More than 10 million men were coerced into being sterilized in the 1970s in an "Emergency Drive" for family planning that saw all kinds of government administrators—from police to teachers and railway inspectors—given monthly quotas to recruit "volunteers" for vasectomy camps. Not surprising, a popular backlash put an end to the sterilization program. Equally important, these efforts did little to curtail population growth.

Demographic changes, however, have come rapidly in some parts of the region in recent years, largely as a result of infrastructural, educational, and economic change. The unevenness of this effect can be seen across the region. **Figure 9.25** shows the average fertility rates by state in India. Notably, many of the states of southern India have rates of roughly two births per woman, effectively a condition of **zero population growth** or ZPG. In a country internationally known for high population and population growth, how is this possible?

In the State of Kerala, often taken to be a case example of demographic success, this impressive achievement appears to have been the result of a combination of important factors. First, rural health care in the state is well developed and well distributed. With access to good maternal and infant care, infant mortality rates are low, which discourages families from births to "insure" against possible deaths. Good health care also means access to birth control, either in the form of prophylactics or sterilization. This is coupled with high levels of literacy in the population, especially of women, whose education is consistently correlated with lower fertility rates. The final factors include high levels of participation in the labor force by women in Kerala, as well as a matrilineal tradition in the state, which gives women significant control over property. States in India where population growth remains high, con-

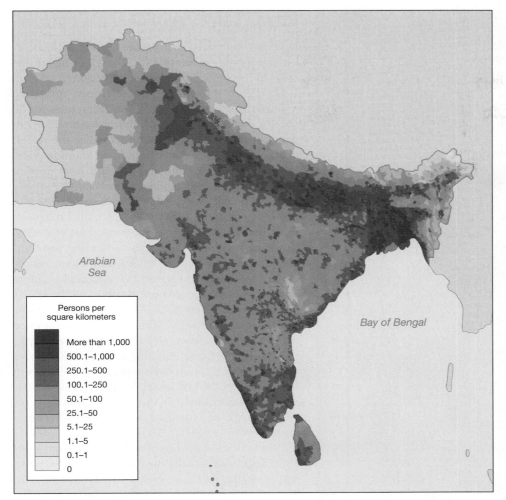

▲ **FIGURE 9.24 Population Density in South Asia 1995** The density of population is very high throughout most of South Asia, but especially so in the Plains and in subregions with good soils and humid climates.

continent, it is possible that the boom in population growth across the subcontinent will halt within a few decades.

**Urbanization**   In comparison with other world regions, South Asia is still very much a land of villages. Indeed, India has the largest rural population in the world. Approximately 63% of Pakistan's population and 70% of India's live in rural settings; in Afghanistan, Bangladesh, and Sri Lanka, the rural population is around 77%; and the tiny state of Bhutan is 63% rural, with Nepal at 82%.

Rural-to-urban migration is shifting the balance toward towns and cities, however. This is a result in part of population pressure in rural areas, where a natural population increase has reduced the amount of cropland per person to half of what it was in 1960. In that year, there were just 9 cities of 1 million or more in South Asia, and only one of these had more than 5 million inhabitants. In 2010 there were 55 cities of 1 million or more, including 11 of 5 million or more. Between 1960 and 2010, Mumbai, the largest metropolis in South Asia, grew from 4.1 million to 20 million; Dhaka, in Bangladesh, grew from fewer than 650,000 to more than 14.8 million; and in Pakistan, Karachi grew from 1.8 million to 13.1 million.

The explosion of cities and middle-sized urban areas and large towns has fundamentally changed the character of South Asian politics, its landscapes, and its culture. Urban areas are characterized by intense mixing of ideas, dress, and traditions, including many relations, classes, and castes. The dynamism of cities extends beyond the signature urban areas of Mumbai and Bangalore to the tens of thousands of middle-sized towns, where housing construction is booming. The competition for land at the edge of cities is intense, as agricultural land prices have risen dramatically through speculation, and formerly quiet rural areas have been quickly incorporated into bustling metropolises.

**The South Asian Diaspora**   The population dynamics of South Asia, like all other characteristics of the region, are highly influenced by their relationships to the rest of the world. The South Asian diaspora—the movement of Indians, Pakistanis, and Bangladeshis abroad—amounts to between five and six million people, most of them located in Europe, Africa, North America, and Southeast Asia (**Figure 9.26**). The origins of this diaspora can be traced to the abolition of slavery in the British Empire in 1833. The consequent demand for cheap labor in the plantations and on the railways of the British Empire was filled in part by emigrants from British India. In the mid-19th century, thousands of Indians left for the plantations of Mauritius (in the Indian Ocean), East Africa, the West Indies, and South Africa.

After World War II the pattern changed significantly. Independence in former British colonies led to the exclusion of South Asian immigrants, and both Burma and Uganda expelled most South Asian immigrants. But a new destination for South Asian emigrants opened as the postwar economic recovery in Europe resulted in a severe shortage of labor on assembly lines and in transportation. Britain received more than

versely, like the states of Rajasthan and Uttar Pradesh in north India, are typically ones with low women's employment, low women's literacy, and poorly developed health infrastructure. In 1998 the Indian government acknowledged the evidence of international experience—that female education is the single most influential determinant of lower birthrates—and finally abandoned targets for sterilization and contraception. Several Indian states are now following Kerala's successful example of emphasizing women's education and better infant and maternal care.

This effect is reproduced across the region. Pakistan and Bangladesh make a useful comparison. Although both countries have high levels of poverty and historically discriminatory cultures in terms of women's access to employment and education, they each have startlingly different population profiles. In 2008, the total fertility rate in Bangladesh was 2.7, while in Pakistan the figure was a far-higher 4.1. The differential availability of health care for women is certainly a factor; ongoing efforts in Bangladesh to provide clinical coverage across the country have meant women can access immunizations for their infants and family planning for themselves, leading to dramatic and encouraging declines in birthrates. If places like Kerala and Bangladesh are any indicator, and to the degree that their successes are reproduced in other parts of the sub-

Persons per square kilometers

More than 1,000
500.1–1,000
250.1–500
100.1–250
50.1–100
25.1–50
5.1–25
1.1–5
0.1–1
0

Arabian Sea

Bay of Bengal

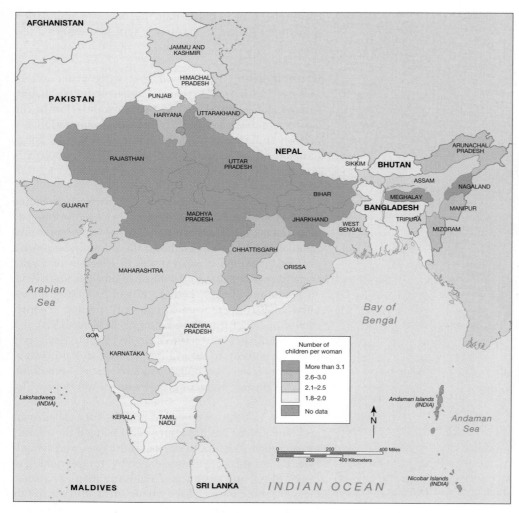

▲ **FIGURE 9.25 Fertility Rates Across the States of India** Although high fertility and large family sizes are the norm in some parts of the country, many states, like Kerala, have achieved nearly zero population growth with a replacement rate of two children per woman.

# CULTURE AND POLITICS

Diversity has to be the key word to describe the cultural geography of South Asia. The whole region has deep cultural roots, but these roots are often tangled. Even where traditions have not been mixed or hybridized, there are significant differences in the degree to which traditional cultures have accommodated or resisted globalization. Notably, the region contains cultures that are the product of influxes and migrations of language and religion from around world, as in the cases of ubiquitous Islam and widely spoken English. At the same time, South Asia is also cradle to several of the world's key global religions, including Buddhism, and the birthplace of several historically critical languages, including Sanskrit. New South Asian cultural traditions are increasingly global, moreover; Bollywood cinema and Bhangra music, for example, are now common on the streets of London and New York. The result is that traditional regional cultures, still strong, are juxtaposed vividly against modern global culture. For example, the tradition of parents seeking marriage partners for their children through newspaper advertisements in India, a long-followed social practice, continues relatively undiminished, but those same advertisements often provide an e-mail address or even a Web site for replies.

Similarly, the territorial and security conditions of the region are equally the outcome of global political struggles. As Pakistan and India, most notably, have internationalized their ongoing enmity by becoming nuclear states, so too has international politics influenced their struggle. The United States has become more deeply invested in securing stability in the region in support of its war in Afghanistan.

1.5 million South Asian immigrants, whose permanent presence not only filled a gap in the labor force but has also served to enrich and diversify British urban culture. About 800,000 South Asians moved to North America, mainly to larger metropolitan areas, where most found employment in service jobs. From the 1970s onward there has also been a steady stream of South Asian immigrants to the oil-rich Persian Gulf states, recruited on temporary visas to fill manual and skilled manual jobs.

South Asia has experienced a "brain drain" of significant proportions over the past several decades. Beginning with the emigration of physicians and scientists to Britain in the 1960s, the brain drain accelerated as South Asian students, having completed their studies in British and American universities, stayed to take better-paying jobs rather than return to South Asia. The idea of living abroad gained popularity among India's cosmopolitan and materialist middle classes as newspaper and television features publicized the global successes of Indian emigrants. In the 1990s the most distinctive aspect of the brain drain from South Asia was the emigration of computer scientists and software engineers from India to the United States and parts of Europe. By 2005, more than 2,000 of the 15,000 employees on Microsoft's Redmond, Washington, campus were South Asian immigrants.

## Tradition, Religion, and Language

**Religion** The two most important religions in South Asia are Hinduism and Islam. Hinduism is the dominant religion in Nepal (where about 90% of the population is Hindu) and India (about 80%). Islam is dominant in Afghanistan (99%), Bangladesh (more than 80%), the Maldives (100%), and Pakistan (about 80%).

Buddhism, though it originated in South Asia, is followed by only about 2% of the region's population. It is the predominant religion in Bhutan and Sri Lanka, and an enclave of Buddhism is found in Ladakh, the section of Kashmir closest to China. Jains are another distinctive religious group whose origins are in South Asia. Jains, like Buddhists, trace their faith to Prince Siddhartha, a religious leader who lived in northern India in the 6th century B.C.E. and who came to be known as Buddha, that is, the Enlightened One. Sikhs, whose religion was founded by Guru Nanak in the 16th century C.E., are concentrated in the Punjab, which straddles the India–Pakistan border (**Figure 9.27**).

The broad regional patterns reflected in Figure 9.27 are much more complex when considered in any detail. Underlying much of this complexity is the fact that Hinduism is not a single organized religion with

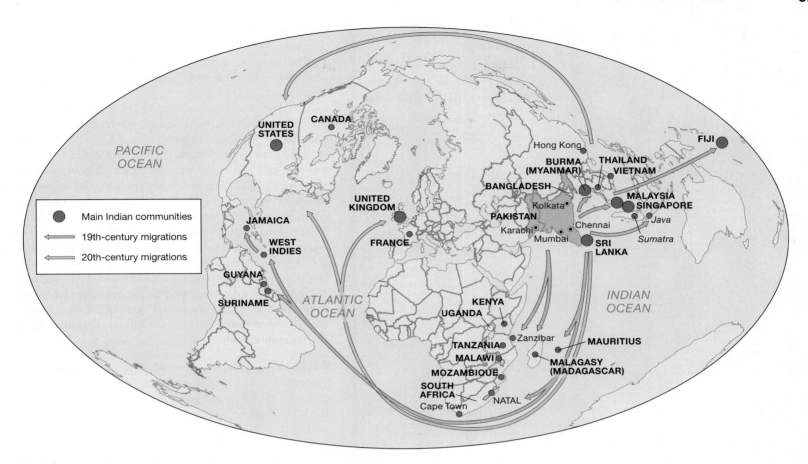

▲ **FIGURE 9.26 The South Asian Diaspora** Nineteenth-century migrations from South Asia were mainly to British colonies in East Africa, South Africa, and Southeast Asia, where there was a demand for cheap labor. In the 20th century the principal flows were to factory and service jobs in Britain and the United States.

one sacred text or doctrine; it has no unifying organizational structure, worship is not congregational, and there is no agreement as to the nature of the divinity. Rather, Hinduism exists in different forms in different communities as a combination of "Great Traditions" and "Little Traditions." The Great Traditions derive from the Rig Veda, a collection of 1028 Vedic poems that date from the 10th century B.C.E. A key aspect of this Great Tradition is the belief that human lives represent an episode of cosmic existence, followed after death by the transmigration of the soul to some other form of life. The "Little Traditions" of Hinduism consist of many local gods, beliefs, rituals, and festivals and the sacred spaces that are associated with them.

In Hinduism, the number seven has special significance. There are seven especially sacred cities in India: Varanasi, associated with the god Shiva (the destroyer, but without whom creation could not occur); Haridwar (where the Ganga enters the Plains from the Himalayas); Ayodhya (birthplace of Rama, one of the incarnations of Vishnu, the preserver); Mathura (birthplace of Krishna, another incarnation of Vishnu, sent to Earth to fight for good and combat evil); Dwarka (legendary capital of Krishna thought to be located off the Gujarat coast); Kanchipuram (the great Shiva temple); and Ujjain (the site every 12 years of the Kumbh Mela, a huge religious fair). There are also seven sacred rivers: the Ganga, the Yamuna, the (mythical) Saraswati, the Narmada, the Indus, the Cauvery, and the Godavari. Hindus visit sacred pilgrimage sites for a variety of reasons, including to seek a cure for

sickness, wash away sins, or fulfill a promise to a deity. The Ganga is India's holiest river, and many sacred sites are located along its banks, including Haridwar and Varanasi. Many important sacred sites, including thousands of temples and shrines, attract Hindu pilgrims.

A sizable minority population adheres to religions other than Hinduism. The most important of these is Islam. Although several million Muslims migrated from India to Pakistan at the time of partition, more than 123 million Muslims still reside in India today. There are also almost 20 million Christians in India. According to legend, Christianity was introduced to South Asia by the Apostle Thomas during the 1st century. Silk traders passing through northwest Pakistan to China during the 2nd century encountered Christians, but the small Christian community did not increase significantly until the arrival of colonial powers. The Portuguese brought Roman Catholicism to the west coast of India in the late 1400s, and Protestant missions, under the protection of the British East India Company, began to work their way through the region in the 1800s. Christianity is most widespread in the State of Kerala, in southwest India, where nearly one-third of the population is Christian. Buddhism prevails in Sri Lanka, although a significant Hindu minority lives there as well (**Figure 9.28**).

**Language** A great diversity of languages is spoken in South Asia. In India alone there are approximately 1,600 different languages, about

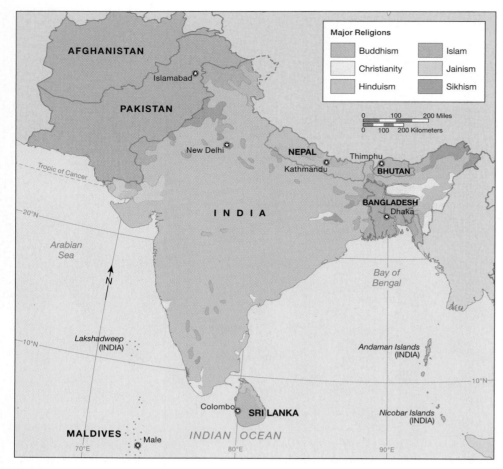

▲ **FIGURE 9.27 The Geography of Religion in South Asia** Partition between India and Pakistan in 1947 resulted in mass migrations of Hindus from Pakistan to India and of Muslims from India to Pakistan. There remain, however, more than 123 million Muslims in India, concentrated in several subregions.

400 of which are spoken by 200,000 or more people. There is, however, a broad regional grouping of four major language families:

1. Indo-European languages, introduced by the Aryan herdsmen who migrated from Central Asia between 1500 and 500 B.C.E., prevail in the northern plains region, Sri Lanka, and the Maldives. This language family includes Hindi, Bengali, Punjabi, Bihari, and Urdu.

2. Munda languages are spoken among the tribal hill people who still inhabit the remoter hill regions of peninsular India.

3. Dravidian languages (which include Tamil, Telegu, Kanarese, and Malayalam) are spoken in southern India and the northern part of Sri Lanka.

4. Tibeto-Burmese languages are scattered across the Himalayan region.

In India the boundaries of many of the country's constituent states were established after partition on the basis of language. No single language is spoken or understood by more than 40% of the people. Since India became independent, there have been efforts to establish Hindi, the most prevalent language, as the national language, but this has been resisted by many states within India, whose political identity is now

closely aligned with a different language. In terms of popular media and literature, there is a thriving Hindi and regional language press, whereas film and television are dominated by Hindi and Tamil, with some Telegu programming.

English, the first language of fewer than 6% of the people, serves as the link language between India's states and regions. As in other former British colonies in South Asia, English is the language of higher education, the professions, and national business and government. Without English-language skills, there is little opportunity for economic or social mobility. A guard, sweeper, cook, or driver who speaks only Hindi or Urdu will likely do the same work all his or her life. In contrast, those who can speak English—by definition, the upper-middle classes—are able to practice their profession or do business in any region of their country or in most parts of the world. English-language South Asian literature has produced many excellent novels. Among the most familiar authors are Anita Desai, V. S. Naipaul, Arundhati Roy, and Salman Rushdie.

**Caste**   A very important—and often misunderstood—aspect of India's cultural traditions is that of caste. **Caste** is a system of kinship groupings, or *jati,* that are reinforced by language, region, and occupation. There are several thousand separate jati in India, most of them confined to a single linguistic region. Many jati are identified by a traditional occupation, from which each derives its name: *jat* (farmer), for example, or *mali* (gardener), or *kumbhar* (potter). Modern occupations such as assembly-line operators, clerks, and computer programmers, of course, do not have a traditional jati, but that does not mean that people doing these jobs cease to be members of the jati into which they were born. People within the same jati tend to sustain accepted norms of behavior, dress, and diet. They are also endogamous, which means

▲ **FIGURE 9.28 Buddhism in Sri Lanka** Four of the Buddha statues in the Seema Malakaya temple complex on Beira Lake, with bustling and modern urban Colombo in the background.

that families are expected to find marriage partners for their children among other members of the jati.

In each village or region, jati exist within a locally understood social hierarchy—the caste system—that determines the accepted norms of interaction between members of different jati. In a normal village caste system, individuals will typically interact on a daily basis with others from about 20 different jati. Each individual person's jati is fixed by birth, but the position of the jati within the local caste system is not. Nevertheless, the broad structure of caste systems always places certain groups at the top and others at the bottom. Caste systems tend to hold in high esteem those who are religious and those who are especially learned. Those who pursue wealth or hold political power are typically less well regarded, and those who perform menial tasks are accorded least status of all. Priestly jatis—known as brahmins—are always at the very top of the caste hierarchy. Brahmins are expected to lead ascetic lives and revere learning.

At the opposite end of all caste systems are the so-called untouchables— jatis whose members dispose of human waste and dead animals. Mohandas Gandhi, the inspirational leader of India before independence, crusaded to dissociate these jatis from the demeaning term *untouchable*. Gandhi called them Harijans, meaning "children of God," but today most people in these jatis prefer to be referred to as Dalits, meaning "the oppressed," and the Indian government refers to them as "Scheduled Castes" (**Figure 9.29**), meaning that these groups sometimes receive set-asides or quotas for government positions. Traditionally, the Dalits were forced to live outside the main community because they were deemed by the brahmins to be capable of contaminating food and water by their touch. They were denied access to water wells used by other jatis, refused education, banned from temples, and subject to violence and abuse. Although these practices were outlawed by India's constitution in 1950, discrimination and violence against Dalits is still routine in many rural areas.

In urban areas, practices relative to caste vary widely. A perusal of any urban Indian newspaper, however, reveals a lengthy "matrimonial" section, where advertisements are made to find "suitable" matches, most often from specific caste groups, to marry.

## Cultural Practices, Social Differences, and Identity

**Marginality and Resistance** Discrimination and marginalization is also common among other groups of society. "Adivasi" or tribal people in India, notably, represent a significant part of the manual labor force in the regions where they predominate and have received government development assistance and infrastructural development. Like Dalit communities, tribals tend to hold the least land, have the poorest job prospects, and live in the most difficult of conditions.

Increasingly, marginal castes and indigenous adivasi communities have become more politically active, and increasingly violent resistance movements are associated with marginal groups in the region. So-called naxalite insurgencies, violent armed uprisings rooted in Maoist ideology, have appeared across the country, particularly in the tribal belt of the country, especially the poorest states: Chhattisgarh, Jharkand, and Bihar. Because they attack landlords and police stations, the naxalites are considered terrorists by the government of India, but they have considerable followings in rural areas and among students.

Within Pakistan, ethnic tensions have developed around linguistic differences. Most indigenous Pakistanis speak Punjabi or Sindhi, but

▲ **FIGURE 9.29 Caste** There are about 138 million Dalits, or "untouchables" in India, most of whom still live on the margins of society.

families who migrated from India at the time of partition—known in Pakistan as *muhajirs*—have tended to retain Urdu as their language. To protect and maintain their distinctive identity, the muhajirs formed a political party, the Mohajir Quami Movement. This attracted considerable resentment among indigenous Pakistanis, and in 1995 more than 1,800 people were killed in riots in Karachi. In 1998 continuing tensions led the government to impose martial law and to exile, in London, the leadership of the Mohajir Quami Movement.

**Religious Identity and Politics** Religious identity and politics have proven equally turbulent in recent years. Historically, the religions of South Asia have displayed a high level of syncretism, meaning that they have co-evolved and have merged with one another over the centuries. Recent changes in identity and geopolitics, however, have caused rising religious antagonisms and increasing social movements oriented around religion.

In Afghanistan and Pakistan, notably, powerful Islamist movements have resisted globalization by attempting to recreate certain aspects of traditional culture as the basis of contemporary social order. There is strong adherence in both countries, for example, to traditional forms of dress and public comportment (**Figure 9.30**). In Afghanistan the ultraorthodox Taliban rulers, who controlled 90% of the country between 1996 and 2002, imposed a harsh version of Islamic law that followed a literal interpretation of the Muslim holy book, the Qu'ran. Under Taliban laws, murderers were publicly executed by the relatives of their victims. Adulterers were stoned to death, and the limbs of thieves were amputated. Lesser crimes were punished by public beatings.

The Taliban government of Afghanistan was toppled in the United States invasion of 2001, but the movement still wields considerable power in Pakistan. In 2009 Taliban insurgents captured territory within 100 kilometers of the capital of Islamabad, resulting in a massive government counteroffensive.

Simultaneously in India, political movements founded in a conservative interpretation of Hinduism have been on the increase since the

▲ **FIGURE 9.30 Traditional Dress** These women at prayer in a Mumbai mosque are wearing the typical everyday attire of Muslim women in the region.

1950s. **Hindutva**, a social and political movement that calls on India to unite as an explicitly Hindu nation, has given birth to a range of political parties over the years, including the Bharatiya Janata Party, or BJP. The BJP quickly attracted popular support in the late 1980s, and between 1999 and 2004 it was the key partner in a 24-party coalition that came to power after national elections (**Figure 9.31**). Given the large populations of non-Hindus in the country and the explicitly secular nature of the national constitution, the rise of Hindu nationalism is extremely contested and viewed with suspicion and fear by many. The BJP failed to capture a place in the 2009 governing coalition.

The case of conservative Islam and Hinduism both suggest that although South Asia's religions have shown a remarkable capacity for integration and coexistence, they also provide cultural symbols and practices that conservative and intolerant groups and parties can rally around and use to organize. The continued success of secular government in India, however, and to a lesser degree in Pakistan, shows that these parties and groups are neither universally embraced nor consistently successful.

**Emerging Popular Cultures**   Contemporary culture provides many sharp contrasts with the deep-rooted traditions of South Asia, although there are places and regions (in Afghanistan, in Bhutan, and in many of the more remote rural areas of South Asia) where contemporary culture finds few expressions. The growth of a large and affluent middle class in India since the country's 1992 economic reforms has brought the sights and sounds of Western-style materialism to India's larger towns and cities: fast-food outlets, ATMs, name-brand leisure wear, consumer appliances, video games, luxury cars, and a fast world of pubs, clubs, and shopping malls (**Figure 9.32**). Between 2001 and 2007, the number of shopping mall developments in India jumped from 3 to 345; and the Indian credit-card industry is growing at a rate of 35% a year. As of 2007, India's luxury goods market was estimated to be $14 billion annually, featuring international brands such as Rolex, Tag Heuer, Hermes, Gucci, Hugo Boss, and Armani. The preeminent sport in both India and Pakistan is cricket, a legacy of British colonialism. The new affluence of the middle classes has taken cricket beyond a popular pastime with a passionate following to a sport that generates huge sums in betting and supports a star system to rival that of baseball in the United States.

Cable television arrived in India in the early 1990s, at about the same time the government initiated its economic reforms. After years without access to popular Western culture, urban middle-class Indians could now watch, via Hong Kong–based Star TV, programming that included MTV, *Baywatch,* and *The Oprah Winfrey Show.* The expectation among many was that such programming would quickly displace Indian culture, at least among the young and the middle classes. The sheer size

▼ **FIGURE 9.32 New Malls** The rapid growth of India's economy has spurred consumer spending for the country's growing middle class. This mall is in Gurgaon, just outside New Delhi.

▲ **FIGURE 9.31 Indian Bharatiya Janata Party (BJP)** At a rally of BJP in New Delhi, supporters listen to a speech given by the party's prime ministerial candidate prior to the 2009 elections. Characterized by its saffron colors and lotus leaf symbolism, BJP iconography merges religious identity and political action.

and market power of India's middle classes, however, has meant that this scenario of an externally imposed global culture has not occurred. Rather, Indian television and cable companies quickly began to produce films, musical shows, sitcoms, and soap operas in Hindi, Tamil, and some other local languages. The only Hollywood-made programs that earn reasonable ratings are those that are dubbed, whereas the domestic Indian television and movie industry, based in Mumbai, has quickly grown to major proportions.

When Mumbai was still known as Bombay, the city developed a huge Hindi-language film industry, which acquired the nickname "Bollywood." Although revenues from the hundreds of movies produced in India each year do not compare to those of Hollywood, they nevertheless represent a significant industry within India. Equally important, they represent a unique cultural element. They provide a popular form of escapism from the harsh realities of daily life for the majority of the population (**Figure 9.33**), and they do so in a form that is culturally distinctive, drawing on classical Hindu mythology and traditional social values. The roots of the Bollywood approach lie in traditions of folk theater and performance that stretch back 2,000 years, with familiar themes: good triumphing over evil, the struggle of the poor, the sins of the big city, and the melodrama of family life. Hindi-language movies have been a potent force in shaping Indian ideas of nationhood. At the same time, many of the Bollywood movies and TV productions address themes such as caste and modernization in ways that relate directly to the lives of Indian viewers.

Although India has produced avant-garde movies that have been recognized for their artistic and dramatic content, most Bollywood products are exuberant, spectacle-driven entertainment: melodramatic fantasies that mix action, violence, romance, music, dance, and moralizing into a distinctive, formulaic form. Bollywood stars such as Salman Khan, Amitabh Bachchan, Priyanka Chopra, Aishwarya Rai, Shilpa and Shamita Shetty, and Twinkle Khanna, have their careers and private lives monitored by adoring fans with an intensity that Hollywood agents would envy. Most Bollywood films have some sort of musical content, and the songs (lip-synched by the actors but sung by "playback artists," who are also stars) dominate Indian pop charts. Sound tracks include sitars, synthesizers, pianos, and violins to provide a score that moves effortlessly from classical Indian ragas to Mozart to hip-hop and rap music. Every taste is catered to, making a bridge between the traditional and the modern, and between East and West.

There is a large market for Bollywood films among the Indian diaspora, earning India more than $225 million in film exports each year. In addition to expatriate Indian markets, Bollywood movies are successful in the Persian Gulf states and in Russia. Nevertheless, Bollywood has its problems. Indian filmgoers, especially those in cities and those with access to satellite television, have become more and more difficult to satisfy with the standard Bollywood recipe. Meanwhile, India has become one of the largest markets for Hollywood films. The collision of Bollywood and Hollywood resulted most recently in the filming of the Oscar Award–winning *Slumdog Millionaire*, a film set and filmed within the slums of Mumbai.

Just as the impact of globalization has been mediated and transformed by India's television and movie industry, other aspects of economic and cultural globalization have found mixed expression amid South Asia's traditional cultural patterns. Thus, for example, it is still common to see people dressed in traditional clothing—saris for women, dhotis (loin cloths) for Hindu men, turbans for Sikh men, and so on—often in combination with Nike or Adidas sneakers or some other nontraditional apparel. Similarly, despite the proliferation of fast-food outlets such as Domino's Pizza and vending machines selling soft drinks such as Pepsi and Coca-Cola, Western-style food retailing has little appeal to affluent households, most of whom still live in neighborhoods where street vendors sell high-quality fruits, vegetables, dairy products, and other basics. Appliances such as washing machines, dishwashers, and power tools are also less prevalent than might be expected among South Asia's affluent middle classes, simply because of the millions of people available to undertake domestic labor at very low wages.

On the other hand, a countermaterialist pattern of culture is celebrated widely in the country of Bhutan. There, the government has devised a measure of the "Gross National Happiness" of the population, which stresses spiritual development and good living as opposed to per capita economic activity. Whether or not the people are happier than others in South Asia is not entirely clear, but this approach to culture contrasts dramatically with that of neighboring states, who have embraced the accumulation of material goods as the key goal of public welfare.

**Global South Asian Culture** Meanwhile, as in other world regions, the cultural shifts involved in globalization flow out as well as in. South Asian mysticism, yoga, and meditation found their way into Western popular culture during the "flower power" era of the 1960s after the Beatles visited India. South Asian cuisine, with its spicy curries and unleavened breads, found its way into Britain at about the same time and has since become established in restaurants and supermarkets in much of Europe and North America. South Asian methods of nonviolent protest such as boycotts and fasting, inspired by the ancient Buddhist concept of dharma and developed in the 20th century by Gandhi, have spread all around the world. Contemporary South Asian literature from writers such as R. K. Narayan, Vikram Seth, Satyajit Ray, and Salman Rushdie has found a global readership. South Asian art and music have been less influential, though Indian singers and musicians are well represented in the "international music" sections of Western record stores, and some artists, such as Sheila Chandra, have crossed over into a broader international audience.

▲ **FIGURE 9.33 Bollywood Escapism** These billboards hint at the escapist themes common to Bollywood movies.

Most recently, the global phenomenon of Bhangra music and dance has taken hold around the world. Originally a regional music style of the Punjab, Bhangra is set to the distinctive rhythmic pulse of the dhol drum and typically involves dancing. Lyrics are typically sung about current or local problems or issues. As Punjabis came to live in greater numbers in the United Kingdom and the United States, their alienation led to a resurgence of this traditional form of music, which soon became increasingly popular among other communities. Bhangra is now heard in nightclubs and on radio stations, as DJs continue to fuse the music with reggae and hip-hop (**Figure 9.34**). Bhangra has long since been untethered from the South Asian diasporic community, and R&B performers and rock musicians around the world incorporate bhangra rhythms and styles into their music.

## Territory and Politics

The colonial imposition of arbitrary borders, as elsewhere in the world, continues to have an enormous impact on governance and conflict in South Asia, where tremendous cultural diversity means that national political boundaries tend to encompass diverse groups in terms of ethnicity, language, religion, and cultural identity, while at the same time dividing some groups, leaving some in one country and some in another. The partition of British India in 1947 demonstrated this in relation to Hindus and Muslims, as did the subsequent secession of Bangladesh from Pakistan in relation to Bengali ethnic and cultural identity. But South Asia's cultural diversity, framed within national boundaries that have been relatively recently imposed, has also given rise to several other cases of regionalism, separatism, and irredentism (**Figure 9.35**) that are a continuing basis for political tension, social unrest, and, occasionally, outright rioting or armed conflict.

**Civil War in Sri Lanka**   Sri Lanka's ethnic tensions involve both language and religion. The majority population is Buddhist and Sinhalese-speaking. In the northeastern part of the country, however, the majority population is an enclave of a Tamil-speaking and Hindu population that represents about 17% of Sri Lanka's total population. Ever since independence from

▲ **FIGURE 9.34 Bhangra Musicians Playing with UB40** Traditional Punjabi bhangra music has become fully global as here, where bhangra music is mixed with English styles at the Live 8 event in Hyde Park, London in 2005.

Britain in 1948, the Sri Lankan government has pursued a nationalistic posture that has resulted in oppression of this Tamil population. The first casualties were 600,000 descendants of Tamil plantation workers, who had been brought to Ceylon (as it was then called) from southern India to work in tea plantations. The deportation of these "plantation Tamils" led to the formation in the 1980s of a militant and bloody Tamil separatist movement that crystallized in 1983 into the "Tamil Tigers"—the Liberation Tigers for Tamil Eelam. In the early 1990s, more than a million Tamil villagers were displaced by fighting between the Tamil Tigers and the Sri Lankan army, becoming refugees in their own land. Since the mid-1990s the level of conflict has diminished, but Sinhalese and Tamil nationalism continues to result in sporadic terrorist attacks and outbreaks of violence. Extensive Sri Lankan army victories by the government in 2009 seemed to tip the balance of the conflict in favor of the government, promising a final end to the decades-old conflict. The status and autonomy of the Tamil minority remains unclear, however, leaving the likely prospect for renewed fighting.

**Kashmir**   Kashmir, whose predominantly Muslim population found itself isolated as a minority within India at partition, has three times been the cause of war between India and Pakistan (in 1948, 1965, and 1971). Kashmir remains a contentious and complex arena. Kashmir's northern border is not an accepted international border—it is a "line of control" established after the 1971 war. Pakistan controls the northwestern portion of what India claims as Kashmir, and China controls the northeastern corner. In 1986 Muslim separatists began a renewed campaign of insurgency in the Indian-controlled portion of Kashmir. Since 1989, more than 30,000 people—separatist guerillas, policemen, Indian army troops, and civilians—have died in a guerilla campaign aimed at incorporating Kashmir into Pakistan as part of a larger Islamic state. The Pakistani-backed campaign culminated in Pakistan sending its own forces across the border into the Kargil Peaks district in 1999. Pakistani troops were withdrawn after India launched a full-scale military offensive to evict them, and U.S. President Clinton put pressure on the Pakistani government, which subsequently fell to a military coup d'état, the fourth such coup since Pakistan became independent in 1947. Pakistan has been governed by an elected government since 2008, when the military leadership stepped down. The problem of Kashmir, however, remains unresolved.

**Indo-Pakistani Conflicts and Tensions**   South Asia occupies a position of critical importance in global geopolitics. South Asia's strategic location meant that it was of great interest to the superpowers during the Cold War. India took full advantage of this, playing both sides against one another in seeking aid, while following a hybrid approach to domestic policy, with a democratic form of governance but a socialist-style approach to economic development. Postcolonial ties to Britain were maintained by Bangladesh, India, Pakistan, and Sri Lanka through the British Commonwealth, a loose association of countries tied together by patterns of trade, a shared language, and a similarity of institutions.

Ironically, South Asia has become even more of a geopolitical hot spot since the end of the Cold War. Both India and Pakistan have developed the capability of producing nuclear weapons. With the territorial dispute over Kashmir still simmering, India announced in May 1998 that five nuclear tests had been carried out in the Thar Desert close to the Pakistani border. This triggered a fervent bout of national pride within India, but it prompted Pakistan to respond within a few weeks with its own series of nuclear tests. India and Pakistan, together with Israel, are among the few states not to have signed the Nuclear Non-Proliferation Treaty that was formulated by the superpowers during the Cold War in 1968.

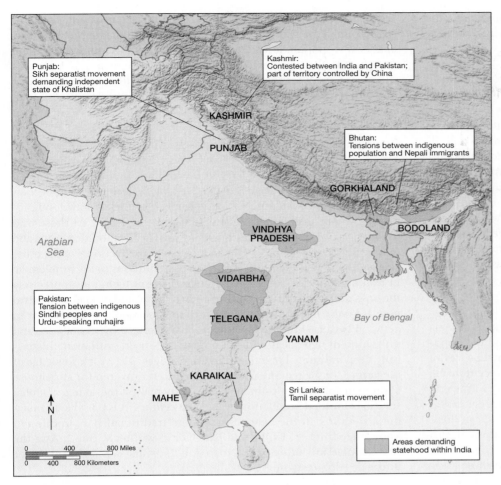

**▲ FIGURE 9.35 Regional and Separatist Movements in South Asia** The imposition of modern political and administrative boundaries on centuries-old patterns of cultural and ethnic differentiation has led to ongoing tensions and several cases of regionalism, separatism, and irredentism.

In the period since, there has been significant détente, with the opening of borders and increased cultural exchange between the two nations. This progress toward peace was brought to a dramatic halt in November of 2008, when roughly a dozen carefully coordinated attacks were carried out in a few hours throughout the dense business and tourist district of Mumbai, India. More than 170 people were killed in the attacks, and hundreds more were wounded. In the aftermath, the sole surviving attacker revealed that the event was staged by a Pakistan-based group of militants, the Lashkar-e-Taiba, who are considered a terrorist organization by both the United States and India. Demands and accusations from India, coupled with recalcitrance on the part of Pakistan, resulted in a disintegration of relations between the two states. As both countries are known nuclear powers, the global implications of tensions in the region are extremely high.

**Geopolitics of Afghanistan and Pakistan** This is exacerbated by the international tensions surrounding Afghanistan, which hold further implications for Indo-Pakistani relations. Afghanistan has long occupied a situation of particular geopolitical significance because it is situated pivotally between Central and South Asia. Control of its mountainous terrain and mountain passes—such as the Khyber Pass—has often been disputed, more often than not leaving its villages and its countryside in ruins. For much of the latter part of the 20th century, Afghanistan had economic and cultural ties to the Soviet Union and began to pursue a Soviet-style program of modernization and industrialization that ran counter to deeply rooted Islamic traditions. This provoked resistance from a zealous group of fundamentalist Islamic tribal leaders called the *mujahideen,* who were armed and trained by Pakistan. A total of more than 120,000 Soviet troops were subsequently sent to Afghanistan, but they were unable to establish authority outside Kabul.

The guerilla war ended in 1989 with the withdrawal of the Soviet Union. With the demise of their common enemy, the militias' ethnic, clan, religious, and personality differences surfaced, and civil war ensued. Eventually, the hard-line Islamist faction of the mujahideen, the Taliban, gained control of Kabul and most of Afghanistan. The Taliban regime not only imposed harsh religious laws and barbaric social practices on the Afghan population but also harbored the source of an entirely new geopolitical factor with worldwide implications: Osama bin Laden and his Al-Qaeda terrorist network, which was responsible for the September 11, 2001, attacks on the Pentagon and the World Trade Center. Consequently, Afghanistan became the focus of a U.S. military operation, "Enduring Freedom," that resulted in the defeat of the Taliban and installation of a U.S.–backed government in Kabul.

Military progress in Afghanistan, however, slowed during the following years. NATO forces, along with the U.S. military, managed to maintain control of Kabul and much of the hinterland, but Taliban attacks intensified in many outlying areas, restricting NATO and U.S. forces from full control of the country. The major goals of the operation, therefore, remain elusive, because Al-Qaeda has managed to survive in the border regions of Afghanistan and Pakistan.

As a result, during 2008 and 2009, this unresolved conflict spilled over into Pakistan. That country's North-West Frontier Province is adjacent to Afghanistan along the remote mountainous border, across which tribal communities have freely moved throughout history. State control has been difficult to achieve in this area over many centuries, including periods of British control in the 1800s. With Taliban forces operating in these areas and Al-Qaeda potentially sheltering there, U.S. forces have conducted military operations in the area. This has raised tensions with Pakistan, who views these actions as transgressions of their territory. Pakistani cooperation in the conflict remains essential for the United States, making the unsettled geopolitical triangle between the three states highly problematic.

# FUTURE GEOGRAPHIES

The shape and character of South Asia over the next 50 years rests on the outcome of several tensions and trends that all link the region to forces and places beyond its boundaries. First, the resolution of the crisis in Kashmir will delineate the future political boundaries of the region and the possibility of ongoing, and potentially cataclysmic, Indo–Pakistani conflict. Second, the speed at which birthrates in the region decline will determine the ultimate total population of the region and the time at which the population will stop growing. Finally, the future of food production techniques and technologies will determine whether South Asia is a global breadbasket or the site of hunger.

**Emerging Political Geographies**   The status of Kashmir is the "other shoe" that South Asia has been awaiting to drop since the time of partition. Should a resolution to the status of the region emerge in the next few decades, possibly in the form of an autonomous or independent region, the most intractable problem in Indo–Pakistani relations would be solved, leading toward considerably higher regional stability and potential opportunities for increased cooperation in areas of critical mutual interest, including increased trade along the enormous shared border and improved interstate cooperation on water allocation and groundwater use. The resolution of this problem, while it rests heavily with the two countries involved, is closely linked to international activities, moreover. As long as the conflict in Afghanistan continues, instability in Pakistan will deter any hope of Indo–Pakistani negotiations. The people of Kashmir themselves may also prove critical, as exhaustion with the decades-old conflict may propel local actions or agreements to catalyze change.

**Emerging Population Geographies**   As we have seen, South Asia is home to more than one-sixth of the world's people, and the population continues to grow. But as is also evident, the growth rates of many parts of the region have declined dramatically in recent years, with some areas actually achieving zero population growth. It is reasonable to assume that overall demographic growth in the region will halt before 2050. How much sooner growth ends than that year, however, depends on several factors, including women's education, health care availability, women's participation in the labor force, and the degree of urbanization of the population. If conditions in these areas improve even modestly, as they have in Bangladesh, where the fertil-

ity rate has fallen dramatically from seven births per woman in 1975 to approximately three births per woman in 2006, the prospect of a smaller final population are good. Conversely, at the time population growth ends, the region will be predominantly composed of older people, presenting a serious economic puzzle for this future generation, who will have to support an extremely large dependent population of elderly people.

**Emerging Landscapes of Food**   South Asia is a food-exporting region and a breadbasket of rice and wheat, among many other crops. Much of this success is leveraged on implementation of the Green Revolution, as we have seen. But with crop yields declining, soils becoming less vital, and water increasingly scarce, it is unclear how food production will be maintained. On one side of the argument are those who call for a "new" Green Revolution, using genetically modified crops and high-technology systems to spur an increase in production. On the other side are those who suggest many of the best solutions to the problem lie in traditional water, soil, and crop management techniques, precisely the ones destroyed during the Green Revolution, and critical to survival now. It is very likely that the answer to the question lies somewhere in between because South Asia remains characterized by its embrace and utilization of new ideas and technology in concert with many historically deep traditional practices and knowledges. The vast knowledge of the farmers of the region on adapting to uncertainty, and the storehouse of drought-tolerant crops in the region, may make it the source of global solutions for food production in the next century. It is possible to imagine a situation in the near future where traditional rice production knowledge from Sri Lanka and Bangladesh will make South Asia the key resource for future global food solutions instead of the target for future development aid.

# Main Points Revisited

■ **South Asia is a region physically distinguished by its ringing arc of mountain ranges and the annual rhythm of its summer monsoon rains.**

The effective human settlement and use of the region's heterogeneous landforms, including mountains, river flood zones, and deltas, ironically make the region vulnerable to the possible effects of climate change, through erosion, flooding, and sea-level rise.

■ **The landscapes of the region are characterized by inventive adaptation to regional climate, especially including highly productive agricultural systems, carefully adapted to the exigencies of drought and flooding.**

The Green Revolution heralded exceptional increases in agricultural productivity that have ironically decreased the diversity of crops and cropping systems, making the region vulnerable to future shocks.

■ **The historical geography of the region is characterized by its presence as a center of global trade, especially in Asian/African trade across the Indian Ocean, which was later exploited, and overridden by, colonial governance.**

British imperialism created dysfunctional monopolies over trade and a divisive two-tiered social system, but also introduced wide-ranging innovations in technology and education.

■ **South Asia has one of the largest and densest populations of all world regions, and its growth is ongoing, but growth rates are dropping rapidly, especially where health care, birth control, and education are available, especially to women.**

Areas with high population growth (northwest India and Pakistan) contrast sharply with areas with far lower growth (southern India and Bangladesh).

■ **The cultural landscape of South Asia is influenced by endemic traditions, including Buddhism and Hinduism, but is also a product of continued adoption and adaptation of other cultural traditions, including religion, artistic traditions, and languages from around the world.**

While a receiver and synthesizer of many global traditions, the region is also a major culture exporter, and South Asian music, literature, and cinema are now global staples.

■   The geopolitics of the region revolve around the relations of India to Pakistan, and the future fate of the region rests on resolution of deep antagonisms between these nuclear states, rooted in the partition of the region after colonialism.

The possibilities for conflict resolution are complicated, and potentially blocked, by the uncertain status of Kashmir and the ongoing war between U.S./NATO forces and the Taliban in Afghanistan.

■   **Economically, South Asia is at a critical juncture with considerable potential in its human and natural resource base and emerging** new economy. Foreign investment has flowed into the region, helping generate exceptionally high economic growth rates and a huge, well-educated, and sophisticated consumer society.

The planned economies of the period following colonialism have given way to free markets and global trade, creating a burgeoning middle class, but also resulting in staggering inequality between the very poor and very rich.

## Key Terms

bonded labor (p. 324)

caste (p. 328)

distributaries (p. 310)

Grameen Bank (p. 322)

Green Revolution (p. 308)

Hindutva (p. 330)

microfinance (p. 322)

monsoon (p. 306)

orographic effect (p. 306)

Raj (p. 317)

Sundarbans (p. 314)

terracing (p. 308)

zero population growth (ZPG) (p. 324)

## Thinking Geographically

1.  How do the Himalayan Mountains and monsoon season help define the character of South Asia? What is the orographic effect?

2.  What challenges face the people of South Asia with regard to the monsoon, and how have they adapted themselves and their landscapes over thousands of years?

3.  How did Britain transform the physical and cultural geographies of South Asia in terms of agriculture, infrastructure, and economy?

4.  Compare poverty in rural and urban South Asia. With malnutrition and illiteracy rates high, how do the poor manage to survive?

5.  India provides outsourcing for core regions, especially in the high-tech industry, where computer programmers create a large amount of computer code for far less money than it would cost in North America, yet for relatively high salaries in South Asia. How does that change the lives of programmers in the United States and in India?

6.  Some parts of South Asia have far lower population growth than others. What accounts for this geographic unevenness?

7.  What qualities of South Asian economies have made them resilient in the face of recent global economic crises?

8.  As South Asians migrated worldwide, which occupations did they take? Which cultural traditions did they bring with them?

9.  Why do India and Pakistan both have a claim to the Kashmir region?

10. How are key external political and military linkages between South Asia and the rest of the world affecting internal relationships within the subcontinent?

Log in to www.mygeoscienceplace.com for Videos, Interactive Maps; RSS feeds; Further Readings; Suggestions for Films, Music and Popular Literature; and Self-Study Quizzes to enhance your study of South Asia.

## SOUTHEAST ASIA KEY FACTS

- Major subregions: Indochina, Malaya Archipelago, Mainland Southeast Asia, Insular Southeast Asia

- Total land area: 4 million square kilometers/ 1.5 million square miles

- Major physiogeographic features: Mekong, Red, Chao Phraya, and Irrawaddy Rivers; more than 20,000 islands, including the world's third largest, Borneo; mostly monsoon climate

- Population (2010): 600 million, including 233 million in Indonesia, 93 million in Philippines, 88 million in Vietnam

- Urbanization (2010): 48%

- GDP PPP (2010) current international dollars: 2.8 trillion, GDP PPP per capita 12,245 (average without Singapore (45,028) and Brunei Darussalam (49,240) is 4,491)

- Debt service (total, % of GDP) (2005): 10.5%

- Population in poverty (% < $2/day; 1990–2005): 27%

- Gender-related development index value: 0.73; highest = Brunei Darussalam, 0.89; lowest = Cambodia, 0.59.

- Average life expectancy/fertility: 70 (M 68, F 72)/2.5

- Population with access to Internet (2008): 23.2%

- Major language families: Austro-Asiatic, Malayo-Polynesian, Papuan, Tibet-Burmese, Tai-Kadai with over 500+ distinct ethnic and linguistic groups

- Major religions: Islam, Theravada Buddhism, Mahayana Buddhism, Catholicism, Hinduism

- $CO_2$ emission per capita, tonnes (2004): 5.3

- Access to improved drinking-water sources/sanitation (2006): 81%/64%

# 10
# SOUTHEAST ASIA

On a cool December morning in 2004, a Moken elder named Saleh Kalathalay, living on Surin Island off the coast of Thailand, began to notice some unusual events.[1] The sea had become eerily calm, and the cicadas—a normally loud insect—grew silent. As a nomadic fisherman, Saleh could sense that a dramatic change was about to occur, as his people had long passed down a tradition of how to read the movements of the sea. He began to warn others in his village to get to higher ground. As he sounded the warning, a massive earthquake registering 9.0 on the Richter scale occurred off the western coast of Indonesia along a line reaching northwest toward the Andaman Islands. The earthquake drove large **tsunami** waves into shallow water. These waves hit the shoreline causing the total destruction of Saleh's village. Unlike the hundreds of thousands of people who died that day in Banda Aceh, Indonesia, Phuket, Thailand, or coastal Burma and Sri Lanka, only one of the Moken villagers and the tourists visiting their island died—a young physically-challenged boy whom the villagers had accidentally left behind.

Over the remainder of the day, thousands of people in coastal communities throughout the region not trained in the ways of the Moken were swept away, and their houses and livelihoods destroyed in the deadliest tsunami in recorded history. Almost everyone was taken by surprise by the enormous waves that surged inland along low-lying coasts around the Indian Ocean. There was no tsunami warning system in the Indian Ocean, in contrast to the Pacific system based in Hawaii, which uses seismic and oceanographic data to issue warnings to at-risk areas.

More than a half million people were displaced by the tsunami at the same time that drinking water and agricultural fields were contaminated by seawater. For the Moken, although resources and aid made their way to their destroyed villages, they now find themselves under increasing government control, which ties them to particular places. This forced settlement has led to a new reliance on food aid, a growing dependence on rice as a staple crop, and the regulation of Moken fishing practices. In fact, the Moken, who have always been nomadic to avoid overfishing, have had their livelihoods limited by these new policies. Ironically, the very knowledge of the sea that saved the Moken from the tsunami of 2004 is being made useless as a result of post-tsunami development efforts. This change is experienced through children, in particular, who are increasingly attending schools, some of which have been built in posttsunami resettlement communities (shown in the image).

For others living in Southeast Asia, the tsunamis illustrate their vulnerability to natural disasters more generally. Women, children, and poorer communities were especially vulnerable. There are indications that the impact of the tsunamis was more serious where protection against waves was reduced by the clearing of mangroves and destruction of coral. What is clear is that the long-term impacts of the tsunami remain, and despite efforts to increase awareness, the geographies of poverty and inequality mean that Southeast Asia's most vulnerable people remain at risk. ■

## MAIN POINTS

■ Southeast Asia is a region that, despite its outward appearance as "tropical," is distinguished by its diverse physical geography and ecological variety. This diversity and variety has led to a wide range of bio-geographical subregions and land use practices.

■ The historical geography of this region is a global geography, as this region has long been in contact with the peoples and places of East and South Asia as well as more recently Europe and the Americas. Through these contacts, Southeast Asia has developed a complex pattern of demographic, religious, and linguistic difference.

■ Southeast Asia is economically enmeshed in the global economy. On the plus side, these connections have enabled this region to foster some of the world's fastest-growing economies. On the minus side, some of these same connections have meant that many of the countries in this region have also been susceptible to the volatility of the global economy.

■ It is challenging to speak of a Southeast Asian set of cultural practices because this is a region that is highly differentiated socially. Southeast Asia is home to hundreds of different languages and ethnic groups, and these differences have led to an array of political tensions and relationships.

■ This region has grown over time to become unified through the development of the Association of Southeast Asian Nations (ASEAN). This transnational organization has institutionalized this region, although conflict and tension across it continue to raise questions about the long-term cohesiveness of ASEAN and the region more generally.

---

[1]Leung, Rebecca, (2007), "Sea Gypsies Saw Signs in the Waves." CBSNews, *60 Minutes*.

# ENVIRONMENT AND SOCIETY

Geographers of Southeast Asia have long understood that the relationship between the region's physical geography and human geography is complex. On the one hand, this region's position in relation to global climatic patterns and massive riverine and oceanic waterscapes has fostered rice cultivation and fishery development for over 5,000 years (**Figure 10.1**). On the other hand, the region's proximity to different global tectonic plate boundaries makes it susceptible to tsunamis, earthquakes, and volcanic activity and gives this region its complex physical geography (**Figure 10.2**). In both cases, the people of Southeast Asia have adapted to these environmental conditions, modifying both physical landscapes—developing agricultural practices that take advantage of monsoon rains—and the biogeographical world around them—producing new species of rice and harnessing biodiversity for medicinal purposes. This section traces some of this complexity, examining how Southeast Asia is situated within a larger global framework of environment–society relations and practices.

## Climate, Adaptation, and Global Change

Despite images of Southeast Asia as a singular tropical region, the climatic map demonstrates that it has a wide variety of climatic subregions (**Figure 10.3**). This map of climatic diversity is partially a function of the region's situation near the equator and the global intertropical convergence zone, or ITCZ (see Chapter 1, p. 4), which helps produce the monsoon rain patterns commonly found throughout the region. As the sun heats inland Asia from May to October, low pressure builds over the continental interior, and winds start to flow in from cooler high-pressure regions over the Indian and Pacific oceans. These winds are laden with moisture and produce heavy rain over land, particularly where air

currents rise and cool to produce orographic precipitation on highlands and south-facing island slopes. The onset of the monsoon is a momentous event in many places. It brings some alleviation from the very high and oppressive temperatures as rains cool the air, and it brings water for crops and drinking. But the monsoons can also cause severe floods, and the constant heavy rain promotes the growth of molds and funguses.

On mainland Southeast Asia, the November-to-March period brings cooler temperatures and higher pressures. The ocean temperatures become warmer relative to land, and lower pressure prevails over the oceans. During this period, the monsoon winds reverse and blow out of the continental interiors and across the ocean. On the mainland, this period is drier, resulting in lower overall annual rainfall totals, but the islands now receive a second sequence of monsoon rainfall, this time on the north-facing slopes. The Indonesian island of Sumatra, for example, receives the bulk of its rain on north-facing slopes during the December-to-February period as winds blow southward from mainland Southeast Asia across the South China Sea but on south-facing slopes between June and August, when monsoon winds sweep northward from the Indian Ocean.

The equatorial location of the Southeast Asian islands brings them within the influence of the ITCZ, where intense sun means high evaporation and a vigorous hydrological cycle associated with daily tropical thunderstorms throughout the year. Warm ocean temperatures from August to October combine with eddies in the trade winds to produce the large rotating storms known as *typhoons* in Pacific Asia, *hurricanes* in the Americas, and *cyclones* in the Indian Ocean. Typhoons most commonly develop east of the Philippines and move eastward into the South China Sea, whereas cyclones begin in the Indian Ocean and sweep into Southeast Asia from the east (**Figure 10.4**). The combined effect of monsoons, the ITCZ, and typhoons/cyclones means that the islands of Southeast Asia are among the wettest regions in the world with lush forest vegetation.

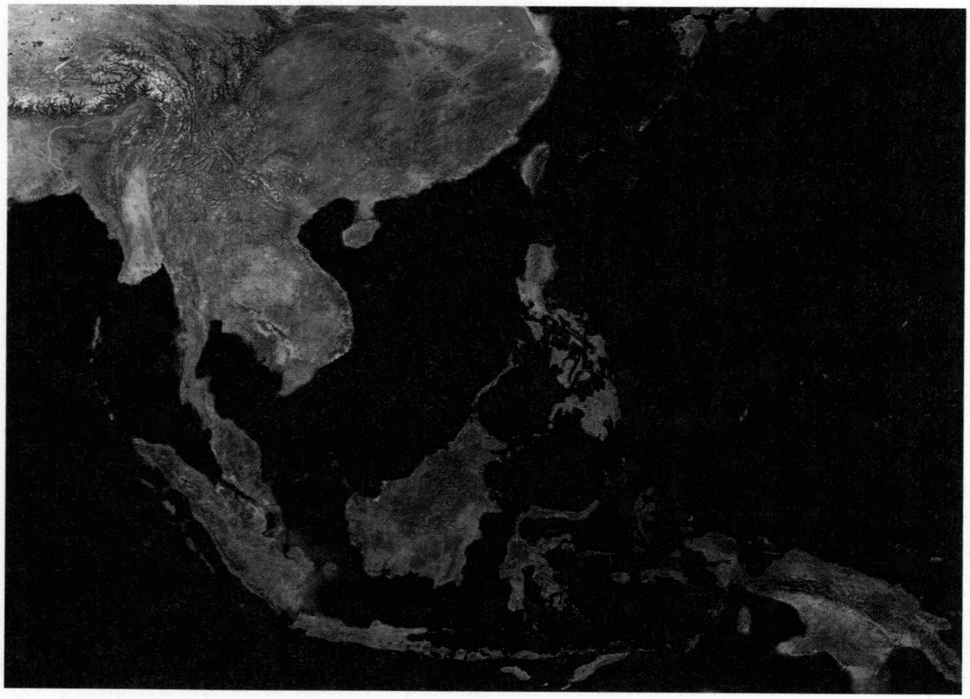

▲ **FIGURE 10.2 Southeast Asia from Space** This image/shows the main islands and major peninsulas that make up Southeast Asia.

**Adapting to Climatic Variability**    Despite the challenges produced by the dynamic climatic forces in this region, people have adapted to these climatic conditions effectively for a very long time. About 5,000 years ago, the selective breeding of a grass with edible seeds produced rice, the basis of human diets in Southeast Asia. Domesticated in several parts of Asia, rice grew where rainfall was evenly distributed through the year and totaled more than 120 centimeters (80 inches). Rice complemented fish and vegetables in the diet, and the plant was also used for fodder and thatch in building and has been served locally with other crops that were probably domesticated in Southeast Asia, including taro, sago, bananas, mango, and sugar.

The region's unique position in relation to global climatic wind patterns has also afforded local people and distanced travelers the opportunity to sail throughout this region and beyond. The summer onshore winds have brought traders and migrants to Southeast Asia, and with them their religious beliefs—first Hinduism and Buddhism and later Islam. The shift of winds in winter carried the traders back home along with spices and other products, which globalized Southeast Asia's contributions to food production and consumption. As a result, tastes changed in parts of Africa, the Middle

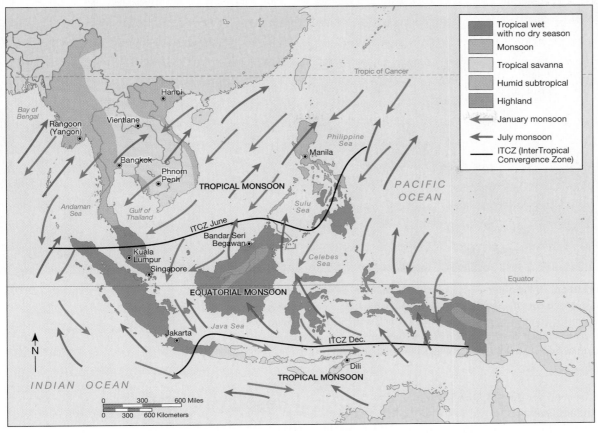

◀ **FIGURE 10.3 Climate Map of Southeast Asia** Southeast Asia has a hot, wet climate with equatorial convergence and seasonal monsoon winds that bring heavy precipitation. Most of the region falls within Koppen's tropical wet with no dry season, tropical monsoon, and tropical savanna climatic classifications, while northern parts of the region also fall within the human subtropical climate region. As a result, the mainland is drier with more seasonal rainfall, and regional climates can vary considerably from year to year as a result of El Niño conditions and the intensity of tropical *cyclones* and *typhoons*.

(a)

(b)

▲ **FIGURE 10.4 Cyclone Strikes Burma** (a) On May 2, 2008, Cyclone Nargis made landfall in Burma. (b) A woman and her children, who survived Cyclone Nargis, are seen where their home had stood in the town of Kyaiklat, southwest of Yangon in the Irrawaddy Delta. The town of around 50,000 inhabitants, where the storm tore rice mills apart and washed away fishing boats, was devastated by the storm. It is clear that such natural disasters differentially affect men and women, as women made up over 60% of the total death rate from this disaster.

▲ **FIGURE 10.5 Terrace Agriculture in Luzon, Philippines** Terracing affords a number of advantages in hilly areas that experience heavy rainfall. It allows people to control the flow of water through irrigation systems between different parts of the field; it protects topsoil from eroding during periods of heavy rainfall, sustaining the nutrient levels of the field; and it allows farmers to divide the land into paddies that can be used to grow rice seedlings, transplant seedlings into rice beds, and grow other basic fruits and vegetables to accompany this staple crop.

East, and Europe as new spices were introduced via the Silk Roads and the transoceanic trade networks of the Indian Ocean.

Adapting to the heavy monsoon winds has also meant modifying the region's physical landscape for rice production. These adaptations have included the construction of terraces, paddies, and irrigation systems (**Figure 10.5**). Terraces cut into steep hillsides provide level surfaces that facilitate water control and reduced erosion. The construction of dikes (ridges) around fields allows them to be flooded, plowed, planted, and drained before harvest in a system called *paddy farming*. Rice, one of the few major crops that can grow in standing water, is suited to the flooding that accompanies heavy monsoon rains. Wet rice, or *sawah*, thus became over time the most important crop in Southeast Asia. Traditional rice production is highly labor intensive, with work throughout the season in preparing and maintaining fields, transplanting seedlings, weeding, and harvesting each stalk by hand. Women perform much of the labor.

Irrigation, although productive, also has the potential to shift in relation to climatic variability in the region. There is the possibility, for example, that global warming will increase the land that might come under rice cultivation in parts of the region, whereas the same processes might cause extended periods of drying and drought, thus reducing rice yields in other areas. Large-scale rice agribusiness, which now keeps rice under cultivation year-round in parts of Southeast Asia, might also affect microclimatic patterns by producing increased moisture in the air around large fields.

Although the people of Southeast Asia have long benefited from the monsoon and the rain it brings, recent global climate trends are presenting new challenges. Increases in global temperature are affecting fisheries in the region as well as the long-term viability of many island communities, which may be subject to problems associated with increasing sea levels. Broader global dynamics, particularly within the wider Indian Ocean region, are also affecting global climate patterns, as deforestation related to increases in agricultural production in sub-Saharan Africa and South Asia may, when combined with similar patterns in Southeast Asia, have a long-term impact on rainfall and the monsoon patterns present throughout the region.

## Geological Resources, Risks, and Land and Water Change

Southeast Asia's geological history is an important part of its geography, as plate tectonics have shaped the physical topography of the region and influenced the configuration of land and oceans, highlands and lowlands, and geological hazards and resources. Topographic changes affect microclimatic differences throughout the region, for example. Topographical as well as geological diversity has also affected the development of the region's dynamic ecosystems—historically, lowland areas in the region have been ideal for rice agriculture, whereas upland areas have been available to the production of cool climate crops, including opium poppies.

Broadly speaking, the region is usually discussed as consisting of two physical geographic subregions—mainland and island (or insular)—although the plate boundaries do not reflect this rather neat geographic division. The mainland of Southeast Asia and the island of Borneo occupy the Sunda shelf of the Eurasian tectonic plate, for example. The Sunda shelf is flooded by the shallow South China Sea between Borneo and the mainland, but it was exposed during the ice ages, forming a **land bridge** to parts of Asia. In fact, humans probably reached Southeast Asia from Eurasia about 60,000 years ago. These people were the ancestors of contemporary indigenous groups in parts of Malaysia and Indonesia, as well as New Guinea, Australia, and the Pacific Islands region. The insular region, in contrast, consists of the Indonesian island arc from Sumatra to East Timor, which lies along the edge of the Eurasian plate. To the south and east, the Australian, Pacific, and Philippine plates are moving toward, and colliding with, the Eurasian plate and are being forced downward in deep-sea trenches along major subduction zones (see Chapter 1, pp. 8–9). The Mariana deep-sea trench, between the Pacific and Philippine plates in the Pacific Ocean southeast of the Mariana Islands, is the deepest in the world at 11,034 meters (36,201 feet). The collision zones are also associated with mountain building and volcanic activity and have created the thousands of islands that form the Indonesian and Philippine archipelagoes (**Figure 10.6**).

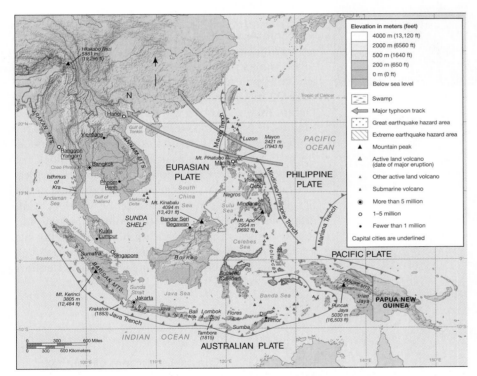

▲ **FIGURE 10.6 Southeast Asia's Physiographic Regions** Southeast Asia, one of the most geologically active regions of the world, experiences many earthquakes and volcanoes because it lies at the intersection of active tectonic plates. The region is very mountainous except for the major valleys of the Irrawaddy, Chao Phraya, Red (or Song Hong), and Mekong rivers and some coastal plains.

In general, the mainland is geologically older, but despite long-term weathering, it is still very mountainous. Peaks in the highlands of northern Burma reach beyond 5,000 meters (16,400 feet), and ridges link to the Himalayas. Between these ridges lie areas of flatter land along the major river valleys and deltas of the Irrawaddy, Chao Phraya, Mekong, and Red (or Song Hong) rivers. The relatively young islands of the region have high mountains, steep slopes, and generally narrow coastal plains. Irian Jaya, on the island of New Guinea, has mountains reaching 5,300 meters (17,300 feet); temperatures are so cold at these high elevations that permanent glaciers cap these mountains. More than a dozen other mountains and volcanoes in insular Southeast Asia reach 3,300 meters (10,800 feet).

Many countries of mainland Southeast Asia have fragmented and elongated geographies that have historically created challenges to national integration, transportation, and economic development. Burma, Thailand, and Vietnam all include long narrow segments that are less than 160 kilometers (100 miles) wide, although the region's rivers cut through these countries providing networks for transportation and large-scale agricultural developments. In the insular region, countries such as Indonesia have more than 13,600 islands (only half of them inhabited), and the Philippines are made up of more than 7,000 islands. The economies and lifestyles of insular Southeast Asia have therefore been oriented to the sea through ocean fishing and maritime trade for centuries (**Figure 10.7a**), and it has often been easier to use ocean or river transport than difficult overland routes. But the complexity of coastlines and the many islands of some nations, together with the economic significance of ocean fishery and oil resources, have created conflicts over maritime territorial jurisdictions and boundaries. Countries within Southeast Asia and outside the region, including the People's Republic of China, lay claim to small island chains in the South China Sea and the ocean spaces and resources that surround those islands (**Figure 10.7b**).

**Riverine Landscapes** Rivers are of central importance to Southeast Asia, particularly for the large urban communities that developed on the

341

mainland. Of all the riverine spaces in Southeast Asia, the Mekong River is the longest waterway, flowing 4,000 kilometers (2,500 miles) from the highlands of Tibet to its delta in Vietnam and Cambodia. It forms the main transportation and settlement corridor for the under-developed and conflict-torn countries of Laos and Cambodia, while also serving as a key trade network between these two countries and China, Burma, and Thailand. It also provides water for irrigation and hydroelectric development and is a productive fishery. One of the most unusual physical features in the Mekong Basin is Tonle Sap in Cambodia. This large lake acts as a safety-valve overflow basin for flood-ing on the Mekong River. During the dry season, water flows out of the lake along the Sab River into the Mekong, and during flooding on the Mekong, water flows back up into the lake, sometimes more than doubling its area.

The Mekong has provided a motive for cooperation among coun-tries of the basin, even during times of tension, because of the potential for water resource development that could provide mutual benefits. The **Mekong River Commission (MRC)**, formally established in 1995, has coordinated the planning of flood control and dam projects and has studied environmental issues within the basin. The 1995 accord pro-vides mechanisms for resolving disputes within the basin between Cambodia, Laos, Thailand, and Vietnam. In 1996, China and Burma joined as "dialogue partners" of the MRC. Several large dams have been constructed on the Mekong tribu-taries—among them the Manwan on the headwaters in China and Nam Ngum in Laos, which generates several million dollars of hydroelectric sales to Thailand—and many others are proposed (**Figure 10.8**). These dams have not been without controversy as local communities struggle to maintain their control over local resources in the wake of post-dam-construction flood-ing and government appropriation of land resources for dam development.

**Volcanic Activity and Soils**   The active volcanoes and earthquake zones of Southeast Asia pose great risks to the human populations of the region. The Indonesian and Philippine volcanoes are part of the Ring of Fire (see page 00 in Chapter 8), which surrounds the Pacific Ocean, linking this region with the seismically active regions of Japan, the western United States, and the Andes Mountains in South America. The island of Borneo sits on the Eurasian Plate and is of older and more stable geological composition. The risks of volca-noes are balanced by the benefits they provide in terms of soil nutrients. As in other parts of the world, volcanic eruptions have deposited ash that contributes to fertile, less-acidic soils that can sustain high crop yields and associated population densities. In Java, the rich vol-canic soils have contributed to productive agriculture

(a)

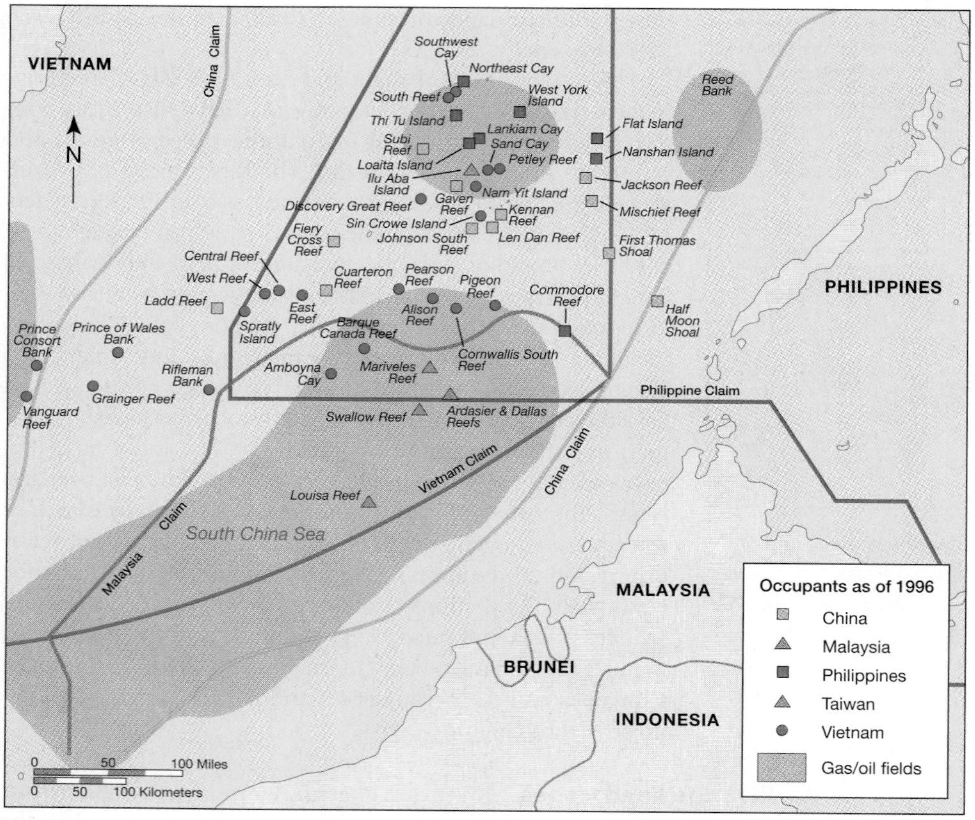

(b)

◀ **FIGURE 10.7  Coastal Landscapes of Southeast Asia**   (a) The many islands and long coastlines in much of Southeast Asia orient human activity toward the sea. These boats (called *junks*) in Halong Bay, Vietnam, are used for transport and fishing, for both subsistence and commercial sales. (b) This map shows the multitude of contradictory claims in the Spratly Islands in the South China Sea.  International law regulates the use and control of ocean space, providing countries with a 200 mile exclusive economic zone around any land claim. Thus, each island in the Spratley's is important not so much for what can be provided for by the land but what the land provides in terms of access to the sea.

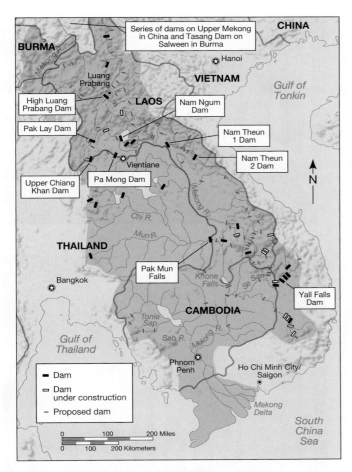

▲ **FIGURE 10.8  The Mekong River**  (a) The Mekong River links the countries of mainland Southeast Asia as it flows from a source on the Tibetan Plateau to the delta in Vietnam and Cambodia. (b) The Mekong River Commission also has a project to collect remotely sensed data to track environmental changes within the river region. This image, taken on a clear day during the dry season, illustrates different land use patterns. The light green color, for example, highlights intense rice agriculture in the river's delta.

and dense populations. Soil fertility is also high in river valleys and deltas that receive regular replenishment of river sediment and nutrients in annual floods. Regions farther from volcanoes and rivers experience soil limitations typical of tropical climates around the world, where heavy rain and warm temperatures wash nutrients through the soil, and organic material is broken down and recycled rapidly into forest vegetation.

**Energy and Mineral Resources**  The geology and tectonics of Southeast Asia have promoted the formation and development of mineral and energy resources, especially tin and oil, but generally the region is not as resource rich as other regions, such as Africa or Latin America. Tin mines, developed by European colonial administrations in the region during the 19th century, employed imported Chinese labor because local people were reluctant to abandon subsistence rice production. This immigration produced an interesting social geography to Southeast Asia, as large Chinese communities were established throughout the region, par-

ticularly in Malaysia, Indonesia, Singapore, and Thailand. Today, oil from Southeast Asia contributes about 5% of global production. Indonesia, Malaysia, and Brunei are the principal producers, exporting oil mainly to Japan from major oil fields off the north coast of Borneo, Sumatra, and Java. With foreign assistance, Vietnam is expanding oil production, and Burma is reinvigorating its energy sector through exploitation of gas reserves for domestic use and for sale to Thailand. From colonial times, tin has been a major export from Malaysia, but exports have dropped in the last 20 years as deposits have become exhausted and competition has lowered prices. Thailand and Indonesia also have significant tin reserves and production. The Philippines has developed a wide range of mineral resources, including copper, nickel, silver, and gold, and Indonesia is developing new gold, silver, and nickel mines in Irian Jaya, despite opposition from environmentalists and indigenous groups. The newest energy sources to be developed in Southeast Asia are **biofuels**, especially palm oil, which can be converted into biodiesel. Indonesia and Malaysia accounted for almost 80% of crude palm oil production in

2007, and Thailand is also starting to increase its production, which complements their already expansive production of sugarcane-based ethanol. These two countries bring a total of 8.4 million hectares (2.07 million acres) under cultivation for palm oil. Concern is growing about the conversion of tropical forests to palm oil plantations across the region, particularly when expansion of this crop continues to take away space for megafauna, such as orangutans (**Figure 10.9**). But as global demand continues to increase, even within the current context of slowing global economic growth, and rising economies, such as India and China, increase their need for palm oil, it is likely to be an important export crop for these countries into the future.

## Ecology, Land, and Environmental Management

Climate change and plate tectonics together have produced a fascinating division in the ecology of the Southeast Asian region. The last ice age, when sea levels dropped as ice sheets locked up moisture, exposed the Sunda Shelf between the mainland and Indonesia until about 16,000 years ago, and many species, including tigers, elephants, and orangutans, migrated across this land bridge to the islands of Indonesia. To the south, animals such as kangaroos and opossums moved north across the Australian Plate to New Guinea and other islands. Between Bali and Lombok is a deep ocean trench that remained ocean, even during ice ages, and prevented these two very different types of species communities from mixing. This created an ecological division called **Wallace's Line**, named after naturalist Alfred Wallace, who traveled extensively in the region in the 19th century.

The vegetation and ecosystems of Southeast Asia reflect the wet tropical climate, with a natural land cover of dense forests originally dominating the region. Indonesia is ranked second in the world in terms of its biodiversity, including at least 10% each of the world's plant, bird, and mammal species. The two major forest types are evergreen (with leaves year-round) tropical forests in the wetter areas and tropical deciduous, or monsoon, forests where rainfall is more seasonal or less intense and trees lose their leaves in the dry season (**Figure 10.10**). In drier regions, these forests are less diverse, and vegetation cover changes to savanna and grasslands where rainfall is less or more seasonal, or at higher elevations. **Mangroves** and marshes are found along the long mainland and island coastlines, together with a rich offshore marine ecosystem that includes coral reefs and fisheries. Indonesia leads the world in mangrove area, with more than 4 million hectares (10 million acres). These mangrove areas serve a vital function, protecting areas from erosion while providing habitat for myriad animal species.

In addition to the naturally occurring vegetation of this region, other ecological systems have been introduced via European colonialism. In the late 19th century, the economy of Malaya, under British control, focused on producing natural rubber. Rubber production grew rapidly after 1876, when plants grown in Britain from seeds smuggled out of Brazil were introduced to Malaya. The area covered by rubber plantations grew from 800 hectares (2,000 acres) in 1898 to 850,000 hectares (2.1 million acres) in 1920, as the explosion of automobiles in North America and Europe increased demand for rubber tires. Even though there was a ban on production by local people in Malaya to protect the profits and monopoly of European planters, rubber quickly became popular with local farmers because of the high price it brought and the low labor its production required.

(a)

(b)

▲ **FIGURE 10.9 Palm Oil Protests** (a) Activists of the Centre for Orangutan Protection (COP) dressed like injured orangutans and lay on the ground as others displayed a banner during a demonstration in Jakarta on May 8, 2008, to voice awareness to protect orangutans. Conservationists believe one of the biggest populations of wild orangutans on Borneo will be extinct in three years without drastic measures to stop the expansion of palm oil plantations. More than 30,000 wild orangutans live in the forests of Indonesia's Central Kalimantan province, or more than half the entire orangutan population of Borneo Island, shared between Indonesia, Malaysia, and Brunei. (b) Forest loss in Southeast Asia, in particular, threatens species such as the orangutan, shown here in East Kalimantan on the island of Borneo.

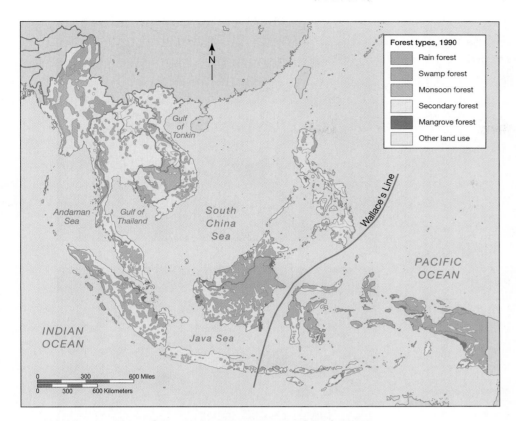

**▲ FIGURE 10.10  Southeast Asian Vegetation and Ecology**  The natural land cover of most of Southeast Asia is forest—evergreen in the wetter areas and tropical deciduous in drier regions. This map shows the actual land cover in 1990 and the extensive conversion of forests to agriculture and other land uses. Wallace's Line, named after 19th-century naturalist Alfred Wallace, indicates the location of a deep-ocean trench. The trench divided two sections of a great *land bridge* that rose above Southeast Asian oceans and seas 16,000 years ago. This line divides the two major biogeographical zones of Asia and Australia.

**Deforestation**  Deforestation is the most significant regionwide environmental problem in Southeast Asia (**Figure 10.11a**). Forests once dominated the region, providing habitat for a diverse ecology and food, medicine, fuel, fiber, and construction materials to local peoples. As trade with countries such as China and later Japan developed, demand for specialized products, such as aromatic sandalwood and teak, grew and resulted in some increased forest exploitation. Whereas **swidden farming** cleared patches of forest for crops, widespread clearance began in the late 1800s with the expansion of rice production and export of tropical hardwoods under European colonial control. Most of the deforestation during this period was in lowland regions such as the Irrawaddy Delta. After World War II, timber extraction expanded in the highlands, especially cutting teak in Thailand and Burma for export to the furniture industry. More recently, the growth of oil palm plantations, the pulp and paper industry, and cutting trees for plywood and veneers has placed even greater pressure on forests.

The driving forces for contemporary deforestation include export agricultural plantations, logging for tropical hardwoods and pulp, and land clearing by frontier migrants and swidden farmers (**Figure 10.11c**). Forest cutting increased as timber prices increased by 50% during the 1990s. By 1996 more than half of the original forests had been cleared—Thailand and Vietnam had about 20% of their forests left, whereas

Cambodia, Malaysia, and Indonesia retained about 65%. Forest loss rates in Indonesia are 1.8 million hectares (almost 7,000 square miles) a year, second only to Brazil. Only 40% of Burma's original forests remain. Timber exports are critical to the Burmese economy, making up 15% of export earnings, including earnings from concession of more than 18,000 square kilometers (6,950 square miles) for virgin teak extraction by Thailand. The timber industry provides thousands of jobs and a significant amount of foreign exchange in Southeast Asia. Swidden clearing, although often blamed for massive forest destruction, has also intensified in areas where population has increased, land is limited, and chain saws are available. The causes of extended swidden practices are tied as much to politics and the economy as they are to local practice. This process of extensive shifting of cultivation was sustainable, for example, when widely scattered villages controlled large areas of land and could use fields for a few years and then leave them for 10 to 30 years to recover. As access to land decreases and population and consumption increase, land may be cleared more frequently, causing long-term declines in environmental productivity.

The impact of deforestation includes loss of species habitat, flooding and soil erosion, and smoke and pollution from forest burning. Local communities have also been marginalized from lands they once occupied as logging concessions are given over to larger corporate or state entities. In the Philippines, floods in Mindanao in 1981 killed 283 people and injured 14,000 below a region of major forest clearing, and deforestation has been associated with the near extinction of the Manobo culture in northern Luzon. Deforestation has destroyed or is threatening the habitat of many species in Southeast Asia, including charismatic animals such as tigers and orangutans. But the deep forests of mainland Southeast Asia are still relatively unexplored by biologists. New species are being discovered, including relatively large animals such as the Vu Quang ox and giant Muntjac deer, identified as recently as the 1990s. But these isolated regions are coming under further attack, as poachers move into once-isolated forest lands to collect prized megafauna, such as tigers (**Figure 10.11b**).

The enormous islands of Borneo, Sulawesi, and Irian Jaya, the Indonesian portion of New Guinea, are frontier regions that have also become a focus for **transmigration** resettlement (see p. 356), mineral development, forest exploitation, and resistance by indigenous groups. Borneo covers more than 750,000 square kilometers (290,000 square miles) and includes territory controlled by Indonesia (Kalimantan), Malaysia (Sarawak and Sabah), and Brunei. Borneo, Sulawesi, and Irian Jaya have more than 25,000 species of plants and 10% of the world's biodiversity. Elephants, tigers, rhinos, and orangutans reside in the mountainous and forested interiors, with human populations concentrated in the coastal plains and river valleys around Hulu Sungai in southern Kalimantan and Pontianak in western Kalimantan.

Governments have responded by setting aside forest reserves and by banning logging in many regions, as well as by insisting on local

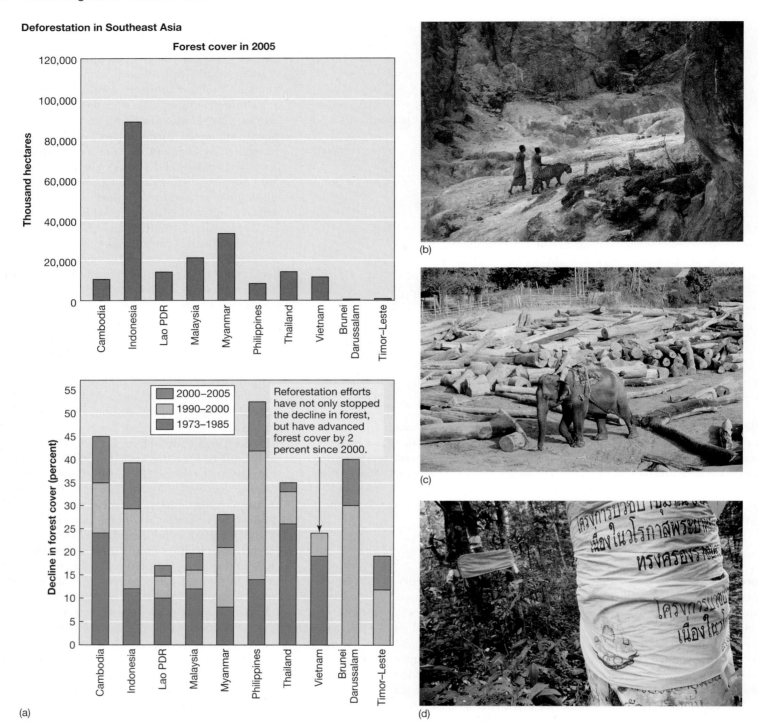

**Deforestation in Southeast Asia**

**Forest cover in 2005**

(a)

(b)

(c)

(d)

▲ **FIGURE 10.11  Deforestation in Southeast Asia**  (a) Forest cover in 2005 (in thousand hectares) and deforestation rates (for 1973–85, 1990–2000, and 2000–2005 in percentage) for selected countries and periods of years in Southeast Asia. (b) Monks walk with a tiger at Wa Pa Luangta Bua, the "Tiger Temple" in northern Thailand. The monks shelter and care for wild tigers that have been injured or abused by poachers. (c) An elephant hauls valuable teakwood near Pak Lay in Laos. The use of elephants for logging and transport in countries such as Laos and Burma has less impact on the environment than mechanized deforestation. (d) Trees have traditionally been robed to make peace with the spirits believed to dwell in them, an example of the Thai blend of animism and Buddhism. In 1988, when loggers threatened to harvest a forest in northern Nan province, a monk named Phra Manasnati Pitak ordained trees to protect them. His action saved the forest, and soon other communities were following his example, such as in this region near Chiang Mai, Thailand.

processing rather than export of raw logs. But earlier patterns of "crony capitalism" (see p. 352) that gave generous timber concessions to friends and relatives of government officials in Indonesia, Malaysia, and the Philippines have not been rescinded. Local people and international environmentalists have responded to deforestation in Southeast Asia by forming social movements and alliances to protect forests. In Thailand, Buddhist monks have helped to protect trees by wrapping them in saffron cloth and ordaining them, thus providing strong religious taboos against deforestation. The environmental movement is one of many organized grassroots efforts that also include action to support women, indigenous groups, factory workers, and religious minorities (**Figure 10.11d**). Deforestation and the fate of indigenous populations on Borneo have garnered the attention of international environmental and human rights groups. By the late 1990s, Borneo was providing more than half the world's tropical hardwoods, mainly through concessions given to multinational timber corporations, such as Weyerhaeuser, Georgia-Pacific, and Mitsubishi. The damage from forest clearing includes loss of biodiversity, erosion and flooding, and loss of forest benefits to indigenous populations. Although bans on export of raw logs were put in place in 2002, the corresponding increase in wood processing on Borneo, including more than 50 plywood mills, keeps the pressure on the forests.

**Postcolonial Agricultural Development** Although agriculture has decreased in economic significance across most of the region, it is still a major employer and is essential to food security, employing more than 40% of the economically active population in Indonesia, Thailand, and Vietnam and almost 70% in Cambodia and Laos. Rice production continues to dominate the land area of Southeast Asia, but the export of spices and plantation crops, such as rubber, tea, coffee, and sugar, are still important. Regional and international demand has increased the area planted with oil palm and pineapples, especially in Malaysia, the Philippines, and Thailand. Multinational corporations such as Del Monte and Dole are heavily involved in producing pineapples and other fruits in Thailand and the Philippines. Southeast Asian agriculture was transformed by two major factors—the Green Revolution and land rice yields reform—in the last 50 years.

The **Green Revolution**—a process of genetic modification that was first developed in Mexico in the 1940s by U.S. food scientists—comprised a technological package of higher yielding seeds, especially wheat, rice, and corn, which in combination with irrigation, fertilizers, pesticides, and farm machinery was able to increase crop production in the developing world. The globalization of Green Revolution technologies contributed to dramatic increases in rice and other cereal production in Indonesia, the Philippines, and Thailand. As an example, Indonesia and the Philippines almost tripled their production of rice per hectare between 1961 and 2005. As in other world regions, the benefits of the Green Revolution in Southeast Asia were unequally distributed. Communal land and rights to the rice harvest were lost in the new orientation to exports and wage labor, and the introduction of mechanical rice harvesting increased rural unemployment and landlessness in countries such as Malaysia. The rate of successful adoption was higher among those with access to irrigation. When the new varieties were planted uniformly across large areas as monocultures, they became vulnerable to diseases and pests. Rather than continue to use pesticides to combat such pests, the Indonesian government sponsored the use of integrated pest management by small-scale farmers, which uses less chemically intense techniques.

# HISTORY, ECONOMY, AND DEMOGRAPHIC CHANGE

The historical geography of Southeast Asia is distinguished by its long-term connections to other world regions, from East Asia (Chapter 8) to South Asia (Chapter 9) to Europe (Chapter 2) and the Americas (Chapters 6 and 7). That Southeast Asia has long been connected through trade networks—both land-based and maritime—to other parts of the world means that the economies of this region are related to a number of other broader global economic processes, including the ebbs and flows of trade in and out of India and China and the changing nature of global financial markets, which has brought Southeast Asian economies into a more direct relationship with the financial markets of Japan, the United States, and Europe. This section examines some of these processes, arguing that Southeast Asia's diversity is a result of its relative location globally by demonstrating how Southeast Asian practices have been adapted to the broader global movements of peoples, ideas, and biological species.

## Historical Legacies and Landscapes

Maritime trade brought merchants from the Middle East, China, and India about 2000 years ago, together with new religions, crops, and technologies. Southeast Asian **entrepôt** (e.g., commercial trading) towns, such as Oceo near the Cambodia–Vietnam border, became points of connection between Southeast Asia and India and China about 1,800 years ago. For most of the region, early interaction was with India, whose Hindu and Buddhist religions and related forms of government were attractive to local chiefs; they saw advantage in the divine privilege they could be granted as god-kings (*deva raja*). Over time, a number of urban-based kingdoms emerged in the region. Two of the most powerful were the mainland Khmer empire and the island Srivijaya culture of Sumatra. The Khmer kingdom emerged in the 9th century near Tonle Sap Lake and built the magnificent 12th-century Hindu and to a lesser extent Buddhist temple compound of Angkor Wat, one of the largest religious structures ever built and a focus of tourism in contemporary Cambodia. Srivijaya ruled Sumatra and the southern Malay Peninsula from the 7th to the 12th centuries by controlling the region's long-distance maritime trade through the Strait of Malacca. This powerful state had a capital at Palembang, which was based in Buddhist religious traditions (**Figure 10.12**).

On the mainland, an increasingly powerful Thai kingdom, centered on the city of Ayutthaya on the Chao Phraya River, conquered the Khmer empire in the 14th century and maintained control of parts of the Malaysian Peninsula for four centuries. The Thai would compete with different empires, including the Burmese and to lesser extent kingdoms found in northern Thailand, Laos, and Cambodia, from the 14th to the 19th centuries, for control of the people and resources of the mainland. These empires privileged Theravada Buddhism, which spread through the mainland—often blending with local animistic beliefs—as these empires were consolidated around a number of very important city–states, such as Bangkok, Thailand, and Pagan, Burma.

By the 11th and 12th centuries, new religious systems, including Islam, also entered insular Southeast Asia through global trade from South Asia. By the 15th century, the political geography of insular

▲ **FIGURE 10.12 Ceremonial and Religious Space in Southeast Asia** By the 10th century there were large kingdoms influenced by Indian culture and religion along the Irrawaddy River in Burma, in the Champa regions in what is now southern Vietnam, and in central Cambodia. These kingdoms relied on state taxation of rice production and power structures based on kinship and religious–royal bureaucracies, and they built spectacular capitals. Examples of these different kingdoms include the complex that includes the 12th-century temple of Angkor Wat and sacred city of Angkor Thom, which was constructed by the Khmer empire just north of the Tonle Sap Lake in Cambodia. The design represents symbols from the Hindu cosmos, including mountains, artificial lakes, and sculptures. After years of conflict, the site is being restored and promoted as a major tourist destination.

Southeast Asia was restructured around a series of Islamic-based sultanates such as Malacca (in present Malaysia) and Brunei (on the island of Borneo). Malacca's strategic location controlling the trading route between India and China gave it great commercial power as a sea-based trading state, and after the rulers enthusiastically adopted Islam, Malacca disseminated Islamic beliefs and institutions in Southeast Asia.

**European Colonialism in Southeast Asia**  Scholars divide the colonial period into two periods. The mercantile (or trade) period spanned from about 1500 to 1800. The industrial (or export-oriented) period lasted from about 1800 to 1945. A number of powers sought to control the valuable markets in the region, including Portugal, Spain, the Netherlands, Britain, and France, whereas the United States played an important colonial role in the region in the 20th century (**Figure 10.13**).

Portugal dominated early mercantile trade in the region. The Portuguese sailed around Africa, established a headquarters in Goa, India, and then moved to control the strategic port of Malacca in 1511 and East Timor in 1515. To obtain commodities such as cloves, nutmeg, and pepper from across the Indonesian archipelago, they relied on indigenous and other Asian producers and merchants. Meanwhile, the Spanish sailed across the Pacific and established the Philippines from several sizable islands and hundreds of smaller ones. As a result, they spread Catholicism and expanded their global empire by gaining access to trade in spices and other commodities in Southeast and East Asia. The city of Manila, founded in 1571, became the center of trade with Latin America, with galleons sailing regularly to Acapulco, Mexico. The third colonial power to move rapidly into Southeast Asia was the Netherlands. The Dutch dominated trade from about 1600 to 1750,

with an initial focus on the Molucca Islands of Indonesia—the famous **Spice Islands**. The formation of the Dutch East India Company in 1602—with its headquarters in Java—consolidated commercial interests to control trade in Indonesia (then called the East Indies) by restricting the production of valuable spices such as pepper and by destroying communities that ignored restrictions or participated in smuggling.

The British colonial effort in Southeast Asia was an extension of their activities in India. It focused on Burma, Malaya, and Borneo and on the control of strategic ports. The British were also led by a trading company, the British East India Company, formed in 1600 and focused on British domination of South Asia (see Chapter 9, pp. 316–18). The British waited until the 19th century, however, to make a move from their empire in India into Southeast Asia, beginning with control of the strategic ports of Penang (1786), Singapore (1819), and Malacca (1824). They fought two wars with the Burmese and used a protectorate system to gain control of the Malay Peninsula; they also reoriented these colonial economies to exports of tin, rubber, and tropical hardwoods and on controlling trade routes between India and China.

The French influence originated in the 18th century, and they consolidated their holdings in the 19th century in response to rivalry with Britain over commercial links to China. The French named their holding *l'francaise indochine* (or Indochina), bringing together Cambodia, Laos, and the districts of Tonkin, Annam, and Cochin China in Vietnam as the Union of French Indochina. They also created a new writing system for Vietnamese, which is still used today, while engaging in a model of colonial rule that instituted French language training and cultural practices for local elites.

The United States was a colonial power in Southeast Asia for less than 50 years, acquiring the Philippines in 1898 after victory in the

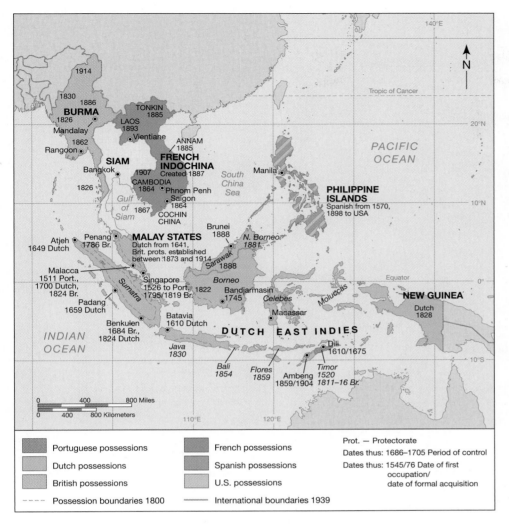

**▲ FIGURE 10.13 European Expansion into Southeast Asia** This map shows the areas controlled by the major colonial powers in Southeast Asia.

Spanish-American War. Spain and the United States wanted to control the Philippines because of their location as a gateway to trade with Asia. Even after the United States granted independence to the Philippines in 1946, it continued to treat the islands as a strategic location, maintaining several military installations such as Clark Air Force Base and Subic Bay Naval Base—closing both after the Mount Pinatubo eruption in 1991.

Unlike other parts of Southeast Asia, Thailand (called Siam until 1939) was able to maintain its political independence throughout the colonial period. It provided a buffer between British and French interests, losing territory to the British in Malaya and Burma and to the French in Cambodia and Laos but maintaining a core independent kingdom with tacit support of Britain. The ability of the Thais to adapt themselves to the changing geopolitical dynamics of the region also benefited their autonomy during this period. Although it remained independent, Siam was linked into colonial trading systems and vulnerable to the policies of the European powers.

During the 19th century, Southeast Asia was integrated into a European-centered global trading system built on longer-standing traditions of trade with the Middle East, India, and China. The colonial economies also altered land use practices in the region. The Dutch intro-duced the **Culture System** into Java in 1830, for example. This required farmers to devote one-fifth of their land and their labor to export-crop production, especially coffee and sugar, with the profits going to the Dutch government. The British and French promoted rice production to feed laborers and growing populations, and expansion was especially dramatic in the Irrawaddy Delta of Burma and the Mekong Delta of Indochina. These colonial systems also created spatial inequalities. Key ports and trading cities such as Singapore, Batavia, and Manila developed as regional cores and gateways to the world, with export agricultural regions oriented to these cores and with remote rural peripheries left in subsistence livelihoods with little investment in education or other services. Most of Southeast Asia remained under colonial control during the first part of the 20th century.

**Postcolonial Southeast Asia**   European colonial power in Southeast Asia was diminished by the Japanese invasion of the region in World War II, which reduced the image of Western racial superiority in Asia (**Figure 10.14**). In addition, the Japanese granted independence to the people of Burma in 1943 and promised independence to Indonesia on their retreat, expanding local desires for absolute autonomy from their former European colonizers. Hastened by postwar global calls for decolonization and the legacy of Japanese influence in the region, independence came almost immediately following the war to most of the region except for Indochina, where extensive French investments made them reluctant to hand over power. The Philippines was granted independence from the United States immediately after the war, in 1946. The British formally granted independence to Burma in 1948 and created the Federation of Malaysia in 1963. Singapore left the federation to become an independent country in 1965. Brunei converted from a British protectorate to an independent nation in 1983. Indonesia was granted independence by the Dutch in 1949 only after a violent struggle, following a declaration of independence in 1945. Western New Guinea, formerly Dutch New Guinea, became part of Indonesia as Irian Jaya in 1963.

In the postcolonial period, new independent countries faced great challenges in forging a sense of national unity within a global framework of bounded nation–states. In Indonesia, the concept of **pancasila**—unity in diversity through belief in one God, nationalism, humanitarianism, democracy, and social justice—as the national ideology and the promotion of a singular Indonesian language were used to try to unify the country across myriad islands and cultures. Fractures soon developed around religious differences and in response to government repression of criticism and political opposition. In 1965 a communist coup against Indonesian leader Sukarno was unsuccessful; at least 500,000 people associated with the communist movement were massacred in the aftermath. Similarly, the Philippines has struggled to develop a unified national identity, as religious and ethnic tensions remain, particularly on the island of Mindanao, where a majority Muslim population struggles against the centralized Catholic-domi-

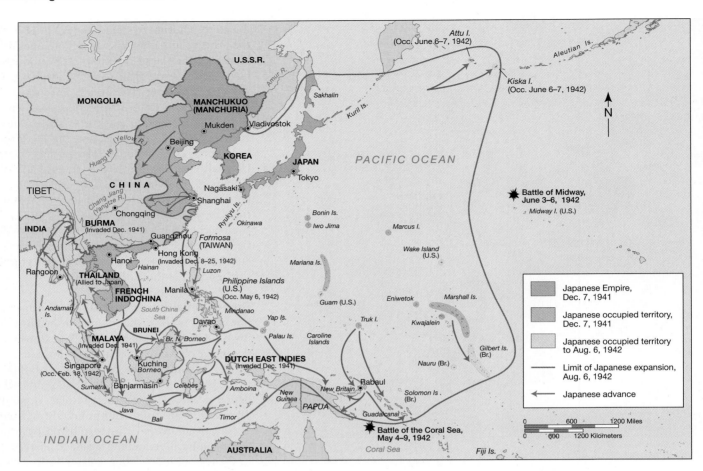

▲ **FIGURE 10.14  Japan's Occupation of Southeast Asia**  The Japanese occupation of Southeast Asia during World War II had severe effects. Japan sought access to the natural resources of the region, claiming legitimacy from a shared set of cultural values with the regions that they captured under the slogan, "Asia for the Asiatics." But the Japanese occupation cut off trade revenues, used forced labor, and diverted resources to Japan at the expense of local economies and food security. The destruction of bridges and roads during the conflict left Southeast Asia's infrastructure in ruins. In Vietnam, the Japanese requisitioned rice and forced farmers to grow jute fiber, resulting in a famine that killed more than two million people in 1944 to 1945. The Japanese were particularly harsh on the Chinese population of Southeast Asia.

nated Philippine state. Less discussed, but no less important, is the struggle in southern Thailand, where a minority Muslim population also fights, sometimes violently, for religious freedom and political autonomy. In Burma, northern-based ethnic groups continue to struggle as well against the Burmese-dominated dictatorship in the central region of that country. Nationalist struggles also intersected with the Cold War in the region, which had the most global and long-lasting impacts in Vietnam, Laos, and Cambodia, where independence movements were simultaneously economic struggles for the development of locally autonomous, planned, socialist economies (see the section "Territory and Politics," later in this chapter). All these struggles demonstrate how the global process of nation building, which has always depended on a rigidly defined set of borders, remains problematic in places where colonial borders and global geopolitical conflicts cut through very ambiguous cultural spaces.

# Economy, Accumulation, and the Production of Inequality

The wars and instability in the countries of Indochina limited their economic development and trade with the world for more than 30 years. In some cases, such as in Laos, the development of a noncapitalist economy intentionally interfered with global market forces by focusing attention on social equality more so than capitalist-led economic development. Meanwhile, other countries in Southeast Asia reoriented their capitalist economies, first to import substitution and later to export manufacturing to take advantage of the postwar global economy. In the mid to late 20th century, Malaysia, Singapore, and Thailand provided examples of how government-led economic development policies brought Southeast Asia into a new relationship with the global capitalist system and the international division of labor (see Chapter 1, p. 21).

**The Shift from Import Substitution to Free Trade** In the early years after independence, governments pursued import substitution industrialization (ISI; see Chapter 7, p. 252) policies to develop domestic manufacturing rather than rely on imported goods. In Malaysia, tariffs on imports such as clothing and plastics were increased dramatically to protect locally produced goods from global competition. Dominated by non-Malay (especially Chinese) investment and producing low-value goods that quickly saturated domestic markets or competed, in the case of steel, with global overcapacity, the economic gains of ISI did not benefit much of the population or the national trade balance. Malaysia, Singapore, and Thailand all shifted to export-oriented industrialization. This meant trying to profit from their competitive advantages in the global economy, especially a low-cost but relatively well-educated workforce. Strong state involvement and incentives and high levels of foreign direct investment (FDI) began to characterize economic development. Over time, more countries have taken advantage of FDI. In 2007, for example, Singapore received over 24 billion USD (United States Dollars) in FDI, up from just under 14 billion USD in 2005, whereas Vietnam has seen increases in FDI nearly triple, from two billion to almost seven billion USD.

To take advantage of foreign capital, Malaysia implemented their New Economic Policy in 1971. It set out to shift the economy from primary commodities to higher value exports and to distribute the benefits of economic development to the ethnic Malay population called **Bumiputra** ("those of the soil"). This program explicitly discriminated against ethnic Chinese populations. Incentives for foreign investment included tax breaks and freedom from customs tariffs when locating within a **free trade zone (FTZ)** or export processing zone (EPZ; see Chapter 7, p. 260). From 1980 onward, foreign investment flowed into a range of Malaysian economic sectors, and the share of manufacturing in exports increased from 20 to 80% by 2003. As a result, the city of Penang shares characteristics with Singapore in that it has become a center for high-technology manufacturing for the computer industry. Government investment in training skilled labor and generous tax and free trade incentives have attracted multinationals such as Intel, Sony, Philips, Motorola, and Hitachi to the island. Kuala Lumpur, located about halfway between Penang and Singapore, originally developed as a center for the colonial tin industry. Now, as the capital of Malaysia, it has a population of about 1.5 million and is the center of a manufacturing region called the Klang Valley Conurbation—an extended urban metropolitan area consisting of more than one city—that includes a free trade zone. As a result, manufacturing sectors that developed or relocated to Southeast Asia included automobile assembly, chemicals, and electronics. Japan was a major source of foreign investment, including the relocation of Japanese-owned firms seeking cheaper labor and land. This "offshore manufacturing" included 250 Japanese firms in Malaysia by 1990, for example.

**The Little Tigers** The rapid growth of a number of Southeast Asian economies, averaging 8% per year in the early 1980s, was seen as part of the larger East Asian economic miracle. The more successful countries—Thailand, Malaysia, and Indonesia—joined the "Asian tigers" (see Chapter 8, p. 289) of Hong Kong, South Korea, Taiwan from East Asia, and Singapore (see Signature Region in Global Context, p. 354). These "**little tigers**" were termed newly industrializing economies (NIEs) rather than "developing" or "underdeveloped" countries. High rates of savings, balanced budgets, and low inflation were all indicators of a successful transition to industrial economies. Foreign investment flowed into Southeast Asia to take advantage of domestic markets for consumer goods, such as automobiles and soft drinks, and of valuable natural resources, such as oil and minerals. In some cases, new industries were developed locally to meet national demand, as is the case with the Proton, Malaysia's "national car," although this industry has benefited tremendously from international trade protection, keeping the cost artificially low (see **Figure 10.15**).

**The Economic Crisis of 1997** The risks of close financial and trade linkages to the global economy became dramatically evident in the 1997 collapse of Southeast Asian economies. The collapse was preceded by a slight decrease in competitiveness, as wages increased in Southeast Asia compared to low-cost labor in China, and by extensive international borrowing by governments and banks at low interest rates. Thailand was the first economy to be affected, as the currency value fell sharply. High interest rates stopped economic development, and the country fell into recession, with massive job layoffs. Furthermore, global currency speculators "attacked" Southeast Asian currencies, as they believed these currencies to be artificially inflated by the protectionist measures of local governments in the region. In the ensuing panic, 12 billion USD was withdrawn from the region. Malaysian markets lost 80% of their value over a two-year period. The crisis in the region was exacerbated by overinflation of stocks, speculative real estate development, and over-lending by banks freed from government oversight.

The effects of the 1997 crisis were most severe in Indonesia, where the rupiah currency lost 80% of its value as the country also faced a severe drought caused by El Niño. Food prices increased; government spending cuts resulted in the removal of subsidies on gas and kerosene. Public resentment focused on the ethnic Chinese, who have long been a target of attack for local people, who saw them as part of the broader colonial project. Many ethnic Chinese fled the

▲ **FIGURE 10.15 Malaysia's National Car Industry** Proton Saga cars rolling out from its assembly line in Shah Alam on the outskirts of Kuala Lumpur on June 27, 2008. Proton intends to fit all its cars with natural gas vehicles (NGV) kits in an attempt to reduce dependency on global oil production. The development of an indigenous car industry has long been the priority of the Malaysian government, which sees such heavy industry production lines as a key aspect of the economic autonomy and national consciousness-building efforts. The signing of the Free-Trade Agreement in ASEAN may have a dramatic impact on the production of the Proton, which has long benefited from strong government protection from outside competition.

country, abandoning their shops and businesses, creating further economic scarcity. Moreover, President Suharto was forced to resign as a result of the crisis. Unemployment reached 40%. Those living below the poverty line increased from 10% to more than 50% of the population.

The International Monetary Fund agreed to help restructure debt and stabilize the economies as long as strict structural-adjustment policies were followed, including the reduction of corruption and removal of tariffs. Indonesia received the third-largest bailout, requiring strict cutbacks in government spending. Thailand and Malaysia responded to the crisis with austerity measures, including reductions in government spending and appeals to the public to accept reduced services and increased prices of basic goods.

Despite the crisis, Southeast Asia remains highly integrated into the global economy. One overall indicator of this linkage is the relationship between the value of exports and the value of GDP—the total value of all materials, foodstuffs, goods, and services produced in a country in a particular year (see Chapter 1, p. 23). Singapore, with exports valued at 170% of GDP in 2004, and Malaysia, at 121%, are the most integrated into the global economy through exports, followed by Indonesia (31%), the Philippines (52%), and Thailand (71%). In terms of key global commodities, Southeast Asia produces more than half of the world's rubber, coconut, tin, palm oil, and hardwoods. This integration means that many of the economies in this region are susceptible to global slow-downs or recessions (**Figure 10.16**).

**The Global Economic Crisis of 2008 in Southeast Asia**    There is no doubt that the global economic crisis of 2008 has also had a significant impact on the economies of Southeast Asia. Ten of the 11 countries in the region will report a negative change in GDP per annum in 2009, and several countries have reported significant increases in the relative rate of inflation between 2007 and 2008 (see Figure 10.16). The economies of Southeast Asia remain intimately linked into the global economy through the production of inexpensive goods and exports. It is likely the global recession will have a further impact on the manufacturing sector in the region as well as the service sector, particularly in the area of tourism, which is the largest foreign exchange earner for some countries in the region, such as Thailand. Although export-led production helped propel Southeast Asia out of the crisis in the late 1990s, it is less likely that exports will provide the boost that it did in the 1990s, as importing countries enact protectionist measures to defend local markets against less-expensive Southeast Asian commodities. The United Nations Economic and Social Commission for Asia and the Pacific has outlined the short-term and long-term impacts of this crisis for Southeast Asia, highlighting the fact that volatility in fuel and food markets coupled with declining demand (both locally and globally) places the region in a precarious global position moving forward. Moreover, the Commission points out the longer-term impacts that global climate change will likely have on the economic and social outlook of the region, suggesting that these changes will increase the vulnerabilities of people whose livelihoods depend on water resources for agriculture and fisheries development, for example.

**Social and Economic Inequality**    Despite the economic gains of the last half of the 20th century in the region, there are large variations in economic and social conditions between Southeast Asia and other world regions and within Southeast Asia itself. Southeast Asia includes countries that have some of the world's highest per capita GDPs (for example, in

2006 the GDPs of Singapore and Brunei were more than 30,000 USD) but also some of the lowest (for example, Timor-Leste's 2007 per capita GDP was only 400 USD, whereas for Laos, the per capita GDP for the same year was 700 USD). The distribution of income tends to be most unequal in the strongest economies. For example, the wealthiest 20% control around half the wealth in Malaysia, the Philippines, Singapore, and Thailand. Wealth concentration in Southeast Asia has been associated with **crony capitalism**, in which leaders allow friends and family to control the economy, and **kleptocracy**, in which leaders divert national resources for their personal gain.

There are wide variations in how poverty is defined in the region, but there do seem to be general improvements in overall conditions. For example, Malaysia reduced the proportion of people living in poverty from 60% in 1970 to 5% by 2002. Geographer Jonathan Rigg suggested that the poorest and most disadvantaged populations include those living in areas of extractive industry, such as timber and mining, indigenous groups, religious and ethnic minorities, guest workers, refugees, and the young and elderly. He pointed out that some are disadvantaged by who they are (ethnicity and religion) and others as a result of their occupation (in an urban informal sector such as sex worker or garbage picker), but that there is considerable variation within the region and within each group. Urban residents tend to be much better off than rural dwellers. In Malaysia and Thailand, for example, about a quarter of rural people live in poverty, compared to less than 10% of urban residents.

In terms of economic and social indicators, women are more equal to men in Southeast Asia than in many other regions. Female literacy averages more than 85%, only 10% less than the male rate, compared to an average for women of less than 50% in sub-Saharan Africa and South Asia. Despite these relatively positive measures, many workers in Southeast Asia are exploited in terms of wages and labor conditions compared to workers producing similar goods in North America and Europe. Many work 12-hour days and seven days a week without benefits or unions, making products such as clothing and electronics for global markets. International companies such as Nike have come under criticism for sweatshop labor practices in countries such as Vietnam and Cambodia, where wages are less than $2 a day, and workers earn less than 5% of what workers earn producing similar goods in the United States.

# Demography, Migration, and Settlement

The historical geography of Southeast Asia is further complicated by the region's demographic diversity. Population distributions remain uneven, as urbanization has taken hold at a faster rate in some parts of the region, whereas rural areas remain less populated, but no less affected by the changes taking place at the national, regional, and global levels. Global and regional economic processes are constantly at work creating new networks of migration in and out of the region, as people from Southeast Asia seek new lives in other regions of the world and people move within the region from places of relatively high instability to those of relative stability. The nature of demographic change in the region is thus fluid. Life expectancies across the region are also quite different, and ageing populations are being asked to take on new responsibilities as the changing realities of family life shift in relation to the distribution of epidemic diseases such as HIV. In particular, in parts of Southeast Asia the HIV epidemic has left some towns and villages with few working-age adults, placing the burden of raising children in the hands of age-

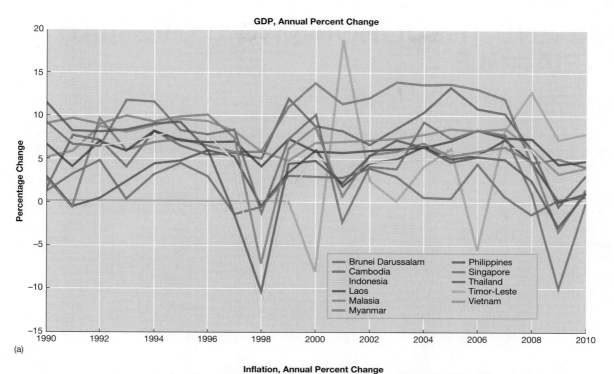

(a)

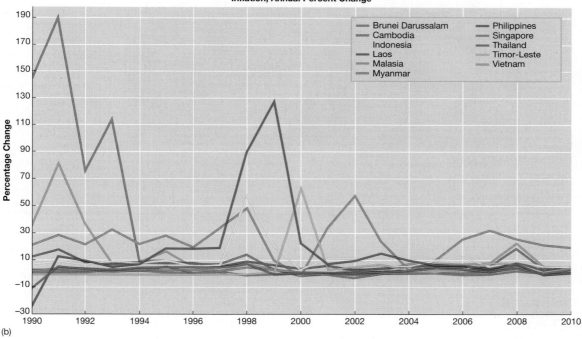

(b)

▲ **FIGURE 10.16  Gross Domestic Product and Inflation Annual Percent Change**  (a) GDP, Annual Percent Change measures the change from one year to the next in total GDP. The 1997 economic crisis clearly had a marked impact on the change in GDP between 1997 and 1999. The crisis of 2008 is also clearly represented on this graph, as all countries in the region have seen fairly distinct decreases in relative GDP since the crisis began. (b) Inflation, Annual Percent Change represents the relative change in the cost of goods and services from one year to the next. Although most of Southeast Asia has seen inflation range between 0 and 10% per year between 1990 and 2010, countries in economic and/or political turmoil, such as Cambodia in the early 1990s and Laos in the late 1990s, have witnessed extreme fluctuations in relative inflation. In both graphs, Timor-Leste has zero percent GDP and inflation change until 1999. This is the year that Timor-Leste became an independent country. Prior to independence, it was part of Indonesia. The experience of independence had a clear and distinct impact on GDP and inflation rates immediately following independence.

# SINGAPORE AND THE JOHOR-SINGAPORE-RIAU GROWTH TRIANGLE

Singapore, a small, flat, marshy island of 4.5 million people that has been drained and developed to become the principal port and business center in Southeast Asia, is one of the 25 wealthiest countries in the world in terms of per capita gross domestic product (GDP) and has the highest standard of living in Asia. The superior infrastructure—especially the excellent port and international airport—has made it the import and transshipment center for the region (often termed an entrepôt) and the busiest port in the world. The international business executive flying into Singapore's modern airport sees towering high-rise offices, luxury hotels, and large-scale industrial facilities (**Figure 1**).

When the British colonized along the shores of the channel between the Malay Peninsula and Indonesia, Singapore became the jewel of their Southeast Asian colonies, with its strategic location on the major route from the India to China. After independence from Malaysia in 1965, the government of Singapore made the attraction of foreign firms and investment a priority. It offered a low-cost, hard-working, and compliant workforce by repressing labor movements and

limiting civil rights, creating an ideal space for labor-intensive manufacturing based in newly designed industrial parks. The country was successful in attracting companies that manufactured textiles and electronic equipment, such as TVs, and the economy began to grow at more than 10% per year.

▲ **FIGURE 1 Singapore** View of the city of Singapore looking across older colonial buildings to the modern high-rise city core with harbor cranes at loading docks in the distance.

As wages increased, labor shortages developed, and trade barriers grew internationally in the late 1970s, these industries became less competitive. The government of Singapore made the astute decision to move to higher-technology activities that required a skilled workforce, such as precision engineering, aerospace, medical instruments, specialized chemicals, and most important, the emerging computer industry. To support this policy, the government invested in education, especially engineering and computing, in high-technology industrial parks, and in some state-owned pilot companies, such as Singapore Aerospace. The strategy was successful. Singapore is now a **world city** and a center for information technology and aerospace, producing many of the world's hard drives and sound cards for companies like Seagate, as well as computer memory chips and software for Hewlett-Packard and Apple. It has also diversified its service sector to include a wider range of tourism, financial, communications, and management activities and has attracted the regional headquarters of many multinational corporations. This diversification, as

ing grandparents, who historically would have been cared for by their own children.

**Population and Fertility** The population of Southeast Asia was estimated at about 600 million in 2010. The country of Indonesia, with a population of 233 million in 2010, is the world's fourth largest. Vietnam and the Philippines will each have about 88 million people, followed by Thailand at 67 million, Burma at 61 million, and Malaysia at 28 million. Population growth and life expectancy in Southeast Asia surged with the eradication of malaria and improved medical care after about 1950. After several decades of growth at more than 2% per year, overall population growth has slowed to 1.6% per year, mainly as a result of significant fertility declines in Indonesia, Thailand, Singapore, and Vietnam. In these countries the total fertility rate (average number of children born to a woman of childbearing age) has fallen from

more than 6 to around 3. In Singapore, where the fertility rate has fallen to 1.4—below the replacement level of 2.1—and population growth is negative, the government is promoting marriage and childbearing, especially among the highly educated.

Fertility rates have remained higher, at 3.3, in the Philippines, partly as a result of opposition to birth control by the Catholic Church, and in Malaysia, where the government encourages the ethnic and Muslim Malay population to have at least 5 children per married couple. Fertility rates are also higher in the poorer countries of the region, including Laos, where the fertility rate is 4.8, and Timor-Leste, where the same rate is almost 6.7. Southeast Asian data support theories that fertility decline is associated with higher income, lower infant mortality, and higher status of women. Cambodia, Laos, and Timor-Leste have lower GDP per capita, higher infant mortality, and lower levels of female literacy and schooling than do other countries in the region. Until recently, however, population well as high consumption levels in the domestic market, allowed Singapore to cope with the 1983 oil crisis and 1997 Asian financial crisis somewhat more easily than other countries in the region.

In recent years Singapore has invested in the wider region. Manufacturing has expanded to a new Growth Triangle (called **SIJORI**) that includes Johor Baharu in Malaysia and the Riau Islands (especially Batam) of Indonesia. It has been promoted as a "borderless" growth area and an example of how to prosper in a globalizing world. Geographer Matt Sparke suggested that the growth triangle has allowed Singaporean corporations to escape the high-wage, resource, and land costs of Singapore (**Figure 2**). Trade restrictions have been eased within the triangle so that labor- and land-intensive activities are carried out in the province of Johor and Batam, with value and business services added in Singapore in a flexible production system. Singapore obtains half of its water supply from Johor and has developed agroprocessing industries in the Riau Islands that supply Singapore with pork, chicken, prawns, and flowers. Singapore investors are also active in Thailand and China, while serving a globalized economy that connects Singapore with investors from around the planet. ■

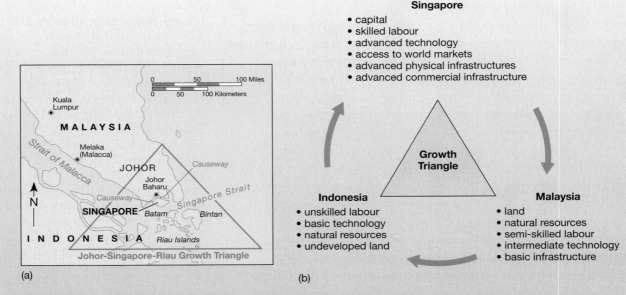

▲ **FIGURE 2  Malay Peninsula and Singapore** (a) The Malay Peninsula and Singapore are a center for economic development in Southeast Asia, especially the Johor-Singapore-Riau growth triangle. (b) The economic complementarities of Singapore; Riau, Indonesia; and Johor, Malaysia.

growth in Cambodia, Laos, and Vietnam was reduced by high death rates from war and famine.

Indonesia's population policy is often promoted as a model for noncoercive family planning. Indonesia promotes two-child families through advertising, grassroots leadership training, and free distribution of birth-control pills and condoms. Population growth rates in Indonesia have fallen from 5.2% between 1970 to 1975 to 2.4% between 2000 and 2005.

The map of population distribution shows that people are concentrated in the river valleys and deltas and on the island of Java (**Figure 10.17**). The highest population densities (people per hectare) are in Singapore, the Philippines, and Vietnam. Levels of urbanization range from 100% in the city–state of Singapore, to more than 50% in Malaysia and the Philippines, to less than 25% in Cambodia, Laos, Thailand, and Vietnam.

**Southeast Asian Diaspora**   Three major types of migration exist within the countries of Southeast Asia. The first is the worldwide phenomenon of people flowing into the cities from rural areas. They are driven by landlessness and agricultural stagnation or pulled by the attractions of urban areas, including job opportunities, education and health services, and access to consumer goods and popular culture. This has commonly resulted in increasing urban primacy, with the development of megacities, such as Bangkok, Thailand, or Manila, Philippines, each of which has significant populations and large slum areas where many recent rural migrants first settle (see further discussion later in chapter).

The second set of flows arises from war and civil unrest within countries, such as the mass evacuations from cities in Cambodia under the Khmer Rouge and from war zones in Vietnam. War, ethnic and religious conflict, and poverty have forced thousands of people to move within the region between 1970 and 1995. Included in the total were

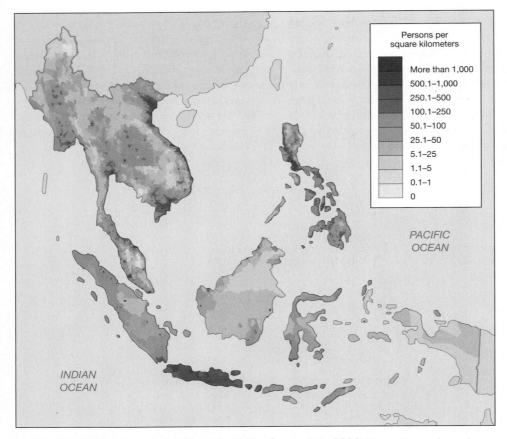

▲ FIGURE 10.17 Population Density in Southeast Asia 2000 The majority of people of Southeast Asia live in or near river valleys and deltas on the islands of Java, the Philippines, and Singapore.

more than 300,000 Laotians and 370,000 Cambodians to Thailand, about 300,000 Vietnamese to China and Hong Kong, 300,000 Muslims from Burma to Bangladesh, 200,000 Muslims from the Philippines to Malaysia, and 110,000 Burmese to Thailand. The UN High Commission for Refugees reported that in 2005 there were 120,000 Burmese in refugee camps in Thailand, 20,000 Burmese refugees in Bangladesh, and 300,000 Vietnamese in China (see **Figure 10.18**). There is also a flow of labor migrants among countries, with more than 500,000 Indonesians working in the Malaysian construction industry and 400,000 Thais working in Singapore and Malaysia (see Signature Region in Global Context Box).

The third, and distinctively Southeast Asian, pattern is the active resettlement of populations from urban to rural areas, especially in Indonesia. In 1950, the Indonesian government initiated a *transmigration* program, which redistributed populations from densely settled Java and the city of Jakarta to reduce civil unrest, increase food production in peripheral regions, and further goals of regional development, national integration, and the spread of the official Indonesian language. It is estimated that more than 4 million people moved to the Moluccas, Sulawesi, Sumatra, and Kalimantan, with 1.7 million of the migrants receiving official government sponsorship in the form of transport, land grants, and social services (**Figure 10.19**). The Philippines followed similar

resettlement programs in subsidizing migration from Luzon to frontier regions such as Mindanao. Geographer Thomas Leinbach identified a number of problems with the transmigration program. He included lack of infrastructure in the new settlements, conflict between the settlers and indigenous groups, and the destruction of forests as peasant farmers accustomed to the fertile soils of Java cleared more rain forests as their crops failed because the soil degraded after deforestation. Malaria, pests, and weed invasion also hindered the success of the program.

**International Migration to and from Southeast Asia** International migration to and from the region has a long history. The most important flow into the region has been the centuries of movement of Chinese into Southeast Asia, beginning as early as the 14th century. Driven by civil wars, famine, and revolution in China, more than 20 million Chinese moved to Southeast Asia during the colonial period to work as contract plantation laborers harvesting rubber and to work in mines and railways. Imported labor was employed to clear forests and build irrigation and drainage systems in these vast deltas, which became the **rice bowls** of Southeast Asia. This persistent stream of migrants from India and China accelerated in the 19th century, a third major process that changed the geography of Southeast Asia. Thousands of workers were brought in to convert the deltas to rice production, to mine tin and other minerals, to manage the services in major ports, and to develop small businesses to serve colonial and local demands for consumer goods. In some cases, a shortage of European administrators was filled when rights to collect taxes and harbor duties, to market opium, or to operate gambling were auctioned by colonial governments and purchased by the Chinese. Such activities were valuable revenue generators. These migrants, who were seen as more entrepreneurial and fit to govern by some colonial administrators, became a core of the colonial economies and are the ancestors of the large populations of South and East Asians, and especially Chinese, that live in Southeast Asia today.

Many of these migrants shifted into jobs in retail and trading and as clerical employees of the colonial trading companies. These so-called **overseas Chinese** became essential to the success of the colonial economy, and on independence they became the entrepreneurs who ran banks, insurance companies, and shipping and agricultural businesses. As a result of this massive in-migration, ethnic Chinese make up a large percentage of the overall population in Singapore (77%), Malaysia (26%), and Thailand (10%). In other countries, such as Indonesia, ethnic Chinese make up a large percentage of the economic elite, even when they are a small percentage of total population. The Chinese generally live in urban areas, in separate neighborhoods often known as Chinatowns, with their own social clubs and schools.

Out-migration from Southeast Asia includes large refugee flows and many thousands of labor migrants (**Figure 10.20a**). Since 1974, 1.5 million people have left Vietnam as refugees from war and commu-

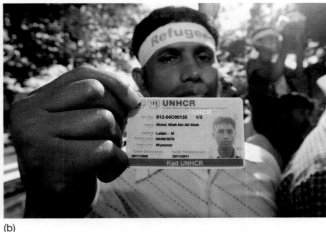

(a)                                        (b)

▲ **FIGURE 10.18  Rohingya Refugees**  (a) In January 2009, 78 ethnic Rohingya, a Muslim ethnic minority population from the Arakan region of western Burma, were released from a Thai prison in Ranong Province after entering Thailand illegally. This is but a small fraction of the estimated 20,000 people of Rohingya descent living in Thailand. This diaspora is a result of the long-term repression of ethnic minorities in Burma by the government. Like many ethnic minority refugees in this region, the Rohingya are "stateless," having been driven from Burma and often not accepted in neighboring states, such as Thailand. (b) The Rohingya are engaging in a campaign to limit the repression by the Burmese state. In this image, a Rohingya man holds up an identity card he received from the UN High Commission on Refugees while at a protest in front of the Burmese Embassy in Kuala Lumpur, Malaysia.

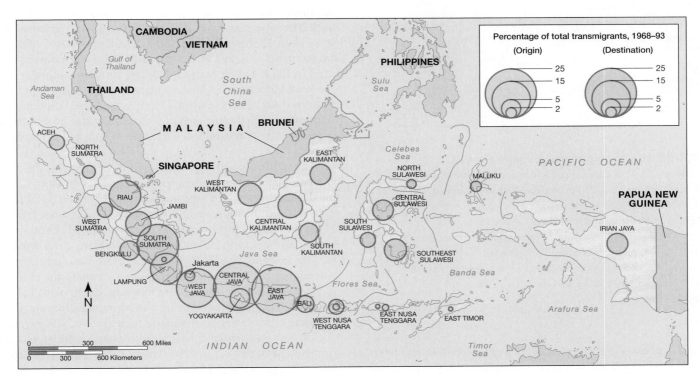

▲ **FIGURE 10.19  Transmigration Flows in Indonesia**  The Indonesian government has relocated thousands of people from urban areas on Java to rural areas of the islands of Moluccas, Sumatra, Sulawesi, and Borneo.

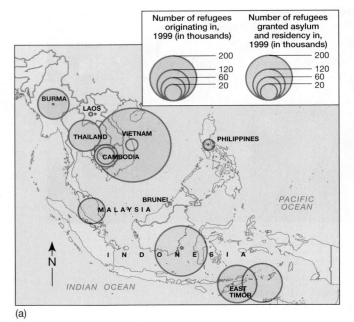

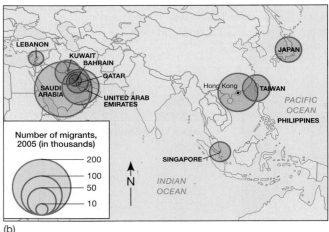

▲ **FIGURE 10.20  Migrations in Southeast Asia**
(a) Southeast Asia has experienced enormous flows within and from the region, including out-migrations from Vietnam, Burma, Laos, and Cambodia to Thailand and beyond the region, from the Philippines to Malaysia, and from Timor-Leste to Indonesia. Between 1999 and 2007, the total refugee population was around 800,000. (b) This map shows where Philippine labor migrants were moving in 2005 and illustrates the significance of remittances from workers in Japan, Hong Kong, and the Middle East to the Philippine economy.

nism. About half of the Vietnamese refugees went to the United States; France, Canada, and Australia accepted others. U.S. cities such as San Francisco have large Vietnamese populations living in distinctive neighborhoods with a strong Vietnamese heritage. Some Vietnamese refugees remained in refugee camps in Hong Kong. After Vietnam invaded Laos in 1975, more than 300,000 Hmong people fled to Thailand, fearing persecution because they had supported the Americans; they were subsequently resettled in the United States. Recently, with peace and

reform, many Vietnamese are returning to Vietnam, and some bring capital for investment.

The largest numbers of labor migrants from Southeast Asia work in the Middle East, especially in Saudi Arabia, Kuwait, and Oman, and in Hong Kong and Japan. Many Southeast Asians in the Middle East are Muslim women working in the service sector as nurses and maids. Philippine men have a tradition of joining the merchant marines of many countries and working as cooks, seamen, and mechanics. Thousands of Thais and Filipinos work in Hong Kong, where their wages are typically much lower than those of local laborers. Many Filipina women work as maids or nannies in North America, Europe, and Singapore.

The money sent back by these labor migrants (remittances) is very important to both national and local economies. For example, it is estimated that $11 billion was sent back to the Philippines in 2005 by the 10% of the population working abroad (**Figure 10.20b**). Thailand's remittances are worth more than $1.5 billion each year.

**Urbanization**    About half of Southeast Asia's people live in cities, most of which have grown rapidly since 1950. Southeast Asia is dominated by several enormous cities with large metropolitan-area populations, among them Bangkok (6.7 m), Ho Chi Minh City (5.4 m), Manila (11.1 m), and Jakarta (9.2 m). These cities have grown so rapidly that they have serious problems of overurbanization (see Chapter 1), with insufficient employment opportunities, an inadequate water supply, sewerage problems, and inadequate housing. With a wider metropolitan population of more than 9 million people, Bangkok, for example, is the current hub of mainland Southeast Asia and is a clear example of primacy. The city dominates Thailand with 32% of the urban population (more than 16 times the size of the next largest city), 90% of the trade and industrial jobs, and 50% of the GDP. There are more than 25,000 factories in the Bangkok metropolitan region, many of them labor-intensive textile producers. Although manufacturing grew rapidly from 25% to more than 70% of national exports between 1980 and 1998, the economic development of Bangkok is now tied to the expansion of its public transportation infrastructure as well as its new international airport.

The unification of North and South Vietnam left the new country with two major urban centers—Hanoi and Ho Chi Minh City (formerly called Saigon). New administrative capitals have been established to promote decentralization in Quezon City (Philippines) and Putrajaya (Malaysia), but their proximity to Manila and Kuala Lumpur creates one large urban region in each case. Key secondary cities with more than 500,000 residents include Palambang, Medan, Bandung, Ujung Pandang, and Surabaya in Indonesia and Cebu and Davao in the Philippines.

Many cities in Southeast Asia mix local design with that of colonial and modern planning, but they are also surrounded by unplanned growth, including desperately poor squatter settlements. Geographer Terence McGee, who has written at length on cities in Southeast Asia, argues that the region has a distinctive urban landscape in the form of extended metropolitan regions (called **desakota**). Cities have extended into intensively farmed agriculture, especially rice paddies; town, industry, and agricultural villages have become intermixed. McGee's work reminds us that boundaries between the city and the countryside in Southeast Asia are often blurred. There are many small industries, such as textiles, in rural areas and considerable agricultural production in the cities, and new suburban developments are taking the place of rural farmland.

# CULTURE AND POLITICS

Southeast Asia is a region marked as much by its cultural and political diversity as its similarity, a legacy of this region's complex relationship to wider global historical and demographic patterns of movement and exchange. This means that any study of Southeast Asia must also appreciate the cultural and political complexity of this region in historical context. For example, although insular Southeast Asia, particularly Indonesia, was long distinguished by its practices of Hinduism and Buddhism, today it is the world's largest Islamic country. At the same time, on the island of Bali, people still practice Hinduism, whereas in parts of Borneo, animistic practices are still maintained. In studying the geography of religion, we must comprehend the changing historical geographies of this region and how these geographies are informed by their interactions with other places and people. This section begins by tracing the modern-day religious and linguistic geographies of the region and how these cultural geographies create complexity and a certain sense of cultural cohesion and cultural disconnect regionally. Next, this section examines local gender, sexual, drug, and ethnic politics in the region. Southeast Asia is a region that has a number of dominant ethnic groups that sometimes suppress minority voices. It is important to examine these social geographies to illustrate how the processes of ethnic identity are interrelated with the politics of nation–states. Finally, this section broadens its scope to investigate the national and regional politics that are both cleaving this region apart and bringing it closer together.

## Tradition, Religion, and Language

To talk about culture in Southeast Asia is to begin with a discussion of religious and linguistic complexity. Although the maps presented in this section create a representation of broad spatial patterns of religious and linguistic differences, these regional maps only provide a partial story of everyday cultural religious and linguistic practice. Take, for example, an average day in northern Thailand, where in a local village a health-care worker might wake and begin her day with a Buddhist prayer. Next, she might attend a meeting with local village healers and participate in a ceremony blessing a local forest, borrowing from a long set of animistic practices. Her day next turns to her work in a hospital, where, on the one hand, she speaks in central Thai (*phasaa klang*) with the head physician, who is from Bangkok, and in northern Thai (*kam muang*) with her patients, who are from the local area. On an unusual day, she might also find herself in a meeting with international visitors, who work through translators and her own English skills to discuss health-care programs for people living with HIV and AIDS in the area.

It is important to keep this complexity in mind and use this as a caveat in the interpretation of the cultural practices discussed here. As was examined earlier, the religious and linguistic geographies of this region are tied to a much longer set of global processes, which has brought Buddhism, Hinduism, Islam, and Catholicism to the region. Moreover, the linguistic geographies of the region are tied to the very long-term migration of Malayo-Polynesian peoples (Malay, Indonesian, or Tagolog) and Austro-Asiatic speakers (Khmer, Mon, and Vietnamese) to more recent migrants of Tai speaking peoples (Thai, Lao, or Shan) to even more recent in-migration of Chinese (e.g.,

Hakka) and Indian (e.g., Tamil) speakers. English is also spoken widely throughout the region, particularly in urban areas and among the region's elite. English is one of the four official languages of Singapore, which also includes Mandarin, Tamil, and Malay as part of its national linguistic tradition. In constructing this regional geography of language and religion, then, it is important to keep in mind that this geography is a partial representation of the broader historical geographies of global interconnection that distinguish the cultural diversity of Southeast Asia today.

**Religion**  The contemporary religious geography of Southeast Asia reflects centuries of evolution under Indian, Chinese, Arab, and European influences (**Figure 10.21**). Generally, Buddhism tends to dominate the mainland region and Islam the islands. Islamic believers in Indonesia outnumber those in the Middle East, although the practices are sometimes seen as more liberal. Islam is also widespread in Malaya and is growing in the Philippines (e.g., Mindanao). Intensification of Islamic belief has resulted in the seclusion and veiling of women; in political conflict over the enforcement of Islamic law, especially in diverse populations; and in terrorism linked to global post–9/11 tensions. In 2002 the bombing of a nightclub in Bali by Islamic terrorists killed 202 people, and further attacks in Indonesia's tourist regions have been linked to radical groups, including the most recent attacks on hotels in Jakarta's tourist district in July 2009.

Hinduism is common in Bali, whereas Christianity is the religion of 85% of the Philippine population as a legacy of Spanish colonialism. Four hundred years of domination by Spain and the United States resulted in several distinctive characteristics, including a majority Catholic religion, mainly Spanish names, highly concentrated land ownership, an agriculture oriented to exports of sugar, tobacco, and pineapples to the Americas, the widespread use of English, and a general orientation to the West, especially the United States. Christianity is also found among the hill tribes of Burma and in Vietnam, converted by French and British missionaries who could not make inroads into the Buddhist beliefs of lowland residents. Remote indigenous groups have maintained animistic beliefs that imbue nature with spiritual meaning, especially in the mountains of Burma, Laos, and Vietnam and in Borneo and Irian Jaya. Religion and cultural tradition intersect in the reverence for the monarchy in countries such as Thailand and Cambodia. The Thai royal family is held in high, almost godlike esteem, with pictures of the king in the majority of homes. The royal family has selectively intervened during times of political crisis and plays a balancing and leadership role.

**Language**  Southeast Asia accommodates an incredibly diverse set of cultures, which is reflected in more than 500 distinct ethnic and language groups. Indonesia, for example, has more than 300 ethnic groups and languages, and the population of Laos speaks more than 90 different languages. No language has unified the region in the way that Arabic has in the Middle East or Spanish has in Latin America. The most common dialects are versions of Malay spoken in Malaysia, Indonesia, and Brunei and a version used as a lingua franca (trade language).

Some national boundaries of Southeast Asia do enclose some dominant cultural and language groups. Thailand (Thai), Burma (Burmese), Cambodia (Khmer), Vietnam (Vietnamese), Laos (Lao), and Malaysia (Malay) all have large majority populations that speak the same lan-

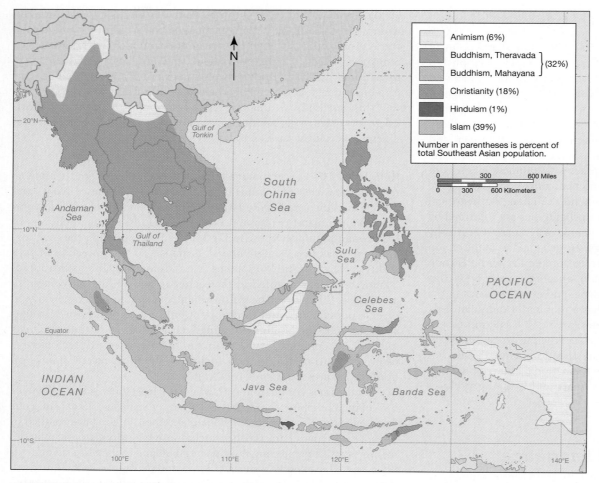

▲ **FIGURE 10.21 Map of Dominant Religions in Southeast Asia** The geography of religious belief in Southeast Asia is dominated by versions of Buddhism on the mainland and Islam on the islands. Buddhism includes both Theravada and Mahayana, the former more conservative and found mostly in Burma, Thailand, and Cambodia, and the latter associated with Vietnam. Catholicism is important in the Philippines. Some ethnic minority groups maintain animist beliefs in remote mountain and island areas.

guage and share cultural and religious traditions (**Figure 10.22**). The adoption of a common language in Indonesia is also an attempt to integrate diverse cultural groups across this wide geographic space. However, there are significant minority populations in many of these countries, including the Karen and Shan in Burma and the Hmong in Laos, as well as large populations of Chinese and Indians in many countries. Even within these groups there are distinct cultural and regional differences. For example, the Indian populations include Bengalis and Sikhs in Burma and Tamils in Malaya.

## Cultural Practices, Social Differences, and Identity

It is a challenge given this region's religious and linguistic diversity to provide an overarching set of regional characteristics on social practice. Indeed, Southeast Asia is a region for example, where one can find both **matrilocality** and **patrilocality** (i.e., where married couples move into the wife's family home on the one or the husband's family home on the other), although much of the region remains marked more generally by patriarchy. Thus, although it is fairly common for married male–female couples to live with the wife's parents, familial authority generally passes from father to son-in-law. This does give women more power because they live with their own families rather than with in-laws. And, in general, women have more authority within Southeast Asian families and societies than in many other world regions. As an example, in many parts of Indonesia women often manage the family money, although authority is not always vested in the sphere of the economy but rather in the sphere of public politics, which men tend to dominate. The value of daughters is reflected in a more equal preference for female compared to male children than in either East or South Asia, where preferences are highly biased toward males. Women also have new employment opportunities in services and manufacturing and are preferred by many high-technology companies because they receive lower wages and are perceived as more careful and docile. This has created some conflict in home communities, where men are "left behind" by female spouses that have more earning potential than their male counterparts.

Nevertheless, the broader cultures of Southeast Asia are patriarchal, and socioeconomic conditions are generally better for men than for women. Ironically, then, this means that women have more access to work, but that work comes with a lower wage. When granted access to work, men tend to receive higher wages and more education. In poorer Theravada Buddhist communities on the mainland, for example, young boys have access to free education and housing through local monasteries, whereas girls do not. This provides boys with a double benefit, as they are granted the resources of temple life, and they are able to increase the merit of themselves and their parents by training as a Buddhist novice.

In recent years families in Southeast Asia have also become more nuclear. And, they have also become more suburban, a phenomenon whereby agricultural land is slowly (or sometimes rapidly) being converted into single-family home housing communities, many of them gated (see **Figure 10.23**). This has changed patterns of obligation from a focus on parents and older people to a focus on children and from a culture of family-centered farm or industrial labor to a focus on education and service-sector work. The expansion of suburbs also means that new spaces serve a growing middle class throughout the region, including malls, restaurants, and vacation resorts. New globalized social identities are emerging in the region that are centered not around production—agricultural or manufacturing—but around consumption—such as high-end consumer goods and luxury items, such as Starbucks coffee.

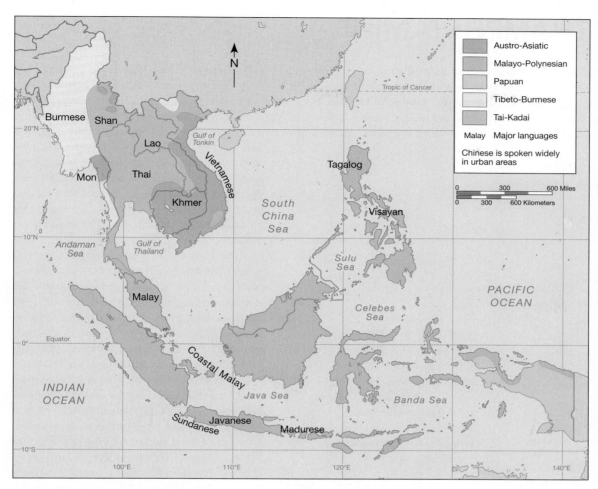

▲ **FIGURE 10.22 Map of Dominant Languages in Southeast Asia** Southeast Asia has hundreds of distinct languages, which fall into five major language families. The major language families are Austronesian (or Malayo-Polynesian), spoken in insular Southeast Asia and the Malay Peninsula, Tibeto-Burmese in Burma, Tai-Kadai focused on Thailand, Papuan in New Guinea, and the Austro-Asiatic languages of Vietnam and Cambodia.

rural and urban areas. This process has increased the desire for the trafficking of women, particularly from neighboring countries in Thailand, including Burma, Laos, Cambodia, and southern China. Some of the sex workers are very young women sold into bondage by rural families or smuggled in slavelike conditions. Studies suggest that for women from poor rural families, sex work may appear as a rational choice for making a living, providing opportunities for them to send money back to their villages and families. The explosion of commercial sex work in the region was also critical to the spread of HIV in the 1990s, as these places were ideal sites for unprotected sex. A campaign to have 100% condom usage in these venues in the late 1990s has been credited with the sharp reductions in HIV rates in Thailand. This campaign, however, failed to address the underlying gender and sexual politics that promote this industry in the first place, ironically leaving intact an industry that is illegal in Thailand. Human trafficking thus continues into this industry.

**HIV/AIDS Politics** Despite massive prevention efforts, according to UN estimates, in 2007 more than 600,000 people in Thailand were known to be living with HIV—

**Sexual Politics** Southeast Asia, particularly Thailand, is well known for its commercial sex work industry. Indeed, Thailand is a country where polygyny—having more than one wife at a time—was legal until the 1930s, when it was formally outlawed by the new constitutional government in an effort to look more "Western." The practice of having more than one wife, although illegal, has persisted for some of Thailand's elite in the informal sexual economy of *mia noi* (minor wife). This practice has long motivated an underground economy of commercial sex in Thailand. The rise of the Vietnam War, however, and Thailand's role in that war as the rest and relaxation (R&R) capital of the region for U.S. soldiers, helped construct the modern-day image of Thailand as a global commercial sex capital. This industry was organized around an image of Asian women as passive and exotic, and by the 1990s tours were being advertised in places such as Japan and Germany.

The explosion of a "high-end" commercial sex work industry serving middle- and upper-class men from Europe and other parts of Asia also promoted new markets for commercial sex work, making commercial sex work more readily available to local men. Today, inexpensive commercial sex work venues exist throughout Thailand, in both

▲ **FIGURE 10.23 Suburbanization in Southeast Asia** In places like Bangkok, Thailand, large tracts of suburban housing have been developed from once-profitable agricultural land. This is creating new upper-middle-class lifestyles and shifting class politics as well. In this image, a worker sweeps the driveway of this expensive suburban home, just one of hundreds in a large-scale suburban development.

# OPIUM AND METHAMPHETAMINE

The mist-shrouded highlands of Burma, Laos, and Thailand contain much of mainland Southeast Asia's remaining forests, but they are also known for the fields of opium poppies that historically have provided as much as half the global supply of the addictive drug known as heroin. More recently, this region, known as the Golden Triangle (see **Figure 1**), has become an important global and regional center for the production of methamphetamines (sometimes called "speed" and in Thai called *yaa baa*). The drug economy has helped local ethnic groups fight off Burmese oppression, funding rebellion against the Burmese state, and it has served as a mythic symbol of the boundary between Burma, Thailand, and Laos, where tourists now visit. It is also an area that continues to know violence, and the recent Thai government-led "war against drugs" has led to an escalation of violence in this area, particularly in rural areas that are cross-border sites for the trafficking of these drugs.

The allure of the opium poppy as a narcotic is centuries old. It was a component of trade and conflict between China and Europe, with the exchange of opium for Chinese tea, silk, and spices, creating thousands of addicts in China (see Chapter 8, p. 287). In 1905, as heroin addiction rose, the U.S. Congress banned opium and eventually prohibited both opium and heroin sales. In Thailand, an active campaign to eradicate opium production was successful, for example, but similar campaigns have largely failed in Burma and Laos. Despite these efforts, often global in scope, large markets for heroin still exist in the United States, Europe, and Australia, as well as in Thailand. The United Nations estimates that there are 11 million heroin and opium addicts worldwide, many unable to work productively and turning to crime to fund their addiction. Addiction can

◀ **FIGURE 1 The Golden Triangle** Map of the region known as the Golden Triangle, highlighting the density of poppy production in the region. Thailand, which engaged in active campaign to reduce poppy production, clearly shows the success of that program. Methamphetamine production takes place across Burma and parts of Laos, while Thailand serves as a key trans-shipment site for the drug.

just under 1% of the total population—whereas an additional 24,000 died in that same year (see **Figure 10.24**). This is by far the highest number of cases in Southeast Asia and is one result of the active sex industry and intravenous drug use (IVDU)—heroin and methamphetamines (speed)—in the country. UNAIDS reports, for example, that around 45% of the IVDU population currently in treatment for drug addiction is HIV positive. This high rate is partially the result of a strong antidrug government policy, which strictly regulates access to clean needles, putting drug users at greater risk for HIV through the reuse and sharing of their syringes. At the same time, even though Thailand is depicted as the epicenter of the HIV epidemic in the region, other countries demonstrate more alarming numbers. The HIV incidence rate for Vietnam in 2007—the rate of cases among the total population in a given year—was 0.5%. Although this rate seems relatively low, it is a significant increase for this country, which has witnessed a doubling of its HIV-positive population since 2000 and now reports HIV cases in all 64 provinces of the country. When comparing these numbers to the adequacy of the health-care systems in the region, it is easy to see how the HIV epidemic could quickly outstrip the capacity of several other Southeast Asian countries, such as Burma and Cambodia, to deal with the epidemic. Unlike Thailand, which has both a long tradition of public health and strong connections to the international HIV and AIDS outreach and care networks, other countries in the region lack the basic infrastructure to handle the growing crisis. Thailand, for example, was able to rely on its strong history of family planning, which dramatically lowered fertility rates in the 1960s and 1970s, as well as on a small cohort of activists and public health officials to expand a free condom campaign throughout the country's commercial sex work spaces in the 1990s. This clearly lowered the HIV rate as well as the rate of other sexually transmitted diseases (STDs).

It is also quite possible that underreporting of HIV cases is common in countries where repression and discrimination against HIV-positive people is widespread. This is perhaps the case in Burma, whose actual incidence rate might be significantly higher than reported. Underreporting of HIV cases is common globally and is partially a reflection of the strong stigma associated with being HIV positive. That is why country-based and regional efforts have turned, since the 1990s, to the development of social support and advocate networks for peo-

▲ FIGURE 2 Opium Harvest A woman of the Wa ethnic group in northern Burma harvests opium poppies. Ethnic minorities, many of whom are poor, have long relied on growing poppies and harvesting opium on small plots in the highlands to bring in cash. While ethnic peoples in the region historically used the drug medicinally. More recently, drug cartels control its production and distribution, using the opium "tar" to produce heroin.

result in overdoses and death from respiratory failure, and the use of shared needles to inject heroin has significantly contributed to the transmission of HIV/AIDS among addicts. Production of opium in Southeast Asia has declined in recent years, from almost 2000 tons in 1990 to only 326 tons, almost all in Burma, in 2005, as new efforts to stop its production have met with some success. Afghanistan is now the largest producer, at more than 4000 tons per year.

The reduction in heroin trafficking has not slowed the production of illegal drugs in Southeast Asia, as local drug warlords turned to new markets and commodities in the wake of the competition from opium production in other global regions. Methamphetamines, for example, are highly sought in the urban and rural markets of the region, where they are used recreationally, and in work contexts, such as in the manufacturing sector, where people want to extend their workday. Youth with disposable income, who participate in an active club scene in Thailand, use these drugs to party for long periods of time. Pierre-Arnaud Chouvy and Joël Meissonnier, in a recent study, examined how the shift from heroin to methamphetamines has created a new drug use landscape in the region. First, the production of methamphetamine is not dependent on the same crop (opium), and therefore it is more mobile. Second, methamphetamines can be used "productively" in work. It thus has a completely different

▲ FIGURE 3 Drugs Seized in the Golden Triangle Heroin (white) and methamphetamine (red), called locally *yaa baa*, are pictured here after they were seized as part of a crackdown on drug trafficking in Thailand. Although heroin is still produced in the region, new drugs, such as *yaa baa*, are now taking their place.

market. Third, the circuits of methamphetamine use have successfully piggybacked on the production of "energy drinks," which were popularized in the region in the 1990s. Estimates by 2000 suggested that hundreds of thousands of people in Thailand were using *yaa baa* regularly in work and leisure, whereas heroin use has been on the decline in the same period. ■

ple living with HIV and AIDS. Thailand was one of the first countries in the region to fund nongovernmental efforts to establish people living with HIV and AIDS (PLWHA) support groups throughout the country. These efforts were particularly strong in the north of the region throughout the 1990s, which has been the region most dramatically affected by the epidemic. Today, these support groups work in conjunction with other community-based groups, such as herbal medicine physicians and public health workers, to provide care and support for both HIV-positive people and their families. Despite these efforts, Thailand and other dramatically affected countries face a crisis in families where one or more of the parents have died from HIV, leaving a legacy of "**AIDS orphans**" throughout the region. The rising number of orphans is placing a strain on both the social service sector and the extended family networks of HIV-positive people, creating new layers of complexity for an already extensive social, political, and economic crisis of prevention and care.

**Minority Politics** Southeast Asia is often envisioned as a region made up of a few distinct cultural groups. But the region is also home to a number of ethnic groups and identities that transgress national boundaries and disrupt any neat and clear reading of this region's ethnicities. In fact, throughout Southeast Asia there are minority groups, some of whom are not considered citizens of any particular nation. This includes refugee groups displaced during the Vietnam War period, such as Hmong, who now live in parts of Thailand, or the Karen, an ethnic group that has long fought against the domination of Burma by the ethnic Burmese military junta. Many of these groups in mainland Southeast Asia are euphemistically called **hill tribes** or *chao kao* in Thai, named for their geographic concentration in the sparsely populated border regions of Thailand, Burma, Laos, and Vietnam. These groups have historically participated in dry farming and slash-and-burn agriculture, supplementing these practices with hunting-and-gathering techniques.

In insular Southeast Asia, ethnic groups often known as **sea gypsies** by the dominant powers in the region rely heavily on fishing and boat travel between islands to sustain their daily caloric intake (see Figure 10.2). These populations predate many of the modern-day dominant ethnic groups in the region, such as the Thai and Burmese, probably arriving over 5,000 years ago as part of the Australesian migrations

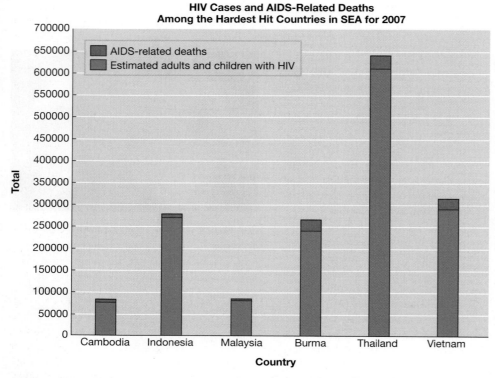

**HIV Cases and AIDS-Related Deaths Among the Hardest Hit Countries in SEA for 2007**

▲ **FIGURE 10.24 HIV Cases and AIDS-Related Deaths in 2007** This graph represents the total number of people living with HIV and AIDS-related deaths in the year 2007. Thailand clearly has the highest rate of HIV cases and AIDS-related deaths. But, the other five countries reported their first HIV cases later than Thailand. This graph represents the 2007 HIV and AIDS-related death incidence rates. It may also suggest where the epidemic is likely to grow in the near future (e.g., Vietnam is witnessing a sharp increase in the number of people diagnosed with HIV).

to the region. As was mentioned at the beginning of this chapter, the Moken, one of these migratory groups, are finding themselves under increasing pressure to settle and participate in the modern-day cash economy. For many of these people, tourism provides a way to bring cash directly to their communities, although these tourism economies are often controlled by people outside the communities. In recent years, advocates of sustainable tourism have pushed for greater autonomy for these groups over their own local tourism economies. In the meantime, many of these people live in limbo; they are not citizens of the states of Southeast Asia, and they do not have the power to fully live their lives as they see fit.

## Territory and Politics

The territorial integrity of Southeast Asia as a region appears fairly stable, although the intersections of local and regional politics and social action challenge some of this region's national governments. In fact, legacies of this region's wider global interconnectivity tied to colonialism, territorial expansionism by some countries after independence, and Cold War conflicts have contributed to continuing instability in the region. The Cold War had the most global and long-lasting impacts in Indochina, where communist groups in Vietnam first led resistance

against Japanese occupation during World War II, then against French recolonization after the war, and then against the United States during the Cold War.

**The Indochina Wars** Led by Ho Chi Minh, the communists established a separate government in the northern Vietnamese city of Hanoi. From this base, they supported a guerilla war against the French in southern Vietnam with assistance from the Soviets and Chinese. When the French withdrew from the region after a devastating loss at Dien Bien Phu in 1954, Laos and Cambodia became permanently independent. At that time, the Geneva accords temporarily divided Vietnam into North Vietnam and South Vietnam. North Vietnam became an independent country, and thousands of refugees, especially Catholics, fled to South Vietnam, which established its capital in Saigon.

While this was taking place, communist rebels in Cambodia and Laos joined guerilla forces in South Vietnam (called the Vietcong), expanding control in several regions of South Vietnam. Subscribing to a **domino theory**, which held that the communist takeover of South Vietnam would lead to the global spread of communism throughout Southeast Asia, the United States sent military advisors to South Vietnam in 1962. This was followed by the bombing of North Vietnam in 1964 and escalation to a full-scale land war, with more than half a million U.S. troops, in 1965 (**Figure 10.25**).

The Vietnam War was probably the most serious global manifestation of Cold War competition; it wrought terrible social and environmental effects in Southeast Asia as well as on U.S. domestic and international politics. More than a million Vietnamese people died, together with 58,000 Americans. U.S. forces sprayed 2 million hectares (5 million acres) of Vietnam with defoliants, such as Agent Orange, that poisoned ecosystems and caused irreparable damage to human health. Cambodia and Laos were also bombed with napalm and defoliated to disrupt communist supply lines and camps. Two million people left South Vietnam fearing repression after unification, many (called "Vietnamese boat people") sailing away in small, fragile boats, creating a global Vietnamese diaspora in Southeast Asia, Europe, and the Americas. The new unified government confiscated farms and factories to create state- and worker-owned enterprises, resettled ethnic minorities into intensive agricultural zones, and moved one million people into new economic development regions. But U.S.-led economic sanctions from 1973 to 1993 limited the potential for exports and restricted some critical imports such as medicines. At the same time, the modern-day pockmarked landscape created by the bombing has promoted the spread of disease, such as malaria because mosquitos can thrive in new pools of standing water in old craters, on the one hand, but has also provided new economic opportunities for some, on the other (**see Figure 10.26**).

During this same period in Cambodia, the Khmer Rouge Maoist revolutionaries overcame the U.S.-backed military government in 1975 and instituted a cruel regime under the leadership of Pol Pot. Their

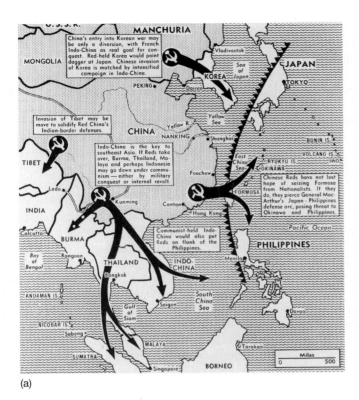

(a)

(b)

interpretation of Mao Tse Tung's revolutionary approach in China led them to suspend formal education, empty the cities, eliminate the rich and educated, and isolate themselves from the world. They renamed the country Kampuchea. Mass murders in the "**killing fields**" and the brutal "death march" out of the capital Phnom Penh in 1975–76 killed at least two million people—a quarter of Cambodia's population, including most intellectuals and professionals—between 1975 and 1979, when Vietnam invaded and installed a new government. Many Cambodians who did escape migrated to countries outside the region, including the United States, where large Cambodian populations now live in cities such as Long Beach, California, and Boston, Massachusetts. Cambodians were not the only peoples affected by the war, as thousands of Hmong refugees (a minority ethnic group living in Laos, Thailand, Burma, and southern China) fled Laos after fighting for the United States during the CIA-led covert war in that country. Today, Hmong people constitute a global diaspora with populations in places as diverse as Merced, California; Paris, France; and Guyana, South America.

**Postcolonial Conflict and Ethnic Tension**    In addition to the obvious legacies of the Cold War in the region, some of the most long-lasting conflicts are those in which ethnic minorities are seeking recognition and independence or fractured physical and cultural geographies have made national unification more difficult (**Figure 10.27**). It is thus hard to examine territory and politics outside the broader politics of social action and resistance. In this context, it is possible to look at three key examples of the relationship between territory, politics, and social action in the region to see how Southeast Asia, as a regional space, is managing its own sociocultural and political–economic difference.

In Indonesia, the outer islands have long resisted the dominance of Java, with the most serious insurrections in East Timor, Irian Jaya, and Aceh. Timor-Leste is mostly Christian, a former Portuguese colony that occupies the eastern portion of the island of Timor, at the eastern end of the Indonesian archipelago. Initial hopes for independence in Timor-Leste were dashed when Indonesia occupied the island after Portugal gave up colonial control in 1976. Twenty years of resistance and more than 200,000 deaths finally resulted in a referendum for independence in 1999. After the vote, however, anti-independence militias went on a rampage, and thousands of refugees fled to the western half Timor, which is still part of Indonesia. In 2000, Timor-Leste was granted status as a newly independent country under initial United Nations administration and with a strong presence from Australia, in particular.

Ever since independence in 1948, the Burmese government has faced rebellion from ethnic and religious minorities, who resist the dom-

◀ **FIGURE 10.25  Domino Theory**  (a) Fear of the expansion of communism in East and Southeast Asia promoted the use of map images to depict the communist threat. This map, which was produced in the 1950s, demonstrates the threat that a communist China and by association a communist Vietnam posed to U.S. national interests in East Asia. Maps such as this one would be used later to justify the extensive, costly, and deadly Vietnam War. (b) U.S. troops are waiting to be evacuated at Khe Sanh, Vietnam, in 1968—a year during which domestic opposition to U.S. involvement in Southeast Asia grew dramatically. Images such as these helped fuel large-scale opposition to the war.

▲ **FIGURE 10.26 The Legacy of U.S. Bombing** The U.S. bombing campaign against the North Vietnamese as well as in Cambodia and Laos created a distinct landscape of bomb craters throughout the region. In many places, these craters remain to this day, and some are used, as they are in Vietnam along the Laos border, as fisheries. Craters have also led to expanses of standing water, which can be ideal breeding grounds for malarial mosquitoes and other pests.

ination of the Burmese-speaking authority. Muslims on the northern Arakan coast have fled into neighboring Bangladesh and more recently Thailand, Malaysia, and Indonesia. The Karen, who live along the border with Thailand and were favored by the British because of their Christian beliefs, have been repressed by the government. With a population of more than three million, the Karen have been able to establish an insurgent state supported by smuggling diamonds and opium. The Shan, with a population of 4.5 million in northern Burma, have similarly used drug profits to establish control of their territory and oppose the Burmese government.

The repressive authority of the Burmese military government is also opposed by Burmese who want a more democratic government. Popular protests in 1988 resulted in martial law and the establishment of the State Law and Order Restoration Council (SLORC—now known as the State Peace and Development Council). Elections were held in 1990 in which an opposition party, the National League for Democracy (NLD), won 60% of the vote, compared to 21% for the existing government party. SLORC cancelled the election results and remained in power. One of the leaders of the NLD is Aung San Suu Kyi, daughter of Burma independence hero General Aung San. Resolute in her quest for democracy and in her opposition to the authoritarian approach of SLORC, she has been kept under house arrest by the government and constitutionally barred from leading the country because she had lived abroad. She was awarded the Nobel Peace Prize in 1991.

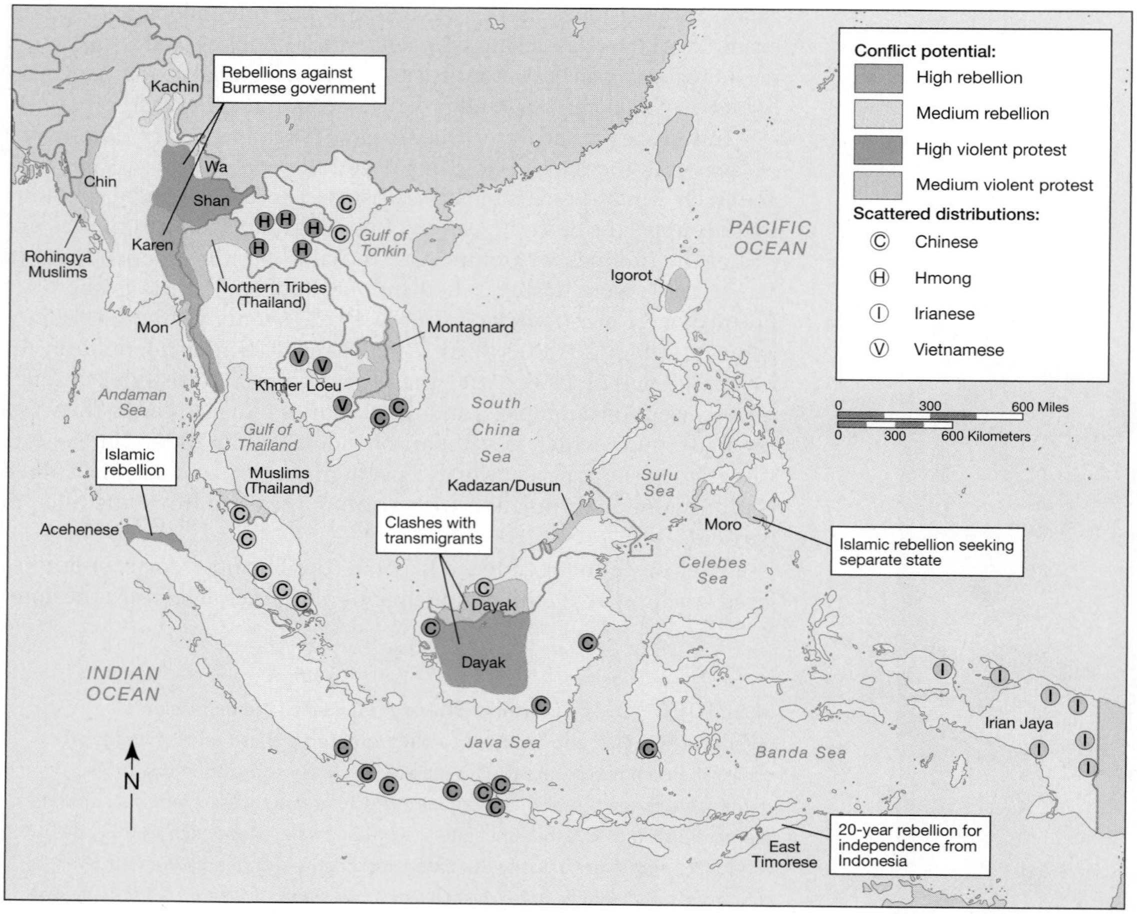

Although Thailand is often described as the "land of smiles" within the literature of the global tourist economy, this country is distinguished by a number of political tensions within its own borders. These include the fact that Thailand has witnessed over 20 military coups since the absolute monarchy was overthrown in 1934. The most recent coup took place in 2006 and resulted in a military dictatorship that has only recently given rise to a newly elected government in that country. Southern Thailand is also a site of conflict between the central government and local separatists, which seek autonomy from the Thai state. This conflict is partially reflective of the tensions that exist in the south between the Islamic minority in that region and the majority Buddhist population. Recently, tensions have emerged around the legitimacy of the most recent postcoup governments, as various social groups and organizations struggle to destabilize and, in some cases, remove the existing government in favor of more open and fair elections. And, Thailand found itself in conflict with its Cambodian neighbor in late 2008, as a border dispute left at least two Cambodian soldiers dead.

**Conflict potential:**

- High rebellion
- Medium rebellion
- High violent protest
- Medium violent protest

**Scattered distributions:**

- Ⓒ  Chinese
- Ⓗ  Hmong
- Ⓘ  Irianese
- Ⓥ  Vietnamese

▲ **FIGURE 10.27 Map of Conflict Zones** This map shows some of the major zones of conflict in Southeast Asia, including separatist movements in the outer islands of Indonesia and the Philippines.

**Regional Values and Cooperation**   Despite internal, regional, and international political conflicts, Southeast Asia provides a model for economic and, to a lesser extent, political cooperation in the form of **ASEAN**—the Association of Southeast Asian Nations formed in 1967. Whereas the initial alliance of Indonesia, Malaysia, Singapore, and Thailand was distinctly anticommunist and formed a defensive regional security group against China, the Asian Free Trade Association (AFTA) was created as part of ASEAN in 1993 to reduce national tariffs within the region. A policy of constructive engagement with both military and socialist governments led to invitations to the rest of the region to join ASEAN (Brunei, 1994; Vietnam, 1995; Burma and Laos, 1997; Cambodia, 1999). Despite some resistance from Malaysia, ASEAN has been open to discussions with other nations and groups, including China and Australia and more recently the United States and India.

More generally, the development of ASEAN signals just how important and "real" the region of Southeast Asia has become over time. Although it remains a disparate and diverse region, global trends toward macroregional organizations, such as free trade associations and supranational political organizations, have helped push Southeast Asia into a coherent regional entity. At the same time, local community groups and nongovernmental organizations have struggled against free trade and extended regional cooperation. In the case of free trade, some argue that reducing trade barriers within Southeast Asia and between ASEAN and larger economies, such as India or the United States, places local economies at a distinct competitive disadvantage. In other cases, activists have argued that ASEAN has done little to intervene in the politics of countries that continue to oppress their populations. ASEAN's reticence to involve itself in the local affairs of countries such as Burma and other member states means that it is losing credibility within and outside the region. It is thus clear that although regional cooperation has produced some integrative regional effects, it has also created new regional and global dialogues that may challenge the stability of Southeast Asia as a region.

# FUTURE GEOGRAPHIES

The region of Southeast Asia has emerged over time to become a relatively coherent economic entity dominated by the Association of Southeast Asian Nations (ASEAN). At the same time, its longer history is tied to its interconnections across a set of global socioeconomic and political–economic processes that have brought the people of this region into contact with people from other parts of the world. Today, this region is even more intimately tied to the practices of distant others, as local environmental practices in one part of the Indian Ocean region are affecting monsoon patterns in Southeast Asia. Additionally, agricultural practices in Southeast Asia, such as deforestation, are changing the patterns of water evaporation globally as well. Economic prosperity in this region is also fragile, as was witnessed in the 1997 economic crash that crippled many of the national economies in this region. The global recession of 2008 to 2009 is further exposing the dependencies that Southeast Asia has on the economies of others.

**Emerging Regional Cooperation and Conflict**   Regional cooperation is also a delicate balancing act between national and regional interests. Southeast Asia, despite the rise of ASEAN as a growing economic cooperative, is still a region that is fragmented across a diversity of cultural and political divisions. This is reflected in the politics of religion, the rise of Islamist practices and conservative Buddhist nationalism, for example, and the tensions that are maintained between countries and subregions that have high levels of inequality. That Singapore remains a dominant economic engine in the region, acting as a net exporter of cheap jobs to labor-rich Malaysia and Indonesia, means that there remain tensions between the economic dominance of one place and the political (and military) power of another place. In the context of all this complexity, ASEAN is now negotiating free trade agreements as a block with other countries and regional entities. What this holds for the future regional geography of Southeast Asia is unclear, as ASEAN currently has a "hands-off" policy when it comes to local politics. Moreover, it is unclear if other partners, such as Papua New Guinea, could be brought into the broader ASEAN umbrella in the future.

**Emerging Environmental Issues and Sustainability**   Future sustainable development in Southeast Asia must also confront the impact of urban growth and land use change in much of the region, especially air pollution, spreading slums, inadequate infrastructure in the cities, and deforestation of the highlands of the mainland and many of the islands such as Borneo. Managing the environment and land use more equitably and ecologically will contribute to maintaining and, it is hoped, improving social and economic conditions in this region. It is clear, however, that Southeast Asia is a key player in the broader global economy, and this region's decisions will continue to affect global environmental, political, and economic relationships into the future.

# Main Points Revisited

■ Southeast Asia is a region that, despite its outward appearance as "tropical," is distinguished by its diverse physical geography and ecological variety. This diversity and variety have led to a wide range of biogeographical subregions and land use practices.

Southeast Asia has lowland with wet-rice agriculture and dry farming practices (rain fed) in the uplands. Monsoon rains are ubiquitous, but rainfall patterns vary dramatically depending on relative location. This creates a wide diversity of flora and fauna.

■ The historical geography of this region is a global geography, as this region has long been in contact with the people and places of East and South Asia as well as more recently Europe and the Americas. Through these contacts, Southeast Asia has developed a complex pattern of demographic, religious, and linguistic difference.

Southeast Asia's relative location to South and East Asia influenced the political, economic, and cultural practices of this region. More recent interaction with Europe and the United States resulted in the establishment of 11 distinct nation–states practicing a variety of global religions. Despite the establishment of these nation–states, the region still hosts hundreds of distinct ethnic and linguistic traditions and practices.

■ Southeast Asia is economically interdependent with the rest of the global economy. As a result of these connections, this region has been home to some of the world's fastest-growing economies. Many of the countries in this region have also been susceptible to the volatility of the global economy.

The 1997 Economic Crisis in the region illustrated Southeast Asia's involvement in the global economy. This was further clarified in 2008 and 2009. The countries of Southeast Asia have continually retooled their economies to meet the needs of the global economy—moving from light industry to high technology. As labor costs increase in the region, Southeast Asia is losing its competitive edge to countries such as China.

■ It is challenging to speak of a Southeast Asian set of cultural practices because this is a region that is highly differentiated socially. Southeast Asia is home to hundreds of different languages and ethnic groups, and these differences lead to an array of political tensions and relationships.

Southeast Asia is a region in which politics and human rights continue to be significant. As stateless peoples are immobilized by state boundaries, questions will continue to emerge about rights to citizenship and basic social services. Human trafficking of women and children continues in this region as well, and this is a concern for global human rights advocates.

■ This region has grown over time to become unified through the development of the Association of Southeast Asian Nations (ASEAN). This regional organization has institutionalized this region, although conflict and tension in and across it continue to raise questions about the long-term stability of this organization and the region more generally.

ASEAN will continue to bring Southeast Asia closer together. But national and international politics within and beyond this region may also destabilize this regional organization and raise questions about its broader efficacy as an advocate for Southeast Asian people.

# Key Terms

AIDS orphans (p. 363)

ASEAN (p. 367)

biofuels (p. 343)

Bumiputra (p. 351)

*chao khao* (hill tribes) (p. 363)

crony capitalism (p. 352)

Culture System (p. 349)

desakota (p. 358)

domino theory (p. 364)

entrepôt (p. 347)

free trade zone (FTZ) (p. 351)

Green Revolution (p. 347)

killing fields (p. 365)

kleptocracy (p. 352)

land bridge (p. 340)

little tigers (p. 351)

mangroves (p. 344)

matrilocality (p. 360)

Mekong River Commission (MRC) (p. 342)

overseas Chinese (p. 356)

pancasila (p. 349)

patrilocality (p. 360)

rice bowl (p. 356)

sea gypsies (p. 363)

SIJORI (p. 355)

Spice Islands (p. 348)

swidden farming (p. 345)

transmigration (p. 345)

tsunami (p. 337)

Wallace's Line (p. 344)

world city (p. 354)

# Thinking Geographically

1. How does the monsoon influence the climate and vegetation of Southeast Asia?

2. What is the main crop in Southeast Asia and the main systems by which it is produced? How and where did the Green Revolution affect this crop?

3. What were the five main European colonial powers in Southeast Asia, and what regions were under their control between about 1500 and 1890? Which country remained independent during the colonial period? Identify four key mineral or agricultural products that the colonial powers exported from Southeast Asia and the regions from which they came.

4. What roles did India and China play in Southeast Asia prior to the colonial period, and how did this influence culture and religion? What political and economic roles did Japan and the United States play in Southeast Asia during the 20th century?

5. What factors led to the Vietnam War? What were some of the effects of the war on the people and environments of the region?

6. What triggered the 1997 financial crisis? What were some of the responses to the crisis? How is the 1997 crisis different from the 2008 global recession in a Southeast Asian context?

7. Who are the "overseas Chinese"? What are their historical roles in the economy?

8. What is ASEAN? What role does it play in the region both politically and economically?

9. What are some of the geographic issues that precipitate the drug economy in Southeast Asia?

10. What have been the consequences of dictatorship in Cambodia and Burma? What are some of the similarities and differences between these two experiences of violence in the region?

Log in to www.mygeoscienceplace.com for Videos, Interactive Maps; RSS feeds; Further Readings; Suggestions for Films, Music and Popular Literature; and Self-Study Quizzes to enhance your study of Southeast Asia.

## ▼ FIGURE 11.1

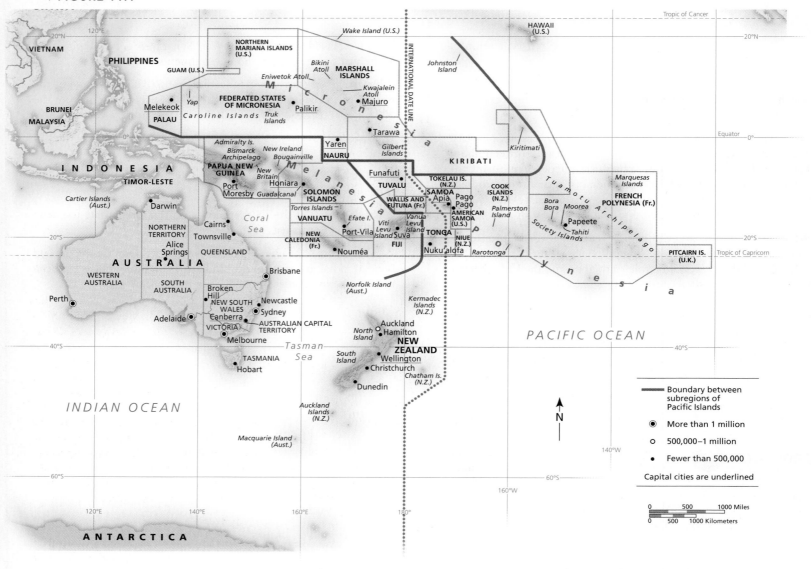

# AUSTRALIA, NEW ZEALAND, AND THE SOUTH PACIFIC KEY FACTS

- Major subregions: Australasia, Melanesia, Micronesia, Polynesia

- Total land area: 8.8 million square kilometers/ 3.3 million square miles

- Major physiogeographic features: Great Barrier Reef, Australian Highlands, Great Artesian Basin, Australian Outback. Much of Oceania lies within the tropics and is dominated by warm seas and moisture-bearing winds. Australia is dominated by very hot and arid conditions.

- Population (2010): 36 million, including 21.9 million in Australia, 4.3 million in New Zealand, and 6.4 million in Papua New Guinea

- Urbanization (2010): 71%

- GDP PPP (2010) current international dollars: 10,459; highest = Australia, 36,583; lowest = Papua New Guinea, 2,197

- Debt service (total, % of GDP) (2005): 2.6%

- Population in poverty (%< $2/day; 1990–2005): No data available

- Gender-related development index value: 0.8; highest = Australia, 0.96; lowest = Papua New Guinea, 0.53

- Average life expectancy/fertility: 76 (M73, F78)/2.4

- Population with access to Internet (2008): 59.5%

- Major language families: Aside from the colonial languages of English, Spanish, and French, Oceania is home to many indigenous languages from the Austronesian family.

- Major religions: Predominantly Christian, with a range of indigenous religious rituals still practiced, including forms of animism in Papua New Guinea as well as Buddhism and Islam among in-migrant communities from other world regions.

- $CO_2$ emission per capita, tonnes (2004): 6.3

- Access to improved drinking-water sources/sanitation (2006): 74%/62%* (*1990–2006)

# 11

# AUSTRALIA, NEW ZEALAND, AND THE SOUTH PACIFIC

In spring 2007, the ambassador to the United Nations from the Pacific island nation of Tuvalu spoke to members of the world community. Although many ambassadors go to the floor of the UN to discuss issues of war and conflict or peace and reconciliation, Ambassador Afelee Pita spoke about the threat of global warming and its impact on his nation. "The world," he argued, "has moved from a global threat once called the Cold War, to what now should be considered the Warming War. Our conflict is not with guns and missiles but with weapons from everyday lives—chimney stacks and exhaust pipes."[1] The ambassador is concerned that if climate change remains unchecked and sea levels continue to rise, island nations such as his own will soon disappear under the ocean.

The ambassador's fears are not unjustified, as recent scientific studies have shown that Tuvalu is being covered at a rate of 5.7 millimeters (.22 inches) a year due to rising sea levels.[2] Although this might not seem like much, it is almost triple the rate of just a decade earlier. This island is expected to be the first to disappear in the South Pacific because, at only 4 meters (13 feet) above sea level at its highest point, the island will soon be inundated with seawater. The process has already begun, as salt water infiltrates freshwater supplies, putting agricultural practices at risk. Funafuti Atoll—one of the nine coral atolls of Tuvalu (shown in the image)—is being inundating by seawater near a local power plant. Tuvalu has a short time to bring attention to this problem and stem the tide toward permanent disappearance. Regionally, the people of Tuvalu are not alone. Many other island nations in the South Pacific are under threat from increased sea level rise and seawater infiltration into their freshwater supplies.

Plans have already been prepared in New Zealand, which agreed in 2003 to take the refugees of global warming into their own country. Migrants have already left Tuvalu for New Zealand since the 1970s for not just environmental reasons but for economic ones also, as the economies of these island communities are often extremely underdeveloped. Population pressures have also exacerbated concerns among those who see the longer-term capacity of these islands as unsustainable. The migration streams have increased even more quickly, however, since 2003. The New Zealand-based Tuvalu population, most of whom now reside in Aukland, accounts for about 20% of the world's entire Tuvalu population. Thus, while regional and global plans to reduce carbon emissions are readied to reduce the long-term impacts of global warming, the people of Tuvalu and other islands in the South Pacific remain at the mercy of SUV drivers in Canada or the United States and cattle ranchers in Brazil. This suggests just how significant the issue of global warming is to the South Pacific, and how climatic changes affect the everyday experiences of people who live in the wide-ranging and interconnected oceanic space. ■

## MAIN POINTS

■ This world region is often discussed as one of the most geographically isolated. This is a result of its unique physical geography of extensive Pacific Ocean waterscapes, which create vast distances between this region and other world regions. At the same time, the region is one of the most vulnerable to global climate change as well as wider geopolitical processes.

■ This region's historical geography is interwoven with a series of human migrations that have brought the region into direct contact with new people as well as new plant and animal species. This has created a region with a dynamic and diverse demography, as well as a distinct set of land use practices.

■ The extensive settlement of this region by people of European ancestry has created a complicated set of cultural and social politics in this region. Although this has rarely led to direct violence between people in this region, there remains tension between indigenous people and colonial people.

■ Although this region is well integrated into the global economy and is highly urbanized, distinct differences remain in economic development across the region. Some smaller islands rely heavily on resource extraction or tourism to sustain their development, whereas the larger economies of the region, such as Australia and New Zealand, have been able to diversity, attracting foreign direct investment and skilled migrant labor.

■ The region is linked by a number of formal political organizations as well as through economic connections. In addition, the region is growing more interconnected through the migration of people from the South Pacific to Australia and New Zealand and through cooperation and the development of supranational organizations.

---

[1]Shapiro, Joseph. (2007). Tuvalu Envoy Takes Up Global Warming Fight. *NPR News.* [accessed 27 May 2009].
[2]Kinver, Mark. (2008). The Ebb and Flow of Sea Level Rise. *BBC News.* [accessed 27 May 2009].

# ENVIRONMENT AND SOCIETY

Commonly described as one of the most isolated world regions in terms of its physical geography (**Figure 11.1**), Australia, New Zealand, and the South Pacific are dynamically interrelated into wider global environmental climatic and tectonic systems. In fact, this region, which is sometimes known as Oceania[3] because of its relation to the vast Pacific Ocean (**Figure 11.2**), is directly connected to El Niño, the periodic change in ocean temperature off the coast of Peru that affects weather worldwide (see Chapter 7). This region is also geologically rich, bringing it into close connection with the wider global economy of resource extraction, as mining of precious minerals makes this region of interest to transnational corporations. The diverse ecosystems of this region are also highly susceptible to human-induced change, and the region is famous for the impact that invasive species—nonnative flora and fauna—have had on the indigenous populations. These invasive species include the earliest introduction of humans as well as the dingo—a small wild dog—and more recently the cane toad from Hawaii. Thus, although it might be easy to discuss this region's dynamic and diverse environments in global isolation, it is important to measure that isolation in relation to the ebb of flow of global interconnection over time.

## Climate, Adaptation, and Global Change

On a global scale, much of Oceania lies within the tropics and is dominated by the warm seas and moisture-bearing winds of these southern latitudes. Australia and New Zealand reach farther south and thus have

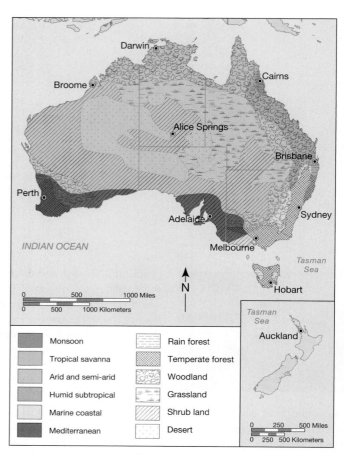

▲ **FIGURE 11.3  The climate of Australia and New Zealand**
The climate of these countries is determined by proximity to the ocean and by latitude, tropical warm and wet conditions, and dry conditions in the interior. As such, Australia's interior is dominated by arid and semi-arid climates, its coastal spaces are much more diverse, and include tropical savanna and some tropical wet with no dry season regions as well as significant humid subtropical and Mediterranean regions. The majority of Australia's population lives in these latter climatic regions. New Zealand is dominated by a marine coastal climate, while much of the South Pacific is tropical wet with no dry season.

climates that range from tropical to the cooler temperate climates of the southern westerly wind belts, with annual average temperatures declining from north to south (**Figure 11.3**). The Australian climate, in particular, is dominated by very dry and hot conditions that parch the interior and define it as one of the most arid regions on Earth, with two-thirds of the country receiving less than 50 centimeters (20 inches) of rainfall a year. This harsh climate limits human activity and has required complex adaptations from both people and ecosystems in the continent's interior. In fact, adaptation to and modification of this environment has been a key for aboriginal peoples as well as European settlers, who developed sophisticated irrigation systems and cattle ranching techniques to adapt this space to their needs.

Most of the coastal perimeter of Australia has higher precipitation; the northeast has heavy rainfall from the southeasterly trade winds that

▲ **FIGURE 11.2  Oceania from space**  This image conveys the importance of the Pacific Ocean to Oceania and the contrast among the large island continent of Australia, the larger islands of New Guinea and New Zealand, and the scattering of smaller islands across the southern Pacific.

---

[3]The convention of naming this region is complex, as Australia and New Zealand are often discussed as the Antipodes or Australasia region. In this case, however, the wider region's connection to maritime space is important, making the term "Oceania" an appropriate shorthand for the entire region.

rise over the eastern uplands with associated orographic rainfall. The rainfall at Cairns in Queensland averages more than 460 centimeters (180 inches) a year. Northern Australia receives most of its rain from monsoon winds, drawn inland by the southward shift of the intertropical convergence zone (ITCZ) and heating of the landmass in the Southern Hemisphere summer from November to February. The rainy season is called the Big Wet, and it sometimes brings tropical cyclones. Southern Australia receives rainfall from storms associated with the westerly wind zone, especially during winter (June to August), when the storm tracks shift northward. The southern coast, specifically the areas around the cities of Adelaide and Perth, has the mild temperatures and winter rainfall associated with the Mediterranean climate type that is also found at this latitude in California, Chile, southern Europe, and South Africa. The southernmost part of New South Wales, Victoria, and the island of Tasmania are wetter as a result of exposure to westerly rain-bearing winds for most of the year. These areas have considerable snowfall in the mountains.

New Zealand sits in the middle of the westerly wind belt, and frequent storms release heavy rain on the west coast as they rise over the high mountain ranges. The eastern coasts of New Zealand are much drier because they lie in the rain shadow to the east of the mountains, and this area can be sunny in summer when subtropical highs move southward and create clear and stable conditions. The North Island is generally much warmer than the south, but mountain climates are cooler and wetter throughout the country, often with heavy snow that favors ski tourism in the June to October period, drawing many skiers from North America.

The Pacific islands have warm temperatures associated with year-round high sun and the warmth of the tropical ocean. Islands with higher elevations experience substantial rainfall as moist winds rise over the land, cooling and releasing their moisture. Islands throughout the region receive rainfall from the towering cumulus clouds that occur at the ITCZ as intense heating creates rising air (convection) and heavy afternoon rains. The lower islands are much drier because they do not benefit from orographic uplift over mountain ranges and are small enough to elude the convective downpours. As a result, many low-lying islands experience near-desert conditions and shortages of freshwater. The 1997 to 1998 El Niño caused severe drought in Papua New Guinea, Australia, and Micronesia, with crop failures and food shortages in New Guinea, expensive shipments of drinking water to smaller islands, and serious livestock loss and wildfires in Australia, whereas the 2004 to 2005 El Niño was much milder in its overall regional affects.

**Adaptation to Climatic Conditions in Interior Australia** The inland areas of Australia are often called the **Outback**, a term generally applied to the remote and drier inland areas of Australia. Although low rainfall and frequent drought are common in inland Australia, the exploitation of underground water from the Great Artesian Basin has allowed the development of scattered homesteads that raise livestock on sheep and **cattle stations**. In these enterprises, cattle are left to fend for themselves for the most part and are only brought into the stations once or twice a year. Sheep are raised where rainfall is higher to the east and west. In interior towns, such as Coober Pedy, which has been a center for the mining of opals valued for jewelry, the temperatures are so intense that much of the town has been built underground (**Figure 11.4a**). Many Aborigines still live in the Outback, and many now work on stations and mines or reside in small settlements known

(a)

(b)

(c)

▲ **FIGURE 11.4 Adaptation to life in Australia** (a) Judy McLean displays the subterranean bedroom of Faye Nayler, an earlier resident of Coober Pedy, who built an underground house in this opal mining town. Half of Coober Pedy's 3,500 residents have dug their homes into the chalky clay rock to escape the harsh climatic conditions. (b) An elderly aboriginal landowner, Lilly Billy, sits with a friend at her Outback camp at Anna Creek Station in Arnhem Land of the Northern Territories. (c) February 7, 2009, is now known as "Black Saturday" in Australia, named for the 400+ separate bush fires that killed 173 people and destroyed over 2,000 homes. Pictured here is Marysville, Australia, just north of Melbourne.

as **outstations** (**Figure 11.4b**). This region is obviously remote from schools, shops, and hospitals. Adaptation to this region thus includes distance education, with children taught through radio broadcasts, whereas the ill receive emergency medical care from the flying doctor service. Changing climatic conditions are further affecting life in this region, however, as Australia is vulnerable to natural hazards, including frequent droughts. Droughts are also associated with severe wildfires that can race across the tinder-dry bush vegetation, especially where oily eucalyptus fuels the fire, putting people with few resources at greater risk (**Figure 11.4c**).

**Climatic and Environmental Change**   Oceania is especially vulnerable to global environmental changes, particularly ozone depletion, global warming, and sea-level rise. Scientists first noticed a dramatic drop in the amount of ozone over the South Pole in the 1980s: the so-called hole in the ozone layer. **Ozone depletion**—the loss of the protective layer of ozone gas—can result in higher levels of ultraviolet radiation and associated increases in skin cancer, cataracts, and damage to marine organisms. Australia, with its southern latitude location, sunny days, and traditions of beach going and sunbathing, is especially vulnerable and has a growing incidence of skin cancer. However, the Montreal Protocol—an international treaty to control chemicals that damage the ozone layer—has succeeded in preventing further ozone depletion.

Human activities have also caused an increase in carbon dioxide associated with global warming (see Chapter 1). In Oceania the impacts of global warming include drier conditions in the already drought-prone interior of Australia, increased risk of forest fires, and the melting of New Zealand's magnificent glaciers. A rise in global temperatures is also likely to produce a significant rise in sea levels, mainly because a warmer ocean takes up slightly more volume than a cooler one and secondarily because global warming may melt glaciers and ice sheets such as those in the Antarctic. Sea-level rise is of urgent concern to many Pacific islands, especially those on low coral atolls, where any increase in sea level may result in the disappearance of the land, saltwater moving into drinking water supplies, and an increased vulnerability to storms. In response to the threat of global warming and sea-level rise, the Pacific islands were early members of the **Alliance of Small Island States (AOSIS)**, which maintains a sustained voice in international negotiations to reduce the threat of global climate change (see introduction to this chapter). In 2007, Australian policy also changed dramatically when the new prime minister, Kevin Rudd, signed the Kyoto Protocol, which calls for targeted reductions in $CO_2$ greenhouse gas emissions for all member states. With Australia's signature, the lone holdout on this historic international regulatory regime is the United States (see Chapter 6).

## Geological Resources, Risks, and Water Management

Geological history, especially tectonic activity, influences the nature of many of the major subregions of Oceania. Australia is a very old and stable landmass that was originally part of the Southern Hemisphere supercontinent called Gondwanaland (see Chapter 1). The Indo-Australian Plate separated about 50 million years ago and began moving northeast until it collided with the Pacific Plate. New Zealand now sits at the boundary where the Pacific Plate is subducting the Indo-Australian Plate and producing high levels of volcanic and earthquake activity.

**Australia**   The Australian landmass lies in the middle of the Indo-Australian Plate and forms a continental shield of ancient stable rock with very little volcanic, earthquake, or other mountain-building activity. Australia, which has an area of 7.7 million square kilometers (3 million square miles), is divided into three major physical regions by a series of interior, low-lying basins that divide the western Australian plateau from the uplands of eastern Australia (**Figure 11.5**). The eastern highlands of Australia are the remnants of an old folded mountain range with a steep escarpment on the eastern flanks. The highland crest is often called the Great Dividing Range because it separates the rivers that flow to the east coast from those flowing inland or to the south.

The interior lowlands were once flooded by a shallow ocean and now contain the Lake Eyre basin, filled only occasionally by inland-draining rivers. A large part of the lowlands is also called the **Great Artesian Basin** because it is underlain by the world's largest groundwater aquifer, a reservoir of underground water in porous rocks. The basin is artesian because overlying rocks have placed pressure on the underground water so that when a well is drilled, the water rises rapidly to the surface and discharges as if from a pressurized tap. The wells that tap the Great Artesian Basin are critical to the human settlements and livestock of arid east-central Australia, although the cost of drilling is high, and the water is often very warm and salty.

Two-thirds of Australia is occupied by the western plateau of old shield rocks with a few low mountains and large areas of flatter desert plains and plateaus. This region has numerous mineral deposits—the basis for Australia's mining industry—and old, weathered soils that are too nutrient poor or salty for agriculture. Mining centers have developed across the interior at Broken Hill in New South Wales and Kalgoorlie in Western Australia and more recently at new finds near Mount Isa (lead, zinc, copper) in Queensland and Pilbara (iron ore) in Western Australia. Uranium is also mined in the Outback. Western Australia and the interior lowlands also contain impressive examples of desert landforms, including wind-shaped undulating ridges of sand dunes, stony plains with varnished rock fragments called *desert pavement*, and dry interior drainage basins called *playas*. Centuries of erosion by wind and water have left erosion-resistant domes of rocks standing above the surrounding landscape. The most famous of these isolated rock domes are Ayres Rock, called *Uluru* by the Aborigines, and the Olgas, called *Kata Tjuta* (**Figure 11.6**).

**New Zealand**   In contrast to Australia, New Zealand is much younger geologically and more tectonically active because it is located where the Pacific Plate is moving under the Indo-Australian Plate. The New Zealand physical landscape includes two major islands spanning 1,600 kilometers (976 miles) from north to south. The South Island has rugged mountains rising to more than 3,500 meters (11,500 feet) in the Southern Alps, dominated by Mount Cook at 3,754 meters (12,316 feet; **Figure 11.7a**). The South Island is far enough south to have extensive permanent snowfields and more than 300 glaciers, some flowing almost to sea level. The southern portion of the west coast is penetrated by

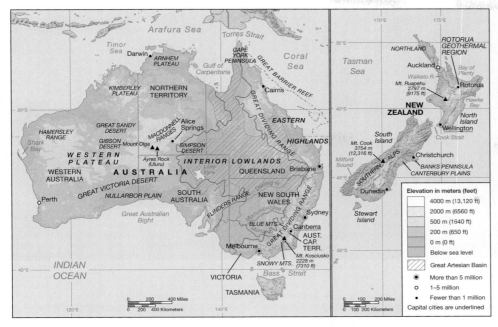

▲ **FIGURE 11.5 The physical landscape of Australia and New Zealand** Major physiographic regions and distinctive landforms of Australia and New Zealand are shown on these maps. For Australia, key features include the eastern highlands, central deserts, Great Artesian Basin, and Great Barrier Reef, and for New Zealand, the Southern Alps and west coast fjords, such as Milford Sound.

**Pacific Islands** The Pacific islands can be classified into the high volcanic islands and the low coral islands called **atolls**. The high islands, which are mostly volcanic in origin, rise steeply from the sea and have very narrow coastal plains and deep narrow valleys (**Figure 11.8a**). Many high islands, such as those of Hawaii and Samoa, are created in linear chains as tectonic plates move over hot spots where molten rock reaches the surface (see Chapter 1). Others, such as the Marianas and Vanuatu islands, form island arcs along the edge of tectonic plates. The heights of the islands promote heavy rainfall and are often capped by clouds, creating spectacular landscapes such as those of Tahiti, where the mountains rise to 2,100 meters (6,900 feet), Bougainville (3,000 meters or 9,840 feet), or Hawaii, where the Mauna Loa volcano reaches 3,900 meters (12,800 feet). The island of New Guinea is the second-largest island in the world (after Greenland, as Australia is a continent) and is much larger than the country of New Zealand at more than 800,000 square kilometers (309,000 square miles). The mountain spine of the island of New Guinea rises to more than 4,000 meters (13,000 feet), with many extinct volcanoes and high isolated basins.

The low islands are mostly atolls created from the buildup of skeletons of coral organisms that grow in shallow tropical waters. Atolls are usually circular, with a series of coral reefs or small islands ringing and sheltering an interior lagoon (**Figure 11.8b**). The lagoon may contain the remnants of earlier islands or a volcanic island that has sunk below the surface. Although in some cases, such as the islands of Nauru and Guam, tectonic activity may uplift coral reefs to create higher-elevation limestone plateaus, many of these islands are very low lying, with most of the land within a meter of sea level, making them vulnerable to storms, tidal waves, and rising seas. Pacific atolls include Kiribati, the Marshall Islands, and Tuvalu. The Kwajalein Atoll in the Marshall Islands is the largest in the world, with

magnificent fjords, created when the sea flooded the deep valleys cut by glaciers. The east coast of the South Island has much gentler relief, with rolling foothills, long valleys with braided rivers and freshwater lakes, and alluvial plains (formed from stream deposits), which are used for agriculture. The North Island has much more volcanic activity than the South Island. Volcanically heated water that emerges from hot springs, geysers, and steam vents is captured in geothermal facilities and used as an energy resource (**Figure 11.7b**). The North Island also has extensive areas of rolling hills and valleys, where the warmer climate and rich volcanic and river-deposited soils nourish a productive agriculture.

▼ **FIGURE 11.6 Australian Highlands** The Olgas are a series of 36 isolated rock domes reaching almost 500 m (1,500 feet) in central Australia; together with Ayres Rock, the Olgas are major tourist destinations. Note the feral camels in the foreground of the picture.

90 small coral islands circling a 650-square-kilometer (251-square-mile) lagoon.

**Mining in the Pacific** One distinctive set of islands contains those on which the economy is almost wholly dependent on exports of mineral resources. Mining has destroyed the landscape and created social tensions over the wealth that flows from exports. Perhaps the most dramatic case is that of Nauru, an oval island with an area of only 21 square kilometers (8 square miles) consisting of an uplifted coral platform about 30 meters (100 feet) above sea level (**Figure 11.9a**).

Centuries of roosting by seabirds covered most of Nauru with deep deposits of **guano** (bird droppings) that have created the highest-quality phosphate rock in the world. Exploitation of the phosphate for use as a fertilizer began in 1906, and Nauru phosphate was especially valuable in making the phosphorus-poor soils of Australia productive for crops and pasture. When Nauru became independent from Australia in 1968, its government took over control of the mining industry. After independence, phosphate dominated the economy and was strip-mined, crushed, and sent by conveyor belts to ships that anchored outside the reef that surrounds the island. The profits were divided among the government, local landowners, and a long-term trust fund. Many locals chose not to work, and temporary migrant contract workers from other islands did the work. The phosphate ran out in the 1990s. The landscape of Nauru is now a desolate wasteland stripped of vegetation and soil, with cavernous pits dotting a rainless rocky plateau. Drinking water comes from an aging desalination plant or is shipped in from Australia. High levels of consumption and sedentary lifestyles among local people have brought the diseases of affluence to the island, including obesity, diabetes, and heart disease and a loss of traditional culture. Nauru invested in real estate in Australia and in other enterprises as alternative sources of revenue after the phosphate income disappeared. Its other alternative income sources included offering its territory as an offshore detention center for asylum seekers trying to enter Australia.

Mining also exacerbates problems on the island of Bougainville, which is controlled by Papua New Guinea but is geographically and culturally part of the Solomon Islands. The giant Panguna copper mine owned by the multinational company Rio Tinto was one of the world's largest open-pit mines and contributed as much as a quarter of Papua New Guinea's export earnings. The mine was developed in the forests without the participation of the resident indigenous Nasioi, and it has polluted several rivers that provided fish and drinking water to other groups. Local people seeking a share of the copper revenues, concerned about the mine's environmental impacts, and demanding independence from Papua New Guinea, have joined a rebel movement. The conflict over mining on indigenous lands in Bougainville parallels that surrounding the Ok Tedi copper and gold mine on the mainland of Papua New

(a)

(b)

▲ **FIGURE 11.7 New Zealand landscapes** (a) Mount Cook (Aoraki) National Park has more than 19 peaks that are over 3,000 m (10,000 ft), with glaciers covering more than one-third of the park area. The region is a popular ski destination. (b) The region around Rotorua, on the North Island of New Zealand, has numerous hot springs, geysers, and steam vents associated with volcanic activity. Geothermal energy sources, such as the Wairakei power plant shown here, have been developed and contribute 10% of New Zealand's electricity production. The region is also popular with tourists attracted by the landscape, spas, and thriving Maori culture.

Guinea, where the pollution of rivers by mine tailings has prompted international environmental concern (**Figure 11.9b**). In the 1990s, the communities downstream sued the BHP Company, who owned the mine until 2002, and received US$28.6 million in an out-of-court settlement, which was the culmination of an enormous public-relations campaign against the company by environmental groups.

**Marine Ecosystems** Oceania is defined by its marine environment. It is not surprising, then, that the region has concerns about territorial claims over ocean boundaries, fisheries, and pollution of marine ecosystems. The declaration of the international 200-nautical-mile

 **FIGURE 11.8  Pacific Islands** (a) The higher islands of the Pacific are mostly volcanic in origin and rise steeply from the sea to forested mountain slopes. The island of Moorea in French Polynesia is a spectacular example of a volcanic high island with surrounding coral reefs that protect shallow lagoons. (b) Low islands and reefs surround a lagoon in Tetiaroa Atoll in French Polynesia.

▲ **FIGURE 11.9  Mining in the Pacific** (a) The landscape of Nauru was devastated by mining phosphate rock that accumulated from the guano of roosting seabirds. This shows the dock where phosphate was loaded onto ships for export. (b) The Ok Tedi mine in Papua New Guinea has polluted the Fly River with mine waste and affected the lives of local indigenous people.

(370-kilometer) *exclusive economic zone* (EEZ), which was formalized in 1982 in the United Nations Convention on the Law of the Sea (**UNCLOS**), was of tremendous significance to Oceania because it allowed countries with a small land area but many scattered islands, such as Tonga and the Cook Islands, to lay claim to immense areas of ocean. The pattern of Pacific islands is such that most of the region is now covered by their EEZs, with relatively little unclaimed ocean (**Figure 11.10**). These island nations can demand licensing fees from the international fleet that seeks to catch tuna, for example, within their zones.

The development of international law governing ocean space is also important to Pacific islanders because they eat more fish per person than any other population, and fishing along with coastal tourism are critical to the majority of smaller island economies. Marine resources include not only fish and shellfish but also valuable exports such as pearls and shell (mostly for shirt buttons), as well as products made

# URANIUM IN OCEANIA

As with oil in the Middle East, the extraction and use of uranium links Oceania to the global hunger for cheap energy and to geopolitical conflicts beyond the region. Uranium is a radioactive element that can be split in a process of nuclear fission to produce a reaction that releases large amounts of energy. Controlled reactions can generate electricity in nuclear power plants, whereas uncontrolled reactions, releasing enormous amounts of energy and radioactivity, can be used in bombs. Uranium became a desirable commodity after the World War II bombings of Hiroshima and Nagasaki, and the potential of nuclear-powered electrical generation became apparent.

This interest in uranium affected several regions of Oceania. The United States, Britain, and France all joined the Cold War arms race and the effort to develop even more powerful weapons based on uranium. They chose to test many of these weapons in the Pacific, with devastating implications for local residents and environments. The United States tested its bombs in the Marshall Islands, relocating the residents of Bikini and Enewetak atolls and exploding several different types of nuclear weapons between 1946 and 1958 (**Figure 1**). Although the prevailing winds were supposed to carry the radioactive fallout from bomb testing away from inhabited islands, in 1954 radioactive ash dusted the island of Rongelap and its almost 100 residents, including several relocated from Bikini.

Radioactive exposure can have serious short- and long-term effects, including acute poisoning, leukemia, and birth defects, so the U.S. government evacuated the residents of Rongelap at short notice with little information about the hazard they had been exposed to or warning that they would not be able to return to their homelands. Years later, in 1968, the residents of Rongelap and Bikini were told it was safe to return; those on Bikini later had to be reevacuated when scientists discovered that dangerous levels of radioactivity persisted in food gathered on the islands. Although the United States has monitored the health of the islanders and established a $90-million trust fund, many residents of the islands are angry and resentful about the experiments that disrupted their lives.

France conducted more than 150 bomb tests on the tiny atolls of Moruroa and Fangataufa in French Polynesia beginning in 1966. The first bombs showered the surrounding regions with radioactivity, reaching as far as Samoa and Tonga hundreds of miles to the west. Opposition from other Pacific islands, including New Zealand and Australia, culminated in boycotts of French products, including wine and cheese, during the 1970s. France moved to underground testing and refused to release information about accidents and monitoring of radioactive pollution or health in French Polynesia. Whereas locals use the bomb tests as a reason to seek independence from France, international activists have tried to stop the tests (**Figure 2**). In 1985 the environmental group Greenpeace planned to protest tests by sailing their ship *Rainbow Warrior* to Moruroa, but French intelligence agents scuttled the ship while it was moored in the harbor of Auckland, New Zealand.

The international scandal prompted New Zealand to take a strong stand against nuclear proliferation, banning all nuclear-powered and nuclear-armed vessels from its harbors, breaking off diplomatic relations with France, and taking a leadership role in the antinuclear movement in the Pacific. This created a long-term strain on relations between New Zealand and the United

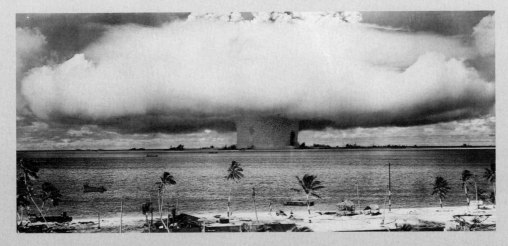

▲ **FIGURE 1 Bikini Atoll** The atomic bomb test at Bikini Atoll in the Marshall Islands on July 25, 1946. Fallout from this and subsequent tests posed serious health risks to Pacific islanders and resulted in the evacuation of several atolls.

from whales, the species that initially attracted many Europeans to the Pacific. At the same time, because of the warm water, the southern Pacific is actually less biologically productive than the colder water that wells up near the continental shelves of Australia and New Zealand. The reefs also limit the harvest of fish and other marine resources because coral organisms use up most of the nutrients, and the coral reefs snag nets. Island societies consume a very wide range of fish species as well as other marine organisms, such as sea cucumbers (exported to China) and giant clams.

## Ecology, Land, and Environmental Management

The relative global isolation of Oceania has contributed to the development of some of the world's most unique, diverse, and vulnerable ecosystems. Many of the species that evolved from the isolated populations have remarkable adaptations to the physical environment and are found only in that locality. Biodiversity in the region is partially defined by the **theory of island biogeography**, which holds that diver-

▲ **FIGURE 2 Greenpeace ship** A Greenpeace ship moored off Tahiti during protests against nuclear testing. This ship replaced the *Rainbow Warrior*, scuttled by French agents in the harbor of Auckland, New Zealand, because of its role in protesting nuclear testing in French Polynesia. New Zealand has declared its ports as nuclear-free zones and has joined Pacific island nations in strongly protesting nuclear activities in Oceania.

▲ **FIGURE 3 The Ranger uranium mine** The Ranger uranium mine is located in Kakadu National Park in Australia's Northern Territory. The area is sacred to the Aborigine population and has striking landscapes and ecosystems. Moves to expand mining activity have resulted in protests by Aborigine groups and environmental activists.

States because U.S. military vessels, which will not admit nuclear capability, were banned from New Zealand. New Zealand's actions contributed to the announcement by France in 1996, after riots in Tahiti and declines in tourism, that it would end nuclear testing.

The consumption of uranium has also threatened Australian Aborigines through mining on or near their lands in northern Australia (**Figure 3**). The Ranger mine began operations in the Northern Territory in 1980 within the boundary of Kakadu National Park, a region of great natural beauty listed as a World Heritage site for both natural and cultural values. The mine has produced more than 16 million tons (35 billion pounds) of radioactive waste and has created serious water-pollution problems in the area. Great controversy has arisen over proposals to open another mine; activists have blockaded mine roads, and protests have occurred around Australia. Australia has 40% of the world's uranium deposits, however, and exports fuel to nuclear power stations in the United States, Japan, Europe, Canada, and South Korea.

Nuclear testing has been halted, but radioactivity persists. Moreover, although uranium prices are low because few new nuclear power stations are under construction in the aftermath of the Chernobyl accident (see Chapter 3, p. 86), many countries are reconsidering the viability of nuclear energy. ■

sity increases with island size (i.e., smaller islands are generally less diverse than the larger ones). This means that biodiversity in the smaller Pacific islands is less dynamic than on the large islands of New Guinea or New Zealand, for example. At the same time, this region has long been in contact globally with other regions and places—tectonically during the period of Gondwanaland through the period of human in-migration to the most recent expansion of global travel and colonization. As such, the region bears the marks of *both* its isolation and global connectivity ecologically.

Within this region, Australia is beloved by biogeographers and conservationists for its great biodiversity and unusual species (**Figure 11.11a**). There are more than 20,000 different plant species, 650 species of birds, and 380 different species of reptiles. Australia has several remarkable species of marsupials and of monotremes, animals descended from mammals that died out on most other continents. A **marsupial** gives birth to a premature offspring that then develops and feeds from nipples in a pouch on the mother's body. Australian marsupials range in size from the large gray kangaroo (with adults reaching 2 meters—6 feet—tall) to

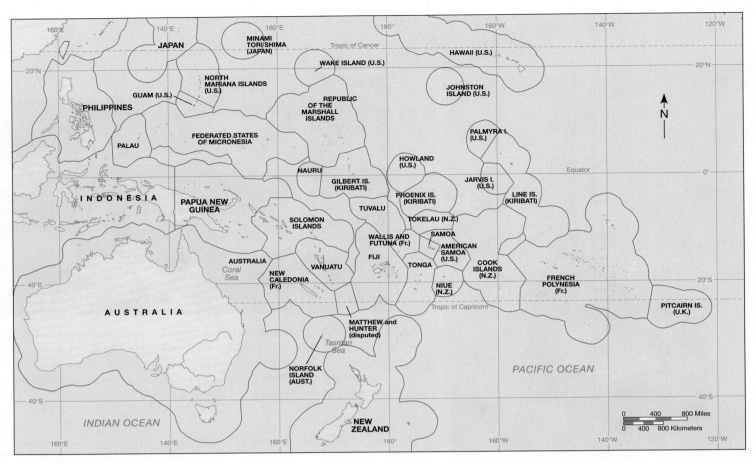

▲ **FIGURE 11.10 Marine territorial claims in the Pacific** The exclusive economic zones (EEZs) of Pacific nations and territories leave very little unclaimed ocean in the South Pacific. The international treaty, which established this zone, was ratified by 119 member states of the UN in 1982 and now consists of more than 130 signatory members. One result of the treaty is that 90% of the world's fisheries are claimed today.

koalas, wombats, and ferocious Tasmanian devils, to small mice and voles. **Monotremes** are very unusual mammals in that they lay eggs rather than gestate their young within the body but then nurture the young with milk from the mother. They include the platypus, with a ducklike bill and a tail like a beaver, and the spiny anteater.

Australian ecosystems are further defined by a number of dominant species, which include several types of forest and shrubland species well adapted to dry climates and frequent fires. The most important type of tree is the eucalyptus, commonly known as the *gum tree*. The leaves contain strong-smelling oils that are used to treat respiratory illness. Eucalyptus has been introduced into many other world regions for reforestation and pulp plantations, including California in North America. Acacia, an important hardwood commonly called *wattle*, is another dominant tree and shrub type, especially in the drier woodlands. The densest forests are found along the wettest portion of the northern east coast, with the largest remnants composed mainly of trees of Southeast Asian origin interspersed with woody vines (*lianas*). The forests of the southeast and southwest are dominated by drought-adapted eucalyptus, with more open woodlands in the transition to the dry interior. The northwestern and southern regions have shrub vegetation called *mallee* that includes

low, multibranching eucalyptus and plains covered with bushes similar to the sagebrush found in North America. The northern interior has extensive but sparse grasslands, and the driest zones have scattered grasses and shrubs characteristic of desert ecosystems (but without cacti).

New Zealand was heavily forested prior to the arrival of humans about 1,000 years ago, with towering *kauri* conifers in the north and beech in the cooler south. The remaining one-third of the land was covered with scrub, with grasses at drier, lower elevations, and with alpine grasslands (tundra) at high altitudes. Faunal evolution in New Zealand produced no predators or carnivores, and several birds remained flightless, including the now extinct moa, which was 3 meters (nearly 10 feet) tall, and the kiwi bird (**Figure 11.11b**). Across the islands of the Pacific, plant species that can be easily transported by ocean (for example, coconuts) or air (for example, fruit seeds eaten and excreted by birds) are more widely distributed. The variety declines as one moves eastward, away from the larger landmasses. Luxuriant rain forests are found on the wetter and higher islands, and marshes and mangroves thrive along the coastal margins. The larger islands, such as New Guinea and Hawaii, also have extensive middle-elevation grasslands. The smaller and drier coral islands have much sparser vegetation, but coconut palms

▲ **FIGURE 11.11 Unusual Australian animals** (a) This road sign on the Nullarbor Plain warns drivers to beware of two of Australia's marsupials—the kangaroo and the wombat—as well as feral camels, an introduced domestic animal that now runs wild. (b) The kiwi is the national emblem of New Zealand. It has given its name to a popular fruit (formerly known as the Chinese gooseberry) and has become a nickname for New Zealanders.

are ubiquitous, are a basis of human subsistence, and grow along many beaches. There are few native mammals on the Pacific islands (with the exception of New Guinea). The richest fauna include the birds that have been able to fly from island to island and marine organisms, especially those of reefs and lagoons, including turtles, shellfish, tuna, sharks, and octopus.

**The Introduction of Exotics** Beginning with the introduction of the dingo from Southeast Asia by Australian Aborigines about 3,500 years ago, foreign species have devastated the native species of Oceania. The dingo, a canine similar to a coyote, probably outcompeted and outhunted the marsupial predators, such as the now-extinct Tasmanian wolf, as well as rodents. But it was **ecological imperialism**—the way in which European organisms were able to take over the ecosystems of

other regions of the world—that led to the endangerment and extinction of numerous other native species through hunting, competition, and habitat destruction. These introduced species are also called "exotics" because they come from elsewhere.

The flightless birds of Oceania were the most vulnerable to exotic predators such as rats, cats, dogs, and snakes. Conservation efforts to protect birds today include the establishment of reserves and the careful monitoring and elimination of predators. On the Pacific islands, there are great efforts to contain the spread of the introduced brown tree snake, the mongoose (a very aggressive small mammal), and a carnivorous snail, all of which prey on local species. Other changes included the introduction of European weeds, pests, and crops and the escape of domesticated livestock. These "feral" animals of Australia include horses, cattle, sheep, goats, and pigs, as well as camels that were introduced to provide transportation across vast deserts (see Figure 11.6). The overall escaped populations are as large as 23 million pigs, 2.6 million goats, 300,000 horses (called brumbies) and 300,000 camels and comprises the largest population of feral domestic animals in the world.

Another ecological disaster was the introduction of the European rabbit to Australia in 1859. Over the next 50 years, the rabbit population exploded to plague proportions that devastated pasturelands. The rabbit population was partially eradicated in the 1950s by the deliberate introduction of a disease called *myxomatosis*. The introduction of the prickly pear cactus in the 1920s to create hedges led to the infestation of more than 20 million hectares (50 million acres) of Australian pasture before the cactus was eradicated by introducing a beetle that fed on the plants. The cane toad, introduced from Hawaii to control pests in the sugarcane district of Queensland, has destroyed frogs, reptiles, and small marsupials; is highly toxic to predators; and has spread to northern Australia. Despite attempts to curb its expansion, it has gone largely unchecked since its ironic introduction—cane toads, it was found, did not actually eat the cane beetle, which it was imported to destroy in the first place.

**Agricultural Production and Land Use in Australia and New Zealand** The most important agricultural region in Oceania is probably the area of southeastern Australia that stretches from the Gold Coast in southeastern Queensland, south along the coast to the cities of Newcastle, Sydney, Wollongong, Melbourne, and Adelaide, and inland to Australia's capital city of Canberra and the agricultural regions of the Darling Downs, Murrumbidgee Irrigation Area, and the Barossa Valley (**Figure 11.12a**). The rich agricultural lands of southeastern Australia host a livestock industry of milk, beef, and lamb production, with animals that graze on pastures improved by fertilizer (especially phosphate) and introduced grasses. Southeastern Australia is also the heart of the Australian wheat industry in a "fertile crescent" that stretches with rolling farmland from the Darling Downs of southern Queensland to central South Australia. Wheat and other grains are often grown in rotation with pasture and sheep-raising on larger farms. Wheat from this area combines with that from Western Australia to constitute an export wheat industry that ranks with that of Canada and France, with only the United States ahead of this group. The irrigated farms of the Murrumbidgee Irrigation Area and the grazing lands of the Riverina are also very important to agriculture. Cotton from New South Wales contributes to Australia's dominance in world cotton exports. Southeastern Australia also includes

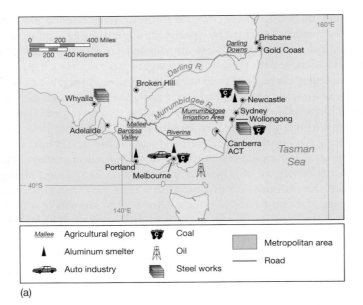

(b)

▲ **FIGURE 11.12 Southeastern Australia** (a) The most economically significant region of Oceania is the agroindustrial region of southeastern Australia with the dynamic cities of Sydney and Melbourne and the federal capital of Canberra. (b) South Australian vineyards may be the perfect spot to grow grapes, as the soil, climate, and afternoon sea breeze are all ideal. But recent droughts, including one in 2007, are forcing winemakers to rethink their commitment to the regional wine economy.

several areas of intensive horticulture (fruits and vegetables) and vineyards (**Figure 11.12b**).

Land use practices in New Zealand are similarly dominated by livestock development, particularly sheep, lamb, and dairy cattle farming as well as wheat and barley production. In similar fashion to Australia, New Zealand has put some of its land into the development of vineyards, while also investing in a reforestation program for productive woods. In 2008, 33,100 hectares (81,791 acres) of forests were replanted on the North Island, whereas an additional 22,500 hectares (55,598 acres) were replanted on the South Island. At the same time, 43,000 hectares (107,800 acres) of exotic forest was harvested for local and global markets in 2008 as well.

**Managing Oceania's Fisheries and Aquatic Life**    Oceania provides many interesting examples of how communities manage renewable resources such as fisheries. Fisheries are often thought of as **common property resources**; these are managed collectively by a community that has rights to the resource, rather than owned by individuals. Strategies for managing common property resources in the Pacific include moratoriums, called *tabu* in the Pacific (periods or places where fishing is not permitted by local communities), and recognition of family or group access based on customary rights to harvest a resource. Recently, some countries have brought fisheries under government control or have regulated harvesting more formally through permits and quotas. Beyond the EEZs, fisheries are open to all and are vulnerable to the so-called tragedy of the commons, in which an open-access common resource is overexploited by individuals who do not recognize how their own use of the resource can aggregate with that of many others to degrade the environment—for example, overfishing a given species to the point of extinction. One of the greatest challenges in the sustainable management of fisheries is the lack of information about fish numbers, movement, and reproduction, especially in the Pacific.

Within this region's massive marine ecosystems, is the Great Barrier Reef, which is a UN **World Heritage Site**. Fringing the northeast coast of Australia, the reef is more than 2,000 kilometers (1,250 miles) long and easily visible from space. The view from the air does not reveal the real beauty of the landscape because the most attractive parts of the reef are underwater. Diving below the surface, the visitor encounters intricate and colorful corals and waving sea grasses that are home to millions of brilliant fish and majestic marine animals such as turtles, whales, and dugong (large seal-like mammals; **Figure 11.13a**).

The Great Barrier Reef consists of 3,400 coral reefs, incorporating more than 300 species of coral and hosting more than 1,500 species of fish and 4,000 different types of mollusks. The reefs were formed over millions of years from the skeletons of marine coral organisms in the warm tropical waters of the Coral Sea. The reef is Australia's second-most-popular foreign tourist destination, after Sydney. Tourists visit the reef and participate in activities that include fishing, diving, snorkeling, and reef walking. Today the reef is under pressure from trawling, chemical and sediment pollution, climate change, and coastal development (**Figure 11.13b**).

# HISTORY, ECONOMY, AND DEMOGRAPHIC CHANGE

As has already been noted, this world region has been through long, extended periods of isolation from human migration and has also witnessed increasing integration into the global economy. This region's history is distinguished by a number of key phases, which includes its early human habitation, the extension of populations into the Pacific,

(a)

(b)

▲ **FIGURE 11.13 The Great Barrier Reef** (a) The undersea landscape of the Great Barrier Reef has become a major destination for tourists who swim among the corals and grasses to view the colorful tropical fish. (b) A Greenpeace diver inspects corals of the Great Barrier Reef, which could be destroyed by global warming. The threat is a global phenomenon known as "coral bleaching," which occurs when water temperature rises, prevents the coral from reproducing fast enough.

European contact and colonization, and a dynamic postcolonial period. In some cases, parts of the Pacific remain colonized—either formally or through association—by Australia, New Zealand, France, the United Kingdom, and the United States. This historical geography has produced

economies that are globally and regionally significant—such as Australia, which plays such an important development role in other regions such as Southeast Asia—and underdeveloped, such as Papua New Guinea, because they have been positioned in a dependent relationship to larger global economic powers. Today, the region remains interconnected to the wider flows of migrants from other regions, including East (Chapter 8) and Southeast Asia (Chapter 10), as well as northern and southern Europe (see Chapter 2). Intraregional migration patterns, based on this region's changing political economy, further suggests the need to examine population dynamics within the context of this region's own interconnectivities.

## Historical Legacies and Landscapes

The early human history of Oceania is divided into two main phases: the migration of humans from Southeast Asia into Australia, New Guinea, and nearby islands about 40,000 years ago, and a second dispersal to the Pacific islands such as Fiji, Tonga, and Samoa about 3,500 years ago.

The early inhabitants of Australia are ancestors of today's Aborigines, who still preserve practices that reflect the early adaptations and modifications of the Australian environment. These practices include a complex spiritual relationship to the land, a nomadic lifestyle, and the use of fire in hunting. Gathering roots, seeds, grubs, insects, and lizards contributed essential calories and proteins to their diets, whereas people in some coastal areas constructed traps for stonefish and eel. As Aborigine populations grew, they may have reduced local populations of major game species such as kangaroos, but the most significant environmental change was the transformation of vegetation through the use of fire to improve grazing for game and to drive animals to hunters. Ecologists believe that over thousands of years, some Australian vegetation became more resistant to these fires. The only domesticated animal was the dingo, a dog that was probably imported from Southeast Asia. The Aborigine worldview linked people to each other, to ancestral beings, and to the land through rituals, art, and taboos. Their worldview is associated with the **Dreamtime**, a concept that joins past and future, people and places, in a continuity that ensures respect for the natural world.

The Maori arrived in New Zealand much more recently but are believed to have caused much more widespread environmental transformations than Australian Aborigines, in part due to the smaller extent of the island environment. They hunted the enormous moa bird to extinction, cleared as much as 40% of the original forests, and practiced agriculture based on shifting cultivation of sweet potatoes—a knowledge system they brought with them when they migrated to the country. The Maori migrated from Polynesia, where subsistence was based on fishing and the cultivation of root crops such as taro and yams and tree crops such as coconuts and breadfruit, all originally domesticated in Southeast Asia.

**European Regional Exploration**   Although Oceania had some early contact with other regions of the world, especially Southeast Asia, it was not until the mid-1700s that European explorers established the area for wider global trade and eventually colonization. Spanish and Portuguese sailors controlled Guam as a stop for galleons traveling from Manila to Acapulco and may have encountered Australia. The Dutch explored the west coast and south coasts of Australia and claimed it as Van Diemen's Land in 1642, later named Tasmania after the Dutch explorer Abel Tasman (**Figure 11.14**). Tasman also made contact with the Maori in 1642, but the encounter was violent, and the Dutch did not land, calling the region *Nieuw Zeeland* (after a region of the Netherlands). It was more than a hundred years later that the most enduring claim on the region was advanced by explorer Captain James Cook, who landed at Botany Bay, Australia, in 1770, claiming the land for Britain and calling the new territory New South Wales. Cook also explored New Zealand at this time and established harmonious relationships with the Maori.

**European Colonization of Australia**    Based on Cook's reports, the British government decided to people New South Wales by using it as a penal colony. The British sent boatloads of convicts to relieve pressure on their prisons but also to reinforce territorial claims and provide cheap labor for economic development. More than 160,000 convicts were eventually transported to Australia by 1868, half to New South Wales and half to Van Diemen's Land (Tasmania). Most were not serious criminals, and many have been identified as petty thieves or Irish political activists. The convicts worked for the government or were assigned to private employers. Many eventually gained freedom through pardons, often prompted by the desperate need for more farmers to produce food. Many other free settlers arrived in Australia. Some were given cash rewards for emigrating and were assigned convicts as laborers. The convicts' forced labor for colonists ended in eastern Australia by 1840 and in the west by 1868.

The initial goal of the settlements was self-sufficiency, but many, including Sydney, were unable to produce an adequate range of foodstuffs because of poor soils, plant disease, and a variable climate, even after convicts were released to increase the number of farmers. As a result, many settlements depended on imports of food and other goods from Britain. The most successful agricultural areas were in coastal valleys, such as the Derwent Valley in Tasmania, which produced wheat. The main exports from the initial settlements were associated with the whaling trade, such as whale oil and seal pelts, with Hobart as the main base of the South Pacific whaling fleet.

A momentous shift took place with the import of the first livestock to Australia, especially the Merino sheep, which thrived in central New South Wales and Tasmania. The first wool shipment to Britain was in 1807, and high prices globally encouraged expansion of the sheep industry in Australia. By 1831 more than 1 million kilograms (2.2 million pounds) of wool were being exported. By 1860 exports totaled 16 million kilograms (35 million pounds), and there were 21 million sheep in Australia. The demand was driven by the success of the British textile industry (see Chapter 2) and tied the pastoral economy of Australia as well as New Zealand to the industrial core. The Australian Agriculture Company was established in 1824 and invested £1 million to promote wool and other agricultural exports. Although the British government was initially reluctant to permit frontier settlements away from the closely supervised port communities, stockmen began to move inland, especially to the grasslands, such as the Bathurst Plain on the western

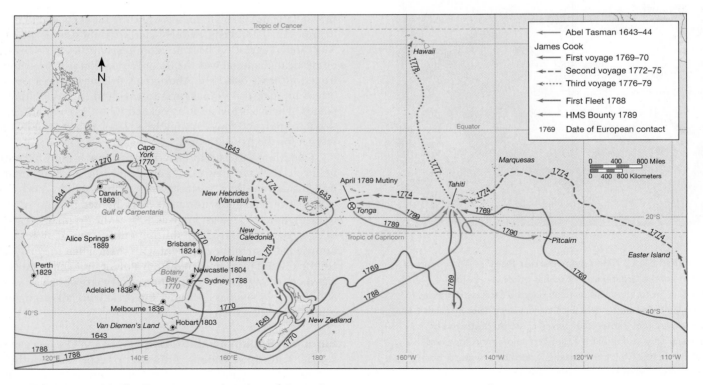

▲ **FIGURE 11.14  The European exploration of Oceania**  This map shows the sequence of European exploration and settlement of Oceania, including the voyages of Abel Tasman, Captain James Cook, and of the HMS *Bounty*. Cook made three voyages to Oceania, landing at Botany Bay and claiming Australia for Britain; he was ultimately killed in Hawaii.

slopes of the Great Dividing Range. As the European frontier expanded, it came into conflict with the Aborigines, who defended their traditional lands and lost their lives in the process.

In addition to sheep farming, commercial wheat production was centered in southeastern Australia, especially on the red-brown soils around Adelaide (see discussion of the fertile crescent earlier in this chapter). The frost-free climate of the central coast of eastern Australia also allowed for sugar plantations around Brisbane beginning in the 1860s, using indentured laborers brought in from Vanuatu and the Solomon Islands. Cattle were also introduced into the warmer and drier regions of central and northern Australia after wells were drilled into the Great Artesian Basin in the 1880s. Cattle grazed more lightly than sheep on the sparse vegetation (**Figure 11.15**). Development was further facilitated by the construction of railroads, which radiated from ports to terminuses at livestock yards, grain elevators, and mines. This transportation pattern increased the importance of the port cities but hindered later national integration, especially when it became evident that three different rail gauges had been selected by different colonies.

Another major transformation of the Australian economy occurred with the discovery of gold in 1851. Gold diggers were drawn by the thousands from all over Australia, England, and China in a gold rush fueled by rumors of giant nuggets weighing 30 kilograms (100 pounds) or more. The town of Broken Hill at the border of New South Wales and South Australia (**Figure 11.15**) became one of Australia's busiest mining communities, producing lead, silver, and zinc that were exported to Europe to sustain industrial development. These economic changes were accompanied by political ones, and the mid-1850s brought a great degree of self-government to Australia, with two-thirds of the legislatures elected by popular vote (and the rest appointed by the British) in the states of New South Wales, Victoria, South Australia, Queensland, and Tasmania. The boundaries between these states were mostly drawn as straight lines, regardless of physical features or indigenous land rights.

### European Colonization of New Zealand

The history of European settlement and economic development in New Zealand is closely tied to that of Australia. A sealing station was established on the South Island of New Zealand in 1792, but British sovereignty and the first official settlers' colonies were not established until the 1840s, initially as part of New South Wales. Small whaling settlements had established trading relationships with the Maori, who through such contacts obtained access to firearms, were exposed to disease, and were brought into a capitalist economy. Missionaries had also established settlements in the early 1800s and contributed to the transformation of Maori culture. The British annexed New Zealand in 1840 through the **Treaty of Waitangi**, a pact with 40 Maori chiefs on the North Island. This treaty purported to protect Maori rights and land ownership if the Maori accepted the British monarch as their sovereign, granted a crown monopoly on land purchases, and became British subjects. At the last minute before the treaty was signed, land agents purchased large areas of land around Cook Strait, often without identifying the true Maori owners. This land was held by the private New Zealand Association and included the sites of the cities of New Plymouth, Wellington, and Nelson.

Alarmed by European settlement, some Maori resisted the British and waged warfare for several years until suppressed in 1847. The introduction of sheep (**Figure 11.16**) prompted further settler expansion in search of pastures and a renewal of hostilities with the Maori during the Maori wars of the 1860s. The discovery of gold in 1861 on the South

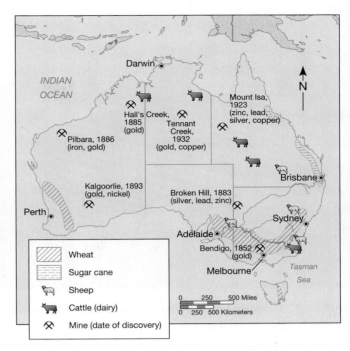

▲ **FIGURE 11.15 The development of livestock, wheat, and mining in Australia** This map shows the main areas of settlement, agriculture, and mining development in 19th-century Australia.

Island in the Otago region made Dunedin the largest settlement in New Zealand (with a population of 60,000) within ten years and fostered development of wheat production on the Canterbury Plains.

The next stage of the integration of Australia and New Zealand into a global economy was driven by a technological innovation, the development of refrigerated shipping, after 1882. This advance allowed both economies to expand or shift from producing nonperishables such as wool, metals, and wheat to the more valuable export of meat and dairy products. New Zealand became Britain's "farm in the South Pacific," trading its high-quality agricultural products for imported manufactured goods from Britain. Trade was facilitated by the opening of the Suez Canal in 1867 and the Panama Canal in 1914, which reduced the time and cost of ocean transport to Europe. Australia and New Zealand, like Canada, became staple economies that depended on the export of natural resources.

### European Colonization of the Pacific Islands

Initially the Pacific islands were of little interest to European explorers, and they were drawn into the world economy more slowly than Australia and New Zealand. During the Age of Discovery (see Chapter 1), Guam, Palau, parts of the Federated States of Micronesia, and the Mariana Islands became Spanish colonies. More generally, explorers brought European diseases to local populations who had no resistance to them, so the most serious impact was mortality.

But as Britain and France rose to power in Europe in the 18th century, a series of explorations set out for the Pacific. British explorer Samuel Wallis and Luis Antoine de Bougainville of France were made welcome by the people of Tahiti in the 1760s, and their reports of

▲ **FIGURE 11.16  Sheep farming**  The landscapes of Australia and New Zealand are still characterized by extensive grazing lands for sheep, initially a colonial enterprise designed to provide Britain's factories with wool. The development of refrigerated shipping in the 1880s boosted sheep farming for meat production. This photograph is from Otago, New Zealand, which has a population of more than six million sheep.

Society was very active in the Pacific, seeking conversions in Tahiti, Samoa, Tuamotu, Tuvalu, and the Cook Islands. The Methodists focused on Tonga and Fiji. The missionaries often worked through native chiefs, who were advised to alter local laws and traditions to conform to European principles, sometimes provoking local rebellions. Missionary activity altered traditional social ties, beliefs, and political structures. Hundreds of whaling ships called regularly at islands such as Tahiti, Fiji, and Samoa for supplies and maintenance. The discovery of sandalwood, a valued aromatic wood, attracted traders to Fiji, Hawaii, and New Caledonia; they were ruthless in their treatment of local people.

Coconut, the staple product of the Pacific, became part of European trade from about 1840 in the form of copra, dried coconut meat used to make coconut oil for soaps, and food. Some Europeans sought to establish plantations on islands such as Fiji and Samoa. A scarcity of cotton during the U.S. Civil War prompted the establishment of cotton plantations on Fiji in the 1860s. The same scarcity increased the need for laborers for the cotton plantations in Queensland, Australia; workers came from the islands. Beginning in the 1840s, people from the Pacific were kidnapped and enslaved in a process called *blackbirding* that brought thousands of laborers, collectively called *kanakas* (because many of them were of Kanak origin from the islands now known as Vanuatu), to Australian cotton and sugar plantations. This practice continued until 1904.

friendly people and abundance cultivated the myth of a tropical paradise (**Figure 11.17**).

At the beginning of the 19th century, the Pacific islands came into contact with missionaries, whalers, and traders. The London Missionary

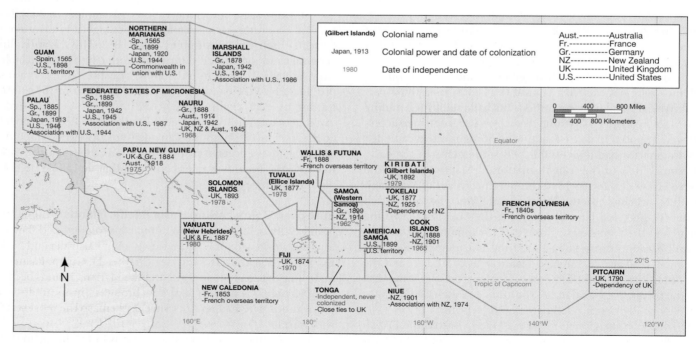

▲ **FIGURE 11.17  The colonization and independence of the Pacific**  Many Pacific islands are now independent from their former colonial rulers. This map shows the pattern and dates of colonial control, dates of independence of Pacific island countries, and current affiliations of territories. Note that only Tonga resisted colonization and that some countries, such as Nauru and Palau, were handed from one European power to another.

The islands were governed according to the different colonial styles of European nations modified to local conditions. Britain ruled through governors who incorporated native leadership into their administrations in a form of indirect rule. The Germans administered their Pacific colonies through commercial companies, and the French practiced direct rule and assimilation into French culture and institutions. The imposition of colonial rule met with resistance in the form of armed uprisings, alternative trading networks, political movements, and defiant behavior. The colonial powers restructured the economies and people of the Pacific islands in ways that left enduring legacies. The British brought large numbers of contract workers from India to work on plantations in Fiji, creating a divided society of ethnic Asians and Pacific islanders that produces significant political tensions even today.

### Political Independence and Global Reorientation

Independence was granted to Australia in 1901. The Commonwealth of Australia was established with a federal structure governing the six states of Western Australia, South Australia, Queensland, New South Wales, Victoria, and Tasmania and administering the federal territories of the Northern Territory and (after 1908) the Australian Capital Territory around Canberra. New Zealand declined to join this new nation and chose dominion status as a self-governing colony of Britain in 1907. Both countries soon gained their own colonies in the Pacific. In 1901, Britain turned over Papua New Guinea to Australia and the Cook Islands to New Zealand. New Zealand was granted a mandate over Samoa in 1920 and jurisdiction over Tokelau in 1925.

By taking over as regional powers, both Australia and New Zealand also established themselves globally. They further exercised this new global authority by fighting with the Allies in World War I in the Australian and New Zealand Army Corps (Anzac), most notably at Gallipoli, Turkey, in 1915, when more than 33,000 of them died. The war contributed to the development of a steel and auto industry in Australia and to increased agricultural exports at higher prices because of difficulties the war created in international trade. This further established these two countries as part of a growing global political–economic core of states.

Importantly, World War II also further influenced Australia's geopolitical orientation. Although Australia fought in defense of Britain in Europe and North Africa, the rapid advance of the Japanese in Asia and the Pacific, including the capture of 15,000 Australians in Singapore and the bombing of the northern Australian city of Darwin, caused Australia to look to the United States more directly as an ally. American and Australian troops fought together in New Guinea and the Pacific islands, forging enduring bonds that were formalized in the ANZUS security treaty between Australia, New Zealand, and the United States, signed in San Francisco in 1951. As the Cold War deepened after 1950, East and Southeast Asia became the focus of U.S. concern about communist expansion from China. As the United States became more involved in East Asia (see Chapter 8) and then Indochina (see Chapter 10), both Australia and New Zealand sent troops to Korea and Vietnam.

### World War II and Independence in the Pacific

Beyond Australia and New Zealand, World War II also marked a critical turning point in the history of the Pacific, with thousands of foreign soldiers fighting across the islands and constructing military bases. In the process, many islanders lost their lives. The most significant impacts were in the western Pacific, where the Japanese advanced from their colonies in Micronesia (such as Palau) to Guam, New Guinea, and the Solomon Islands and then attacked the U.S. base at Pearl Harbor in Hawaii in 1941. Three years of intense and bitter warfare on land, air, and sea included many famous battles such as those of Guadalcanal and Saipan, in which the Allies, especially the United States, retook the islands from Japan. The damage from bombing was extensive. After the war, the United States was determined to maintain military bases in the Pacific, especially in response to the Cold War. Despite this, self-government began with elected governments and small independence movements. Western Samoa became independent from New Zealand in 1962 (and dropped the "Western" from its name in 1997), Nauru from Australia in 1968, Fiji and Tonga from the United Kingdom in 1970, Papua New Guinea from Australia in 1975, and the Solomons, Kiribati, Tuvalu (from the U.K.) and Vanuatu (from the U.K. and France) by 1980 (see Figure 11.17).

## Economy, Accumulation, and the Production of Inequality

Within the context of this region's historical geography there have been numerous attempts to regulate economic production and accumulation in relation to the changing demands of the global economy. The geopolitical positioning of certain countries, such as Australia and New Zealand, has meant that some have benefited more from global economic integration, whereas other countries, which remain dependent on primary commodities, for example, have found themselves in a position of underdevelopment. Like other world regions, however, the countries of Oceania have often tried to position themselves in relation to a wider global economy either through the development of protectionist measures or through resistance to global capitalist development. The results of these attempts have been mixed, as protectionism has been broken down. In other cases, self-sustaining local economies offer hope that global economic integration is not a fait accompli for everyone.

### Australia and New Zealand: Isolation and Integration

Beginning in the 1920s, policies to substitute expensive imports with domestic production resulted in heavy subsidies of manufacturing and high tariffs on imported goods, especially in Australia. The goal was to create new jobs, diversify the economy, and reduce the sensitivity to global demand for wool and minerals. As in other regions, these import substitution policies had mixed success. Although Australia and New Zealand had middle-class populations with a demand for manufactured goods, the overall market was small, and labor costs were high as a result of a strong tradition of labor union activism. The new industries were often inefficient because they were protected from competition in the world market. New Zealand had a particularly high level of government involvement in the economy, with state-run marketing boards controlling the export of commodities, such as wool, meat, dairy products, and fruit, and government ownership of banking, telecommunications, energy, rail, steel, and forest enterprises.

Although Australia protected manufacturing and subsidized agriculture, there were few barriers to foreign investment. Many sectors had high levels of foreign ownership, including minerals and land, especially by British firms. During the 1970s, investment patterns began to change, with Asian, especially Japanese, capital starting to flow into Australia. Another major change occurred when the U.K. decided to enter the European Community in 1973 and was forced to end preferential trade relationships with British Commonwealth nations, including Australia and New Zealand. New Zealand was particularly hard hit by the loss of guaranteed markets, but the shock provided an impetus to seek new markets in Asia.

Both Australia and New Zealand decided to reduce government intervention in and regulation of the economy in the 1980s. The Australian government deregulated banking, reduced subsidies to industry and agriculture, and sold off public-sector energy industries beginning in about 1983. The move to the neoliberal policies of free trade and reduced government was even more dramatic in New Zealand. There, sweeping policy reversals eliminated agricultural subsidies, removed trade tariffs, reduced welfare spending, and privatized government-owned enterprises, including airlines, postal services, and forests. Although these policies succeeded in reducing the national debt, they also increased economic inequality and unemployment.

The most rapidly growing sector during the 1980s and 1990s was services, with employment in this sector increasing from 58% in 1980 to 71% in 2006 in Australia and from 58% to 66% in New Zealand. Finance, tourism, and business services expanded the most and were associated with an increase in female employment and considerable foreign investment. The region has become a major international tourist destination. Of the more than 6.4 million visitors in 2003, many came from Asia, especially Japan, China, and Singapore, as well as the United Kingdom. A majority of visitors to the region in 2007 traveled to Australia (**Table 11.1**).

The travel numbers to Australia, although robust, illustrate a slight downturn in the total number of visitors from just over 5.5 million in 2005. The slowing in tourism volume is tied, in part, to the global economic recession, which hit Oceania more generally, and Australia and New Zealand in particular, in 2008 and 2009. In fact, although some predicted that countries such as Australia could weather the global recession because of its strong ties to China—considered one of the world's fastest-growing economies (see Chapter 8)—recent trends have shown that China's vulnerability is affecting Australia as well. In New Zealand, shifting global patterns of consumption led to drops in the sales of key international commodities, such as milk, lamb, and timber in 2008, demonstrating how this country is also subject to shifts in the wider global economy.

**The Changing Face of the Agricultural Sector**  The future of agriculture in Australia and New Zealand in the wake of economic restructuring and global economic change is unclear, as competition from Latin America and Asia, the high cost of inputs, the loss of subsidies, and the changing structure of demand create a new geography of international agricultural trade. That said, Australian wine has gained an excellent reputation in world markets, especially the wines of the Barossa Valley north of Adelaide and the Hunter Valley of New South Wales. Australian wine exports are the fourth largest in the world (behind France, Italy, and Spain). Wine is produced with the most advanced technology and innovative marketing and supports a thriving tourist industry similar to that of the Napa Valley in the U.S. state of California.

New Zealand agriculture has had to change significantly in response to the restructuring of the global food system over the last three decades. Agricultural policies in 1984 abruptly removed most price supports, trade protections, and farm subsidies and required farms to pay for extension services, water, and quality inspections. New Zealand agriculture was thrown into a global free market, whereas most other devel-

### TABLE 11.1  Top 10 Tourism Source Countries for Visitors to Australia, 2007

| Country of residence | Number of visitors |
| --- | --- |
| New Zealand | 975,504 |
| United Kingdom | 690,895 |
| Japan | 581,710 |
| USA | 429,258 |
| China (excluding SAR/Taiwan) | 318,856 |
| South Korea | 244,002 |
| Singapore | 224,454 |
| Germany | 144,469 |
| Hong Kong | 138,761 |
| Malaysia | 137,682 |
| **Total top 10** | **3,885,591** |
| **Total of all arrivals** | **5,156,650** |

(*Source*: Tourism Australia, 2007.
http://www.tourism.australia.com/content/Research/Factsheets/Top_10_Markets_Mar_07_20(2).pdf [accessed 8 June 2009].)

oped countries, including the United States, Canada, and Europe, maintained considerable state regulation and support for their agricultural systems. Although New Zealand farmers coped by adjusting their herd sizes and changing their crop mix, some farmers went out of business, and their properties were horizontally integrated into larger farms. The landscapes of agricultural regions such as the Hawkes Bay region of the North Island began to change, as farms switched into fruit (and wine) production (**Figure 11.18**), often associated with increases in pesticide use. At the same time, New Zealand was also one of the first countries to adopt certification for organic agricultural products, and there is a thriving domestic market for sustainably grown foods. And, the transnational agribusiness firm H. J. Heinz has purchased New Zealand agricultural processing enterprises with the goal of supplying growing Asian markets, and the international corporation ConAgra owns grain companies, feedlots, meat-processing, and wholesale distribution in Australia.

**The Pacific Islands in the Contemporary Global Economy**  Many Pacific islands are now integrated into the world system through their dependence on imported goods, transfer payments, improvements in transportation, and the emergence of international tourism as the major source of foreign exchange. With their long experience of global trade, investment, and migration, the Pacific islands have in many ways been less affected by more recent globalization of trade, capital flows, culture, and labor than have other world regions. They are sometimes called MIRAB (migration, remittances, aid, and bureaucracy) economies because of their dependence on labor migration, jobs with foreign governments, and foreign aid.

▲ **FIGURE 11.18 Kiwi production**  These kiwi orchards on the North Island of New Zealand in the Bay of Plenty are surrounded by lines of trees that protect the delicate kiwi fruit from strong winds.

At the same time, most of the nations in Polynesia spend twice as much on imports as they gain from exports, and they must finance the deficit through borrowing, foreign aid, or remittances from citizens working elsewhere. However, compared to countries in Latin America and the Caribbean or Southeast Asia, total international debt is relatively low in Oceania, with annual interest payments averaging less than 5% of GDP. This low level of debt has been maintained despite the trade deficits because of the large flows of official aid, remittances from overseas workers, and interest on savings held outside the country. Kiribati and Nauru finance their imports from offshore financial assets, including interest on profits saved from phosphate mining.

**Tourism and the Pacific**   The widespread image of the Pacific islands as a vacation paradise has origins in the accounts of the first European visitors, who described tropical abundance and peaceful locals, including the romantic depiction of island women as exotic and available partners (**Figure 11.19**). The notion of the Pacific as a tourist destination grew as international air and cruise routes included stops at island groups such as Hawaii and Fiji, often en route to Australia or Asia. But the big boom in Pacific tourism occurred from about 1980 onward as air travel became more accessible, and increased numbers of North Americans and Asians (especially Japanese) sought luxury and exotic vacations. The most popular Pacific island destinations, after the U.S. state of Hawaii, are Guam, Fiji, and Tahiti; the total number of tourists to the Pacific islands (Hawaii excepted) reached more than

three million a year in the late 1990s. The significance of international tourism to economies and employment in individual countries is tremendous and is the major source of foreign exchange for the Cook Islands, Fiji, French Polynesia, Samoa, Tonga, Tuvalu, and Vanuatu. Challenges faced by these tourism-dependent economies include vulnerability to international trends in tourism, political unrest that dissuades tourists, the need to ensure that the benefits of tourism reach throughout the population, and the need to minimize negative effects on local cultures and environments.

**Poverty and Inequality**   Although Oceania has generally higher incomes and better living conditions than many other world regions, there are significant differences between and within countries of the region. Australia and New Zealand are distinctive for their very high average incomes as expressed by the per capita annual gross domestic product (GDP) at almost $36,600 (in PPP) for Australia and more than $26,400 for New Zealand in 2010. Some Pacific islands have an average GDP in PPP near or above $10,000 per person as a result of their associations with the United States (Guam) or France (French Polynesia, New Caledonia). Others, especially Kiribati and Vanuatu, register less than $6,000 in GDP per capita per year. Although there is little published information on the distribution of incomes within most of the smaller countries, and although inequality is apparently less than in many other world regions, there is persistent poverty throughout the region, including in Australia and New Zealand, where the Aborigine and Maori populations are particularly disadvantaged. Australia and New Zealand were for many years reputed to have strong welfare systems and equitable societies,

▲ **FIGURE 11.19 Bora-Bora**  Luxury tourist resorts of Bora-Bora in French Polynesia attract tourists by describing this location as the "islands of dreams," an "emerald in a setting of turquoise, encircled by a necklace of pearls." Tourists are invited to purchase crafts and to view the traditional dances of "beautiful Polynesian women." Visitors to these resorts bring valuable foreign exchange to the economy, but the resorts put pressure on the local resource base and encourage local people to modify their livelihoods and rituals to please foreign visitors.

at least for the nonindigenous populations. However, income inequality has increased in the last two decades, and the government has reduced or privatized social services, especially in New Zealand. Larger populations of single parents, the elderly, refugees, and workers in low-paid service-sector jobs are also diminishing the overall ranking of Australia and New Zealand as places where everyone can make a good living.

Monetary measures such as GDP and PPP are of limited use where many people are living in economies based on exchanges and barter or on subsistence. The concept of **subsistence affluence** has been used to describe Pacific island societies: Monetary incomes may be low, but local resources such as coconut and fish provide a reasonable diet, and extended family and reciprocal support prevent serious deprivation. Adequate diets and relatively effective health and education systems contribute to comparatively high life expectancies and literacy and low infant mortality throughout the Pacific. Literacy is above 90% for both men and women in much of the region. Papua New Guinea, Kiribati, and Vanuatu, on the other hand, have higher infant mortality and lower literacy than other parts of Oceania.

## Demography, Migration, and Settlement

Oceania is one of the least-populated world regions, with just over 35 million people, of which 21.9 million live in Australia, 6.4 million in Papua New Guinea, and 4.3 million in New Zealand. Most people live along the coasts of the larger land areas, especially along the coast of southeast Australia and coastal Papua New Guinea. Overall population densities of the larger countries are very low: 2.5 people per square kilometer (6 per square mile) in Australia, 12 per square kilometer (31 per square mile) in Papua New Guinea, and 15 per square kilometer (39 per square mile) in New Zealand. The smaller islands, in contrast, can have fairly high population densities, reaching more than 150 per square kilometer (300 people per square mile) in Guam, the Marshall Islands, Nauru, and Tonga (**Figure 11.20**).

The overall population of the region is growing quite slowly (at 1% per year), and it will take 65 years for the population to double at current growth rates. This slow growth is in large part due to the very low growth and fertility rates in Australia and New Zealand. Women in these two countries bear an average of less than two children during their reproductive years. The low birthrate, combined with long life expectancy, is creating a growing proportion of older residents and raising concerns about the provision of services such as Social Security, health care, and pensions for the aging population. Fertility rates are higher in the Pacific islands, ranging from 2 children per woman during her childbearing years in French Polynesia to 4.1 children per woman in Kiribati.

**Immigration and Ethnicity** There were at least half a million Australian Aborigines and perhaps a quarter of a million New Zealand Maori when Europeans arrived. The introduction of European diseases and the violence of some colonial encounters reduced native populations significantly and resulted in majority European populations in Australia and New Zealand within a century or so of initial settlement. The contemporary ethnic composition of Australia and New Zealand is strongly influenced by a larger history of immigration.

Almost a quarter of Australia's current residents were born elsewhere. As noted earlier, the first European settlers were mostly English. Some Chinese arrived in Australia during the gold rush of the 19th century. Migrants of Irish and Scottish origins, as well as some Germans, were also common.

Australia's immigration policy sought to maintain a European "look" through the adoption of the **White Australia policy** after independence in 1901. This policy essentially restricted immigration to people from northern Europe through a ranking that placed British and Scandinavians as the highest priority, followed by southern Europeans. By 1936 more people were coming from southern Europe, especially Italy and Greece. After World War II, the government made extra efforts to attract immigrants as labor for manufacturing, as a source of military volunteers, and in response to humanitarian pressure to resettle refugees from eastern Europe.

In 1973, the racist restrictions on immigration were removed and replaced with skills criteria, and a new wave of immigration from Asia began, with large numbers of migrants from Vietnam, Hong Kong, and the Philippines. By the end of the 20th century, Australia's population was much more diverse, with about 73% of the population of British or Irish heritage, 20% from elsewhere in Europe, 5% from Asia, and only 2% Aborigine. That all said, the economic crisis of 2008 to 2009 has slowed Australian in-migration, as the new Labor Party government has had to limit migration as unemployment levels increased throughout the country.

New Zealand also practiced an immigration policy designed to attract white immigrants, and prior to World War II this included a bias not just against nonwhites but also against Irish Catholics. After the war, labor shortages were met by temporary labor migration from the Pacific islands. More recent changes in immigration policy have opened New Zealand to more permanent settlers from the Pacific as well as to Asian immigrants who can bring capital with them. In 2005, New Zealand was still dominated by people of English and Scottish ancestry, but with a growing percentage of people of Pacific island (around 6.5%) or Asian (around 6.6%) heritage and a significant group (more than 15% who are full or part) of Maori population.

The Pacific islands have a much higher proportion of indigenous people, averaging more than 80%, with Asians (14%) and Europeans (6%) representing the remainder. The exceptions are Fiji, which has a very large Asian Indian population associated with the importation of contract labor during colonial times; New Caledonia, where about 40% of the population are indigenous, 40% French, and 20% descendants of Southeast Asian contract workers; and Guam, where there are significant American and Filipino populations.

**Oceanic Diasporas** One of the largest diasporas is internal to Oceania and is the emigration of Pacific islanders to work and live in New Zealand and Australia. This emigration has been encouraged in the last 50 years by labor needs, refugee movements, and lenient rules allowing migration from current and former colonies. It has also been encouraged more recently by global climate change (see chapter introduction). More people from Niue, Tokelau, and the Cook Islands now live in the capital of New Zealand, Auckland, than remain on the islands, as well as about one-eighth of all Samoans and migrants from Tonga and Fiji. Australia has accepted thousands of refugees from East Timor in Indonesia and from Fiji.

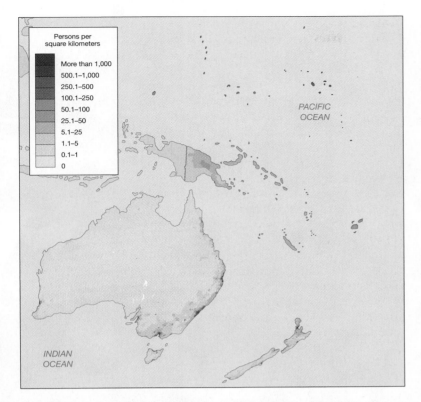

▲ **FIGURE 11.20 Population density map of Oceania, 2000** Although most of Oceania appears thinly populated on this map, several small islands have high population density. Australian population is concentrated in the cities and in the southeastern region of the country.

A second diaspora is from the Pacific islands to North America, especially to Hawaii and California. Many Samoans and Tongans have moved to the United States, and recent unrest in Fiji sent a wave of several thousand Asian-Fijians to the United States and Canada. There is a small emigration flow of Australians and New Zealanders to Europe and North America, including people of British heritage returning to retire in Britain and young people seeking educational and work opportunities in England, Canada, or the United States. Within Australia there is an internal migration flow from the population concentration in southeastern Australia northward along the east coast and to western Australia.

**Urbanization** According to United Nations projections, Australia is about 94% urban, whereas New Zealand is just over 86%, and the Micronesian region is almost 80% urban. This contrasts with other parts of the region, such as the wider Polynesian region, which is about 45% urban, and Papua New Guinea, which is just over 13% urban. Despite these differences, Oceania is highly urbanized in contrast to other world regions, including sub-Saharan Africa or Southeast Asia, with an average percent urban of around 73%. These high levels of urbanization are due, in part, to the settlement patterns of European descendents and the harsh climatic conditions of much of the region, which has limited large-scale rural development. In fact, despite mythic images of Outback Australia, most Australians

live close to the coast. One exception is the capitol of Australia, Canberra, which is found inland between Melbourne and Sydney, Australia's two largest cities. This planned city was chosen as a site for the Australian capitol because of its proximate relationship between the country's two major cities—it gave no distinct advantage to either Melbourne or Sydney. Even so, Canberra is only a few hours drive from the coast.

Within the region, Sydney is the largest urban metropolitan space. With a population of almost 4.4 million people, one in five Australians lives in and around Sydney. It is a place with high levels of car ownership, and the population sprawls into surrounding suburbs. The wealthier areas of the city are to the east, with poorer residents concentrated in the western suburbs (**Figure 11.21**). As immigrants from different regions settled in groups of similar origin, Sydney has developed several ethnic neighborhoods, including Greek, Italian, and Vietnamese. Many Aborigines who migrated to the city settled in one suburb, and poverty and discrimination has made their neighborhood a focus of indigenous action and social programs. A number of older downtown neighborhoods, such as Paddington, have been renovated by higher-income groups in a process of gentrification (older working-class neighborhoods are converted to serve higher-income households). Some of the older warehouse and manufacturing districts near the harbor have been redeveloped into shopping, museum, convention, and entertainment structures, such as the Darling Harbor district. Sydney is also surrounded by parks and protected areas and by magnificent beaches that encourage swimming, surfing, and sailing.

In contrast to Sydney is the city of Melbourne, which is the historic manufacturing center of southeastern Australia. Located on a sheltered harbor on Port Phillip Bay, Melbourne has a local population of almost four million people. The city first developed as the transport hub for the 19th-century gold rush; it grew further when thousands of refugees and migrants were sponsored to come to Australia after World War II and were sent to work in Melbourne's industrial sector, which included textiles, clothing, and metal processing. Contemporary industry includes chemicals, food processing, automobiles, and computers. The Latrobe Valley, east of Melbourne, has the world's largest reserves of brown coal (or lignite) used to fuel Australia's thermoelectric power stations and for export. Australia is the world's biggest exporter of hard coal—one of the reasons the current government is reluctant to ratify the Kyoto Protocol to combat climate change.

In New Zealand, the city of Auckland remains the largest, and plans are in place to extend the boundary of the city to create an administrative "super city" in the area surrounding Auckland. In fact, the plan titled "Making Auckland Greater" would subsume seven cities into one Auckland urban space. This would bring together a population of 2.5 million people, making this new "super city" the third-largest metropolitan area and the largest single administrative urban space in the region. The push to create this super city is tied to the needs of the population to better integrate transportation networks as well as provide better services more generally to the growing population. The political establishment of this super city is not without controversy, not the least of which is tied to the politics of representation and who will be given seats on the new super city council (see "Multiculturalism and Indigenous Social Movements" section later in this chapter).

▲ **FIGURE 11.21 Sydney harbor** Two architectural symbols dominate the landscape—the Sydney Harbor Bridge, completed in 1932 as one of the longest steel arch bridges in the world, and the Sydney Opera House, with its brilliant white sail- or shell-shaped roof. These images became familiar to many worldwide during the broadcasts of the 2000 Summer Olympics, which brought thousands of visitors to the city. Although Sydney has some manufacturing, the city economy is overwhelmingly service oriented, focusing on trade, banking, and tourism.

# CULTURE AND POLITICS

The culture and politics of this region are, as we have already seen, intimately interconnected to the wider global processes of migration, colonialism, conflict, postcolonial development, and environmental change, to name just a few. For example, the linguistic and religious practices of the people in this region are partially related to the region's colonial experience, whereas the culture of migration in the region is tied to both the long-term experience of movement and recent environmental and economic changes that are creating push and pull factors within and beyond the region. Political tensions and the emergence of social movements in Oceania are also tied into the complex politics of race, particularly in Australia and New Zealand, where indigenous rights are organized quite differently in relation to white settler majorities.

## Tradition, Religion, and Language

British settlement in Australia and New Zealand resulted in English being the most widely spoken and official language. Of an estimated 260 interrelated languages spoken by Australian Aborigines, and unique to the continent, many have become extinct or have only a few surviving speakers. Only Mabuiag (the language of the Western Torres Strait islands) and the Australian Western Desert languages are spoken by more than a few hundred people. Many Aborigines no longer speak anything but English, although that English can be modified locally.

The British heritage of Australia and New Zealand is reflected in the dominance of the Christian religion—mostly Protestant in New Zealand and about half Protestant and half Catholic in Australia—although many people report themselves as having no strong religious beliefs at all. Catholicism was brought to the region by the Irish as well as by southern European groups, particularly Italians, whereas the Greek Orthodox religion can be found in key ethnic enclaves in cities, such as Sydney and Melbourne. Non-Christian, local beliefs are still important in more remote islands and regions, especially in Papua New Guinea, and some communities have fused local and Christian practices by combining harvest rituals with the celebration of Christmas, for example. As a result of Asian immigration, both Buddhism and Hinduism are growing in significance, and many Aboriginal communities maintain their own locally situated religions, despite a loss of local linguistic diversity in the wake of English-language domination in Australia. In New

Zealand, Maori communities abandoned many local beliefs when they converted to Christianity and lost their land in the 19th century, but there is now a revival of religion, ritual, and the teaching of the Maori language, which was made an official language of New Zealand in 1987, creating New Zealand's modern bilingual society. Almost all Maori speak English, and about 100,000 also speak or understand Maori, a language related to those of Polynesia and especially to Hawaiian and Samoan.

The Pacific islands have an enormous number and variety of languages, which can be divided into two general families—Austranesian and Papuan. It is believed that almost 20% of all living languages are spoken on the island of New Guinea and that most are not understood by even the people in the next valley—creating a fascinating linguistic geography. To compensate for Papua New Guinea's 820 distinct living languages, a widespread trade language (or *lingua franca*) called *Tok Pisin* is used that combines local and English words into pidgin. Some island groups also have a large number of languages. For example, Vanuatu has 110 living languages, many spoken by at least 500 people, and the Solomon Islands have 70 different languages, most with at least 500 speakers. This diversity of languages has been explained by the fragmentation and isolation of New Guinea's physical geography and Vanuatu's many islands. Linguistic diversity generally decreases toward

the eastern Pacific. The missionaries were successful in converting most of the Pacific islanders to Christianity, with a range of Protestant denominations and Catholicism prevalent in French Polynesia. Hinduism is important in Fiji among the Asian Indian population. Exposure to the United States and its allies during WWII also transformed some of the values and goods found on the islands. For example, canned foods such as Spam, as well as U.S.-style music, clothing, and sports, became popular.

## Cultural Practices, Social Difference, and Identity

If the story of Oceania is partially about the hegemony of white settler life and experience, the politics of social difference and identity in this region today is tied to the diversity of cultural practice as well. As is already clear, the indigenous cultures of Australia, New Zealand, and the Pacific islands have been disrupted by contact with the rest of the world. They have sometimes become transformed into societies oriented to cultures based on the demands of tourism and the need to strengthen political identity. In Australia, Aboriginal cultures, based on strong spiritual ties to land, were marginalized and homogenized when Aborigines were removed from their ancestral lands and resettled on reserves. Aboriginal art, often based on rock art designs in dotted forms, silhouettes, and so-called X-ray styles, has become very popular in contemporary markets (**Figure 11.22a**), and native dances and songs, such as those that are part of the social gathering known as corroboree, are performed for tourists. Aboriginal symbols have been appropriated for major events, such as the Sydney 2000 Olympic Games, whereas Aboriginal land rights have remained contested (see Figure 11.25). In some ways, it is easy to see how Aboriginal culture has been appropriated for a global consumer and tourism market, whereas these representations fail to fully appreciate the complexity of Aboriginal life in Australia.

Maori tradition—such as the welcome ceremonies that include the *hongi* (pressing noses together) and the *haka* war dance (**Figure 11.22b**) now performed at international sporting events—is celebrated as integral to New Zealand's official bicultural identity. Maori architecture includes distinctive carved and decorated meeting houses called *whare*. Artistic expressions include intricate carvings such as those found on war canoes and decorative masks and tattoos (*moko*), which are also found on other Pacific islands.

Overall, society in Australia and New Zealand is still influenced by British legacies, including the significance of sports such as cricket and rugby, where the national teams, such as the New Zealand All Blacks rugby team (see Figure 11.22b), have gained international renown and have enthusiastic, if not fanatic, local support. Australia and New Zealand have also produced a series of award-winning films and novels, as well as two of the world's most popular opera singers—Australian Joan Sutherland and New Zealander and part-Maori Kiri Te Kanawa. The echoes of the colonial relationship with Britain provoke considerable ambivalence, and there have been strong attempts to establish distinct identities by embracing indigenous traditions, new immigrant cultures such as those from Asia, or the particular livelihoods and landscapes of the Outback frontier or the surfing beach.

**Regional Identities in the South Pacific** The islands of the Pacific form a distinctive image in the minds of most of the world—tropical paradises where local people fish, collect coconuts, and make crafts while tourists relax on beautiful beaches and swim in peaceful lagoons fringed with coral reefs. The Pacific islands are also marked by many differences and are often divided into three broad groups, defined by geography and ethnicity (see Figure 11.1). **Melanesia** is the region of the western Pacific that includes the largest islands of New Guinea, the Solomon Islands, Fiji, Vanuatu, and New Caledonia. Only the eastern half of New Guinea (the country of Papua New Guinea) is included in Oceania because the western half is the Indonesian province of Irian Jaya and is usually included in Southeast Asia (see Chapter 10). **Micronesia** (meaning "small islands") is the group of islands at, or north of, the equator, including the independent countries of Nauru, Kiribati, Palau, the Marshall Islands, and the Federated States of Micronesia; the U.S. territory of Guam; and the Commonwealth of the Northern Mariana Islands in association with the United States. **Polynesia** (meaning "many islands") comprises the eastern Pacific islands. The U.S. state of Hawaii is sometimes considered part of Polynesia because of ethnic and linguistic links between the indigenous Maori, Hawaiiana, and people of the eastern Pacific islands.

**Gender Roles, Sexuality, and Identity in Oceania** Gender roles in Oceania are influenced by stereotypes, but roles are as rapidly changing and complex as in any other world region. In Australia, the image of the frontier rancher or miner is associated with heavy drinking, gambling, male camaraderie, and a tough, laconic attitude epitomized by movie characters such as those in *Crocodile Dundee* or the lone outlaw

(a)

(b)

▲ **FIGURE 11.22 Aborigine and Maori culture** (a) A Sotheby's auction house employee walks past the Aboriginal artwork titled "Wayampajarti area," a collaborative work by Jukuna Mona Chuguna and Ngarta Jinny Bent in Sydney. Sotheby's international clientele bid on this and other pieces of Aboriginal art at an international auction in July 2009. (b) The New Zealand national rugby team—the All Blacks—perform a traditional Maori *haka* dance prior to an international match against South Africa.

portrayed in *Mad Max* (**Figure 11.23a**). But other films, such as *Priscilla, Queen of the Desert*, celebrate an alternative image of Australia as a safe space for gay and transgendered identities (**Figure 11.23b**). Sydney's Mardi Gras parade has become a celebration of gay culture and a major tourist attraction. Many women in Australia and New Zealand have shifted from roles as traditional housewives into a multitude of careers and to senior political positions (such as Prime Minister Helen Clark of New Zealand) supported by a strong feminist movement.

It is nearly impossible to generalize about the cultures of the Pacific islands, which are as varied as the many languages. Some of the most distinctive social and material forms include the strict separation of the male and female in much of New Guinea and Melanesia; the tradition of ritual warfare and reciprocal gift exchange; the importance of local leaders, or "big men"; and close links with kin and extended families. Pigs are still considered a measure of wealth on many islands, including New Guinea, and a traditional plant, kava, is consumed as a recreational and ritual relaxant on many islands. Polynesian cultures have a strong orientation to the ocean. Natives have been stereotyped by some explorers, anthropologists, and tourists as sexually promiscuous and living an easy life of tropical abundance. Contemporary cultures on the Pacific islands reflect the tensions between the maintenance and revival of traditional cultures, their selective construction for the tourist industry, and the widespread penetration of global culture, especially formal education, television, and processed foods.

Some of the more serious social problems in the region include alcohol abuse and high levels of domestic violence. Papua New Guinea has some of the highest levels of violence against women in the world. UNICEF estimates that one million children may be directly affected by domestic violence, but women's organizations are emerging in the region to combat what is considered a widespread problem in the country. Through the development of safe spaces—shelters and support groups—women in Papua New Guinea are building "safe houses" for women who may be subject to violence.[4]

## Territory and Politics

Oceania has seen regional changes in recent years, including the shift in alignment from Europe to North America and Asia and the challenges of coping with its relative geographic isolation within a global economy. The stability of some independent democracies and dependencies in the Pacific has been threatened by internal tensions, whereas political and economic integration has been sought through regional cooperation agreements. Moreover, glaring inequalities within Australia and New Zealand have highlighted the fate of indigenous groups, while at the same time these countries have embraced multicultural identities. Oceania's political geography is thus a complex, demanding assessment of the relationship between broader geopolitical forces and locally mediated conflicts and tensions.

**Political Stability**   Oceania is relatively stable compared to many other regions, whereas a number of smaller or resource-poor islands have maintained close associations with or are still under the control of the United States or New Zealand. U.S. dependencies receive financial

(a)

(b)

▲ **FIGURE 11.23 Gender and sexual politics in Australia**
(a) The international symbol of Australian Outback masculinity is typified in the Hollywood representation of Crocodile Dundee. Although only a small percentage of the Australian population lives in the Outback, images of Australian masculinity are often incorrectly identified with life in this region. (b) Cast members from the stage musical version of *Priscilla, Queen of the Desert* attend an Australian awards show in full regalia. This international stage production follows the film of the same name. The film and show have brought international attention to the complicated politics of sexuality in Australia.

subsidies, called *transfer payments*, in return for military base sites and security control. France has maintained its control over the several islands, insisting that they are integral parts of the French state. The continued rule in New Caledonia is supported by the descendants of French immigrants but opposed by the indigenous Melanesian Kanaks. New Caledonia remains an important producer of nickel for the French,

[4]UNICEF. (2009). "At a Glance: Papua New Guinea." [accessed 25 May 2009].

# ANTARCTICA IN THE GLOBAL IMAGINATION

It is always difficult to categorize Antarctica within the normal groupings of world regions, but in some ways it fits well within Oceania because of the importance of the marine environment, its relationship with New Zealand and Australia, and its relative experience as a simultaneously globally interconnected and isolated regional space. Its relative isolation is tied largely to its climate. Antarctica temperatures average –51°C (–60°F) during the six-month winter, when the sun does not rise above the horizon and the continent is in perpetual twilight. By September each year, half the surrounding ocean is frozen, creating a vast mantle of Antarctic pack ice with an area of 20 million square kilometers (32 million square miles) and a thickness of more than 2 meters (6.6 feet).

Although scarcely inhabited by humans, there is a great deal of interest in Antarctica, both from the point of view of scientific research and from the point of view of the potential exploitation of reserves of natural resources such as deposits of iron ore, coal, gas, and oil that may lie beneath Antarctica and the seas around it. International relations on the continent are governed by the **Antarctic Treaty**, which covers the area south of 60° S. The treaty, created in 1958 and now signed by 45 countries, bans nuclear tests and the disposal of radioactive waste and ensures that the continent can be used only for peaceful purposes and mainly for scientific research. In 1991 the treaty added a 50-year ban on mineral and oil exploration. Nevertheless, several countries—Australia, Argentina, Chile, France, New Zealand, Norway, and Great Britain—still claim specific slices of the Antarctic pie, hoping to be able to assert rights to offshore fisheries and onshore resource exploitation (**Figure 1**).

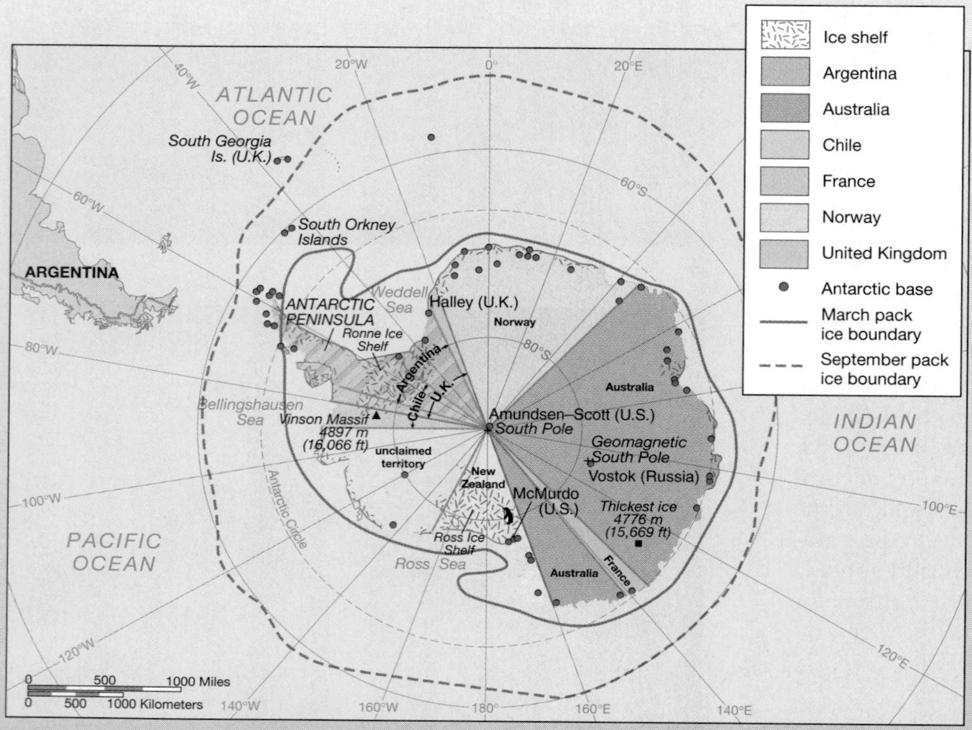

Antarctica also figures in a wider global imagination, and each summer about 30,000 travelers visit it. This is because visually it is one of the most striking landscapes in the world. First, there is Antarctica's sheer scale. In the clear, bright light of unpolluted air, the unbounded snow and ice fields seem endless. The distances are indeed immense, and the atmosphere is so clear that one can easily be deceived: Mountains that seem no more than 20 kilometers (11.4 miles) away may in fact be 80 kilometers (49.7 miles) distant. In detail, the snowy, icy landscape is subject to constant change

▲ **FIGURE 1 Territorial claims in Antarctica** Seven countries have claims in Antarctica, but more than 40 have signed the Antarctic Treaty. The treaty bans nuclear tests and the disposal of radioactive waste, ensures that the continent can be used only for peaceful purposes and scientific research, and includes a ban on mineral and oil exploration. Sixteen countries also maintain scientific research bases on the continent.

however, and thus the island nation holds value for the European power. In the context of this region's colonial experience, artificial lines were drawn across otherwise coherent geographic and ethnic regions in Oceania, leaving a number of ethnic groups "cut up" through the process of independence

The most serious recent political conflicts in Oceania has involved encounters between ethnic groups in Fiji and demonstrations by independence or secessionist movements in New Caledonia and Papua New Guinea. But the region has also seen renewed attempts at regional integration and peacekeeping through a variety of regional and subregional cooperation agreements. The conflict in Fiji is a legacy of British colonial policies that brought Asian Indians as indentured workers for sugar plantations from 1879 to 1920. By the 1960s, Indo-Fijians almost outnumbered the ethnic Fijian population, dominating commerce and urban life and maintaining a separate existence with little intermarriage and continued cultural and religious differences. The indigenous Fijians took over government at independence in 1970, but subsequent elections have produced contested wins for Indo-Fijian parties and coups in 1987 and 2000. As of 2006 a coalition dominated by indigenous Fijians is in power. In the nickel-rich islands of New Caledonia, the indigenous Melanesian popula-

A.

B.

▲ **FIGURE 2 Antarctic landscape A.** Antartica is an ice-covered landscape populated by penguins and other species adapted to cold temperatures. The penguins feed on abundant fish in the oceans around the icecaps.
**B.** Norwegian Prime Minister Jens Stoltenberg visits the Norwegian Antarctic research station in Troll 19 January 2008. Norway's government said 17 January 2008 the country would dramatically slash its carbon dioxide emissions by 2020 and aim to be completely carbon neutral by 2030—20 years ahead of schedule.

as ice features move and new snow blankets old features (**Figure 2**). Along parts of the coast, stark granite promontories provide fixed landmarks, but much of the coastline of Antarctica is ephemeral. The latest maps of Antarctica are always out of date because of the changing configuration of glaciers as they reach the sea. In summer, the Antarctic pack ice breaks up, and the coastal glaciers carve huge icebergs that shift, drift, and change shape by the hour.

Though these numbers are minuscule on the overall scale of world tourism, there may already be environmental impact. Bird species such as petrels, penguins, and albatrosses are declining in number. Petrels are long-lived birds—the oldest on record survived 50 years—making them an ideal Antarctic "indicator species." Many ecologists believe that the increased human presence in Antarctica may be disturbing these sensitive birds so much that they fail to breed. Another possible culprit is commercial fishing, which is booming, often illegally, throughout the southern oceans. Meanwhile, the efficiency, persistence, and greed of Japanese, Icelandic, and Norwegian commercial and "scientific" whaling fleets continue hunting whales, particular minke, despite international protests. There is also growing concern that global warming is threatening Antarctica, especially the Antarctic Peninsula, where temperatures are warming and changing the ecology, and ice cover is disappearing. In 2002 a 3250-square-kilometer (1225-square-mile) chunk of the Larsen B ice shelf broke away from the mainland. If Antarctic ice continues to melt as a result of climate change, it may contribute to a worldwide rise in sea levels, although recent studies suggest that the process of warming and cooling in Antarctica is even more complex than originally suggested—as the western half appears to be warming whereas recent studies show the eastern half may be cooling and the ice getting thicker. ■

tion, known as *Kanaks*, has been militantly pressing for independence for years but has been outvoted by those of French descent (called *demis*), who prefer to remain part of France. The Nouméa Accord of 1998 promises to grant independence by 2018. In the Solomon Islands, residents of Bougainville are trying to secede from Papua New Guinea, claiming ethnic affiliation with the other Solomon Islands that are independent and complaining that they do not receive a fair share of the profits from mining. Within the Solomon Islands there are conflicts between ethnic groups over land, such as those between longtime residents of Guadalcanal and immigrants from the neighboring island of Malaita.

Regional cooperation agreements include the South Pacific Commission, founded in 1947, which focuses on social and economic development and includes 21 island nations and territories, Australia, New Zealand, the United States, France, and the United Kingdom. The **South Pacific Forum**, established in 1971, excludes France, the United Kingdom, the United States, and their colonies and promotes discussion and cooperation on trade, fisheries, and tourism among all the independent and self-governing states of Oceania. It has supported maritime territorial rights and a nuclear-free Pacific as well as the independence goals of French Polynesia and New Caledonia. A summit in 2000 legitimized peacekeeping military operations led by New Zealand and Australia in the Solomon Islands and Papua New Guinea. There are also dozens of nongovernmental organizations and intergovernmental agencies that operate regionwide, especially among the smaller Pacific islands. For example, the University of the South Pacific fosters higher education across 12 countries through distance education and three main campuses in Fiji, Samoa, and Vanuatu.

Australia and New Zealand are members of larger economic and political alliances such as the Asia-Pacific Economic Cooperation group (APEC), which also includes Papua New Guinea and focuses on improving transportation links and liberalizing regional trade. Both Australia and New Zealand have been able to take advantage of APEC to increase exports to Asia, especially to Japan. In attempts to foster regional markets as global trade liberalizes and restructures around them, Australia and New Zealand created the free trade focused **Closer Economic Relations (CER) agreement** in 1983, which built on an earlier New Zealand-Australia Free Trade Agreement (NZAFTA). CER set out to remove all tariffs and restrictions on trade between the two countries. The resulting increased trade has been especially beneficial to New Zealand, with its small domestic market, which has doubled its exports to Australia. Many manufacturing firms now operate in both countries.

**Multiculturalism and Indigenous Social Movements**    Among the most passionately debated issues in contemporary Australia and New Zealand are those relating to the rights of their indigenous peoples and the creation of a multicultural society and national identity. The countries share a history of British colonialism and dispossession of indigenous lands and cultures. They have distinctly different contexts and contemporary approaches to intercultural relationships.

In New Zealand, Maori rights are framed by the 1840 Treaty of Waitangi. Although the Maori interpreted the treaty as guaranteeing their land and rights, the century that followed saw large-scale dispossession of Maori land and disrespect for Maori culture. Maori landholdings were reduced from 27 million hectares (100,000 square miles) to only 1.3 million hectares (5000 square miles), or 3% of the total area of New Zealand. A series of protests, court cases, and reawakening to Maori tradition led in 1975 to the establishment of the Waitangi tribunal, which eventually reinterpreted the Treaty of Waitangi as more favorable to the Maori and investigated a series of Maori land and fishery claims. The Maori were established as *tangata whenua* (the "people of the land"), and Maori was recognized as an official language of New Zealand. Some land claims were settled or compensated through money or grants of government land. Others are too large or threatening to private interests to be easily recognized. A bicultural Maori and *Pakeha* (a Maori term for whites) society has been adopted rather than a multicultural policy that would encompass other immigrant groups, such as Pacific islanders and Asians, or recognize the differences within Maori and other cultures.

New Zealand's recognition of Maori rights and culture as part of a national identity has not solved some of the deeper problems of racism toward the Maori or of their poverty and alienation. Maori unemployment is twice that of white residents; average incomes, home ownership, and educational levels are less than half; and welfare dependence is much higher. There also remain consistent problems of political representation, as was witnessed in the 2009 protests over Maori representation on the new Auckland "super city" council (**Figure 11.24**).

Australian Aborigines, in contrast, have had no recourse to a treaty to assert their rights. The European colonists saw the indigenous peoples as primitive and their land as *terra nullius*, owned by no one, and therefore freely available to settlers. Only in the 1930s were reserves set aside for Aboriginal populations, mostly in very marginal environments with little autonomy or access to services. In many ways the Aboriginal population had been made "invisible," not counted in the census or allowed to vote until the 1960s. It was also stereotyped as a primitive and homogeneous nomadic culture, when in fact there were many different cultures. One of the most misguided programs set out to assimilate the Aboriginal population by forcibly removing their children from their families and communities and placing them in white foster homes and institutions from 1928 to 1964 (as represented in the film *Rabbit-Proof Fence*). This **stolen generation** of as many as 100,000 Aboriginal children was given voice and officially acknowledged by the Australian government in a national inquiry in the 1990s.

Growing awareness and regret at the treatment of Aborigines has led to efforts at apology and reconciliation by many white Australians. In 2000, more than 300,000 people marched across the Sydney Harbor Bridge in a walk of reconciliation, and more than 1 million Australians signed "Sorry Books" that stated, "We stole your land, stole your children, stole your lives. Sorry," as a way of apologizing for the treatment of Aboriginal peoples (**Figure 11.25a**). In February 2008, Prime Minister of Australia, Kevin Rudd, formally apologized for these atrocities for the first time, stating that:

> We apologise for the laws and policies of successive parliaments and governments that have inflicted profound grief, suffering and loss on these our fellow Australians. We apologise especially for the removal of Aboriginal and Torres Strait Islander children from their families, their communities and their country. For the pain, suffering and hurt of these Stolen Generations, their descendants and for their families left behind, we say sorry. To the mothers and the fathers, the brothers and the sisters, for the breaking up of families and communities, we say sorry. And for the indignity and degradation thus inflicted on a proud people and a proud culture, we say sorry.[5]

Despite the apologies, indigenous Australians remain disadvantaged on almost all economic and social indicators, with unemployment at four times the national average, much lower average incomes, housing quality, and educational levels, and higher levels of suicide, substance abuse, disease, and violence. The Aboriginal Land Rights Act of 1976 gave Aborigines title to almost 20% of the Northern Territory and opened government land to claims through regional land councils. The states of South Australia and Western Australia have also handed over land to Aboriginal ownership or leases. In 1992 the Australian High Court effectively overruled the doctrine of *terra nullius*, which has had the effect of encouraging Aboriginal claims for land and compensation. Aboriginal control now extends over about 15% of Australia, with claims to at least another 20% (**Figure 11.25b**). The more contentious claims surround land with valuable mineral resources, especially uranium, or where development threatens spiritual sites.

Australian Aborigines still have much less power and recognition than the Maori of New Zealand, and this is reflected in Australia's adoption of a multicultural rather than bicultural policy of national integration. Multiculturalism emerged in the 1970s and set out to embrace the distinctive cultures of many different ethnic and immigrant groups. The

[5]N.A. (2008). "Kevin Rudd's National Apology to Stolen Generations," News.com.au. [accessed 30 May 2009].

◀ **FIGURE 11.24 Maoris protest for representation in Auckland** Campaigners take part in a *Hikoi* or Maori protest march, to protest the dropping of Maori seats from the new Auckland "super city" proposal, at Queen Street on May 25, 2009, in Auckland, New Zealand. A Royal Commission proposal suggested having 3 Maori seats on a 23-member council, 2 elected and 1 appointed by local *iwi* (Maori tribes) but this was rejected by the government in favor of having only 20 councillors, none directly Maori-elected. The Auckland "super city" would create a political bureaucracy that incorporated all the cities in the surrounding Auckland area.

National Agenda for a Multicultural Australia (published in 1989) set out to promote tolerance and cultural rights and to reduce discrimination, while maintaining English as the official language and avoiding special treatment for any one group. In contrast to New Zealand, where Maori language and culture is an essential component of the national bicultural identity, Aborigines are just one of many ethnic groups in a multicultural society. Some have resented this status. There has also been considerable opposition to immigration, Aboriginal rights, and multiculturalism in the last decade, much of this exacerbated by recent global economic conditions. In the 1990s a new political party, the One Nation party, emerged with a platform that opposed immigration, multiculturalism, and any special preferences for Aborigines, but the party has now lost favor. The relatively new Kevin Rudd government's policies are still being rolled out, but it is clear that land rights and reconciliation for Aboriginal peoples will remain contested for the Australian government, which is balancing competing interests and needs.

(a)

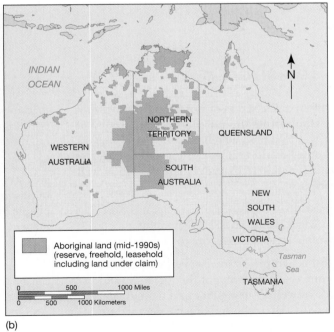

(b)

▲ **FIGURE 11.25 Aboriginal issues in Australia** (a) Millions of Australians have apologized to the Aborigines for discrimination and the damage to the "stolen generation" of Aboriginal children, who were taken forcibly from their homes. This apology was flown over the Sydney Opera House. (b) This map shows the current status of Aboriginal lands in Australia, with the majority in the Northern Territory. Key legal settlements such as the Mabo and Wik judgments have prompted Aborigine land claims to more than 35% of the country.

# FUTURE GEOGRAPHIES

Often discussed as a relatively isolated world region globally, Oceania, which stretches across a vast oceanic waterscape, is dynamically interconnected to the rest of the world in terms of its physical and human geographies. The future of this region is uncertain, however, as global climatic change continues to dramatically affect the livelihoods and cultural practices of those living in the region. It is clear, for example, that there will continue to be shifts in the demographics of the region as rising sea levels and increased droughts affect those living in the South Pacific, on the one hand, and the interior agricultural spaces of Australia, on the other. This suggests that it remains vitally important to examine this region as a barometer for future environmental change globally—studying how it might be possible to mitigate some of these challenges as they become more demonstrable will be the challenge facing this region.

**Emerging Demographic Challenges**   Despite these challenges, Australia's and New Zealand's relatively high standards of living continue to attract in-migrants, and debates are underway in both countries on how to manage growing migrant populations. In fact, the management of migration into both countries has become an even more acute political issue in the context of the recent global economic recession. Regional cooperation may become even more important, as interdependencies increase between the economies of the South Pacific and Australasia.

**Emerging Global Development Leaders**   It is clear that Australia and New Zealand will continue to play an important role as global actors, not only in the context of their own wider region, but also in the region's Southeast Asia, just to the north. Organizations, such as **AusAID**, Australia's International Development Assistance Program, for example, continue to offer development assistance to a diverse set of countries, from Vanuatu in the South Pacific to Indonesia and Timor-Leste in Southeast Asia. New Zealand also continues to be a global leader in terms of antinuclear proliferation. That this country stands at the forefront against the further expansion of nuclear power and nuclear violence is important not only symbolically but also geopolitically.

**Emerging Trans-Regional Cooperation**   It is clear, as well, that global changes may bring the various actors and diverse cultures of this region closer together, as they create new cooperative arrangements to meet the challenges facing them on both land and sea. What these new cooperative organizations will look like is not yet apparent, and the actors in this region may find themselves further integrated into organizations, such as ASEAN in Southeast Asia (Chapter 10), as economic and political relations bring these two regions closer. Furthermore, the states in Oceania may find themselves further aligned with organizations in the Americas (see Chapters 6 and 7), sharing a common concern about changing sea-surface temperatures, for example, in the southern hemisphere. There is no doubt, however, that the dynamic interrelationships between this world region and other world regions will continue to create new opportunities and challenges for the peoples and environments of this very diverse geographic space.

# Main Points Revisited

■   This world region is often discussed as one of the most geographically isolated. This is a result of its unique physical geography of extensive Pacific Ocean waterscapes, which create vast distances between this region and other world regions. At the same time, the region is one of the most vulnerable to global climate change as well as wider geopolitical processes.

Despite being relatively isolated by its physical geography, it is clear that Oceania is integrated into wider global physical geographic processes, demonstrating how vulnerable local ecologies are to global change.

■   This region's historical geography is interwoven with a series of human migrations over time that have brought the region into direct contact with new peoples as well as new plant and animal species. This has created a region with a dynamic and diverse demography as well as a distinct set of land use practices.

Oceania has been dramatically changed through the long-term migrations of people from outside this region. From the earliest human migrants to the more modern-day movements, the cultural landscapes of this region bear a remarkable set of imprints based on these ongoing migrations.

■   The extensive settlement of this region by peoples of European ancestry has created a complicated set of cultural and social politics in this region. Although this has rarely led to direct violence between people in this region, there remains tension between indigenous people and colonial people.

Despite a relative lack of violence, the cultural violence associated with historical white settlement and colonization still creates tension. This can be seen in Australia and New Zealand as well as in Fiji and New Caledonia.

■   Although this region is well integrated into the global economy and is highly urbanized, there remain distinct differences in economic development across the region. Some smaller islands rely heavily on resource extraction or tourism to sustain their development, whereas the larger economies of the region, such as Australia and New Zealand, have been able to diversity their economies, attracting foreign direct investment and skilled migrant labor.

Economic and social development remains geographically differentiated across the region and within the region's member states. Although it is difficult to measure these differences—due to a lack of data—evidence suggests that the expanse between rich and poor is growing.

■   The region is linked by a number of formal political organizations as well as through economic connections. In addition, the region is growing more interconnected through the migration of people from the South Pacific to Australia and New Zealand and through cooperation and the development of supranational organizations.

Regional cooperation is taking place at different scales. It is unclear how that cooperation will affect long-term connections in the region or the relationship of this world region to other world regions, such as Southeast Asia or Latin America.

# Key Terms

Alliance of Small Island States (AOSIS) (p. 374)

Antarctic Treaty (p. 396)

atolls (p. 375)

AusAID (p. 400)

cattle stations (p. 373)

Closer Economic Relations (CER) agreement (p. 398)

common property resources (p. 382)

Dreamtime (p. 383)

ecological imperialism (p. 381)

Great Artesian Basin (p. 374)

guano (p. 376)

marsupial (p. 379)

Melanesia (p. 394)

Micronesia (p. 394)

monotremes (p. 380)

Outback (p. 373)

ozone depletion (p. 374)

outstations (p. 374)

Polynesia (p. 394)

South Pacific Forum (p. 397)

stolen generation (p. 398)

subsistence affluence (p. 390)

theory of island biogeography (p. 378)

Treaty of Waitangi (p. 385)

UNCLOS (p. 377)

White Australia policy (p. 390)

World Heritage Site (p. 382)

# Thinking Geographically

1. What are some ways in which the large expanse of ocean has defined culture and economy in Oceania?

2. Describe some of the physical differences between high volcanic islands and low coral atolls. (Discuss how they formed and what they look like, whether they receive a lot of rainfall, what the soils and vegetation are like, and so forth.)

3. Why is Oceania particularly vulnerable to global environmental changes, and what is the name of the organization that helps some of the smaller countries respond to these changes?

4. How did the perceptions of early colonial explorers influence the image of the Pacific in other parts of the world?

5. How did World War II affect the Pacific? Which regions of the Pacific are still strongly influenced by international geopolitical activities?

6. Using specific islands as examples, how do resource extraction and tourism pose regional development challenges for Pacific island states?

7. How, why, and when did Australia and New Zealand start to shift their economic and political orientations away from Great Britain toward Asia and North America?

8. In what ways do livestock, mining, and indigenous livelihoods make the Outback a distinctive but contested landscape?

9. What are the differences between the ways in which Australia and New Zealand have dealt with their indigenous peoples, immigration, and multiculturalism in general?

10. What is ecological imperialism, and how did it transform the environment of Oceania?

Log in to www.mygeoscienceplace.com for Videos, Interactive Maps; RSS feeds; Further Readings; Suggestions for Films, Music and Popular Literature; and Self-Study Quizzes to enhance your study of Australia, New Zealand, and the South Pacific.

# MAPS AND GEOGRAPHIC INFORMATION SYSTEMS

Maps are representations of the world. They are usually two-dimensional, graphic representations that use lines and symbols to convey information or ideas about spatial relationships. Maps express particular interpretations of the world, and they affect how we understand the world and see ourselves in relation to others. As such, all maps are social products. In general, maps reflect the power of the people who draw them. Just including things on a map—literally "putting something on the map"—can be empowering. The design of maps—what they include, what they omit, and how their content is portrayed—inevitably reflects the experiences, priorities, interpretations, and intentions of their authors. The most widely understood and accepted maps—"normal" maps— reflect the view of the world that is dominant in universities and government agencies.

Maps that are designed to represent the *form* of Earth's surface and to show permanent (or at least long-standing) features such as buildings, highways, field boundaries, and political boundaries are called *topographic maps* (see, for example, **Figure A.1**). The usual device for representing the form of Earth's surface is the *contour*, a line that connects points of equal vertical distance above or below a zero data point, usually sea level.

Maps that are designed to represent the spatial dimensions of particular conditions, processes, or events are called *thematic maps*. These can be based on any of a number of devices that allow cartographers or map makers to portray spatial variations or spatial relationships. One device is the *isoline*, a line (similar to a contour) that connects places of equal data value (for example, air pollution, as in **Figure A.2**). Maps based on isolines are known as *isopleth maps*.

Another common device used in thematic maps is the *proportional symbol*. For example, circles, squares, spheres, cubes, or some other shape can be drawn in proportion to the frequency of occurrence of some phenomenon or event at a given location. Figure 9.19, p. 453 shows an example using proportional circles. Symbols such as arrows or lines can also be drawn proportionally to portray flows between particular places. Simple distributions can be effectively portrayed through *dot maps*, in which a single dot represents a specified number of occurrences of some phenomenon or event.

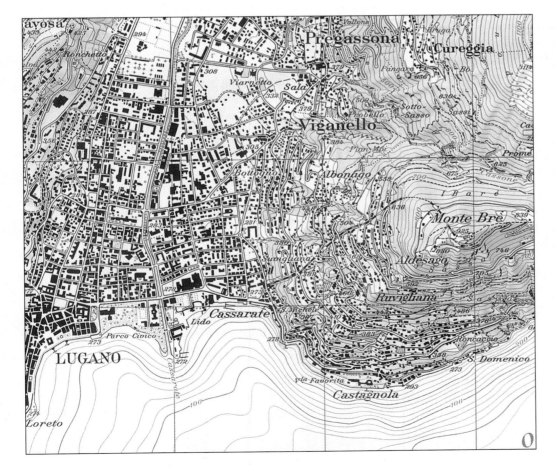

◀ **FIGURE A.1 Topographic maps** Topographic maps represent the form of Earth's surface in both horizontal and vertical dimensions. This extract is from a Swiss map of Lugano at the scale of 1: 25,000. The height of landforms is represented by contours (lines that connect points of equal vertical distance above sea level), which on this map are drawn every 20 meters. Features such as roads, power lines, built-up areas, and so on are shown by stylized symbols. Note how the closely spaced contours of the hill slopes represent the shape and form of the land.

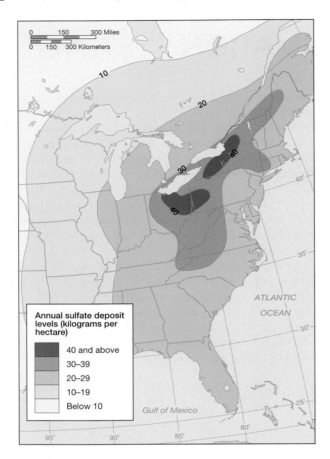

▲ **FIGURE A.2  Isoline maps** Isoline maps portray spatial information by connecting points of equal data value. Contours on topographic maps (see Figure A.1) are a type of isoline. This map shows one type of air pollution in the eastern United States.

Yet another type of map is the *choropleth map*, in which tonal shadings are graduated to reflect area variations in numbers, frequencies, or densities (see, for example, Figure 1.22, p. 24). Finally, thematic maps can be based on *located charts*, in which graphs or charts are located by place or region. In this way, a tremendous amount of information can be conveyed in one map (see Figure 1.25, p. 27).

## MAP SCALES

A *map scale* is simply the ratio between linear distance on a map and linear distance on Earth's surface. It is usually expressed in terms of corresponding lengths, as in "one centimeter equals one kilometer," or as a *representative fraction* (in this case, 1/100,000) or ratio (1:100,000). *Small-scale* maps are based on small representative fractions (for example, 1/1,000,000 or 1/10,000,000). They cover a large part of

Earth's surface on the printed page. A map drawn on this page to the scale of 1:10,000,000 would cover about half of the United States; a map drawn to the scale of 1:16,000,000 would easily cover the whole of Europe. *Large-scale* maps are based on larger representative fractions (e.g., 1/25,000 or 1/10,000). A map drawn on this page to the scale 1:10,000 would cover a typical suburban subdivision; a map drawn to the scale of 1:1,000 would cover just a block or two.

## MAP PROJECTIONS

A **map projection** is a systematic rendering on a flat surface of the geographic coordinates of the features found on Earth's surface. Because Earth's surface is curved and not a perfect sphere, it is impossible to represent on a flat plane, sheet of paper, or monitor screen without some distortion. Cartographers have devised a number of techniques for projecting **latitude** and **longitude** (**Figure A.3**) onto a flat surface, and the resulting representations each have advantages and disadvantages. None can represent distance correctly in all directions, though many can represent compass bearings or area without distortion. The choice of map projection depends largely on the purpose of the map.

Projections that allow distance to be represented as accurately as possible are called **equidistant projections**. These can represent distance accurately in only one direction (usually north–south), although they usually provide accurate scale in the perpendicular direction (which in most cases is the equator). Equidistant projections are often more aesthetically pleasing for representing Earth as a whole, or large portions of it. An example is the Polyconic projection (**Figure A.4**).

Projections on which compass directions are rendered accurately are known as **conformal projections**. On the Mercator projection (see Figure A.4), for example, a compass bearing between any two points is plotted as a straight line. As a result, the Mercator projection has been widely used in navigation for hundreds of years. The Mercator projection was also widely used as the standard classroom wall map of the world for many years, and its image of the world has entered deeply into general consciousness. As a result, many Europeans and North Americans have an exaggerated sense of the size of the northern continents and underestimate the size of Africa.

Some projections are designed such that compass directions are correct from only one central point. These are known as **Azimuthal projections**. They can be equidistant, as in the Azimuthal Equidistant projection (see Figure A.4), which is sometimes used to show air-route distances from a specific location.

Projections that portray areas on Earth's surface in their true proportions are known as **equal-area** or **equivalent projections**. Such projections are used when the cartographer wishes to compare and contrast distributions on Earth's surface—the relative area of different types of land use, for example. Examples of equal-area projections include the Eckert IV projection, Bartholomew's Nordic projection (used in Figure 1.25) and the Mollweide projection (see Figure A.4). Equal-area projections such as the Mollweide are especially useful for thematic maps showing economic, demographic, or cultural data. Unfortunately, preserving accuracy in terms of area tends to result in world maps on which many locations appear squashed.

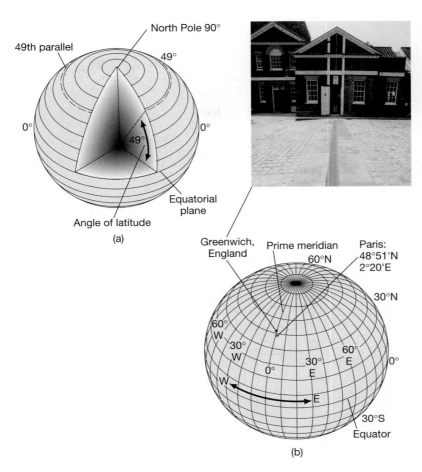

The prime meridian at the Royal Observatory in Greenwich, England. The observatory was founded by Charles II in 1675 with the task of setting standards for time, distance, latitude, and longitude—the key components of navigation.

▲ **FIGURE A.3  Latitude and longitude**  Lines of latitude and longitude provide a grid that covers Earth, allowing any point on Earth's surface to be accurately referenced. Latitude is measured in terms of angular distance (that is, degrees and minutes) north or south of the equator, as shown in (a). Longitude is measured in the same way, but east and west from the prime meridian, a line around Earth's surface that passes through both poles (North and South) and the Royal Observatory in Greenwich, just to the east of central London, in England. Locations are always stated with latitudinal measurements first. The location of Paris, France, for example, is 48°51′ N and 2°20′ E, as shown in (b).

For some applications, aesthetic appearance is more important than conformality, equivalence, or equidistance, so cartographers have devised a number of other projections. Examples include the Times projection, which is used in many world atlases, and the Robinson projection, which is used by the National Geographic Society in many of its publications. The Robinson projection (**Figure A.5**) is a compromise projection that distorts both area and directional relationships but provides a general-purpose world map.

There are also political considerations. Countries may appear larger and so more important on one projection than on another. The Peters projection, for example (**Figure A.6**), is a deliberate attempt to give prominence to the underdeveloped countries of the equatorial regions and the Southern Hemisphere. As such, it has been officially adopted by the World Council of Churches and numerous agencies of the United Nations and other international institutions. Its unusual shapes give it a shock value that gets people's attention. For some, however, those shapes are ugly.

In this book we have occasionally used another striking projection, the Dymaxion projection devised by Buckminster Fuller (**Figure A.7**). Fuller was a prominent modernist architect and industrial designer who wanted to produce a map of the world with no significant distortion to any of the major land masses. The Dymaxion projection does this, though it produces a world that at first may seem disorienting. This is not necessarily a bad thing, for it can force us to take a fresh look at the world and at the relationships among places. Because Europe, North America, and Japan are all located toward the center of this map projection, it is particularly useful for illustrating two central themes of this book: the relationships among these prosperous regions and the relationships between this prosperous core group and the less prosperous, peripheral countries of the world. Fuller's projection shows the economically peripheral countries of the world as cartographically peripheral, too.

One kind of projection that is sometimes used in small-scale thematic maps is the *cartogram*. In this projection, space is transformed

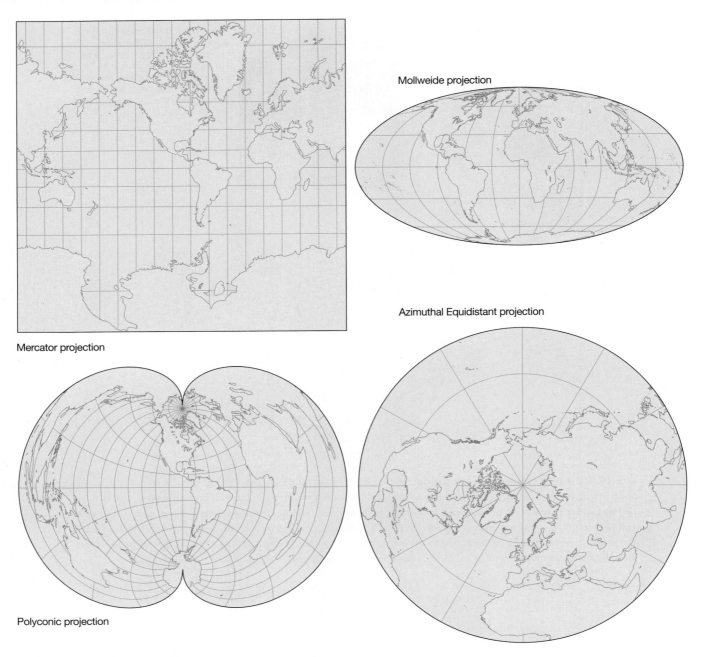

Mollweide projection

Azimuthal Equidistant projection

Mercator projection

Polyconic projection

▲ **FIGURE A.4 Comparison of map projections** Different map projections have different properties. The Polyconic projection is true to scale along each east–west parallel and along the central north–south meridian. It is free of distortion only along the central meridian. On the Mercator projection, compass directions between any two points are true, and the shapes of land masses are true, but their relative size is distorted. On the Azimuthal Equidistant projection, distances measured from the center of the map are true, but direction, area, and shape are increasingly distorted with distance from the center point. On the Mollweide projection, relative sizes are true but shapes are distorted.

according to statistical factors, with the largest mapping units representing the greatest statistical values. **Figure A.8a** shows a cartogram of the world in which countries are represented in proportion to their population. This sort of projection is particularly effective in helping us visualize relative inequalities among the world's populations. **Figure A.8b** shows a cartogram of the world in which the cost of telephone calls made from the United States has been substituted for linear distance as the basis of the map. The deliberate distortion of the shapes of the continents in this sort of projection dramatically emphasizes spatial variations.

Finally, the advent of computer graphics has made it possible for cartographers to move beyond two-dimensional representations of

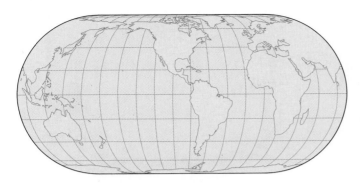

◀ **FIGURE A.5  The Robinson projection**  On the Robinson projection, distance, direction, area, and shape are all distorted in an attempt to balance the properties of the map. It is designed purely for appearance, and is best used for thematic and reference maps at the world scale.

Earth's surface. Computer software that renders three-dimensional statistical data onto the flat surface of a monitor screen or a piece of paper facilitates the **visualization** of many aspects of human geography in innovative and provocative ways (**Figure A.9**).

# GEOGRAPHIC INFORMATION SYSTEMS

**Geographic information systems (GIS)**—organized collections of computer hardware, software, and geographic data that are designed to capture, store, update, manipulate, and display geographically referenced information—have rapidly grown to become a predominant method of geographic analysis, particularly in the military and commercial worlds. The software in GIS incorporates programs to store and access spatial data, to manipulate those data, and to draw maps.

Between 1999 and 2005, GIS services grew at a rate of around 10 percent per year. In 2005, estimates of global investment in GIS technologies ranged from $5 billion to more than $8 billion. In the United States, employment in GIS is now one of the 10 fastest-growing technical fields in the private sector. The U.S. government spends more than $4 billion per year on geographic data acquisition, and the total annual market for GIS services in North America is valued at around $3 billion.

The primary requirement for data to be used in GIS is that the locations for the variables are known. Location may be annotated by x, y, and z coordinates of longitude, latitude, and elevation, or by such systems as zip codes or highway mile markers. Any variable that can be located spatially can be fed into a GIS. Data capture—putting the information into the system—is the most time-consuming component of GIS work. Different sources of data, using different systems of measurement, scales, and systems of representation, must be integrated with one another; changes must be tracked and updated. Many GIS operations in the United States, Europe, Japan, and Australia have begun to contract out such work to firms in countries where labor is cheaper; India has emerged as a major data conversion center for GIS.

## Applications of GIS

The most important aspect of GIS, from an analytical point of view, is that they it allows the merger of data from several different sources, on different topics, and at different scales. This capability allows analysts to emphasize the spatial relationships among the objects being mapped. A geographic information system makes it possible to link, or integrate, information that is difficult to associate through any other means. For example, using GIS technology and water-company billing information, it is possible to simulate the discharge of materials in the septic systems in a neighborhood upstream from a wetland. The bills show how much water is used at each address. Because the amount of

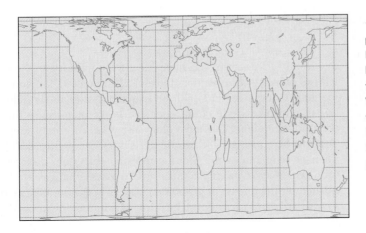

◀ **FIGURE A.6  The Peters projection**  This equal-area projection offers an alternative to traditional projections which, Arno Peters argued, exaggerate the size and apparent importance of the higher latitudes—that is, the world's core regions—and so promote the "europeanization" of Earth. While it has been adopted by the World Council of Churches, the Lutheran Church of America, and various agencies of the United Nations and other international institutions, it has been criticized by cartographers in the United States on the grounds of aesthetics: One consequence of equal-area projections is that they distort the shape of land masses.

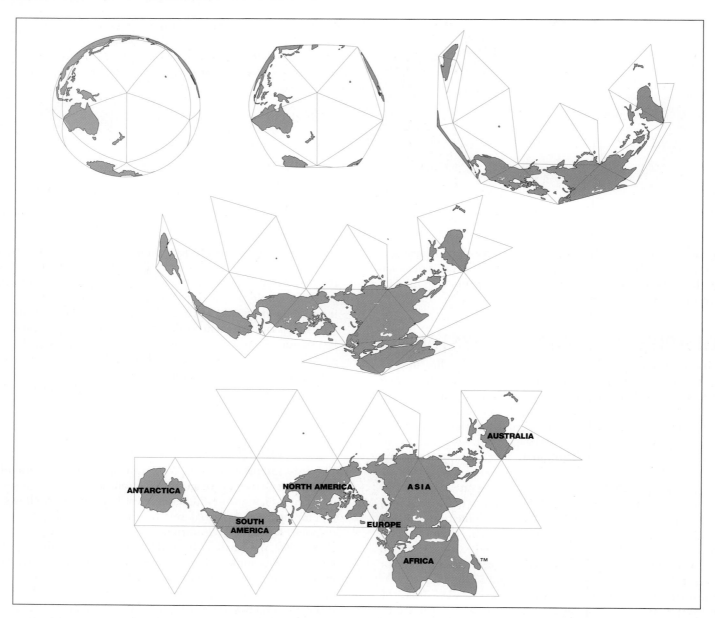

▲ **FIGURE A.7  Fuller's Dymaxion projection**  In this striking map projection, Buckminster Fuller (1895–1983) created a map with the minimum of distortion to the shape of the world's major land masses by dividing the globe into triangular areas. Those areas not encompassing major land masses were cut away, allowing the remainder of the globe to be "unfolded" into a flat projection.

water a customer uses will roughly predict the amount of material that will be discharged into the septic systems, areas of heavy septic discharge can be located using a GIS.

GIS technology can render visible many aspects of geography that were previously unseen. For example, GIS can produce incredibly detailed maps based on millions of pieces of information—maps that could never have been drawn by human hands. At the other extreme of spatial scale, GIS can put places under the microscope, creating detailed new insights using huge databases and effortlessly browsable media (**Figure A.10**).

Many advances in GIS have come from military applications. GIS allows infantry commanders to calculate line of sight from tanks and defensive emplacements, allows cruise missiles to fly below enemy radar, and provides a comprehensive basis for military intelligence. Beyond the military, GIS technology allows an enormous range of problems to be addressed. It can be used, for example, to decide how to manage farmland; to monitor the spread of infectious diseases; to monitor tree cover in metropolitan areas; to assess changes in ecosystems; to analyze the impact of proposed changes in the boundaries of legislative districts; to identify the location of potential business customers;

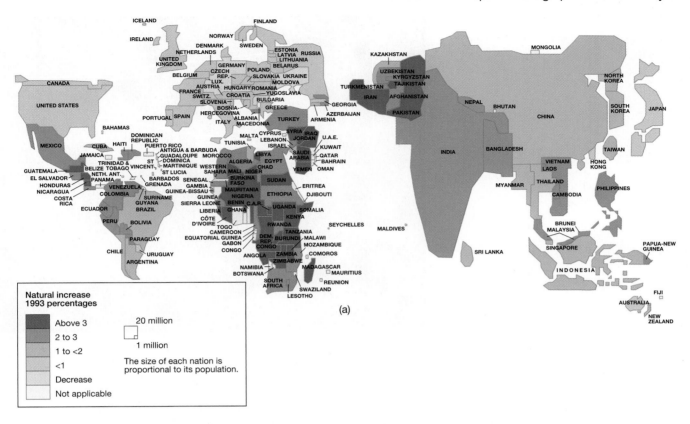

Natural increase
1993 percentages

| Above 3 |
| 2 to 3 |
| 1 to <2 |
| <1 |
| Decrease |
| Not applicable |

20 million

1 million

The size of each nation is
proportional to its population.

(a)

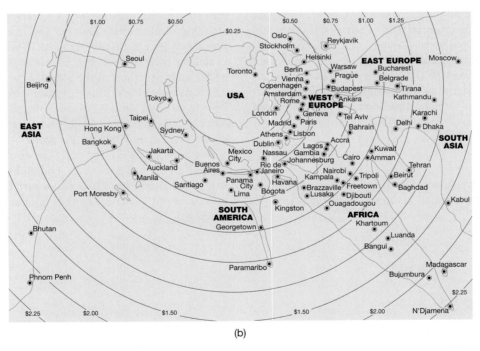

(b)

▲ **FIGURE A.8 Cartograms** In a cartogram, space is distorted to emphasize a particular attribute of places or regions. (a) This example shows the relative size of countries based on their population rather than their area; the cartographers have maintained the shape of each country as closely as possible to make the map easier to read. As you can see, population-based cartograms are very effective in demonstrating spatial inequality. (b) In this example, the cost of telephone calls is substituted for linear distance as the basis of the map, thus deliberately distorting the shapes of the continents to dramatic effect. Countries are arranged around the United States according to the cost per minute of calls made from the United States in 1998.

▶ **FIGURE A.9 Visualization** This example shows the spatial structure of the Internet backbone and associated traffic flows within the United States.

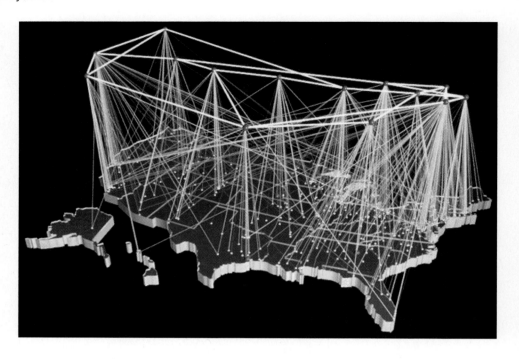

to identify the location of potential criminals; and to provide a basis for urban and regional planning. Some of the most influential applications of GIS have resulted from geodemographic research. **Geodemographic research** uses census data and commercial data (such as sales data and property records) about the populations of small districts in creating profiles of those populations for market research. The digital media used by GIS make such applications very flexible. With GIS it is possible to zoom in and out, evaluating spatial relationships at different spatial scales. Similarly, it is possible to vary the appearance and presentation of maps, using different colors and rendering techniques.

## Critiques of GIS

Within the past five years, GIS has resulted in the creation of more maps than were created in all previous human history. One result is that, as maps have become more commonplace, more people and more businesses have become more spatially aware. Nevertheless, some critics have argued that GIS represents no real advances in geographers' understanding of places and regions. The results of GIS, they argue, may be useful but are essentially mundane. This misses the point that, however routine their subjects may be, all maps constitute powerful and influential ways to represent the world.

A more telling critique, perhaps, is that the real impact of GIS has been to increase the level of surveillance of the population by those who already possess power and control. The fear is that GIS may be helping create a world in which people are not treated and judged by who they are and what they do, but more by where they live. People's credit ratings, ability to buy insurance, and ability to secure a mortgage, for example, are all routinely judged, in part, by GIS-based analyses that take into account the attributes and characteristics of their neighbors.

## Further Reading

Dorling, D., and D. Fairbairn, *Mapping: Ways of Seeing the World*. London: Addison Wesley Longman, 1977.

Greene, R. W., *GIS in Public Policy*. Redlands, CA: ESRI Press, 2000.

O'Looney, J., *Beyond Maps: GIS and Decision Making in Local Government*. Redlands, CA: ESRI Press, 2000.

Thrower, N. J. W., *Maps & Civilization: Cartography in Culture and Society*, 2nd edition. Chicago: University of Chicago Press, 1999.

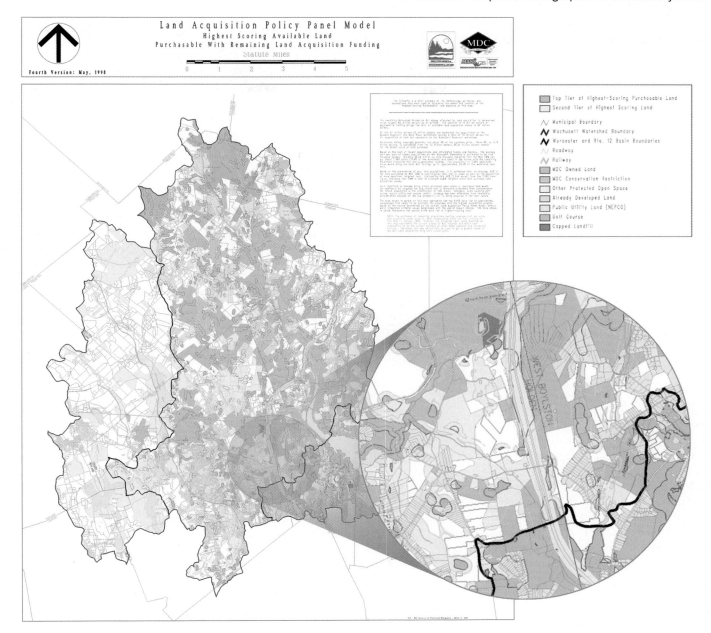

▲ **FIGURE A.10  GIS-derived planning map**  This map of the Wachusett Reservoir watershed in Massachusetts was prepared from a composite of 12 layers of GIS maps, each depicting a key aspect of land use and topography. The map is used to show in detail the parcels of land that the Commonwealth of Massachusetts needs to buy in order to keep development from damaging the watershed and the quality of the reservoir water.

# GLOSSARY

**Aborigines:** indigenous peoples of Australia.

**acid rain:** precipitation that has mixed with air pollution to produce rain that contains levels of acidity—often in the form of sulfuric acid—that are harmful to vegetation and aquatic life.

**afforestation:** converting previously unforested land to forest by planting trees or seeds.

**agglomeration:** clustering together of economic activities at the scale of metropolitan areas or industrial subregions.

**agglomeration economies:** cost advantages that accrue to individual firms because of their location among functionally related activities.

**AIDS orphans:** children who have lost one or more parents to an AIDS-related death.

**airshed:** a unit for air quality measurement.

**Alliance of Small Island States (AOSIS):** association of more than 40 low-lying, mostly island, countries that have formed an alliance to combat global warming, which threatens their existence through sea-level rise.

**altiplano:** high-elevation plateaus and basins that lie within even higher mountains, especially in Bolivia and Peru, at more than 3000 meters (9500 feet) in the Andes of Latin America.

**altitudinal zonation:** vertical classification of environment and land use according to elevation based mainly on changes in climate and vegetation from lower (warmer) to higher (cooler) elevations.

**americanization:** process by which a generation of individuals born elsewhere felt less loyalty and fewer cultural ties to their countries of origin and developed a new indigenous ethos.

**Anime:** a distinctive stylized form of Japanese animated film.

**Antarctic Treaty:** international agreement to demilitarize the Antarctic continent, delay mineral exploration, and preserve it for scientific research.

**anti-globalization movement:** groups that stand in opposition to the unregulated political power of large, multinational corporations and to the powers exercised through trade agreements and deregulated financial markets.

**apartheid:** South Africa's policy of racial separation that, prior to 1994, structured space and society to keep separate black, white, and colored populations.

**archipelago:** group of islands or expanse of water with many islands.

**aridity:** climate with insufficient moisture to support trees or woody plants.

**Asian Tigers:** newly industrialized territories of Hong Kong, Taiwan, South Korea, and Singapore that have experienced rapid economic growth and become semiperipheral within the world-system.

**aspect:** the direction in which a sloping piece of land faces.

**assimilation:** process by which peoples of different cultural backgrounds who occupy a common territory achieve sufficient cultural solidarity to sustain a national existence.

**Association of Southeast Asian Nations (ASEAN):** international organization of the nations of Southeast Asia established to promote economic growth and regional security.

**atoll:** low-lying island landform consisting of a circle of coral reefs around a lagoon, often associated with the rim of a submerged volcano or mountain.

**AusAID:** Australia's International Development Assistance program, which programs training and support for development programs in countries outside Australia.

**Azimuthal projection:** map projection on which compass directions are correct only from one central point.

**backwash effects:** negative impacts on a region (or regions)—including outmigration and the loss of capital—due to the economic growth of some other region.

**Balfour Declaration:** 1917 British mandate that required the establishment of a Jewish national homeland.

**balkanization:** division of a territory into smaller and often mutually hostile units.

**banana republics:** term used to describe small tropical countries, often run by a dictator, dependent on the export of a few crops such as bananas.

**barchan:** crescent-shaped sand dune, concave on the side sheltered from the prevailing wind.

**Berlin Conference:** meeting convened by German chancellor Bismark in 1884–85 to divide Africa among European colonial powers.

**biodiversity:** variety in the types and numbers of species in particular regions of the world.

**biofuels:** fuels that are produced from raw, often renewable, biological resources, such as plant biomass.

**biogeography:** study of the spatial distribution of vegetation, animals, and other organisms.

**biome:** largest geographic biotic unit, a major community of plants and animals or similar ecosystems.

**bioprospecting:** search for plants, animals, and other organisms that may be of medicinal value or may have other commercial use.

**blood diamonds:** diamonds that have been mined in war zones and used to finance conflicts.

**bonded labor:** labor that is pledged against an outstanding debt.

**bracero:** guest worker from Mexico given temporary permit to work as a farm laborer in the United States between 1942 and 1964.

**British Commonwealth:** group of former British colonies and other countries allied with Britain that cooperate on political, economic, and cultural activities.

**buffer zone:** group of smaller or less powerful countries situated between larger or more powerful countries that are geopolitical rivals.

**bumiputra:** a term used to define the ethnic Malay population in Malaysia; literally translated as "those of the soil," it intentionally excludes non-Malays from the definition (particularly people of Chinese ancestry living in Malaysia).

**bush fallow:** modification of shifting cultivation where crops are rotated around a village and fallow periods are shortened.

**cap-and-trade:** a three part regulatory system in which the "cap" is a government-imposed limit on carbon emissions, where emission permits or quotas are then given or sold to states (or firms), and the "trade" occurs when unused permits are sold by those who have been able to reduce their emissions beyond their quota to those who are unable to meet theirs.

**capitalism:** form of economic and social organization characterized by the profit motive and the control of the means of production, distribution, and exchange of goods by private ownership.

**carbon footprint:** a way of understanding how much an individual, a region, or a country is contributing to global climate change through their greenhouse gas emissions and where their impacts might be reduced.

**cargo cult:** Pacific island religious movements in which the dawn of a coming new age was associated with the arrival of goods brought by spiritual beings or foreigners.

**cartogram:** map projection that is transformed in order to promote legibility or to reveal patterns not readily apparent on a traditional base map.

**caste:** system of kinship groupings that is reinforced by language, religion, and occupation.

**cattle station:** livestock enterprises where cattle (or sheep) are raised on large grazing leases in the remote regions of Australia.

**chador:** loose, usually black, robe worn by Muslim women that covers the body, including the face, from head to toe.

**chaebol:** South Korean term for the very large corporations in that country that, with government help, control numerous businesses and dominate the national economy.

**Chao Khao (hill tribes):** those ethnic minorities living in mainland Southeast Asia, particularly, but not exclusively, in the northern more mountainous parts of the region.

**chernozem:** thick, dark grassland soils (also called Black Earths) that are neutral in terms of acidic content and rich in humus.

**circle of poison:** use of imported pesticides on export crops in developing countries that then export back the contaminated crops to the regions where the pesticides were manufactured.

**circular migration:** traditional and long-standing population movements that respond to seasonal availability of pasture, droughts, and wage employment.

**civil society:** network of social groups and cultural traditions that operate independently of the state and its political institutions.

**climate:** typical conditions of the weather expected at a place, often measured by long-term averages of temperature and precipitation (e.g., a rainy place).

**Closer Economic Relations (CER) Agreement:** agreement in 1983 that built upon an earlier New Zealand–Australia Free Trade Agreement (NZAFTA) and set out to remove all tariffs and restrictions on trade between the two countries.

**cognitive image (mental map):** psychological representation of locations that are made up from people's individual ideas and impressions of these locations.

**Cold War:** The period between World War II and 1990, in which the United States and the Soviet Union established their role as global hegemons and in which conflicts between these two powers took place through a variety of proxy conflicts.

**colonialism:** establishment and maintenance of political and legal domination by a state over a separate and alien society.

**colonization:** establishment of a settlement in a place or region.

**Columbian Exchange:** interchange of crops, animals, people, and diseases between the Old World of Europe and Africa and the New World of the Americas beginning with the voyages of Christopher Columbus in 1492.

**command economy:** national economy in which all aspects of production and distribution are centrally controlled by government agencies.

**commodity:** anything useful that can be bought or sold.

**commodity chains:** networks of labor and production processes that originate in the extraction or production of raw materials. The end result is the delivery and consumption of a finished commodity.

**common market:** market in which internal restrictions on the movement of capital, labor, goods and services are removed within the basic framework of a customs union.

**common property resources:** resources such as fish or forests that are managed collectively by a community that has rights to the resource rather than being owned by individuals.

**communism:** form of economic and social organization characterized by the common ownership of the means of production, distribution, and exchange.

**comparative advantage:** principle whereby places and regions specialize in activities for which they have the greatest advantage in productivity relative to other regions—or for which they have the least disadvantage.

**conformal projection:** map projection on which compass bearings are rendered accurately.

**Confucianism:** a spiritual and philosophical tradition native to China expression in the philosophy that places emphasis on ethics and principles of good governance and on the importance of education as well as family and hard work.

**continental drift:** slow movement of the continents over long periods of time across Earth's surface (*see* plate tectonics).

**core regions:** regions that dominate trade, control the most advanced technologies, and have high levels of productivity within diversified economies.

**counterurbanization:** net loss of population from cities to smaller towns and rural areas.

**creative destruction:** withdrawal of investments from activities (and regions) that yield low rates of profit, in order to reinvest in new activities (and new places).

**crony capitalism:** a system of capitalism by which leaders and the friends of leaders are privileged in the economy.

**cultural hearth areas:** centers or regions of sociocultural and political–economic production from which major, distinct cultures—ways of life, modes of agriculture, forms of craft and industry, language, and religions—historically emerged.

**culture:** shared set of meanings that are lived through the material and symbolic practices of everyday life.

**Culture System:** Dutch colonial policy from 1830 to 1870 that required farmers in Java to devote one-fifth of their land and their labor to production of an export crop.

**deforestation:** the clearing, thinning, or elimination of tree-cover in historically forested areas, most typically referring to human-caused tree-cover loss.

**deindustrialization:** decline in industrial employment in core regions as firms scale back their activities in response to lower levels of profitability.

**demographic collapse:** after about 1500, the rapid die-off of the indigenous populations of the Americas as a result of diseases introduced by the Europeans to which residents of the Americas had no immunity.

**demographic transition:** replacement of high birth and death rates by low birth and death rates.

**dependency school (of development theory):** school of thought based on the premise that the economic progress of peripheral or less-developed countries is constrained by the economic and political power of affluent, core countries.

**Desakota:** a "village-town" that is conceptualized as part of a wider metropolitan region in Southeast Asia wherein urban and rural characteristics of economy, communication, transportation, and livelihood are blended.

**desertification:** process by which arid and semiarid lands become degraded and less productive, leading to more desertlike conditions.

**development theory:** analysis of social change that assesses the economic progress of individual countries in an evolutionary way.

**diaspora:** spatial dispersion of a previously homogeneous group.

**disinvestment:** sale of assets such as factories and equipment.

**distributary:** river branch that flows away from the main stream.

**division of labor:** separation of productive processes into individual operations, each performed by different workers or groups of workers.

**domestication:** adaptation of wild plants and animals through selective breeding by humans for preferred characteristics into cultivated or tamed forms.

**domino theory:** (also known as the domino effect) view that political unrest in one country can destabilize neighbors and start a chain of events like the fall of a stack of dominoes.

**Dreamtime:** aboriginal worldview that links past and future, people and places, in a continuity that ensures respect for the natural world.

**dry farming:** arable farming techniques that allow the cultivation of crops without irrigation in regions of limited moisture (50 centimeters, or 20 inches per year).

**Earth system science:** integrated approach to the study of Earth that stresses investigation of the interactions among Earth's components in order to explain Earth and atmosphere dynamics, Earth evolution and ecosystems, and global change.

**ecological imperialism:** concept developed by historian Alfred Crosby to describe the way in which European organisms, including diseases, pests, and domestic animals, were able to take over the ecosystems of other regions of the world, often with devastating impacts on local peoples, flora, and fauna.

**ecosystem:** complex of living organisms, their physical environment, and all their relationships in a particular place.

**ecotourism:** environmentally oriented tourism designed to protect the environment and often to provide economic opportunities for local people.

**egalitarian society:** society based on belief in equal social, political, and economic rights and privileges.

**El Niño:** periodic warming of sea surface temperatures in the tropical Pacific off the coast of Peru that results in worldwide changes in climate, including droughts and floods.

**enclave:** culturally distinct territory that is encompassed by a different cultural group or groups.

**encomienda:** system by which groups of indigenous people were "entrusted" to Spanish colonists who could demand tribute in the form of labor, crops, or goods and in turn were responsible for the indigenous groups' conversion to the Catholic faith and for teaching them Spanish.

**Enlightenment:** eighteenth-century movement marked by a belief in the sovereignty of reason and empirical research in the sciences.

**entrepôt:** seaport that is an intermediary center of trade and transshipment.

**Environmental Determinism:** the idea that the natural environment is the main influence on economic development and human potential.

**equal-area (equivalent) projection:** map projection that portrays areas on Earth's surface in their true proportions.

**equidistant projection:** map projection that allows distance to be represented as accurately as possible.

**ethnic cleansing:** systematic and forced removal of members of an ethnic group from their communities in order to change the ethnic composition of a region.

**ethnic group:** group of people whose members share cultural characteristics.

**ethnicity:** a socially-created system of rules about who belongs to a particular group based on actual or perceived commonalities, such as language or religion.

**ethnocentrism:** attitude that one's own race and culture is superior to that of others.

**europeanization:** highly selective process involving the mixing of native and imported practices that eventually created distinct colonial cultures and societies.

**exclave:** portion of a country or a cultural group's territory that lies outside its contiguous land area.

**failed state:** a nation in which the central government is so weak that it has little control over its territory.

**farm crisis:** financial failure and foreclosure of thousands of family farms across the U.S. Midwest.

**fascism:** political philosophy characterized by a centralized, autocratic government that values nation and race over the individual.

**fast food:** edibles that can be prepared and served very quickly, sold in a restaurant, and served to customers in packaged form.

**favela:** Brazilian term for informal settlements lacking good housing and services that grow up around the urban core.

**federal state:** form of government in which power is allocated to units of local government within the country.

**feminization of poverty:** likelihood that women will be poor, malnourished, and otherwise disadvantaged because of inequalities within the household, the community, and the country.

**feng shui:** application of a collection of ancient principles of geomancy that are believed by adherents to ensure health, wealth, happiness, long life, and healthy offspring through the spatial organization of cities, buildings, and furniture.

**feudal systems:** regional hierarchies composed, at the bottom, of local nobles and, at the top, of lords or monarchs owning immense stretches of land. In these systems, landowners delegated smaller parcels of land to others in return for political allegiance and economic obligations in the form of money dues or labor.

**fjord:** steep-sided, narrow inlet of the sea, formed when deeply glaciated valleys are flooded by the sea.

**flexible production region:** region within which there is a concentration of small- and medium-sized firms whose production and distribution practices take advantage of computerized control systems and local subcontractors in order to quickly exploit new market niches for new product lines.

**formal region:** region with a high degree of homogeneity in terms of particular distinguishing features.

**free trade association:** association whose member countries eliminate tariff and quota barriers against trade from other member states but continue to charge regular duties on materials and products coming from outside the association.

**free trade zone:** area within which goods may be manufactured or traded without customs duties.

**functional region:** area characterized by a coherent functional organization of human occupancy.

**G8:** Group of Eight countries (Canada, France, Germany, Italy, Japan, Russia, the United Kingdom, and the United States) whose heads of state meet each year to discuss issues of mutual and global concern.

**gender:** social differences between men and women rather than the anatomical differences related to sex.

**gender and development (GAD):** approach to development that links women's productive and reproductive roles and an approach to understanding the gender-related differences and barriers to better lives of both men and women.

**gender division of labor:** separation of productive processes based on gender.

**gender-related development index:** a composite indicator of gender equality assessing the standard of living in a country that aims to show the inequalities between men and women in the following areas: long and healthy life, knowledge, and a decent standard of living.

**gentrification:** invasion of older, centrally located, working-class neighborhoods by higher-income households.

**geodemographic research:** use of census data and commercial data (such as sales data and property records) for small districts in creating profiles of the populations of those districts for market research.

**geographic information systems (GIS):** integrated computer tools for the handling, processing, and analyzing of geographical data.

**geomancy:** belief that the physical attributes of places can be analyzed and manipulated in order to improve the flow of cosmic energy, or *ch'i.*

**geomorphology:** study of landforms.

**global civil society:** the worldwide array of voluntary civic and social organizations and institutions—also called nongovernmental organizations, or NGOs—that operate to represent the interests of citizens against the power of formal states and markets.

**global north:** those parts of the globe experiencing the highest levels of economic development, which are typically—though by no means exclusively—found in northern latitudes.

**global south:** those parts of the globe experiencing the lowest levels of economic development, which are typically—though by no means exclusively—found in southern latitudes.

**global warming:** increase in world temperatures and change in climate associated with increasing levels of carbon dioxide and other gases resulting from human activities such as deforestation and fossil-fuel burning.

**globalization:** increasing interconnectedness of different parts of the world through common processes of economic, environmental, political, and cultural change.

**GMO (genetically modified organism):** any organism that has had its DNA modified in a laboratory rather than through cross-pollination or other forms of evolution.

**Grameen Bank:** a nongovernmental grassroots organization formed to provide small loans—microfinance—to the rural poor in Bangladesh, which has grown to an international institution that lends worldwide to poor borrowers who have no credit.

**Great Artesian Basin:** world's largest reserve of underground water located in central Australia and under pressure so that water rises to the surface when wells are bored.

**Great Leap Forward:** a Chinese policy scheme launched in 1958 to accelerate the pace of economic growth by merging land into huge communes and seeking to industrialize the countryside. Largely viewed as a massive failure and partly responsible for starvation in the period following.

**Green Revolution:** technological package of higher yielding seeds, especially wheat, rice, and corn, that in combination with irrigation, fertilizers, pesticides, and farm machinery was able to increase crop production in the developing world after about 1950.

**greenhouse effect:** trapping of heat within the atmosphere by water vapor and gases, such as carbon dioxide, resulting in the warming of the atmosphere and surface.

**gross domestic product (GDP):** estimate of the total value of all materials, foodstuffs, goods, and services that are produced in a country in a particular year.

**gross national product (GNP):** similar to GDP, but also includes the value of income from abroad.

**guano:** bird dropping that can create high-quality phosphate rock, which is used in fertilizers.

**guest worker:** foreigner who is permitted to work in another country on a temporary basis.

**hacienda:** large agricultural estate in Latin America and Spain that grows crops mainly for domestic consumption (for example, for mines, missions, and cities) rather than for export.

**hajj:** pilgrimage to Mecca required of all Muslims.

**harmattan:** hot, dry wind that blows out of inland Africa.

**hate crime:** act of violence committed because of prejudice against women; homosexuals; and ethnic, racial, and religious minorities.

**heathland:** open, uncultivated land with poor soil and scrub vegetation.

**hegemony:** domination over the world economy, exercised through a combination of economic, military, financial, and cultural means, by one national state in a particular historical epoch.

**Hindutva:** a social and political movement that calls on India to unite as an explicitly Hindu nation, which has given birth to a range of political parties over the years, including the Bharatiya Janata Party, or BJP.

**homelands:** areas set aside in South Africa for black residents as tribal territories where they were given limited self-government but no vote and limited rights in the general politics of South Africa.

**Hydraulic civilizations:** large state societies hypothesized to have arisen from the needs of organizing massive irrigation systems.

**Ideograph:** a linguistic symbol representing a single idea or object, rather than a sound.

**imperialism:** extension of the power of a nation through direct or indirect control of the economic and political life of other territories.

**import substitution:** process by which domestic producers provide goods or services that were formerly bought from foreign producers.

**indentured servant:** individual bound by contract to the service of another for a specified term.

**informal economy:** economic activities that take place beyond official record and not subject to formalized systems of regulation or remuneration (e.g., street selling, petty crime).

**intermontane:** lying between or among mountains.

**internal migration:** movement of populations within a national territory.

**internally displaced persons:** individuals who are uprooted within their own countries due to civil conflict or human rights violations, sometimes by their own governments.

**international division of labor:** specialization of different people, regions, and countries in certain kinds of economic activities.

**International Monetary Fund (IMF):** international organization that monitors the international financial system and provides loans to governments throughout the world.

**international regime:** orientation of contemporary politics around the international arena rather than the national one.

**intertropical convergence zone (ITCZ):** region where air flows together and rises vertically as a result of intense solar heating at the equator, often with heavy rainfall, and shifting north and south with the seasons.

**intifada:** the uprising of Palestinians against the rule of Israel in the Occupied Territories.

**IPCC:** the Intergovernmental Panel on Climate Change, an international group of scientists that, since 1990, have prepared regular assessments on climate change.

**irredentism:** assertion by the government of a country that a minority living outside its borders belongs to it historically and culturally.

**Islam:** religion that is based on submission to God's will according to the Qur'an.

**Islamism:** Political movement or political identity that promotes Islamic law, pan-Islamic unity, and rejection of western influences in the Muslim world.

**jihad:** sacred struggle or striving to carry out God's will according to the tenets of Islam.

**keiretsu:** Japanese business networks facilitated after World War II by the Japanese government in order to promote national recovery.

**killing fields:** fields found throughout Cambodia where millions of people were buried during the reign of the Khmer Rouge (1975–1979).

**kinship:** shared notion of relationship among members of a group often, but not necessarily, based on blood, marriage, or adoption.

**kleptocracy:** a form of government whereby leaders divert national resources for their personal gain.

**Kyoto Protocol:** an international treaty providing legally binding rules, which seeks to limit and reduce greenhouse gas emissions in signatory countries, in order to avert dangerous anthropogenic climate change.

**La Niña:** periodic abnormal cooling of sea surface temperatures in the tropical Pacific off the coast of Peru that results in worldwide changes in climate, including droughts and floods that contrast with those produced by El Niño.

**land bridge:** dry land connection between two continents or islands, exposed, for example, when sea level falls during an ice age.

**land reform:** change in the way land is held or distributed, such as the division of large private estates into small private farms or communally held properties.

**latitude:** angular distance of a point on Earth's surface, measured north or south from the equator, which is 0°.

**law of diminishing returns:** tendency for productivity to decline, after a certain point, with the continued addition of capital and/or labor to a given resource base.

**leadership cycles:** periods of international power established by individual states through economic, political, and military competition.

**liberation theology:** Catholic movement, originating in Latin America, focused on social justice and on helping the poor and oppressed.

*Lingua franca:* common language used to communicate between people of different backgrounds and languages, often for trading purposes.

**little tigers:** this represents the economies of Thailand, Malaysia, and Indonesia, which have had relatively high rates of economic growth since the early 1980s.

**local food:** usually organically grown, that it is produced within a fairly limited distance from where it is consumed.

**loess:** surface cover of fine-grained silt and clay deposited by wind action and usually resulting in deep layers of yellowish, loamy soils.

**longitude:** angular distance of a point on Earth's surface, measured east or west from the prime meridian (the line that passes through both poles and through Greenwich, England, and is given the value of 0°).

**Main Street:** dominant urban corridor of Canada, extending from Québec City to Windsor, Ontario.

**mandate:** delegation of political power over a region, province, or state.

**Manga:** Japanese print cartoons books (or komikku), which date at least to the last century.

**mangroves:** ecosystem of trees or shrubs growing in saline coastal regions in the tropics.

**map projection:** systematic rendering on a flat surface of the geographic coordinates of the features found on Earth's surface.

**maquiladora:** industrial plant in Mexico, originally within the border zone with the United States and often owned or built with foreign capital, that assembles components for export as finished products free from customs duties.

**market economies:** economies in which goods and services are produced and are distributed through free markets.

**Marshall Plan:** strategy (named after U.S. Secretary of State George Marshall) of quickly rebuilding the West German economy after World War II with U.S. funds in order to prevent the spread of socialism or a recurrence of fascism.

**marsupial:** Australian mammal such as the kangaroo, koala, and wombat that gives birth to premature infants that then develop and feed from nipples in a pouch on the mother's body.

**massif:** mountainous block of Earth's crust bounded by faults or folds and displaced as a unit.

**matrilocality:** the cultural practice by which a married male-female couple lives with the family of the woman.

**megalopolis:** continuous chain of metropolitan development such as the urban corridor that extends from Boston to Washington DC on the Eastern seaboard of the U.S. (also used for any megacity of more than 10 million people).

**Mekong River Commission (MRC):** an intergovernmental organization that coordinates the management of the Mekong River Basin, with members including: Cambodia, Laos, Thailand, and Vietnam as well as "dialogue partners" Burma and the People's Republic of China.

**Melanesia:** region of the western Pacific that includes the westerly and largest islands of Papua New Guinea, the Solomon Islands, Fiji, Vanuatu, and New Caledonia.

**mercantilism:** Economic system typical of European states from the 16th to 18th centuries, to increase national wealth and power through regulation of the economy and expansion of trade, especially through colonies and accumulating precious metals such as gold.

**merchant capitalism:** form of capitalism characterized by trade in commodities and a highly organized system of banking, credit, stock, and insurance services.

**mestizo:** term used in Latin America to identify a person of mixed white (European) and American Indian ancestry.

**microfinance:** programs that provide credit and savings to the self-employed poor, including those in the informal sector, who cannot borrow money from commercial banks.

**Micronesia:** a region of island states in the South Pacific, which includes Guamn, Kiribiti, Marshall Islands, the federate states of Micronesia, Nauru, Northern Mariana Islands, Palau.

**mikrorayon:** neighborhood-scale planning unit of the Soviet era.

**Millennium Development Goals (MDGs):** eight goals to be met by the year 2015, agreed to by members of the United Nations, that include the eradication of poverty, universal primary education, gender equality, the reduction of child mortality, the improvement of maternal health, the combating of disease, environmental sustainability, and the creation of global partnerships.

**minifundia:** very small parcels of land farmed by tenant farmers or peasant farmers in Latin America.

**minisystem:** society with a single cultural base and a reciprocal social economy.

**Modernity:** forward-looking view of the world that emphasizes reason, scientific rationality, creativity, novelty, and progress.

**modernization theory:** economic development occurs when investment rates enable higher levels of industrialization, thus raising labor productivity and increasing the GDP per capita levels.

**monoculture:** agricultural practice in which one crop is grown intensively over a large area of land.

**monotremes:** a biological group of egg-laying mammals most often found in Australia and New Guinea.

**Monroe Doctrine:** proclamation of U.S. President James Monroe in 1823 stating that European military interference in the Western Hemisphere, including the Caribbean and Latin America, would no longer be acceptable.

**monsoon:** seasonal reversal of wind flows in parts of the lower to middle latitudes. During the cool season, a dry monsoon occurs as dry offshore winds prevail; in hot summer months a wet monsoon occurs as onshore winds bring large amounts of rainfall.

**moraine:** accumulation of rock and soil carried forward by a glacier and eventually deposited at its frontal edge or along its sides.

**multiculturalism:** process of immigrant incorporation in which each ethnic group has the right to enjoy and protect their officially recognized "native" culture.

**municipal housing:** rental housing that is owned and managed by a local government or municipality; a form of *social housing*.

**Muslim:** member of the Islamic religion.

**nation:** group of people often sharing common elements of culture, such as religion, language, a history, or political identity.

**nationalism:** feeling of belonging to a nation as well as the belief that a nation has a natural right to determine its own affairs.

**nationalist movement:** organized groups of people, sharing common elements of culture, such as language, religion, or history, who wish to determine their own political affairs.

**nationalization:** process of converting key industries from private to governmental organization and control.

**nation-state:** ideal form consisting of a homogeneous group of people governed by their own state.

**Near Abroad:** independent states that were formerly republics of the Soviet Union.

**neocolonialism:** economic and political strategies by which powerful states in core economies indirectly maintain or extend their influence over other areas or people.

**neoliberal policies:** economic policies that are predicated on a minimalist role for the state and the reduction in the role and budget of government, including reduced subsidies and the privatization of formerly publicly owned and operated concerns, such as utilities.

**neoliberalism:** economic doctrine based on a belief in a minimalist role for the state, assuming the desirability of free markets as the ideal condition not only for economic organization but also for social and political life.

**neotropics:** ecological region of the tropics of the Americas.

**new international division of labor:** decentralization of manufacturing production from core regions to some peripheral and semiperipheral countries.

**nontraditional agricultural exports (NTAEs):** new export crops, such as vegetables and flowers, that contrast with the traditional exports such as sugar and coffee, and often require fast, refrigerated transport to market.

**North American Free Trade Agreement (NAFTA):** 1994 agreement among the United States, Canada, and Mexico to reduce barriers to trade among the three countries, through, for example, reducing customs tariffs and quotas.

**oasis:** spot in the desert made fertile by the availability of surface water.

**offshore financial services:** provision of banking, investment, and other services to foreign nationals and companies who wish to avoid taxes, oversight, or other regulations in their own countries.

**One-child policy:** a Chinese policy involving rewards for families giving birth to only one child, including work bonuses and priority in housing.

**orographic effect:** influence of hills and mountains in lifting airstreams, cooling the air, and thereby inducing precipitation.

**Outback:** dry and thinly populated interior of Australia.

**Outstations:** a remotely located and often sparsely populated settlement, often based around cattle and sheep or aboriginal communities, associated with the Australian interior or "Outback."

**overseas Chinese:** migrants from China who settled in Southeast Asia as early as the fourteenth century, but mainly during the period of European colonialism, when they arrived as contract plantation, mine, and rail workers and then moved into clerical and business roles.

**ozone depletion:** loss of the protective layer of ozone gas that prevents harmful ultraviolet radiation from reaching Earth's surface and causing increases in skin cancer and other ecological damage.

**Pacific Rim:** loosely defined region of countries that border the Pacific Ocean.

**Pancasila:** the Indonesian, post-colonial nation-building ideology whereby all Indonesians are connected through unity in diversity through belief in one god, nationalism, humanitarianism, democracy, and social justice.

**pastoralism:** system of farming and way of life depending on the environment and based on keeping herds of grazing animals—cattle, sheep, goats, horses, camels, yaks, and so on.

**pastoralists:** people who gains a significant portion of their subsistence from the raising of livestock, typically in large herds.

**patrilocality:** the cultural practice by which a married male-female couple lives with the family of the man.

**peripheral regions:** regions that are characterized by dependent and disadvantageous trading relationships, by inadequate or obsolescent technologies, and by undeveloped or narrowly specialized economies with low levels of productivity.

**permafrost:** permanently frozen subsoil, which may extend for several meters below the surface layer and may defrost up to a depth of a meter or so during summer months.

**petrodollar:** revenues generated by the sale of oil.

**physiographic region:** broad region within which there is a coherence of geology, relief, landforms, soils, and vegetation.

**pinyin:** system of writing Chinese language using the Roman alphabet.

**place:** specific geographic setting with distinctive physical, social, and cultural attributes.

**plantation:** large agricultural estate that is usually tropical or semitropical, monocultural (one crop), and commercial- or export-oriented, most of which were established in the colonial period.

**plantation economies:** the economic systems typical of colonial trade, made up of extensive, European-owned, operated, and financed enterprises where single crops were produced by local or imported labor for a world market.

**plate tectonics:** theory that Earth's crust is divided into large solid plates that move relative to each other and cause mountain building, and volcanic and earthquake activity when they separate or meet.

**pluralist democracy:** society in which members of a diverse group continue to participate in their traditional cultures and special interests.

**polder:** area of low land reclaimed from a body of water by building dikes and draining the water.

**Polynesia:** central and southern Pacific islands that include the independent countries of Samoa, Tonga, the Cook Islands, Niue, and Tuvalu; the U.S. territory of American Samoa; the French territories of Wallis, Fortuna, and French Polynesia (including the island of Tahiti, the Society Islands, the Tuamotu archipelago, and the Marquesas Islands); the New Zealand territory of Tokelau; and the British territory of the Pitcairn Islands. Polynesia sometimes includes New Zealand and the Hawaiian islands.

**primacy:** condition in which the population of the largest city in an urban system is disproportionately large in relation to the second- and third-largest cities in that system.

**primary activity:** economic activity that is concerned directly with natural resources of any kind.

**primary commodities:** economic goods derived from organic or geological processes, like agricultural products, minerals, or fish.

**pristine myth:** erroneous belief that the Americas were mostly wild and untouched by humans prior to European arrival. In fact, large areas had been cultivated and deforested by indigenous populations.

**privatization:** turnover or sale of state-owned industries and enterprises to private interests.

**purchasing power parity (PPP):** a measure of how much of a common "market basket" of goods and services a currency can purchase locally, including goods and services that are not traded internationally; PPP makes it possible to compare levels of economic prosperity between countries where the price of goods might be relatively much higher or lower.

**quaternary activity:** economic activity that deals with the handling and processing of knowledge and information.

**race:** a problematic and illusory classification of human beings based on skin color and other physical characteristics; biologically speaking, no such thing as race exists within the human species.

**racialization:** The practice of creating unequal castes where whiteness is considered the norm, despite the biological reality that no such thing as race exists within the human species.

**Raj:** rule of the British in India.

**region:** large territory that encompasses many places, all or most of which share similar attributes in comparison with the attributes of places elsewhere.

**regional geography:** study of the ways in which unique combinations of environmental and human factors produce territories with distinctive landscapes and cultural attributes.

**regionalism:** strong feelings of collective identity shared by religious or ethnic groups that are concentrated within a particular region.

**regionalization:** geographer's classification of individual places or areal units.

**remittances:** money sent home to family or friends by people working temporarily or permanently in other countries.

**remote sensing:** collection of information about parts of Earth's surface by means of aerial photography or satellite imagery designed to record data on visible, infrared, and microwave sensor systems.

**Rice bowl:** commonly refers to regions in Asia where large-scale, wet-rice agriculture production takes place, providing a continuous source of this staple crop.

**Richter scale:** logarithmic scale ranging from 1 to 10 used to measure the amount of energy released by an earthquake.

**rift valley:** block of land that drops between two others, forming a steep-sided trough, often at faults on a divergent plate boundary.

**Ring of Fire:** chain of seismic instability and volcanic activity that stretches from Southeast Asia through the Philippines, the Japanese archipelago, the Kamchatka Peninsula, and down the Pacific coast of the Americas to the southern Andes in Chile. It is caused by the tension built up by moving tectonic plates.

**salinization:** salt deposits caused when water evaporates from the surface of the land and leaves behind salt that it has drawn up from the subsoil.

**satellite state:** national state that is economically dependent and politically and militarily subservient to another—in its orbit, figuratively speaking.

**savanna:** grassland vegetation found in tropical areas with a pronounced dry season and periodic fires.

**sawah:** Indonesian term used in Southeast Asia that means irrigated or wet rice cultivation.

**sea gypsies:** nomadic fisherpeoples and ethnic minorities who commonly make their living and homes in the coastal waters of South and Southeast Asia.

**secondary activity:** economic activity involving the processing, transformation, fabrication, or assembly of raw materials, or the reassembly, refinishing, or packaging of manufactured goods.

**secondary commodities:** economic goods derived from industrial processes that add value to primary goods or raw materials by combining or refining them.

**sectionalism:** extreme devotion to local interests and customs.

**semiperipheral regions:** regions that are able to exploit peripheral regions but are themselves exploited and dominated by the core regions.

**sense of place:** feelings evoked among people as a result of the experiences and memories that they associate with a place and the symbolism that they attach to it.

**shifting cultivation:** agricultural system that preserves soil fertility by moving crops from one plot to another.

**Shinto:** a Japanese indigenous religious culture, which stresses a belief in the nature of sacred powers that can be recognized in every individual existing thing, and which include practices entailing ritual purification, the offering of food to sacred powers, sacred music and dance, solemn worship, and joyous celebration.

**short twentieth century:** expression often used to describe the period from the outbreak of World War I in 1914 until the collapse of the Soviet Union in 1991, a time of distinctive economic and geopolitical development.

**SIJORI:** an economically integrated growth region that includes Singapore, Johor Baharu (Malaysia), and Riau Islands (Indonesia).

**Silk Road:** ancient east–west trade route between Europe and China.

**site:** physical attributes of a location that could include terrain, soil, vegetation, and water sources.

**situation:** location of a place relative to other places and human activities.

**slow food:** an attempt to resist fast food by preserving the cultural cuisine and the associated food and farming of an ecoregion.

**slash and burn:** agricultural system often used in tropical forests that involves cutting trees and brush and burning them so that crops can benefit from cleared ground and nutrients in the ash.

**social capital:** networks and relationships that encourage trust, reciprocity, and cooperation.

**social housing:** rental housing that is owned and managed by a public institution or nonprofit organization.

**social movements:** organized movements of people with an agenda of political opposition and activism.

**South Pacific Forum:** institution that promotes discussion and cooperation on trade, fisheries, and tourism between all of the independent and self-governing states of Oceania.

**sovereign state:** a political unit that exercises power over a territory and people and is recognized by other states with its independent power codified in international law.

**sovereignty:** exercise of state power over people and territory, recognized by other states and codified by international law.

**spatial diffusion:** way that things spread through geographic space over time.

**spatial justice:** fairness of the distribution of society's burdens and benefits, taking into account spatial variations in people's needs and in their contribution to the production of wealth and social well-being.

**Special Economic Zones (SEZs):** carefully segregated export-processing areas in China that offer cheap labor and land, along with tax breaks, to transnational corporations.

**Spice Islands:** island chain in Southeast Asia (modern-day eastern Indonesia in particular) where spices, such as pepper and mustard, were domesticated and globally traded to other world regions, particularly beginning with the European colonial period, when Portuguese and the Dutch occupied the region.

**staple crops:** food crops that are eaten regularly and which dominate the basic diet of a population, and so form the basis of regional or national subsistence.

**staples economy:** economy based on natural resources that are unprocessed or minimally processed before they are exported to other areas where they are manufactured into end products.

**state:** independent political unit with territorial boundaries that are internationally recognized by other states.

**state socialism:** form of economy based on principles of collective ownership and administration of the means of production and distribution of goods, dominated and directed by state bureaucracies.

**steppe:** semiarid, treeless, grassland plains.

**stolen generation:** Aboriginal children that were forcibly removed from their homes in Australia and placed in white foster homes or institutions.

**structural adjustment policies:** economic policies, mostly associated with the International Monetary Fund, that required governments to cut budgets and liberalize trade in return for debt relief.

**subsistence affluence:** achievement of a good standard of living through reliance on self-sufficiency in local foods and with little cash income.

**suburbanization:** growth of population along the fringes of large metropolitan areas.

**Sundarbans:** a distinctive ecology of untouched mangrove and tropical swamp forest stretching from West Bengal in India into coastal Bangladesh; it is home to crocodiles, Chital deer, and one of the largest single populations of Bengal tigers in the world.

**superfund site:** locations in the United States officially deemed by the federal government as extremely polluted and requiring extensive, supervised, and subsidized cleanup.

**supranational organization:** collection of individual states with a common economic and/or political goal that diminishes, to some extent, individual state sovereignty in favor of the collective interests of the membership.

**sustainable development:** vision of development that seeks a balance among economic growth, environmental impacts, and social equity.

**swidden farming:** a form of agriculture in which land is cleared for cultivation by cutting and burning shrubs or trees, allowing multiple years of planting until forest regrowth occurs; also pejoratively called "slash-and-burn."

**taiga:** ecological zone of boreal coniferous forest.

**technology system:** cluster of interrelated energy, transportation, and production technologies that dominates economic activity for several decades.

**terms of trade:** relationship between the prices a country pays for imports compared to the prices for its exports where poor terms of trade are when import prices are much higher than export prices.

**Terracing:** the creation of a distinctive landscape of stepped and reinforced flat agricultural fields cut into steep slopes in order to stabilize the land for cropping in the face of potentially catastrophic soil erosion, which typically allows sustained agricultural yields amid torrential rainfall in mountainous areas.

**territorial production complex:** regional groupings of production facilities based on local resources that were suited to clusters of interdependent industries.

**tertiary activity:** economic activity involving the sale and exchange of goods and services.

**theory of island biogeography:** theory that smaller islands will generally be less biologically diverse than larger ones.

**Three Gorges Dam:** the largest electricity-generating facility in the world, based on an enormous hydroelectric river dam spanning the Yangtze River in central China and several dozen associated generators.

**time-space convergence:** rate at which places move closer together in travel or communication time or costs.

**total fertility rate:** average number of children a woman will bear throughout her childbearing years, approximately ages 15 through 49.

**trade creation effects:** positive economic effects of transnational integration.

**trade diversion effects:** negative economic effects of transnational integration.

**transform boundary:** process of plates sliding past each other horizontally.

**transgenic crops:** crops containing genes that have been artificially inserted rather than acquired through pollination, typically derived from other species or breeds; also known as GM, or genetically modified crops.

**transhumance:** movement of herds according to seasonal rhythms: to warmer, lowland areas in the winter, and cooler, highland areas in the summer.

**transmigration:** policy of resettling people from densely populated areas to less populated, often frontier regions.

**transnational corporation:** corporation that has investments and activities that span international boundaries, with subsidiary companies, factories, offices, or facilities in several countries.

**Treaty of Tordesillas:** agreement made by Pope Alexander VI in 1494 to divide the world between Spain and Portugal along a north–south line 370 leagues (about 1800 kilometers, or 1100 miles) west of the Cape Verde Islands. Portugal received the area east of the line, including much of Brazil and parts of Africa, and Spain received the area to the west.

**Treaty of Waitangi:** 1840 agreement in which 40 Maori chiefs gave the Queen of England governance over their land and the right to purchase it in exchange for protection and citizenship. Reinterpreted by the Waitangi tribunal in the 1990s, it provides the basis for Maori land rights and New Zealand's bicultural society.

**treaty ports:** ports in Asia, especially China and Japan, that were opened to foreign trade and residence in the mid-nineteenth century because of pressure from powers such as Britain, France, Germany, and the United States.

**tribe:** form of social identity created by groups who share a common set of ideas about collective loyalty and political action.

**tsunami:** sometimes-catastrophic coastal waves created by offshore seismic (earthquake) activity.

**tundra:** arctic wilderness where the climate precludes any agriculture or forestry. Permafrost and very short summers mean that the natural vegetation consists of mosses, lichens, and certain hardy grasses.

**UNCLOS:** United Nations Convention on the Law of the Sea, which governs territorial seas and sea use.

**unitary state:** form of government in which power is concentrated in the central government.

**urban bias:** the tendency of governments and others to concentrate investment and attention in urban rather than rural areas.

**viceroyalty:** largest scale of Spanish colonial administration leading to the emergence of Mexico City and Lima as the headquarters of the viceroyalties of New Spain and Peru, respectively.

**visualization:** computer-assisted representation of spatial data, often involving three-dimensional images and innovative perspectives, in order to reveal spatial patterns and relationships more effectively.

**Wallace's Line:** division between species associated with the deep-ocean trench between the islands of Bali and Lombok in Indonesia that could not be crossed even during the low sea levels of the ice ages.

**watershed:** drainage area of a particular river or river system.

**weather:** instantaneous or immediate state of the atmosphere (e.g., it is raining).

**welfare state:** institution with the aim of distributing income and resources to the poorer members of society.

**White Australia policy:** Australian policy, until 1975, that restricted immigration to people from northern Europe through a ranking that placed British and Scandinavians as the highest priority, followed by southern Europeans, with the goal of attaining a homogenized-looking and culturally similar population.

**World Bank:** development bank and the largest source of development assistance in the world.

**world city:** city in which a disproportionate share of the world's most important business—economic, political, and cultural—is transacted.

**World Heritage Site:** place that has been formally identified as a protected site by the United Nations Educational, Scientific, and Cultural Organization (UNESCO).

**world region:** large-scale geographic division based on continental and physiographic settings that contain major clusters of humankind with broadly similar cultural attributes.

**world religion:** belief system with worldwide adherents.

**world-system:** interdependent system of countries linked by political and economic competition.

**xenophobia:** hate and/or fear of foreigners.

**yurt:** circular tent used by nomadic groups in Mongolia, northwest China, and Central Asia, constructed from collapsible wooden frames and felt made from sheep's wool.

**zaibatsu:** large Japanese conglomerate corporation.

**Zero population growth (ZPG):** a demographic state where the number of births match the number of deaths in a population, such that no natural population growth occurs.

**zionism:** movement whose chief objective has been the establishment for the Jewish people of a legally recognized home in Palestine.

# PHOTO AND ILLUSTRATION CREDITS

## CONTENTS

TOC.01: Christopher Herwig/Getty Images

TOC.02: Paulo Whitaker/Corbis

TOC.03: Chen Yehua/XinHua/Xinhua Press/Corbis

TOC.04: Jose Fuste Raga/Corbis

## CHAPTER ONE

1.2: E. Aguado and J. E. Burt, *Understanding Weather and Climate*, 2nd ed. Upper Saddle River, NJ: Prentice Hall, 2001.

1.3: R. W. Christopherson, *Geosystems: An Introduction to Physical Geography*, 6th ed. Upper Saddle River, NJ: Prentice Hall, 2005.

1.4b: Ed Darack/Science Faction/Corbis

1.5 (Left): Modified from T.L. McKnight, *Physical Geography: A Landscape Appreciation*, 6th ed. Upper Saddle River, NJ: Prentice Hall, 1999, pp. 97, 160-61

1.5 (Right): Modified from R.W. Christopherson, *Geosystems*, 4th ed. Upper Saddle River, NJ: Prentice Hall, 2000, p. 268.

1.7: 10.8., 2007 IPCC reports for (a) Working Group 1 and (b) Working Group 2.

1.9: Alison Wright/Corbis

1.10: T. L. McKnight, *Physical Geography: A Landscape Appreciation*, 6th ed. Upper Saddle River, NJ: Prentice Hall, 1999.

1.12a: Redrawn from Armesto, *The World: A History*, 2010.

1.12b: Barry Lewis/Corbis

1.13: Redrawn from Armesto, *The World: A History*, 2010.

1.14: Redrawn from Armesto, *The World: A History*, 2010.

1.15: Redrawn from Armesto, *The World: A History*, 2010, pp. 526-527.

1.16: Redrawn form B. Crow and A. Thomas, *Third World Atlas*. Milton Keynes: Open University Press, 1982, pp. 37, 41.

1.17: Adapted from B. Crow and A. Thomas, *Third World Atlas*. Milton Keynes: Open University Press, 1982, p. 27.

1.18: After P. Hugill, *World Trade Since 1431*. Baltimore: Johns Hopkins University Press, 1993, p. 136.

1.19: Redrawn from Armesto, *The World: A History*, 2010.

1.20: Yang Liu/Corbis

1.21: M. Zook, http://www.zookNIC.com/.

1.27: Richard Cummins/Corbis

1.28: Knox/Marston, *Human Geography*, 5th ed.

1.29: Knox/Marston, *Human Geography*, 5th ed.

1.30: Adapted from E.F. Bergman, *Human Geography: Cultures, Connections, and Landscapes*, Upper Saddle River, NJ: Prentice Hall, 1994; Western Hemisphere after J.H. Greenberg, *Language in the Americas*. Stanford, CA: Stanford University Press, 1987; Eastern Hemisphere after D. Crystal, *The Cambridge Encyclopedia of Language*. Cambridge: Cambridge University Press, 1997).

1.31: Knox/Marston, *Human Geography*, 5th ed.

1.32: Catherine Karnow/Corbis

1.33: Seppe Van Grieken/Getty Images

1.34: Jason Lee/Corbis

1.35: Ashley Cooper/Corbis

1.36: Vincent West/Corbis

1.39: History: Energy Information Administration, *International Energy Annual 2003*, and Projections: Energy Information Administration, *Systems for the Analysis of Global Energy Markets, 2006*. Available at http://www.eia.doe.gov/iea/.

1.40: Millennium Ecosystem Assessment Board, *Living Beyond Our Means: Natural Assets and Human Well-Being*. Washington, DC: Island Press, 2006, p. 13.)

1.B1.2: Getty Images

## CHAPTER 2

2.CO: Atlantide Phototravel/Corbis

2.2: NASA/http://visibleearth.nasa.gov/

2.4: Adapted from R. Mellor and E. A. Smith, *Europe: A Geographical Survey of the Continent*. London: Macmillan, 1979.

2.5: Jean-Pierre Lescourret/Corbis

2.6: Bo Braennhage/Corbis

2.7: Michael Busselle/Corbis

2.8: Adam Woolfitt/Corbis

2.9: Ljupco Smokovski/Shutterstock

2.10: Redrawn from R. King et al., *The Mediterranean*. London: Arnold, 1997, pp. 59 and 64.

2.11: Redrawn from P. Hugill, *World Trade Since 1431*. Baltimore: Johns Hopkins University Press, 1994, p. 50.

2.13: Hulton-Deutsch Collection/Corbis

2.14: Alessandra Benedetti/Corbis

2.15: Sean Gallup/Getty Images

2.17: EU Inforegio. European Commission, European Regional Development Fund and Cohesion Fund, 2002. Available at http://europa.eu.int/comm/regional_policy/funds/prord/guide/euro2000–2006_en.htm.

2.18: Radius Images/Corbis

2.19: Sandy Sockwell/Corbis

2.20: Paul Knox

2.21: John Harper/Corbis

2.22: Kurt Krieger/Corbis

2.23: Wolfram Steinberg/dpa/Corbis

2.24: Peter Adams/Corbis

2.25: D. Pinder [ed.], *The New Europe: Economy, Society, and Environment*. New York: John Wiley & Sons, 1998, p. 106.)

2.26: Center for International Earth Science Information Network [CIESIN], Columbia University, *Gridded Population of the World [GPW]*, Version 3. Palisades, NY: CIESIN, Columbia University, 2005. Available at http://sedac.ciesin.columbia.edu/gpw/.

2.27: J. McFalls, Jr., "Population: A Lively Introduction," *Population Bulletin*, 46[2], 1991, p. 1.

2.28: D. Pinder [ed.], *The New Europe: Economy, Society, and Environment*. New York: John Wiley & Sons, 1998, p. 265.)

2.29: Redrawn from R. Mellor and E. A. Smith, *Europe: A Geographical Survey of the Continent*. London: Macmillan, 1979, p. 22.

2.30: Francis Dean/The Image Works

2.31: Thomas Coex/AFP/Getty Images

2.32: Christian Hartmann/Reuters

2.33: Howard Davies/Corbis

2.34: Euratlas—Nüssli, Rue du Milieu 30, 1400 Yverdon-les-Bains, Switzerland. http://www.euratlas.com/history_europe/europe_map_1500.html.

2.35: Euratlas—Nüssli, Rue du Milieu 30, 1400 Yverdon-les-Bains, Switzerland. http://www.euratlas.com/history_europe/europe_map_1900.html.)

2.36: P. Knox, *Geography of Western Europe*. London: Croom Helm, 1984, p. 69.)

2.B1.1: Grand Tour/Corbis

2.B1.2: Adapted from T. Unwin, *Wine and the Vine*. New York: Routledge, 1996, pp. 35 and 219.

2.B2.1: Jason Hawkes/Corbis

2.B2.2: Corbis

2.B2.3: Corbis

## CHAPTER 3

3.CO: Mladen Antonov/AFP/Getty Images

3.2: NASA/http://visibleearth.nasa.gov/

3.5: Kevin Schafer/Corbis

3.6: Pavel Filatov/Alamy

3.7: Corbis

3.8: Patrick Landmann/Getty Images

3.9: Corbis/Sygma

3.10: Kazuyoshi Nomachi/Corbis

3.11: Yapanchintsev Yevgeny/Corbis

3.12: Wolfgang Kaehler/Corbis

3.13: Dean Conger/Getty Images

3.14: Arctic-Images/Corbis

3.15: David J. Cross/Peter Arnold, Inc.

3.16: Sovfoto/Eastfoto

3.17: Adapted from J. H. Bater, *Russia and the Post-Soviet Scene*. London: Arnold, 1996, p. 314.

3.18: Christopher Herwig/Getty Images

3.19: Redrawn from D. J. B. Shaw, *Russia in the Modern World*. Oxford: Blackwell, 1999, p. 7.)

3.20: *Atlas of Twentieth Century World History*. New York: HarperCollins Cartographic, 1991, pp. 86–87.

3.21: Redrawn from P. L. Knox and J. Agnew, *The Geography of the World Economy*, 3rd ed. London: Arnold, 1998, p. 168.

3.23: Vladimir Pcholkin/Getty Images

3.24: Alexander Demianchukai/Reuters/Corbis

3.25: Arctic-Images /Getty Images

3.26a: Center for International Earth Science Information Network [CIESIN], Columbia University; International Food Policy Research Institute [IFPRI]; and World Resources Institute [WRI], 2000. *Gridded Population of the World [GPW]*, Version 3. Palisades, NY: CIESIN, Columbia University, 2005. Available at http://sedac.ciesin.columbia.edu/gpw/.

3.26b: Updated from J. H. Bater, *Russia and the Post-Soviet Scene*. London: Arnold, 1996, p. 94.

3.27: Updated from J. H. Bater, *Russia and the Post-Soviet Scene*. London: Arnold, 1996, p. 101.

3.28: Redrawn from R. Millner-Gulland and N. Dejevsky, *Cultural Atlas of Russia*, rev. ed. New York: Checkmark Books, 1998, pp. 26–27.

3.29: Redrawn from G. Smith, *The Post-Soviet States: Mapping the Politics of Transition*. London: Arnold, 1999, p. 129.

3.30: Remi Benali/Corbis

3.B1.1: RPG /Roman Poderny/Corbis

3.B1.2: Nancy Kaszerman/ZUMA/Corbis

3.B2.1: Lystseva Marina/Corbis

## CHAPTER 4

4.CO: Paul Gilham/Getty Image

4.2: NASA/http://visibleearth.nasa.gov/

4.4: Luis Orteo/Corbis

4.5: Reza/Corbis

4.7: Redrawn from C. C. Held, *Middle East Patterns: Places, Peoples and Politics*, 3rd ed. Boulder: Westview Press, 2000, p. 38.

4.8: Ed Darack/Corbis

4.9: Redrawn from P. English, *City and Village in Iran*. Madison: University of Wisconsin Press, 1966, p. 31.

4.10: Chris North/Corbis

4.11: Fred Friberg/Getty Images

4.12: George Steinmetz/Corbis

4.13: Redrawn from C. C. Held, *Middle East Patterns: Places, Peoples and Politics*, 3rd ed. Boulder: Westview Press, 2000, p. 23.

4.14: Guenter Rossenbach/Corbis

4.15: Redrawn from D. Hiro, *Holy Wars*. London: Routledge, 1989, frontispiece.

4.16: Brian Lawrence/Getty Images

4.18: Redrawn from *Hammond Times Concise Atlas of World History*. Maplewood, NJ: Hammond, 1994, pp. 100–1.

4.19: The British Library, Image # 013028;

4.20: Jospeh Barrack/Getty Images

4.21: Courtesy of the Jumeirah Resort

4.22: Patrice Latron/Corbis

4.23: Socioeconomic Data and Applications Center. Columbia University, Center for International Earth Science Information Network [CIESIN]. *Gridded Population of the World [GPWv3]*. Palisades, NY: CIESN, Columbia University. Available at: http://sedac.ciesin.org/plue/gpw

4.24: Peter Adams/Corbis

4.25: Stephanie Kuykendal/Corbis

4.27: Redrawn and modified from A. Segal, *An Atlas of International Migration*. London: Hans Zell, 1993, pp. 95 and 103.

4.28: Adapted from *The Guardian*, October 14, 2000, p. 5.

4.29: Corbis/Bettmann

4.30: Suhaib Salem/Corbis

4.31: Stephanie Cardinale/Corbis

4.32: Sandro Vannini/Corbis

4.33: Arco Images GmbH/Alamy

4.35: University of Texas, http://www.lib.utexas.edu/maps/middle_east_and_asia/kurdish_lands_92.jpg, accessed September 23, 2006.

4.36: Ian G.R. Shaw, Ph.D. student, *School of Geography and Development*, University of Arizona

4.37: J. M. Rubenstein, 2008, *The Cultural Landscape: An Introduction to Human Geography*, 9th ed. Upper Saddle River, NJ: Prentice Hall, p. 211.

4.38: BBC News, Web edition; http://news.bbc.co.uk/1/hi/world/middle_east/3111159.stm; accessed June 19, 2009.

4.39: http://dissidentvoice.org/deathtoll/; accessed June 19, 2009).

4.B1.1: Tim Keirn

4.B1.2: Martin Harvey/Corbis

4.B2.1: Adapted from M. Janofsky, "Forecast of $250 Oil Finds Few Believers," *Arizona Daily Star*, June 22, 2008. www.azstarnet.com/sn/fromcomments/244867.php (accessed 5 June 2009.)

4.B2.2: Yann Arthus-Bertrand/Corbis

4.B2.3: Adapted *from Energy Watch Group, Crude Oil: The Supply Outlook*, 2007, p. 12. http/www.energywatchgroup.org/fileadmin/global/pdf/EWG_Oilreport_10-2007.pdf, accessed June 18, 2009.

## CHAPTER 5

5.CO: P. Deloche/International Labor Organization

5.2: NASA/http://visibleearth.nasa.gov/

5.4a: http://soils.usda.gov/use/worldsoils/papers/desert-africa-fig2.gif.

5.4b: FAO Photo/R. Faidutti

5.5: http://maps.grida.no/go/graphic/climate_change_vulnerability_in_africa

5.6: S. Aryeetey-Attoh (ed.), The Geography of Sub-Saharan Africa (Upper Saddle River, NJ: Prentice Hall, 1997), p. 5.

5.7a: Blaine Harrington III/Corbis

5.7b: Panoramic Images/Getty Images

5.8: Africa Program. The Woods Hole Research Center. www.whrc.org. Laporte, N., J. Stabach, R. Grosch, T. Lin, and S. Goetz, 2007, Expansion of Industrial Logging in Central Africa. *Science,* 316, 1451. Baccini, A., Laporte, N., Goetz, S.J., Sun, M., and Dong, H., 2008. A first map of tropical Africa's above-ground biomass derived from satellite imagery. *Environmental Research Letters*, 3: 045011.doi10.1088/1748-9326/3/4/045011.

5.9: Wendy Stone/Corbis

5.10a: Peter Johnson/Corbis

5.10b: Renee Lynn/Corbis

5.10c: Martin Harvey/Corbis

5.11: Modified from I. L. L. Griffiths, *An Atlas of African Affairs*. New York: Methuen, 1985, pp. 20–21.

5.12: Rob Howard/Corbis

5.13b: Hulton-Deutsch Collection/Corbis

5.14: Adapted from I. L. L. Griffiths, *An Atlas of African Affairs*. New York: Methuen, 1985; D. L. Clawson and J. S. Fisher (eds.), *World Regional Geography: A Developmental Approach*. Upper Saddle River, NJ: Prentice Hall, 1998, Fig. 23.2; and C. McEvedy, *The Penguin Atlas of African History*. New York: Penguin Books, 1995.

5.15a: Sandro Vannini/Corbis

5.15b: Nik Wheeler/Corbis

5.16: Adapted from A. Thomas and B. Crow [eds.], *Third World Atlas*. Buckingham, UK: Open University Press, 1994, p. 28; and J. F. Ade, A. Crowder, M. Crowder, P. Richards, E. Dunstan, and A. Newman [eds.], *Historical Atlas of Africa*. Harlow, Essex, UK: Longman, 1985, p. 67.

5.17a: Hulton-Deutsch Collection/Corbis

5.17b: Hulton-Deutsch Collection/Corbis

5.18: Adapted from A. Thomas and B. Crow [eds.], *Third World Atlas*. Buckingham, UK: Open University Press, 1994, p. 35.

5.19a: Michael S. Lewis/Corbis

5.19b: Bernard and Catherine Desjeux/Corbis

5.19c: serrv.org

5.20: Adapted from I.L.L. Griffiths, *An Atlas of African Affairs*. New York: Methuen, 1985.

5.21: UNCTAD. http://www.fao.org/docrep/007/y5419e/y5419e02.htm

5.22: Darren Staples/Corbis

5.23: Jon Hrusa/epa/Corbis

5.24a: Adapted from International Telecommunication Union (www.itu.int).

5.24b: Strauss/Curtis/Corbis

5.25: Center for International Earth Science Information Network [CIESIN], Columbia University; International Food Policy Research Institute [IFPRI]; and World Resources Institute [WRI]. *Gridded Population of the World [GPW]*, Version 2. Palisades, NY: CIESIN, Columbia University, 2000. Available at http://sedac.ciesin.org/plue/gpw/index.html?main.html&2.)

5.26: Demographic and Health Survey statistics, http://www.statcompiler.com/

5.27: From the World Bank Group, Intensifying Action Against HIV/AIDS in Africa: Responding to a Development Crisis, 1998. Available at http://www.worldbank.org/html/extdr/offrep/afr/aidstrat.pdf

5.28: Gideon Mendel/ActionAid/Corbis

5.29: Liba Taylor/Corbis

5.30: Redrawn from S. Aryeetey-Attoh [ed.], *The Geography of Sub-Saharan Africa*. Upper Saddle River, NJ: Prentice Hall, 1997, Fig. 4.6.

5.31a: Seylouu/Getty Images

5.31b: Ben Radford/Corbis

5.31c: Thierry Orban/Corbis

5.32a: "One Woam's Day in Sierra Leone," Food and Agricultural Organization of the United Nations, 1997. Available at http://www.fao.org/NEWS/FACTFILE/FF9719-E.HTM.

5.32b: Earl & Nazima Kowall/Corbis; c) Vince Streano/Corbis

5.34a: Aaron Ufemeli/Corbis

5.34b: Ed Kashi/Corbis

5.34c: Urs Flueeler/Corbis

5.B1.1: David Turnley/Corbis

5.B2.1 : Jon Hicks/Corbis

5.B2.2: Radhika Chalasani/Sipa Press

## CHAPTER 6

6.CO: Paul Chinn/Corbis

6.3: NASA/http://visibleearth.nasa.gov/

6.4: The Vulcan Project/Dr. Kevin R. Gurney/Purdue University/NASA

6.5: Western Governors' Association/Black & Veatch

6.7: Michele Falzone/JAI/Corbis

6.8: Redrawn and adapted from E. Homberger, *The Historical Atlas of North America.* London: Penguin, 1995, p. 21.

6.9: Dave Reede/Corbis

6.10: Omar Torres/Getty Images

6.11: Eastcott/Momatiuk/Valan Photos

6.13: All Canada Photos/Alamy

6.14: James Quine/Alamy

6.15: U.S. Census Bureau, Economic Census 2007.

6.16: P. L. Knox, *Urbanization.* Upper Saddle River, NJ: Prentice Hall, 1994, p. 295.

6.17: *Demographic Profiles: 100 Percent and Sample Data.* Washington, DC: U.S. Census Bureau, 2008.

6.18: Center for International Earth Science Information Network [CIESIN], Columbia University; International Food Policy Research Institute [IFPRI]; and World Resources Institute [WRI]. 2000. *Gridded Population of the World [GPW]*, Version 2. Palisades, NY: CIESIN, Columbia University. Available at http://sedac.ciesin.org/plue/gpw).

6.19: Modified from J. M. Rubenstein, *The Cultural Landscape: An Introduction to Human Geography*, 8th ed. Upper Saddle River, NJ: Prentice Hall, 2004.

6.20: John Zich/Corbis

6.21: E. Homberger, *The Historical Atlas of North America.* London: Penguin, 1995, pp. 88–89.

6.22: Redrawn and adapted from E. Homberger, *The Historical Atlas of North America.* London: Penguin, 1995, pp. 106–7.

6.23: Barry Lewis/Corbis

6.24: Jan Butchofsky-Houser/Corbis

6.26: Getty Images/Time Life Pictures

6.28: fotoVoyager/Istockphoto.com

6.29: Jay Syverson/Corbis

6.31: Walter Bibikow/JAI/Corbis

6.32: Bo Zaunders/Corbis

6.32: Source for casualty figures: http://www.antiwar.com/casualties/.

6.33: mseidelch/Istockphoto.com

6.B1.1: Chris Hellier/Corbis

6.B1.2: UN Food and Agricultural Organization, "FAOStat," 2007, http://faostat.fao.org/site/339/default.aspx accessed July 6, 2009).

6.B2.1: David E Myers/Getty Images

6.B2.2: MODIS/NASA

## CHAPTER 7

7.CO: Keith Dannemiller/Corbis

7.2: Greg Probst/Corbis

7.5: Pensa Nicarabua/Corbis

7.6: Eberhard Hummel/Corbis

7.8: NASA/http://visibleearth.nasa.gov/

7.9a: Kevin West/Getty Images

7.9b: Owen Franken/Corbis

7.10: Paulo Fridman/Corbis

7.11: Joseph Sohm/Corbis

7.12a: Charles O'Rear/Corbis

7.12b: Kit Houghton Photography/Corbis

7.12c: Tim Wright/Corbis

7.13b: Ted Spiegel/Corbis

7.13c: International Potato Center

7.15a: Adapted from G. Knapp and C. Caviedes, *South America.* Englewood Cliffs, NJ: Prentice Hall, 1995, p. 233, and from "Controversial Infrastructure Projects Proposed or Underway in the Amazon Basin." Available at http://www.amazonwatch.org

7.15b: Photo Researchers

7.15c: Herve Collart/Corbis

7.16: Bob Krist/Corbis

7.17: Adapted from P. L. Knox and S. A. Marston, *Human Geography: Places and Regions in Global Context,* 3rd ed. Upper Saddle River, NJ: Prentice Hall, 2004.

7.18: http://chicago.grubstreet.com/2008/10/tribune_food_digested_a_fresh.html

7.19: Keith Dannemiller/Corbis

7.20: Mariana Bazo/Corbis

7.21a: UN Food and Agricultural Organization. Available at http://faostat.fao.org/

7.21b: Charles O'Rear/Corbis

7.21c: The Fairtrade Foundation

7.22: Oxfam. Economic Commission for Latin America and the Caribbean (ECLAC), on the basis of special tabulations of data from household surveys conducted in the relevant countries.

7.23: Center for International Earth Science Information Network [CIESIN], Columbia University; International Food Policy Research Institute [IFPRI]; and World Resources Institute [WRI]. *Gridded Population of the World [GPW]*, Version 2. Palisades, NY: CIESIN, Columbia University, 2000. Available at http://sedac.ciesin.org/plue/gpw

7.25: Gabriel M. Covian/Getty Images

7.26: Paulo Whitaker/Corbis

9.29: Antonie Serra/Corbis

9.30: Peter Adams/Corbis

9.31: Aninito Mukherjee/Corbis

9.32: Robert Nickelsberg/Getty Images

9.33: Robert Holmes/Corbis

9.34: Mick Hutson/Getty Images

9.B1.2: JJ Travel Photography/Getty Images

9.B1.3: Daniel O'Leary/Panos Pictures

9.B2.1: Martin Puddy/Getty Images

## CHAPTER 10

10.CO: Craig Lovell/Corbis

10.2: MODIS/NASA

10.3: Redrawn from R. Ulack and G. Pauer, *Atlas of Southeast Asia*. New York: Macmillan, 1988, p. 6.

10.4a: NASA/http://visibleearth.nasa.gov/

10.4b: Reuters/Corbis

10.5: Corbis

10.6: Map is based on information from T. R. Leinbach and R. Ulack, *Southeast Asia: Diversity and Development*. Upper Saddle River, NJ: Prentice Hall, 2000, Map 2.1; and H. C. Brookfield and Y. Byron, *South-East Asia's Environmental Future: The Search for Sustainability*. New York: United Nations University Press, 1993, Figure 13.1.

10.7a: Nik Wheeler/Corbis

10.7b: Redrawn from M.J. Valencia, J.M. Van Dyke, and N.A. Ludwig, *Sharing the Resources of the South China Sea*. Honolulu: University of Hawaii Press, 1999, Plate I.

10.8a: Redrawn from Mekong River Commission http://www.mr-cmekong.org/img/programmes/hydro/LMBhydroPowerProject080911draft.jpg.

10.8b: Mekong from Space Website, http://www.mrcmekong.org/MfS/html/monitoring_the_mekong_basin.html.

10.9a: Jewel Samad/AFP/Getty Images

10.9b: Patrick Robert/Corbis

10.10: Based on maps in R. Ulack and G. Pauer, *Atlas of Southeast Asia*. New York: Macmillan, 1988, p. 11; and T. R. Leinbach and R. Ulack, *Southeast Asia: Diversity and Development*. Upper Saddle River, NJ: Prentice Hall, 2000, Map 2.6b.

10.11a: 1990–2000, 2000–2005 percent forest loss and 2005 forest cover: FAO State of the World's Forests. Rome, FAO, 2009; 1973–85 percent forest loss: Rain Forest Report Card, East Lansing, MI: Michigan State University, 2000.

10.11b: Michael S. Yamashita/Corbis

10.11c: Guido Cozzi/Corbis

10.11d: Patrick Brown/Panos Pictures

10.12: Simon Podgorsek/Istockphoto.com

10.13: Redrawn from B. Crow and Thomas [eds.], *Third World Atlas*. Philadelphia: Open University Press, 1984, p. 39.

10.14: Redrawn from M. Dockrill, *Atlas of 20th Century World History*. New York: Harper, 1991, pp. 74–75.

10.15: Kamarul Akhir/AFP/Getty Images)

10.16: International Monetary Fund, World Economic Outlook Database, April 2009, http://www.imf.org/external/pubs/ft/weo/2009/01/weodata/index.aspx).

10.17: Socioeconomic Data and Applications Center. Columbia University, Center for International Earth Science Information Network [CIESIN]; International Center for Tropical Agriculture [CIAT]. 2005. *Gridded Population of the World* [GPWv3]. Palisades, NY: CIESN, Columbia University. Available at: http://sedac.ciesin.org/plue/gpw.

10.18a: epa/Corbis

10.18b: Shamshahrin Shamsudin/Corbis

10.19: Redrawn from T. R. Leinbach and R. Ulack, *Southeast Asia: Diversity and Development*. Upper Saddle River, NJ: Prentice Hall, 2000, Map 12.7.

10.20a: UN High Commission on Refugees *Statistical Online Population Database*, 2007. Available at: http://www.unhcr.org/statistics/45c063a82.html.

10.20b: Philippines Overseas Employment Administration, 2000-2005. Available at: http://www.poea.gov.ph/html/statistics.html.

10.21: Redrawn from R. Ulack and G. Pauer, *Atlas of Southeast Asia*. New York: Macmillan, 1988, p. 29.

10.22: Redrawn from R. Ulack and G. Pauer, *Atlas of Southeast Asia*. New York: Macmillan, 1988, p. 27.

10.23: James Marshall/Corbis

10.24: UNAIDS, 2008, *Report on the Global AIDS Epidemic*. http://www.unaids.org/en/KnowledgeCentre/HIVData/GlobalReport/2008/2008_Global_report.asp.

10.25a: Corbis

10.25b: Bettmann/Corbis

10.26: Les Stone/Corbis

10.27: Adapted from T. R. Leinbach and R. Ulack, *Southeast Asia: Diversity and Development*. Upper Saddle River, NJ: Prentice Hall, 2000, Map 10.2.

10.B1.1: Jose Fuste Raga/Corbis

10.B1.2a: Adapted from T. R. Leinbach and R. Ulack. *Southeast Asia: Diversity and Development*. Upper Saddle River, NJ: Prentice Hall, 2000, Map. 14.1 and 14.6.

10.B1.2b: M. Sparke, T. Bunnell, and C. Grundy-Warr; "Triangulating the Borderless World: Geographies of Power in the Indonesia-Malaysia-Singapore Growth Triangle." *Transactions of the Institute of British Geographers*, NS29, 485-398.

10.B2.1: Redrawn from "The Opium Kings," PBS Frontline, 1998. Available at: http://www.pbs.org/wgbh/pages/frontline/showsherion/maps/shan.html/.

10.B2.2: Christophe Loviny/Corbis

10.B2.3: Barbara Walton/Corbis

## CHAPTER 11

11.CO: Ashley Cooper/Corbis

11.2: NASA/http://visibleearth.nasa.gov/

11.3: Based on T. McKnight, *Oceania: The Geography of Australia, New Zealand and the Pacific Islands*.Englewood Cliffs, NJ: Prentice Hall, 1995, Figure 2.2e; and G. M. Robinson, R. J. Loughran, and P. J.

Tranter, *Australia and New Zealand: Economy, Society and Environment*. New York: Oxford University Press, 2000, Figure 3.1.

11.4a: Torsten Blackwood/AFP/Getty Images

11.4b: Penny Tweedie/Corbis

11.4c: Lucas Dawson/Getty Images

11.6: Tim Wimborne/Corbis

11.7a: Karl-Heinz Haenel/Corbis

11.7b: Geoff Renner/Corbis

11.8a: Bill Ross/Corbis

11.8b: Douglas Peebles/Corbis

11.9a: Heldur Netocny/Panos Pictures

11.9b: Wayne Lawler/Corbis

11.10: Redrawn from M. Rapaport (ed.), *The Pacific Islands: Environment and Society*. Honolulu: The Bess Press, 1999, 30.3.

11.11a: Howard Davies/Corbis

11.11b: Geoff Moon/Corbis

11.12b: Madeleine Coorey/AFP/Getty Images

11.13a: Getty Images

11.13b: AFP/Getty Images

11.14: Based on C. McEvedy, *The Penguin Historical Atlas of the Pacific*. New York: Penguin Books, 1998, pp. 49, 63, 65, 90.

11.15: Based partly on T. McKnight, *Oceania: The Geography of Australia, New Zealand and the Pacific Islands*. Englewood Cliffs, NJ: Prentice Hall, 1995, Figures 4.1 and 4.2; J. M. Powell, *An Historical Geography of Modern Australia: The Restive Fringe*. New York: Cambridge University Press, 1988, pp. 3–55; and G. M. Robinson, R. J. Loughran, and P. J. Tranter, *Australia and New Zealand: Economy, Society and Environment*. New York: Oxford University Press, 2000, 6.5.

11.16: Jose Fuste Raga/Corbis

11.17: Data from Australian Broadcasting Company. Available at http://www.abc.net.au/ra/carvingout/maps/statistics.htm.au

11.18: Kevin Fleming/Corbis

11.19: Douglas Peebles/Corbis

11.20: Columbia University, Center for International Earth Science Information Network [CIESIN]; International Center for Tropical Agriculture [CIAT]. 2005. *Gridded Population of the World* [GPWv3]. Palisades, NY: CIESN, Columbia University. Available at: http://sedac.ciesin.org/plue/gpw.

11.21: Panoramic Images/Getty Images

11.22a: Greg Wood/Getty Images

11.22b: Corbis

11.23a: Yoshikazu Tsuno/Getty Images

11.23b: Kristian Dowling/Getty Images

11.24: Hannah Johnston/Getty Images

11.25a: AP Wide World Photos

11.25b: Adapted from G. M. Robinson, R. J. Loughran, and P. J. Tranter, *Australia and New Zealand: Economy, Society and Environment*. New York: Oxford University Press, 2000, 5.5.

11.B1.1: Bettmann/Corbis

11.B1.2: Jacques Langevin/Corbis

11.B1.3: John Carnemolla/Corbis

11.B2.1: Based on G. Lean and D. Hinrichsen (eds.), *Atlas of the Environment*. Santa Barbara, CA: ABC-CLIO, 1994, pp. 182–83; and Terraquest, *Virtual Antarctica Expedition*. Available at http://www.terraquest.com/va/expedition/maps/cont.map.html.

11.B2.2: Theo Alofs/Corbis

## APPENDIX

A.1: Extract from Carta Nazionale Della Svizzera, Sheet 1353, 1:25,000 series, Ufficio Federale di Topografia, 3084 Wabern, Switzerland. Edizione 1998.

A.2: Reprinted with permission of Prentice Hall, from J. M. Rubenstein, *The Cultural Landscape: An Introduction to Human Geography*, 1996, p. 584. Adapted from William K. Stevens, "Study of Acid Rain Uncovers Threat to Far Wider Area," *New York Times*, January 16, 1990, p. 21, map.

A.5: Reprinted with permission of Prentice Hall from E. F. Bergman, *Human Geography: Cultures, Connections, and Landscapes*, © 1995, p. 12.

A.6: Reprinted with permission of Prentice Hall from E. F. Bergman, *Human Geography: Cultures, Connections, and Landscapes*, © 1995, p. 13.

A.7: Buckminster Fuller Institute and Dymaxion Map Design, Santa Barbara, CA. The word Dymaxion and the Fuller Projection Dymaxion™ Map design are trademarks of the Buckminster Fuller Institute, Santa Barbara, California, © 1938, 1967 & 1992. All rights reserved.

A.8: (a) M. Kidron and R. Segal (eds.), *The State of the World Atlas*, rev. 5th ed. London: Penguin Reference, 1995, pp. 28–29; (b) G. C. Staple (ed.), TeleGeography 1999. Washington, DC: TeleGeography, 1999, p. 82.

A.9: Donna Cox and Robert Patterson, http://www.ncsa.uiuc.edu/SCMS/DigLib/text/technology/Visualization-Study-NSFNET-Cox.html.

A.10: R. W. Greene, *GIS in Public Policy*. Redlands, CA: ESRI Press, 2000, p. 67.

# INDEX

# N